AF761535

POUNDER'S MARINE DIESEL ENGINES AND GAS TURBINES

POUNDER'S MARINE DIESEL ENGINES AND GAS TURBINES

Tenth Edition

MALCOLM LATARCHE

ELSEVIER

Butterworth-Heinemann
An imprint of Elsevier

Butterworth-Heinemann is an imprint of Elsevier
The Boulevard, Langford Lane, Kidlington, Oxford OX5 1GB, United Kingdom
50 Hampshire Street, 5th Floor, Cambridge, MA 02139, United States

Library of Congress Cataloging-in-Publication Data
A catalog record for this book is available from the Library of Congress

British Library Cataloguing-in-Publication Data
A catalogue record for this book is available from the British Library

ISBN: 978-0-08-102748-6

For information on all Butterworth-Heinemann publications
visit our website at https://www.elsevier.com/books-and-journals

Publisher: Matthew Deans
Acquisitions Editor: Carrie Bolger
Editorial Project Manager: Charlotte Rowley
Production Project Manager: Prasanna Kalyanaraman
Cover Designer: Victoria Pearson

Typeset by SPi Global, India

Contents

Preface

It has been more than a decade since the last (ninth) edition of Pounder's Marine Diesel Engines was authored by Doug Woodyard. Sadly, in the intervening period, Doug has passed away and I would like to dedicate this new edition to his memory.

In the last edition, Doug added the details of 5 years of intense development of marine diesels that occurred when the world fleet of commercial was growing at its fastest rate ever. The economic crash that occurred in 2008 barely had time to impact ship construction and ordering was still continuing at a high level when Doug's work was published.

It is fair to say that some of the events that have happened since 2009 could not have been envisaged at the time. The IMO's regulations controlling NOx and SOx had been put in place, but the strictest controls were still in the future then. Today, the SOx regulation has reached its final limits and so has NOx, although in the case of the latter new areas are being added and there are plans for perhaps even stricter limits.

In 2009, the EEDI rules that effectively limit CO_2 emissions for new ships were under discussion, but not adopted. The rules are not only now in place having come into force in 2013, but the rolling programme of increasing reductions has been accelerated and again there are voices calling for even more action.

Arguably, the biggest change has been the IMO announcing in 2018 that it has ambitions to decarbonize shipping altogether by the end of the century and to halve emissions compared to 2008 by 2050 or even earlier. As of August 2020, these are still just ambitions, but if the will is there among IMO member states, it could become mandatory within the next 5 years.

Operating diesel engines on fuels containing no carbon is not impossible. However, it will take some intense research and that is underway. Engine makers must be applauded for grasping the nettle, but constant moving of the goalposts by regulators is not ideal and almost certainly is restricting the efficiency improvements that were being pursued.

The engine designers and manufacturers may have a hard task, but so do ship operators. A ship is not a short-term asset, nor is it cheap. Planning, design, and construction can take two years or more and ships are generally considered as having a 20–30 year lifespan. Almost certainly, none of the ships built in the next few years will satisfy the IMO's decarbonization ambitions, but some of the more innovative ones will be seen as a step on the way.

This edition of Pounder's contains the detail of the new regulations, and in various chapters, looks at the advances that have been made and those being pursued to ensure that the diesel engine retains its dominance in maritime propulsion. It does still contain much of the material from earlier editions that are still relevant.

I make no apology for retaining older material—the history of the marine diesel engine is still being written and it is important to know where we have come from as well as perhaps getting some idea of where we are headed.

Just as Doug and previous authors have done, I would also like to thank those within the industry with whom it has been my pleasure to come in contact with over the years who have given their time and provided information. I hope this edition meets the expectations of readers and achieves the status that earlier versions deservedly did.

Malcolm Latarche

Introduction

Marine diesels—Powering ships for over a century

Destined to dominate ship propulsion

Exactly which ship was the first to be powered by a diesel engine is a matter of some debate. Most reference works point to the 7400 dwt *Selandia* of 1912, but four-stroke diesels had been fitted to many more vessels before the *Selandia* was built. Both France and the UK had diesel-powered submarines in operation and the Russian navy also had some diesel-powered craft.

Several inland waterway crafts, as well as some small coasting vessels, also had been powered by diesel engines. If *Selandia* is remembered most, it is perhaps because it served for more than 30 years using the same two engines it was delivered with. Probably undisputed in being the first two-stroke diesel-engined vessel, *Monte Penedo* was delivered six months after *Selandia.*

Regardless of which vessel was first, it is a fact that in the intervening years, the Diesel engine has come to dominate the marine power sector. Diesel engines did not immediately displace steam as their use in reciprocating engines continued for several decades through to the middle of the 20th century, and by way of turbines, is still in use today in many LNG carriers including in newbuildings.

The remorseless rise of the diesel engine at the expense of steam reciprocating and turbine installations was symbolized in 1987 by the steam-to-diesel conversion of Cunard's prestigious cruise liner *Queen Elizabeth 2*. Her turbine and boiler rooms were ignominiously gutted to allow the installation of a 95,600 kW diesel–electric plant.

The revitalized *QE2*'s propulsion plant was based on nine 9-cylinder L58/64 medium-speed four-stroke engines from MAN B&W Diesel which provided a link with the pioneering *Selandia:* the 1912-built twin-screw 7400 dwt cargo/passenger ship was powered by two Burmeister and Wain eight-cylinder four-stroke engines (530 mm bore/730 mm stroke), each developing 920 kW at 140 rev/min. An important feature was the effective and reliable direct-reversing system. Progress in raising specific output over the intervening 70 years was underlined by the 580 mm bore/640 mm stroke design specified for the *QE2* retrofit: each cylinder has a maximum continuous rating of 1213 kW.

Selandia was built by the Burmeister and Wain yard in Copenhagen for Denmark's East Asiatic Company and, after trials in February 1912, successfully completed a 20,000 mile round voyage between the Danish capital and the Far East. The significance of the propulsion plant was well-appreciated at the time. On her first arrival in London, the ship was inspected by Sir Winston Churchill, then First Lord of the Admiralty; and *Fiona*, a sister ship delivered four months later by the same yard, so much impressed the German Emperor that it was immediately arranged for the Hamburg Amerika Line to buy her.

A third vessel in the series, *Jutlandia*, was built by Barclay, Curle, in Scotland and handed over to East Asiatic in May 1912. The Danish company's oceangoing motor ship fleet numbered 16 by 1920, the largest being the 13,275 dwt *Afrika* with twin six-cylinder B&W engines of 740 mm bore/1150 mm stroke developing a combined 3300 kW at 115 rev/min.

It is not difficult to understand the attraction of the new diesel engines to owners of steam-powered vessels. Oil fuel for diesel engines can be stored in almost inaccessible tanks and pumped to the engine. Coal by contrast needed to be shifted manually by gangs of men wheelbarrowing coal from bunkers to the engines. Those men, their wages, and the space needed onboard to accommodate them could be swept away by adopting diesel.

Early steam-to-diesel conversions included three 4950 dwt vessels built in 1909 and repowered in 1914/1915 by the B&W Oil Engine Co. of Glasgow, each with a single six-cylinder 676 mm bore/1000 mm stroke engine developing 865 kW at 110 rev/min.

Selandia operated successfully for many years (latterly as *Norseman*) and maintained throughout a fully loaded service speed of 10.5 knots before being lost off Japan in 1942. The propulsion plant of the second *Selandia*, which entered service in 1938, demonstrated the advances made in diesel technology since the pioneering installation. The single, double-acting two-stroke, five-cylinder engine of the new 8300 dwt vessel delivered 5370 kW at 120 rev/min: three times the output of the twin-engined machinery powering the predecessor.

In 1914, there were fewer than 300 diesel-powered vessels in service with an aggregate tonnage of 2,35,000 grt; a decade later, the fleet had grown to some 2000 ships of almost two million gross register tonnage; and by 1940, the total tonnage had risen to 18 million gross register tonnage embracing 8000 motor ships.

Between the two world wars, the proportion of oil-engined tonnage in service thus expanded from 1.3% to 25% of the overall oceangoing fleet.

By 1939, an estimated 60% of the total tonnage completed in world yards comprised of motor ships, compared with only 4% in 1920.

In outlining the foundations of the diesel engine's present dominance in shipping, other claimants to pioneering fame should be mentioned. In 1903, two diesel-powered vessels entered service in quick succession: the Russian naphtha carrier *Vandal*, which was deployed on the Volga, and the French canal boat *Petit Pierre*. By the end of 1910, there were 34 trading vessels over 30 m long worldwide with diesel propulsion, and an unknown number of naval vessels, especially submarines.

The earliest seagoing motor vessel was the twin-screw 678 ton *Romagna*, built in 1910 by Cantieri Navali Riuniti with twin four-cylinder port-scavenged trunk piston engines supplied by Sulzer. Each 310 mm bore/460 mm stroke engine delivered 280 kW at 250 rev/min. The year 1910 also saw the single-screw 1179 dwt Anglo-Saxon tanker *Vulcanus* enter service powered by a 370 kW Werkspoor six-cylinder four-stroke crosshead engine with a 400 mm bore/600 mm stroke. The Dutch-built vessel was reportedly the first oceangoing motor ship to receive classification from Lloyd's Register.

In 1911, the Swan Hunter-built 2600 dwt Great Lakes vessel *Toiler* crossed the Atlantic propelled by two 132 kW Swedish Polar engines. Krupp's first marine diesel engines, six-cylinder 450 mm bore/800 mm stroke units developing 920 kW at 140 rev/min apiece, were installed the same year in the twin-screw 8000 dwt tankers *Hagen* and *Loki* built for the German subsidiary of the Standard Oil Co. of New Jersey.

Each of the twin four-cylinder Sulzer 4S47 crosshead units (470 mm bore/680 mm stroke) delivered 625 kW at 160 rev/min. (The adoption of the two-stroke cycle by Sulzer in 1905 greatly increased power output and fostered a simpler engine. Port-scavenging, introduced in 1910, eliminated the gas exchange valves in the cylinder cover to create a simple valveless concept that characterized the Sulzer two-stroke engine for 70 years: the change to uniflow scavenging only came with the RTA-series engines of 1982 because their very long stroke—required for the lower speeds dictated for high propeller efficiency—was unsuitable for valveless port scavenging.)

Another important delivery in 1912 was the 3150 dwt Furness Withy cargo ship *Eavestone*, powered by a single four-cylinder Carels two-stroke crosshead engine with a rating of 590 kW at 95 rev/min. The 508 mm bore/914 mm stroke design was built in England under licence by Richardsons Westgarth of Middlesbrough.

There were, inevitably, some failures among the pioneers. For example, a pair of Junkers opposed-piston two-stroke engines installed in a 6000 dwt

Hamburg-Amerika Line cargo ship was replaced by triple-expansion steam engines even before the vessel was delivered. The Junkers engines were of an unusual vertical tandem design, effectively double-acting, with three pairs of cylinders of 400 mm bore and 800 mm combined stroke to yield 735 kW at 120 rev/min. More successful was Hapag's second motor ship, *Secundus*, delivered in 1914 with twin Blohm + Voss-MAN four-cylinder two-stroke single-acting engines, each developing 990 kW at 120 rev/min.

After the First World War, diesel engines were specified for increasingly powerful cargo ship installations and a breakthrough made in large passenger vessels. The first geared motor ships appeared in 1921, and in the following year, the Union Steamship Co. of New Zealand ordered a 17,490 grt quadruple-screw liner from the UK's Fairfield yard. The four Sulzer six-cylinder ST70 two-stroke single-acting engines (700 mm bore/990 mm stroke) developed a total of 9560 kW at 127 rev/min—far higher than any contemporary motor ship—and gave *Aorangi* a speed of 18 knots when she entered service in December 1924.

Positive experience with these engines and those in other contemporary motor ships helped to dispel the remaining prejudices against using diesel propulsion in large vessels. Swedish America Line's 18,134 grt *Gripsholm*—the first transatlantic diesel passenger liner—was delivered in 1925; an output of 9930 kW was yielded by a pair of B&W six-cylinder four-stroke double-acting 840 mm bore engines. Soon after, the Union Castle Line ordered the first of its large fleet of motor passenger liners, headed by the 20,000 grt *Caernarvon Castle* powered by 11,000 kW Harland and Wolff-B&W double-acting four-stroke machinery.

Another power milestone was logged in 1925 when the 30,000 grt liner *Augustus* was specified with a 20,600 kW propulsion plant based on four MAN six-cylinder double-acting two-stroke engines of 700 mm bore/1200 mm stroke.

It was now that the double-acting two-stroke engine began to make headway against the single-acting four-stroke design, which had enjoyed favour up to 1930. Two-stroke designs in single- and double-acting forms, more suitable for higher outputs, took a strong lead as ships became larger and faster. Bigger bore sizes and an increased number of cylinders were exploited. The 20,000 grt *Oranje,* built in 1939, remained the most powerful merchant motor ship for many years, thanks to her three 12-cylinder Sulzer 760 mm bore SDT76 single-acting engines aggregating 27,600 kW.

The groundwork for large bore engines was laid early on. Sulzer, for example, in 1912 tested a single-cylinder experimental engine with a

1000 mm bore/1100 mm stroke. This two-stroke crosshead type 1S100 design developed up to 1470 kW at 150 rev/min and confirmed the effectiveness of Sulzer's valveless cross-scavenging system, paving the way for a range of engines with bores varying between 600 mm and 820 mm. (Its bore size was not exceeded by another Sulzer engine until 1968.)

At the end of the 1920s, the largest engines were Sulzer five-cylinder 900 mm bore models (3420 kW at 80 rev/min) built under licence by John Brown in the UK. These S90 engines were specified for three twin-screw Rangitiki-class vessels of 1929.

Goodbye to blast injection

It was towards the end of the 1920s that most designers concluded that the blast air–fuel injection diesel engine—with its need for large, often troublesome and energy-consuming high-pressure compressors—should be displaced by the airless (or compressor-less) type.

Air-blast fuel injection called for compressed air from a pressure bottle to entrain the fuel and introduce it in a finely atomized state via a valve needle into the combustion chamber. The air-blast pressure, which was only just slightly above the ignition pressure in the cylinder, was produced by a water-cooled compressor driven off the engine-connecting rod by means of a rocking lever.

Rudolf Diesel himself was never quite satisfied with this concept (which he called self-blast injection) since it was complicated, and hence, susceptible to failure—and also because the 'air pump' tapped as much as 15% of the engine output. Diesel had filed a patent as early as 1905 covering a concept for the solid injection of fuel, with a delivery pressure of several hundred atmospheres. A key feature was the conjoining of pump and nozzle and their shared accommodation in the cylinder head. One reason advanced for the lack of follow-up was that few of the many engine licensees showed any interest.

A renewed thrust came in 1910 when Vickers' technical director McKechnie (independently of Diesel, and six months after a similar patent from Deutz in Germany) proposed in an English patent an 'accumulator system for airless direct fuel injection' at pressures between 140 bar and 420 bar. By 1915, he had developed and tested an 'operational' diesel engine with direct injection and is thus regarded as the main inventor of high-intensity direct fuel injection. Eight years later, it had become possible to manufacture reliable production injection pumps for high pressures, considerably expanding the range of applications.

The required replacement fuel injection technology thus had its roots in the pioneering days (a Doxford experimental engine was converted to airless fuel injection in 1911), but suitable materials and manufacturing techniques had to be evolved for the highly stressed camshaft drives and pump and injector components. The refinement of direct fuel injection systems was also significant for the development of smaller high-speed diesel engines.

A boost from turbocharging

A major boost to engine output and reductions in size and weight resulted from the adoption of turbochargers. Pressure charging by various methods was applied by most engine builders in the 1920s and 1930s to ensure an adequate scavenge air supply: crankshaft-driven reciprocating air pumps, side-mounted pumps driven by levers off the crossheads, attached Roots-type blowers, or independently driven pumps and blowers. The pumping effect from the piston underside was also used for pressure charging in some designs.

The Swiss engineer Alfred Büchi, considered the inventor of exhaust gas turbocharging, was granted a patent in 1905 and undertook his initial turbocharging experiments at Sulzer Brothers in 1911/1915. It was, however, almost 50 years after that first patent before the principle could be applied satisfactorily to large marine two-stroke engines.

The first turbocharged marine engines were 10-cylinder Vulcan-MAN four-stroke single-acting models in the twin-screw Preussen and Hansestadt vessel *Danzig*, commissioned in 1927. Turbocharging under a constant pressure system by Brown Boveri turboblowers increased the output of these 540 mm bore/600 mm stroke engines from 1250 kW at 240 rev/min to 1765 kW continuously at 275 rev/min, with a maximum of 2960 kW at 317 rev/min. Büchi turbocharging was keenly exploited by large four-stroke engine designers, and in 1929, some 79 engines totalling 162,000 kW were in service or contracted with the system.

In 1950/1951, MAN was the forerunner in testing and introducing high-pressure turbocharging for medium-speed four-stroke engines for which boost pressures of 2.3 bar were demanded and attained.

Progressive advances in the efficiency of turbochargers and systems development made it possible by the mid-1950s for the major two-stroke engine builders to introduce turbocharged designs.

A more recent contribution of turbochargers, with overall efficiencies now topping 70%, is to allow some exhaust gas to be diverted to a power recovery turbine and supplement the main engine effort or drive a generator. Turbocharging is also playing a key role in reducing emissions from ships as first variable and later two-stage turbochargers have been developed to increase power output and reduce NOx emissions.

Heavy fuel oils

Another important step in strengthening the status of the diesel engine in marine propulsion was R&D, enabling it to burn cheaper, heavier fuel oils. Progress was spurred in the mid-1950s by the availability of cylinder lubricants able to neutralize acid combustion products, and hence, reduce wear rates to levels experienced with diesel oil-burning. All low-speed two-stroke and many medium-speed four-stroke engines are now released for operation on low-grade fuels of up to 700 cSt/50°C viscosity, and development work is extending the capability to higher speed designs.

Combating the deterioration in bunker quality is just one example of how diesel engine developers—in association with lube oil technologists and fuel treatment specialists—have managed successfully to adapt designs to contemporary market demands.

The ability to burn heavy oil has reduced the bunker cost burden for shipowners which at times has reached levels equivalent to almost half of all ship running costs. However, heavy fuel oil is considered by many environmental NGOs to be a dangerous substance and calls for it to be banned from use in some areas, especially polar zones.

Environmental pressures

A continuing effort to reduce exhaust gas pollutants is another challenge for engine designers who face tightening international controls in the years ahead on nitrogen oxide, sulphur oxide, carbon dioxide, and particulate emissions.

In-engine measures (e.g. retarded fuel injection) can cope with the IMO's NOx requirements, while direct water injection, fuel emulsification, and charge air humidification can affect greater curbs. Selective catalytic reduction (SCR) systems, however, may be dictated to meet the toughest regional limits not least the IMO Tier III NOx levels and potentially even tighter controls demanded in future.

Demands for 'smokeless' engines, particularly from cruise operators in pollution-sensitive arenas, have been successfully addressed—common rail fuel injection systems playing a significant role—but the development of engines with lower airborne sound levels remains a challenge.

Environmental pressures are also stimulating the development and wider application of dual-fuel diesel and gas engines, which have earned breakthroughs in offshore support vessel, ferry, and LNG carrier propulsion.

Although engine developers and shipowners have been able to successfully comply with all of the NOx and SOx rules in force as of 2020, the reduction of CO_2 emissions envisaged and announced by the IMO in 2018 may mean that fossil fuels–including LNG–are coming to an end as far as ship propulsion goes.

That emphatically does not end the role of the diesel engine for ship propulsion as work is already advanced to produce engines running on alternative fuels such as ammonia and hydrogen. In addition, biofuels and synthetic fuels which do produce almost identical levels of CO_2 as fossil fuels are allowed as replacements because regulators consider that such fuels come from renewable sources and the CO_2 produced is reabsorbed almost immediately.

Lower speeds, larger bores, and longer strokes

Specific output thresholds of the low-speed two-stroke engines found in the largest of ships and accounting for around 80% of installed ship power have been boosted over time by various developments.

Engine development has also focused on greater fuel economy achieved by a combination of lower rotational speeds, higher maximum combustion pressures, and more efficient turbocharging. Engine thermal efficiency has been raised to over 54% and specific fuel consumptions can be as low as 155 g/kWh. At the same time, propeller efficiencies have been considerably improved due to minimum engine speeds reduced to as low as 55 rev/min.

The pace and expense of development in the low-speed engine sector have been such that only three designer/licensors remain active in the international market. The roll call of past contenders includes names either long forgotten or living on in other fields: AEG-Hesselman, Deutsche Werft, Fullagar, Krupp, McIntosh and Seymour, Neptune, Nobel, North British, Polar, Richardsons Westgarth, Still, Tosi, Vickers, Werkspoor, and Worthington. The most recent casualties were Doxford, Götaverken, Stork, Paxman, and Allen, some of whose products remain at sea in dwindling numbers.

These pioneering designers displayed individual flair within generic classifications which offered two- or four-stroke, single- or double-acting, and single- or opposed-piston arrangements. The Still concept even combined the Diesel principle with a steam engine: heat in the exhaust gases and cooling water was used to raise steam which was then supplied to the underside of the working piston.

Evolution has decreed that the surviving trio of low-speed engine designers (MAN Energy Solutions, WinGD and Japan Engine—all of which names have changed since the last edition of this book) should all settle—for the present at least—on a common basic philosophy: uniflow-scavenged, single hydraulically actuated exhaust valve in the head, constant pressure turbocharged, two-stroke crosshead engines exploiting increasingly high stroke/bore ratios, and low operating speeds for direct coupling to the propeller.

Another thing that has changed from the last edition is that the trend towards larger bore sizes has ended and even reversed. The largest bore that ever appeared in the MAN engine portfolio was 1080 mm, but none were ever built and the latest range of that designer extends from 300 to 950 mm bore sizes. The last 980 mm bore engines from MAN (the largest that ever entered service) were installed in ships delivered up to 2015.

As with MAN, Wärtsilä has been now superseded by WinGD. The largest engines under this designer reached a peak with 960 mm bore engines. The last engines in this size date back to 2014 and have been replaced by the 920 mm engines of the X Generation.

The increase in bore size may have peaked, but the newer engines do have a longer stroke. An engine cylinder with a longer stroke-to-bore ratio will have a smaller surface area exposed to the combustion chamber gasses compared to a cylinder with a shorter stroke-to-bore ratio, leading directly to reduced in-cylinder heat transfer, increased energy transfer to the crankshaft, and therefore, higher efficiency.

As the stroke-to-bore ratio increases, so does the distance fresh air has to travel between a cylinder's intake and exhaust ports. This increased distance results in higher scavenging efficiency and lower pumping work.

Through to 2010, most of the large container ships in operation were powered by MAN B&W K98 engines. The K indicating a short stroke of 2400 mm and the 98 referring to the 980 mm bore. Thereafter, it was the S90 engines that became standard. The S referring to Super-Long stroke with these engines has a stroke of 3260 mm and the 90 indicating a bore of 900 mm.

MAN Energy Solutions' G-type programme entered the market in October 2010. The 'G' prefix before an engine means it has a design that follows the principles of the large-bore, Mark 9 engine series that MAN Diesel and Turbo introduced in 2006 with an ultra-long stroke that reduces engine speed, thereby paving the way for ship designs with unprecedented high-efficiency and allowing for the G for 'Green' prefix.

G-type engines' longer stroke results in a lower rpm for the engine driving the propeller. This lower optimum engine speed allows the use of a larger propeller and is, ultimately, significantly more efficient in terms of engine propulsion. The largest of the G engines is the G95, which has a stroke of 3460 mm and a bore of 950 mm.

In contrast to the limited number of makers in the low-speed sector, the high-/medium-speed engine market is served by numerous companies offering portfolios embracing four-stroke, trunk piston, uniflow- or loop-scavenged designs, and rotating piston types, with bore sizes up to 640 mm. Recent years have seen the development of advanced medium- and high-speed designs with high power-to-weight ratios and compact configurations up to V20 cylinders for fast commercial vessel and warship propulsion.

There are plenty of defunct builders in this sector also, but new names occasionally appear. HiMSEN, which is an in-house brand developed by Hyundai Heavy Industries, has a history that dates back only to 2001.

The future

It is difficult to envisage the diesel engine being seriously troubled by alternative prime movers in the short-to-medium term, but any regulation- or market-driven shift to much cleaner fuels (liquid or gas) could open the door to thwarted rivals, such as gas engines and gas turbines. Fuel cells will seek a shipboard foothold, initially in low auxiliary power applications.

Besides stifling coal- and oil-fired steam plant in its rise to dominance in commercial propulsion, the diesel engine shrugged off challenges from nuclear (steam) propulsion and gas turbines. Both modes found favour in warships, however, and aero-derived gas turbines have carved niches in fast ferry and cruise tonnage. A sustained challenge from gas turbines has been undermined by fuel price rises (although combined diesel and gas turbine solutions remain an option for high-powered installations) and diminishing fossil fuel acceptability may yet see nuclear propulsion revived in the longer term.

However, diesel engine makers appear untroubled by the challenges that appear from time to time and this confidence can be put down to the inherent flexibility of the diesel engine. A little over a decade ago, engine designers were involved in adapting engines to run on heavy fuel oil as a response to soaring oil prices, but in 2014 with a crash in crude oil prices, this became less pressing.

In 2018, the IMO unveiled ambitious plans to decarbonise shipping with a target of a 50% reduction in CO_2 by 2050 and phasing out of all CO_2 emissions by the end of the century. It should be stressed that these are ambitions and may never receive sufficient backing to become mandatory. Nevertheless, engine designers have been obliged to rethink strategies for the future and have begun research into alternative fuels.

There is no carbonless fuel yet ready to replace fossil fuels, but ammonia and hydrogen are the most promising options. Both MAN Energy Solutions and Wärtsilä have embarked upon researching and testing and it is expected that a vessel may be running on ammonia in two or three years' time. Hydrogen is expected to take longer to be accepted not least because it needs to be stored at extreme cryogenic temperature, whereas ammonia is more akin to LPG and LNG which are already in common use.

Bio- and synthetic fuels also hold considerable promise as these are drop in replacements for fossil fuels, but are deemed to be carbon-neutral as all emissions are absorbed by replacement sources for the next supplies of fuel. Therefore, it will be highly likely that, beyond 2050, ships will either be fitted with a dual-fuel engine that can burn a variety of 'nonpolluting' fuels or, in extreme cases, run on whatever fossil fuels are still marketed at that time.

Fossil fuels may even have a longer lifespan than some envisage as there are projects underway seeking to develop carbon capture and storage systems that could be fitted to large ships. It has to be said that there is a degree of scepticism over whether or not such a system would be practical, but the idea has attracted research funding and it could well be that a successful system is devised in the same way that exhaust gas cleaning systems that remove SOx from the exhaust stream have proven popular with around 4000 having been fitted to the largest types of vessels.

CHAPTER ONE

International regulations

Since the first marine diesel engine was fitted to a ship in 1912, any regulation has mostly been the province of the flag state and any classification society involved with the vessel. International regulation is a relatively new development and falls under the auspices of the International Maritime Organization (IMO).

There are two conventions involved: The International Convention for the Safety of Life at Sea (SOLAS) and The International Convention for the Prevention of Pollution from Ships (MARPOL). Both conventions are regularly amended and published by the IMO with appropriate date but for the purpose of this volume, it is sufficient to refer to them merely by name.

The structure of the IMO includes two main committees: the Maritime Safety Committee (MSC), which covers safety and security related matters; and the Marine Environment Protection Committee (MEPC), which addresses environmental issues under IMO's remit. Both committees have a number of subcommittees which meet at least annually and do most of the groundwork on matters which are later adopted by the two main committees.

Although engine development has been heavily influenced by both conventions, the texts themselves do not concern themselves overly with how the requirements might be achieved but merely with what the goal of the regulation is. Regulations relating to engines in SOLAS cover aspects such as ensuring availability of essential services, fire and explosion prevention, and control and safety. Examples include a requirement for double walled fuel lines and for some engines to be fitted with crankcase oil mist detectors or engine bearing temperature monitors.

SOLAS requires ships to undergo a regular cycle of special surveys at five yearly intervals with an interim survey taking place between two and 3 years after each special survey. It is usual at these surveys for major repairs and overhauls to take place, but the SOLAS rules only require that the main machinery should be in a satisfactory condition and fit for purpose.

It is the MARPOL convention which impacts most on engine development and operation. The convention has six annexes distinguished by

Pounder's Marine Diesel Engines and Gas Turbines
https://doi.org/10.1016/B978-0-08-102748-6.00001-3

Roman numerals. Annex VI is devoted to emissions to air from engines and boilers, and it is this where most regulation of engines emanates from.

Exhaust gas emissions from marine diesel engines largely comprise nitrogen, oxygen, carbon dioxide, and water vapour, with smaller quantities of carbon monoxide, oxides of sulphur (SOx) and nitrogen (NOx), partially reacted and noncombusted hydrocarbons (HC), and particulate material.

For a vessel with a typical two-stroke engine using fuel with a sulphur content of 3%, the results of the combustion process contain around 600 ppm of SOx around 1500 ppm of NOx. In addition for every tonne of fuel consumed more than 3 tonnes of CO_2 is released. There are also unburnt hydrocarbons and small amount of particulates which are considered harmful to health. After January 2020, fuel with a sulphur content above 0.5% (0.1% in ECAs) will be prohibited unless the vessel is fitted with an exhaust gas cleaning system. In such a vessel, the SOx, is converted to sulphates which are a normal component of seawater.

Nitrogen oxides (NOx)—generated thermally from nitrogen and oxygen at high combustion temperatures in the cylinder—are of special concern since they are believed to be carcinogenic and contribute to photochemical smog formation over cities and acid rain (and hence excess acidification of the soil). Internal combustion engines primarily generate nitrogen oxide but $<10\%$ of that oxidizes to nitrogen dioxide the moment it escapes as exhaust gas.

Sulphur oxides (SOx)—produced by the oxidation of sulphur in the fuel—have an unpleasant odour, irritate the mucous membrane, and are a major source of acid rain (reacting with water to form sulphurous acid). Acid deposition is a transboundary pollution problem: once emitted, SOx can be carried over hundreds of miles in the atmosphere before being deposited in lakes and streams, reducing their alkalinity.

Sulphur deposition can also lead to increased sulphate levels in soils, fostering the formation of insoluble aluminium phosphates which can cause a phosphorus deficiency. Groundwater acidification has been observed in many areas of Europe; this can lead to corrosion of drinking water supply systems and health hazards due to dissolved metals in those systems. Forest soils can also become contaminated with higher than normal levels of toxic metals, and historic buildings and monuments damaged.

HC—created by the incomplete combustion of fuel and lube oil, and the evaporation of fuel—have an unpleasant odour, are partially carcinogenic and smog forming, and irritate the mucous membrane (emissions, however, are typically low for modern diesel engines).

Carbon monoxide (CO)—resulting from the incomplete combustion due to a local shortage of air and the dissociation of carbon dioxide—is highly toxic but only in high concentrations.

Particulate matter (PM) is a complex mixture of inorganic and organic compounds resulting from incomplete combustion, partly unburned lube oil, thermal splitting of HC from the fuel and lube oil, ash in the fuel and lube oil, sulphates, and water. More than half of the total particulate mass is soot (inorganic carbonaceous particles), whose visible evidence is smoke. Soot particles (unburned elemental carbon) are not themselves toxic, but they can cause the build-up of aqueous HC, and some of them are believed to be carcinogens. Particulates constitute no more than around 0.003% of the engine exhaust gases.

The IMO has marked out Black Carbon (BC) as one of the more problematic components of PM. BC is strongly light absorbing carbonaceous material emitted as solid particulate matter created through incomplete combustion of carbon-based fuels. BC contains more than 80% carbon by mass, a high fraction of which is sp2-bonded carbon, and when emitted forms aggregates of primary spherules between 20 and 50 nm in aerodynamic diameter. BC is a potent climate forcer creating significant atmospheric warming. It absorbs solar radiation across all visible wavelengths, and freshly emitted BC has a mass absorption efficiency of $5\,m^2\,g^{-1}$ at the mid-visible wavelength of 550 nm. The strength of this light absorption varies with the composition, shape, size distribution, and mixing state of the particle. BC emissions from the international commercial shipping industry are thought to contribute about 1%–2% of global BC.

MARPOL Annex VI mentions almost all of the above exhaust components, but as of 2019 only SOx and NOx emissions are regulated. The various means for controlling the emissions are detailed in Chapter 4 of this volume.

Regulating NOx emissions

The global approach to NOx emission controls was taken by the IMO, whose Annex VI (Regulations for the Prevention of Air Pollution from Ships) to MARPOL 73/78 was adopted at a diplomatic conference in 1997. Ships burning marine diesel oil and heavy fuel oil at that time were reportedly responsible for around 7% of global NOx emissions, around 4% of global sulphur dioxide emissions, and 2% of global carbon dioxide emissions.

Ships constructed after 1 January 2000 (date of keel laying) were required to comply with the annex, which addresses engines with a power output of more than 130 kW (other than those used only for emergency purposes) installed in new ships constructed from that date and engines in existing ships that undergo a major modification.

Engines have to fulfil the NOx emission limits set by the IMO curve, which is related to engine speed (discussed later). To show compliance, an engine has to be certified according to the NOx technical code and delivered with an Engine International Air Pollution Prevention (EIAPP) certificate of compliance. The certification process includes NOx measurement for the engine type concerned, stamping of components thatvaffect NOx formation, and a technical file that is delivered with the engine.

NOx technical code-certified engines have a technical file, which includes the applicable survey regime, termed the on-board NOx verification procedure. The associated parameter check method effectively stipulates the engine components and range of settings to be adopted to ensure that NOx emissions from the given engine, under reference conditions, will be maintained within the certified value. These NOx emission critical components are broadly divided into three groups: the combustion chamber (including piston, cylinder cover, and liner); the charge air system (turbocharger and charge air cooler); and the fuel injection system (fuel pump, injection nozzle, and timing cam). The key settings are either the maximum permitted combustion pressures across the load range or the fuel injection timing.

Consequently, although this NOx certification still allows for an engine to be operated at any load point within the available range, including that corresponding to slow steaming, it does not severely restrict the ability to have the engine reoptimized for lower powers than used for the initial certification. This is particularly apparent in the case of the fuel oil injection nozzle, a significant NOx-critical component but one which previously would have been readily changed for a 'slow steaming' version with no wider implications.

For a NOx-certified engine, however, such alternative components can be fitted only if approved within the requirements of the NOx technical code. In the case of alternative fuel injection nozzles, the process would require running a full emissions test, normally calling for input from an engine builder. Ship operators must therefore ensure that any changes to

NOx-critical components or settings are duly approved prior to installation to avoid invalidating an engine's certification.

An alternative route is to change from the parameter check method for the on-board NOx verification procedure to the direct measurement and monitoring method, which checks inputs of fuel and air to the engine and its outputs of power and exhaust emissions. Under such an arrangement, the NOx-critical components and settings may be changed, provided that the engine is shown by the direct method to remain compliant with the relevant NOx emission limit.

IMO's current maximum allowable NOx emission levels depend on the speed category of the engine and range from 17 g/kW h for engines of speed <130 rev/min to 9.84 g/kW h for engines of speed >2000 rev/min (Fig. 1.1). Much tougher curbs on NOx and other emissions are set by regional authorities such as California's Air Resources Board, and Sweden has introduced a system of differentiated port and fairway dues, making ships with higher NOx emissions pay higher fees than more environment-friendly tonnage of a similar size.

Amendments agreed by the IMO in 2008 set progressively tighter NOx emission standards for new engines, depending on the date of their installation and engine speed:

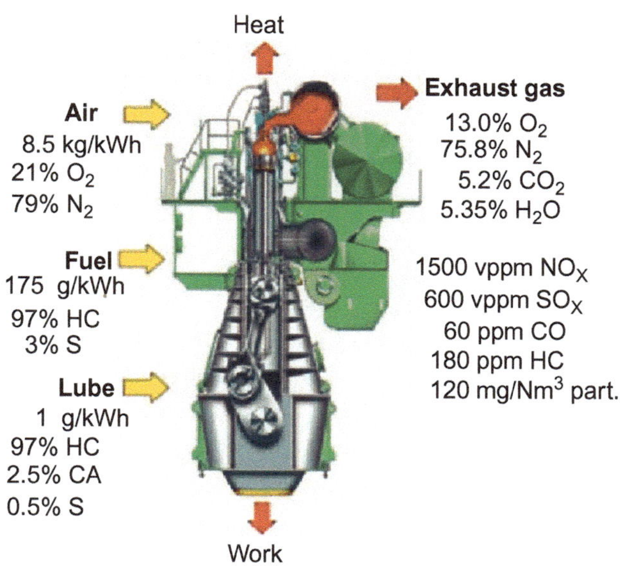

Fig. 1.1 Typical two-stroke engine combustion products.

MARPOL Annex VI NOx emission limits

		NOx limit (g/kWh)		
Tier	**Date**	$n<130$	$130\le n<2000$	$n\ge 2000$
Tier I	2000	17.0	$45\times n^{-0.2}$	9.8
Tier II	2011	14.4	$44\times n^{-0.23}$	7.7
Tier III	2016[a]	3.4	$9\times n^{-0.2}$	1.96

[a]In NOx emission control areas (Tier II standards apply outside ECAs). Exact date depends on ECA establishment.
n = engine rpm.

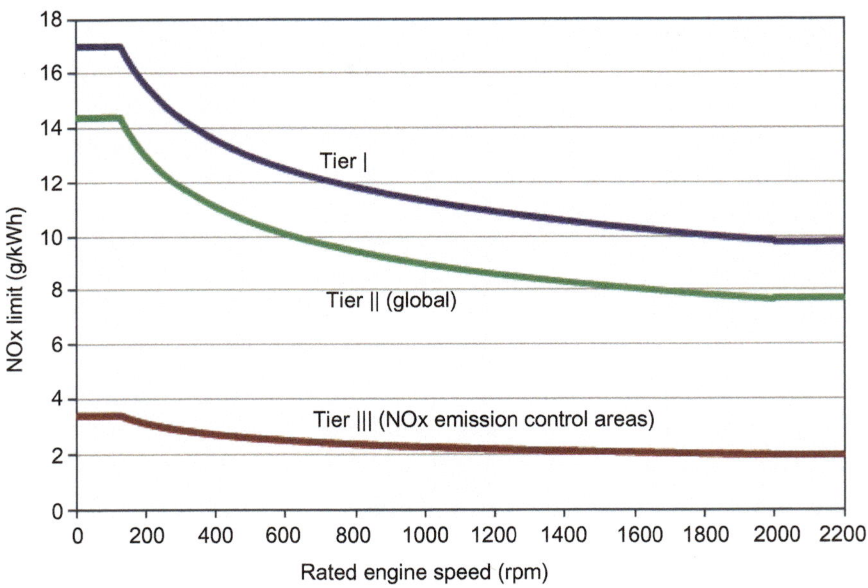

When the amendments above were adopted, there were no designated NOx emission control areas defined. However, in 2010, the North American ECA covering most of the US and Canadian waters was established becoming effective in 2012. In 2011, the US Caribbean ECA, including Puerto Rico and the US Virgin Islands, was established to take effect in 2014. Both of these ECAs regulate emissions of NOx and SOx.

Before the establishment of the American ECAs, two sulphur emission control areas (SECAs) had been established in Northern Europe. The Baltic Sea SECA established in 1997 entered into force in 2005, and the North Sea SECA established in 2005 came into effect in 2006.

In 2017, the IMO agreed that these two areas would also in future include the Tier III limits on NOx. Accordingly, the Tier III levels will apply to ships constructed (keel laying) on or after 1 January 2021, or existing ships which replace an engine with "nonidentical" engines or instal an "additional" engine.

There are moves to introduce a new ECA in the Mediterranean Sea with the possibility that this will be debated at the IMO in 2020. It is expected that the EU will be a strong supporter, but the final decision may require a unanimous acceptance by the 22 states that are signatories to the Barcelona Convention aimed at protecting the Mediterranean Sea environment. A future ECA in Chinese waters has also been proposed, but no dates have yet been put forward as to when this might be requested.

With stricter controls planned by the IMO, the reduction of NOx emissions remains a priority for engine designers whose concern is to secure environmental acceptability without compromising the impressive gains in engine fuel economy and reliability achieved in recent years.

Regulating SOx emissions

After setting NOx reduction targets, the IMO next set about tackling pollution by SOx.

Compared with land-based power installations, the fuel oil burned by much of shipping has a high sulphur content and contributes significantly to the overall amount of global SOx emissions at sea and in port areas. Studies on sulphur pollution showed that in 1990, SOx emissions from ships contributed around 4% to the total in Europe. In 2001, such emissions represented around 12% of the total.

SOx emissions in diesel engine exhaust gas—which mostly comprise sulphur dioxide with a small amount of sulphur trioxide—are a function of the amount of sulphur in the fuel and cannot be controlled by the combustion process. If the fuel contains 3% sulphur, for example, the volume of SOx generated is around 64 kg per tonne of fuel burned; if fuel with 1% sulphur content is used, SOx emissions amount to around 21 kg per tonne of fuel burned.

Since the formation of SOx depends on the fuel rather than the engine, the IMO regulations place limits on the sulphur content of the fuel. First introduced in 2005, the limits were subject to a gradual tightening in the same way as the NOx regulations. Two limits on permissible sulphur levels were set, one for ECAs and another for areas outside of any ECA.

In 2005, when the rules first took effect, the limits were:

- a global heavy fuel oil sulphur content cap of 4.5% and
- a fuel sulphur limit of 1.5% in SECAs such as the Baltic Sea, North Sea, and English Channel.

In 2008, the IMO approved further amendments to curb SOx emissions:

- The fuel sulphur limit applicable in ECAs from 1 March 2010 would be 1% (10,000 ppm), reduced from the existing 1.5% content (15,000 ppm).
- The global fuel sulphur cap would be reduced to 3.5% (35,000 ppm), reduced from the existing 4.5% (45,000 ppm), effective from 1 January 2012.
- The fuel sulphur limit applicable in ECAs from 1 January 2015 would be 0.1% (1000 ppm).
- The global fuel sulphur cap would be reduced to 0.5% (5000 ppm) effective from 1 January 2020, subject to a feasibility review to be completed no later than 2018, otherwise 1 January 2025.

The review was duly carried out in 2016 and despite some dissenting arguments, the 1 January 2020 date was selected for the reduction in the global cap to 0.5%.

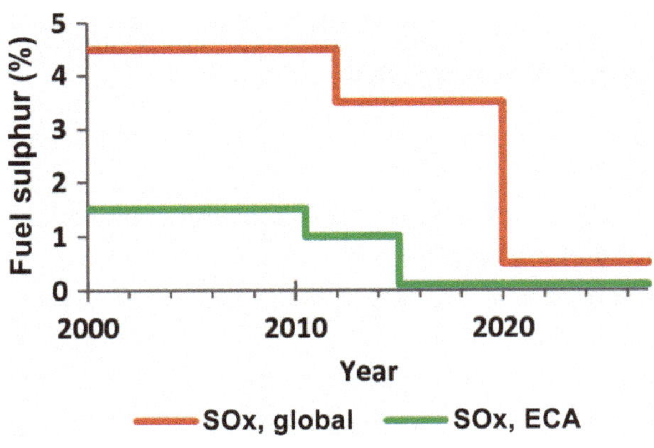

Under pressure from the shipping industry, the IMO agreed before introducing the sulphur limits to allow for alternative means to meet the SOx reduction targets. Although there were at that time no commercially available means to do this, the idea was that technology from shore power stations might be able to be adapted to remove the SOx from the exhaust stream.

The IMO devised an equivalence formula based on monitoring the level of SOx and CO_2 in the exhaust stream. The original table showing the equivalence figures has now been made obsolete, and, as from 1 January 2020, only two of the original six equivalence figures are relevant. In an

ECA, the ratio of SO_2 measured in ppm to CO_2 measured as %v/v must not exceed 4.3 and outside an ECA must not exceed 21.7.

In systems where the exhaust gas cleaning process may affect the amount of CO_2 in the exhaust gases, the CO_2 concentration is to be measured before the scrubber, and the SO_2 concentration after it, to calculate the ratio correctly.

The exhaust gas cleaning systems devised to remove SOx are commonly referred to as scrubbers and are marinized versions of systems commonly found in power plants and other industries ashore.

Early take-up of scrubbers was low and mainly confined to ferries and cruise ships operating for long periods in ECAs. However, with the introduction of the new 0.5% global sulphur cap in 2020, many more ship types have been fitted with them. It is estimated that there are now more than 3000 scrubber fitted ships, and this is expected to grow in the future.

There are two types of scrubber—wet or dry. In a wet scrubber, the exhaust gas flow is brought into intensive contact with an alkaline fluid (both raw seawater and fresh water mixed with caustic soda can be used for this purpose). In a dry system, the exhaust gas is made to come into contact with either hydrated lime or sodium bicarbonate where it reacts chemically to form new chemical compounds that can be disposed of ashore.

Exhaust gas scrubbers, however, are heavy and require significant installation space, and if of the wet type they produce large quantities of wastewater and sludge. The water has to be properly treated before discharge into the sea, and the sludge must be disposed of at reception facilities ashore.

The simplest approach to reducing SOx emissions is to burn bunkers with a low sulphur content. (If a selective catalytic reduction [SCR] system is installed to achieve the lowest NOx emission levels—see the next section—then lower sulphur fuels may be dictated anyway to avoid premature fouling of the system's catalyst package.) The drastic reduction or complete elimination of sulphur from diesel fuels also contributes to reduced formation of PM.

Regulating other exhaust emissions

As of 1 January 2020, NOx and SOx are the only regulated elements of the exhaust stream. The IMO has been debating the issues of PM and BC without yet coming to anything like consensus on whether regulation is desirable or even possible.

Because much of the PM in exhausts is related to the sulphur content, the lower permitted level of sulphur in fuels goes someway to reducing it. Ships equipped with wet scrubber systems will release far less PM into the atmosphere because much of it will be trapped in the wash water and form the sludge that is disposed of later.

Some 6% of the exhaust gas emissions from marine engines is carbon dioxide. Although not itself toxic, carbon dioxide contributes to the greenhouse effect (global warming and climate change). This gas is an inevitable product of combustion of all fossil fuels, but emissions from diesel engines—thanks to their thermal efficiency—are the lowest of all heat engines. A lower fuel consumption translates to reduced carbon dioxide emissions since the amount produced is directly proportional to the volume of fuel used and, therefore, to the engine or plant efficiency. As a rough guide, burning 1 t of diesel fuel produces approximately 3.25 t of carbon dioxide.

International concern over the atmospheric effect of carbon dioxide has stimulated measures and plans to curb the growth of such emissions, and the marine industry has not been immune.

Under international agreements, such as the Kyoto Protocol and the European Union's accord on greenhouse gases, many governments are committed to substantial reductions in total emissions of carbon dioxide. (A switch from other transport modes—air, road, and rail—to shipping would nevertheless yield a substantial overall reduction in emissions of the greenhouse gas because of the higher efficiency of diesel engines.)

There are currently no mandatory limits on carbon dioxide emissions from engines, although the IMO does have a mandatory system in place for measurement of the amount produced by ships. There is also a mandatory system in place whereby newbuild ships are obliged to improve their efficiency relative to a determined baseline. This is known as the Energy Efficiency Design Index (EEDI).

Under EEDI rules, ships are measured to a complex formula which attempts to calculate the amount of carbon dioxide produced per tonne/mile of work done using the carrying capacity of the ship. While a ship's EEDI rating must be at or below a fixed figure, how this is achieved is left open and the ship can take advantage of energy saving measures such as battery energy storage systems, hull air lubrication systems, and more.

In 2018, The IMO adopted an ambitious target of decarbonization of the shipping industry. The aim is to reduce total emissions from shipping by 50% in 2050, and to reduce the average carbon intensity by 40% in 2030 and 70% in 2050, compared to 2008.

The scope for cutting carbon dioxide emissions by raising the already high-efficiency level of modern diesel engines is limited and other routes have to be pursued: for example, operating the engines at a fuel-saving service point; adopting diesel–electric propulsion (the engines can be run continuously at the highest efficiency) and exploiting advanced waste heat recovery systems and diesel combined cycles incorporating a steam turbine; using alternative fuels with lower carbon content.

The highest reduction targets set by the IMO will not be reachable by using fuels such as LNG or methanol where these are derived from fossil sources. However, regulators in some countries or regions are prepared to consider some so-called biofuels derived from plant, animal, and waste matter as being carbon neutral. There are biofuel equivalents of methane and different types of fuel oils. In practice, these fuels emit similar or identical amounts of carbon dioxide as do fossil fuels, but because the carbon dioxide released is expected to be reabsorbed by the next season's crops and by future generations of animals, it is not considered as adding to the greenhouse gas effect. Whether or not this dispensation for biofuels is accepted internationally remains to be seen. There are some objections to using valuable food resources for fuel production purposes.

Research is being done into use of alternative fuels for diesel engines. One which seems to offer potential is ammonia which has a high level of hydrogen but no carbon. Other projects are exploring the manufacture of synthetic fuels such as methanation using carbon dioxide captured from the air and hydrogen to create methane.

The intensity or R&D being put into these projects bodes well for diesel engines as a means of providing motive power to ships. If they are not successful and if the direction of regulation is not reversed or slowed, then the future of the diesel engine could be limited.

CHAPTER TWO

Theory and general principles

Theoretical heat cycle

In the original patent by Rudolf Diesel, the diesel engine operated on the diesel cycle in which heat was added at constant pressure. This was achieved by the blast injection principle. Today 'diesel' is universally used to describe any reciprocating engine in which the heat induced by compressing air in the cylinders ignites a finely atomized spray of fuel. This means that the theoretical cycle on which the modern diesel engine works is better represented by the dual or mixed cycle, which is diagrammatically illustrated in Fig. 2.1. The area of the diagram, to a suitable scale, represents the work done on the piston during one cycle.

Starting from point C, the air is compressed adiabatically to point D. Fuel injection begins at D, and heat is added to the cycle partly at constant volume as shown by vertical line DP, and partly at constant pressure, as shown by horizontal line PE. At point E expansion begins. This proceeds adiabatically

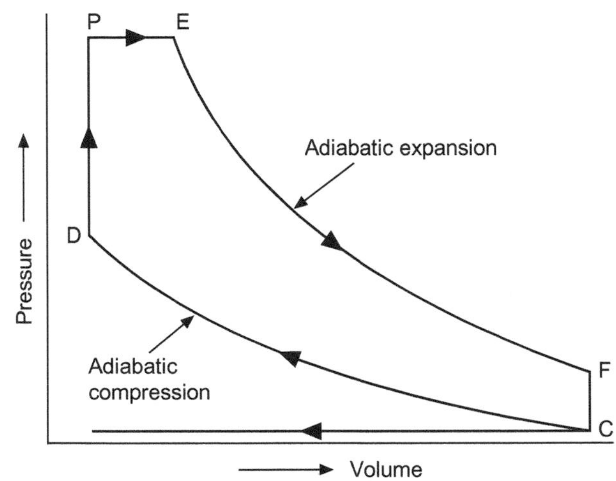

Fig. 2.1 Theoretical heat cycle of true diesel engine.

https://doi.org/10.1016/B978-0-08-102748-6.00002-5

to point F when the heat is rejected to exhaust at constant volume as shown by vertical line FC.

The ideal efficiency of this cycle (i.e. the hypothetical indicator diagram) is about 55%–60%: that is to say, about 40%–45% of the heat supplied is lost to the exhaust. Since the compression and expansion strokes are assumed to be adiabatic, and friction is disregarded, there is no loss to coolant or ambient. For a four-stroke engine, the exhaust and suction strokes are shown by the horizontal line at C, and this has no effect on the cycle.

Practical cycles

While the theoretical cycle facilitates simple calculation, it does not exactly represent the true state of affairs. This is because:

1. The manner in which, and the rate at which, heat is added to the compressed air (the heat release rate) is a complex function of the hydraulics of the fuel injection equipment and the characteristic of its operating mechanism; of the way the spray is atomized and distributed in the combustion space; of the air movement at and after top dead centre (ATDC) and to a degree also of the qualities of the fuel.
2. The compression and expansion strokes are not truly adiabatic. Heat is lost to the cylinder walls to an extent, which is influenced by the coolant temperature and by the design of the heat paths to the coolant.
3. The exhaust and suction strokes on a four-stroke engine (and the appropriate phases of a two-stroke cycle) do create pressure differences which the crankshaft feels as 'pumping work'.

It is the designer's objective to minimize all of these losses without prejudicing the first cost or reliability, and also to minimize the cycle loss: that is, the heat rejected to exhaust. It is beyond the scope of this book to derive the formulae used in the theoretical cycle, and in practice, designers have at their disposal sophisticated computer techniques, which are capable of representing the actual events in the cylinder with a high degree of accuracy. But broadly speaking, the cycle efficiency is a function of the compression ratio (or more correctly the effective expansion ratio of the gas–air mixture after combustion).

The theoretical cycle (Fig. 2.1) may be compared with a typical actual diesel indicator diagram such as that shown in Fig. 2.2. Note that in higher speed engines, combustion events are often represented on a crank angle, rather than a stroke basis, in order to achieve better accuracy in portraying events at the top dead centre (TDC) as shown in Fig. 2.3. The actual

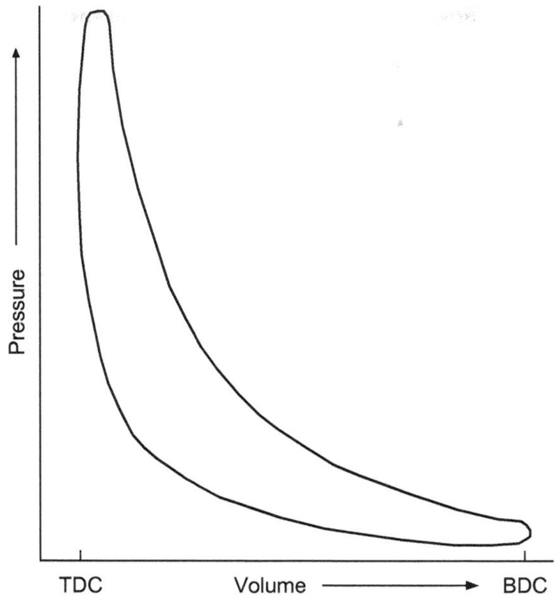

Fig. 2.2 Typical indicator diagram (stroke-based).

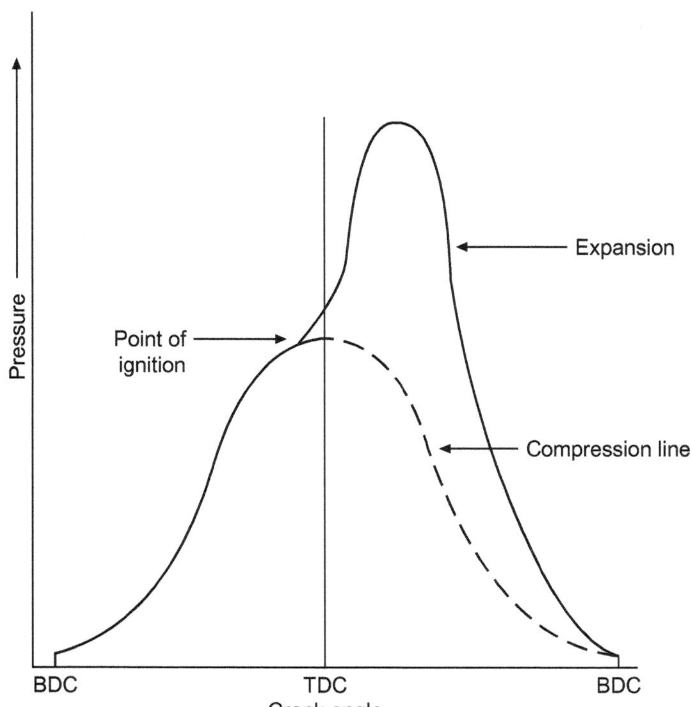

Fig. 2.3 Typical indicator diagram (crank angle-based).

indicator diagram is derived from it by transposition. This form of diagram is useful too when setting injection timing. If electronic indicators are used, it is possible to choose either form of a diagram.

An approximation to a crank angle-based diagram can be made with mechanical indicators by disconnecting the phasing and taking a card quickly, pulling it by hand: this is termed a 'draw card'.

Efficiency

The only reason a practical engineer wants to run an engine at all is to achieve the desired output of useful work, which is, for our present purposes, to drive a ship at a prescribed speed, and/or to provide electricity at a prescribed kilowattage.

To determine this power, he or she must, therefore, allow not only for the cycle losses mentioned earlier but also for the friction losses in the cylinders, bearings, and gearing (if any) together with the power consumed by engine-driven pumps and other auxiliary machines. He or she must also allow for such things as windage. The reckoning is further complicated by the fact that the heat rejected from the cylinder to exhaust is not necessarily totally lost, as practically all modern engines use up to 25% of that heat to drive a turbocharger. Many use much of the remaining high-temperature heat to raise steam and use low-temperature heat for other purposes.

The detail is beyond the scope of this book but a typical diagram (usually known as a Sankey diagram), representing the various energy flows through a modern diesel engine, is reproduced in Fig. 2.4. The right-hand side represents a turbocharged engine, and an indication is given of the kind of interaction between the various heat paths as they leave the cylinders after combustion.

Note that the heat released from the fuel in the cylinder is augmented by the heat value of the work done by the turbocharger in compressing the intake air. This is apart from the turbocharger's function in introducing the extra air needed to burn an augmented quantity of fuel in a given cylinder, compared with what the naturally aspirated system could achieve, as in the left-hand side of the diagram.

It is the objective of the marine engineer to keep the injection settings, the airflow, and coolant temperatures (not to mention the general mechanical condition) at those values, which give the best fuel consumption for the power developed.

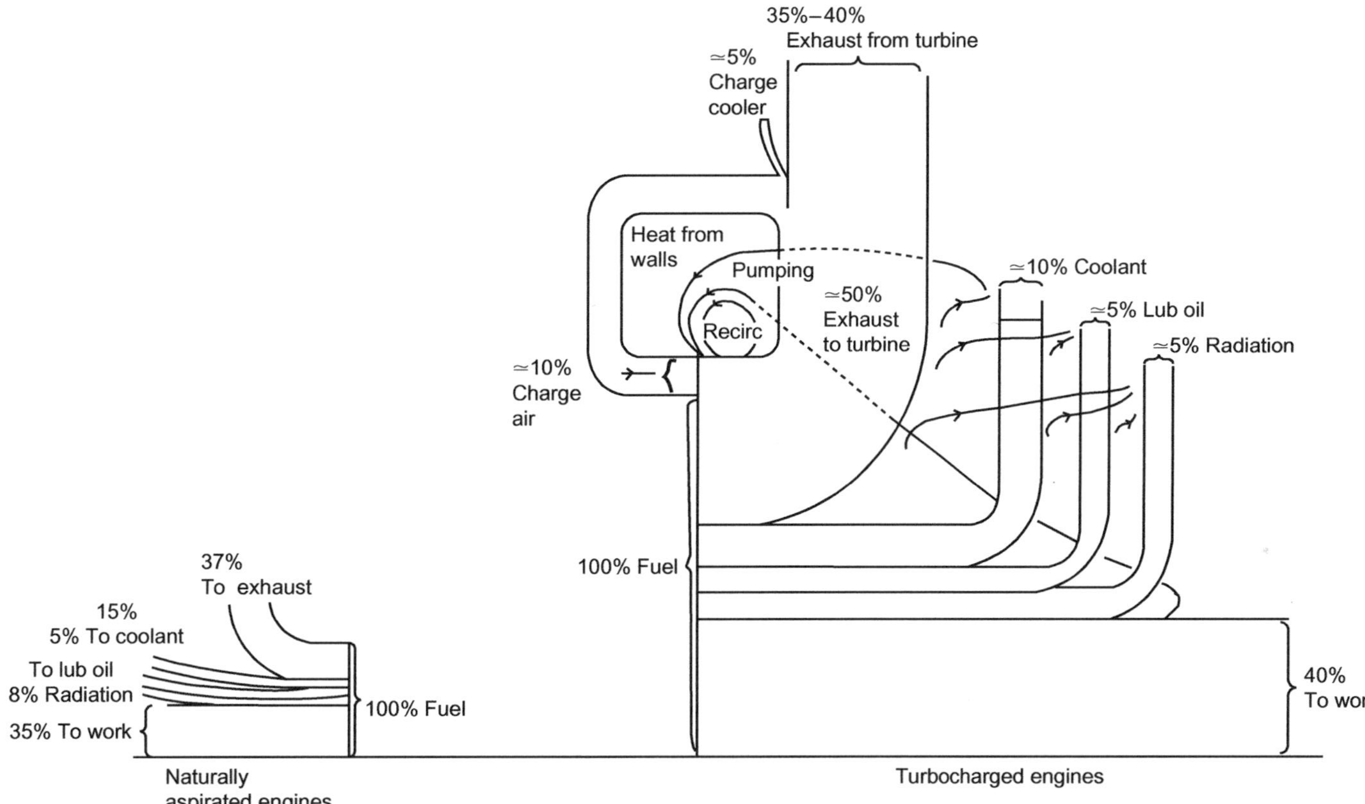

Fig. 2.4 Typical Sankey diagrams.

Note also that, whereas the fuel consumption is not difficult to measure in tonnes per day, kilograms per hour, or in other units, there are many difficulties in measuring work done in propelling a ship. This is because the propeller efficiency is influenced by the entry conditions created by the shape of the afterbody of the hull, by cavitation, and so on and also critically influenced by the pitch setting of a controllable pitch propeller. The resulting speed of the ship is dependent, of course, on hull cleanliness, wind and sea conditions, draught, and so on. Even when driving a generator, it is necessary to allow for generator efficiency and instrument accuracy.

It is normal when defining efficiency to base the work done on that transmitted to the driven machinery by the crankshaft. In a propulsion system, this can be measured by a torsionmeter; in a generator, it can be measured electrically. Allowing for measurement error, these can be compared with figures measured on a brake in the test shop.

Thermal efficiency

Thermal efficiency (Thη) is the overall measure of performance. In absolute terms, it is equal to:

$$\frac{\text{Heat converted into useful work}}{\text{Total heat supplied}} \tag{2.1}$$

As long as the units used to agree, it does not matter whether the heat or work is expressed in pounds feet, kilograms metres, BTU, calories, kilowatt hour, or joules. The recommended units to use now are those of the SI system.

$$\begin{aligned}\text{Heat converted into work per hour} &= \text{NkWh}\\ &= 3600\,\text{NkJ}\end{aligned}$$

where N is the power output in kilowatts

$$\text{Heat supplied} = M \times K$$

where M is the mass of fuel used per hour in kilograms and K the calorific value of the fuel in kilojoules per kilograms

$$Th\eta = \frac{3600N}{M \times K} \tag{2.2}$$

It is now necessary to decide where the work is to be measured. If it is to be measured in the cylinders, as is usually done in slow-running machinery, by means of an indicator (though electronic techniques now make this possible directly and reliably even in high-speed engines), the work measured (and hence power) is that indicated within the cylinder, and the calculation leads to the *indicated* thermal efficiency.

If the work is measured at the crankshaft output flange, it is net of friction, auxiliary drives, etc., and is what would be measured by a brake, whence the term *brake* thermal efficiency. (Manufacturers in some countries do include as output the power absorbed by essential auxiliary drives, but some consider this to give a misleading impression of the power available.)

Additionally, the fuel is reckoned to have a higher (or gross) and a lower (or net) calorific value (LCV), according to whether one calculates the heat recoverable if the exhaust products are cooled back to standard atmospheric conditions, or assessed at the exhaust outlet. The essential difference is that in the latter case the water produced in combustion is released as steam and retains its latent heat of vaporization. This is the more representative case—and more desirable as water in the exhaust flow is likely to be corrosive. Today the net calorific value or LCV is more widely used.

Returning to our formula (Eq. 2.2), if we take the case of an engine producing a (brake) output of 10,000 kW for an hour using 2000 kg of fuel per hour having an LCV of 42,000 kJ/kg,

$$
\begin{aligned}
(\text{Brake})\ Th\eta &= \frac{3600 \times 10,000}{2000 \times 42,000} \times 100\% \\
&= 42.9\% (\text{based on LCV})
\end{aligned}
$$

Mechanical efficiency

$$
\text{Mechanical efficiency} = \frac{\text{output at crankshaft}}{\text{output at cylinders}} \tag{2.3}
$$

$$
= \frac{\text{bhp}}{\text{ihp}} = \frac{\text{kW (brake)}}{\text{kW (indicated)}} \tag{2.4}
$$

The reasons for the difference are listed earlier. The brake power is normally measured with high accuracy (98% or so) by coupling the engine to a dynamometer at the builder's works. If it is measured in the ship by torsionmeter, it is difficult to match this accuracy and, if the torsionmeter cannot be installed between the output flange and the thrust block or the gearbox input, additional losses have to be reckoned due to the friction entailed by these components.

The indicated power can only be measured from diagrams where these are feasible, and they are also subject to significant measurement errors.

Fortunately for our attempts to reckon the mechanical efficiency, testbed experience shows that the 'friction' torque (i.e. in fact, *all* the losses reckoned to influence the difference between indicated and brake torques) is not very greatly affected by the engine's torque output, nor by the speed. This means that the friction power loss is roughly proportional to speed, and fairly constant at fixed speed over the output range. Mechanical efficiency, therefore, falls more and more rapidly as brake output falls. It is one of the reasons why it is undesirable to let an engine run for prolonged periods at less than about 30% torque.

Working cycles

A diesel engine may be designed to work on the two-stroke or four-stroke cycle—both of these are explained below. They should not be confused with the terms 'single-acting' or 'double-acting', which relate to whether the working fluid (the combustion gases) acts on one or both sides of the piston. (Note, incidentally, that the few opposed-piston two-stroke engines in service today are *single-acting*.)

The four-stroke cycle

Fig. 2.5 shows diagrammatically the sequence of events throughout the typical four-stroke cycle of two revolutions. It is usual to draw such diagrams starting at TDC (firing), but the explanation will start at TDC (scavenge). TDC is sometimes referred to as an inner dead centre (IDC).

Proceeding clockwise around the diagram, both inlet (or suction) and exhaust valves are initially open. (All modern four-stroke engines have poppet valves.) If the engine is naturally aspirated or is a small high-speed type with a centripetal turbocharger, the period of valve overlap (i.e. when both valves are open) will be short, and the exhaust valve will close some 10° ATDC.

Propulsion engines and the vast majority of auxiliary generator engines running at speeds below 1000 rev/min will almost certainly be turbocharged and will be designed to allow a generous throughflow of scavenge air at this point in order to control the turbine blade temperature (see also Chapter 10). In this case, the exhaust valve will remain open until exhaust valve closure (EVC) at 50–60° ATDC. As the piston descends to the outer or bottom dead centre (BDC) on the suction stroke, it will inhale a fresh charge of air.

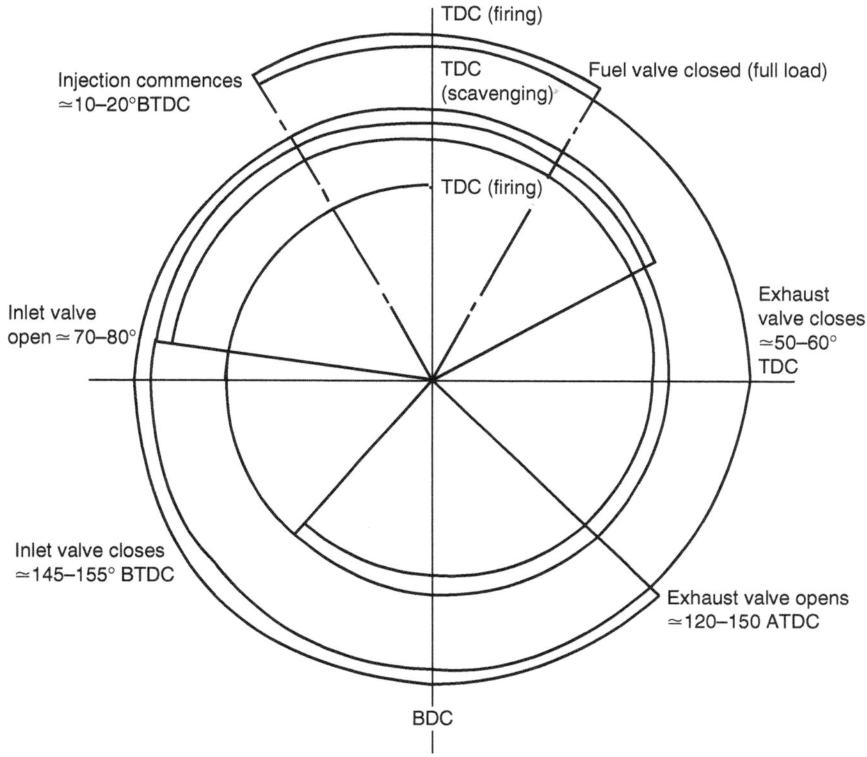

Fig. 2.5 Four-stroke cycle.

To maximize this, balancing the reduced opening as the valve seats against the slight ram or inertia effect of the incoming charge, the inlet (suction valve) will normally be held open until about 25–35° after BDC (ABDC) (145–155° BTDC). This event is called inlet valve closure (IVC).

The charge is then compressed by the rising piston until it has attained a temperature of some 550°C. At about 10–20° BTDC (before top dead centre) (firing), depending on the type and speed of the engine, the injector admits finely atomized fuel which ignites within 2–7° (depending on type again) and the fuel burns for 30–50° while the piston begins to descend on the expansion stroke, the piston movement usually helping to induce air movement to assist combustion.

At about 120–150° ATDC the exhaust valve opens (EVO), the timing is chosen to promote a very rapid blow-down of the cylinder gases to exhaust. This is done to (a) preserve as much energy as is practicable to drive the turbocharger and (b) reduce the cylinder pressure to a minimum by BDC to reduce pumping work on the 'exhaust' stroke. The rising piston expels the remaining exhaust gas and at about 70–80° BTDC, the inlet valve opens (IVO) so that the inertia of the outflowing gas, and the positive pressure difference, which usually exists across the cylinder by now, produces a throughflow of air to the exhaust to 'scavenge' the cylinder.

If the engine is naturally aspirated, the IVO is about 10° BTDC. The cycle now repeats.

The two-stroke cycle

Fig. 2.6 shows the sequence of events in a typical two-stroke cycle, which, as the name implies, is accomplished in one complete revolution of the crank. Two-stroke engines invariably have ports to admit air when uncovered by the descending piston (or air piston where the engine has two pistons per cylinder). The exhaust may be via ports adjacent to the airports and controlled by the same piston (loop scavenge) or via piston-controlled exhaust ports or poppet exhaust valves at the other end of the cylinder (uniflow scavenge). The principles of the cycle apply in all cases.

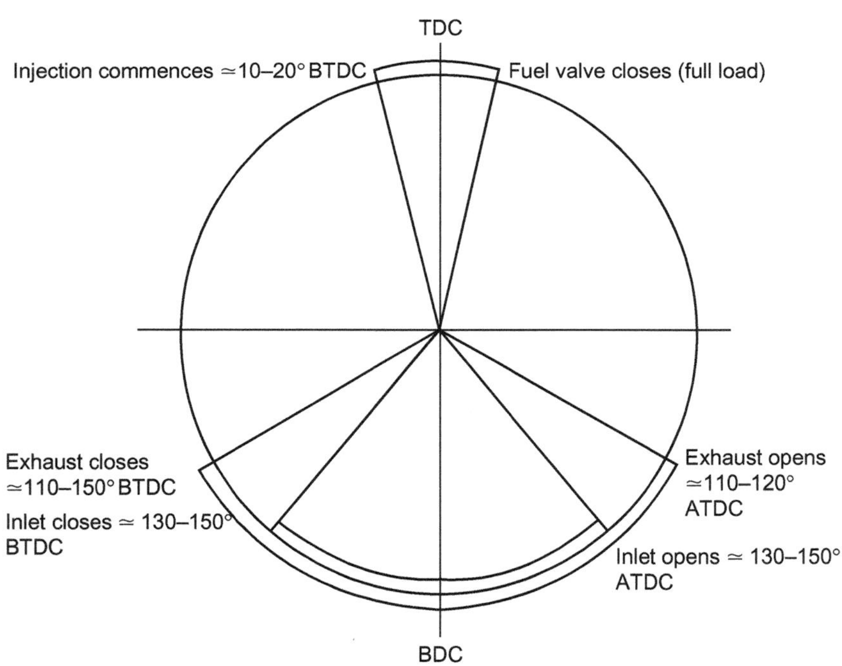

Fig. 2.6 Two-stroke cycle.

Starting at TDC, combustion is already underway and the exhaust opens (EO) at 110–120° ATDC to promote a rapid blow-down before the inlet opens (IO) about 20–30° later (130–150° ATDC). In this way, the inertia of the exhaust gases—moving at about the speed of sound—is contrived to encourage the incoming air to flow quickly through the cylinder with a minimum of mixing, because any unexpelled exhaust gas detracts from the weight of air entrained for the next stroke.

The exhaust should close before the inlet on the compression stroke to maximize the charge, but the geometry of the engine may prevent this if the two events are piston controlled. It can be done in an engine with exhaust valves, but otherwise, the inlet closure (IC) and exhaust closure (EC) in a single-piston engine will mirror their opening. The IC may be retarded relative to EC in an opposed-piston engine to a degree which depends on the ability of the designer and users to accept greater out-of-balance forces.

At all events, the inlet ports will be closed as many degrees ABDC as they opened before it (i.e. again 130–150° BTDC) and the exhaust in the same region. Where there are two cranks and they are not in phase, the timing is usually related to that coupled to the piston controlling the airports. The dephasing is described as 'exhaust lead'.

Injection commences at about 10–20° BTDC depending on speed, and combustion occurs over 30–50°, as with the four-stroke engine.

Horsepower

Despite the introduction of the SI system, in which power is measured in kilowatts, horsepower cannot yet be discarded altogether. Power is the rate of doing work. In a linear measure, it is the mean force acting on a piston multiplied by the distance it moves in a given time. The force here is the mean pressure acting on the piston. This is obtained by averaging the difference in pressure in the cylinder between corresponding points during the compression and expansion strokes. It can be derived by measuring the area of an indicator diagram and dividing it by its length. This gives naturally the indicated mean effective pressure (imep), also known as mean indicated pressure (mip). Let this be denoted by 'p'.

Mean effective pressure is mainly useful as a design shorthand for the severity of the loading imposed on the working parts by combustion. In that context, it is usually derived from the horsepower. If the latter is 'brake' horsepower (bhp), the mep derived is the brake mean effective pressure (bmep); but it should be remembered that it then has no direct physical significance of its own.

To obtain the total force, the mep must be multiplied by the area on which it acts. This, in turn, comprises the area of one piston, $\alpha = \pi d^2/4$, multiplied by the number of cylinders in the engine, denoted by N. The distance moved per cycle by the force is the working stroke (l), and for the chosen time unit, the total distance moved is the product of $l \times n$, where n is the number of *working* strokes in one cylinder in the specified time. Gathering all these factors gives the well-known 'plan' formula:

$$\text{Power} = \frac{p \times l \times a \times n \times N}{k} \tag{2.5}$$

The value of the constant k depends on the units used. The units must be consistent as regards force, length, and time.

If, for instance, SI units are used (newtons, metres, and seconds), k will be 1000 and the power will be given in kilowatts.

If imperial units are used (pounds, feet, and minutes), k will be 33,000 and the result is in imperial horsepower.

Onboard ship the marine engineer's interest in the above formula will usually be to relate mep and power for the engine with which he or she is directly concerned. In that case l, a, n become constants as well as k and

$$\text{Power} = p \times C \times N \tag{2.6}$$

where $C = (l \times a \times n)/k$.

Note that in an opposed-piston engine 'l' totals the sum of the strokes of the two pistons in each cylinder. Applying the formulae to double-acting engines is somewhat more complex since, for instance, allowance must be made for the piston rod diameter. Where double-acting engines are used, it would be advisable to seek the builder's advice about the constants to be used.

Torque

The formula for the power given in Eqs (2.5) and (2.6) is based on the movement of the point of application of the force on the piston in a straight line. The inclusion of the engine speed in the formula arises in order to take into account the total distance moved by the force(s) along the cylinder(s): that is, the number of repetitions of the cycle in unit time. Alternatively, power can be defined in terms of rotation.

If F is the effective resulting single force assumed to act tangentially at a given radius from the axis of the shaft about which it acts, r the nominated radius at which F is reckoned, and n revolutions per unit time of the shaft specified, then the circumferential distance moved by the tangential force in unit time is $2\pi rn$.

Hence,

$$\begin{aligned}\text{Power} &= \frac{F \times 2\pi rn}{K} \\ &= \frac{Fr \times 2\pi n}{K}\end{aligned} \tag{2.7}$$

The value of K depends on the system of units used, which, as before, must be consistent. In this expression $F \times r = T$, the torque acting on the shaft, and is measured (in SI units) in newton metres.

Note that T is constant irrespective of the radius at which it acts. If n is in revolutions per minute, and power is in kilowatts, the constant $K = 1000$, so that

$$\begin{aligned}T &= \frac{\text{power} \times 60 \times 1000}{2\pi n} \\ &= \frac{30{,}000 \times \text{power}}{\pi n}\ (\text{Nm})\end{aligned} \tag{2.8}$$

If the drive to the propeller is taken through gearing, the torque acts at the pitch circle diameter of each of the meshing gears. If the pitch circle diameters of the input and output gears are, respectively, d_1 and d_2 and the speeds of the two shafts are n_1 and n_2, the circumferential distance travelled by a tooth on either of these gears must be $\pi d_1 \times n_1$ and $\pi d_2 = n_2$, respectively. But since they have meshed $\pi d_1 n_1 = \pi d_2 n_2$.

Therefore

$$\frac{n_1}{n_2} = \frac{d_2}{d_1}$$

The tangential force F on two meshing teeth must also be equal. Therefore, the torque on the input wheel

$$T_1 = \frac{Fd_1}{2}$$

and the torque on the output wheel

$$T_2 = \frac{Fd_2}{2}$$

or

$$\frac{T_1}{T_2} = \frac{d_1}{d_2}$$

If there is more than one gear train, the same considerations apply to each. In practice, there is a small loss of torque at each train due to friction, etc. This is usually of the order of 1.5%–2% at each train.

Mean piston speed

This parameter is sometimes used as an indication of how highly rated an engine is. However, although in principle a higher piston speed can imply a greater degree of stress and wear, in modern practice, the lubrication of piston rings and liner, as well as other rubbing surfaces, has become much more scientific. It no longer follows that a 'high' piston speed is of itself more detrimental than a lower one in a well-designed engine.

$$\text{Mean piston speed is simply} \frac{l \times n}{30} \tag{2.9}$$

This is given in metres per second if l = stroke in metres and n = revolutions per minute.

Fuel consumption in 24H

In SI units:

$$W = \frac{w \times \text{kW} \times 24}{1000} \tag{2.10}$$

$$w = \frac{1000\,W}{\text{kW} \times 24} \tag{2.11}$$

where w is the fuel consumption rate (kg/kWh)

W = total fuel consumed per day in tonnes

(1 t = 1000 kg).

Vibration

Many problems have their roots in, or manifest themselves as, vibration. Vibration may be in any linear direction, and it may be rotational (torsional). Vibration may be resonant, at one of its natural frequencies, or forced. It may affect any group of components or any single one. It can occur at any frequency up to those which are more commonly called noise.

That vibration failures are less dramatic now than formerly is due to the advances in our understanding of vibration. It can be controlled, once recognized, by design and correct maintenance, by minimizing at source, by damping, and by arranging to avoid exciting resonance. Vibration is a very complex subject and all that will be attempted here is a brief outline.

Any elastically coupled shaft or other systems will have one or more natural frequencies, which, if excited, can build up to an amplitude which is perfectly capable of breaking crankshafts. 'Elastic' in this sense means that displacement or a twist from rest creates a force or torque tending to return the system to its position of rest, and which is proportional to the displacement. An elastic system, once set in motion in this way, will go on swinging, or vibrating, about its equilibrium position forever, in the theoretical absence of any damping influence. The resulting time/amplitude curve is exactly represented by a sine wave; that is, it is sinusoidal.

In general, therefore, the frequency of torsional vibration of a single mass will be as follows:

$$f = \frac{1}{2\pi}\sqrt{\frac{q}{I}\text{cycles per second}} \quad (2.12)$$

where q is the stiffness in newton metres per radian and I the moment of inertia of the attached mass in kilograms square metres.

For a transverse or axial vibration

$$f = \frac{1}{2\pi}\sqrt{\frac{s}{m}\text{cycles per second}} \quad (2.13)$$

where s is the stiffness in newtons per metre of deflection and m is the mass attached in kilograms.

The essence of control is to adjust these two parameters, q and I (or s and m), to achieve a frequency which does not coincide with any of the forcing frequencies.

Potentially the most damaging form of vibration is the torsional mode, affecting the crankshaft and propeller shafting (or generator shafting). Consider a typical diesel propulsion system, say a six-cylinder two-stroke engine with a flywheel directly coupled to a fixed pitch propeller. There will be as many 'modes' in which the shaft can be induced to vibrate naturally as there are shaft elements: seven in this case. For the sake of simplicity, let us consider the two lowest: the one-node mode and the two-node mode (Fig. 2.7A and B).

In the one-node case, when the masses forward of the node swing clockwise, those aft of it swing anticlockwise and vice versa. In the two-node case,

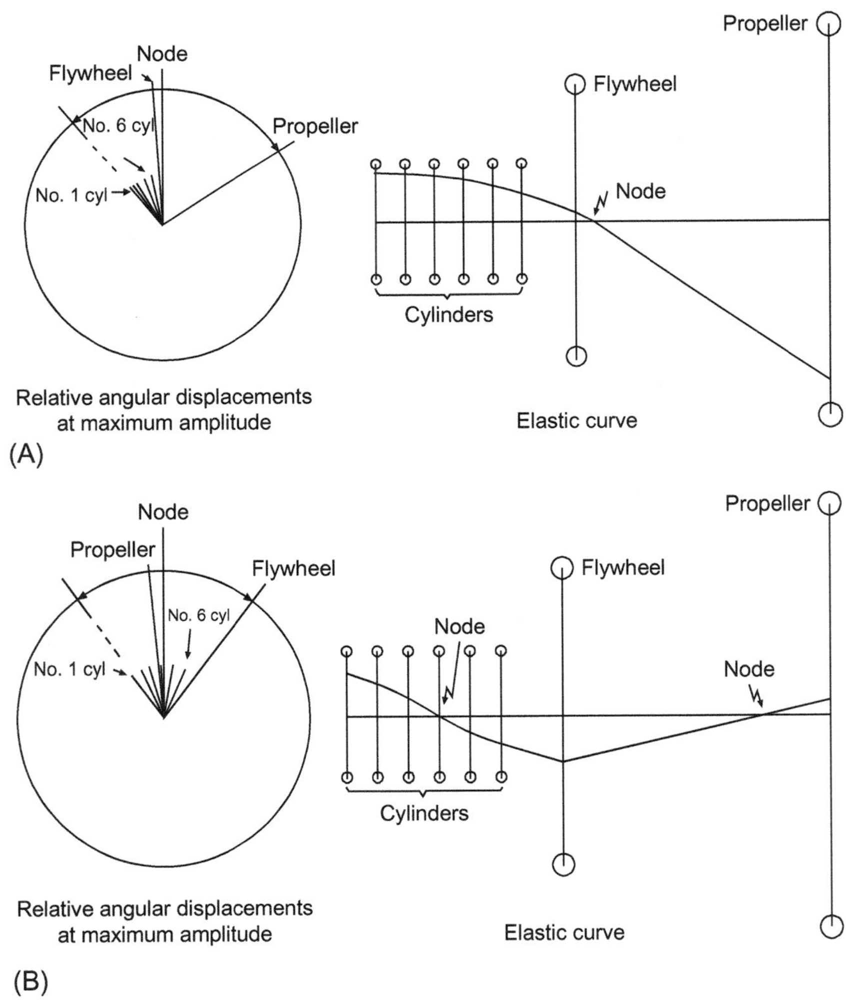

Fig. 2.7 (A) One-node mode and (B) two-node mode.

when those masses forward of the first node swing clockwise, so do those aft of the second node, while those between the two nodes swing anticlockwise, and vice versa.

The diagrams in Fig. 2.7 show (exaggerated) at left the angular displacements of the masses at maximum amplitude in one direction. At right they plot the corresponding circumferential deflections from the mean or unstressed condition of the shaft when vibrating in that mode. The line in the right-hand diagrams connecting the maximum amplitudes reached simultaneously by each mass on the shaft system is called the 'elastic curve'.

A node is found where the deflection is zero and the amplitude changes sign. The more nodes that are present, the higher the corresponding natural frequency.

The problem arises when the forcing frequencies of the externally applied, or input, vibration coincide with, or approach closely, one of these natural frequencies. A lower frequency risks exciting the one-node mode; a higher frequency will possibly excite the two-node mode and so on. Unfortunately, the input frequencies or—to give them their correct name—the 'forcing frequencies' are not simple.

As far as the crankshaft is concerned, the forcing frequencies are caused by the firing impulses in the cylinders. But the firing impulse put into the crankshaft at any loading by one-cylinder firing is not a single sinusoidal frequency at one per cycle. It is a complex waveform which has to be represented for calculation purposes by a component at 13 cycle frequency; another, usually lower in amplitude, at 23 cycle frequency; another at 33, and so on, up to at least 10 before the components become small enough to ignore. These components are called the first, second, third up to the tenth orders or harmonics of the firing impulse. For four-stroke engines, whose cycle speed is half the running speed, the convention has been adopted of basing the calculation on running speed. There will, therefore, be 'half' orders as well, for example, 0.5, 1.5, 2.5, and so on.

Unfortunately, from the point of view of complexity, but fortunately from the point of view of control, these corresponding impulses have to be combined from all the cylinders according to the firing order. For the first order, the interval between successive impulses is the same as the crank angle between successive firing impulses. For most engines, therefore, and for our six-cylinder engine, in particular, the one-node first order would tend to cancel out, as shown in the vector summation in the centre of Fig. 2.8. The length of each vector shown in the diagram is scaled from the corresponding deflection for that cylinder shown on the elastic curve such as is in Fig. 2.7.

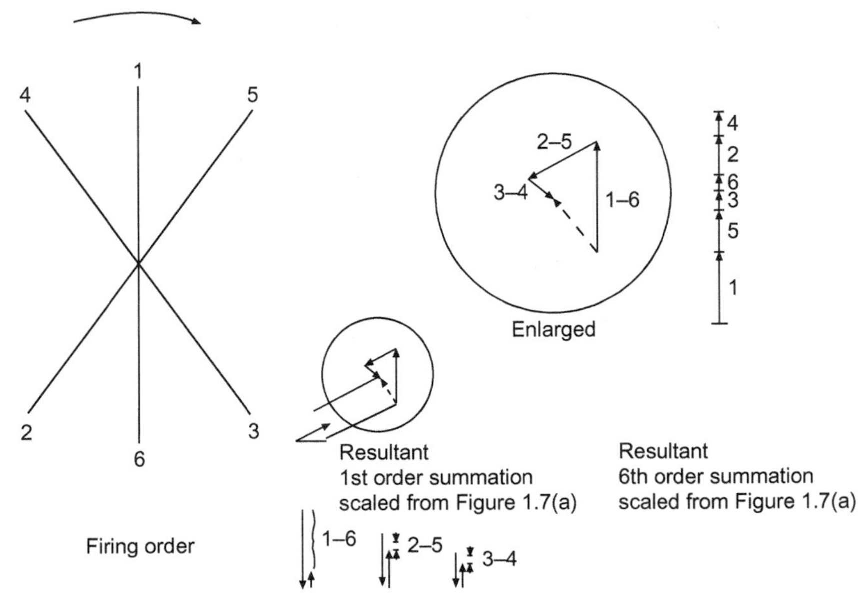

Fig. 2.8 Vector summations based on identical behaviour in all the cylinders.

On the other hand, in the case of our six-cylinder engine, for the sixth order, where the frequency is 6 times that of the first order (or fundamental order), to draw the vector diagram (right of Fig. 2.8), all the first-order phase angles have to be multiplied by 6. Therefore, all the cylinder vectors will combine linearly and become much more damaging.

If, say, the natural frequency in the one-node mode is 300 vibrations per minute (vpm) and our six-cylinder engine is run at 50 rev/min, the sixth harmonic (6 × =0 = 300) would coincide with the one-node frequency, and the engine would probably suffer major damage. Fifty revolutions per minute would be termed the 'sixth-order critical speed' and the sixth order, in this case, is termed a 'major critical'.

Not only the engine could achieve this. The resistance felt by a propeller blade varies periodically with depth while it rotates in the water and with the periodic passage of the blade tip past the stern post, or the point of closest proximity to the hull in the case of a multi-screw vessel. If a three-bladed propeller were used and its shaft run at 100 rev/min, the third order of propeller-excited vibration could also risk damage to the crankshaft (or whichever part of the shaft system was most vulnerable in the one-node mode).

The most significant masses in any mode of vibration are those with the greatest amplitude on the corresponding elastic curve. That is to say, changing them would have the greatest effect on frequency. The most vulnerable shaft sections are those whose combination of torque and diameter induce in them the greatest stress. The most significant shaft sections are those with the steepest change of amplitude on the elastic curve and therefore the highest torque. These are usually near the nodes, but this depends on the relative shaft diameter. Changing the diameter of such a section of the shaft will also have a greater effect on the frequency.

The two-node mode is usually of a much higher frequency than the one-node mode in propulsion systems, and in fact, usually, only the first two or three modes are significant. That is to say that beyond the three-node mode, the frequency components of the firing impulse that could resonate in the running speed range will be small enough to ignore.

The classification society chosen by the owners will invariably make its own assessment of the conditions presented by the vessel's machinery and will judge by criteria based on experience.

Designers can nowadays adjust the frequency of resonance, the forcing impulses, and the resultant stresses by adjusting shaft sizes, number of propeller blades, crankshaft balance weights, and firing orders, as well as by using viscous or other dampers, detuning couplings, and so on. One of the effects of energy efficiency regulations that have been applied to ships under the International Maritime Organization's Energy Efficiency Design Index (EEDI) regulations introduced in 2011 is that far more attention is given to the power and propulsion system of a ship as a whole rather than as individual components.

Gearing, of course, creates further complications—and possibilities. Branched systems, involving twin input or multiple output gearboxes, introduce complications in solving them; but the principles remain the same.

The marine engineer needs to be aware, however, that designers tend to rely on reasonably correct balance among cylinders. It is important to realize that an engine with one cylinder cut out for any reason, or one with a serious imbalance between cylinder loads or timings, may inadvertently be aggravating a summation of vectors which the designer, expecting it to be small, had allowed to remain near the running speed range.

If an engine were run at or near a major critical speed, it would sound rough because, at mid-stroke, the torsional oscillation of the cranks with the biggest amplitude would cause a longitudinal vibration of the connecting rod. This would set up, in turn, lateral vibration of the piston and hence of the entablature. Gearing, if on a shaft section with a high amplitude, would also probably be distinctly noisy.

The remedy, if the engine appears to be running on a torsional critical speed, would be to run at a different and quieter speed while an investigation is made. Unfortunately, noise is not always distinct enough to be relied upon as a warning.

It is usually difficult, and sometimes impossible, to control all the possible critical parameters so that in a variable speed propulsion engine, it is sometimes necessary to 'bar' a range of speeds where vibration is considered too dangerous for continuous operation.

Torsional vibrations can sometimes affect camshafts also. Linear vibrations usually have simpler modes, except for those which are known as axial vibrations of the crankshaft. These arise because firing causes the crankpin to deflect, and this causes the crankwebs to bend. This in turn leads to the setting up of a complex pattern of axial vibration of the journals in the main bearings.

The vibration of smaller items, such as bracket-holding components, or pipework, can often be controlled either by using a very soft mounting whose natural frequency is *below* that of the lowest exciting frequency or by stiffening.

Balancing

The reciprocating motion of the piston in an engine cylinder creates out-of-balance forces acting along with the cylinder, while the centrifugal force associated with the crankpin rotating about the main bearing centres creates a rotating out-of-balance force. These forces, if not in themselves necessarily damaging, create objectionable vibration and noise in the engine foundations, and through them to the ship (or building) in which the engine is operating.

Balancing is a way of controlling vibrations by arranging that the overall summation of the out-of-balance forces and couples cancels out or is reduced to a more acceptable amount.

The disturbing elements are in each case forces, each of which acts in its own plane, usually including the cylinder axis. The essence of balancing is that a force can be exactly replaced by a parallel force acting in a reference plane (chosen to suit the calculation) and a couple whose arm is the distance perpendicularly between the planes in which these two forces act (Fig. 2.9). Inasmuch as the balance is usually considered at and about convenient reference planes, all balancing involves consideration of forces and couples.

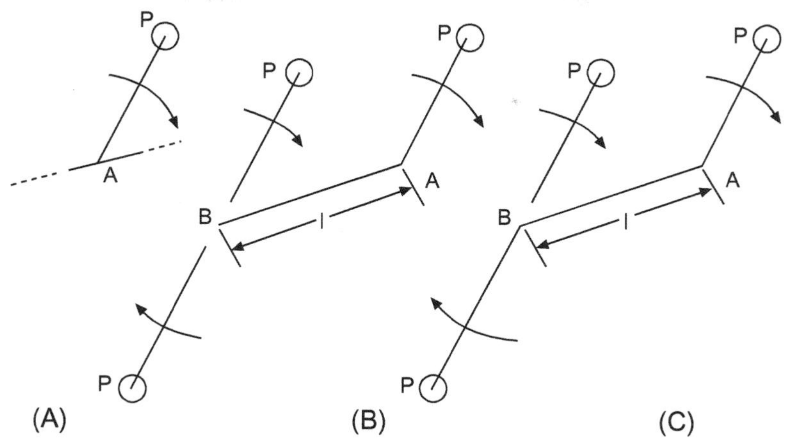

Fig. 2.9 Principle of balancing: (A) force (*P*) to be balanced acting at A (B) equal and opposite forces of magnitude (*P*) assumed to act at B, a distance (*l*) arbitrarily chosen to suit calculation (C) system is equivalent to couple (*P* 3 *l*) plus force (*P*) now acting at B.

In multi-cylinder engines, couples are present because the cylinder(s) coupled to each crank throw act(s) in a different plane.

There are two groups of forces and couples. These relate to the revolving and reciprocating masses. (This section of the book is not concerned with balancing the power outputs of the cylinders. In fact, cylinder output balance, while it affects vibration levels, has no effect on the balance that we are about to discuss.)

Revolving masses are concentrated at a radius from the crankshaft, usually at the crankpin, but are presumed to include a proportion of the connecting rod shank mass adjacent to the crankpin. This is done to simplify the calculations. Designers can usually obtain rotating balance quite easily by the choice of crank sequence and balance weights.

Reciprocating masses are concentrated at the piston/crosshead and similarly are usually assumed to include the rest of the connecting rod shank mass.

Revolving masses turning at a uniform speed give rise to a single system of revolving forces and couples, each system comprising the related effects of each crank and of any balance weights in their correct phase relationship.

Reciprocating masses are more complicated, because of the effect of the obliquity of the connecting rod. This factor, particularly as the connecting rod length is reduced in relation to crank radius, causes the motion of the piston to depart from the simple sinusoidal time-displacement pattern along the cylinder axis. It has instead to be represented by a primary (sinusoidal) component at a frequency corresponding to engine speed in revolutions per minute, a secondary component at 2 × engine speed, a tertiary component at 3 × engine speed, and so on. (It is seldom necessary to go beyond secondary.)

In practical terms, the reversal of direction involves greater acceleration at TDC than at BDC. This can be thought of as an average reversal force (the primary) at both dead centres, on which is superimposed an extra *inward* force at twice the frequency, which adds to the primary at TDC and detracts from it at BDC. Reciprocating forces and couples always act in the plane of the cylinder. For V-cylinder engines, they are combined vectorially. (The effect of side-by-side design for V-engine big ends is usually ignored.)

Most cylinder combinations of two- and four-stroke engines lend themselves to complete balance of either primary forces, primary couples, secondary forces, or secondary couples; some to any two, some to any three, and some to all of them.

Where one or more systems do not balance, the designer can contain the resulting vibration by low-frequency antivibration mountings (as in the four-stroke four-cylinder automotive engine, which normally has severe unbalanced secondary forces); or he or she can ameliorate it by adding extra rotational balance weights on a different shaft or shafts running at crankshaft speed (for primary unbalance) or at twice that speed (for secondary unbalance).

Note that a reciprocating unbalance force can be cancelled out if two shafts, each carrying a balance weight equal to half the offending unbalanced weight and phased suitably, are mounted in the frame, running in opposite directions. The lateral effects of these two contra-rotating weights cancel out. This is the principle of Lanchester balancing, which is used to cope with the heavy unbalanced secondary forces of four inline and V8-cylinder four-stroke engines with a uniform crank sequence.

In general, it is nowadays dangerous to consider only the crankshaft system as a whole, because of the major effect that line-by-line balance has on oil film thickness in bearings. It is often necessary to balance internal couples much more closely; that is, for each half of the engine or even line-by-line.

Noise

Noise is unwanted sound. It is also a pervading nuisance and a hazard to hearing, if not to health itself. Noise is basically a form of vibration, and it is a complex subject, but a few brief notes will be given here.

To the scientist, noise is reckoned according to its sound pressure level, which can be measured fairly easily on a microphone. Since this topic is essentially subjective, that is, to say, it is of interest because of its effect on people via the human ear, the mere measurement of sound pressure level is of very limited use.

Noises are calibrated with reference to a sound pressure level of 0.0002 dynes/cm^2 for a pure 1000 Hz sine wave. A dyne is a force that gives 1 g acceleration of 1 cm/s^2. It is 10^{-5} N. A linear scale gives misleading comparisons, so noise levels are measured on a logarithmic scale of bels. (1 bel means 10 times the reference level, 3 bel means 1000 times the reference level, i.e. 10^3.) For convenience, the bel is divided into 10 parts, hence the decibel or db. This means that 80 db (8 bel) is 10^8 times the reference pressure level.

The human ear hears high and low frequencies (over say 1 kHz or below 50 Hz, respectively) as louder than the middle frequencies, even when sound pressure levels are the same. For practical purposes, sounds are measured according to a frequency scale weighted to correspond to the response of the human ear. This is the 'A' scale and the readings are quoted in dB(A).

The logarithmic-based scale is sometimes found to be confusing, so it is worth remembering that:

30 db corresponds to a gentle breeze in a meadow.

70 db to an open office.

100 db to a generator room.

Any prolonged exposure to levels of 85 db or above is likely to lead to hearing loss in the absence of ear protection; 140 db or above is likely to be physically painful.

The scope for reducing the source noise emitted by a diesel engine is limited, without fundamental and very expensive changes of design principle. However, all the measures which lead to efficiency and economy—higher pressures, faster pressure rise and higher speeds—unfortunately lead also to greater noise, and to noises emitted at higher frequencies and therefore more objectionably.

The noise from several point sources is also added to the logarithmic scale. Two-point sources (e.g. cylinders) are twice as noisy as one. $\text{Log}_{10} 2 = 0.3010$, that is, 0.301 bel or 3.01 db louder than one. Six cylinders, since $\log_{10} 6 = 0.788$, are 7.88 db louder than one.

If there is no echo, sound diminishes according to the square of the distance from the source (usually measured at a reference distance of 0.7 m in the case of diesel engines). So, moving 10 times further away reduces the noise to 1/100th, or by 20 db.

Unfortunately, a ship's engine room has plenty of reflecting surfaces so there is little benefit to be had from moving away. The whole space tends to fill with noise only a few decibels less than the source. Lining as much as possible with sound-absorbing materials does the following two things:

1. It reduces the echo so that moving away from the engine gives a greater reduction in perceived noise.
2. It tends to reduce sympathetic (resonant) vibration of parts of the ship's structure, which, by drumming, add to the vibration and noise which is transmitted into the rest of the ship.

Antivibration mounts also help in the latter case but are not always practical. The only other measure which can successfully reduce noise is to put weight, particularly if a suitable cavity can be incorporated (or a vacuum), between the source and the observer.

A screen, of almost any material, weighing 5 kg/m^2 will affect a reduction of 10 db in perceived noise, and *pro rata*. Where weight is increased by increasing thickness, it is more effective to do so in porous/flexible material than in rigid material. For the latter, the decibel reduction in transmitted noise is proportional to the $\log_{10}$ of the weight, but for the former, it is proportional to the thickness. A 10-fold increase in the thickness of the rigid material of 5 kg/m^2 would double the attenuation; in porous material, it would be 10-fold.

The screen must be totally effective. A relatively small aperture will destroy much of the benefit. For example, most ships' engine rooms now incorporate a control room to provide some noise protection. The benefit will be appreciated every time the door is opened.

Achieving quieter engine rooms

In 2012, the IMO adopted Resolution MSC 337(91)—generally known as the Noise Code which entered into force on 1 July 2014. The provisions of the code are mandatory for ships built after it became effective, but flag states are also encouraged to adopt the provisions for older vessels.

On the basis of engine noise measurements and frequency analyses, the main source of marine diesel engines, MAN Energy Solutions, says that noise emissions from two-stroke engines primarily originate from:

- The turbocharger, air, and gas pulsations
- Exhaust valves

- Fuel oil injection systems.

The chain drive also contributes to a certain extent. The best way of reducing engine-related noise is, naturally, to reduce the vibrational energy at the source or, if this is neither feasible nor adequate, to attenuate the noise as close to its source as possible (Fig. 2.10).

National and international standards for noise levels in ships have, in general, resulted in a considerable reduction of noise in newly built tonnage, especially in accommodation spaces, but engine rooms remain a problem. Limits for maximum sound pressure levels in areas of a newbuilding are either defined specifically between shipowner/yard and engine builder or indirectly by referring to relevant national and international legislation.

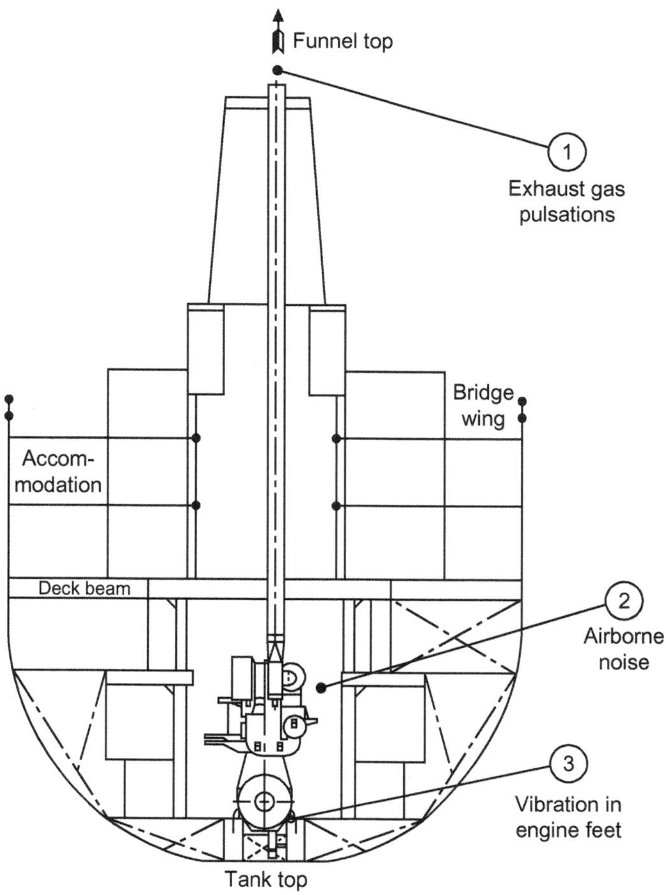

1. Exhaust gas noise (on bridge wing)
2. Airborne noise (in engine room)
3. Structure-borne noise (in accommodation)

Fig. 2.10 Typical sources of engine noise (MAN Diesel). *From MAN Energy Solutions.*

Table 2.1 IMO noise limits (sound pressure level) in dB(A).

Workspaces	
Machinery spaces (continuously unmanned)[a]	90
Machinery spaces (not continuously manned)[a]	110
Machinery control rooms	75
Workshops	85
Unspecified workspaces[a]	90
Navigation spaces	
Navigating bridge and chartrooms	65
Listening posts (including bridge wings and windows)	70
Radio rooms (with radio equipment operating but not producing audio signals)	60
Radar rooms	65
Accommodation spaces	
Cabins and hospital	60
Mess rooms	65
Recreation rooms	65
Open recreation areas	75
Offices	65

[a]Ear protectors should be worn when the noise level is 0.85 dB(A), and no individual's daily exposure duration should exceed 4 h continuously or 8 h in total.

Many owners refer to the German SBG (see Berufsgenossenschaft) specifications or the newer IMO Noise Code limits (see Table 2.1). On the bridge wing, where exhaust gas noise predominates, there are certain limitations as this area is regarded as a listening post. The requirement, depending on the noise standard to be met, is a maximum of 60–70 dB (A), which can always be satisfied by installing a suitable exhaust gas silencer. Ships built and registered in the Far East are often not equipped with exhaust gas silencers, while European-built tonnage is normally fitted with silencers in accordance with rules and regulations prevailing in Europe. In saving the cost of silencers, owners often do not specify compliance with European rules for ships built in the Far East; and even if such tonnage is later transferred to Europe, compliance with the rules is not required.

Noise reductions have not been achieved in the engine room, however, where airborne noise from the main engine dominates. The reason, MAN Energy Solutions suggests is that the acceptable noise limits for periodically manned engine rooms have been set at around 110 dB(A) for many years, and the introduction of stricter requirements has not been realistic as noise emissions from engines have increased over the years with rising power ratings. The unchanged noise limit in itself seems to have constituted a serious limitation for engine builders. It is recognized, however, that a noise level of over 110 dB(A) can, in the long term, cause permanent damage to hearing, and any easing of this limit cannot be expected. Engine designers must, therefore, pay particular attention to reducing the airborne sound levels emitted from their future models.

Noise emitted by the main diesel engine's exhaust gas and the structure-borne noise excited by the engine is generally so low that keeping within the noise level requirements for the bridge wing and accommodation does not pose a problem. Airborne noise emitted from the engine, on the other hand, is so high that in some cases there is a risk that the noise limits for the engine room cannot be met unless additional noise reduction measures are introduced. These measures may include:

- A Helmholtz resonator lining in the scavenge air pipe
- External insulation of the scavenge air receiver
- External insulation of the scavenge air cooler
- Additional absorption material at the engine and/or at the engine room walls (shipyard's responsibility)
- Additional turbocharger intake silencer attenuation (turbocharger maker's responsibility)
- Additional attenuation material at the turbocharger inspection cover
- Low noise diffuser for turbocharger compressor (if available)

Such measures may reduce the maximum noise levels by 3–5 dB(A) or more, depending on their extent. It would nevertheless be extremely difficult to meet stricter engine room noise level requirements of, say, 105 dB(A) instead of 110 dB(A), especially in view of the influence of sound reverberations and the noise emitted by other machinery. The possibility of reducing the noise from an existing engine is greatly limited because the noise stems from many different sources, and because the noise transmission paths (through which vibrational energy is transferred from one area to another through the engine) are numerous. In principle, however, the transmission of airborne noise from the engine room to other locations (e.g. accommodation quarters) normally has no influence on the actual noise level in those locations.

Antivibration mountings

Over the years, the use of elastomer-based mounting systems to suppress or attenuate noise and vibration in ships has advanced in both technical sophistication and growth of the application. Rubber-to-metal bonded systems are the most commonly applied antivibration mountings, and they deliver the highest performance rating for a given size. Shipboard installations mainly benefit propulsion engines, gensets, and diverse ancillary machineries such as ventilation fans, compressors, pumps, sewage treatment units, refrigeration plants, and instrument panels.

Any item of machinery with moving parts is subject to out-of-balance forces and couples created by the interaction of those parts. The magnitude of the forces and couples will vary considerably, depending on the configuration of the machine design. In theory, some machines will be totally balanced, and there will be no external forces or couples (e.g. an electric motor or a six-cylinder four-stroke engine). In practice, however, manufacturing tolerances may create out-of-balance excitations that can become significant in large elements of the machinery. The weights of pistons and connecting rods in large diesel engines may vary substantially, although to some extent this is allowed for by selective assembly and balancing. Uneven firing in large engines, due to variation in combustion between the cylinders, can also cause significant excitation.

Either the various excitations in a machinery element will have discrete frequencies, based on constant running speed, or there will be a frequency range to contend with (as in the case of a variable speed machine). In practical terms, it may not be possible or cost effective to eliminate vibration at source, and the role of antivibration mountings is therefore to reduce the level of transmitted vibration.

In an active isolation system, the use of antivibration mountings beneath a machine will isolate the surrounding structure from the vibrating machine and equipment; in a passive system, the mountings may be used to protect an item of sensitive equipment, such as an instrument panel, from external sources of vibration or shock. The amplitude of movement of the equipment installed on a flexible mounting system can be reduced by increasing the inertia of the vibrating machine, typically by adding mass in the form of an inertia block. Unfortunately, adding mass is not generally acceptable for marine and offshore applications because of the penalties of weight and cost.

A number of standard ranges of rubber/metal-bonded antivibration and suspension components are offered by specialist companies, but most installations differ and may call for custom-designed solutions. Computer analysis is exploited to study individual applications, predict the behaviour of various isolation systems, and recommend an optimum solution. It is usually possible to design an antivibration system of at least 85% efficiency against the worst possible condition, reports Sweden-based specialist Trelleborg.

Active mountings

Most ships are built for carriage of cargo or related work purposes and usually, crew comfort is low on the list of priorities. In the passenger and luxury yacht sectors, reducing noise and vibration is high on the list of design drivers.

Active, rather than passive, mountings can enhance noise insulation many times over by exploiting antivibration to counteract vibration from the engine. The technique makes use of the fact that sound emissions take the form of a wave. An active mounting system developed in 2008 jointly by high-speed engine builder MTU Friedrichshafen and French specialists Paulstra and Stop-Choc is said to significantly diminish vibration noise. The compact active mountings function on the antivibration principle and apply noise reduction techniques used in the automotive and aircraft industries.

Structure-borne noise and vibration are produced by an engine's moving parts and the combustion process; airborne noise is generated from the air intake, the structure, and the exhaust. An MTU Series 4000 engine, for example, deflects the needle on a sound level meter to around the 100 dB (A) mark. (Because decibel noise levels are calculated logarithmically, an increase or reduction by 10 dB(A) is, respectively, equivalent to doubling or halving the noise output.)

All MTU engines are supplied as standard with single-resilient mountings made of an elastomer material, which absorb a proportion of the engine vibrations and reduce the amount transmitted to the hull. This measure is adequate for vessels including ferries and workboats. Higher quality mountings with even better elastic properties are applied to vessels with greater demands on noise reduction and comforts, such as mega yachts and cruise ships. Here, the engine sub-frame and structure are also optimized for noise reduction, enabling vibration to be reduced by a further 12–20 dB over standard mountings.

Even stricter requirements for smoothness and comfort are met by MTU by placing the resiliently mounted engine on a heavy baseframe, which, in turn, stands on another set of resilient mountings. Thanks to the inertia of the frame and the double-resilient mountings, the engine vibration can reportedly be reduced by more than 30 dB compared with standard mountings. The greater the mass of the frame, the greater the degree of attenuation achieved. Halving the vibration transmitted from the engine to the hull means doubling the mass of the passive mountings.

MTU's active mounting system electronics react to the engine's oscillations by controlling actuators and producing antivibrations with a phase shift of 180°. The result of active noise emission management is that antivibration patterns are superimposed over the original vibrations, thereby largely neutralizing them. Combining passive and active mountings can deaden individual noise peaks and reduce noise levels by up to 30 dB, MTU reports.

All passive noise reduction solutions can be combined with active mountings either to further improve attenuation levels or to save weight and space. The passive mountings sit on top of the active mountings, which contain sensors and vibrating coils (actuators) that act in all three planes. Other sensors can also be attached outside the mountings to further enhance system performance. Special software running on the associated control unit analyses the sensor signals in real time and sends control signals to the actuators to produce the appropriate antivibration. The speed of signal analysis is a fundamental factor in the successful functioning of the active mounting system.

For even higher performance, engines can be housed in an acoustic enclosure to minimize the transmission of sound through the air—a solution mainly adopted for mega yachts and naval vessels. An enclosure usually makes sense with good-quality engine mountings as high levels of airborne noise would be superimposed over the reduced levels of structure-borne vibration so that the mountings would not be used to best effect. Only by simultaneously lowering airborne and structure-borne noise can the best possible reduction be achieved. The key means of attenuating airborne noise are intake mufflers in the air filters, exhaust silencers, or complete enclosure of the engine in an insulated housing.

Low-speed engine vibration: Characteristics and cures

Ship machinery installations have two principal sources of excitation: the main engine(s) and the propeller(s). The two components are essentially linked by elastic shaft systems and may also embrace gearboxes and elastic couplings. The whole system is supported in flexible hull structures, and the forms of vibration possible are therefore diverse.

Problems in analysing ship machinery vibration have been compounded in recent years by the introduction of new fuel-efficient engine designs exploiting lower running speeds, longer stroke/bore ratios, and higher combustion pressures. Another factor is the wider use of more complex machinery arrangements including power take-offs (PTOs), shaft alternators, exhaust gas power turbines, and multiple geared engines. The wider popularity of four- and five-cylinder low-speed two-stroke engines for propulsion plant—reflecting attractive installation and operating costs—has also stimulated efforts by designers to counteract adverse vibration characteristics.

The greater complexity of vibration problems dictates a larger number of calculations to ensure satisfactory vibration levels from projected installations. For optimum results, the vibration performance of the plant has to be investigated for all anticipated operational modes.

Sophisticated tools such as computer software and measuring systems are now available to yield a more detailed and accurate analysis of complex vibration problems. Even so, inaccuracies in determining shaft stiffness, damping effects, coupling performance, and hull structure response make it advisable in certain borderline cases to allow, at the design stage, for suitable countermeasures to be applied after vibration measurements have been taken during sea trials.

All major engine producers stress that proper consideration should be given to the vibration aspects of a projected installation at the earliest possible stage in the ship design process. The available countermeasures provide a good safety margin against potential vibration problems. Close collaboration is desirable between naval architects, machinery installation designers, engine builders, and specialist component suppliers.

The key factor is the interaction with the ship and not the mere magnitude of the excitation source. Excitations generated by the engine can be divided into two categories:

1. *Primary excitations:* forces and moments originating from the combustion pressure and the inertia forces of the rotating and reciprocating masses. These are characteristics of the given engine, which can be calculated in advance and stated as part of the engine specification with reference to a certain speed and power.

2. *Secondary excitations:* forces and moments stemming from a forced vibratory response in a ship substructure. The vibration characteristics of substructures are almost independent of the remaining ship structure.

Examples of secondary excitation sources from substructures could be anything from transverse vibration of the engine structure to longitudinal vibration of a radar or light mast on top of the deckhouse. Such substructures of the complete ship might have resonance or be close to resonance conditions, resulting in considerable dynamically magnified reaction forces at their interface with the rest of the ship. Secondary excitation sources cannot be directly quantified for a certain engine type but must be calculated at the design stage of the specific propulsion plant.

The vibration characteristics of low-speed two-stroke engines, for practical purposes, can be split into four categories that may influence the hull (Fig. 2.11A and B):

1. External unbalanced moments: these can be classified as unbalanced first- and second-order external moments, which need to be considered only for engines with certain cylinder numbers.
2. Guide force moments.
3. Axial vibrations in the shaft system.
4. Torsional vibrations in the shaft system.

The influence of the excitation forces can be minimized or fully compensated if adequate countermeasures are considered from the early project stage. The firing angles can be customized to the specific project for 9-, 10-, 11-, and 12-cylinder engines.

External unbalanced moments

The inertia forces originating from the unbalanced rotating and reciprocating masses of the engine create unbalanced external moments although the external forces are zero. Of these moments, only the first order (producing one cycle per revolution) and the second-order (two cycles per revolution) need to be considered, and then only for engines with a low number of cylinders. The inertia forces on engines with more than six cylinders tend, more or less, to neutralize themselves.

Countermeasures have to be taken if hull resonance occurs in the operating speed range, and if the vibration level leads to higher accelerations and/or velocities than the guidance values given by international standards or recommendations (e.g. with reference to a special agreement between the

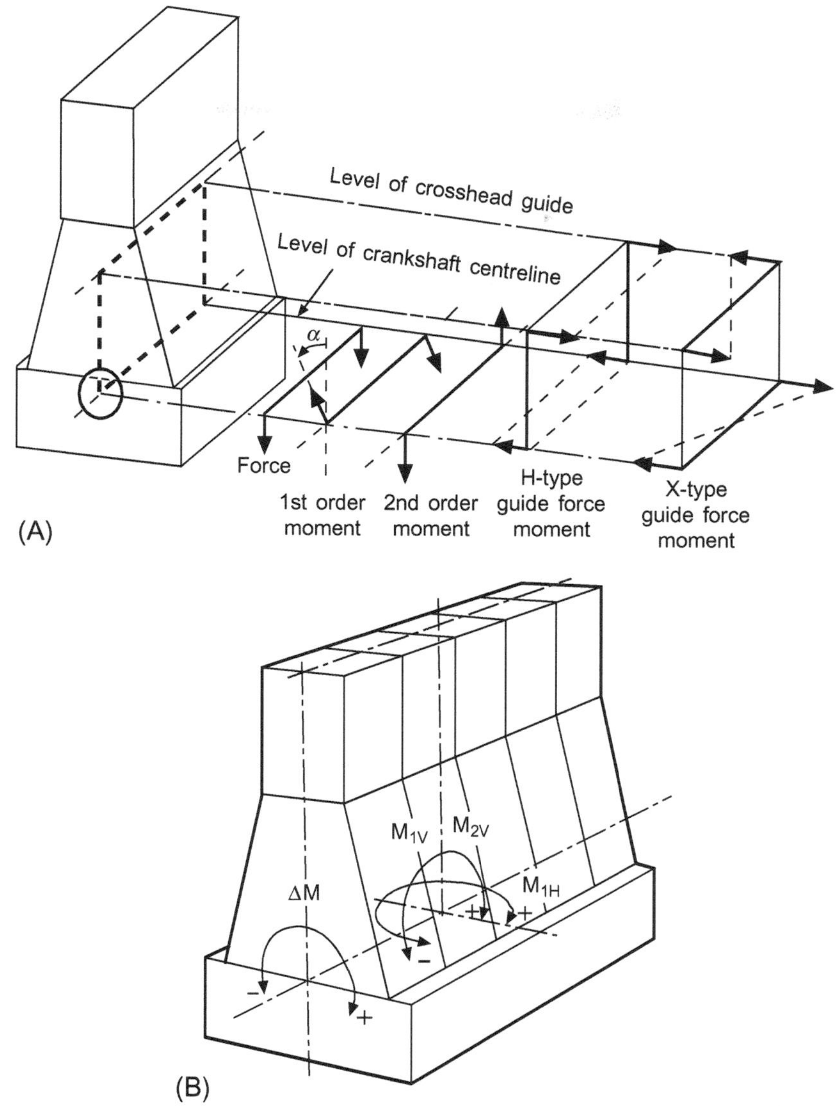

Fig. 2.11 (A) Forces and moments of a multi-cylinder, low-speed, two-stroke engine. The firing order will determine the vectorial sum of the forces and moments from the individual cylinders. A distinction should be made between external forces and moments, and internal forces and moments. The external forces and moments will act as resultants on the engine and thereby also on the ship through the foundation and top bracing of the engine. The internal forces and moments will tend to deflect the engine as such (MAN Diesel). (B) Free couples of mass forces and the torque variation about the centre lines of the engine and crankshaft (Wärtsilä): M_{1V} is the first-order couple having a vertical component, M_{1H} is the first-order couple having a horizontal component, M_{2V} is the second-order couple having a vertical component, and ΔM is the reaction to variations in the nominal torque. Reduction of the first-order couples is achieved by counterweights installed at both ends of the crankshaft. *(A) From MAN Energy Solutions.*

ship owner and shipyard). The natural frequency of the hull depends on its rigidity and distribution of masses, while the vibration level at resonance depends mainly on the magnitude of the external moment and the engine's position in relation to the vibration nodes of the ship.

First-order moments

These moments act in both vertical and horizontal directions and are of the same magnitude for MAN B&W two-stroke engines with standard balancing. For engines with five cylinders or more, the first-order moment is rarely of any significance to the ship, but it can be of a disturbing magnitude in four-cylinder engines.

Resonance with a first-order moment may occur for hull vibrations with two and/or three nodes. This resonance can be calculated with reasonable accuracy, and the calculation will show whether a compensator is necessary on four-cylinder engines. A resonance with the vertical moment for the two-node hull vibration can often be critical, whereas the resonance with the horizontal moment occurs at a higher speed than the nominal because of the higher natural frequency of the horizontal hull vibrations.

Four-cylinder MAN B&W MC two-stroke engines with bores from 500 to 980 mm are fitted as standard with adjustable counterweights (Fig. 2.12). These can reduce the vertical moment to an insignificant value (although

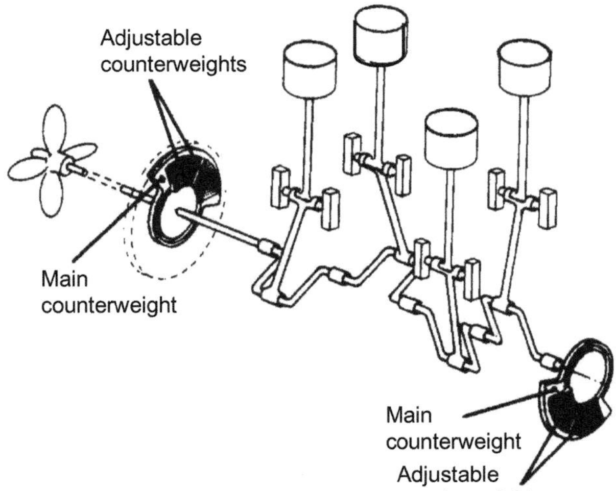

Fig. 2.12 Four-cylinder MAN B&W 500–980 mm bore MC engines are fitted with adjustable counterweights.

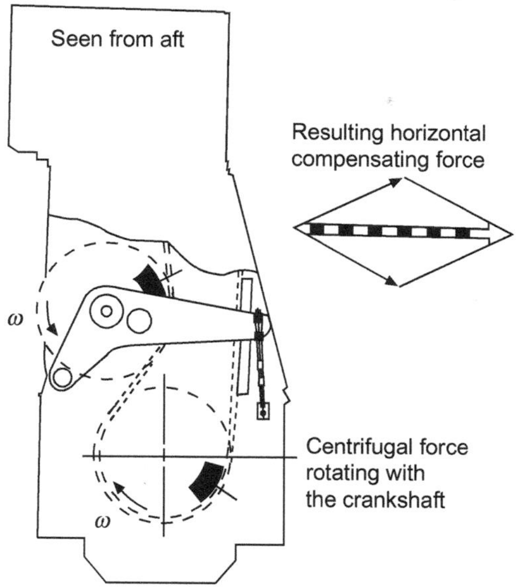

Fig. 2.13 First-order moment compensator (MAN Diesel). *From MAN Energy Solutions.*

increasing, correspondingly, the horizontal moment), so this resonance is easily handled. A solution with zero horizontal moments is also available.

In rare cases, where the first-order moment will cause resonance with both the vertical and horizontal hull vibration modes in the normal speed range of the engine, a first-order compensator (Fig. 2.13) can be introduced in the chain-tightener wheel, reducing the first-order moment to a harmless value. The compensator is an option and comprises two counterrotating masses rotating at the same speed as the crankshaft.

With a first-order moment compensator fitted aft, the horizontal moment will decrease to between 0% and 30% of the value stated in MAN Diesel tables, depending on the position of the node. The first-order vertical moment will decrease to around 30% of the value stated in the tables.

Since resonance with both the vertical and horizontal hull vibration modes is rare, the standard engine is not prepared for the fitting of such compensators.

Second-order moments

The second-order moment acts only in the vertical direction and precautions need to be considered only for four-, five-, and six-cylinder

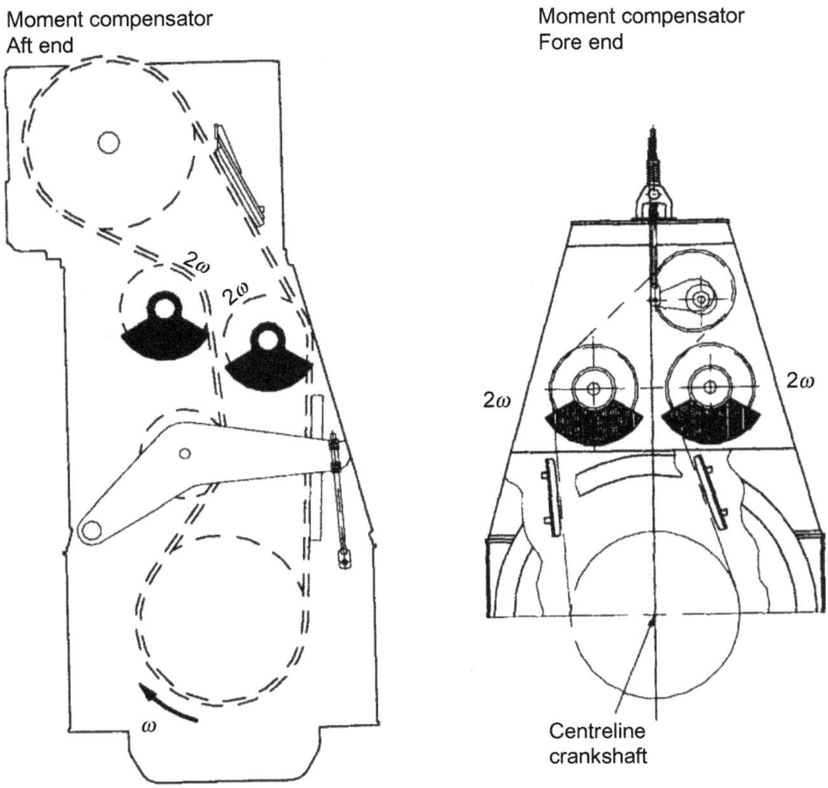

(A) (B)

Fig. 2.14 Second-order moment compensators (MAN Diesel). *From MAN Energy Solutions.*

engines. Resonance with the second-order moment may occur at hull vibrations with more than three nodes.

A second-order moment compensator comprises two counterrotating masses running at twice the engine speed. Such compensators are not included in the basic extent of MAN B&W engine deliveries.

Several solutions are available to cope with the second-order moment (Fig. 2.14) from which the most efficient can be selected for the individual case:

- No compensators, if considered unnecessary on the basis of natural frequency, nodal point, and size of the second-order moment
- A compensator mounted on the aft end of the engine, driven by the main chain drive
- A compensator mounted on the fore end, driven from the crankshaft through a separate chain drive

- Compensators on both aft and fore end, completely eliminating the external second-order moment

A decision regarding the vibrational aspects and the possible use of compensators must be taken at the contract stage. If no experience is available from sister ships (which would be the best basis for deciding whether compensators are necessary or not), calculations should be made to determine which of the recommended solutions should be applied.

If compensator(s) is omitted, the engine can be delivered prepared for their fitting at a later date. The decision for such preparation must also be taken at the contract stage. Measurements were taken during the sea trial, or later in service with a fully loaded ship, which will show whether compensator(s) have to be fitted or not.

If no calculations are available at the contract stage, MAN Diesel advises the supply of the engine with a second-order moment compensator on the aft end and provision for fitting a compensator on the fore end.

If a decision is made not to use compensators and, furthermore, not to prepare the engine for their later fitting, an electrically driven compensator can be specified if annoying vibrations occur. Such a compensator is synchronized to the correct phase relative to the external force or moment and can neutralize the excitation. The compensator requires extra seating to be fitted—preferably in the steering gear compartment where deflections are larger, and the effect of the compensator will therefore be greater.

The electrically driven compensator will not give rise to distorting stresses in the hull, but it is more expensive than the engine-mounted compensators mentioned earlier. Good results are reported from the numerous compensators of this type in service (Fig. 2.15).

Power-related unbalance

To evaluate if there is a risk that first- and second-order external moments will excite disturbing hull vibrations, the concept of 'power-related unbalance' (PRU) can be used as a guide, where

$$\text{PRU} = \frac{\text{External moment}}{\text{Engine power}} \text{Nm/kW}$$

With the PRU value, stating the external moment relative to the engine power, it is possible to give an estimate of the risk of hull vibrations for a specific engine. Actual values for different models and cylinder numbers rated at layout point L1 are provided in MAN B&W engine charts.

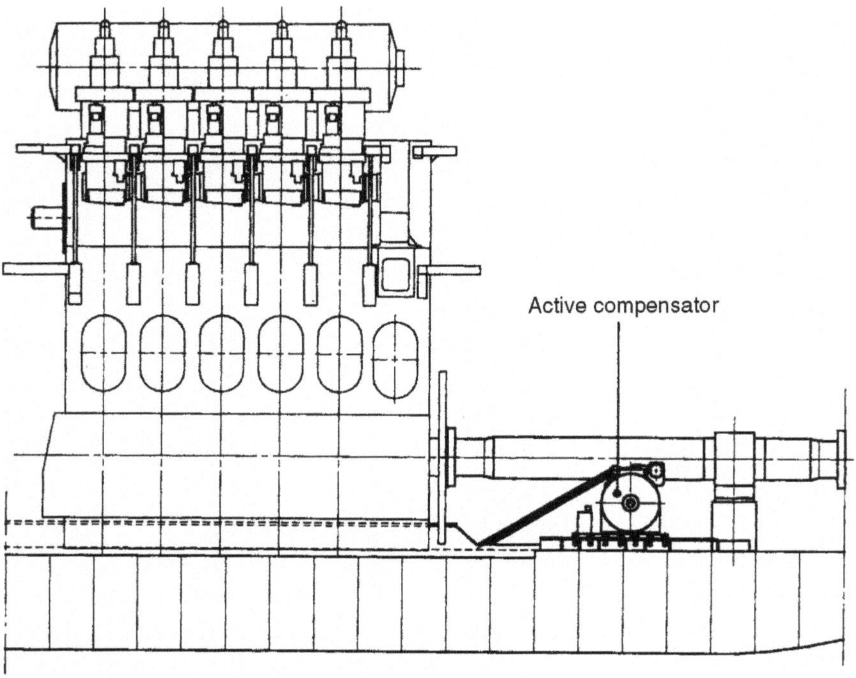

Fig. 2.15 Active vibration compensator used as a thrust pulse compensator (MAN Diesel). *From MAN Energy Solutions.*

Guide force moments

The so-called guide force moments are caused by the transverse reaction forces acting on the crossheads due to the connecting rod/crankshaft mechanism. These moments may excite engine vibrations, moving the engine top athwart ships and causing a rocking (excited by H moment) or twisting (excited by X-moment) movement of the engine (Fig. 2.16).

Guide force moments are harmless except when resonance vibrations occur in the engine/double bottom system. As this system is very difficult to calculate with the necessary accuracy, MAN Energy Solutions strongly recommends as standard that top bracing is installed between the engine's upper platform brackets and the casing side for all of its two-stroke models except the S26MC and L35MC types.

The top bracing comprises stiff connections (links) with either friction plates which allow adjustment to the loading conditions of the ship or, alternatively, a hydraulic top bracing (Fig. 2.17). With both types of top bracing, the above-mentioned natural frequency will increase to a level where resonance will occur above the normal engine speed.

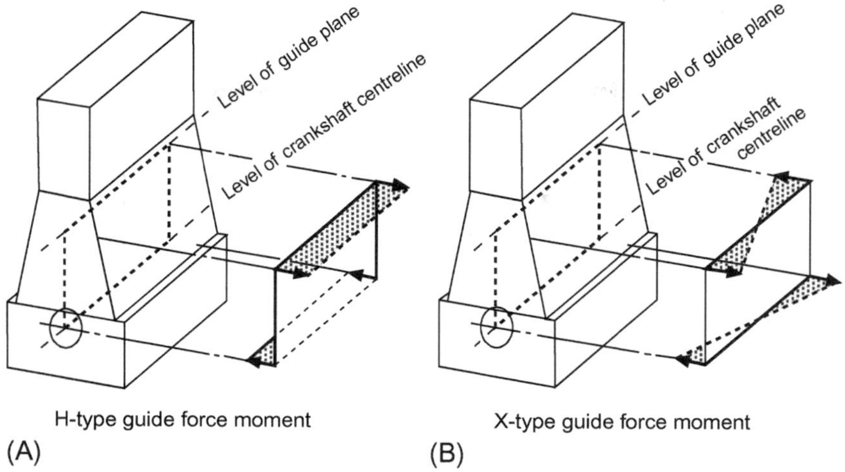

Fig. 2.16 H- and X-type guide force moments (MAN Diesel). *From MAN Energy Solutions.*

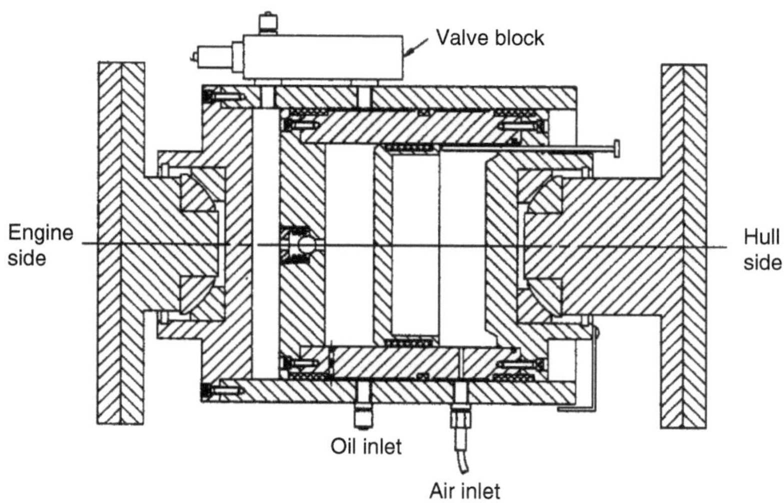

Fig. 2.17 Hydraulically adjustable top bracing, modified type (MAN Diesel). *From MAN Energy Solutions.*

Axial vibrations

The calculation of axial vibration characteristics is only necessary for low-speed two-stroke engines. When the crank throw is loaded by the gas pressure through the connecting rod mechanism, the arms of the crank throw deflect in the axial direction of the crankshaft, exciting axial vibrations. These vibrations may be transferred to the ship's hull through the thrust bearing.

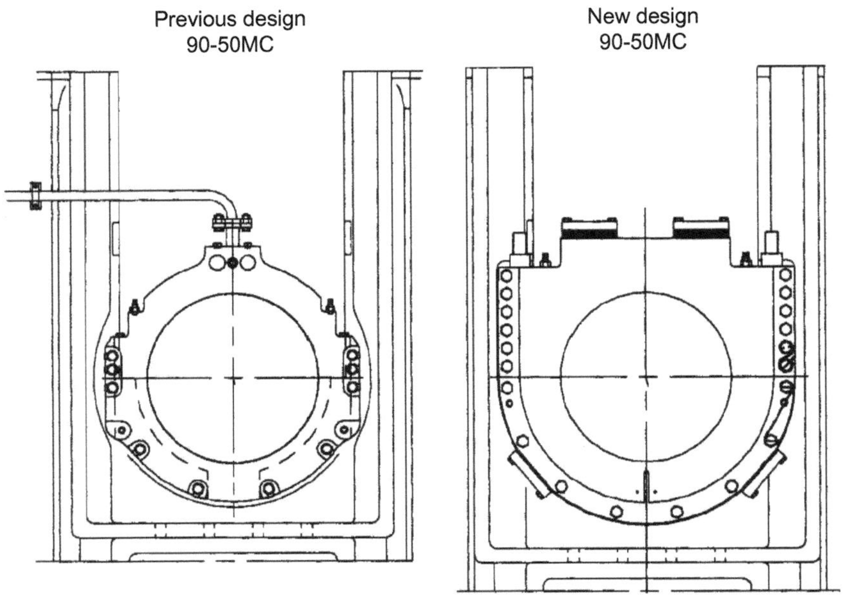

Fig. 2.18 Mounting of previous and improved axial vibration dampers for MAN B&W 500–900 mm bore MC engines.

Generally, only zero-node axial vibrations are of interest. Thus, the effect of the additional bending stresses in the crankshaft and possible vibrations of the ship's structure due to the reaction force in the thrust bearing is to be considered.

An appropriate axial damper is fitted to all engines to minimize the effects of the axial vibrations (Fig. 2.18). Some engine types in five- and six-cylinder form with a PTO mounted at the fore end require an axial vibration monitor for alarm and slowdown. For the crankshaft itself, however, such a damper is only necessary for engines with large cylinder numbers.

The examination of axial vibration (sometimes termed longitudinal vibration) can be confined to the main shaftline embracing crankshaft, intermediate shaft, and propeller shaft because the axial excitation is not transmitted through PTO gearboxes to branch shaftlines. The vibration system can be plotted in equivalent constant masses and stiffnesses, just as with torsional vibration.

The masses are defined by the rotating components as a whole and can be calculated very accurately with the aid of CAD programs. Axial stiffnesses of the crankshaft are derived, as with the torsional values, from an empirical formula and, if necessary, corrected by comparing measured and calculated axial natural frequencies. Alternatively, axial stiffnesses can be calculated directly by finite element analysis. Damping in the crankshaft is derived from measurements of axial vibration.

The dominating order of the axial vibration is equivalent to the number of cylinders for engines with less than seven cylinders. For engines with more than six cylinders, the dominating order is equal to half the cylinder numbers. For engines with odd cylinder numbers, the dominating orders are mostly the two orders closest to half the cylinder number.

Torsional vibrations

Torsional vibration involves the whole shaft system of the propulsion plant, embracing engine crankshaft, intermediate shafts, and propeller shaft, as well as engine running gear, flywheel, propeller and (where appropriate) reduction gearing, flexible couplings, clutches, and PTO drives.

The varying gas pressure in the cylinders during the working cycle and the crankshaft/connecting rod mechanism creates a varying torque in the crankshaft. It is these variations that cause the excitation of the torsional vibration of the shaft system. Torsional excitation also comes from the propeller through its interaction with the non-uniform wakefield. Like other excitation sources, the varying torque is cyclic and thus subject to harmonic analysis. Such an analysis makes it possible to represent the varying torque as a sum of torques acting with different frequencies, which are multiples of the engine's rotational frequency.

Torsional vibration causes extra stresses, which may be detrimental to the shaft system. The stresses will show peak values at resonances: that is, where the number of revolutions multiplied by the order of excitation corresponds to the natural frequency.

Limiting torsional vibration is vitally important to avoid damage or even fracture of the crankshaft or other propulsion system elements. Classification societies, therefore, require torsional vibration characteristics of the engine/shafting system to be calculated, with verification by actual shipboard measurements. Two limits are laid down for the additional torsional stresses. The lower limit T1 determines a stress level which may only be exceeded for a short time; this dictates a 'barred' speed range of revolutions in which continuous operation is prohibited. The upper-stress limit T2 must never be exceeded.

Taking a shaftline of a certain length, it is possible to modify its natural frequency of torsional vibration by adjusting the diameter: a small diameter results in a low natural frequency, a larger diameter in a high natural frequency. The introduction of a tuning wheel will also lower the natural frequency. Classification societies have also laid down rules determining the shaft diameter. An increase in the diameter is permissible, but a reduction must be accompanied by the use of a material with higher ultimate strength.

Based on its experience, MAN Diesel offers the following guidelines for low-speed engines:

Four-cylinder engines normally have the main critical resonance (fourth order) positioned earlier but close to normal revolutions. Thus, in worst cases, these engines require an increased diameter of the shaftline relative to the diameters stipulated by classification society rules in order to increase the natural frequency, taking it 40%–45% above normal running range.

For five-cylinder engines, the main critical (fifth order) resonance is also positioned close to—but below—normal revolutions. If the diameter of the shafting is chosen according to class rules, the resonance with the main critical will be positioned quite close to normal service speed, thus introducing a barred speed range. The usual and correct way to tackle this unacceptable position of a barred speed range is to mount a tuning wheel on the front end of the crankshaft, design the intermediate shaft with a reduced diameter relative to the class diameter, and use better material with a higher ultimate tensile strength. In some cases, however, an intermediate shaft of large diameter is installed in order to increase the resonance to above the maximum continuous rating (mcr) speed level. This is termed 'overcritical running' (see section Overcritical running).

Only torsional vibrations with one node generally need to be considered. The main critical order, causing the largest extra stresses in the shaftline, is normally the vibration with order equal to the number of cylinders: that is, five cycles per revolution on a five-cylinder engine. This resonance is positioned at the engine speed corresponding to the natural torsional frequency divided by the number of cylinders.

The torsional vibration conditions may, for certain installations, require a torsional vibration damper which can be fitted when necessary at additional cost. Based on MAN Energy Solutions' statistics, this need may arise for the following types of installation:

- Plants with a controllable pitch propeller
- Plants with an unusual shafting layout and for special owner/yard requirements
- Plants with 8-, 10-, 11-, or 12-cylinder engines.

Four-, five-, and six-cylinder engines require special attention. To avoid the effect of the heavy excitation, the natural frequency of the system with one-node vibration should be situated away from the normal operating speed range. This can be achieved by changing the masses and/or the stiffness of the system so as to give a much higher or much lower natural frequency (respectively, termed 'undercritical' or 'overcritical' running).

Owing to the very large variety of possible shafting arrangements that may be used in combination with a specific engine, only detailed torsional vibration calculations for the individual plant can determine whether a torsional vibration damper is necessary.

Undercritical running

The natural frequency of the one-node vibration is so adjusted that resonance with the main critical order occurs about 35%–45% above the engine speed at specified mcr. The characteristics of an undercritical system are normally the following:

- Relatively short shafting system
- Probably no tuning wheel
- Turning wheel with relatively low inertia
- Large diameter shafting, enabling the use of shafting material with a moderate ultimate tensile strength, but requiring careful shaft alignment (due to relatively high bending stiffness)
- Without barred speed range.

When running undercritical, significant varying torque at mcr conditions of about 100%–150% of the mean torque is to be expected. This torque (propeller torsional amplitude) induces a significant varying propeller thrust, which, under adverse conditions, might excite annoying longitudinal vibrations on engine/double bottom and/or deckhouse. Shipyards should be aware of this and ensure that the complete aft body structure of the ship, including the double bottom in the engine room, is designed to cope with the described phenomena.

Overcritical running

The natural frequency of the one-node vibration is so adjusted that resonance with the main critical order occurs about 30%–70% below the engine speed at specified mcr.

Such overcritical conditions can be realized by choosing an elastic shaft system, leading to a relatively low natural frequency. The characteristics of overcritical conditions are the following:

- Tuning wheel may be necessary on crankshaft fore end.
- Turning wheel with relatively high inertia.
- Shafts with relatively small diameters, requiring shafting material with a relatively high ultimate tensile strength.
- With barred speed range of about ±10% with respect to the critical engine speed.

Torsional vibrations in overcritical conditions may, in special cases, have to be eliminated by the use of a torsional vibration damper, which can be fitted when necessary at extra cost.

For six-cylinder engines, the normal procedure is to adopt a shaftline with a diameter according to the class rules and, consequently, a barred speed range. Excitations associated with engines of seven or more cylinders are smaller, and a barred speed range is not normally necessary.

Hull vibration

The natural frequencies of ships' hulls are relatively difficult to predict accurately and are also influenced by the loading condition. MAN B&W and Wärtsilä (Sulzer) two-stroke engines do not generate any free forces; and free moments are generated only for certain numbers of cylinders.

Hull vibration can only be excited by a Wärtsilä RTA low-speed engine if it is located on or near a node of critical hull vibration, and if the frequency of this vibration mode coincides with the first or second harmonic of the engine excitation. Furthermore, the magnitude of the troublesome free moment has to exceed the stabilizing influence of the natural damping in the hull structure.

If the resonance is foreseen, Wärtsilä notes the following solutions (Figs 2.19 and 2.20):

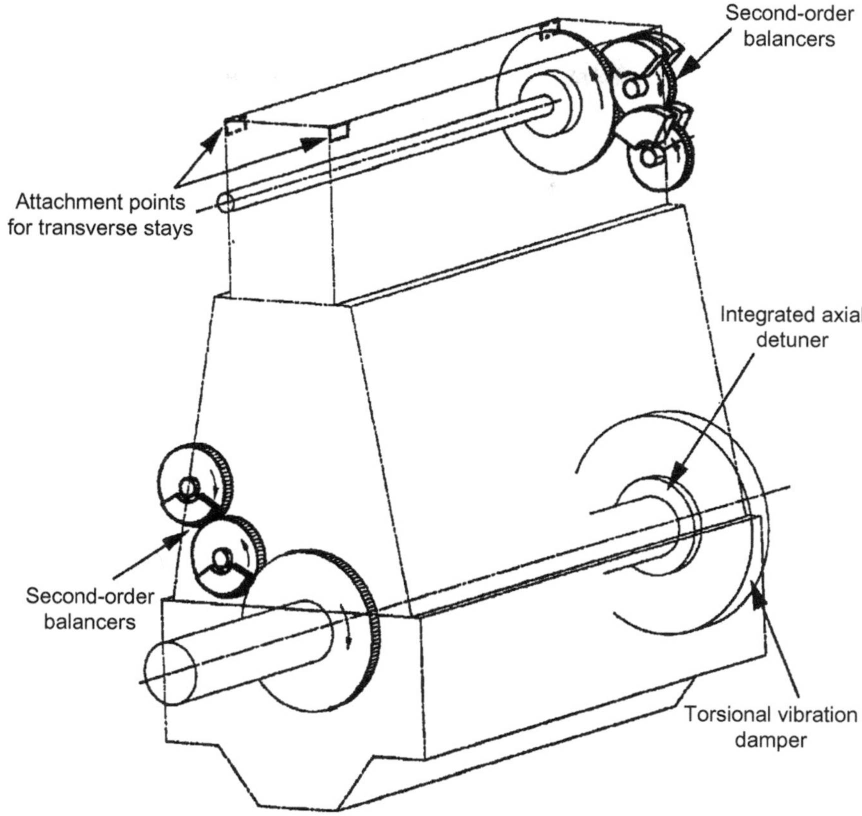

Fig. 2.19 Provisions for vibration control and balancing on Wärtsilä RTA low-speed engine.

- Lanchester balancers, either on the engine or electrically driven units usually located on the steering gear flat, compensate for ship vibration caused by the second-order vertical moment.
- Counterweights on the crankshaft often represent a simple and effective solution for primary unbalance, which is only relevant to four-cylinder engines.
- Combined primary/secondary balancers are available for RTA engines to counteract entirely both primary and secondary unbalances.
- Side stays can be fitted between the engine top and hull structure to reduce vibration caused by lateral moments—a situation mainly arising for 4-, 8-, and 12-cylinder engines. Hydraulic-type side stays are preferable.

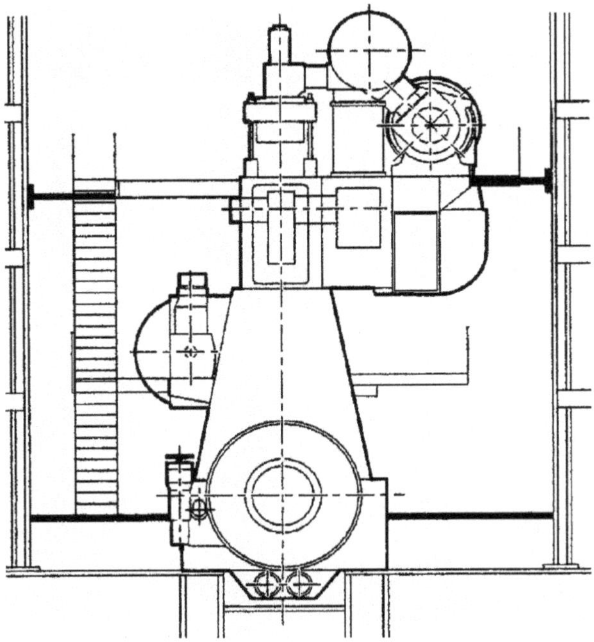

Fig. 2.20 Typical attachment points for transverse stays of Wärtsilä RTA low-speed engine.

Acknowledgements

MAN Energy Solutions and Wärtsilä Corporation. Both designers have produced valuable papers, available on request, detailing the theory of vibration in shipboard machinery installations, analysis techniques, and practical countermeasures.

CHAPTER THREE

Dual-fuel and gas engines

To call any diesel engine a dual-fuel (DF) engine is perhaps something of a misnomer since all engines of the type are capable of burning many different types and grades of fuel. The term came into regular use in the marine community around the turn of the century and refers specifically to an engine designed to run on either oil fuels or gas—particularly liquefied natural gas (LNG).

Natural gas is well established as a major contributor to the world's energy needs. It is derived from the raw gas from onshore and offshore fields as the dry, light fraction, and mainly comprises methane and some ethane. It is available directly at the gas field itself, in pipeline systems, condensed into liquid as LNG or compressed as CNG.

Operation on natural gas results in very low emissions thanks to the clean-burning properties of the fuel and its low content of pollutants. Methane, the main constituent, is the most efficient hydrocarbon fuel in terms of energy content per amount of carbon, containing the highest amount of hydrogen per unit of energy of all fossil fuels. In gas operating mode, the specific carbon dioxide emissions are thus typically reduced by between 15% and 20% compared with heavy fuel oil (HFO)/marine diesel oil operation. The corresponding reduction in nitrogen oxide (NOx) emissions is 85%–90%, while sulphur oxide (SOx) and particulate emissions are almost eliminated. Furthermore, there is no visible smoke, no sludge deposits, and no lead emissions; and benzene emissions are reduced by around 97%.

Growing opportunities for dual-fuel and gas engines in land and marine power markets have stimulated designs from leading medium-speed and low-speed engine builders. Development is driven by the increasing availability of gaseous fuels, the much lower level of noxious exhaust emissions associated with such fuels, reduced engine maintenance and longer intervals between overhauls for power plants. A healthy market is targeted from projects for floating oil production vessels and storage units, rigs, shuttle tankers, offshore support vessels, and LNG carriers.

Initially, there were two driving factors behind the use of dual-fuel engines. The first was the introduction by Norway of a levy on NOx

Pounder's Marine Diesel Engines and Gas Turbines
https://doi.org/10.1016/B978-0-08-102748-6.00003-7

emissions of vessels operating in its waters and the second was a desire to improve the efficiency of LNG carriers which were at the time all powered by steam turbine propulsion systems utilizing boil-off gas from the cargo. When the first sulphur emission control areas allowed under MARPOL Annex VI regulations were established in 2005 (Baltic Sea) and 2006 (North Sea) there was increased interest in LNG as a means of meeting the SOx emission limits as the fuel contains virtually no sulphur.

Valuable breakthroughs in mainstream markets have been made since 2000 when LNG-burning engines were specified for propelling *Glutra* a small Norwegian double-ended ferry (Mitsubishi high-speed engines). By May 2019 there were 435 ships capable of operating on LNG with a further 270 on order. These ships are in addition to those using LNG-fired steam turbine power systems. Most ship types are represented including ferries, offshore vessels, car carriers, LNG carriers, tankers, and cruise ships.

Since 1 January 2017, new ships that use LNG as a fuel are obliged to meet the requirements of the IMO's International Code of Safety for Ships using Gases or other Low-flashpoint Fuels (IGF Code). While the IMO had published earlier guidelines, the rules governing the installation of gas-fired engines, fuel storage systems, training, and bunkering requirements were left for the Flag States to devise. This was an obstacle to wider acceptance and since the advent of the IGF Code which sets universal requirements and the adoption by the IMO of new lower sulphur limitations on fuels effective in 2020, an increase in the number of dual-fuel vessels has been observed.

The adoption of the IGF Code was the culmination of over 10 years of work by several IMO bodies, starting with the approval by MSC78 (May 2004) of a work item on 'Development of provisions for gas-fuelled ships'. Following the adoption by MSC86 (June 2009) of the Interim Guidelines on safety for natural gas-fuelled engine installations in ships [resolution MSC.285(86)], MSC 87 approved the expansion of the scope of the work on the development of provisions for gas-fuelled ships to include ships fuelled by low-flashpoint liquid fuels. The present version of the IGF Code includes regulations to meet the functional requirements for natural gas fuel. Regulations for other low-flashpoint fuels will be added as and when they are developed by the IMO.

Natural gas has very good combustion characteristics in an engine and as it is lighter than air and has a high ignition temperature, is also considered a very safe fuel. The thermal efficiencies of various prime movers as a function of load are illustrated in Fig. 3.1.

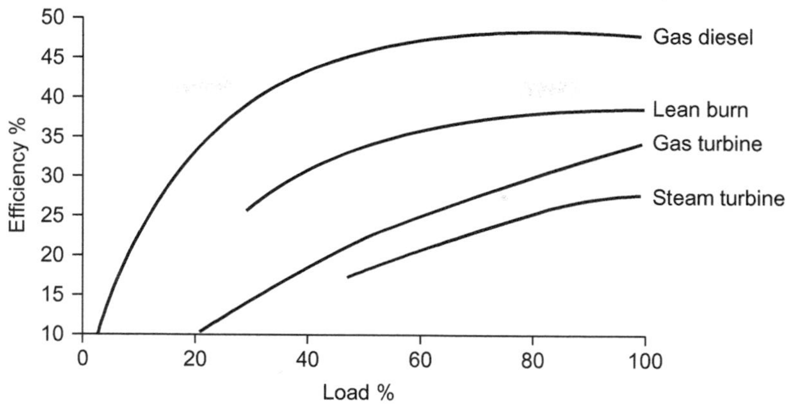

Fig. 3.1 Thermal efficiency of diesel and GD engines as a function of load (Wärtsilä).

Following on from the *Glutra* with its Mitsubishi GS12R-MPTK engines, the majority of dual-fuel engines installed were Wärtsilä medium-speed models. The success of these prompted other makers to enter the market and today there is competition in the medium-speed sector from MAN Energy Solutions, MaK, and many other makers.

Development of dual-fuel two strokes followed with MAN Energy Solutions (then trading as MAN Diesel & Turbo) announcing a new range of engines designated ME-GI in 2011. These engines were a development of the maker's MC-GI engines used in shore power production. Wärtsilä came slightly later to the two-stroke market with its first two-stroke dual-fuel engine being the X62DF introduced in late 2013.

The Swiss company, Wärtsilä Switzerland Ltd., responsible for the low-speed, two-stroke engine within the Wärtsilä group, was merged with China State Shipbuilding Corporation (CSSC) in early 2015 and renamed Winterthur Gas & Diesel Ltd. (WinGD). In 2016, Wärtsilä Corporation transferred its remaining shares of WinGD to CSSC making WinGD 100% owned by CSSC. The engine brand was then changed from 'Wärtsilä' to 'WinGD'. The new company has been successful in capturing a good share of the market in diesel and dual-fuel engines evolved from the earlier Wärtsilä models. The low-pressure X-DF technology is based on the lean-burn Otto-cycle combustion principle in which fuel and air are premixed and burned at a relatively high air-to-fuel ratio; a concept already used widely on medium-speed engines. Since LNG is mixed with the scavenge air before compression, the required gas pressure is maximum 13 bar or lower when operating at lower loads. As a result, the fuel supply system is relatively simple.

Wärtsilä Corporation is still considered as the largest manufacturer in the DF four-stroke sector. Its engines can be run in either gas mode or liquid-fuelled diesel mode. In gas mode, the engines work according to the lean-burn Otto principle, with a lean premixed air–gas mixture in the combustion chamber. (Lean burn means the mixture of air and gas in the cylinder has more air than is needed for complete combustion, reducing peak temperatures.) Less NOx is produced and efficiency increases during leaner combustion because of the higher compression ratio and optimized injection timing. A lean mixture is also necessary to avoid knocking (self-ignition).

The gas is fed into the cylinder in the air inlet channel during the intake stroke (Fig. 3.2). Instead of a spark plug for ignition—normally used in lean-burn gas engines—the lean air–gas mixture is ignited by a small amount of diesel fuel injected into the combustion chamber. This high energy source ensures a reliable and powerful ignition of the mixture, which is needed when running with a high specific cylinder output and lean air–gas mixture. To secure low NOx emissions, it is essential that the amount of injected diesel fuel is very small. The Wärtsilä DF engines, therefore, use a 'micro-pilot' injection, with less than 1% diesel fuel injected at nominal load, to achieve NOx emissions approximately 1/10th those of a standard diesel engine.

When the DF engine is running in gas mode with a premixed air–gas mixture, the combustion must be closely controlled to prevent knocking and misfiring. Wärtsilä says that the only reliable way to effect this is to use complete electronic control of both the pilot fuel injection and the gas admission on every cylinder head. The global air–fuel ratio is controlled by a wastegate valve, which allows some of the exhaust gases to bypass the

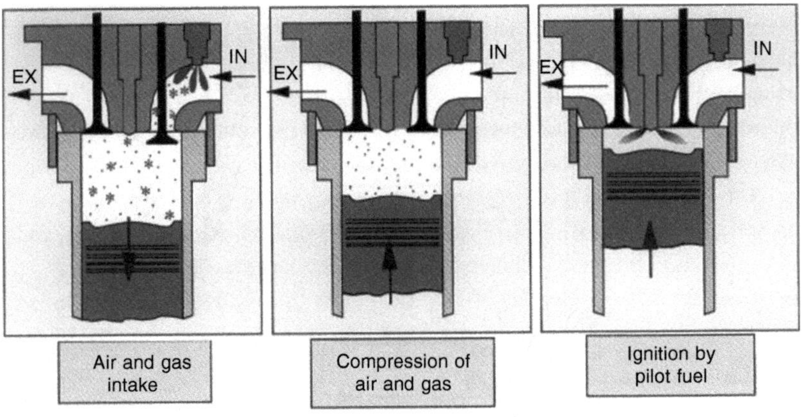

Fig. 3.2 The lean-burn DF diesel operating system (Wärtsilä).

turbine of the turbocharger. This ensures that the ratio is of the correct value independent of changing ambient conditions such as the temperature (Fig. 3.3).

The quantity and timing of the injected pilot fuel are adjusted individually together with the cylinder-specific and global air–fuel ratio to keep every cylinder at the correct operating point and within the operating window between the knock and misfire limits. This, according to Wärtsilä, is the key factor for securing reliable operation in gas mode, automatically tuning the engine according to varying conditions.

In diesel mode, the engine works according to the normal diesel concept using a traditional jerk pump fuel injection system: diesel fuel is injected at high pressure into the combustion chamber just before the top dead centre (TDC). Gas admission is deactivated but the pilot fuel remains activated to ensure reliable pilot ignition when the engine is transferred to gas operation.

The gas pressure in the engine is <4 bar at full load, making a single-wall pipe design acceptable if the engine room is arranged with proper ventilation and gas detectors. The gas valve on every cylinder head is a simple and robust, electronically controlled solenoid valve, promising high reliability with long maintenance intervals.

The pilot fuel system is a common rail (CR) system with one engine-mounted high-pressure pump supplying the pilot fuel to every injection valve at a constant pressure of 900 bar. Due to the high pressure, a double-walled supply system is used with leak fuel detection and alarms. The injection valve is of twin-needle design, with the pilot fuel needle

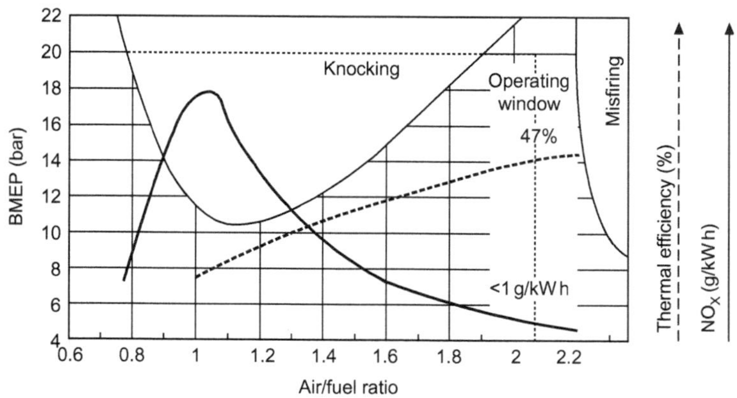

Fig. 3.3 A special electronic system for the Wärtsilä DF engine controls combustion in each cylinder, and optimizes efficiency and emissions under all conditions by keeping each cylinder within the operating window.

electronically controlled by the engine control system. It is important that the injection system can reliably handle the small amount of pilot fuel with a minimum cycle-to-cycle variation. The main needle design is a conventional system in which mechanically controlled pumps control the injection.

Other main components and systems are similar to designs well proven over a decade on Wärtsilä standard diesel engines, further underwriting high reliability of the DF engines.

Wärtsilä DF engines can be run on gas or light fuel (marine diesel or gas oil) and HFO. When running in gas mode, the engine instantly switches over to diesel operation if the gas feed is interrupted or component failure occurs. The switchover takes less than 1 s and has no effect on the engine speed and load during the process. Transferring from diesel to gas operation, however, is a gradual process: the diesel fuel supply is slowly reduced while the amount of gas admitted is increased. The effect on the engine speed and load fluctuations during transfer to gas is minimal, Wärtsilä reports.

The DF engine is normally started only with the pilot fuel injection enabled to a speed of approximately 60% of the nominal engine speed. When the ignition is detected in all cylinders, the gas admission is activated and the engine speed increased to the nominal speed; this safety measure ensures that no excess unburnt fuel is fed directly into the exhaust gas system.

Wärtsilä supplemented its portfolio with DF versions of the 320- and 500-mm bore medium-speed designs, which have, respectively, earned contracts for offshore supply vessel and LNG carrier propulsion projects, both burning natural gas. These 32DF and 50DF engines—covering a power band from 2010 to 17,100 kW with six-cylinder to V18-cylinder models—extended a programme which also embraced gas–diesel (GD) variants of the W32 and W46 engines as well as smaller spark-ignited gas engines.

The Wärtsilä 32DF engine was introduced for marine applications in 2000 to meet the requirements of a new safety class for installations with a gas pressure of <10 bar in a single-pipe arrangement. It provided an alternative to the 32GD engine (see later), which had been successful in offshore markets. An early application called for shipsets of four 6L32DF LNG-burning genset engines to power diesel-electric offshore supply vessels; each vessel has capacity for 220 m^3 of LNG stored at a pressure of 5.5 bar. The larger 50DF engine was officially launched in the following year, four six-cylinder models (each developing 5700 kW) being specified to drive the main gensets of a diesel-electric LNG carrier.

The 32DF was dropped when in 2008 the Wärtsilä 34DF was introduced as the successor for the company's first dual-fuel engine. The Wärtsilä 34DF is based on the Wärtsilä 32 diesel engine platform. It offers the operators the same fuel flexibility that was pioneered in the Wärtsilä 50DF but in a smaller package. This popular engine has seen strong sales in markets where complete fuel flexibility is paramount and has been upgraded to develop 11% more power than the original version.

In 2015 Wärtsilä introduced a new engine range with a slightly smaller bore than its 32 models. The 31 engines are available as conventional diesel (31D), dual-fuel (31DF), and spark-ignited gas engines (31SG)—the latter intended mainly for shore power production. The engine was primarily intended for the offshore sector but by the time of its introduction, offshore ships sales had collapsed along with the price of crude oil.

The Wärtsilä 31 engine was developed taking into account increased interest in new fuel types, reduced lifecycle cost and increased uptime, reduced environmental impact in combination with increased fuel and operational flexibility, and with guaranteed reliability and safe operation.

At the time of its introduction, the engine was awarded a Guinness World Records title as the world's most efficient four-stroke diesel engine. With the Wärtsilä 31, diesel fuel consumption can run as low as 165 g/kWh—putting it significantly ahead of any other four-stroke diesel engines available in the market. Its exceptional efficiency in excess of 50% is made possible through the use of technologies, such as two-stage turbocharging, a high-pressure fuel injection system, and adjustable valve actuation, in combination with the next-generation engine control system. By utilizing the new technologies and optimizing the engine parameters, efficiency is improved for all areas related to combustion, as well as thermal, mechanical (including friction and pump losses), and gas exchange efficiency.

The Wärtsilä 31 can be used both as a main propulsion engine and in diesel-electric configurations, or as an auxiliary engine, and it can be optimized for running either at a constant speed, along a propeller curve, or with constant torque. It is available in 8–16 cylinder configurations and has a power output ranging from 4.2 to 9.8 MW, at 720 and 750 rpm.

Designed to serve a wide selection of vessel types, the Wärtsilä 31's application areas include Offshore, Cruise & Ferry, and other marine segments. The multi-fuel capabilities that the Wärtsilä 31 brings to the market extend the possibilities for operators to utilize different qualities of fuels, from very light low sulphur fuel to very heavy diesel, and a range of different qualities of gas.

Wärtsilä claimed that the engine marks the first time that an engine platform has been developed concurrently for all its fuel variants. While previous engines were typically developed to run on diesel and subsequently adapted for gas, they were never fully optimized for the gas or dual-fuel modes. In contrast, the Wärtsilä 31 delivers maximum fuel efficiency and performance across all its fuel variants. Early sales of 31 series engines were all of the conventional diesel types but in 2018 contracts were agreed for the first 31DF engines which should debut in 2019/2020.

The older and larger bore Wärtsilä 50DF has proven popular for LNG carriers and some other ship types with no less than 168 vessels in operation using various combinations of six-cylinder, eight-cylinder, nine-cylinder, and V12-cylinder models in various combinations.

Wärtsilä DF engine data

Model	32DF	34DF	31DF	50DF
Bore	320 mm	340 mm	310 mm	500 mm
Stroke	350 mm	400 mm	430 mm	580 mm
Engine speed (rev/min)	720/750	720/750	720/750	500/514
Mean piston speed (m/s)	8.4/8.75	9.6/10.0	10.32/10.75	9.7/9.9
Mean effective pressure (bar)	19.9	22	27.2	20/19.5
Output/cylinder (kW)	335/350	480/500	530/550	950
Cylinder numbers	6, 9 L 12, 18 V	6, 8, 9 L 12, 16 V	8, 10, 12, 14, 16 V	6, 8, 9 L 12, 16, 18 V
Power band	2010–6300 kW	2880–8000 kW	4240–8800 kW	5700–17,100 kW

Wärtsilä 50DF engine

Based on the well-proven Wärtsilä 46 diesel engine, but with special systems for DF operation, the 50DF is produced in six-cylinder in-line to V18-cylinder configurations, yielding 950 kW per cylinder at a speed of 500 or 514 rev/min. A thermal efficiency of 47% (higher than any other gas engine at the time of its introduction) is claimed, and the engine delivers the same output whether running on natural gas, light fuel oil, or HFO.

The air–fuel ratio is very high, typically 2.2:1. Since the same specific heat quantity released by combustion is used to heat up a larger mass of air, the maximum temperature, and hence NOx formation, is lower. The mixture is uniform throughout the cylinder as the fuel and air are premixed before introduction into the cylinders, which helps to avoid local NOx formation points within the cylinder.

Engine starting is always in diesel mode, using both main diesel and pilot fuel. At 300 rev/min the main diesel injection is disabled and the engine transferred to gas mode. Gas admission is activated only when combustion is stable in all cylinders, ensuring safe and reliable starting. In gas running mode, the pilot fuel amounts to, 1% of the full load fuel consumption, the quantity of pilot fuel being controlled by the Wärtsilä engine control system (WECS).

Natural gas is fed to the engine through a valve station, the gas first filtered to ensure a clean supply. The gas pressure—dependent on the engine load—is controlled by a valve located near the engine driving end. The full load pressure is 3.6 bar (g) for a lower heating value (LHV) of 36 MJ/m^3 N; for a lower LHV, the pressure has to be increased. The system incorporates the necessary shut-off and venting valves to secure a safe and trouble-free gas supply. Gas is supplied through common rail (CR) pipes running along the engine, each cylinder then served by an individual feed pipe to the gas admission valve on the cylinder head.

The diesel fuel oil supply on the engine is divided into two systems: one for the pilot fuel and one for the backup fuel (Fig. 3.4). The pilot fuel is raised to the required pressure by a pump unit incorporating duplex filters, pressure regulator, and engine-driven radial piston-type pump. The high-pressure pilot fuel is then distributed through a CR pipe to the injection valves at each cylinder; pilot fuel is injected at ~900 bar pressure, and the timing and duration are electronically controlled.

Backup fuel is fed to a normal camshaft-driven injection pump by which it is pumped at high pressure to a spring-loaded injection valve of standard design for a diesel engine. The larger needle of the twin-needle injection valve is used in diesel engine mode and the smaller needle for pilot fuel oil when the engine is running in gas mode. Pilot injection is electronically controlled; main diesel injection is hydro-mechanically controlled.

The individually controlled solenoid valve allows optimum timing and duration of pilot fuel injection into every cylinder when the engine is running in gas mode. Since NOx formation depends greatly on the amount of pilot fuel, the system secures a very low NOx creation while retaining a

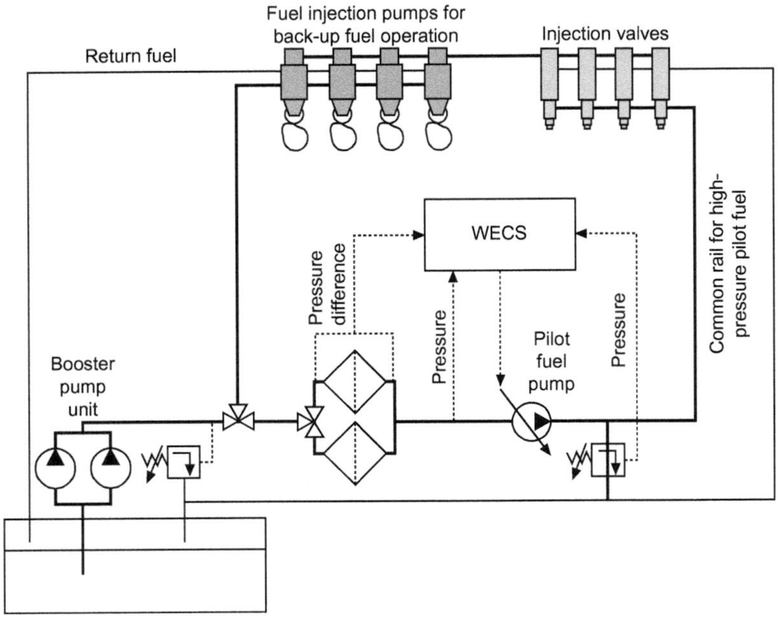

Fig. 3.4 Diesel fuel oil supply system for the Wärtsilä DF engine.

stable and reliable ignition source for the lean air–gas mixture in the combustion chamber.

Gas is admitted to the cylinders just before the air inlet valve. The gas admission valves are electronically actuated and controlled by the engine control system to deliver exactly the correct amount of gas to each cylinder. Combustion in each cylinder can thus be fully and individually controlled. Since the admission valve can be timed independently of the inlet valves, the cylinder can be scavenged without the risk of gas being fed directly to the exhaust system. Independent gas admission ensures the correct air–fuel ratio and optimized operating point with respect to efficiency and emissions; it also fosters reliable performance without shutdowns, knocking, and misfiring. The gas admission valves have a short stroke and embody special materials to underwrite low wear and long maintenance intervals.

A well-proven Wärtsilä-developed monobloc fuel injection pump is specified to withstand the high pressures involved, with a constant pressure relief valve provided to avoid cavitation and a plunger featuring a wear-resistant coating. The fuel pump is ready for operation at all times.

A pilot fuel pump is a stand-alone unit comprising a radial piston pump and necessary filters, valves, and control system. It receives the signal and required pressure level from the engine control unit and independently sets and maintains the pressure at that level, transmitting the prevailing fuel pressure to the engine control system. High-pressure fuel is delivered to each injection valve via a CR pipe, which acts as a pressure accumulator and damper against pressure pulses in the system. The fuel system is of double-wall design with alarm for leakage.

The 50DF engine (Fig. 3.5) can be switched automatically from fuel oil-to-gas operation at loads below 80% of the full load. The transfer takes place automatically after the operator's command, without load changes. During the switchover—which lasts for about 1 min—the fuel oil is gradually substituted by gas. A trip from gas-to-fuel oil operation occurs instantaneously and automatically (e.g. in the event of gas supply interruption). A correct air–fuel ratio under any operating conditions is essential for optimum performance and emissions—a status addressed by an exhaust gas wastegate valve. Part of the exhaust gases bypass the turbocharger through this valve, which adjusts the air–fuel ratio to the correct value independently of the varying conditions under high engine loads. The wastegate is actuated electropneumatically.

A flexible cooling system, optimized for different cooling applications, has two separate circuits: high temperature and low temperature. The HT circuit controls the cylinder liner and cylinder head temperatures, while

Fig. 3.5 Wärtsilä 50DF engine in V12 cylinder form for LNG carrier propulsion.

the LT circuit serves the lubricating oil cooler. Both circuits are also connected to the respective parts of the two-stage charge air cooler, and both HT and LT water pumps are engine driven as standard.

All engine functions are controlled by the WECS, a microprocessor-based distributed control system mounted on the engine. The various WECS modules are dedicated and optimized for certain functions and communicate with each other via a databus. Diagnostics are built in for easy troubleshooting. The core of the WECS is the main control module, which is responsible for keeping the engine at optimum performance in all operating conditions such as varying ambient temperature and methane number. The main module reads the information sent by all the other modules and applies it in adjusting the engine speed and load control by determining reference values for the main gas admission, air–fuel ratio and pilot fuel amount and timing. It also automatically controls the start and stop sequences of the engine and the safety system and communicates with the plant control system.

Each cylinder control module is arranged to monitor and control three cylinders; it controls the cylinder-specific air–fuel ratio by adjusting the gas admission individually for each cylinder. The module measures the knock intensity (uncontrolled combustion in the cylinder) and this information is applied in adjusting the cylinder-specific pilot fuel timing and gas admission. Light knocking leads to automatic adjustment of the pilot fuel timing and cylinder-specific air–fuel ratio; heavy knocking leads to load reduction or a gas trip.

MAN 51/60DF engine

MAN Energy Solution's 510-mm-bore 51/60DF medium-speed engine introduced in 2011 (Fig. 3.6) was designed for LNG carrier propulsion with the capability to operate in both gaseous and liquid fuel modes. In the former mode, the engine runs according to the lean-burn Otto combustion process: the premixed lean gas–air mixture is ignited by the compression ignition of a small quantity of marine diesel pilot fuel injected into the main combustion space. A pilot injection of $<1\%$ of the normal liquid fuel quantity ensures very low NOx emissions in gas mode. In liquid fuel operating mode, the fuel is injected via conventional main pumps adopted from the MAN 48/60B heavy fuel engine, and the thermodynamic working process is the diesel combustion cycle. Fuels ranging from marine diesel oil to HFO

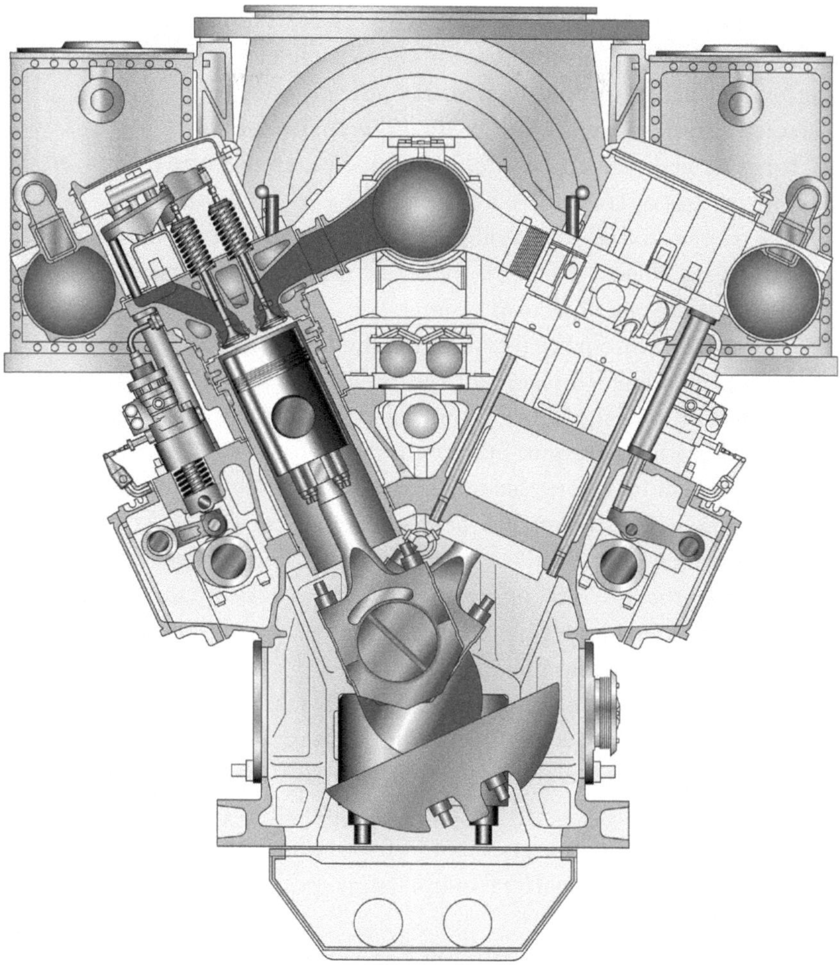

Fig. 3.6 Cross-section of MAN Diesel 51/60DF medium-speed engine.

can be used, with swift and smooth switching between liquid and gaseous fuels with the engine running.

Development was mainly influenced by the need to challenge steam turbine and gas turbine propulsion in the LNG carrier market by offering: high power output and high thermal efficiency; fuel flexibility (vaporized LNG or DMA, DMB, and HFOs); high availability in liquid and gaseous fuel modes; short start-up and loading times; low emission levels and easy maintenance.

MAN 51/60DF engines are offered in 6-to-9 in-line and V12 to 18-cylinder models covering a power range up to 18,900 kW at

500/514 rev/min with a mean effective pressure rating of 20.0–20.6 bar. The design was based on the proven 48/60B diesel engine, the following main features being adapted to facilitate operation on gas:

- Gas admission to each cylinder by individual valves
- Charge air and exhaust manifolds equipped with pressure relief valves to avoid any gas operation risks
- A CR system for pilot fuel oil injection integrated in the cylinder head; although well below 1% of the main fuel supply, the pilot fuel quantity is controllable by the injection system
- Separate conventional fuel system for the liquid fuel mode
- SaCoS DF engine management system for the whole operating range.

The cylinder bore of the base engine was enlarged from 480 to 510 mm to secure a higher power output, but the moderate brake means effective pressure of 19 bar—more favourable in gas mode than in liquid fuel operation—was maintained. The bmep has been increased up to 20.6 in the 2019 versions of the engine. Design work mainly involved adapting gas-related components to the 48/60B engine, the key elements being the air–fuel ratio control, individual cylinder gas admission valves, knock detection system, and CR pilot fuel injection system.

A compressor bypass controls the air–fuel ratio of the cylinders, the optimum ratio fostering stable operation in gas mode with high efficiency and low emissions. The electronically driven gas valves feed each cylinder individually with its required amount of gas, and an electrical actuator controls the main injection pumps. An additional phase pickup is necessary to deliver the required signals of crank angle position in the liquid fuel operating mode.

Knocking control plays an important role in reliable gas operation, a system detecting knock behaviour for each cylinder unit individually via an acoustic sensor mounted on the cylinder head. The independent acoustic signals are fed into the main sensor unit where the knocking level is detected. The overall engine control system reacts to the knocking level and takes appropriate measures by adjusting the engine operating parameters.

Pilot fuel injection is executed by an electrically driven CR pump and a solenoid-activated pilot fuel oil injector. For safety reasons, the pressure relief valves are mounted on the charge air manifold and exhaust gas manifold. Installing the solenoid-driven CR pilot fuel injector separately from the existing main fuel injection system eases changing of the injectors (which have different times-between-overhauls) compared with combining them in a central injection system, MAN Energy Solutions asserts.

In gas operating mode the required amount of gas is metered for each cylinder individually by the gas admission valves integrated into the rocker arm housing. A flow control pipe ensures a homogeneous mixture of air and gas entering the combustion space. Double-wall piping with insulation and ventilation of the main gas valve itself follows the gas-free engine room concept, which allows the installation of several engines in single machinery space.

The SaCoS DF engine management system was evolved from the group's established SaCoS safety and control system to cover DF engine operation. It undertakes control in both liquid and gas operating modes, with the air–fuel ratio controlled by a compressor bypass installed after the charge air cooler. The SaCoS system controls the ratio via the compressor bypass flap.

SaCoS system architecture is dominated by two independent governors: one controls the main injection fuel oil pumps and the individual fuel gas admission valves for each cylinder unit; the other controls the CR pilot fuel injectors. The individual knocking levels, collected from each cylinder by the knocking detection unit, are processed in the SaCoS system. In combination with the individual cylinder control of pilot injection and gas admission, the system control secures stable operation in gas mode with a sufficient margin for knocking and misfiring. In addition, the control system establishes connections with the ship's overall alarm and safety networks.

Gas-diesel engines

Wärtsilä GD technology differs from the DF diesel concept in using direct, high-pressure injection of gas and 3% pilot fuel for its ignition. GD medium-speed engines can be operated on heavy fuel and diesel oils, crude oil directly from the well or natural gas; supply can be switched instantly and automatically from one fuel to another without shutdown. Conversions of existing engines from normal heavy fuel mode to natural gas–diesel oil operation can be executed with small modifications. Containerized engine room packages have been delivered along with the associated gas compressors and other ancillaries.

The GD engine was developed in the late 1980s mainly as a prime mover for the offshore industry. High-pressure direct injection for the gaseous fuel maintains the diesel process, making the concept insensitive to the methane content of the fuel. Such a feature makes the GD engine especially suitable for mobile oil production processing plants, where the gas composition may change with the location of the vessel and the stage in the production process.

Initially, the gas injection was controlled by a mechanical/hydraulic system using a cam-actuated jerk pump to generate the control pressure for the gas needle. Although reliable, Wärtsilä explains, this system did not offer the required flexibility and was later replaced by an electrohydraulic system based on a constant rail pressure to control the gas injection via solenoid valves. Gas-specific monitoring and safety functions were all integrated, and the system allowed adjustment of individual cylinders. This represented the first step towards fully computerized systems for engine control, and valuable experience was gained in both software and hardware requirements.

Gas engines

Rolls-Royce is a specialist in marine gas engine technology, its Norwegian subsidiary Bergen Diesel having started the development of a lean-burn gas-fuelled version of the 250 mm bore K-series medium-speed diesel engine in the mid-1980s. More than 400 engines of this type have been sold for land power and cogeneration installations, reportedly demonstrating high efficiency and very low emissions. The company has also been keen to demonstrate the potential in natural gas-fuelled ferries in an era of concern over atmospheric coastal pollution and the increasing availability of gas from domestic pipeline terminals in Norway.

Rolls-Royce Bergen spark-ignition lean-burn gas engine technology has matured through several generations of engines and thousands of operating hours in shoreside power installations, running on various gas fuels including natural gas and biogas derived from landfill. Its lean-burn gas engine operates on the Otto cycle with mixture compression and external ignition. By providing a strong ignition source, the prechamber and the gas–air mixture in the cylinder can be leaned out, giving much- improved engine performance. Efficiency is increased, emissions reduced (particularly NOx), and the specific power of the engine can be significantly raised because the knock limit is extended.

Combustion air is supplied by the turbocharger to the individual cylinders via the intercooler and the charge air manifold, and a timed mechanical valve injects gas into the inlet air stream. A special inlet port design and engine control system ensures a homogeneous and lean mixture of air and gas.

During compression, the lean mixture in the cylinder is partially pushed into the prechamber, where it mixes with pure gas to form a rich mixture that is easily ignited by the spark plug. Fast and complete combustion of the main mixture in the cylinder is then fostered by a powerful ignition discharge from the prechamber and an optimized combustion chamber configuration. A responsive and flexible ABB turbocharger with variable turbine geometry reduces 'turbo lag', while Lamda-control avoids 'over-fuelling' during hard manoeuvring.

An advanced electronic control system ensures that the operating parameters of the engine are adjusted and optimized in relation to each other. The system sets the optimum main and prechamber gas pressures, the air–fuel ratio, the fuel rack position, the ignition timing, and air throttle position.

Extensive test bed running on various load profiles has reportedly demonstrated an ability to handle propeller law operation and all the load conditions imposed by dynamic interactions between the hull, mechanical transmission, and the prime mover itself. No limits are placed on low load running; optimum combinatory profiles can be used with CP propellers, and crash stops and rapidly changing loads are said to present no problems.

Bergen's KV-G4 gas engine was derived from the company's long-established 250 mm-bore K-series medium-speed diesel design. Introduced for genset applications in 2002, it was released in 2004 as a variable speed engine for direct mechanical propulsion. Electric propulsion installations with the engine were pioneered in 2007 with V12- and V16-cylinder LNG-burning versions driving the main gensets of Norwegian double-ended coastal ferries.

Design work on the spark-ignited 350 mm-bore B-gas (B35:40) engine began in spring 2002, tapping experience with the 320 mm-bore B-series medium-speed diesel engine launched in the late 1980s. A V12-cylinder prototype developing 5100 kW made its commercial debut in a Danish combined heat and power station in 2003 after a comprehensive test programme in Bergen.

The B-gas engine is also offered in V16- and V20-cylinder versions, taking the power band up to 9000 kW. Complementing the K-series and B-series portfolios, Rolls-Royce also planned a gas variant of Bergen's 250 mm-bore C-series (C25/33) diesel engine in six-, eight-, and nine-cylinder models.

MAN Energy Solution's ME-GI engine family

Medium- and low-speed gas-burning engine designs are offered by MAN Energy Solutions building on early experience gained with a Holeby 16V28/32-GI (gas injection) four-stroke model (Fig. 3.7), and a 6L35MC two-stroke model.

A high-pressure gas-injection system for its MC low-speed engines was jointly developed by MAN Energy Solutions and its key Japanese licensee Mitsui, which delivered a 41,000kW 12-cylinder K80MC-GI engine for power generation in an environmentally sensitive region (Figs 3.8 and 3.9). Extensive land-based experience from the early 1990s is reported from this installation in peak load service on high-pressure gas.

A new generation of large LNG carriers stimulated refinement of the ME-GI engine design to burn cargo boil-off gas in almost any ratio with fuel oil. Different natural gas qualities, depending on the source field, maybe transported as LNG, but the ME-GI system is claimed to be easily adaptable to a new shipping route with a different gas composition. An ME-GI system can be retrofitted to existing ME/ME-C HFO-burning engines should fuel

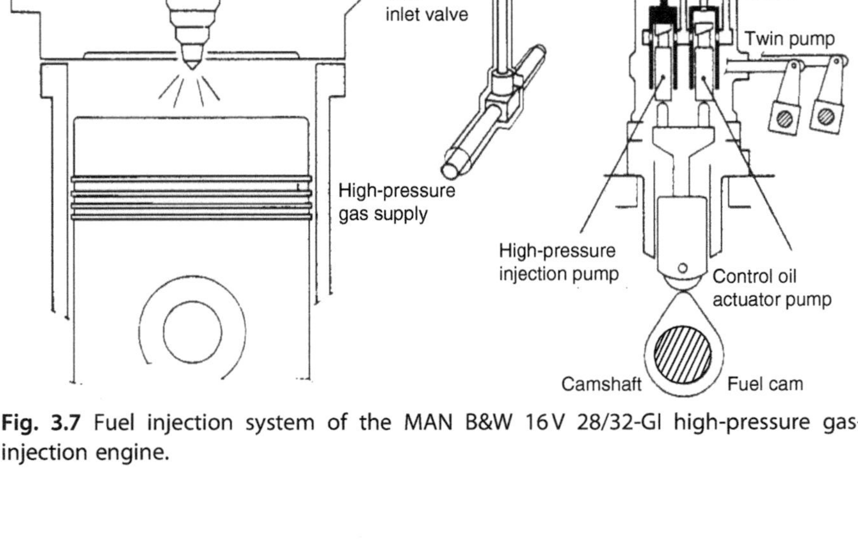

Fig. 3.7 Fuel injection system of the MAN B&W 16V 28/32-GI high-pressure gas-injection engine.

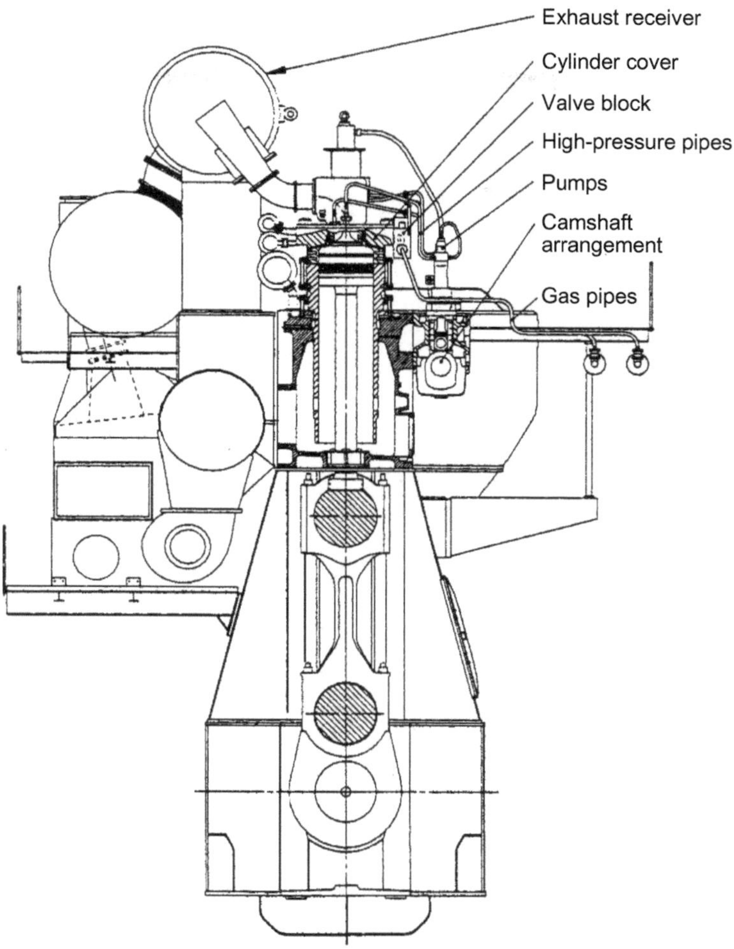

Fig. 3.8 MAN Diesel's MC-GI low-speed DF engine, indicating new or modified components compared with the standard diesel version.

prices or environmental regulation make such projects economically desirable.

The working principle of the GI engine is similar to that of its traditional two-stroke diesel counterpart but with the combustion process based on a high air surplus and a high-pressure gas injection system. There is no derating or knocking due to gas pressure or airflow, and the exhaust gas boiler design is also unchanged. Power, exhaust gas volume, and temperature values are similar for GI and HFO-burning engines; and performance conditions, choice of cylinder liner, cover, piston, and cooling system are the

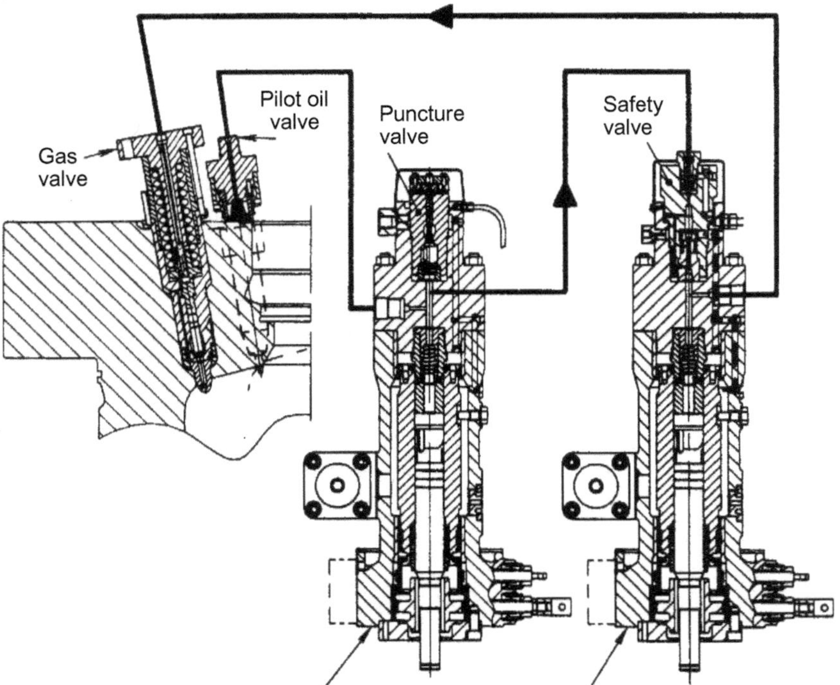

Fig. 3.9 Gas injection is stopped immediately on the first failure to inject the pilot oil by the patented safety valve incorporated in the control oil pump of the MAN B&W MC-GI engine.

same for the two types. The cylinder cover is slightly changed for the GI engine, however, to accommodate the gas valves and a block for handling the fuel gas. Pressurized fuel gas (at a maximum of 250–300 bar) is required but the thermodynamic condition in the combustion chamber is similar to that associated with heavy fuel burning. There is one limitation: condensation in the system is not allowed.

GI engine control and safety are handled by an add-on unit to the proven ME engine control system. The main modified parts in a GI engine are double-walled high-pressure gas pipes; a gas valve control block with internal accumulator on the (slightly modified) cylinder cover; gas injection valves and electronic gas injection (ELGI) valves to control the amount of gas injected. Small modifications to the exhaust gas receiver and the control and manoeuvring system were also dictated. The new units required were a ventilation system for the space between the inner and outer pipes of the double-wall piping; a sealing oil system, delivering oil to the gas valves separating the control oil and the gas: and an inert gas system to enable purging of the engine's gas arrangements.

Operational safety in DF mode is addressed by incorporating a control and safety system with an analyser checking the hydrocarbon content of the air in the gas pipes. The scavenge air pressure, exhaust gas temperature and compression, combustion and expansion pressure is continuously monitored. The GI control and safety system is designed to 'fail to safe' condition. All failures detected during gas–fuel operation (including failures of the control system itself) will result in a gas–fuel stop/shutdown and changeover to HFO operation. Changeover from LNG to HFO—which will reportedly never trigger any engine power loss—is followed by blowout and purging of the high-pressure gas pipes and the complete gas supply system.

Since the gas supply piping is of CR configuration, the gas injection valve must be controlled by an auxiliary control oil system. In principle, this consists of the ME engine hydraulic control oil system and an ELGI valve. High-pressure control oil is supplied to the gas injection valve, thereby controlling the timing and opening of the valve. DF operation requires the injection of both pilot and gas fuels into the combustion chamber for which different types of valves are used; two are fitted for gas injection and two for pilot fuel. The auxiliary media for both fuel and gas operation call for a high-pressure gas supply; a fuel oil supply (pilot oil); control of oil supply for activating the gas injection valves and sealing oil supply.

Under normal operation, with no malfunctioning of the fuel oil valve, the fuel gas valve is opened at the correct crank angle position and gas is injected. The gas is fed directly into an ongoing combustion process, so the chance of having unburnt gas eventually slipping past the piston rings and into the scavenge air receiver is considered very low. Monitoring the scavenge air receiver pressure safeguards against such a situation, and in the event of high pressure the gas mode is stopped and the engine returned to burning only fuel oil.

Gas flow to each cylinder during one cycle is detected by measuring the pressure drop in the accumulator. Any abnormal gas flow, whether due to seized gas injection valves or blocked valves, will be detected immediately. The gas supply will then be discontinued and the gas lines purged with inert gas. Also in such an event, the engine will continue running on fuel oil alone without any power loss.

At the outset, the prime marine industry target for large gas-fuelled engines was LNG carrier propulsion, where the cargo boil-off gas can be tapped. However, increasingly stringent emission regulations have stimulated interest in other sectors. Today there are container ships, tankers, and bulk carriers fitted with dual-fuel two-stroke engines as well as LNG carriers.

Development of the ME-GI engine has seen variants that will run on other fuels as well as HFO and LNG. In June 2016, MAN, in conjunction with its licensee Mitsui Engineering & Shipbuilding Co., Ltd. announced the completion of the world's first ME-GIE ethane-operated two-stroke diesel engine referring to it as the first triple-fuel engine. The Mitsui-MAN B&W 7G50ME-C9.5-GIE was one of three for installation in LEG (liquefied ethylene gas) carriers of 36,000 m^3 built for Hartmann Schiffahrt of Germany and Ocean Yield of Norway at Sinopacific Offshore Engineering in China.

While ME-GI engines had previously been designed for use by methane, operation on ethane was a new development. After methane, ethane is the second largest component of natural gas with a content that varies from <1% to >6% by volume. However, before the natural gas can be used as a commercial fuel, it must be processed to remove impurities, including ethane. Like many hydrocarbons, ethane is isolated on an industrial scale from natural gas, usually by liquefaction at cryogenic temperatures. Its chief use is within ethylene production.

In 2012, MAN decided to expand the engine portfolio for use with low-flashpoint fuels and, as a result, the ME-LGI engine series was introduced. The ME-LGI engine is the dual-fuel solution for low-flashpoint liquid fuels, unlike the ME-GI engine where the fuel is injected in the gaseous state.

One fuel which was envisaged for this engine from the start was methanol. There had been an interest in using methanol as a marine fuel for some time not least because it is a liquid and therefore handling and storage is less problematic than with gas fuels that need to be either refrigerated or else stored under high pressures. Methanol is characterized by a low cetane number and the self-ignition quality is therefore poor and similar to fuel oil. However, it has a much lower flashpoint of just 12°C.

To be able to use methanol fuel on the ME-LGI, the cylinder covers are equipped with the fuel booster injector valves designed specifically for methanol operation. For a 50 bore engine, this means that each cylinder cover will be equipped with two additional methanol booster injectors. A liquid gas injection (LGI) block will be fitted on the cylinder cover.

This block contains a control valve for methanol fuel injection, a sealing booster activation valve, a forced suction activation valve, an LGI purge valve, and methanol fuel inlet/outlet valves. All pipes for hydraulic oil and fuel are double-walled. Furthermore, the double-walled pipes for methanol fuel are vented with ventilation air.

Principle of the BFIV for methanol ME-LGI engine

Principle of the BFIV – Booster Fuel Injection Valve.

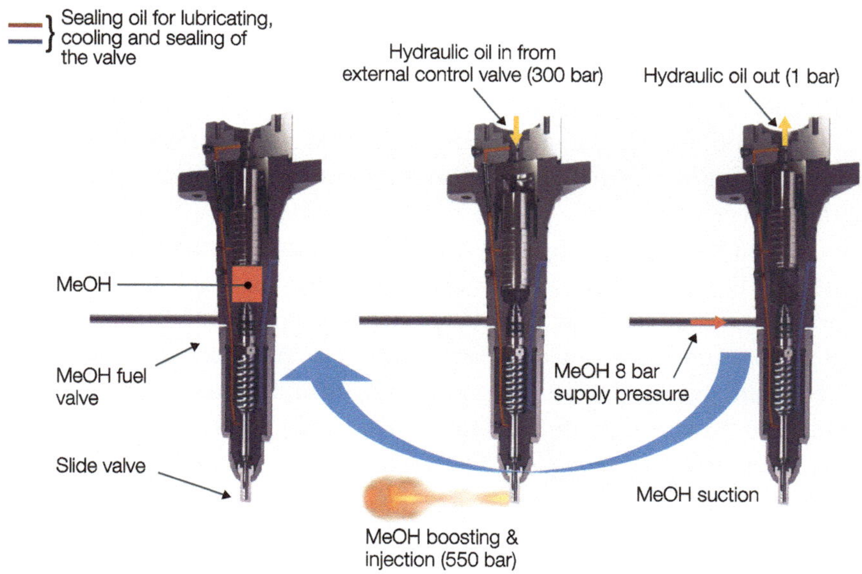

The methanol booster injector valve must be cooled and running surfaces must be lubricated. For this purpose, a combined sealing and cooling oil system delivering a 50 bar system oil pressure has been integrated on the engine, and the system both lubricates all running surfaces and controls that the temperature in the booster valve is lower than max. 60°C. The sealing oil pressure is generated internally in the booster valve in order to avoid contamination of the hydraulic oil operating the valve.

The sealing oil has other advantages. It avoids methanol from going into the umbrella system and further down into the drain oil system. The cooling oil and sealing oil system are fully integrated into the engine design, including equipment for continuous monitoring of methanol contamination in the oil system. If methanol is detected in the system, the engine will switch to fuel oil mode, and the methanol will be purged from the engine.

At the same time, the cooling oil pump supply side will be switched to clean system oil, and the oil circuit will be flushed with clean oil. Then, the clean oil will be collected together with the contaminated oil in the cooling oil tank, and the system will only be able to continue operation when no methanol is detected in the tank.

To ensure the correct temperature of the BFIV, the oil is cooled in a heat exchanger, which is connected to the low-temperature cooling system. When the fuel is injected, the combustion condition is monitored with pressure mean indicator (PMI) sensors located in each of the cylinder covers. The injection pressure is approximately 500 bar.

Three combustion conditions are monitored. The compression pressures, combustion pressures, and expansion pressures. The 8 bar pressurized methanol is delivered to the engine via double-walled pipes, ventilated with dry air taken from the starting air system. To suck in air, a ventilation system is fitted at the outlet. All methanol fuel equipment is made in a double-walled design, and any methanol leakage will develop into methanol vapour. This is monitored by HC sensors located close to the outlet of the double-walled piping system. In case of a too high methanol vapour content in the ventilation, the safety system will shut down the operation on methanol and return to operate on sole fuel oil. This switch is done smoothly and without any power loss.

A methanol control and safety system is integrated into the engine. The main operating panel (MOP) is equipped with a user-friendly interface for methanol operation. At this panel, the LGI system monitors and indicates the relevant pressure, temperatures, and position of the different valves.

The first ships with ME-LGIM engines entered service in 2016 and all 12 vessels have been built. All are methanol carriers. One other ship is currently running on methanol—the RoPax ferry *Stena Germanica* but this is a conversion project undertaken by Wärtsilä in 2015. The ship's four Sulzer 8ZAL40S four-stroke medium speed engines were adapted to run on methanol.

The other development of the ME-GI engine is the ME-LGIP introduced in 2018. This variant is designed to run on liquid petroleum gas (LPG). The primary characteristics of ME-LGIP engines include:

- a low-pressure supply system
- a fuel-injection system similar to that most recently developed for MAN Energy Solutions' conventional MDO/HFO engines—the FBIV (Fuel Booster Injection Valve)
- an injection pressure of 500–600 bar

- the ability to handle low-sulphur/low-flashpoint fuel types: methanol, ethanol, LPG, and dimethyl ether (DME).

LPG is considered an environmentally friendly fuel, in much the same class as liquefied natural gas (LNG), and an LPG-fuelled engine will significantly reduce emissions, enabling vessels to meet the stringent IMO SOx emission regulations. LPG's future as a viable fuel for general marine transportation looks promising as it will not require as large an investment in infrastructure. As a widespread energy source, availability is high and LPG is easier to store and handle, compared with cryogenic gaseous fuels.

Furthermore, LPG is traditionally a cheaper fuel than MGO yet delivers the same performance and efficiency. Importantly, the ability to use LPG cargo as a supplemental fuel source provides significant cost savings for LPGC owners or charterers, including reduced time and fees for fuel bunkering. MAN Energy Solutions expects strong demand for the ME-LGIP engine from very large gas carriers (VLGCs) and coastal vessels from its introduction.

VOC as a fuel supplement

Oil vapours or VOCs are light components of crude oil evaporated mainly during tanker loading and unloading but also during a voyage caused by cargo movement within tanks.

VOC discharge represents a significant loss of energy as well as an environmental problem (the nonmethane part of the VOC released to the atmosphere reacts in sunlight with NOx and may create a toxic ground-level ozone and smog layer). It is estimated that around 5 million tonnes of VOC are released annually into the atmosphere from oil cargoes carried at sea.

Both MAN Energy Solutions and Wärtsilä have explored the potential of capturing and liquefying VOC released from cargoes and using it to supplement the main fuel of the vessel's main and auxiliary engines or boilers.

The VOC-burning system was first applied at sea to the twin 6L55GUCA main engines of the 125,000 dwt shuttle tanker *Navion Viking* in 2000—an on board recovery system collecting the discharges and liquefying the product by compressing and cooling. The resulting condensate is stored and burned as fuel for propulsion, yielding substantial savings in HFO consumption (Fig. 3.10).

Engines for such applications must be able to operate on ordinary HFO if VOC fuel is unavailable, and they must also accept any possible composition

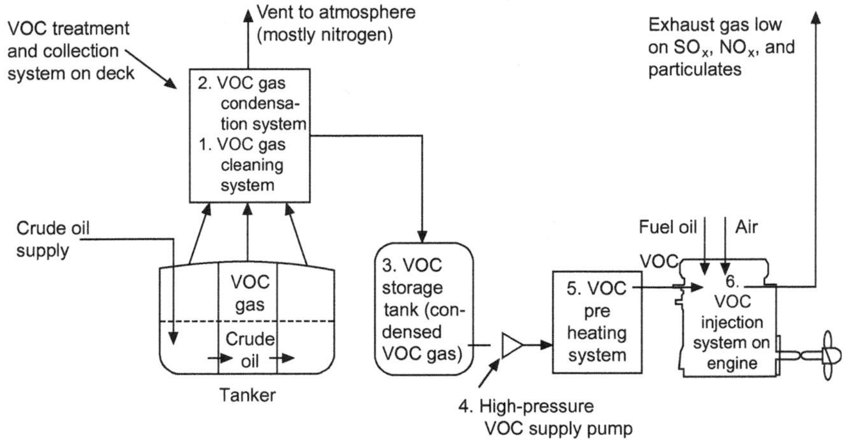

Fig. 3.10 Schematic representation of system for burning VOC discharges in the main engines of a shuttle tanker (MAN Diesel). *From MAN Energy Solutions.*

of VOC condensate since it is not possible to apply any fuel specification: whatever is collected must be burned. Shoreside tests by MAN Diesel confirmed that its DF two-stroke engines could use any VOC/HFO ratio between 92/8% and 0/100%, as well as any relevant VOC quality.

Wärtsilä has developed a VOC recovery unit which is also finding use on board vessels. The Wärtsilä Hamworthy VOC recovery system captures the VOC emissions and removes the heavier hydrocarbon fractions (mainly C3+) of the gas by a two-stage condensation process before the non-condensable hydrocarbons in surplus gas (mainly methane and ethane) is fed to power generating module. Surplus gas constitutes up to 90% of the energy fed into the power generating module, the rest can be supplied from the liquefied VOC (LVOC) stored in a storage tank. LVOC is a light hydrocarbon fuel which can be applied as a clean fuel. The system has been successfully utilized on board several shuttle tankers and floating storage units (FSU) in the North Sea. The ships concerned will use the LVOC in either the two-stroke WinGD XDF main engines or in Wärtsilä DF four-stroke engines.

Design features of gas-burning engines

The fuel type will depend on the demands of the plant and fuel availability. For large engines, the DF system with a liquid pilot fuel is necessarily applied, providing the plant with redundancy to operate on liquid fuel alone in the event of shortage or failure of the gas supply. In smaller plants, the engine may exploit the same principles or be optimized for using a single gaseous fuel or even a number of different gases.

The combustion chamber design may be the single chamber configuration used on all large engines, or it may be a divided chamber design featuring a prechamber. In the latter case the prechamber normally operates with an overrich fuel–air mixture, provided either by a separate gas supply to the chamber or by pilot fuel injection. The aim in both cases is to secure stable ignition in the prechamber and to minimize NOx formation (due to the lack of oxygen). Combustion in the main chamber may then use a lean mixture effectively ignited by the hot combustion products emerging from the prechamber. The lean mixture burns at a lower temperature which, in turn, minimizes the formation of NOx in the main chamber.

Charge exchange may be with or without supercharging and the medium may be pure air or a gas–air mixture. Small, simple plants are normally naturally aspirated with a gas–air mixture while larger plants, for cost reasons, are turbocharged. Four-stroke engines may use air or a gas–air mixture for the charge exchange but two-stroke engines are restricted to air.

The air–fuel ratio may be selected according to engine type and application. A rich mixture can be exploited to secure a high output and stable running from small engines without emission restraints. Such applications will become scarce in the future: a stoichiometric air–fuel ratio controlled by a sensor and combined with a three-way catalyst to comply with emission regulations is considered a more suitable choice.

Except for such small engines, lean-burn or stratified charge is the principle being applied in most new developments, and this may be used without a prechamber. The reasons are the favourable emission characteristics and high engine efficiency. Mixture formation, defined as the mixing process of air and gas, may take place in the cylinder by direct gas injection at low pressure (after valve closure but before compression start) or by high-pressure gas injection near TDC. Two-stroke engines need to apply one of these alternatives to avoid excessive gas losses, which would promote high HC-emissions and reduced efficiency.

With low-pressure direct injection, the gas–air mixture is exposed to the compression process and hence increasing pressure and temperature. This may lead to self-ignition and severe knocking, excluding its use in new engine developments.

In the case of late-cycle high-pressure gas injection, the gas is only admitted to the combustion chamber at the desired time for combustion. This is initiated either by a pilot fuel injection or by directing the gas jets to a glow plug.

The high-pressure gas injection system is relatively costly, but it renders the engine practically insensitive to fuel gas self-ignition properties. For this reason, MAN Energy Solutions suggests that the system can be considered as underwriting true multi-fuel engines, provided that the auxiliary systems for fuel supply have been designed accordingly.

For most four-stroke gas-burning engines, the mixture formation occurs outside the cylinder (an exception is the high-pressure gas-injection engine). The gas may be mixed with air before the turbocharger, yielding a very uniform mixture but also a rather large volume of ignitable gas in the intake system of the engine. Alternatively, it may be added to the intake air in the cylinder head or the intake air manifold, close to the intake valve, by a combined intake valve/gas valve or a separate gas valve.

Ignition sources take various forms but are all, basically, just high temperature sources. MAN Energy Solutionsss explains that the highest ignition energy is provided by pilot injection, which also makes it possible to locate multiple ignition points deep inside the gas–air mixture: beneficial for the fast and efficient combustion of lean mixtures. This solution is reliable and yields extended times-between-overhaul, but it is more costly than a spark plug or glow plug arrangement. Pilot injection, by its nature, dictates the presence of a liquid fuel, which, in some cases, is considered a disadvantage. In other cases, this is viewed positively as a means of fuel redundancy.

CHAPTER FOUR

Exhaust emissions and control

Marine engine designers in recent years have had to address the challenge of tightening controls on noxious exhaust gas emissions imposed by national, regional, and international authorities responding to concern over atmospheric pollution and its impact on human health and climate change.

As discussed in the first chapter, the only substances regulated are SOx and NOx. Since SOx formation is dependent on the quantity of sulphur in the fuel being used, engine-based emission controls have mainly been focused on NOx reduction.

Curbing emissions has stimulated unprecedented cooperation between rival marine engine builders/designers, most notably in the European Union-sponsored Hercules Projects, which aim at developing new technologies to significantly reduce gaseous and particulate emissions from engines while raising their efficiency and reliability. Among the R&D tasks are thermodynamics and mechanics of 'extreme' parameter engines, advanced combustion concepts, multistage 'intelligent' turbocharging, 'hot' engines with energy recovery and compounding, internal emission reduction methods and advanced after-treatment techniques, new sensors for emissions and performance monitoring, and adaptive control for electronically controlled engines.

Noxious emissions amount to 0.25%–0.4% by volume of the exhaust gas, depending on the amount of sulphur in the fuel and its lower heat value, and the engine type, speed, and efficiency. After 1 January 2020, when the IMO's global cap on sulphur levels in fuels is reduced from a maximum 3.5% (in practice the average is around 2.8%) to 0.5%, the noxious emissions will likely be at or below the 0.25% level.

Controlling NOx emissions

Dominating influences in the formation of NOx are temperature and oxygen concentration: the higher the temperature and the higher the residence time at high temperature in the cylinder, the greater the amount of

Pounder's Marine Diesel Engines and Gas Turbines
https://doi.org/10.1016/B978-0-08-102748-6.00004-9

thermal NOx that will be created. A longer combustion time span means that low-speed two-stroke engines, therefore, generate higher NOx emissions than medium- and high-speed four-stroke engines of equivalent output. Apart from the use of alternative fuels, such as methanol, two main approaches can be pursued in reducing NOx emission levels.

Primary (in-engine) measures aimed at reducing the amount of NOx formed during combustion by optimizing engine parameters with respect to emissions (valve timing, fuel injection, and turbocharging). Emission levels can be reduced by 30%–60%.

Secondary measures designed to remove NOx from exhaust gas by downstream cleaning techniques such as selective catalytic reduction (SCR). Emission reductions of over 95% can be achieved.

Primary NOx reduction measures can be categorized as follows:

- *Water addition*: either by direct injection into the cylinder or by emulsified fuel
- *Optimized combustion-chamber geometry:* for superior combustion and avoidance of temperature peaks that produce NOx formation
- *Altered fuel injection*: retarded injection, rate-modulated injection and a NOx-optimized fuel spray pattern, Miller cycle
- *Combustion air treatment*: supercharging, turbocooling, intake air humidification, exhaust gas recirculation (EGR), and selective noncatalytic reduction (supercharging and turbocooling are discussed in Chapter 10).
- *Change of engine process*: compression ratio and boost pressure

The basic aim of most of these measures is to lower the maximum temperature in the cylinder since this result is inherently combined with a lower NOx emission. A combustion chamber geometry which is designed to optimize mixing of the fuel and air in the cylinder achieves more complete and homogeneous combustion, avoiding the temperature peaks that cause over 90% of NOx formation. New low-swirl cylinder heads and high-compression re-entrant pistons contribute to more favourable gas flow and hence a decrease in NOx formation. New design tools such as computational fluid dynamics can be very useful in optimizing the geometry of the combustion chamber.

The temperature of combustion—and hence NOx formation—can also be decreased by retarding fuel injection, although this measure increases the specific fuel consumption (the so-called diesel dilemma). Common-rail fuel injection systems enable precise and flexible control of injection pressure, timing, and duration, helping performance, emissions, and fuel consumption to be optimized over the entire engine load and speed ranges.

The Miller cycle involves early closure of the inlet valve, causing the intake air to expand and cool; the lower intake air temperature reduces the combustion temperature peaks responsible for most NOx formation. Higher emissions of PM at part load are suffered, but this PM penalty can be eliminated by adopting a variable valve timing system with the Miller process. High-efficiency turbochargers with increased pressure ratios compensate for the shorter inlet valve opening times associated with the Miller cycle, ensuring that the quantity of combustion air entering the cylinder—and thus the engine performance and efficiency—remains unaffected. Two-stage turbocharging systems are also under development to support Miller cycle applications.

New generations of medium-speed engines with longer strokes, higher compression ratios, and increased firing pressures addressed the NOx emission challenge (Fig. 4.1). The low NOx combustion system exploited by Wärtsilä throughout its medium-speed engine programme, for example, is based on an optimized combination of compression ratio, injection timing, and injection rate. The engine parameters affecting the combustion process are manipulated to secure a higher cylinder pressure by increasing the compression ratio. The fuel injection equipment is optimized for late injection with a short and distinct injection period. NOx reductions of up to 50% are reported without compromising thermal efficiency, achieving

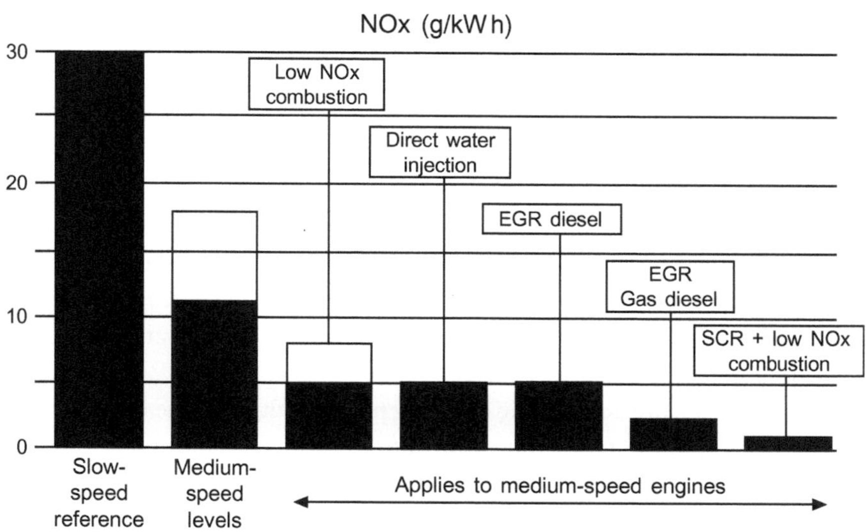

Fig. 4.1 Emission control measures and their potentials in reducing NOx output. EGR = exhaust gas recirculation; SCR = selective catalytic reduction (Wärtsilä).

a NOx rate of 5–8 g/kW h compared with the 15 g/kW h of a typical conventional engine with virtually no effect on fuel consumption.

Low NOx combustion is based on.

- A higher combustion air temperature at the start of injection, which significantly reduces the ignition delay.
- A late start of injection and shorter injection duration to place combustion at the optimum point of the cycle with respect to efficiency.
- Improved fuel atomization and matching of combustion space with fuel sprays to facilitate air and fuel mixing.

Water-based NOx reduction techniques

When the IMO introduced NOx reduction regulation, engine makers responded firstly by making adjustments to the running parameters of the engines. In almost every case, only small adjustments were needed to allow existing engine models to meet the IMO Tier I limits.

Slightly more work was needed to achieve Tier II compliance in 2011, but the additional reduction from a maximum 17.0 to 14.4 g/kWh was again achieved quite easily. However, the Tier III maximum of 3.4 g/kWh represents a much harder target for engine makers. The promise of water-based NOx reduction was a compelling choice for research and was actively pursued by the leading engine makers.

Using water in an engine may at first seem to counter-intuitive not least because most of the fuel-conditioning processes on a modern ship are aimed at removing water from fuel. Engine makers have traditionally recommended that water in HFO should be removed entirely by separation before entering the engine.

There are two reasons for this. Firstly, seawater in fuel can be a source of sodium which along with ash and vanadium should be avoided wherever possible because compounds of the chemicals tend to promote mechanical wear, high-temperature corrosion, and the formation of deposits in the turbocharger and on the exhaust valves.

Secondly, it is due to catalytic cracking process which is one of the modern crude oil refining techniques. This process uses aluminium and silicon oxide and small particles of these are often contained in residual fuels. These so-called catalytic fines or cat fines are extremely hard and abrasive causing damage to injectors, piston rings, and cylinder liners. The presence of water makes the cat fines more likely to pass through the separation process.

Emulsified fuels work by using a quantity of water in the fuel which has the effect of reducing the size of oil droplets compared to conventional fuels. This results in more complete combustion of the oil and so increasing the energy delivered from a given quantity of fuel. Because the oil droplets surround a water core, the heat in the combustion chamber also causes the water to vaporize which breaks the oil down into even finer droplets. The water vapour itself adds energy much as it would in a steam engine.

Water–fuel emulsions injected via the fuel valve can achieve a significant reduction in NOx production. The influence varies with the engine type, but generally 1% of water reduces NOx emissions by 1%. Fuel water emulsion (FWE) methods mix the fuel with freshwater onboard the ship to form an emulsion suitable for injection into the combustion chamber. In the case of heavy fuel oil, up to 30% water can be emulsified into the fuel, resulting in a NOx reduction of approximately 30%.

Both MAN Energy Solution and Wärtsilä have used water-based reduction techniques in a small number of marine engines. Although NOx reduction was greater than that required under Tier II of the NOx Code, reaching the much lower levels needed for Tier III was not considered practically by this method. This was because the required amount of water in the emulsion resulted in ignition problems before the needed NOx reduction could be reached.

Mixing large amounts of water into diesel can result in a prolonged ignition delay which may result in diesel knocking and increased HC emissions due to poor combustion quality. Increasing the water content further and the engine may eventually suffer from complete misfiring.

Nevertheless, the following accounts of attempts before the engine makers opted to pursue other options are interesting. Adoption of the techniques mention can allow ships to go beyond Tier II which may be beneficial in cases—such as in Norway—where some form of the levy is made based on NOx levels.

FWE can be produced continuously online during engine operation (the preferred method) or offline in batches. MAN Energy Solutions' continuous method is installed in a number of cruise ships and cargo vessels, reportedly demonstrating very satisfactory long-term performance. Moderate capital costs, compactness, and minimal impact on fuel consumption are cited as merits.

Wärtsilä noted that the proportion of water added for the water–fuel emulsion is mainly limited by the maximum delivery capacity of the fuel injection pumps. In practice, therefore, the engine has either to be derated or the maximum achievable NOx reduction limited to around 10% or 20%. Obtaining the maximum NOx reduction at full load also would dictate redesign not only of the injection system but also of the camshaft and its drive. The proportion of water added is also limited by the viscosity of the emulsion and the degree of heating required to reduce the viscosity for injection: this is a property of the water–fuel emulsion and cannot be addressed by the engine or system design. Another aspect is that the injection nozzle design (such as hole diameter) has to be adapted to the increased quantity of liquid injected.

Wärtsilä says that with such a modified nozzle design, both fuel consumption and component temperatures may be penalized when the engine is running without water. Although camshaft-controlled engines are restricted in the timing of injection, electronically controlled designs enjoy flexibility that is beneficial when burning water–fuel emulsions. The electronic control system can be readily adapted to different optimized injection characteristics for operating on emulsions or fuel alone. Furthermore, engine settings can be optimized according to engine load and speed. The electronic control system would also be extended to provide load-dependent mixing of water and fuel in the emulsifier.

Water can also be introduced to the combustion space through separate nozzles or by the stratified segregated injection of water and fuel from the same nozzle. Unlike other techniques, direct water injection (DWI) enables the water to be injected at the right time and place to obtain the greatest NOx reduction. It thus represents one of the most efficient means of reducing NOx emissions through internal measures.

DWI is effective in reducing NOx by adding mass and stealing heat from the combustion process when the water is evaporated. Wärtsilä developed DWI for its medium-speed engines after initially investigating the merits of injecting ammonia into the cylinder during the expansion stroke. It was found that injecting water into the combustion chamber during the compression stroke could achieve the same degree of NOx reduction as ammonia.

Wärtsilä medium-speed engines with DWI systems feature a combined injection valve and nozzle for injecting water and fuel oil into the cylinder (Fig. 4.2). The nozzle has two needles that are controlled separately, such that neither mode (water on/water off) will affect the operation of the engine. The engine can be transferred to nonwater operational mode at any load, the transfer in alarm situations being automatic and instantaneous. Water injection takes place before fuel injection, resulting in a cooler

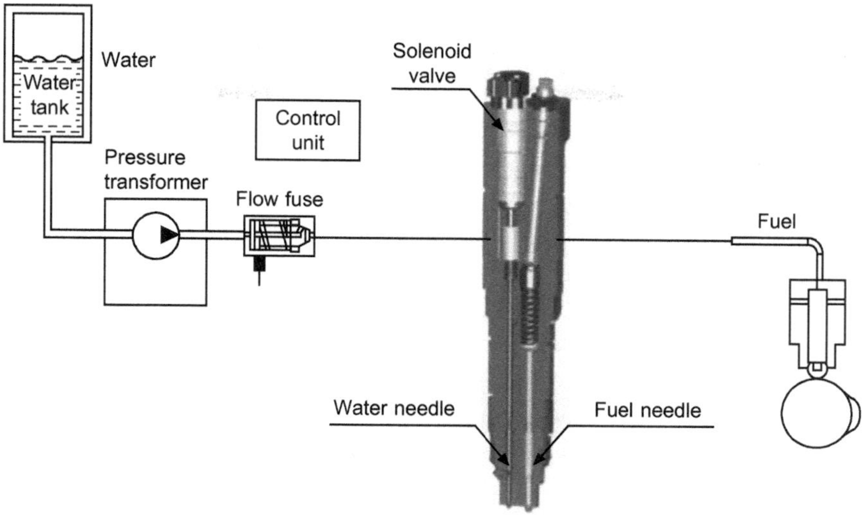

Fig. 4.2 Wärtsilä's DWI system features a combined injection valve and nozzle for water and fuel oil.

combustion space and hence lower NOx generation; water injection stops before fuel oil is injected into the cylinder so that the ignition and combustion process is not disturbed.

Clean water is fed to the cylinder at a pressure of 210–400 bar (depending on the engine type) by a common-rail system, the pressure generated in a high-pressure pump module; a low-pressure pump is also necessary to ensure sufficiently stable water flow to the high-pressure pump. The water is filtered before the low-pressure pump to remove all solid particles; the pumps and filters are built into modules to ease installation.

A flow fuse installed on the cylinder head side acts as a safety device, shutting off the water flow into the cylinder if the water needle gets stuck; immediate water shut-off is initiated in the event of excessive water flow or leakage. Water injection timing and duration are electronically controlled by the control unit, which receives its input from the engine output; timing and duration can be optimized conveniently via a keyboard for different applications. Space requirements for the equipment are minimal, facilitating retrofit applications.

Water and fuel are injected in a typical water-to-fuel ratio of 0.4–0.7:1, reducing NOx emissions by 50%–60% without adversely affecting power output or engine components. NOx emissions are typically 4–6 g/kWh when the engine is running on marine diesel oil, and 5–7 g/kWh when heavy fuel oil is being burned. NOx reduction is most efficient from 40% load and higher of nominal engine output.

Numerous applications of the above-mentioned DWI system have benefited Wärtsilä 32 and 46 medium-speed engines powering diverse ship types. Some W46 'EnviroEngines' combine electronically controlled common-rail fuel injection with DWI, and then the water is injected into the combustion chamber separately from the side of the cylinder head through a dedicated valve.

A DWI system developed and tested by Wärtsilä reportedly confirmed the potential of the technology for two-stroke engines, the water is handled by a fully independent, common-rail delivery system under electronic control. The system offers the possibility of injecting very large amounts of water and fuel with different timings. Water can be injected in parallel with the fuel and/or during the compression stroke so that optimizing injection timing with respect to fuel and water consumptions, NOx and other emissions is possible without undermining engine reliability.

When DWI is applied to a Wärtsilä RT-flex common-rail two-stroke engine, the fuel injection system can be provided with different engine settings so that it can run with the water injection turned on or off without affecting the fuel injection behaviour. There is no restriction on the quantity of water injected. With around 70% water/fuel ratio, DWI has shown the capability to reduce NOx emissions down to around 8 g/kWh or some 50% below the IMO Tier I limit.

Wärtsilä notes that whichever system of injecting water is applied, consideration must be given to the logistics of providing sufficient freshwater onboard. Substantial quantities will be required when 20%–50% water addition is used, and there must be sufficient tank capacity for the water and an associated handling system.

Mitsubishi Heavy Industries developed a stratified fuel–water injection (SFWI) system using a common valve to inject 'slugs' of fuel/water/fuel sequentially into the combustion chamber. The system reportedly worked in a stable condition throughout extensive trials on a ship powered by a Mitsubishi UEC 52/105D two-stroke engine, and NOx emission reduction was proportional to the amount of water injected.

Charge air humidification

Another way of introducing water into the combustion zone is by humidifying the scavenge air: warm water is injected and evaporated in the air intake, whose absolute humidity is thereby increased. An early drawback was that too much water in the air can be harmful to the cylinder condition but the introduction of the antipolishing ring in the liner (see Chapter 18) has allowed a much higher humidity to be accepted.

In the combustion air saturation system (CASS), which resulted from cooperation between Wärtsilä and the Finnish company Marioff Oy, special Hi-Fog nozzles introduced water directly into the charge air stream after the turbocharger in the form of very small droplets. Only a few micrometres in size, the droplets evaporate very quickly in an environment of more than 200°C and 75 m/s air velocity. Further heat for evaporation is provided by the air cooler (but acting now as a heater), resulting in combustion air with a humidity of around 60 g/kg of air. Wärtsilä reports that with such an amount of water, experiments demonstrated that it was possible to secure NOx levels of 3 g/kWh (assuming the starting value is 10–15 g/kWh).

Wärtsilä's WetPac H system promised NOx emission reductions of 30%–60% from its four-stroke engines, applying 1.5:2 water/fuel ratios, with no substantial increase in fuel consumption and no rise in material temperatures. Some of these systems were used on auxiliary engines in ships.

MAN Energy Solutions has also pursued NOx reduction by increasing the humidity of the charge air with water vapour. Based on the humid air motor (HAM) system developed by the German company Munters Euroform (Fig. 4.3), the process can reduce NOx formation by up to

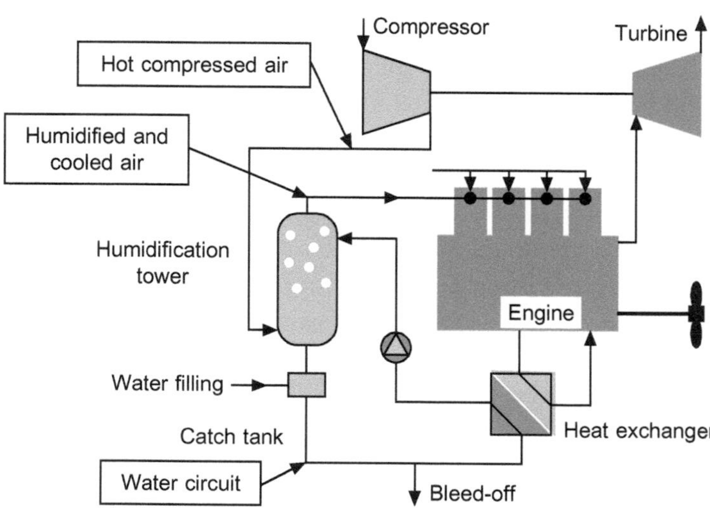

Fig. 4.3 Schematic diagram of HAM system for NOx emission reduction.

65%. Turbocharged combustion air is saturated with water vapour produced onboard from raw seawater using engine heat sources. Compressed and heated air from the turbocharger is passed through a cell that humidifies and cools the air with evaporated water, and the distillation process makes it possible to use seawater rather than freshwater.

Above 90% of NOx formation results from combustion temperature peaks. The HAM principle is used to humidify the inlet air in order to lower these peaks. The ability of water to decrease NOx formation is exploited in the same way as with fuel water emulsification. However, the quantity of water added is much higher, and the heat for vaporization is taken from the compressed air after the turbocharger or from other engine-related heat sources.

When the water vapour is mixed with the compressed charge air, two mechanisms can be identified: an increase in the specific heat capacity of the mixture and dilution of the charge air (water vapour replaces air). The quantity of water (in gram per kilogram dry air) which can be injected into the inlet air depends on the temperature and the pressure of the mixture. When the air temperature rises, so does the quantity of water it is possible to vaporize.

MAN Energy Solutions says that HAM's outstanding advantage is the use of the heat of the engine to raise the sea-water temperature; no external energy source is needed. In addition to the heat of the charge air after the turbocharger, in many applications, heat from the engine coolant and exhaust gases can be introduced into the charge air to increase its capacity to absorb moisture. An NOx reduction level of 40% is said to be achievable without additional heating of the intake air, but 65% reduction is possible when additional heat is introduced from the engine coolant or exhaust gases.

Using untreated seawater fosters very low operating costs. Further benefits cited are very low maintenance costs, decreased lube oil consumption, and high availability, with only minimal impact on fuel oil consumption. The presence of water vapour in the combustion chamber also yields cleaner combustion, reducing deposits in the chamber, on the turbine side of the turbocharger and the rest of the exhaust gas tract.

Test results from a SEMT-Pielstick 3PC2.6 medium-speed engine equipped with a prototype HAM system showed a NOx reduction from 13.5 to 3.5 g/kWh. There was no significant influence on specific fuel consumption, no significant increase in carbon monoxide and HC emissions, and no smoke deterioration. These results—along with no trace of water in the lube oil, no corrosion, and cleaner engine internals—were repeated in subsequent seagoing installations. In retrofit projects, the HAM system was replaced the original intercoolers.

Water use research continuing

For practical reasons—the need to reach Tier III levels before 2016—development of water-based NOx reduction was scaled back but not abandoned altogether. At the CIMAC Conference in 2019, a paper was presented by Stefan Mayer of MAN Energy Solutions which detailed research work being carried out by MAN Energy Solutions and Mitsui E&S.

The paper was titled *NOx Reduction to Tier III Levels on MAN B&W LGI Low-Speed Engines by Water Addition to Either Methanol or Conventional Diesel Fuel Oil.* It detailed novel tests on the MAN B&W 4S50ME-LGI test engine at Mitsui, Tamano, Japan, where two different types of water mixtures have been tested: firstly, water/methanol blends and secondly, water/diesel emulsions. In both the case of water/methanol blends and water/fuel emulsions, it is demonstrated that very satisfactory engine performance can be achieved while keeping NOx emissions well below the IMO Tier III limits, without the need of either EGR or SCR systems.

The paper concluded among other things that during the tests, no issues were seen with regard to the conditions of the cylinder liner, piston ring pack, or for the components in the fuel system. However, the operating hours (80 and 60 h on water/methanol and water/diesel, respectively) are far too few to say anything conclusive about possible operational or component lifetime problems that could occur if this technology is implemented on an engine for an extended period of time. Other issues, which will have to be investigated over time, include fuel and emulsifier compatibility, water quality influences, and fuel supply system compatibility. The other critical question in regard to technology implementation, besides the technology's effect on component lifetime, is of course whether this new technology can be made commercially attractive compared to the established alternatives. In this connection, one must consider both installation and operational cost.

Exhaust gas recirculation

Exhaust gas recirculation (EGR) is a method of modifying the inlet air to reduce NOx emissions at source, an approach widely and successfully used in automotive applications. Some of the exhaust gas is cooled and cleaned before recirculation to the scavenge air side. Its effect on NOx formation is partly due to a reduction of the oxygen concentration in the combustion zone, and partly due to the content of water and carbon dioxide in the exhaust gas. The higher molar heat capacities of water and carbon dioxide lower the peak combustion temperature, which, in turn, curbs the formation of NOx.

Some of the early work on EGR considered it more practical for engines burning cleaner bunkers such as low sulphur and low ash fuels, alcohol, and gas. Engines operating on high sulphur fuel might invite corrosion of turbochargers, intercoolers, and scavenging pipes. However, EGR has emerged as the favoured method for reaching Tier III levels albeit in conjunction with SCR on some engine types.

For ships with a dual-fuel engine operating on oil fuels, Tier II levels can be reached but to reach Tier III would require SCR. However, the same ships operating on LNG or methanol can reach Tier III without SCR. Any ship planned for operation outside of an ECA or built before any ECA it operates in was established need only ever reach Tier II standards.

The main components of an exhaust gas recirculation system comprise:

- a high-pressure exhaust gas scrubber fitted before the engine turbocharger
- a cooler to further reduce the temperature of the recirculated gas
- a water mist catcher (WMC) to remove entrained water droplets
- a high-pressure blower to increase recirculated gas pressure before reintroduction to the engine scavenge air
- automated valves for isolation of the system

A Tier III engine has two emission cycle operating modes: Tier II for operation outside NOX emission control areas and Tier III for operation inside NOX emission control areas.

EGR is a suitable NOx reduction method for ships using fuels with almost any level of sulphur likely to be supplied for marine engine use. However, to prevent sulphur and particles from damaging the engine, cleaning of the recirculated exhaust gas is required. This is performed in a combined cooling and cleaning process by a prespray and an EGR cooler spray in the EGR string, using recirculated water. In order to maintain the ability of the water to clean, cool, and neutralize the exhaust gas, a water handling system (WHS) is needed.

The system must ensure the removal of accumulated particles and neutralization of sulphuric acid in the water and ensure the delivery of water at a sufficient pressure and supply rate to the EGR unit. In addition, the WHS handles the bleed-off water, which is the surplus of water accumulated in the system from the combustion process.

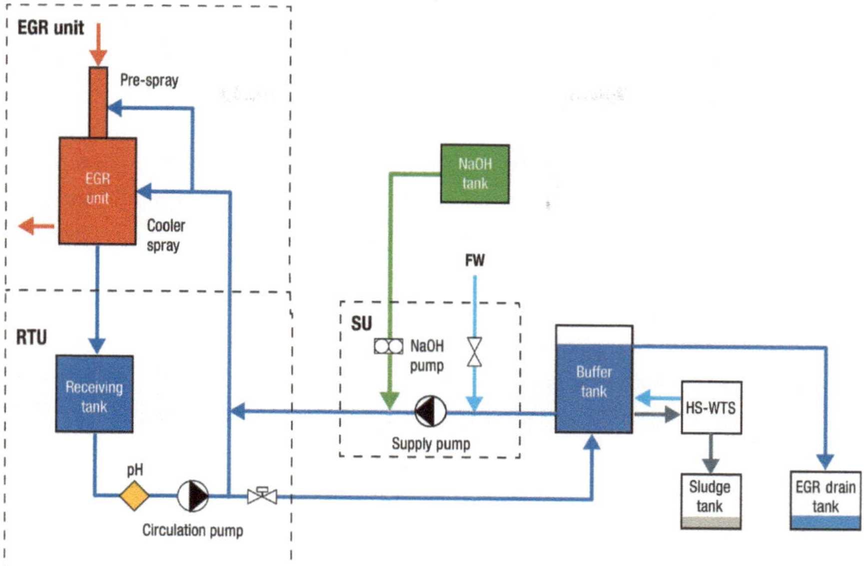

Fig. 4.4 Layout of the WHS system.

MAN Energy Solutions similar to other engine makers has designed a system capable of dealing with different levels of sulphur that may be encountered. The principle of the WHS is shown in Fig. 4.4.

The receiving tank, circulation pump, and control valve are part of the receiving tank unit (RTU), placed on the engine below the EGR unit. The supply pump, the NaOH pump, and the supply of freshwater are part of the supply unit (SU) installed in the engine room. The water from the EGR unit is drained to the receiving tank and recirculated to the EGR unit by the circulation pump. Part of the water is directed to the buffer tank by the control valve and returned to the EGR unit by the supply pump. In this string, an appropriate amount of NaOH is supplied to neutralize the sulphuric acid in the system. The bleed-off water is discharged to the drain tank by an overflow pipe on the buffer tank. The water accumulated in the drain tank may be discharged to the sea provided the specified discharge criteria set by the IMO are met. Any sludge that is collected must be disposed of ashore.

The WHS is designed according to the specified sulphur limits for the fuel oil used, when the engine is running in EGR mode:

- LS-WHS for EGR designed for low-sulphur fuels (max 0.5% S): In case the EGR system is designed for fuels of max 0.5% sulphur, the production of particles is ignorable, and cleaning of the recirculation water will not be required. However, the neutralization of the recirculated water will still be required. For LS-WHS, the flow of water in the SU loop and the amount of NaOH is relatively low.

- HS-WHS for EGR designed for high-sulphur fuels (noncompliant fuels): In case the EGR system is designed for fuels not complying with the SOX requirements, the production of particles is significant, and the recirculated water will need cleaning to prevent damage to the engine. A water treatment system (HS-WTS) is installed in the buffer tank loop to reduce particulate matter in the water. For HS-WHS, the flow of water in the SU loop and the amount of NaOH is relatively high.

The buffer tank is part of the WHS and required for control of the amount and quality of water in the EGR process. The tank size is defined by the volume of water needed in the EGR system during start-up. The buffer tank is arranged with an overflow pipe for automatic discharge of bleed-off water, which accumulates in the system due to condensation of exhaust gas.

- LS buffer tank (max 0.5% S)

In case the EGR system is designed for low-sulphur fuel, cleaning of the recirculated water will not be needed. However, to prevent the accumulation of particulate matter of high density at the tank bottom, the overflow pipe (low overflow) is connected to the lowest possible point of the tank. In addition, to avoid trapping of foam, particulate matter, and similar low-density substance in the top of the tank, an overflow pipe (high overflow) with a slightly higher discharge level is also arranged. Due to the ship movements at sea, bleed-off water will occasionally be discharged this way.

- HS buffer tank (noncompliant fuels)

In case the EGR system is designed for high-sulphur fuel, the recirculated water will need cleaning by a water treatment system, HS-WTS. The outlet for the WTS is taken from the lowest possible point of the buffer tank to ensure that particulate matter of a high density is removed from the system. The treated water is returned to the buffer tank at a level similar to the water inlet from the EGR circuit. Bleed-off water including foam, particulate matter, and other low-density substance is discharged by an overflow pipe, similar to the high overflow at LS buffer tank.

MAN Energy Solutions now incorporates EGR into all of its two-stroke engines. Three different matching methods are used for the EGR systems:

- EGR with bypass, configured with only one turbocharger and used for engines of bore 70 or less.
- EGR with TC cut-out matching, configured with two or more turbochargers and used for engines of bore 80 or greater.

- EcoEGR, configured as EGR with bypass. EcoEGR is a fuel optimized engine, which lowers the SFOC in both Tier II and Tier III mode using EGR to meet the relevant NOx limit.

An EGR system configured with bypass matching is shown in Fig. 4.5. Two strings, the main string and an EGR string, are available to direct the scavenge air into the scavenge air receiver:

- The main string, with the capacity to lead all the scavenge air through the turbocharger compressor and the scavenge air cooler.
- The EGR string, with the capacity to lead up to 40% of the exhaust gas through the prespray and the EGR unit (EGR cooler and WMC) to a mixing point in the main string.

Two modes are available for bypass matching: Tier II and Tier III modes.

In Tier II mode, only the main string is in operation. The valves in the EGR string (SOV/BTV) and the cylinder bypass (CBV) is kept closed. In this mode, the exhaust gas bypass (EGB) is fully open at high loads and partly open at low loads to balance the turbocharger. However, on engines with a

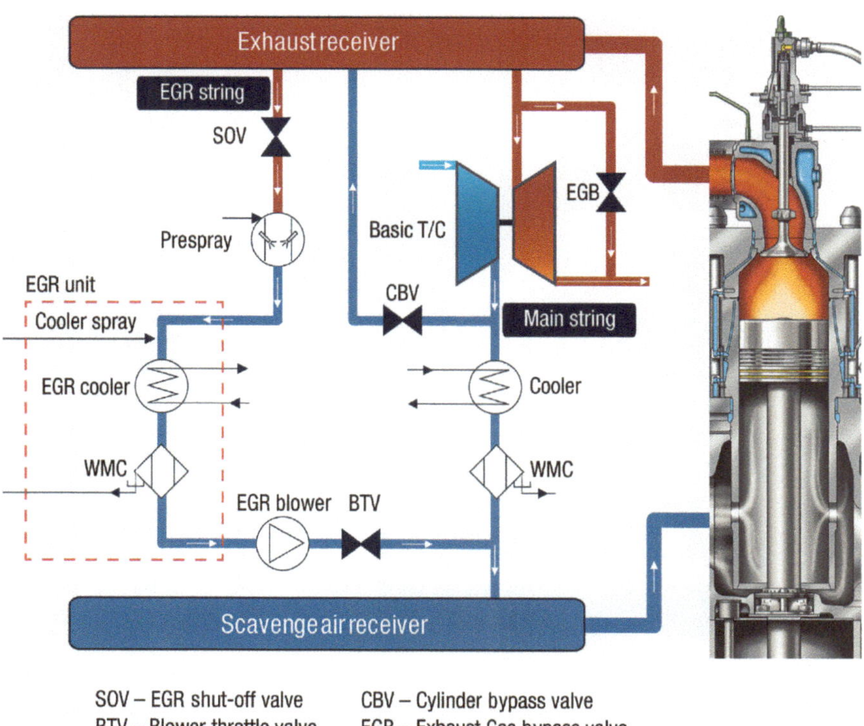

Fig. 4.5 An EGR system configured with bypass matching.

bore of 40 or less, the exhaust gas bypass will be closed at high loads and the EGR string open, to obtain sufficient scavenge air pressure while meeting restrictions on the turbocharger speed.

In Tier III mode, the EGR string is activated by opening the EGR shut-off valve and the blower throttle valve (SOV/BTV). The exhaust gas is led through the prespray and the EGR unit to the mixing point and scavenge air receiver, forced by the EGR blower. The EGR ratio is controlled by changing the flow of the EGR blower. The cylinder bypass (CBV) is active in this mode to increase the scavenge air pressure and thereby reduce the SFOC. The exhaust gas bypass (EGB) is closed.

An EGR system with TC cut-out matching is shown in the diagram in Fig. 4.6. Three strings, the main string, a cut-out string, and an EGR string, are available in the system to direct the scavenge air into the scavenge air receiver:

- The main string leads up to 70% of the scavenge air through the basic turbocharger and the scavenge air cooler.

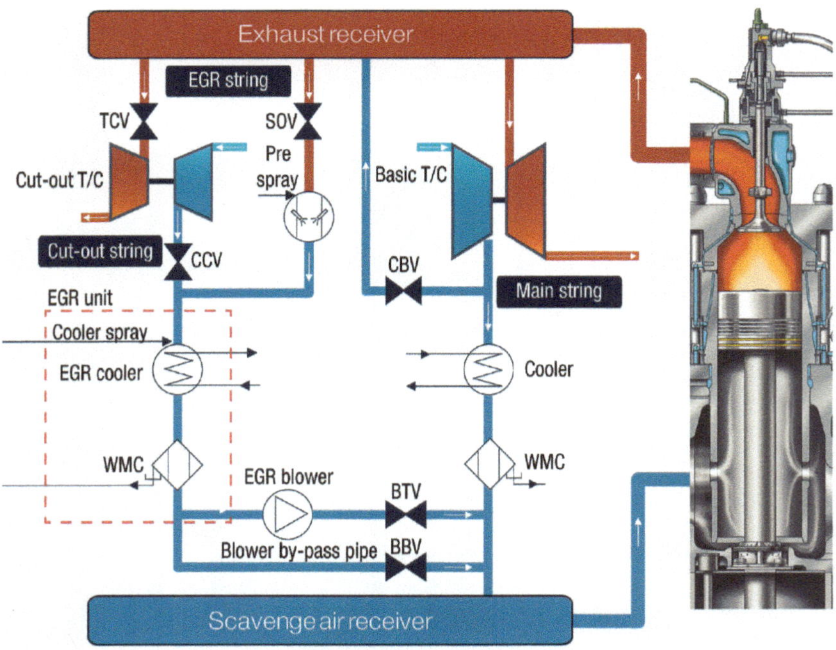

SOV – EGR shut-off valve
BTV – Blower throttle valve
TCV – Turbine cut/out valve
CCV – Compressor cut/out valve
BBV – Blower system bypass valve
CBV – Cylinder bypass valve

Fig. 4.6 An EGR system with TC cut-out matching.

- The cut-out string leads up to 40% of the scavenge air through the cut-out turbocharger and the EGR unit (EGR cooler and WMC) before entering the scavenge air receiver through the balance pipe.
- The EGR string leads up to 40% of the exhaust gas through a prespray and EGR unit to a mixing point in the main string, forced by one or more EGR blowers. In this case, the cut-out string is closed.

On some larger engines, a configuration with more than two turbochargers will be needed. The principle is unchanged although the number of turbochargers and EGR units is increased.

Three modes are available for TC cut-out matching:

- Tier II mode

In Tier II mode the main string and the cut-out string are in operation. The TC cut-out valves (TCV/CCV) and the blower by-pass valves (BBV) are open, while the EGR string is kept closed by the EGR shut-off valve and the blower throttle valve (SOV/BTV). In this mode, the EGR cooler works as a normal scavenge air cooler. About 40% of the scavenge air is passed through the cut-out string, the remaining 60% through the main string. The cylinder bypass (CBV) is kept close in this mode.

- Tier II mode—TC cut-out

The cut-out string gives an opportunity to run the engine in Tier II mode at low loads with a TC cut-out and the SFOC could thereby be reduced. In this case, only the main string will be open, while the cylinder bypass (CBV) is kept closed.

- Tier III mode

In Tier III mode the cut-out string is closed (TCV/CCV). The EGR string is opened by the EGR shut-off valve and the blower throttle valve (SOV/BTV). The exhaust gas is led through the prespray and the EGR unit to the mixing point and the scavenge air receiver, forced by the EGR blowers. The EGR ratio is controlled by changing the flow of the EGR blower. The cylinder bypass (CBV) is partly active in this mode to increase the scavenge air pressure and thereby reduce the SFOC.

The EcoEGR system operates permanently in EGR state regardless of the sailing area. The EGR rate is adapted to the NOx limits in the area i.e. Tier III mode inside NECA and Tier II mode outside NECA. The SFOC reduction is around 2.5% in Tier II mode and 1.0% in Tier III mode.

An additional Tier II fall back mode is available to maintain Tier II compliance in case a serious failure to the EGR system occurs. The EcoEGR system is configured similar to the EGR bypass concept and may be installed with one or two turbochargers, as shown in Figs 4.6 and 4.7.

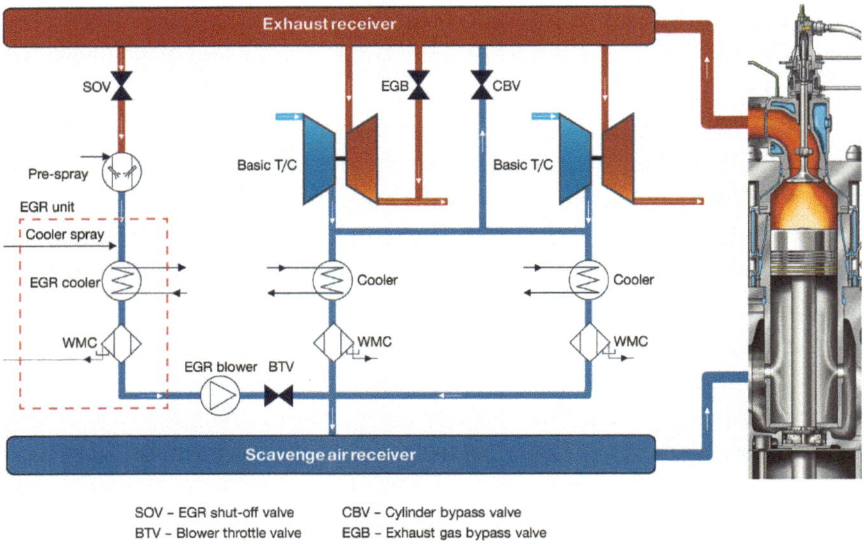

Fig. 4.7 EGR process diagram—EcoEGR configured with two turbochargers.

Three modes are available for EcoEGR:

- EcoEGR Tier II mode

In EcoEGR Tier II mode the engine operates with a low EGR rate, around 10%–15%. The EGR shut-off valve (SOV) and the blower throttle valve (BTV) are fully opened, while the cylinder bypass valve (CBV) is closed. Except at high loads, the exhaust gas bypass (EGB) is also closed.

- EcoEGR Tier III mode

In EcoEGR Tier III mode the engine operates with an EGR rate 25%–45%, similar to EGR Bypass matching. The EGR shut-off valve (SOV) and the blower throttle valve (BTV) are fully opened. The cylinder bypass valve (CBV) is opened except at low loads, while the exhaust gas bypass valve (EGB) is closed.

- EcoEGR Tier II fall back mode

The fall back mode, which is only relevant in case of EGR break down, operates without EGR. Accordingly, the EGR shut-off valve (SOV) and the blower throttle valve (BTV) are closed. Except at high loads, the exhaust gas bypass (EGB) is also closed.

EGR units cannot be described as small, but they do fit well within the envelope of the main engine and in themselves are not excessively space consuming as can be seen for the image below combining an EGR unit, a basic turbocharger, and a cut-out turbocharger (Fig. 4.8).

Fig. 4.8 Side-mounted turbocharger and side-mounted EGR unit.

However, the peripheral components will add to the space requirements of the system. These can be positioned where best suited although the NaOH supply will need to be at or above the supply unit and drain tanks should naturally be below to allow for gravity flow (Fig. 4.9).

Selective catalytic reduction

Primary methods are generally adequate for the IMO NOx emission limits, but tougher international and regional controls may dictate the use of secondary methods—exhaust gas treatment techniques—either alone or in combination with engine modifications. Here the focus is on SCR systems, developed from land-based power station installations for shipboard applications, which can cut NOx reductions by over 90%. An SCR installation thus gives engine designers greater scope to pursue primary in-cylinder measures without increasing fuel consumption.

In an SCR system, the exhaust gas is mixed with ammonia (preferably in the form of a 40% solution of urea in water) before passing through a layer of special catalyst at a temperature between 290°C and 450°C. The lower limit is mainly determined by the sulphur content of the fuel: at temperatures below 270°C ammonia and SOx will react and deposit as ammonium sulphate. At excessive temperatures, the catalyst will be degraded depending on

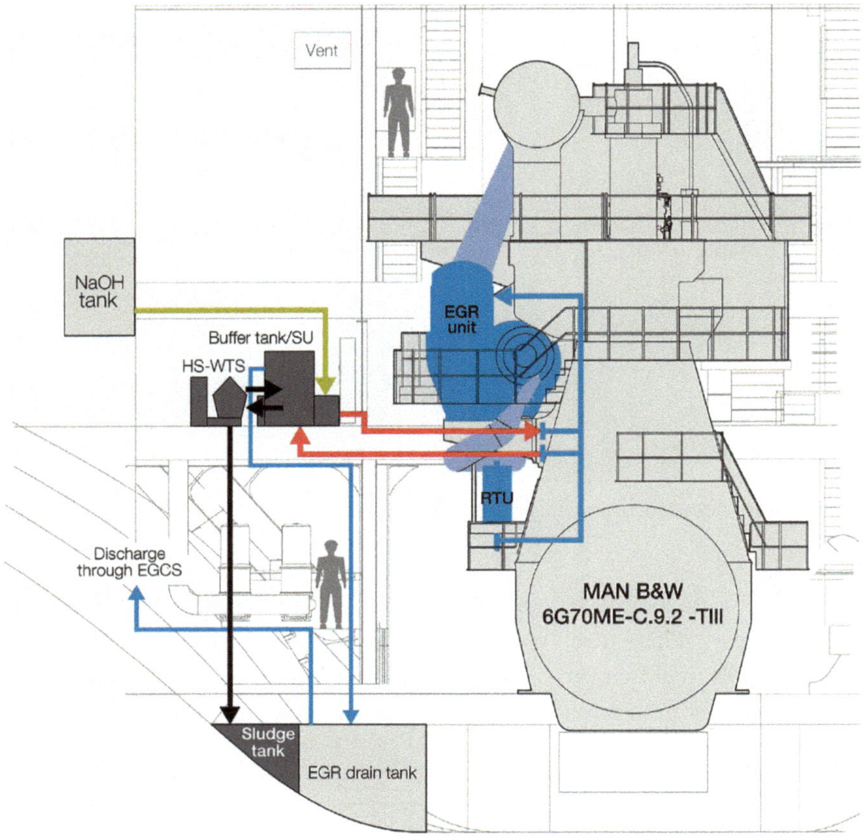

Fig. 4.9 Typical EGR full system layout suited to high sulphur fuels.

the exhaust gas temperature; therefore, reliable SCR operation calls for fuels with low sulphur contents.

Urea decomposes into ammonia and carbon dioxide on injection into the hot exhaust gas stream. The ammonia reduces NOx to harmless gaseous waste products—water and nitrogen; and parts of the soot and HC in the exhaust are also removed by oxidation in the SCR process reactor.

Among the desirable qualities of a catalyst are low inertia, which means that the ammonia slip (the quantity slipping past the catalyst) is extremely low, even during transient operations; a low-pressure drop and a short heating-up time; and a low fouling tendency, ensuring minimal deterioration in performance over time.

The catalytic conversion rate of SCR systems is highly dependent on the amount of urea dose: increased dosage yields increased conversion.

However, excessive urea causes ammonia to slip downstream of the reactor, which is detrimental to both the process and the operational economy. Urea dosing must, therefore, be very accurate under different load conditions. The set point for dosing is derived primarily from the engine speed and load; in addition, control is adjusted by measuring the residual NOx level after the reactor. The NOx measurement data are used for tuning the stoichiometric ammonia/NOx ratio and maintaining the ammonia slip at a constant level.

The temperature requirement of the process generally dictates that the SCR reactor serving a low-speed two-stroke engine is installed before the turbocharger; the postturbocharger exhausts gas temperature of a four-stroke engine, however, is sufficient for the catalytic process. If an exhaust gas boiler is specified, it should be installed after the SCR unit. The temperature window has been the subject of catalyst development.

The impressive de-NOxing efficiency of SCR technology has been demonstrated in deep-sea and coastal vessel installations since early 1990. NOx emission reductions of up to 95% are yielded by pioneering marine SCR plant serving the MAN B&W 6S50MC low-speed engines of bulk carriers on a dedicated trade between Korea and California where strict environmental regulations have to be satisfied. The amount of ammonia injected into the exhaust gas duct is controlled by a process computer, which arranges dosing in proportion to the NOx produced by the engine as a function of engine load (Fig. 4.10).

Typically, a ship on a regular trade to California will bypass the SCR reactor on the ocean leg of the voyage, complying with the IMO emission rules by using primary methods for controlling NOx. Approaching the Californian-regulated waters, the engine feed is switched from heavy fuel to fuel complying with California Air Resources Board rules (gas oil). The exhaust gas is then gradually passed through the SCR reactor, and, when the temperature has been raised to the right level, ammonia dosing is started to affect near-complete NOx control.

An early SCR installation serving medium-speed marine engines was commissioned in 1992 on a diesel–electric ferry plying short crossings between Denmark and Sweden. Its initial NOx reduction efficiency of 96% was increased to 98% after 12,000 running hours.

The number of SCR systems in service has proliferated as ship owners respond to tougher regulations and seek NOx emission levels of 2 g/kWh or lower from the main and auxiliary engines of all ship types. Applications in newbuilding and retrofit projects have been facilitated by the efforts of system designers to reduce reactor space requirements and offer compact

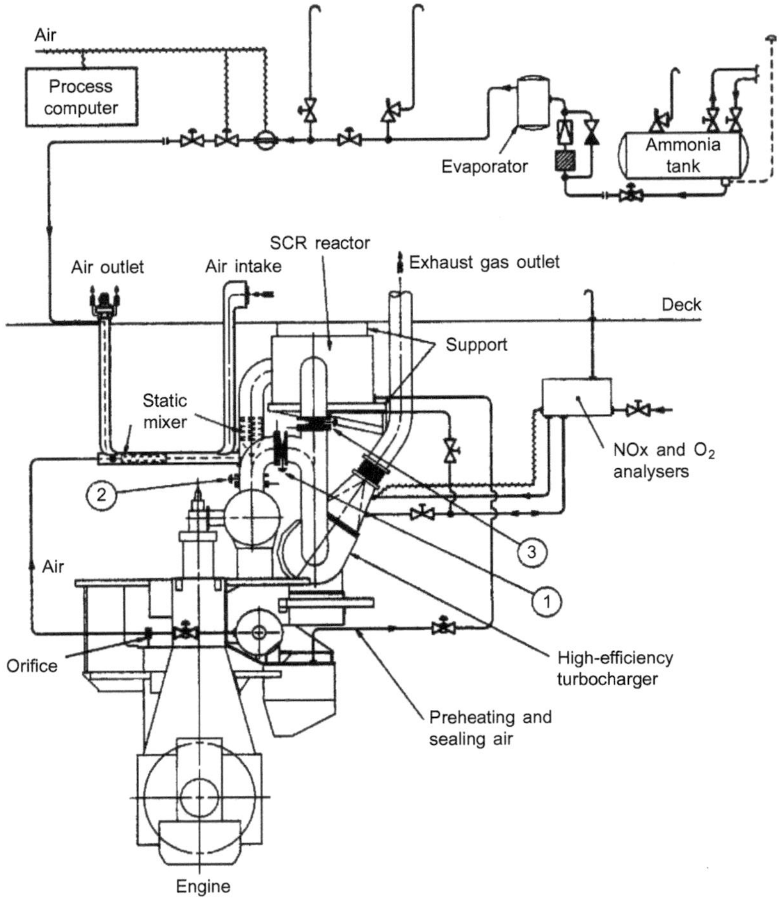

Fig. 4.10 Schematic layout of SCR system for a low-speed diesel (MAN Energy Solutions).

solutions. The reactors can now be installed separately in the engine room or integrated with the exhaust manifolds of both four-stroke and two-stroke engines, doubling as an efficient silencer.

R&D by SCR system designers targets increased compactness, low weight, no waste products, extended catalyst lifetimes, and lower operating and maintenance costs. Catalysts themselves remain the subject of continuous development to improve performance and longevity (Fig. 4.11).

WinGD, the successor of Wärtsilä's two-stroke engine programme prefers to use SCR rather than EGR for its engines which employ the Otto rather than the diesel cycle. The company claims this offers several benefits not least a lower capital outlay, lower fuel consumption, and no residue to

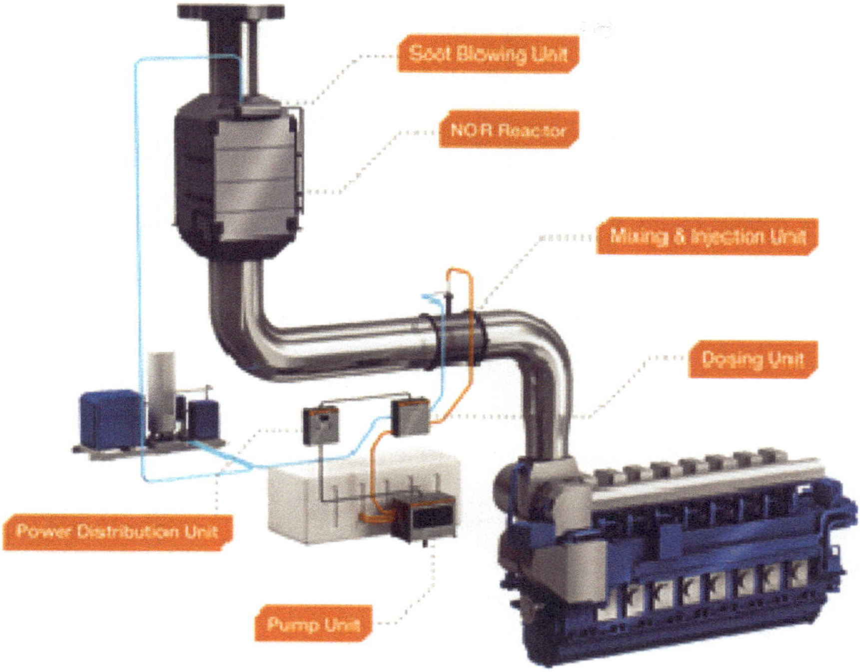

Fig. 4.11 Wartsila NOR SCR system.

dispose of ashore. The company also points out that SCR has no impact on steam production.

With SCR in operation, it is still possible to generate steam using the exhaust gas economizer which produces the same steam production power. With EGR, steam production is reduced by about 40% due to exhaust gas recirculation.

WinGD employs two variants of SCR: high pressure and low pressure. In a high-pressure SCR system, the reactor is put on the high-pressure side, before the turbine. Integrating the SCR reactor before the turbine allows the reactor to be designed in the most compact way due to the higher density of the exhaust gas. WinGD has developed and is systematically deploying high-pressure SCR solutions for the low-speed diesel engine portfolio with single and multi-turbocharger applications. It allows third parties to supply high-pressure SCR systems in accordance with the interface specifications (Fig. 4.12).

In the low-pressure SCR, the reactor is put on the low-pressure side, after the turbine. WinGD has developed an interface specification for low-pressure SCR applications for all low-speed engines which comply

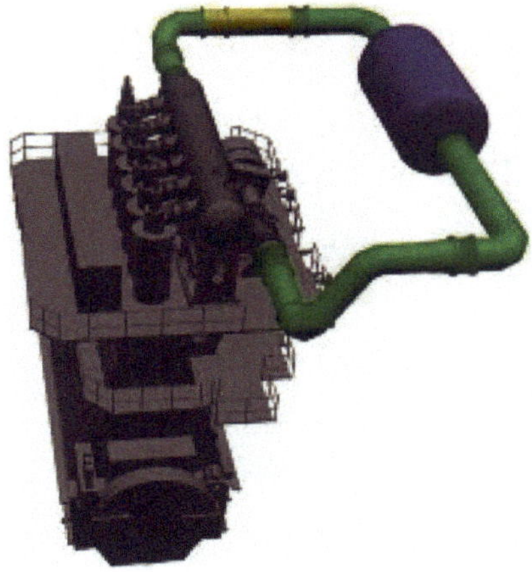

Fig. 4.12 Typical high-pressure SCR arrangement.

with the known low-pressure SCR system providers. A low-pressure SCR is typically larger in volume and has the advantage of being less complicated to integrate into the exhaust stream.

Particulates, soot, and smoke

Particulate emissions are considered a contributory factor in causing asthma, allergies, and various other human health problems. Studies have shown that smaller particulates show a stronger correlation between ambient concentrations and health symptoms than larger particulates. Smaller particulates are thought more likely to penetrate deep into the human lungs, the most minute of them perhaps even moving into the blood stream, and the chemical composition of the particulate may significantly contribute to the biological effect.

Particulates are made up of randomly agglomerated carbonaceous spherules, which build up into a highly branched three-dimensional structure. Various HC, ash, and sulphur compounds are all associated with the structure. The complexity of the particulate makes it impossible to give a satisfactory general definition because the characterization depends largely on the measurement method. It can be defined, for example, in terms of opacity, filter blackening, particle number, size, or mass.

A study carried out by Wärtsilä found that typically between 50% and 70% of the particulate composition comprises compounds that are related directly to the quality of the residual fuel oil (notably its sulphur and ash contents) and cannot be reduced by improved combustion. Consequently, only around 30%–50% of the particulate composition can be affected. Even a significant improvement in engine combustion will thus not necessarily result in any major reduction of particulate emissions.

After formation in the cylinder, the nanometre-sized primary particles coagulate to create larger particulates to which HC and sulphate are attached as the exhaust gas cools. Secondary particulates can be formed outside the combustion chamber as a result of the absorption and condensation processes. This means that the particulate composition and size distribution in the engine exhaust duct is completely different from that reaching the human respiratory tract.

Wärtsilä advises that abatement measures for diesel particulates can be divided into three categories: improvement of fuel and lube oil quality, improvement of the engine combustion process, and exhaust gas cleaning. Examples of engine measures for improving combustion include advanced fuel injection properties (such as rate shaping with common-rail technology) and improved combustion chamber geometry (including swirl and squish).

Traps and oxidation catalysts, used for exhaust gas cleaning in truck engines, are unsuitable for applications involving large residual fuel-operated diesel engines due to the high sulphur and ash content of the fuel. For engines running on such fuels, however, an electrostatic precipitator is a viable option, but its size makes marine applications impractical.

Exhaust gas scrubbers are among potential options for the future and their ability to remove particulates from the exhaust stream has been used as a point in their favour by proponents of their use. Alfa Laval makes the point that its PureSOx exhaust gas cleaning system typically removes 80% of particulate matter.

Whatever measures are favoured for lowering particulate emissions, measurement systems are an inevitable legislative requirement. Various methods are approved by different administrative bodies, but the results are often not comparable because regulators define particulates in different ways.

The ISO 8178 method (with dilution) is the standard applied method to most marine emission regulations, although it should not be used in conjunction with high sulphur fuel. According to ISO, the upper sulphur limit for applying this method is 0.8%, whereas a CIMAC recommendation is 0.05%. Dilution conditions are crucial to the measurement result.

Wärtsilä reports that the reproducibility of measurement results with the ISO 8178 standard when operating on typical marine fuel is often poor. Furthermore, an improper choice of dilution and measurement settings within the permitted requirements of the standard makes it possible to manipulate the results significantly. To achieve repeatable results and enable comparison with land-based industrial sources, Wärtsilä recommends using direct measurement methods.

MAN Energy Solutions explains that particulate generation in diesel engines is a complex process depending on numerous factors, such as engine type, speed, engine setting, operating mode, load, fuel, and even weather conditions.

Particulates comprise all solid and liquid exhaust gas components, which, after cooling by exhaust gas dilution with filtered particulate-free ambient air to a temperature below 51.7°C, are collected on specified filters (dilution tube sampling). The so-called PM fraction represents a broad mixture of partly burned or unburned HC, sulphate bound water, sulphates, ash, and elemental carbon (soot).

Soot and ash are therefore only part of the total particulates. At high loads, ash and soot might contribute <20% to the particulates, but at low loads and idling, this percentage can be much higher.

Much of the elemental carbon formed in a diesel engine is oxidized during combustion, and only the remainder leaves the combustion space with the exhaust gas as soot, which becomes visible as a dark smoke plume emanating from the funnel. There is a clear correlation between the level of soot formation and the type of fuel used; heavy fuel oil combustion generates substantially greater volumes of particles (and soot) than the burning of cleaner fuels, such as marine diesel and marine gas oils. MAN Energy Solutions notes that although soot particles themselves are not toxic, the potential hazard posed by the build-up of liquid HC onto them is viewed critically by many.

At very high engine loads, combustion in a state of the art medium-speed diesel engine can be modelled to give invisible smoke (IS). At low service loads, however, and especially during rapid start-up manoeuvres and load changes, the turbochargers deliver less intake air than the engine needs for complete combustion and the engine 'smokes'. This can be overcome to a great extent by modern turbocharger developments such as variable turbine geometry and two-stage turbocharging.

Smoke as the visible manifestation of soot production is highly undesirable in all types of ships, but particularly in passenger ferries and cruise liners.

In sensitive waters, such as the glacier regions and bays of Alaska or the Galapagos Islands, a vessel producing excessive amounts of visible soot (smoke) can even be banned from cruising there. Achieving zero emissions may dictate shutting down all engines and boilers during a port stay and plugging into a land-based electrical supply system.

MAN Energy Solutions designed and successfully field tested a package of measures to suppress the formation of soot in highly turbocharged medium-speed engines, even during long periods of slow steaming. The full package for these IS near-zero soot engines with low NOx emissions comprises the following:

- Turbocharger optimized for part load, with a waste gate
- Charge air bypass below 65% engine load
- Charge air preheating (80°C) below 20% load
- Smaller injection bores
- Nozzles with short sac holes
- Auxiliary blower for operation below 20% load
- Fuel–water emulsification (15%–20% water)
- Retarded fuel injection below 80% engine load

With smoke readings of 0.3 maximum on the Bosch scale, at very low engine (MAN L/V 48/60IS) loads and even idling, soot production at steady operating conditions is drastically reduced and kept at an invisible level over the entire operating range. The smoke reduction was by a factor of seven at idling and by a factor of five at 10% load, respectively. The mass of soot produced during diesel combustion is reduced by even higher factors than that due to the nonlinear correlation between smoke emission and soot mass.

Elements of the IS package have become standard for all MAN medium-speed engines; the other measures are specified only when extremely strict requirements on soot and NOx emissions have to be met. The package is available for new engines and by retrofit to existing engines.

A different approach to smokeless medium-speed engines was adopted by Wärtsilä, which explains that smoke in a large diesel engine can be formed in two different ways: if the fuel spray touches a metal surface and there is not enough remaining combustion time to burn away the soot formed (this is consequently a low load problem), and when more fuel is injected compared to that can be burned in the amount of air in the cylinder (this can easily happen at load variations with a conventional injection system).

Wärtsilä asserts that a common-rail fuel system can cure both of these problems because it is possible to maintain high injection pressures independently of engine load and thus it ensures good spray atomization even at very low loads. If the combustion space is also optimized, there will be no risk of low load smoke. The risk of over-injection of fuel can also be avoided by a common-rail system because its computer—supplied with air temperature, pressure, and rotational speed data—can calculate the amount of air trapped in the cylinder. Based on empirical maps, the computer can decide how much fuel can be injected. A common-rail fuel-injected engine can have a faster load response than a conventionally injected engine without smoke. (The common-rail system for Wärtsilä medium-speed engines is detailed in Chapter 8.)

Wärtsilä's EnviroEngine concept for its W46 and W32 medium-speed designs exploits an electronically controlled common-rail fuel injection system to achieve 'smokeless' performance (Fig. 4.13). The package embraces the following:

- Camshaft-driven high-pressure pumps
- Accumulators for eliminating pressure waves
- Hot box enclosure of all elements for maximum safety
- Engine-driven control oil pump for easy 'black' start
- Suction control of fuel flow for high efficiency
- Low cam load for high reliability

Such a system offers the freedom to choose the fuel injection pressure and timing completely independent of the engine load, whereas computerized control allows several key engine parameters to be considered and the injection and combustion optimized for each load condition. The ability to

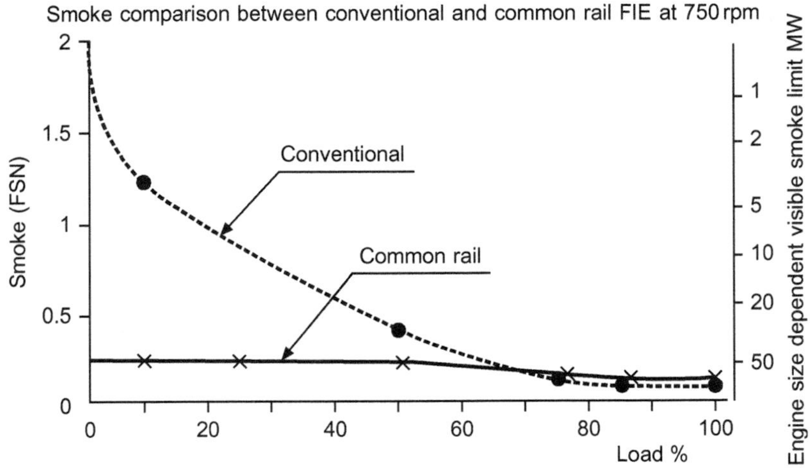

Fig. 4.13 Smoke test results from a Wärtsilä 32 medium-speed engine with and without common-rail fuel injection.

maintain fuel injection pressures sufficiently high at all engine loads and speeds (even at the lowest levels and during starting and transient load changes) contributes to clean combustion with no visible smoke emissions. Four 16V46 EnviroEngines were specified as the diesel element of the CODAG-electric propulsion plant of Cunard's *Queen Mary 2*.

Wärtsilä notes that there is limited scope to significantly reduce particulate emissions by improving combustion, but methods of removing particulates from the exhaust gas are available. Compact ceramic filters can be used with some success on smaller engines but appear not to work when operating with heavy fuel-burning engines because they become clogged by the metal matter. Bag filters work with certain efficiency, but most applications are too bulky and hence impracticable. Electrostatic precipitators are efficient in removing particulates, but they are also extremely bulky and too expensive.

Conventional scrubbers have proved to be quite inefficient for the task because the surface contact between the exhaust gas and the water is limited. Promising results were reported by Wärtsilä, however, for a scrubber based on ultra-fine water droplets created by Hi-Fog nozzles, which yield a much greater surface area between gas and liquid.

Acknowledgements

The authors would like to acknowledge MAN Energy Solutions, Wärtsilä Corporation, and WinGD.

CHAPTER FIVE

Oil fuels chemistry and treatment

Fuel remains one of the highest single cost factors in running a ship and also the source of the most disruptive operating problems. For most of the history of the diesel-engined motor ship, there have been no regulations on any aspect of the fuel used beyond a minimum flashpoint of 60°C.

In 2005, the IMO adopted the regulation adopting the first sulphur emission control area (SECA) in the Baltic Sea. When the regulation took full effect in 2010, the sulphur content of fuels used within the SECA zone was limited to 1.5% by mass. Since then, the sulphur limits in fuels have been gradually reduced so that as from 1 January 2020, the limit in controlled zones is 0.1% by mass and everywhere else 0.5%. There is one exception to the sulphur limit and that is granted to ships which have been fitted with an exhaust gas cleaning system or scrubber.

Early motor ships ran on marine distillate fuels with later ships able to burn intermediate fuel oil (IFO) of 180 cSt. It was not until the oil crises of the 1970s and onwards that heavy fuel oil (HFO) became a fuel of choice for shipping although it had been in use for some years previously. HFO is defined either by a density of greater than 900 kg/m^3 at 15°C or a kinematic viscosity of more than 180 mm^2/s at 50°C. The increasing use of HFO in two-stroke engines was made possible by new types of cylinder lubricants which could effectively neutralize the high-sulphur content of HFO.

New refining techniques—fluid catalytic cracking (FCC) and visbreaking, introduced as a result of political developments in the Middle East in 1973/1974—yielded a more concentrated residual fuel of very poor quality. This residual fuel is the HFO traditionally supplied to ships as bunkers and used in the main engines of most deep-sea and some coastal vessels. Despite the high cost of these poor-quality residual fuels, owners generally had no alternative but to burn them, though some still prefer to use more expensive intermediate grades produced as a result of mixing residual fuel oil with distillate.

Residual fuel or bunker oil is essentially a refinery by-product blended to satisfy market demand for a cheap source of energy, some of which is used in shipping. The main driver in the refining industry is the production of light

Pounder's Marine Diesel Engines and Gas Turbines
https://doi.org/10.1016/B978-0-08-102748-6.00005-0

and middle distillate grades, which are used to formulate motor gasoline, jet fuel, automotive diesel fuel, and chemical feedstock. Over the past 20 years, the worldwide demand for distillate grades has continued to rise. Refineries have thus incorporated more sophisticated processes to extract as much distillate feedstock as possible from a barrel of crude oil.

The net effect is a reduction in the percentage of the crude oil barrel ending up as residual fuel oil. In 1990, for example, approximately 20% of a barrel was processed into residual fuel compared with 14% in early 2006. The downward trend is expected to continue.

Refinery processes

Crude oil is a complex mixture of hydrocarbons that must be processed to provide both the products and the quantity of each product required to satisfy worldwide demand. No two crude oils are alike: there are significant variations in density, viscosity, sulphur, and vanadium contents, as well as in the yield of each product that can be produced by a refinery.

Originally, crude oil was distilled in an atmospheric distillation process to derive the required grades in the required quantity. The vacuum distillation process was developed to enable further refinement of atmospheric process residue. Both atmospheric distillation and vacuum distillation are refinery processes based on the physical separation of crude oil components into distillate fuels. Because these processes do not chemically alter the structure of the oil, the original crude oil could be reproduced by mixing together all of the output streams and components in the appropriate proportions.

These two simple distillation processes, however, do not produce the amount of distillate oil products consistent with increasing worldwide demand. Other, more complex refinery processes—known as secondary conversion processes—are therefore required to satisfy the demand for distillate fuel products, taking some of the product streams from the distillation processes and altering the chemical structure of the oil. This reduces the amount of residual fuel oil and increases the amount of distillate fuel (Fig. 5.1).

Among the secondary conversion, processes are ‘cracking’ processes that break the long molecules of the heavier fuel fractions into shorter molecules, which can be more easily processed into fuel oil products in demand. There are two basic types of cracking processes: thermal cracking and catalytic cracking.

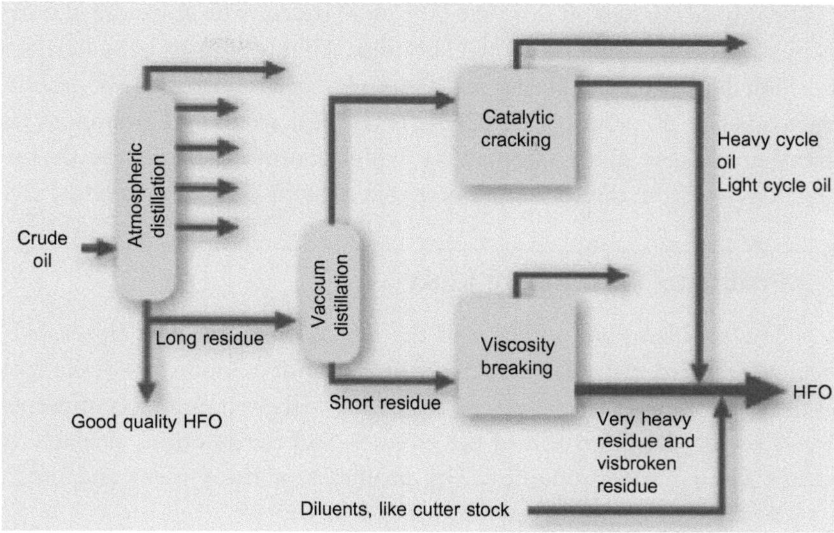

Fig. 5.1 Simplified configuration of a typical complex fuel refinery.

Thermal cracking uses temperature, pressure, and time to provoke a chemical reaction that alters the structure of the oil and can be applied to both distillate and residual fuels. Typical thermal cracking processes include *visbreaking*, which significantly lowers the viscosity of a heavy residue to enable blending with other fuels, and *coking*, which destroys the fuel oil, producing only distillate fuel and coke (a form of carbon).

Catalytic cracking processes also alter the chemical composition of residual fuel oil, using a catalyst rather than high pressure (HP) to break down complex hydrocarbons into simpler molecules. The catalyst is a substance that assists the process of the chemical reaction but does not change its own properties.

FCC, the most common of these processes, can convert gas oil and residual oil into high-octane gasoline and diesel fuel. In this case the catalyst—in the form of fine particles approximately 20–100 μm in diameter—circulates between a reactor and a regenerator in a fluidized-bed process. Large catalytic crackers contain around 500 t of expensive catalysts.

Continual recycling of the catalyst causes its structure to break up, primarily due to attrition, and some of the resulting small particles are carried over into the fractionator. Although refiners attempt to minimize the loss of catalyst from the catalytic cracking process, some carryover of the expensive catalytic fines from the unit is inevitable.

The product drawn off from the bottom of the fractionator is called slurry oil, also known as decant oil or FCC bottoms. Highly aromatic, the slurry oil has a high density (generally around 1000 kg/m^3 at 15°C) and a low viscosity (approximately 30–60 cSt at 50°C). It is an ideal blending component for residual fuels due to its high aromaticity, which contributes stability to the finished fuel. It is from this refinery source that catalytic fines enter residual fuel.

Problems with heavy fuels

The problems are reflected in the effects on the engine in terms of wear and tear and corrosion resulting from harmful components in the fuel. It is the duty of the ship's engineer to be aware of these harmful constituents, their effect on the operation of the engines, and the solutions available to counter the harmful properties. The problems of the present and future heavy residual fuels can be categorized as follows:

1. Storage and handling
2. Combustion quality and burnability
3. Contaminants, resulting in corrosion and/or damage to engine components, for example: burnt-out exhaust valves.

Storage problems

The problems of storage in tanks of bunker fuel result from the build-up of sludge, leading to difficulties in handling. The reason for the increase in sludge build-up is that heavy fuels are generally blended from a cracked heavy residual using a lighter cutter stock, resulting in a problem of incompatibility. This occurs when the asphaltene or high-molecular-weight compound suspended in the fuel is precipitated by the addition of the cutter stock or other diluents. The sludge which settles in the bunker tanks or finds its way to the fuel lines tends to overload the fuel separators with a resultant loss of burnable fuel, and perhaps problems with fuel injectors and wear of the engine through abrasive particles.

To minimize the problems of sludging, the ship operator has a number of options. He or she may ask the fuel supplier to perform stability checks on the fuel being provided. Bunkers of different origins should be kept segregated wherever possible and water contamination kept to a minimum. Proper operation of the settling tanks and the fuel-treatment plant is essential to prevent sludge from entering the engine itself. A detergent-type chemical additive can be used to reduce the formation of sludge in the bunker tanks (Table 5.1).

Table 5.1 Effects of heavy fuels.

Properties	Present HO	Future HO	Effect on engine
Viscosity (Red 1 at 37°C heating temperature)	3500	5200	Increased fuel heating required
Pumping	50	65	
Centrifuging	95	98	
Injection	110–120	115–130	
Density at 15°C	0.98	0.99	Water elimination becomes more difficult
Pour point (°C)	30	30	
Noxious element Carbon residue (%)	6–12	15–22	Fouling risk of components/Increased combustion delay
Asphaltenes (%)	4–8	10–13	Hard asphaltene producing hard particles Soft asphaltene giving sticky deposits at low output Increased combustion delay with defective combustion and pressure gradient increases
Cetane number	30–55	25–40	High-pressure gradients and starting problems
Sulphur (%)	2–4	5	Wear of components due to corrosion below dew point of sulphuric acid (about 150°C)
Vanadium (ppm)	100–400	120–500	Burning of exhaust valves at about 500°C
Sodium (ppm)	18–25	35–80	Lower temperature in case of high Na content
Silicon and aluminium (CCF slurries)			Wear of liners, piston grooves, rings, fuel pump and injectors

Water in the fuel

Water, whether fresh or saline, has always been a problem because it finds its way into the fuel during transport and onboard the ship from various sources including condensation in the fuel storage tanks. Free water can seriously damage fuel-injection equipment, cause poor combustion, and lead to excessive cylinder liner wear. If it is seawater, it contains sodium, which will contribute to corrosion when combined with vanadium and sulphur during combustion.

Water can normally be removed from the bunkers by proper operation of separators and properly designed settling and daily service tanks. However, where the specific gravity of the fuel is the same or greater than water, removal of water is difficult—or indeed not possible—and for this reason, the maximum specific gravity of fuel supplied for ship's bunkers has generally been set at 0.99.

Burnability

The problems that are related to poor or incomplete combustion are many and complex and can vary with individual engines and even cylinders. The most significant problem, however, is the fouling of fuel injectors, exhaust ports and passages, and the turbocharger gas side due to failure to burn the fuel completely. Fuels that are blends of cracked residual are much higher in aromatics and have a high carbon-to-hydrogen ratio, which means they do not burn as well.

Other problems arising from these heavier fuels are engine knocking, afterburning, uneven burning, variation in ignition delay, and a steeper ignition pressure gradient. These factors contribute to increased fatigue of engine components, excessive thermal loading, increased exhaust emissions, and critical piston ring and liner wear. The long-term effects on the engine are a significant increase in fuel consumption and component damage. The greatest fouling and deposit build-up will occur when the engine is operated at reduced or very low loads.

The fuel qualities used to indicate a fuel's burnability are Conradson carbon residue, asphaltenes, cetane value, and carbon-to-hydrogen ratio.

To ensure proper fuel atomization, effective use of the centrifugal separators, settling tanks, and filters are essential, and the correct fuel viscosity must be maintained by heating, adequate injection pressure, and correct injection timing. The maintenance of correct running temperatures according to the engine manufacturer's recommendations is also important, particularly at low loads. Additives which employ a reactive combustion catalyst can also be used to reduce the products of incomplete combustion.

High-temperature corrosion

Vanadium is the major fuel constituent influencing high-temperature corrosion. It cannot be removed in the pretreatment process, and it combines with sodium and sulphur during the combustion process to form eutectic compounds with melting points as low as 530°C. Such molten compounds are very corrosive and attack the protective oxide layers on steel, exposing it to corrosion.

Exhaust valves and piston crowns are very susceptible to high-temperature corrosion. One severe form creates mineral ash deposits on the valve seats, which, with constant pounding, cause dents and eventually a small channel through which the hot gases can pass. The compounds become heated and then attack the metal of the valve seat.

As well as their capacity for creating corrosion, vanadium, sulphur, and sodium deposit out during combustion to foul the engine components and, being abrasive, lead to increased liner and ring wear. The main defence against high-temperature corrosion has been to reduce the running temperatures of engine components, particularly exhaust valves, to levels below that at which the vanadium compounds are melted. Intensively cooled cylinder covers, liners, and valves, as well as rotators fitted to valves, have considerably reduced these problems. Special corrosion-resistant coatings such as stellite and plasma materials have also been applied to valves.

Low-temperature corrosion

Sulphur is generally the cause of low-temperature corrosion. In the combustion process, the sulphur in the fuel combines with oxygen to form sulphur dioxide (SO_2). Some of the sulphur dioxides further combines to form sulphur trioxide (SO_3). The sulphur trioxide formed during combustion reacts with moisture to form sulphuric acid vapours, and where the metal temperatures are below the acid dew point (160°C), these vapours condense as sulphuric acid, resulting in corrosion.

The obvious method of reducing this problem is to maintain temperatures in the engine above the acid dew point through good distribution and control of the cooling water. There is always the danger that an increase in temperatures to avoid low-temperature corrosion may lead to increased high-temperature corrosion. Attack on cylinder liners and piston rings as a result of high-sulphur content fuels has been effectively reduced by the controlled temperature of the cylinder liner walls and alkaline cylinder lubricating oils.

Abrasive impurities

The normal abrasive impurities in fuel are ash and sediment compounds. Solid metals, such as sodium, nickel, vanadium, calcium, and silica, can result in significant wear to fuel-injection equipment, cylinder liners, piston rings, and ring grooves. A comparatively new contaminant—metallic catalyst fines comprising very hard and abrasive alumina and silica particles—is cause for much concern. These particles carry over in the catalytic cracking refinery process and remain suspended in the residual bottom fuel for extended periods. It has been known for brand new fuel pumps to be worn out in a matter of days to the point where an engine fails to start through insufficient injection pressure, as a result of catalyst fines in the fuel.

Catalytic fines formed through the break-up of the catalyst as it is recycled through a cat-cracker plant are variable in size, with a range from submicron to approximately 30 μm and occasionally up to 100 μm. While considered to be spherical particles, this is not necessarily the case.

The composition of cat fines varies, depending on the type of feedstock and on whether the main unit of the cracker is intended to optimize gasoline (light) or diesel (heavier) grades. Exact catalyst compositions are not quoted today, but all of them contain some form of synthetic crystalline zeolite material. The zeolite is dispersed in a matrix comprising an active matrix, clay, and binder. Both the zeolite and the active matrix are forms of aluminium oxide and silicon oxide. The ratio of silicon to aluminium in the catalyst varies considerably, ranging from 0.65:1 to 2:1.

In the mid-1990s, Det Norske Veritas determined that the annual worldwide average ratio of silicon to aluminium was 1.64:1, based on analyses of fuel oil samples. Recent analyses indicate that the worldwide average ratio was around 1.2:1 and that this is likely to fall because modern catalysts tend to contain a higher concentration of aluminium.

The only effective method of combating abrasive particles is the correct fuel pretreatment. Separator manufacturers recommend that the separators should be operated in series (a purifier followed by a clarifier) at throughputs as low as 20% of the rated value (see Fuel Oil Treatment later in this chapter).

Properties of fuel oil

The quality of fuel oil is generally determined by a number of specific parameters or proportions of metals or impurities in a given sample of the particular fuel. Such parameters include viscosity, specific gravity, flash point, Conradson carbon, asphaltenes content, sulphur content, water content, vanadium content, and sodium content. Two parameters of traditional importance have been the calorific value and viscosity. Viscosity, once the best pointer to a fuel's quality or degree of heaviness, is now considered as being only partially a major quality criterion because of the possible effects of constituents of a fuel.

ISO 8217

Even today, obtaining bunkers of consistent quality is not guaranteed. Different refining processes, different crude stocks, and differences in blending techniques often result in very different fuels being supplied even when specifications are quoted. This problem was much worse prior to 1987 when the first ISO standard for Marine Fuels was drawn up. ISO 8217 first published in 1987 has been revised several times since 1996, 2005, 2010, 2012, 2017, and now under review awaiting a new edition.

Most shipowners now order fuels according to one version of the ISO standard or another. Usually, the latest version is used but where changes have occurred that the shipowner is unhappy with, they may continue to specify the earlier version.

The ISO 8217 standard includes two tables that detail the characteristics of both residual and distillate fuels. The residual fuel table contains details of 11 different fuels ranging from RMA 10 with a viscosity at 50°C of 10 cSt through to RMK 700 with a viscosity of 700 cSt. The distillate fuels table list four types of fuels.

Calorific value

The calorific value or heat of combustion of fuel oil is a measure of the amount of heat released during complete combustion of a unit mass of the fuel, expressed in kilojoules per kilogram. Calorific value is usually determined by a calorimeter, but a theoretical value can be calculated from the following equation:

$$\text{Calorific value (kcal/kg of fuel)} = \frac{8100\text{C} + 34{,}000\left(\text{H} - \frac{0}{8}\right)}{100}$$

where C, H, and O are percentages of these elements in 1 kg of fuel. Carbon yields 8080 kcal/kg when completely burned, hydrogen 34,000 kcal/kg; oxygen is assumed to be already attached to its proportion of hydrogen: that is, an amount of hydrogen equal to one-eighth the weight of the oxygen is nullified. The sulphur compounds are assumed to have their combustion heat nullified by the oxy-nitrogen ones.

Another variant of the formula is as follows:

$$\text{Calorific value (kcal/kg of fuel)} = 7500\text{C} + 33{,}800\left(\text{H} - \frac{0}{8}\right)$$

C, H, and O being fractions of a kilogram per kilogram of fuel.

To convert the kilocalories per kilogram values to SI units as widely used:

$$1\,\text{kcal/kg} = 4.187\,\text{kJ/kg}$$

For example, a fuel of 10,500 kcal/kg in SI units:

$$10{,}500\,\text{kcal/kg} = 10{,}500 \times 4.187\,\text{kJ/kg} = 44{,}000\,\text{kJ/kg}$$

The calorific value as determined by a bomb calorimeter is the gross or 'higher' value, which includes the latent heat of water vapour formed by the combustion of the hydrogen. The net or 'lower' calorific value is that obtained from subtracting this latent heat. The difference between the gross and net values is usually about 600–700 kcal/kg, depending upon the hydrogen percentage.

A formula that can be used to calculate the net value is:

$$\text{CV}_\text{n} = \text{CV}_\text{g} - 25\,(f + w)\ \text{kJ/kg}$$

where CV_n is the net calorific value in kilojoules per kilogram, CV_g the gross calorific value in kilojoules per kilogram, f the water content of the fuel in percentage by weight, and w the water percentage by weight generated by the combustion of the hydrogen in the fuel.

Typical gross calorific values for different fuels are as follows:

Diesel oil	10,750 kcal/kg
Gas oil	10,900 kcal/kg
Boiler fuel	10,300 kcal/kg

When test data are not available, the specific gravity of fuel will provide an approximate guide to its calorific value:

Specific gravity at 15°C	0.85	0.87	0.91	0.93
Gross CV (kcal/kg)	10,900	10,800	10,700	10,500

Viscosity

The viscosity of an oil is a measure of its resistance to flow, which decreases rapidly with an increase in temperature. Heating is necessary to thin the heavy fuels of high viscosity in current common use and ease their handling. Viscosity is not actually a direct measure of HFO quality, but it determines the complexity of the shipboard fuel heating and handling system. HFO must be heated to reach the correct injection viscosity to ensure optimized combustion and engine performance. Furthermore, HFOs are typically classed and marketed according to their viscosity.

The viscosity of the fuel determines the storage and handling temperature, particularly if the pour point of the fuel is low. The typical maximum fuel viscosity for transfer is 800–1000 cSt. The temperature for atomization of the fuel also depends on its viscosity.

Over the years different units have been used for viscosity (Engler Degrees, Saybolt Universal Seconds, Saybolt Furol Seconds, and Redwood No. 1 Seconds). The majority of marine fuels today are internationally traded on the basis of viscosity measured in centistokes ($1\,\text{cSt} = 1\,\text{mm}^2/\text{s}$). This is a measurement named after the physicist George Gabriel Stokes.

When quoting a viscosity, it must be accompanied by the temperature at which it is determined. Accepted temperatures for the viscosity determination of marine fuels are 40°C for distillate fuels and 100°C for residuals. If fuel contains an appreciable amount of water, testing for viscosity determination at 100°C becomes impossible, and therefore many laboratories may routinely test fuels at a lower temperature (80–90°C) and calculate the viscosity at 100°C.

Cetane number

The cetane number of a fuel is a measure of the ignition quality of the oil under the conditions in a diesel engine. The higher the cetane number, the shorter the time between fuel-injection and rapid pressure rise. The ignition quality of heavy fuel oils varies considerably. Low ignition quality may cause trouble at engine start-up and during low load operation, particularly if the engine is not sufficiently preheated; it may also result in a long ignition delay and also cause a fast pressure rise and very high maximum pressures. Deposits on the piston top and exhaust valves, in the exhaust system and on the turbine nozzle ring and blades can also be expected, with turbocharger fouling leading to reduced turbocharger efficiency and higher thermal load on the engine. Modern diesel engines with a higher compression ratio and optimized combustion process, however, are not so sensitive to the ignition properties of HFOs compared with earlier designs.

A more usable pointer of ignition quality is the diesel index, expressed as:

$$\text{Diesel index number} = \frac{G \times A}{100}$$

where G is the specific gravity at 60°F on the API scale and A the aniline point in °F, which is the lowest temperature at which equal parts by volume of freshly distilled aniline and the fuel oil are fully miscible.

Calculated carbon aromaticity index (CCAI)

The ignition quality of residual fuels is more difficult to predict than distillate fuels because they comprise blends of many different components, but the ignition quality of such fuels may be ranked by determining the calculated carbon aromaticity index from density and viscosity measurements. It should be noted, however, that the ignition performance of residual fuels is mainly related to engine design and operational factors. Formulae and nomograms for CCAI determination are published by major fuel suppliers and engine builders.

The formula for calculating the CCAI is

$$\text{CCAI} = D - 140.7\log(\log(V + 0.85)) - 80.6 - 483.5\ \log(t + 273/323)$$

where D is the density in kg/m^3 at 15°C, V the viscosity in cSt (mm^2/s) at T°C; and t the temperature at which V is determined.

The CCAI gives numbers in the range 800–870. A low figure is an indication of a better fuel whereas fuels with a figure greater than 870 are considered problematic.

CCAI can be considered a rough tool for estimating the ignition properties of an HFO, but more sophisticated devices, such as the fuel ignition analyser (FIA), have been introduced in recent years for more accurate determination.

Carbon residue

The carbon residue of a fuel is the tendency to form carbon deposits under high-temperature conditions in an inert atmosphere. It is considered to give an approximate indication of the combustibility of the fuel and may be expressed as a Ramsbottom carbon residue (RCR—ASTM D524), a Conradson carbon residue (CCR—ASTM D189), or a micro carbon residue

(MCR—ASTM D5430). Numerically the CCR value is the same as that of MCR. Reporting units are percentage by mass (%m/m).

The carbon residue value of fuel depends on the refinery processes employed in its manufacture. Four straight run fuels the typical value is 10%–12% by mass, while for fuels derived from secondary conversion processing the value depends on the severity of the processes applied. On a global basis, it is typically 15%–16% by mass but it can be as high as 20% by mass.

Ash content

The ash content is a measure of inorganic impurities in the fuel (typically sand, nickel, aluminium, silicon, sodium, and vanadium), which can cause different kinds of problems. Typically the ash value is in the range of 0.03%–0.07% by mass. The ISO 8217 standard sets a maximum limit for ash but also includes maximum limits for vanadium, aluminium silicon, and sodium. The most troublesome inorganic impurities are sodium and vanadium, which form a mixture of sodium sulphate and vanadium pentoxide, which melt and adhere to engine components, particularly exhaust valves.

Oxides of vanadium and sodium, mainly sodium vanadyl vanadates, are formed during combustion and will mix or react with oxides and vanadates of other ash constituents (e.g. nickel, calcium, silicon, and sulphur). The sticking temperature of the mixture may cause a deposit to be formed on the exhaust valve, in the exhaust system or in the turbocharger. Highly corrosive in molten salt, this deposit destroys the protective oxide layer (e.g. on an exhaust valve) and leads to hot corrosion and a burnt valve.

Deposits and hot corrosion in the turbocharger, especially on the nozzle ring and turbine blades, will reduce turbocharger efficiency. The gas exchange will also be disturbed; less air flows through the engine and hence the thermal load on the engine increases. Deposit formation rises with increased temperatures and engine outputs.

Several measures are dictated to avoid these problems when running on high-ash HFOs. It is important, for example, to operate an efficient fuel separation system, regularly clean the turbocharger, and ensure strict quality control of the bunkered fuel (to check that the amounts of ash and dangerous ash constituents remain low). It is also essential to ensure clean air filters and charge air coolers by regular cleaning based on pressure drop monitoring.

Catalytic fines content

The catalytic cracking process in refineries requires a catalyst which is usually an aluminium and silicon compound. The majority of the catalyst is recovered for subsequent operations but a small quantity of remnants of the catalyst remains in the refined fuels. These fragments are referred to as catalytic fines or more commonly as cat fines.

Low-sulphur fuel oils typically have a higher content of cat fines than high-sulphur fuels, as fines end up in the low-sulphur by-products from refining processes and the by-products are blended with residual fuels to reduce their sulphur content.

ISO Standard 8217:2012 introduced a maximum permissible 60 ppm level of cat fines, expressed as Aluminium + Silicon, for marine residual fuels, a reduction from the 80 ppm levels in ISO 8217:2005. The level of 60 ppm Al + Si is maintained in the latest published ISO 8217:2017 Fuel Standard.

However, it should be noted that engine manufacturers generally recommend a maximum of 15 ppm level of cat fines in the fuel entering the engines. As this level is significantly lower than the levels specified in the ISO Standards, it is essential to ensure adequate fuel handling and purification equipment and procedures are in place onboard to effectively bring the levels of cat fines in the fuel below 15 ppm.

As cat fines are very hard, they embed in the softer metal surfaces of cylinder liners, piston grooves, and rings. Cat fines are particularly damaging to cylinder liners because their surfaces are not polished or smooth. The presence of aluminium and silicon in fuel samples is an indication of the presence of cat fines.

Asphaltene content

Asphaltenes are complex, highly aromatic compounds with a high-molecular weight that usually contain sulphur, nitrogen, and oxygen as well as metals such as vanadium, nickel, and iron. A high asphaltene content indicates that fuel may be difficult to ignite and will burn slowly and may also contribute to deposit formation in the combustion chamber and exhaust system, especially at low engine loads.

If the HFO is unstable, the asphaltenes will precipitate from the fuel and block filters and/or cause deposits in the fuel system, as well as lead to excessive sludge formation in the fuel separator. Furthermore, when operating on HFOs with high asphaltene content, good performance of the lubricating oil should be emphasized. It is important that the lube oil is able to bind the combustion residues containing asphaltenes from blow-by gases entering the crankcase and thus prevent deposit formation on the engine component surfaces.

Sulphur content

This has no influence on combustion, but high-sulphur levels can be dangerous because of acid formation, mentioned earlier in this chapter. In recent years, there has been a tendency to equate sulphur content with cylinder liner wear, but opinions differ on this matter. Sulphur in the fuel may cause cold corrosion and corrosive wear, especially during the low load operation. It also contributes to deposit formation in the exhaust system, normally together with vanadium and/or sodium in the form of sulphates. These deposits can also cause high-temperature corrosion.

Sulphur content has come under scrutiny in recent years, with the IMO and European Union among other national authorities setting limits to reduce emissions of SOx in the exhaust gases. Among the options are to refine low-sulphur crude oils, equip refineries with desulphurization units, fit shipboard scrubbers, or to use alternative fuels such as LNG, LPG, methanol, and many more that have appeared in recent years or which are under development.

As from January 2020, only scrubber equipped ships may carry fuel oil with a sulphur content above 0.5% by mass. In ECAs, fuels should have a sulphur content not exceeding 0.1% by mass.

Water content

This is the amount of water in a given sample of the oil and is usually determined by centrifuging or distillation. Water content in fuel is usually very low with a value of a maximum of 0.2% by volume. Water can come from a variety of sources including tank condensation, tank leakage, or deliberate adulteration, and where steam is used to a tank heating purposes, heating coil leakage is another potential source. In a fuel treatment system, a further possible source is the purifier itself if the gravity disc is incorrect for the density of fuel being treated.

Sediment content

All HFOs contain a certain amount of sediment, which can be organic or inorganic. The total sediment content (TSP analysis) describes the fuel's cleanliness (presence of sand, dust, dirt, catalyst fines, and other solid/inorganic contaminants); stability (resistance to breakdown and precipitation of asphaltenes); and compatibility with another fuel quality.

Cloud point

The cloud point of an oil is the temperature at which the crystallization of paraffin wax begins to be observed when the oil is being cooled down.

Pour point

This is the lowest temperature at which oil remains fluid and thus is important to know for onboard handling purposes. An alternative is a solidifying point or the highest temperature at which the oil remains solid.

Because residual fuels are heated in use and distillates are generally not, Pour Point is more critical in distillates. ISO 8217 sets limits on the Pour Points of distillate fuels but problems can still be experienced. Blocked filters and wax build up on heating coils and tank surfaces are also encountered and can be problematic because when the wax may not redissolve even if heat is applied.

Flash point

The flashpoint is defined as the lowest temperature at which an oil gives off combustible vapours or the point at which an air/oil vapour mixture can be ignited by a flame or spark. A low flash point will not affect combustion, but the fuel can be dangerous to handle and store; this is especially so if the pour point is high, making it necessary to heat the HFO close to its flashpoint. Special explosion-proof equipment and fuel separators can be used in extreme cases.

A high vapour pressure (often indicating a low flash point) can also cause cavitation and gas pockets in the fuel pipes; these can be avoided by using an elevated pressure in the fuel handling system. SOLAS sets a minimum flash point of 60°C and it should be noted that some insurance companies demand the use of fuels with a flashpoint higher than 60°C.

Stability

This is normally defined as the ability of a fuel to remain in an unchanged state despite circumstances which could lead to a change. A stable fuel will, therefore, remain mostly as expected but an unstable fuel will begin to breakdown.

An unstable fuel on the other hand will breakdown to form a sludge. This is the result of precipitation of the high-molecular-weight hydrocarbons known as asphaltenes (see above). Whether a fuel will precipitate depends on the nature of the asphaltenes and also—and to a greater extent—on the paraffinic nature of the bulk hydrocarbon content. Asphaltenes are soluble in aromatic hydrocarbons which contain benzene type ring structures but less so in paraffinic hydrocarbons which have a linear structure.

If fuel is thought to be unstable because of filter blockage, this can be confirmed by taking a sample of material blocking a filter and placing it in an open container of water at 60–70°C. If the material is wax, it will melt but an asphaltenic sludge will not.

Compatibility

Usually, residual fuels are generally stable over the short to medium terms that they would be stored until used onboard. However, it is normal practice wherever possible to keep the fuels from different bunkering operations separate providing tank space allows.

In recent years as the use of blended fuels has increased, the problem of compatibility has become more common. This is a problem that is expected to become even more of an issue because most of the fuels produced to meet the 2020 sulphur rules are quite different in characteristics between producers and refining runs.

There is a high risk of incompatibility where one fuel has a high density and the other a low density. Even with two fuels of similar densities, there can be issues if the density values are high. The issue between tow low-density fuels are less common.

Compatibility issues are normally evidenced by heavy sludge formation and blockages in the bunker tanks, pipe runs, and filters. Some cases have required manual removal of the sludge build up.

Specific gravity

This is normally expressed in kilograms per cubic metre or grams per cubic centimetre at 15°C. As the density of the fuel depends on the density of the individual components, fuels can have identical densities but widely varying individual component densities. Apart from being an indicator of the 'heaviness' of fuel, when measured by a hydrometer, the specific gravity can be used to calculate the quantity of fuel by weight in a tank of given dimensions. Typical values for standard fuels are given in Table 5.2.

Table 5.2 Typical values for standard fuels.

Fuel	SG (g/cm^3)	Flash point (°C)	Lower CV (kJ/kg)
Gas oil	0.82–0.86	65–85	44,000–45,000
Diesel oil	0.85	65	44,000
Heavy fuel	0.9–0.99	65	40,000–42,000

200s Redwood No. 1–3500s Redwood No. 1.

Density mainly affects fuel separation. Traditional centrifugal separators can remove water and, to some extent, solid particles from HFOs with densities up to 991 kg/m^3 at 15°C. The latest separator designs can clean heavy fuels with densities up to 1010 kg/m^3 at 15°C. It is important to ensure that the correct separation flow rate and temperature are applied to achieve an efficient reduction of water, catalyst fines, sodium, and sediment from the fuel.

The following is the fuel oil specification of a leading low-speed diesel engine manufacturer. The properties are considered the worst in each case that can be burnt in a particular engine.

Density at 15°C	Maximum 991 kg/m^3
Kinetic viscosity	
At 50°C	Maximum 700 mm^2/s (cSt)
At 100°C	Maximum 55 mm^2/s (cSt)
Carbon residue (CCR)	Maximum 22 m/m (%)
Sulphur	Maximum 5.0 m/m (%)
Ash content	Maximum 0.2 m/m (%)
Vanadium	Maximum 600 mg/kg (ppm)
Sodium	Maximum 100 mg/kg (ppm)
Aluminium	Maximum 30 mg/kg (ppm)
Silicon	Maximum 50 mg/kg (ppm)
Sediment (SHF)	Maximum 0.10 m/m (%)
Water content	Maximum 1.0 v/v (%)
Flashpoint	Minimum 60°C
Pour point	Maximum 30°C

Combustion equations

Table 5.3 gives a list of the symbols designating the atoms and molecules of the elements and compounds in liquid fuels. For the atomic weights given, the suffix denotes the number of atoms accepted as constituting one molecule of the substance.

Table 5.3 Elements and compounds in liquid fuels and products of combustion.

Name	Nature	Atomic symbol	Atomic weight	Molecular symbol	Molecular weight
Carbon	Element	C	12.00	C	12.00
Hydrogen	Element	H	1.008	H_2	2.016
Oxygen	Element	O	16.00	O_2	32.00
Nitrogen	Element	N	14.01	N_2	28.02
Sulphur	Element	S	32.07	S_2	64.14
Carbon monoxide	Compound	–	–	CO	28.00
Carbon dioxide	Compound	–	–	CO_2	44.00
Sulphur dioxide	Compound	–	–	SO_2	64.07
Sulphur trioxide	Compound	–	–	SO_3	80.07
Sulphurous acid	Compound	–	–	H_2SO_3	82.086
Sulphuric acid	Compound	–	–	H_2SO_4	98.086
Water	Compound	–	–	H_2O	18.016

Tables 5.4–5.6 summarize the equations of the chemical reactions taking place during combustion. Nitrogen compounds are ignored. The weight of oxygen needed is evaluated, from the appropriate equation, for every element in the fuel, and the results are added together. The chemically correct quantity of air required to burn the fuel exactly to produce carbon dioxide and water vapour and unburnt nitrogen is approximately

Table 5.4 Combustion equations.

Nature of reaction	Equation	Gravimetric meaning
Carbon burned to carbon dioxide	$C+O_2=CO_2$	12 kg C + 32 kg O = 44 kg CO_2
Carbon burned to carbon monoxide	$2C+O_2=2(CO)$	24 kg C + 32 kg O = 56 kg CO
Carbon monoxide burned to carbon dioxide	$2(CO)+O_2=2(CO_2)$	56 kg CO + 32 kg O = 88 kg CO_2
Hydrogen oxidized to steam	$2H_2+O_2=2(H_2O)$	4 kg H + 32 kg O = 36 kg steam (or water)
Sulphur burned to sulphur dioxide	$S_2+2O_2=2(SO_2)$	64 kg S + 64 kg O = 128 kg SO_2
Sulphur dioxide burned to sulphurous acid	$SO_2+H_2O=H_2SO_3$	64 kg SO_2 + 18 kg H_2O = 82 kg H_2SO_3
Sulphur dioxide burned to sulphur trioxide	$O_2+2(SO_2)=2(SO_3)$	32 kg O + 128 kg SO_2 = 160 kg SO_3
Sulphur trioxide and water to form sulphuric acid	$SO_3+H_2O=H_2SO_4$	80 kg SO_3 + 18 kg H_2O = 98 kg H_2SO_4

Table 5.5 Volumetric meaning of equations.

Nature of reaction	Volumetric result
Carbon burned to carbon dioxide	1 vol C + 1 vol O = 1 vol CO_2
Carbon burned to carbon monoxide	2 vols C + 1 vol O = 2 vols CO
Carbon monoxide burned to carbon dioxide	2 vols CO + 1 vol O = 2 vols CO_2
Hydrogen oxidized to steam	2 vols H + 1 vol O = 2 vols H_2O (steam not water)
Sulphur burned to sulphur dioxide	1 vol S + 2 vols O = 2 vols SO_2
Sulphur dioxide burned to sulphurous acid	–
Sulphur dioxide burned to sulphur trioxide	–
Sulphur trioxide and water to form sulphuric acid	–

Table 5.6 Heat evolved by combustion.

Nature of reaction	Thermo-chemical equation	Heat evolved per kilogram of element burned (kcal/kg)
Carbon burned to carbon dioxide	$C + O_2 = CO_2 + 96{,}900$	8080
Carbon burned to carbon monoxide	$2C + O_2 = 2(CO) + 58{,}900$	2450
Carbon monoxide burned to carbon dioxide	$2(CO) + O_2 = 2(CO_2) + 315{,}000$	5630 (per kg CO burned)
Hydrogen oxidized to steam	$2H_2 + O_2 = 2(H_2O) + 136{,}000$	33,900[a] 29,000[b]
Sulphur burned to sulphur dioxide	$S_2 + 2O_2 = 2(SO_2) + 144{,}000$	2260
Sulphur dioxide burned to sulphurous acid	–	–
Sulphur dioxide burned to sulphur trioxide	–	–
Sulphur trioxide and water to form sulphuric acid	–	–

[a]Value if the latent heat of the steam formed is to be *included*.
[b]Value if the latent heat of the steam formed is to be *excluded*.

15:1, termed the stoichiometric ratio. The atmosphere may be assumed to contain 23.15% of oxygen by weight and 20.96% by volume. At standard atmospheric pressure, 1 kg of air occupies a volume of 0.83 m^3 at 20°C and 0.77 m^3 at 0°C. Average diesel fuel should comprise in percentages, carbon 86–87; hydrogen 11.0–13.5; sulphur 0.5–2; oxygen and nitrogen 0.5–1.0.

Table 5.7 Hydrogen: carbon ratios of typical fuels.

Fuel	H:C ratio	Calorific value (kJ/kg)
Methane	4:1	55,500
Ethane	3:1	51,900
Propane	2.7:1	50,400
Kerosene	1.9:1	43,300
Heavy fuel	1.5:1	42,500
Benzene	1:1	42,300
Coal	0.8:1	33,800

The greatest energy output in the combustion of fuels is from the hydrogen content, and fuels having high hydrogen:carbon ratio generally liberate greater quantities of heat. Typical examples are given in Table 5.7.

Low-sulphur fuels

Environmental and human health concerns over the emission of sulphur oxides have been driving IMO regulations under MARPOL Annex VI since the late 1970s. In 2005, the global limit for sulphur levels in fuels was set at 4.5% by mass and in 2005 when the first Emission Control Area was established, the limit set for these areas was 1.5% sulphur by mass.

On 1 July 2010, the limit in ECAs was reduced to 1.0% and the global cap was reduced from 1 January 2012 to 3.5%. After 1 January 2015, the limit in ECAs—of which four were then in operation—was reduced to 0.1% and as from 1 January 2020 the global cap has been reduced to 0.5%. Ships which are fitted with exhaust gas cleaning systems are permitted to continue to burn fuel with any sulphur content as long as the SOx component of the exhaust gases is equivalent to that of ships burning compliant fuels.

From the point when the limit in ECAs was reduced to 0.1%, some bunker suppliers began offering ultra-low-sulphur fuel oils (ULSFO) which complied with this very low-sulphur limit. Many ship operators chose instead to operate in these areas using distillate fuels. A very small number of ships with dual-fuel engines were also able to comply using LNG as a fuel. Generally, the switch to ULSFO when necessary was achieved without too many problems but there were cases of engine failures and fires caused by the incorrect switch over procedures.

Around this time, it was reckoned that of all bunker fuel sold, some 75% was residual fuel oil and the balance being distillate fuels. The number of ships using LNG and a small number of other alternatives was so small as to barely register.

When the IMO decided in 2016, to reduce the global cap to 0.5% in 2020, there was considerable alarm among the ship operating fraternity. Considering the issues that had been experienced in the ECAs, it was expected that a switch over to very low-sulphur fuel oils (VLSFO) with a sulphur content of just 0.5% would prove problematical for refiners and suppliers.

This alarm was caused partly by few refineries showing much inclination to produce the compliant fuel and also by expert opinions expressed over how VLSFO could actually be produced to a consistent quality. As a blended fuel, VLSFO was expected to display many of the undesirable characteristics around stability and compatibility mentioned above.

The concerns over practical issues thus revolved around the following points:

- VLSFO and ULSFO are both more expensive than HFO. Typically by around $200–$300 per tonne.
- Procedures need to be established for masters and chief engineers to know when and where they must burn specific fuels.
- Better records must be kept, showing exactly what fuel is burned when and where, and providing proof to authorities.
- Fuel must be tested on bunkering to ensure that it meets sulphur content limits.
- Shipowners must ensure that operation teams and bunker buyers are aware of the fleet's needs for low-sulphur fuel in specific areas, and, when chartering out a ship, the owner will need to negotiate a modification to the charter party to ensure that the charterer is required to supply low-sulphur fuel if trading to SECAs.
- Owners will need to consider modifying their existing tonnage to permit better segregation of bunkers, and shipbuilders and designers will have to create ships with much better means of segregating fuels and for switching from one fuel grade to another.
- Lubrication requirements for VLSFO and ULSFO are quite different for two-stroke engines than when operating on conventional HFO.

In addition to the practical concerns over the use of low-sulphur fuels, many ship owners were convinced that unscrupulous competitors whose ships were not fitted with scrubbers would continue to make use of HFO and so gain a competitive advantage. To allay these fears, the IMO amended the regulations making it illegal for ships to carry any noncompliant fuel onboard unless the ship was fitted with a scrubber or the fuel being carried was as cargo and not for the ship's operational use.

Fuel additives

While it is common practice in automotive applications to include additives in distillate fuel oils to enhance performance aspects, the use of additives in the marine sector to improve combustion or secure other benefits is not yet widespread. Many ship operators in the past have succumbed to the promises of disreputable suppliers, the disappointing results leading to a general mistrust by the shipping of fuel additives. The potential advantages for operators from using additives in both marine distillate and residual fuels are becoming more appreciated, however, particularly as solutions for targeting specific problems. Nevertheless, the cost of an additive on top of the fuel price can only be justified when the additive contributes to more reliable engine operation and a reduction in the ship's total operational costs.

The combustion phase is one of the areas where marine residual fuel additives can be used to advantage. Using an iron-based component such as ferrocene as an oxidation catalyst to accelerate the main combustion and enhance the complete combustion of residual fuel oils is a proven technology. Chevron's Marine Combustion Improver is also iron based, but its chemistry is different from conventional ferrocene technology and reportedly offers improved environmental, health, and safety properties. Available in liquid form for blending into the HFO, the combustion improver is formulated to enhance engine operational efficiency and reliability by minimizing postcombustion fouling, maintaining the cleanliness and efficiency of the exhaust system by controlling and removing soot build-up and fostering a reduction in smoke emissions.

Iron-based combustion improvers are offered by Infineum to reduce exhaust tract deposits and ensure that turbochargers maintain their efficiency, as well as to reduce the risk and consequences of exhaust tract fires. Magnesium-based additives can counter corrosion that may result from using high vanadium/sodium content residual fuels. Biocide additives can minimize and control the proliferation of fuel-borne bacteria and algae, together with the fuel filter-plugging waste products that can result from the presence of such species.

Fuel oil treatment

Low-, medium-, and high-speed diesel engines are all designed to burn marine distillate fuels while low-speed and many medium-speed engines can also accept commercially available HFO with a viscosity up to 700 cSt. In practice, most engines today are run on HFO either 180 or 380 cSt. The preparation and treatment of fuel oil begins at the bunkering stage when the fuel is taken onboard.

There are many opportunities for the fuel to be contaminated between the refinery and the delivery of the bunkers onboard, via any intermediary storage that might be required: sometimes so badly that it may damage the ship's machinery and associated systems. Especially threatening is the presence in the fuel of cat fines resulting from the catalytic cracking stage at the refinery in which the oil is heated over aluminium silicate catalyst, a substance harder than metal.

Immediate and longer-term risks to ship and machinery are associated with poor fuel quality, including fire and explosion, loss of main engine power and/or auxiliary power black-out, and difficulty in fulfilling charter party clauses with regard to ship speed, engine power, and specific fuel consumption.

Bunkering practice

Little legislation exists to govern bunker fuel, but the parameters of the international standard ISO 8217 have been widened in recent years. Since the most risk associated with bunkers impacts on the ship and owner even if the fuel is bought by and belongs to any time charterer, the best protection is for all parties to know exactly what fuel is needed for the vessel, and then to test that the delivered bunkers meet these requirements. Most problems and disputes can be traced back to either the ordering or the sampling of the fuel. The traditional way of ordering bunkers has been by quoting IFO grades, which refer only to viscosity at 50°C, but this can be technically inaccurate. It is more important in the first instance to make sure that the fuel is suitable for the machinery and fuel system in which it will be used as well as meeting any environmental regulations as might be in use.

The best starting point is the fuel specification provided by the engine builder. It makes commercial sense for ship operators to follow a recognized specification such as ISO 8217. If operating under a time charter, the charter party clause should clearly state what fuel is required and may also include other aspects such as segregation, testing, and the services of a surveyor for quantity determination. The charter party clause should also be kept as simple as possible. Ship operators are encouraged to stipulate that the bunker supplied is 'fit for its intended use' on the delivery note.

Among the key parameters of a fuel specification are the following:

- A maximum viscosity: this is required to be within the engine builder's specified limit; it is also a guide to the storage and handling properties of the fuel.
- A maximum density or specific gravity: dictated by the ship's fuel-treatment plant.
- A minimum flashpoint: dictated by safety regulations for the storage of fuel at sea.
- A maximum pour point: this limits the amount of heating required to store and handle the fuel.
- A maximum water content.
- A maximum sulphur content: this may be necessary to satisfy regional environmental regulations but is also desirable to limit the corrosive products of combustion and to ensure that the lubricants used in the engine can cope with these products.

Bunker fuel short delivery and quality disputes can be extremely contentious. In the event of a dispute, the conduct of measurement and sampling techniques will usually come under close scrutiny, and it will be necessary for the ship's interests to demonstrate that proper procedures were carried out during the bunkering.

Diligent participation of both ship and supplier during measurement and sampling is very important in preventing or resolving disputes, and the importance of maintaining proper records and documentation before, during, and after bunkering cannot be overstated.

It is essential that prior to bunkering, the chief engineer and any attending surveyor should first check the bunker receipt to ensure that the fuel on the barge complies with that ordered. The tanks of the receiving ship and the bunker barge should then be carefully measured to ascertain the exact quantities on both vessels. If no surveyor is in attendance, the task should be performed by either the chief or first engineer assisted by an engine room rating. The engineer should preferably use his or her own sounding tape, thermometers, hydrometers, and temperature/volume correction tables. Water-finding paste should be used on the measuring tape or sounding rod to check for free water. If found, the quantity of free water should be deducted from the volume of fluids in the tank to arrive at the correct oil volume.

On the receiving vessel, when sufficient empty tanks are available, many chief engineers prefer to segregate fuels from different stemmings: both to ease the calculation of the quantity loaded and to eliminate as far as possible any problems with incompatibilities of fuels produced from different crude feedstocks.

Once the measuring has taken place, the quantity onboard the barge should be calculated using its calibration tables. As the supply of fuel is often under customs control, bunker tankers have officially approved calibration tables adorned with various stamps verifying their authenticity. The receiving ship's representatives must, therefore, ensure that the tables offered to them are suitably approved and declined if they are at all suspected. Poor quality, unapproved tables may be encountered and even cases when no tables are available. If no tables can be provided, measures should be taken to obtain them. If no tables can be found, the fuel on the barge should be rejected, even if there is a delivery meter onboard. The veracity of fuel meters should be regarded as a suspect even when meter correction factors and calibration certificates are present.

Transfer can commence provided that the quantity in the barge tanks is the same as that on the bunker delivery receipt. The agreement must be reached on the rate of delivery and signals for the starting, stopping, and slowing down of delivery.

It is now mandatory for vessels to have clearly laid down procedures for bunkering. For bunkering, these rules required that a senior engineer officer be nominated as 'in charge' and the ranks of his or her assistants to be given. Before bunkering, a detailed plan should be prepared showing how much of each type of oil is to be loaded or transferred, which tanks were to be used and anticipated finishing ullages. A deck officer should be nominated to maintain the ship's mooring and other arrangements during bunkering. All deck scuppers are to be sealed and sawdust or other absorbent material provided. All bunker points should have a saveall fitted to catch any small leaks.

Receptacles for spilled oil and oil-contaminated materials must be at hand. Fire extinguishers and hoses deployed nearby and the fire main pressurized. The method of communication between the various participants to be given, with a back-up mode provided.

After all necessary pollution prevention precautions have been taken, the valves of the receiving ship's bunkering lines can be set and delivery started. When the integrity of the delivery hose and its connection to the ship's system has been proven, the rate of delivery can be slowly increased to the agreed level.

During the delivery, a set of three samples of the fuel should be taken using a small-bore valve fitted to the ship's loading connection and retained for any necessary future analysis. Proprietary sampling devices which can be bolted to the loading flange are available. The best of these has a small bore pipe directed against the flow, which leads outside to three branches, each fitted with a valve. It is therefore possible to obtain three continuous drip samples during loading to give an 'average' of the fuel quality.

The bunker barge normally provides samples of the fuel delivered, and the officer in charge of the operation should always attend to observe the samples being taken and sealed. It is customary for the ship's representative to sign the labels for the sample cans. Both barge and receiving vessel then retain samples. The ship's samples should be retained for at least 6 months.

When the transfer has been completed and the hose blown through and disconnected, the tanks of both vessels should be remeasured to enable the quantity transferred to be calculated. Any discrepancy between the ship and barge figures then becomes a matter for negotiation or protest, depending on the magnitude involved.

Vigilance on the part of personnel at the time of stemming fuel, during its subsequent correct storage and centrifuge treatment, is essential in preventing or mitigating future problems.

Before the fuel loaded into the bunker tank is used in the engine, any testing process should be put in hand so as to ascertain that the fuel is fit for use before it is actually used.

All fuel delivered in a bunkering operation should be fully detailed on the bunker delivery note (BDN). This document will form a key part of the oil record keeping that is mandatory on all ships. If a ship is suspected of burning bunkers that contravene and emission regulations, the first point of checking undertaken by port authorities will be the BDNs.

After 1 March 2020, when it became illegal for ships to carry fuel oil with a sulphur content above 0.5% unless the ship was scrubber fitted, a new document was designed for the cases where no MARPOL compliant bunker fuel was available. This document is known as the Fuel Oil Non-Availability Report (FONAR) and is completed by the ship detailing the efforts made to acquire compliant bunkers and the reason why this was not successful. There is some doubt still existing as to how this will be enforced.

The main stumbling point is whether the ship will be allowed to continue using noncompliant bunkers if complaint fuels are available at a subsequent port and if not what must be done with any noncompliant bunkers remaining on board. This issue was to be discussed at the IMO but the meetings have been indefinitely postponed because of the actions needed in the COVID-19 Pandemic declared in March 2020.

Bunker quality testing

Onboard test kits are available to allow ship's staff to carry out a range of tests during or after bunkering or during storage (if fuel stratification is suspected), for checking purifier or clarifier efficiency, for determining the cause of excessive sludging at the separators, if water ingress into the fuel is suspected, or when transferring bunkers to a new tank. The most comprehensive kits can detect catalytic fines and the presence of microbes, and test comparative viscosity, compatibility/stability, density, the content of insolubles, pour point, saltwater, sludge, or wax, and water content.

Fuel testing services—such as those offered worldwide by specialist divisions of classification societies or independent laboratories—can perform analysis and give advice on the handling and use of bunkers based on the samples supplied. Under ISO 8217, even if a fuel meets all the usual specification criteria, it should still not contain harmful chemicals or chemical wastes that would cause machinery damage or pose a health hazard to personnel.

Any ship operator doubting the need to test bunkers should be reminded that some 3%–5% of stems are critically off-spec. Among the diverse contaminants (mixed inadvertently with the fuel) and adulterants (added wilfully) found over the years have been acids, polypropylene, organic chlorides, and hydrogen sulphite. A number of ships have suffered serious machinery damage in recent years from contaminated or adulterated fuels, the culprit components identified as:

- Petroleum-based solvents, with the potential to destroy lubricant oil film and cause damage to the rubbing parts of fuel pumps and engine cylinders.
- Organic chlorides, which not only damage machinery but are also hazardous to ship staff.
- Methyl esters, while environmentally safe and friendly, are nevertheless solvents, which can remove the oil film and destroy lubrication between rubbing parts.

The damage potential is very high. A very high wear rate can be expected between fuel pump plungers and barrels and also between piston rings and cylinder liners. Additionally, the solvents are hazardous to personnel and exposure, even at low ppm levels, is not recommended. If they are mixed with bunker fuels, there is a real risk of engine room staff coming into contact with the solvent fumes, particularly in the purifier room. In the absence of any method of removing such contaminants from onboard, it is better to offload the bunkers.

Fuel storage

After fuel is loaded into the bunker tank and satisfactory test results have been obtained it will be heated and pumped into a settling tank. In the settling tank, there is a small amount of natural separation as sludge migrates to the bottom and any entrained air escapes. Before the fuel is ready to be used, it will be pumped through a filter to remove large-size contaminants.

Most fuel treatment systems incorporate two separators to aid cleaning the fuel before it reaches the engine but in some older vessels, only one separator may be fitted. After the initial filtration, the fuel will be heated before being routed to the first separator where more water, sludge, and the inevitable cat fines will be removed.

The role of filtration systems has been strengthened by higher fuel-injection pressures and the wider use of common rail fuel systems. Solids remaining after previous treatment stages are removed from the fuel circuit by manual cleaning in a simplex filter or by an automatic back-flushing filter. Common rail fuel systems call for stricter filtration regimes.

Separators are usually type approved by classification societies but there are no mandatory performance standards and shipbuilders are free to instal any size of separator even if it is not best suited to the flow rates of fuels for a specific ship. Some separator manufacturers, including Alfa Laval and GEA Westfalia, have developed a voluntary separation standard with the aid of class societies and engine-makers, but it is for the potential owner of a new-building to insist on applying it (Fig. 5.2). The standard aims to ensure maximum removal of cat fines and other contaminants by using a slightly oversized separator that will not be overwhelmed by any but the worst fuels.

The effective separator throughput should accord with the maximum consumption of the diesel engine plus a margin of about 18% to ensure that separated fuel oil flows back from the daily tank to the settling tank. The separators should be in continuous operation from port to port. To maintain a constant flow through the separators, individual positive displacement-type pumps operating at constant capacity should be installed.

After separation, the fuel then moves on to the service tank in readiness to be pumped to the engine. A vessel operating on HFO normally has two service tanks: one for high-sulphur HFO for use outside ECAs and one for low-sulphur HFO for use inside ECAs. When a service tank is full, there is generally a return line allowing overflow back to the settling tank.

Fig. 5.2 An Alfa Laval centrifugal separator module.

From the service tank, the fuel is pumped to the engine via a series of filters. More heaters are needed at this stage to ensure that the viscosity at the engine is in accordance with the engine maker's operating standards and booster pumps to ensure fuel that pressure is maintained. While it is possible to construct a fuel treatment system from individually sourced components, the leading specialists in the field produce modular units that can be customized to specific requirements.

Apart from the core components such as pumps, heat exchangers, and separators, many fuel treatment systems will incorporate additional components. Counted among the most useful of these is a homogenizer. In very heavy fuels, molecules of carbon often form long chains that will never combust entirely in the engine. The homogenizer overcomes this by grinding the long chains and cutting them into smaller particles of submicron sizes and finely distributes any water present as small droplets. Such a device will greatly reduce the amount of sludge and improve the combustion characteristics of the fuel. Some of these devices also have a modification that allows the introduction of controlled amounts of water to form an emulsion. Emulsified fuels are known to produce less NOx and are predicted to become highly beneficial in helping ship operators meet the most stringent NOx emission regulations.

There is some disagreement, however, over the positioning of a homogenizer in the fuel-treatment system (before or after the separator). Alfa Laval argues that installing such a unit prior to the separator reduces the ability of the treatment system to remove those particles that can cause damage to the fuel-injection system and the main engine and should therefore be avoided.

Another potentially very useful device was first presented at SMM in 2012 by Chris Marine. The Fuel Analyser monitors the cat fine level continuously in the fuel treatment system at different sample points. If a preset alarm level is reached at the engine inlet, it is possible to switch to cleaner fuel from the second-day tank as an emergency action.

The device also permits an analysis of the cause of the rise in cat fines and implements countermeasures preventing similar events. With the typical configuration, it is possible to trace a problem to the settling tank, the purifier, or the service tank by comparing the measurements from the different sampling locations. A similar device called the Catguard was launched at the following SMM in 2014 by NanoNord.

Cat-fines have a higher density than fuel oil and therefore tend to settle at the bottom of the tanks. In rough seas, accumulated cat fines could stir up and suddenly generate a high concentration at the tank's outlet. This may be circumvented by cleaning the tank as soon as the measurement device picks up an elevated cat fine level at the tank's outlet (purifier inlet or engine inlet). The device monitors up to seven automatic sampling points and one off-line point at 1–10 bar and 25–140°C.

After passing through the separator, the fuel will be maintained at around 80–85°C in the service tank. The temperatures required for treating heavy residual fuels exceed the flashpoint of the distillates and therefore a different treatment system is required. There is still a need to remove impurities particularly scale and water so filtration and separation are needed. However, to achieve the viscosity requirements of some engines it may be necessary to cool rather than heat the fuel in the final stages of its journey to the combustion chamber.

The fuel oil heaters used in a fuel treatment system may be of the shell and tube or plate heat exchanger type, with electricity, steam, or thermal oil as the heating medium. The required heating temperature for different oil viscosities is derived from a fuel oil heating chart (Fig. 5.3), based on the information from oil suppliers with respect to typical marine fuels with a viscosity index (VI) of 70–80. Since the viscosity after the heater is the

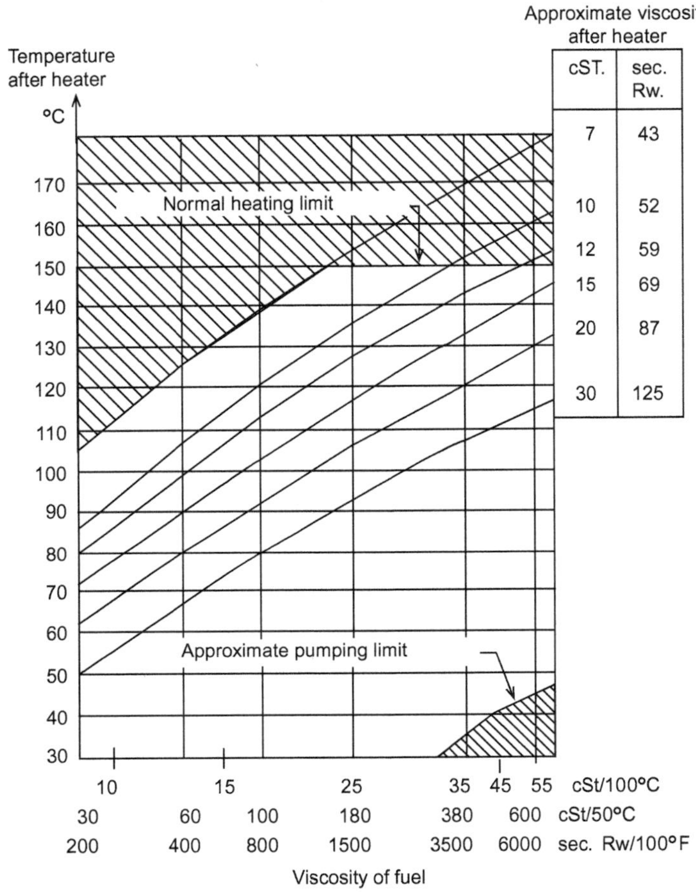

Fig. 5.3 Fuel oil heating chart (MAN Diesel). *From MAN Energy Solutions.*

controlled parameter, the heating temperature may vary, depending on the viscosity and viscosity index of the fuel. The viscosity meter setting, reflecting the desired fuel-injection viscosity recommended for an engine by the engine builder, is typically 10–15 cSt.

Dealing with waste

Waste fuel oil from the settling and service tank drainages, leakages, filters, and purifiers is usually collected in the waste oil tank along with waste lubricating oil and subsequently landed or incinerated. The disposal can be expensive and so adds to the original cost of the fuel. Alfa Laval has developed a device called PureDry which is an advanced separator that can recover a high percentage of usable fuel oil from the waste oil tank.

Because the waste lubricating oil is undesirable in fuel oil, there is a need to have separate waste tanks for fuel and lubricating oils. Although the waste fuel oil tank appears to contain just black oil, it is actually oil-polluted water containing 20%–30% energy in the form of recoverable fuel oil. The remainder is oil-polluted water 70%–80% and, accumulating at the bottom, suspended solids which make up approximately 1%.

The PureDry separator recovers the fuel oil from the oily water in the waste fuel oil tank and it is returned to the fuel oil bunker tank for reuse after normal treatment. For the shipowner, the result is a reduction of up to 2% in the total volume of fuel oil consumed and a corresponding reduction in the ship's fuel bill.

The process reduces the volume of waste oil by 99%, producing typically 5–15 kg per day of nonpumpable "super-dry" solids that can be landed as dry waste and disposed of in the same way as oily rags and used filter cartridges. There are no oil losses and no additional wastes are generated. The separated water, now with an oil content of less than 1000 ppm, is pumped to the bilge water system. The PureDry can also be used to treat waste lubricating oils but there will only be a reduction in the waste product to be disposed of and no usable product.

CHAPTER SIX

Alternative fuels and treatment

Alternative fuels and decarbonization

As mentioned in Chapter 3, the original diesel engine concept was for an engine which able to burn many fuel types. The ready availability of oil fuels around the globe has been the key factor in the decision of shipowners to make either distillates or residual oil fuels derived from crude as the fuels of choice.

Were it not for the environmental movement to end the use of fossil fuels wherever possible, it is highly doubtful that marine diesel engine builders would have followed the path that they have followed over the last two decades. Arguably, the use of LNG would still have been a natural evolution as, despite the problems associated with its use, it is undoubtedly a cleaner fuel because it reduces maintenance time and expenses.

Under most circumstances, the use of oil and LNG along with some exhaust emission limitations would have been more than sufficient to meet all regulatory limits. However, the movement to reduce CO_2 output from all human activities and the IMO's declared ambition of cutting CO_2 from shipping.

Since 2011, the IMO has put in place mandatory efficiency improvements known as the Energy Efficiency Design Index (EEDI) aimed at reducing the CO_2 per tonne/mile from a baseline established using data from existing ships. The reductions are to be done on a phased in basis and will result in reductions between 30% and 50% for specific ship types by 2025 or 2030.

In April 2018, IMO's Marine Environment Protection Committee (MEPC) adopted an initial strategy on the reduction of greenhouse gas emissions from ships beyond that required under EEDI, setting out a vision to reduce GHG emissions from international shipping and phase them out, as soon as possible in this century. The vision confirms IMO's commitment to reducing GHG emissions from international shipping and, as a matter of urgency, to phasing them out as soon as possible.

Pounder's Marine Diesel Engines and Gas Turbines
https://doi.org/10.1016/B978-0-08-102748-6.00006-2

More specifically, under the identified 'levels of ambition', the initial strategy envisages for the first time a reduction in total GHG emissions from international shipping which, it says, should peak as soon as possible and reduce the total annual GHG emissions by at least 50% by 2050 compared to 2008, while, at the same time, pursuing efforts towards phasing them out entirely. The strategy includes a specific reference to 'a pathway of CO_2 emissions reduction consistent with the Paris Agreement temperature goals'.

It follows that for a complete phase out of CO_2 emissions, there must either be a means of extracting CO_2 from the exhaust stream—this is currently considered as not being technically feasible, or ships must burn fuels containing less carbon. The latter would of course only be a temporary step if CO_2 emissions are to be phased out altogether. It would also mean that fuels such as LNG would also be disallowed.

There is a further option which some consider as 'cheating', but which is currently permitted by some national and supranational regulation and that is the use of so-called biofuels. Biofuels are produced from animal or vegetable material either grown specifically for conversion into fuels or the waste product of animal or vegetable material produced for other reasons.

The argument in favour of biofuels is that the CO_2 released during combustion as a fuel will be reabsorbed by the growth of new material. Thus, there will be no release of additional CO_2 into the atmosphere. The argument may have some merit but there are objections based on the fact that land used for the production of biofuels cannot be used for other purposes including food production. Another argument is that for reasons such as drought, floods, or other loss of land, the CO_2 released in one season may not be reabsorbed the following.

Biofuels

The production and use of biofuels have grown significantly in recent years, the focus initially being mainly on bioethanol blended into fossil motor gasoline (petrol) or used directly, and biodiesel or fatty acid methyl ester (FAME) diesel blended into fossil diesel or also used directly. While biofuels in land-based transport are commonplace, usage at sea is still limited but promises some technical merits. Marine engines are generally of lower speed and more tolerant of burning alternative fuels than smaller, high-speed automotive engines.

Biofuel development has been quite rapid aided by the fact that engines used in shore situations have proved good experience builders. Today biofuels are sometimes categorized as first, second, and third generation, based on the technology and/or the raw materials used in production.

In first-generation biofuels, the carbon source comes from sugar, lipid, or starch which is directly extracted from a plant. In this category, the following are included: biodiesel, vegetable fats, biogas, bio-alcohols, and synthetic gas. First-generation biofuels are mostly used for nonmarine purposes although some do end up in marine fuels. The main disadvantage of first-generation biofuels is the food-versus-fuel debate and one of the major reasons for rising food prices is due to the increase in the production of these fuels. In addition, biodiesel is proven to be a noncost-efficient emission abatement technology.

Second-generation biofuels can be broadly grouped into those produced either biochemically or thermochemically, either route using nonfood crops, especially from lignocellulosic feedstocks sourced from the crop, forest, or wood process residues, or purpose-grown perennial grasses or trees. Such crops are likely to be more productive than most crops used for first-generation biofuels in terms of energy levels for comparable growing areas.

Second-generation biofuels include methyl esters derived from used cooking oils, bio-oil, butanol, mixed alcohols, lignocellulosic alcohol, and hydrogenated vegetable oils (HVOs).

Another advantage of second-generation biofuels is that they do not compete with food crops and some types of them can offer better engine performance based on early trials.

Third-generation biofuels are also promising both in terms of energy and economically. These are specially engineered energy crops such as algae that is cultured to act as a low-cost, high-energy, and entirely renewable feedstock. Algae can be grown using land and water unsuitable for food production, therefore, reducing the strain on already depleted water sources.

Until the 2017 version, the fuel standard for marine applications, ISO 8217 did not recognize biofuels. However, the 2017 version introduced new grades for biofuels for the first time. DF (distillate FAME) grades DFA, DFZ, and DFB which allow up to 7% FAME content by volume. Other than the 7% FAME allowance, these grades are identical to the traditional DMA, DMZ, and DMB grades for all other parameters.

The standard also permits up to 0.5% of FAME in other distillate grades. This is because the working group developing the standard was confident that this level does not cause any operational issues based on field experience. The limit is still low enough to prevent the intentional blending of FAME into marine fuel, but high enough to allow for accidental trace amounts caused by shared supply chain infrastructure with fuels for the automotive sector and should broaden the scope for fuel availability. The ISO standard should ensure that blended fuels containing biofuel are to a consistent standard and safe for use by most vessels.

Some owners are participating in trials using other biofuels that have been developed outside of the ISO standard. The owners participating in the trials should be well aware of the risks and issues to be expected. A leading player in the development of biofuels is Netherlands-based GoodFuels. It has been involved in a number of projects including one with Wärtsilä and Dutch shipowner Boksalis and others with Danish operator Norden, French container ship operator CMA CGM, and heavy-lift ship operator Jumbo.

The latter will use GoodFuels second-generation Bio-Fuel Oil MD1-100 which is completely derived from forest residues and waste cooking oil products. GoodFuels describe this as a drop in solution and say it is expected to deliver 80%–90% well-to-propeller CO_2 reduction versus fossil equivalents and contains no sulphur. It also promises NOx reduction of 37% and PM cuts of up to 60%. The fuel can be used instead of fossil fuels in most ship engines. While promising, current production is insufficient to meet all but the smallest demand.

As with mineral oil-based fuels, biofuels have their own specific issues. Biodiesel blended prior to delivery to the ship can be affected by shelf life. Fuel ageing and oxidation can lead to high acid number, high viscosity, and the formation of gums and sediments. Fuel management will, therefore, become more challenging. Supplies of biofuel and diesel for on-board mixing enable the operator to dictate the exact blend of biofuel depending on the conditions, but that would require new technology installed on-board and extra complexity for the crew to handle.

Some operational problems can occur from using biodiesel, related to the cold filter plugging point (CFPP) at which the fuel starts to turn to gel. Considered to be the indication of low-temperature operability, the CFPP is in a range between 0°C and 15°C for different types of biodiesels and can cause problems with filter clogging, which can only be overcome by carefully monitoring the fuel tank temperature. Such a problem could well affect ships operating in cold climates, where additional tank heating coils might be required to prevent it from happening.

Another challenge in using biodiesels is corrosion caused in the fuel injection and treatment systems as biodiesels are hygroscopic and maintain 1200–1500 ppm water. It is essential, therefore, that the fuel is thoroughly conditioned prior to injection and also important that the fuel acid number is monitored to ensure no rancid, and acidic fuel is introduced into the injection system. A typical fuel-treatment system should incorporate separators to ensure water is removed, as well as heaters at various stages to ensure the fuel is at the correct temperature for introducing in the engine.

Acidity could be a particular problem for crew used to matching lubricants to the sulphur content of conventional fuels in two-stroke engines. Although biofuels contain little or no sulphur, their unique formulation can include other types of acid which would need to be assessed and matched to an appropriate lubricant.

As well as biofuel versions of oil fuels, there is also biomethane which can substitute for LNG. Biomethane is produced from the biogas derived from decomposing plant and animal material including human sewage waste. Biogas contains a relatively low level of methane, but the level can be increased through treatment to a similar level to LNG. Chemically, biomethane has exactly the same formula as methane from fossil sources but is considered as being carbon neutral because the material it is made from is considered as naturally renewable.

Methanol

Methanol is a simple alcohol fuel which in terms of mass has less than half the carbon content of fuel oil and around half of the energy value. While this would make its use difficult for long ocean voyages because of a doubling of bunker space required, it can be considered a good fuel for vessels with shorter voyages or those that carry methanol as cargo as this would increase the range of the ship. Methanol and LNG are similar in terms of energy density.

Methanol has already made the transition to marine use from land applications where it has been used successfully for some time. The number of ships making use of methanol is currently very small and includes a ro-pax ferry and several methanol carriers. Details of the engine adaptations necessary to use methanol in MAN two-stroke engines were covered in Chapter 3.

As an alternative to fuel oils, methanol has several advantages in meeting environmental and regulatory demands. Methanol, in common with other alcohols, provides clean burning in the engine and produces low levels of soot in combustion compared with diesel oil or HFO (less than 0.01 g/kWh for methanol in heavy-duty engines compared to more than 0.1 g/kWh for best diesel).

In practical use in the operation and in tests of methanol fuel in marine diesel engines, emissions of nitrogen oxides and particulates have been found very low and, being sulphur-free, methanol does not produce SOx emissions. Nitrogen oxide levels are low, in line with Tier III NOx emissions (2–4 g/kWh). When using alcohol fuels, formaldehyde is sometimes formed. Emission measurements for methanol in use do not show any measurable formaldehyde formation. While running on methanol, engine efficiency is as high or even higher than for traditional fuels.

When used in a diesel engine, methanol requires an ignition enhancer, which may be a small amount of diesel oil. In all tests performed, methanol shows good combustion properties and energy efficiency as well as low emissions from combustion.

Methanol is a low flashpoint fuel, meaning that it can vaporize and mix with air to form a flammable mixture at a relatively low temperature. Having a low flashpoint is a characteristic that methanol shares with LNG. However, unlike LNG, methanol is a liquid at ambient temperature and pressure, meaning that it can be stored in ordinary tanks with few modifications. Therefore, methanol shares many characteristics with HFO.

From a safety viewpoint, one of the defining features of methanol is that it is a low-flashpoint fuel. Methanol has a flashpoint of 11°C and a boiling point of 65°C. The flashpoint of HFO is 60°C while LNG's flashpoint ranges from −188°C to −135°C, with the boiling point standing at −163°C.

Since the fuel storage and treatment of methanol would normally be very similar to HFO, its low flashpoint brings into focus the hazard of fire. Methanol is not yet covered by the IMO's IGF Code which currently only applies to LNG, but its characteristics have been accounted for in the IMO's IGC Code when carried at ambient temperatures as a cargo. The IGC Code thus provides the basis for regulation when methanol is carried as fuel.

Most microorganisms have the ability to oxidize methanol in an enzymatic reaction to formic acid, which is converted to carbon dioxide in the presence of folic acid. This means that methanol that is released into the environment would be biodegraded rapidly. A large spill would have very local effects, but rapid degradation and dilution can be expected.

The general health hazards of methanol have been well known for a long time, as is the treatment to prevent intoxication after exposure. Poisoning through drinking methanol was first reported in literature 150 years ago.

Uptake of methanol is possible through ingestion, but also through the skin and by inhalation. Human beings, in contrast to most other species, have a very limited ability to degrade methanol into carbon dioxide. The enzymatic degradation occurring in the liver will instead result in an increasing level of formic acid, causing intoxication. Ethanol may inhibit the reaction, by being the preferred reagent in the conversion. This means that the effect may be delayed by the ingestion of ethanol. Ethanol conversion in man proceeds through acetic aldehyde to acetic acid and further to carbon dioxide and water.

A typical fuel handling system would make contact with methanol unlikely but when the IGF Code is extended to cover other fuels, appropriate safety measures will be made mandatory both in regard to the construction of the vessel and its equipment and through changes to STCW, the training and certification of the crew will be determined.

LNG

Details of engines able to burn LNG as a marine fuel were extensively covered along with some of its characteristics in Chapter 3. As a fuel, LNG has the benefits of being clean burning reducing maintenance and reductions in emissions of NOx, SOx, CO_2, and particulates. Its use in the marine sphere has been restricted by an absence of fuel storage and bunkering infrastructure and increased capital expenditure on engines and fuel treatment systems onboard.

As a consequence of the IMO 2020 sulphur cap on fuels, the bunkering infrastructure is now improving rapidly in some parts of the world sufficiently enough for owners of large container ships, tankers car carriers, and other ship types to have committed to its use.

Regulations with regard to ship construction and the use of LNG as a marine fuel was initially determined by individual flag states usually in consultation with classification societies and other specialists. This piecemeal approach ended on 1 January 2017 when the IMO's International Code of Safety for Ships using Gases or other Low-flashpoint Fuels (IGF Code) came into effect. With an international standard covering the storage and use of LNG, there was an increase in the number of ships fitted with dual-fuel engines that have continued to advance each year since.

LNG—and for that matter LPG—as a marine fuel requires several changes of thinking by ship designers and ships' crews. This is because the storage arrangement and characteristics are very different from oil fuels requiring attention to new circumstances that previously did not exist.

LNG has approximately 600 times the energy density of natural gas at atmospheric conditions and is typically stored at a pressure between 1 and 4 bar, whereby the equilibrium temperatures are approximately −160°C to −155°C. In order to minimize the risks related to both property and life, it is vital that the material used for the LNG system has been certified for cryogenic temperatures and that the system has built-in pressure relief functionality.

The placement of the LNG fuel tank and processing equipment must be decided at the design stage as must arrangements for ventilation ducts, pressure relief masts, and piping arrangements. Access to hazardous areas must be arranged in a safe manner and great effort must be put into developing a complete and consistent safety philosophy from the beginning of any vessel design. All of the leading classification societies will cover these elements within their construction rules.

Engine makers that produce dual-fuel engines almost always also provide fuel containment and treatment systems. The following is a description of the Wärtsilä LNGPac system that has been installed on some vessels. Other makers' systems will be essentially similar.

As with any type of fuel, the first component in the fuel supply system onboard is the bunkering station that provides the connection to the LNG bunkering barge. Each station includes one bunkering line (LNG line), one return line, and one nitrogen purging line with respective control/thermal relief valves (pressure safety valves) and flanges.

A return line is necessary in case the bunkering operation takes place with two hoses connected, and the evaporated gas is returned to the bunkering terminal or barge. During bunkering operations, LNG could evaporate due to heat leakages in the piping and/or due to the higher temperature in the storage tank onboard compared to the refilling tank.

From the bunkering station, LNG travels to the storage tank via insulated pipes. Vacuum insulation is selected for its excellent insulation properties, and to minimize LNG evaporation during bunkering. The pressurized storage tank is cylindrically shaped with dished ends and is designed in accordance with the IMO IGC Code and EN 13458-2—cryogenic vessels. LNGPac tanks are insulated with perlite/vacuum. The tank consists of a stainless-steel inner pressure vessel and an outer vessel that acts as a secondary barrier. The outer vessel can be made of either stainless steel or carbon steel.

According to the current IMO Guidelines, the LNG fuel tanks have to be selected from among the 'Independent Types A, B, or C'. The LNGPac is designed according to Type C requirements. The pressure vessel allows easy handling of the boil-off, since the tank is designed to withstand a significant pressure increase and the pressure relief valves are set at 9 bar(g).

In practice, dual-fuel vessels can operate for a long time in liquid fuel mode (HFO or MDO) before having to take care of the pressure increase in the tank. The handling of the boil-off is done very simply by a temporary switch over of the engines to gas mode, and the gas is taken from the vapour phase in the upper part of the tank.

As an indication, a 200 m^3 pressurized type C tank, filled at 50% could hold LNG for about 25 days, even without any gas consumption from the tank. A Wärtsilä dual-fuel engine requires approximately 4–5 bar(g) at the inlet of the gas valve unit. In case LNG is stored at atmospheric pressure (Type A and Type B tanks), the fuel system should include either compressors or cryogenic pumps to deliver the fuel at the correct pressure.

The process skid includes all the connections and valves between the tank and the pressure build-up unit (PBU) and the product evaporator, together with the evaporators themselves. The PBU consists of an insulated pipe, an evaporator, valves, a single wall pipe, and sensors. The mission of the PBU is to build up the pressure in the tank after bunkering LNG and to maintain the required pressure in the tank [around 5 bar(g)], during normal operation.

Maintaining the correct pressure in the tank ensures that the Wärtsilä dual-fuel engines are able to achieve 100% MCR at any time. Since the LNGPac system has no cryogenic pump or compressor, the engine gas inlet pressure requirements are met by achieving the correct storage pressure inside the LNG tank.

The circulation of LNG to the PBU evaporator is achieved by the hydrostatic pressure difference between the top and bottom of the tank, with LNG from the bottom of the tank being fed to the evaporator. The evaporated gas is then returned to the top of the tank. The natural circulation through the PBU continues until the required pressure in the tank is achieved.

The Product Evaporator circuit consists of an insulated pipe, an evaporator, valves, a single wall pipe, and sensors. The task of the product evaporator is to evaporate the LNG into gas and heat it to a minimum of 0°C as per engine specifications. The gas is then fed to the gas valve unit before the engines. Both the PBU and Product Evaporator are heated by a water/glycol mixture, which is recirculated to an external cooler. Here, the mixture is heated by the waste heat from the low-temperature engine cooling water circuit.

The process skid has a modular design, making it easy to be selected and assembled for the entire LNGPac product range. The key parameters influencing the modularization of the process skid are the sizes and volume of the tank, and the engines (model and number) connected to the tank.

The tank room is a stainless-steel barrier welded to the outer vessel of the tank. The structure contains the process skid and all the pipe penetrations to the tank. In the unlikely event of an LNG leakage, the tank room acts as a barrier that avoids damage to the external compartments and facilitates the quick ventilation of the evaporated gas. The tank room and ventilation system are to be fire protected to A-60/A-0 insulation class, depending on the safety designation of the adjacent space.

The LNGPac control system is based on Wärtsilä's vessel automation system platform. When combined with the Wärtsilä Integrated Automation System (IAS), the same hardware and human machine interface can be used throughout the vessel to operate the LNGPac, the dual-fuel engines, and the propulsion system. In addition, separately delivered features typical of DF-engine applications, have been incorporated into the IAS. These include the Wärtsilä Operator's Interface System, Condition Based Maintenance, and monitoring of IMO Tier III compliance in Emission Control Areas.

The core of the control system is a PLC cabinet placed in a safe area near the tank room. All LNGPac transmitters and intrinsically safe sensors, as well as all interfaces to external systems such as fire and alarm, gas detection, etc., originate from the cabinet. The pneumatic valves are controlled by solenoids placed in a safe area adjacent to the tank room and bunkering station(s). In the case of ships retrofitted with LNGPac, the control system can operate as a separate system with monitors on the bridge and engine control room, or by being integrated into the existing system.

Future fuels

All of the fuel types currently in use: oil fuels, biofuels, LNG, methanol, and LPG are able to be used in two- and four-stroke diesel engines without much difficulty albeit with some modifications. The quest for shipping to decarbonize does however make all of these fuels—with the possible exception of biofuels and biomethane—as transition fuels, lowering the CO_2 output but not able to meet the extreme low carbon target.

Two potential fuels are being considered for use in combustion engines that contain no carbon in their composition and thus theoretically meet the decarbonization targets. These are hydrogen and ammonia.

Most would consider that hydrogen is best used in producing power from a fuel cell which is quite different from an internal combustion engine. However, it is possible to use hydrogen in a diesel engine although very few trials have been made on the scale needed for marine purposes.

Hydrogen has a very high energy content by mass, but it is a very light gas with 1 kg occupying 5.4 M^3 at standard temperature and pressure. Thus, a large amount of storage space would be needed for gaseous hydrogen making it unsuitable for practical purposes. Liquid hydrogen is a viable alternative but requires extremely low temperatures (−253°C) and pressures. This is much more challenging even than LNG. The density of liquid hydrogen is only 70.99 g/L at −253°C, a relative density of just 0.07. Although the specific energy of liquid hydrogen is around twice that of other fuels, this gives it a remarkably low volumetric energy density.

As a fuel hydrogen also has other undesirable properties. The hydrogen atom is very small and can penetrate into almost all materials including steel and aluminium. When this happens, the affected material can become brittle and prone to damage. Since hydrogen has a very low flash point and is highly explosive/flammable, the method of storage on a ship which is subject to many stresses is something that will pose problems.

There is a great deal of momentum behind the future use of hydrogen and it is very likely that it will be used at some point. However, to consider it as the best method to decarbonize shipping in a very short space of time is probably premature. Its first uses will doubtless be in small craft and ferries where some initial breakthroughs have already been made.

A leading proponent of hydrogen as a fuel for diesel engines is the Belgian ship operator CMB. It has already built a small shuttle ferry fuelled by hydrogen, and in 2018, it entered into a joint venture with engine maker ABC under the name BeHydro.

The *CMB Hydroville* a small 14 m shuttle ferry built in 2019 is powered by two H2ICED engines which are standard diesel engines modified to run on hydrogen. The engines have been in use in road vehicles for more than a decade and are similar to those used in the BMW Hydrogen 7 concept car of which 100 were built in 2006 for BMW executives.

The BeHydro project was expected to produce a medium speed marine diesel running on hydrogen as early as 2020 but no announcements about its launch have yet been made beyond that in August 2019 which intimated the engines currently under development would have a power range between 0.8 and 2.8 MW and be available in 6-, 8-, 12-, and 16-cylinder configurations. Two initial applications have been publicized. A workboat catamaran for use in offshore wind and a harbour tug for the port of Antwerp.

The second carbon-free fuel being considered is ammonia which has a chemical formula of NH3. This would mean that combustion in a diesel engine would produce mostly water vapour and NOx as exhaust emissions. The NOx would be higher than in an engine burning fuel oil but could be reduced using the same methods as for engines burning fuel particularly selective catalytic reduction.

Ammonia has been trialled in diesel engines but has numerous problems. Although containing a high ratio of hydrogen which is normally easy to combust, ammonia burns slowly and requires a temperature of 650°C to begin burning. The slow-burning characteristic means that engines running on ammonia react badly to load changes which are commonly experienced by marine engines.

It is considered that fuel comprised of a mixture of ammonia and liquid hydrogen could possibly overcome the effects of load on the engine. Several academic studies have been done in the regard. In early 2020, Wärtsilä announced that it had initiated combustion trials using ammonia.

As part of the tests, ammonia was injected into a combustion research unit to better understand its properties. Based on initial results, the tests will be continued on both dual-fuel and spark-ignited gas engines. These will be followed by field tests in collaboration with ship owners from 2022, and potentially also with energy customers in the future.

The tests are just the latest step as Wärtsilä aims to develop a complete ammonia fuel solution comprising engines, fuel supply, and storage. The company is working with ship owners, shipbuilders, classification societies, and fuel suppliers to learn more about system and safety requirements, as well as fuel composition, emissions, and efficiency.

Wärtsilä is developing ammonia storage and supply systems as part of the EU project ShipFC to instal ammonia fuel cells on Eidesvik Offshore's supply vessel *Viking Energy* by 2023. The company has also gained significant experience with ammonia from designing cargo handling systems for liquid petroleum gas carriers, many of which are used to transport ammonia.

As a fuel, ammonia does not present any storage difficulties being similar in characteristics to LPG. However, it is not an energy-rich fuel having an energy density by mass less than half that of MGO and by volume only a third of MGO.

The main hazard with ammonia is the toxicity. Ammonia is a gas at atmospheric pressure at room temperature which is lighter than air. So, depending on the situation different methods are applied to cope with leakages and unwanted pressure build in ammonia storage systems.

To prevent hazardous concentrations of ammonia an enclosed space with a system containing ammonia is continuously vented and monitored by ammonia detectors. Furthermore, in case of an actual leakage, additional air ventilation is applied to keep the ammonia concentration as low as possible. At 4–5 ppm, ammonia has a distinctly unpleasant smell but when concentrations rise to 300 ppm it becomes practically odourless.

Ammonia has two distinct different health hazards compared to conventional fuels causing severe skin burns and eye damage and being toxic if inhaled. At concentration as low as 110 ppm, ammonia can cause irreversible health problems and exposure at that level for 4 hours would render an impaired ability to escape. At 220 ppm, the same effects can occur in as little as 10 minutes. At 2700 ppm death could occur after an exposure of 10 minutes.

CHAPTER SEVEN

Lubes: Choices and testing

Lubricating oils

Tribology, the science of interacting surfaces and the friction, wear, and lubrication necessary to ensure the machine works efficiently and safely, is something that few may have studied but all can suffer the effects when things go wrong. Two-stroke and four-stroke engines have differing requirements for lubrication.

The low-speed two-stroke engine requires a system oil to lubricate the lower half of the engine containing crankshaft, bearings and potentially also the turbochargers and camshafts and an upper cylinder lubricant for the pistons and cylinder liners. This upper cylinder lubricant is actually mostly consumed during the combustion process so requires constant renewal. The four-stroke engine on the other hand is more like a road engine and generally requires just a single lubricant type for all purposes.

Significantly improved thermal efficiency, fuel economy, and ability to burn poor-quality bunkers have resulted from the intensive development of low-speed cross head and high-/medium-speed trunk piston engine designs in the past 20 years. Further progress in performance and enhanced lifetimes can be expected from the exploitation of higher firing pressures and better combustion characteristics.

Such advances pose a continuing challenge to engine lubricant formulators; however, higher pressures and temperatures make it more difficult to provide an oil film in the critical zones, and longer strokes (another design trend) lead to spreadability problems. The success of new generations of engines will, therefore, continue to depend on the quality of their lubricants.

Lubricants are complex blends of base oils and chemical additives. Base oils are derived from the refining of specific crudes and their properties—such as stability at high temperature and viscosity—depend on the nature of the crude and the production process. Additives either improve the basic properties of oils or underwrite new properties. Marine lubricants exploit many types of additives:

Pounder's Marine Diesel Engines and Gas Turbines
https://doi.org/10.1016/B978-0-08-102748-6.00007-4

- Detergents hold the acid-neutralizing compound in solution. They are alkaline and salts of alkaline metals such as magnesium and calcium. These salts react with strong acids produced during combustion to neutralize their corrosive effect. Some detergents are neutral and help in giving anticorrosion, antiwear, and high-pressure properties to the lubricant.
- Dispersants help to keep soot and combustion products in suspension in the lubricant so preventing lacquering or sludge deposition.
- Antiwear and extreme pressure additives contribute to the maintenance of satisfactory lubrication under the most severe wear conditions. They are usually zinc and phosphorus based.
- Antioxidant additives retard damage due to thermal and oxidation phenomena at high lubricant pressures and hence extend the life of the lubricant. Oxidation results in the formation of gums, lacquers, and sludge so prevention are considered an essential function of the lubricant.
- Antifoam additives prevent air entrapment which can cause oil starvation due to the air bubbles at contacting surfaces. They are thus essential in preventing damage to engine components.
- Anticorrosion additives help prevent atmospheric corrosion when engines are not in use.
- Pour point depressants help prevent increases in viscosity at low temperatures and crystallization of paraffin waxes.

Environmental legislation has also impacted the composition of lubricants. A reduction in the sulphur content of fuels, for example, influences TBN levels and calls for a new balance of detergent and dispersant additives. Measures taken to reduce NOx emissions from engines—such as delayed fuel timing, the injection of water into the cylinder, emulsified fuel, and selective catalytic reduction (SCR) systems based on ammonia—also dictate attention to lubricant detergency, resistance to hydrolysis, and compatibility with catalysts.

All additives are created from a number of ingredients performing diverse functions. One part of the chemical molecule of 'overbased' detergents, for example, cleans the engine and another neutralizes acids produced by combustion. Additives can interact with each other either positively (synergy) or negatively (antagonism). The art of the lubricant formulator is to select the best base oils and optimize the proportions of the various additives for the specific duty.

Whether or not the engine is a two-stroke or a four-stroke design, operating on distillate or residual fuel, the function of the cylinder lubricant is the same as follows:

- To assist in providing a gas seal between the piston rings and the cylinder liner
- To eliminate or minimize metal-to-metal contact between piston rings, piston, and liner
- To act as a carrier fluid for the functional alkaline additive systems, particularly that which neutralizes the corrosive acids generated during the combustion process
- To provide a medium by which combustion deposits can be transported away from the piston ring pack to keep rings free in grooves
- To minimize deposit build-up on all piston and liner surfaces.

Formulating an 'ideal' lubricant to perform all these functions for all engines on the market is difficult and will always be subject to experience and judgement. First, there are varying requirements depending on engine type (single or dual level lubrication, oil injection position in the liner, and oil feed rate applied), operating condition (power and rotational speed set points), and fuel quality used. Secondly, in many cases the requirements are contradictory; for example, the high viscosity necessary to yield good load-carrying performance will adversely affect spreadability.

Synthetic lubricants

The vast majority of marine lubricants are still formulated with mineral bases, but the development of synthetic lubricants targets enhanced machine running safety and durability and the extension of oil change intervals. Such products are particularly valued for harsh duties and are applied widely in aviation, high-speed trains, and Formula 1 racing cars.

Synthetic lubricants are made from synthetic base oils derived from the chemical reaction of one or more products, supplemented by special additives incorporated to boost performance properties. These base oils may be formulated from synthetic hydrocarbons, organic esters, polyalkylene glycols, phosphatic esters, silicates, silicones and fluorosilicones, polyphenyl esters, and fluorocarbons.

Marine industry applications currently embrace reciprocating and rotary auxiliary machinery such as turbochargers, air compressors, refrigeration compressors, LPG/LNG compressors, and centrifugal separators.

Some high-speed ferry engines also exploit synthetic lubricants. Most of the major lubricant producers now have a range of synthetic lubricants in their product portfolios.

While offering improved performance in engines when used as system oils, synthetic oils are rarely used for cylinder lubrication in two-stroke engines. This is because synthetic lubricants carry a higher price and cylinder lubricant in two-stroke engines is considered as a consumable product with a very short life.

Two-stroke cylinder oils

The cylinder conditions in the latest designs of low-speed cross head engines impose a higher maximum pressure, higher mean effective pressure, and higher cycle temperature than those of previous generations. In addition, the combined effect of a longer piston stroke and lower engine rotational speed exposes the cylinder oil film to higher temperatures for a longer period of time.

The diversity of engine designs, operating conditions, and worldwide bunker fuel qualities means that there can never be a single truly optimum cylinder oil grade. In certain special circumstances, other grades may perform better or at least more cost-effectively. Attempts have been made to develop and sell 'universal' oils suited to all fuels and operations and some success has been achieved in sales as owners attempt to minimize oil change operations when different fuels are involved. However, the general feeling among most engine makers and chief engineers is that whenever possible the cylinder lubricant should closely match the fuel type. It is possible to use a mismatched lubricant and fuel if the fuel is only to be used for a very short duration—such as switching from HFO to MDO while entering a port. In such cases, a simple adjustment to the fuel feed rate is usually considered sufficient.

Modern cross head engine cylinder oils are formulated from good quality base oils with an additive package, which performs various functions within the cylinder. The product must possess adequate viscosity at high working temperatures, yet be sufficiently fluid to spread rapidly over the working surfaces; it must have good metal wetting properties and form an effective seal between rings and liners, and it must burn cleanly and leave a deposit which is as soft as possible.

It is generally accepted that a 70-BN/SAE-50 cylinder oil, formulated with the right technology, is the most cost-effective way of providing cylinder lubrication for the majority of low-speed engines. Alternative cylinder lubricants can be considered for special applications or to counter specific problems, or if the ship involved is engaged in a trading pattern where the sulphur level of the bunkers is consistently high or consistently low.

Decades of accepted practices for selecting the appropriate cylinder lubricants ended on 1 January when the global sulphur cap of 0.5% under MARPOL Annex VI came into effect. Previously there had been a maximum permitted level of sulphur in fuels outside ECAs of 3.5% although the average level found in practice was around 2.8%.

After 1 January 2020, only scrubber equipped ships are permitted to use fuels with a sulphur content above 0.5%. With some 4000 vessels equipped with scrubbers and more on order or scheduled for retrofit, bunker suppliers are quite happy to continue producing conventional HFO. Most of the scrubber equipped ships are large vessels and thus account for a significant proportion of all marine fuels used. It has been suggested that in future scrubber fitted vessels will be consuming as much as 25% of all marine fuel.

Scrubbers are able to handle fuels with a higher sulphur content than the former cap of 3.5% by mass and it is thought that the sulphur content of future supplies of HFO will be above the 2.8% average prior to 2020.

For nonscrubber equipped ships, the choice of fuels will normally be between those with 0.5% sulphur levels for use outside ECAs and for 0.1% fuels inside ECAs. However, some consideration should be given to the fact that in the early days of the new regulation and occasionally at other times later, ULSFO and VLSFO may not be available and the ship may be obliged to either run only on distillates or use conventional HFO with higher sulphur content.

Cylinder lubricants and fuel sulphur levels

Cylinder lubricants in two-stroke engines have a number of functions: they are designed not only to achieve good lubrication and internal cleanliness but also to neutralize the sulphuric acid formed in the combustion chamber. The total amount of base introduced into an engine by the cylinder lubricant should match the amount of acid produced, which depends on the fuel sulphur content.

If the engine is burning high-sulphur fuel, the amount of sulphuric acid produced will be high as well. If the total amount of base provided to the engine is low compared with the amount of acid produced, corrosive wear will increase and steps will have to be taken to raise the cylinder oil feed rate so that more base becomes available.

When the engine is burning low-sulphur fuel, it will produce less acid. At equal oil feed rate, the amount of base fed into the engine will be much higher than that required to neutralize the acids produced. As a result, the amount of instantaneous wear will be low. At the same time, however, the additives responsible for this excess in the base will be burnt in the combustion chamber, resulting in the formation of a thick layer of ash on the piston top land. This could eventually interrupt the oil film and promote abrasive polishing of the liner surface and ring/liner scuffing.

High oil feed rates were traditionally viewed as beneficial because they were thought to provide a greater margin of safety against excessive wear and sticking rings, as well as improving piston cleanliness. It is now accepted, however, that the optimum feed rate is the minimum consistent with acceptable wear in the engine. Apart from adding to operating costs—cylinder lube oils are expensive—oil feed rates that are higher than necessary can lead to excessive deposits and should be avoided.

To achieve the correct balance, a cylinder lubricant with a lower BN can be selected or the oil feed rate adjusted to the sulphur content of the fuel. The BN offering has consolidated to a 70 BN product for high-sulphur fuel and a 40 or 50 BN products for use with fuel having a lower sulphur content.

There is a caveat to this general rule of thumb in that the operational profiles of engines have changed over the past 10–15 years from vessels running at high speed, high load to running at low speed, low load. Feed rates have reduced, and lubrications system have become more efficient. This has created a significant impact on optimal cylinder oil use and in some cases with lower liner surface temperatures, allowing cold corrosion to form. More recently, 100–140 BN cylinder oils have been needed to protect newer engines against cold corrosion under part-load conditions for those burning higher sulphur fuels.

So far there has been little data produced on the sulphur content of HFO supplied since 1 January 2020. Lubricant makers have highlighted that continued HSFO use does not necessarily mean continued use of current cylinder oil products without further consideration. Higher than previous sulphur levels carry the risk of an increased corrosive regime occurring in the engine unless the cylinder oil BN or the feed rates are increased. There is significant risk associated with using a too low BN cylinder oil with a high-sulphur content fuel. If the BN level of the lubricant is too low and the feed rate is not optimized, then an engine will be experiencing severe corrosive wear.

Optimizing feed rate is essential for any ship changing fuels or cylinder oil lubricants. It is not always possible to operate an engine at or close to the

OEM's minimum feed rate without entering a corrosive regime, although this may be overcome with a move to a higher BN cylinder lubricant providing additional neutralization and enabling the engine to operate on an optimized feed rate. This can often not only reduce engine wear but also reduce the overall cost of operation.

Low BN cylinder lubricants must provide adequate detergency, dispersancy, and thermo-oxidative stability in combination with a low BN. All the major engine builders have issued guidelines for operating on low-sulphur fuel, giving recommendations on cylinder lubricant BN and oil feed rates. Below a certain minimum limit prescribed by the engine builder, the oil feed rate becomes too low to secure lubrication and engine cleanliness. If the engine is scheduled to run on low-sulphur fuel for an extended period, therefore, the use of a 40 or 50 BN cylinder lubricant is strongly recommended, especially in modern high output engines. Such engines will obviously be more sensitive to the mechanism of piston crown deposit build-up than older designs having much lower mechanical and thermal load levels.

For ships which operate full time in ECAs using fuel with 0.1% or running on fuels such as LNG, LPG, or ethane and for future fuels such as ammonia all of which have negligible sulphur levels, a lubricant with a BN of 25 may be more appropriate. Such lubricants are produced by several of the leading suppliers.

The following table produced by Chevron Marine Lubricants shows preferred lubricants for different fuels after 2020.

Fuel strategy	**Optimum lubricant**
Up to 3.5%, or even higher HSFO with scrubber	100–140 BN
Distillates	25 BN
LNG	25 BN
Intermittent use of LSFO and distillates	40 BN
Ultra-low-sulphur alternatives	25 BN
Methanol	40 BN

Cylinder lubricant feed rates

Cylinder oil feed rates should be set with a reasonable margin of safety on the recommendations of the engine builder. A high oil feed rate, far greater than 150% nominal, is seldom required other than for a short period during running-in or when special circumstances justify it. There have been reports from both MAN Energy Solutions and Wärtsilä engines that over-lubrication can in fact be harmful in certain adverse conditions.

MAN Energy Solutions advises that unless there are abnormal circumstances, it is pointless to pursue prolonged operation with an excess of 2g/bhph (2.72g/kWh), especially in conjunction with low-sulphur fuel, 70 BN cylinder oil and an engine series well designed against corrosive wear and with no need for excessive BN. Even for a highly detergent oil, over-lubrication can lead to excessive calcium carbonate deposits in the remote top edge of the hot piston crown land. (The BN of all commercial cylinder oils today is derived largely from calcium carbonate.)

In adverse conditions, it is sometimes not the oil detergency that cannot cope but rather there is an enormous generation of calcium carbonate deposit. Engine builders are also concerned that an engine well designed against corrosive wear can become sensitive to excessive feed rate on 70 BN oil when burning fuel with normal sulphur content. This is because if the BN is not used, the unused BN can result in hard calcium carbonate deposits (used BN results in a softer calcium sulphate). Wärtsilä adds that a high cylinder oil feed rate can lead to a disturbed pressure drop across the ring pack and negatively affect piston running. Excessive cylinder oil also tends to increase the formation of deposits on the top land of a piston. A solid crust is formed, which is softer or harder depending on the additives used in the oil, the type of fuel, and its sulphur content, as well as its combustion characteristics. An excessive feed rate can, in fact, also lead to the destruction of a scuffing liner within hours.

Another reason for caution over excessive feed rates is that modern engines are significantly more powerful, burn more fuel, and use more cylinder oil per cylinder than before. The cylinder oil ash level could be more than 5 times higher than in engines of the 1970s and 1980s.

'Intelligent' cylinder lubrication

Demands placed on cylinder oil in two-stroke engines vary widely and depend on the operational conditions. The cylinder of a large-bore engine is typically fed with just 1 g of lubricant during each stroke, and this small amount must be spread correctly to fulfil all requirements. Studies of the relationship between wear and lube oil dosage reveal that interaction between operational factors—such as load variations, fuel and lube oil qualities, and atmospheric humidity—exerts a strong influence on the wear rates of piston rings, ring grooves, and cylinder liners. It has also become clear that over-lubrication has a considerable negative effect on the tribological conditions of liners and rings.

Traditional cylinder lubrication systems, whereby the lube oil is injected at a fixed rate proportional to the engine revolutions per minute or mean effective pressure and perhaps adjusted occasionally based on a scavenge port inspection, may yield a reasonable average condition. Such a condition, however, will result from long periods of excessive lubrication and shorter periods of oil starvation because the sum of influencing factors will change on a daily or sometimes even on an hourly basis. Furthermore, there is a risk of ruining the cylinder condition if unfavourable operational factors are not detected and properly counteracted by the engine operators.

Operational conditions beyond the norm are unavoidable, and many ship owners simply over-lubricate their engines under the false assumption that they will thus always be 'on the safe side'. But over-lubrication is not only expensive, it may even be counterproductive in promoting scuffing through excessive carbon deposits and/or 'bore polished' running surfaces. A wear study has proved that the optimal basic setting of cylinder lubricators should be proportional to the engine load and the fuel oil sulphur content. Feed rate control proportional to the load is one of the two standard options of MAN Energy Solutions's Alpha Lubricator system; the other is to control lubrication in proportion to the mean effective pressure. Lubricator control in relation to the fuel oil sulphur content may be carried out either automatically—based on a feed-forward signal from the fuel inlet line—or manually, based on the sulphur content from the bunker receipt or fuel oil analysis data.

The electronically controlled Alpha Lubricator was developed to inject the oil into the cylinder directly on the piston ring pack at the exact time that the effect is optimal. The system features a number of injectors that inject a specific amount of lubricant into the cylinder every four (five, six, etc.) revolutions of the engine. The lubricator has a small piston for each quill in the cylinder liner, with power for injecting the oil provided from system pressure generated by a pump station (Figs. 7.1 and 7.2).

The properties of cylinder oil scraped from the liner wall reflect the chemical environment in the cylinders as well as the physical condition of rings and liner, and there is a direct relationship between some of the key parameters in the scrape-down oil and the actual cylinder condition. A lubrication algorithm—based on scrape-down analysis data, cylinder oil dosage, engine load, and cylinder wear rate—can thus be created. Automatic optimization of lube oil dosage and cylinder lubrication efficiency is facilitated by online monitoring of the scrape-down oil composition from each

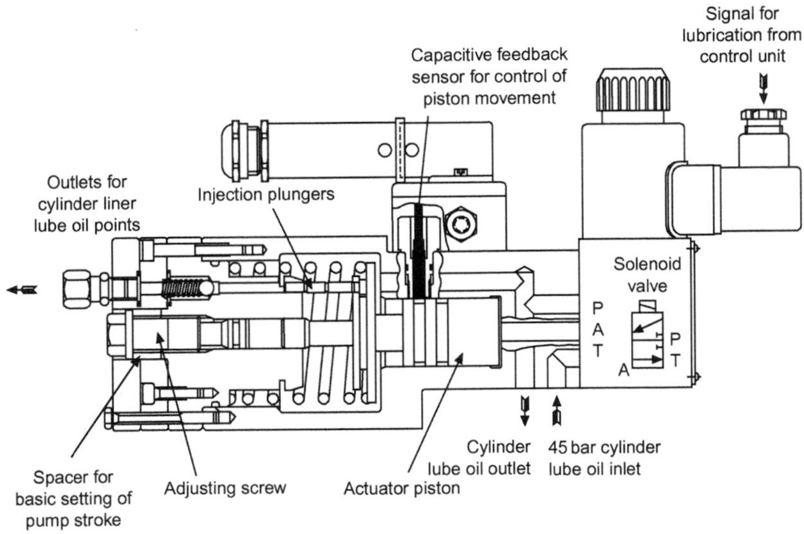

Fig. 7.1 Alpha cylinder lubricator unit (MAN Diesel). *From MAN Energy Solutions.*

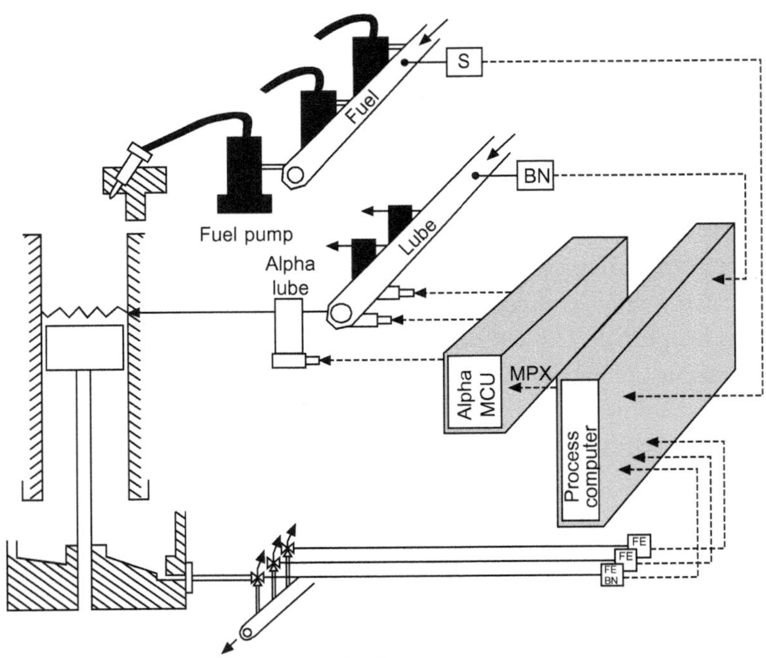

Fig. 7.2 Arrangement of alpha electronic cylinder lubricant system (MAN Diesel). *From MAN Energy Solutions.*

cylinder, feeding the results into a computer (along with the above algorithm), and sending signals to each Alpha Lubricator. Corrosive wear control can be based on either feed rate control or control of the cylinder oil BN; the latter approach calls for two or more lube oil tanks or blending facilities onboard.

Two basic requirements have to be addressed: the cylinder oil dosage must not be lower than the minimum needed for lubrication, and the additive amount supplied through the BN (the alkalinity throughput) must be sufficient only for neutralization and for keeping the piston ring pack clean. As the second criterion usually overrides the first, the following determine the control: the cylinder oil dosage shall be proportional to the engine load (i.e. the amount of fuel entering the cylinders) and the sulphur percentage in the fuel. A minimum cylinder oil dosage is set to address the other duties of the lubricant, such as securing sufficient oil film detergency.

Specialists stress that the lubricant feed rate is critical and should be carefully balanced between cylinders. Any downward departure from the engine builder's recommended feed rate is potentially dangerous, while an excessive rate is generally not cost effective.

Types of cylinder wear

Cylinder lubricants are an essential component in the control of liner and piston ring wear in low-speed cross head engines. High wear rates can be caused by corrosive, abrasive, and adhesive wear, the conditions prevailing in the combustion space being the key influence on the type of wear experienced. In older engine designs the wear was predominantly corrosive, but in the latest engine designs, it is now largely adhesive wear.

Traditionally, cylinder oils had high alkalinity to protect against corrosive wear: 70 BN was the industry norm, with higher BN oils available to secure even greater protection. The increase in engine performance, and in particular the rise in cylinder liner temperature, effectively engineered out high corrosive wear. Engines designed and built in the late 1980s and early 1990s had liner temperatures sufficiently high enough to prevent corrosive wear; and the use of a 70 BN lubricant was often the most economical means of providing cylinder lubrication and protection.

Once liner temperatures are raised above 250°C, the performance of standard cylinder oils becomes marginal in protecting increased liner and piston ring wear. Under these conditions, the type of wear is largely adhesive, and the performance of cylinder oils has to be assessed by their ability to minimize such wear. (Corrosive wear, however, is a problem that may not have been completely eliminated. Theory, as well as data published on test engines with very high peak pressures, points to an increased corrosion potential as more of the sulphur is converted to sulphur trioxide. If the engine charge air contains enough water, then its combination with sulphur trioxide leads to the formation of sulphuric acid in sufficient quantities to tip the wear balance towards corrosion.)

Adhesive wear, sometimes referred to as 'scuffing', occurs when the surface asperities on the liner and piston ring have metal-to-metal contact. When contact takes place, local welding, and metal deformation result, with the amount of contact determining the rate of wear. The rougher the surfaces, the greater the likelihood of adhesive wear occurring (which is why it is usually experienced during the engine running-in period).

Basic separation of the contact surfaces is provided by the lubricant film, and the higher the viscosity the thicker the oil film. Because of the high liner temperatures experienced in modern engines, the oil viscosity on the liner surface can be as low as 2 cSt near top centre, the resultant film thickness being 1 μm or less. Even under these conditions, adhesive wear will not take place, provided the surfaces are not rough. The roughness of the liner and ring surfaces has a significant influence on the load at which adhesive wear starts, and the rougher the surfaces, the less the load-carrying ability. Rougher surfaces also require thicker oil films (or higher viscosity oils) to carry the same load.

The term 'boundary lubrication' describes the condition where the peaks of one rough surface (asperities) come into contact with those of another rough surface, and the lubricant film between the surfaces is not thick enough to prevent metal-to-metal contact. Full separation is achieved only when the ratio of film thickness over the mean roughness value (or mean height of asperities) is greater than three. A typical surface finish for the cylinder liner of a modern engine in service is around 0.5 μm, while the oil film thickness of three times this value (1.5 μm)—required to prevent asperity contact—may not be established until some distance down the stroke from top dead centre. The conditions for adhesive wear are therefore present at around the top dead centre.

Control of the onset of adhesive wear calls for an adequate oil film thickness, which can be enhanced by chemical films from lubricant additives. Such films have lower shear strengths than the metal surface and hence are removed, partially or fully, during sliding. Additives that are effective at reducing friction chemically react with the metal surfaces to form relatively thin planar layers of low shear strength. Replacement of these films during the out-of-contact time of the surfaces is therefore essential to the control of wear.

Additive effectiveness is strongly temperature dependent: as the metal surface temperatures increase, the chemical bond between additive and surface is broken, and the additive de-absorbs from the metal surface. Only additive systems that absorb at elevated temperatures (250°C) are effective in providing the chemical film that can give added protection against the onset of adhesive wear. Additionally, the calcium detergents used to provide the alkalinity of the cylinder oil have the effect of lowering the coefficient of friction. The long-chain natural acid present in such detergents is slowly released during neutralization and migrates to the metal surface to form a friction-modified surface. Consequently, higher base number cylinder oils will provide higher load-carrying performance before the onset of adhesive wear and then provide some antiwear performance to reduce the amount of adhesive wear.

Chemically active surface components give further protection against the onset of adhesive wear by providing a low shear film between the surface asperities. In terms of wear protection, therefore, this addition is equivalent to the use of a higher viscosity cylinder oil: a 19 cSt oil of 80 BN and containing high-temperature antiwear additives is equivalent to a 23 cSt standard 70 BN oil.

Once the conditions for adhesive wear are established then measurable and rapid wear will take place. If the conditions are particularly severe the rate of wear can accelerate, leading to severe damage and seizure. When the metal surface asperities come into contact, local temperatures are generated that are much higher than the bulk temperature of the oil or the metal surface temperatures. The local temperatures (or flash temperatures) for cylinder lubrication are typically between 400 °C and 500 °C. The only way to control the rate of adhesive wear is to limit the local welding and deformation of the asperities, and this can be achieved by using high-temperature high-performance additives.

Special tests were conducted on a modified test engine in which the liner temperatures were raised to almost 300°C. Under these conditions, severe adhesive wear took place within 100 h with a standard 70 BN cylinder oil. Using an 80 BN oil incorporating high-temperature antiwear additives, however, minimized the amount of adhesive wear. Other tests showed that this combination of high BN and antiwear can reduce the amount of adhesive wear by up to 50%.

Severe adhesive wear will increase the surface roughness on the rings and liners, with the result that the load to the onset of adhesive wear is reduced. For example, if an engine which experiences adhesive wear at a certain operating load is shut down (perhaps to examine the wear), then the increased roughness of the surfaces means that when the engine is run again adhesive wear will start at a lower load. If the conditions that resulted in adhesive wear are now removed (by operating at a much lower load), then it may be possible that the surface damage to liners and rings may be polished out to give acceptable running surfaces.

Antiwear performance summary

- Viscosity has a major influence on the load at which adhesive wear starts; higher viscosity index (VI) cylinder oils provide more protection than lower VI oils.
- The rougher the surfaces, the lower the onset threshold for adhesive wear or the higher the viscosity required to prevent adhesive wear.
- The load at which adhesive wear starts is also influenced by the antiwear performance of the alkaline detergent and the use of high-temperature antiwear additives.
- Only a standard BN cylinder oil with a viscosity of 23 cSt at 100°C will provide similar protection against the onset of adhesive wear as an 80 BN oil with a high-temperature antiwear performance at 19 cSt at 100°C.
- The practical limit on the viscosity of cylinder oils (monogrades) for reasons of pumpability, handling, and distribution over the liner surface is 21 cSt at 100°C with a VI of 100.
- The load at the onset of adhesive wear is 30% higher from an 80 BN cylinder oil with high-temperature antiwear performance compared with a 70 BN oil.
- True antiwear performance is a measure of the ability to reduce the rate of wear once the conditions for adhesive wear are established. A reduction of up to 50% can be achieved by using an 80 BN cylinder oil with high-temperature antiwear performance compared with a standard 70 BN oil.

Medium-speed engine system oils

Lubricating medium-speed trunk piston engines is considered much more of an exact science than lubricating low-speed cross head engine cylinders. Sulphurous by-products from the combustion of high-sulphur fuel are countered by increasing the amount of over-based additives in the formulation to give an oil of higher alkalinity, as measured by TBN. But this is not quite as simple as merely adding alkaline additives. Research is carried out to establish a balanced blend of the many functionally different additives, for example, antioxidants, antiwear agents, corrosion inhibitors, detergents, dispersants, antifoam agents, and pour-point depressants.

The performance of lubricating oil in a marine medium-speed engine tends to be measured in terms of its ability to maintain an acceptable level of TBN in service. TBN retention is influenced by factors such as fuel sulphur content, combustion quality, lubricating oil make-up, piston ring blow-by, and water contamination. With regard to the oil itself, however, retention is directly related to additive balance and the quality of the additives used.

Although marine lubricants will continue to rely largely on the types of additives that have ensured success so far, formulators suggest that more sophisticated and effective versions can be expected. The search for more powerful synergisms between additives of different species will intensify, and lube oils that can outlast the life of an engine could be possible in the future.

Liner lacquering

In recent years, a number of highly rated medium-speed engines of certain types have experienced problems with lacquer formation on cylinder liners. This build-up of lacquer in the honing grooves can result in a smoothed or glazed liner surface that leads to a significant increase in the rate of oil consumption. If left unchecked, lacquering can also be accompanied by hard carbonaceous deposits, which lead to scoring or polishing of the liners and increased engine operating costs.

A number of common factors link examples of engines in which liner lacquering has been found: large variations in load (e.g. frequent and long periods of idle followed by full load operation); high mean effective pressure medium-speed designs; and low-sulphur, mainly distillate, fuels. Lacquering can typically occur in the propulsion engines of offshore supply vessels and short-sea ships. Lacquer characteristically contains both organic and inorganic (metal salts) materials, the colour varies from very light amber to dark brown, and there is uneven distribution over the liner.

Modern, highly rated engines exploit much higher fuel-injection pressures and shorter injection times. The total injection, ignition, and combustion process have to take place in a few degrees of crank angle under conditions in which, besides complete combustion, thermal cracking of fuel components can be expected. If these cracked components reach the relatively cool cylinder liner surface, they will condense, concentrate, and start the formation of resinous lacquer by a process of polymerization and evaporation of the light ends (see Fig. 7.3).

Fuel quality has not suffered a step change in recent years that would account for the incidences of liner lacquering problems. However, the complexity of refining processes has increased during the past decade, and therefore, the complexity of components in both distillate and residual fuels has also increased. Experiments have shown that different gas oils produce different distributions of deposits between pistons and liners under conditions known to promote lacquer formation. While the fuel clearly influences liner lacquer formation, however, the link to the sources of gas oil is not clear.

An insight into how fuel contributes to liner lacquering was derived from service experience with a ship running on residual fuel rather than distillate fuel. The ship bunkered a highly cracked, high-sulphur residual fuel before each outward voyage leg and a straight-run, low-sulphur residual fuel before each return leg on a regular sailing schedule. This practice always resulted in

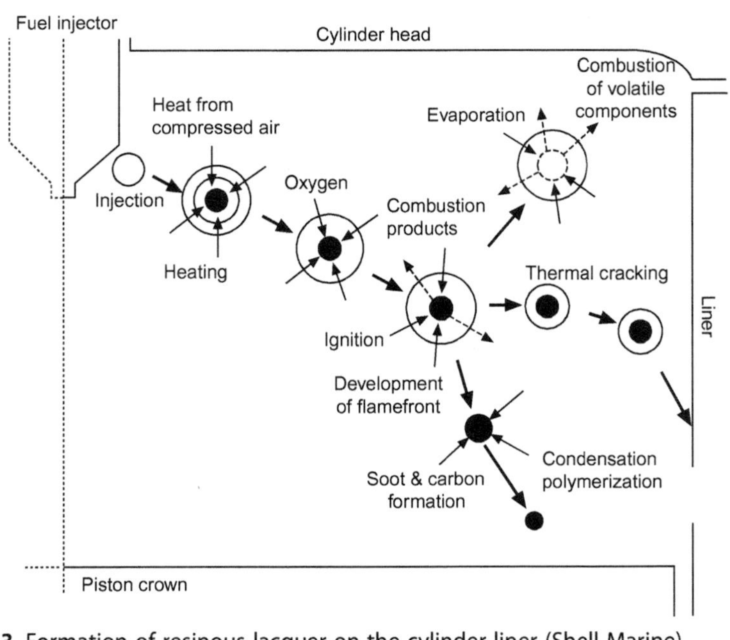

Fig. 7.3 Formation of resinous lacquer on the cylinder liner (Shell Marine).

an increase in sump oil top-up and building of liner lacquer during the return voyage. Although distillate fuels do not show the same sensitivity as this example, it does show that the fuel composition is a factor in lacquer formation.

Evidence from liner lacquering in the field as well as in laboratory engines suggests the following mechanism for lacquer formation, characterized by the following steps:

- Fuel droplets injected at HP begin to burn rapidly.
- Rapid burning leads to incomplete combustion of some larger fuel droplets but intense heating of droplet cores.
- Thermal cracking of some fuel components occurs within larger droplets.
- Some droplets impinge on the cylinder liner where thermal cracking reactions continue to form unsaturated reactive hydrocarbon products.
- Condensation and polymerization of unsaturated products on the liner forms a resinous organic polymeric deposit.
- Sticky resinous deposit is spread over the liner by the piston ring movement.
- Lubricant, including metal-containing additives, becomes entrained in the deposit.
- Continual exposure of the deposit to combustion, evaporation of light ends, and rubbing by piston movement forms a hardened lacquer deposit.
- The build-up of lacquer fills honing grooves and leads to an increased rate of oil consumption.

Current understanding suggests that any fuel will contribute to liner lacquer formation under engine conditions that promote incomplete combustion and thermal breakdown of the fuel. However, the combination of particular fuel compositions combined with such engine conditions is the principal cause of liner lacquer formation in medium-speed engines.

The lubricant can affect lacquer formation. Research by Shell Marine has shown that a lubricant formulation must have the correct balance of detergency, dispersancy, alkalinity, and antioxidant characteristics to prevent lacquering from occurring. Lacquering will occur most frequently when the engine is running under very high load conditions. Good maintenance is essential, and the following are dictated to prevent lacquer formation: injection timing should be correct; injectors should not be worn, and liner and inlet air temperatures should not be too low. Shell's investigations showed that high-performance lubricants can contribute to minimizing lacquering and thus reducing lube oil consumption.

Black sludge formation

A problem in medium-speed engines may arise from the contamination of lube oil by the combustion blow-by gases and/or by direct raw fuel dilution. Higher fuel pump pressures in modern engines have contributed to increased fuel leakage: analyses of used lube oils from engines running on heavy fuel oil have indicated an average HFO contamination of 2%; in some cases, levels of up to 15% were detected.

Most heavy fuel oils supplied today emerge from cracking installations. The cracked asphaltenes, an inherent part of such bunkers, do not dissolve in engine lubricants but instead coagulate and form floating asphalt particles, 2–5 μm in size. These particles are very sticky and form black deposits on all metal surfaces of the engine, including the cambox and the crankcase. The admixture of residual fuels and some lubricants may lead to coagulation of asphaltenic particles causing sludge formation in engines, clogging the oil scraper ring, and increasing piston ring groove deposits. Sludge may overload the lube oil separators and the filters may become clogged.

This problem posed by so-called black sludge or black paint—a sticky, viscous mixture of fuel and combustion-derived deposits—stimulated leading lube oil suppliers to develop new system lubricants able to disperse high concentrations of cracked asphaltenes. Products were developed to meet the challenge posed by engine blackening, undercrown deposits, piston head corrosion, purifier heater fouling, increased lube oil consumption, base number depletion, oil scraper ring clogging, and increased piston deposits. Additives prevent the contaminants causing black sludge from depositing in the engine, the lube oil holding the impurities in suspension so that they can be removed by centrifugal separation. The lubricants are claimed to disperse the highest level of cracked asphaltenes, resulting in cleaner engines (especially excellent piston cleanliness), reduced lube oil purifier cleaning intervals and filter consumption, and the near elimination of deposit formation in purifier heaters.

Lube oil stress

Is the term used to describe the effect of different parameters that cause the lubricant to degrade and it's protective properties to be reduced. There are three basic categories.

Acid stress is related to the sulphur content of the fuel. If the additives in the lube oil cannot neutralize the products of combustion, then the acidic attack can occur on cylinder liners, piston rings, and grooves; the resulting wear reduces component life and reliability. The presence of acid stress is easily noticed from the depletion of the oil BN, which shortens the oil life.

Thermal stress is caused by exposure of the oil to high temperatures. This leads to a breakdown of the oil molecules and oxidation, which in turn fosters sludge and lacquer formation in the cooler areas of the engine. In addition, thermal stress can cause deposits in piston ring grooves and on the internal surface of the piston crown. Deposits in ring grooves lead to a decrease in the pressure build-up behind the ring, causing a reduction in the sealing pressure of the ring against the liner. Deposits on the internal surface of the piston crown can lead to overheating of the crown as the heat transfer is reduced.

Piston undercrown deposits reduce the life of the crown; ring groove deposits shorten the life of liners and piston rings and reduce reliability, and an increase in the viscosity of the oil shortens its life.

Asphaltenes stress: HFO contains asphaltenes (very large hydrocarbon molecules), which become mixed with the lubricant: fuel contamination of the lubrication system is virtually impossible to prevent, especially in trunk piston engines. As the asphaltenes do not dissolve in the lubricant, they settle out in the form of sludge in the crankcase and cam case and can also lead to frequent clogging of the oil filter. Fuel pump sticking can also be attributed to the precipitation of asphaltenes from the lubricant if the latter cannot cope with asphaltene stress.

To summarize, oil stress in whatever form can result in the shortening of lubricant life, the lower lifetime of key components, reduced reliability, and increased downtime. All of these effects raise operational costs with respect to spare parts consumption, lube oil renewal, and labour costs but can be avoided.

It is the responsibility of the shipowner/manager's technical department to ensure that ships' engineers are aware of oil stress, the symptoms to look for and how to prevent it occurring. If the lube oil is operating under highly stressed conditions, its properties change and deteriorate more quickly than oils operating under less stressed conditions. Minimizing oil stress-related failures or preventing them entirely dictates maintaining the lubricant in the best possible condition, which can only be ensured by well-trained and motivated shipboard staff.

Lubricant testing

Lubricants are a valuable indicator of the overall functioning and wear of machinery. Under normal engine operating conditions, the deterioration of a lubricant takes place slowly, but severe conditions and engine malfunctions promote faster degradation. Monitoring lubricants in service and analysing the results determine not only their condition but also that of the engine and auxiliary machinery that they serve. All the major marine lubricant suppliers offer used lube oil testing services to customers, with results and advice transmitted by chosen communication channels to owner and ship from specialist laboratories. The range of analyses undertaken on engine lubricant samples submitted typically embraces checks on:

- *Water content*: even in small quantities, water is always undesirable in a lubricant because it can act as a contaminant. A check on the water content will indicate defects in the engine, purifier, and other auxiliary machinery. Where water is present, testing determines whether it is fresh or seawater.
- *Insoluble products*: lubricants in an engine are contaminated by a number of insoluble products. Monitoring their presence provides a very good indication of the engine's operating condition or the effectiveness of the purifier or filter; it also safeguards against breakdowns since a sudden increase in the insoluble products present indicates a malfunction.
- *Viscosity*: a measure of the ability of a lubricant to flow, viscosity cannot be used in isolation to assess a lubricant's condition, but it yields useful information in conjunction with other determining factors, such as the level of oxidation or contamination by fuel, water, or other elements.
- *Flashpoint*: the temperature at which the application of a flame will ignite the vapours produced by an oil sample under standard conditions. Below a certain value, there is a high risk of very serious incidents in service, including a crankcase explosion. The flashpoint must, therefore, be monitored very carefully.
- *Base number (BN) or alkalinity reserve*: during the combustion process, the sulphur contained in the fuel can produce acidic products that can damage the engine; this acid must be neutralized to avoid extensive corrosive wear of the cylinder liner or piston rings. Neutralization is the role of specific additives in the lubricants, the amount, and type of which are measured by the BN, which represents the alkalinity reserve. Close monitoring of the BN in service is essential to ensure that the lubricant still has sufficient alkalinity reserve to perform correctly.
- *Wear metal elements*: monitoring changes in metal elements (measured in ppm) provides vital information on how patterns of wear develop inside the machinery. Regular monitoring detects any sudden increase in a

wear element, such as iron or copper, which would indicate that excessive wear was occurring.

The correct interpretation of results dictates analyses on a regular basis so that curves can be plotted and extrapolated to indicate trends. Any measurement that deviates significantly from the curves must be investigated further; adopting a historical approach to analyses can minimize the risk of machinery malfunction and breakdown. When machinery manufacturers offer no specific recommendations, the following can be considered as average sampling periods:

Main engine	3–6 months
Diesel genset	6–12 months
Hydraulic system	12 months
Turbocharger	6 months
Gearing	12 months
Air and gas compressors	12 months

Microbial contamination of fuels and lubricants

Shipboard microbial fouling and corrosion have become more commonplace and will probably continue to increase due to fuel oil formulation and handling trends, polluted harbour waters, and regulatory pressures.

Naval architects and shipbuilders are advised to design systems that are less conducive to harbouring microorganisms and more readily sampled and decontaminated if problems are experienced. Comprehensive onboard test kits are available for detecting and quantifying microbial contamination and for monitoring the success of antimicrobial measures.

Microbial contamination can manifest itself in the infection of fuel, lubricants, cooling water, and bilge and ballast waters. The results can be engine damage (pistons and cylinder liners may wear faster, and filters and fuel-injection nozzles become plugged, sometimes within a few hours); fuel tank pitting corrosion and accelerated localized pitting; and general wastage corrosion of bilge pipes, tank bottoms, and hulls. In some cases, the internal and external microbial attack has led to rapid ship structural perforation. Of special concern is the possibility that certain strains of bacteria may endanger the health of the crew, as would the incorrect handling and application of hazardous chemical biocides in remedial treatment operations.

Various kinds of microbes are encountered: bacteria, moulds, and fungi, and yeasts. Bacteria are classified into two types: aerobic, which use oxygen to oxidize their food, and anaerobic, which cannot tolerate oxygen. One anaerobic strain, termed sulphate-reducing bacteria (SRB), is particularly virulent and produces corrosive sulphide.

Bacteria produce biosurfactants, which break down the oil/water interface, promoting emulsions that are detectable when the fuel is tested for water separation. If the infection is severe and longstanding, the aerobic bacteria consume all the oxygen, and problems with SRB may result.

Microbes need water and a food source to survive, ingredients provided by the hydrocarbons and chemical additives in the oil. A temperature between 15°C and 40°C is also required. Warm engine rooms offer an ideal breeding ground. Laid-up tonnage or ships in intermittent service are most vulnerable to attack because microbes dislike movement. Dormant fuel systems may be aggravated by condensation and water leakage.

Various strategies are available for reducing the problems, but solutions must be safe and environmentally acceptable. The details needed to build these strategies into full working protocols will depend on individual circumstances, the time available, and access to antimicrobial agents. There are many different microbiological problems which need to be met with specific tailored solutions but there are certain common principles of good practice which can be implemented:

- *Physical prevention*: preventing ingress of inoculating microbes, particularly those already adapted to growth in relevant environments. Avoid spreading contamination from passing clean fluids through contaminated pipes and filters and into dirty tanks. Minimize the conditions which encourage microbial growth and water.
- *Physical decontamination*: settling, heat, filtration, and centrifugal procedures all aid decontamination, the choice of process depending on equipment and time constraints.
- *Chemical prevention*: protecting against minor contamination, coupled with good housekeeping to prevent rather than cure infection.
- *Chemical decontamination*: a wide range of chemical biocides is available. Only a few are appropriate for each specific application (there is no universal eradication fluid). All are toxic and must only be used with due regard to health and safety and environmental impact.

Microbial problems in the shipping industry were originally mainly confined to distillate fuels and lubricants, but areas of attack have recently spread to residual fuels and bilge, ballast, and potable waters. Specialists cite the following contributory factors:

- Lower levels of shipboard manning and less experienced personnel have undermined stringent housekeeping regimes.
- Adverse trading conditions have led to ships being laid up or deployed intermittently, fostering long and undisturbed incubation periods for opportunistic microorganisms.
- Marine pollution legislation under the MARPOL 73/78 regulation which restricts the pumping of bilges has led to water laying stagnant for long periods.
- Environmental restrictions on the use of toxic biocidal chemicals within bilges and fuels exacerbate the problems associated with microbial contamination.
- Lack of knowledge of the factors which cause microbial contamination and accurate diagnosis of the operational problems being experienced.
- Shortcomings in the design of tanks, bilges, and pipe systems whose configuration should foster effective water draining and subsequent treatment.

There will always be an initial source of shipboard fuel contamination, such as a shore tank, dirty pipeline, road tanker, or polluted tank wash water. Airborne contamination via tank breathers is less likely. Once the infestation has occurred, particular circumstances may encourage microbial proliferation onboard. Further inoculation is then immaterial as the key aggravating factors are water and warmth.

Water accumulates in tanks when there is no drain or water scavenge system, where the drain is not at the lowest point in the tank, and where draining procedures are not enforced. Tanks in the engine room or other warm locations and tanks receiving recirculated distillate fuel from injectors are ideal for incubating microbes. Double bottom tanks, due to their lower temperatures, are less prone to microbial proliferation.

The following visual symptoms of fuel microbial contamination are given as a guide:

- Aggregation of microbes into biomass is observed as discolouration, turbidity, and fouling.
- Biosurfactants produced by bacteria promote stable water hazes and encourage particulate dispersion.
- Purifiers and coalescers, which rely on a clean fuel/water interface, may malfunction.
- Tank pitting.

Operational indications of fuel microbial contamination are as follows:

- Bacterial polymers may completely plug filters and orifices within a few hours.
- Filters, pumps, and injectors foul and fail.
- Nonuniform fuel flow and variations in combustion may accelerate piston ring and cylinder liner wear rates and affect camshaft torque.

When heavily infected fuel is used, all or some of these phenomena will confront the ship's engineers within a few hours. Initially, they will probably be faced with filter plugging, fuel starvation, injector fouling, and malfunctioning of any coalescer in use.

The extent of the contamination must first be established by the following procedure:

1. Clear glass sample bottles rinsed with boiling water should be used to take drain or bottom samples from service, header, and storage tanks.
2. Microbial contamination in a sample will be apparent as a haze in the fuel from the presence of sludge. This sludge readily disperses in the fuel when the sample is swirled; at this time sticky 'cling film' flakes may be seen adhering to the wall of the bottle. The water will be turbid and there may be some bottom sludge; if this is black, SRB is present and they will be an added corrosion hazard.
3. The engineer can now evaluate the various fuel locations and instigate an emergency strategy by using the cleanest fuel. If only heavily contaminated fuel is available, it should first be allowed to settle for as long as possible and then drawn off from the top to a clean tank, preferably via a filter, purifier, centrifuge, or coalescer. Using a biocide at this stage (see the next section) is usually not advisable due to a tendency to block filters with the dislodged biofilms.
4. At the earliest opportunity, drain or bottom samples should be taken and the supplier's retained sample forwarded for microbiological examination. This will identify, by sophisticated 'fingerprinting', whether or not the bunkers supplied were contaminated.

The standard methods for curing heavily contaminated fuels involve the use of biocides, which may be toxic and incur disposal problems. Careful handling is necessary. Once a biocide has been added, the dead biomass must be allowed to settle before drainage and filtration. The careful monitoring of filters is also required to prevent blockage by the dead microbes.

In expert hands, biocides eliminate the microbial menace, but using an incorrect biocide can exacerbate the problem rather than effect a remedy.

The choice of biocide is therefore important. Prevention is better than cure, and the source of the infections should be identified to prevent their recurrence.

Lube oil contamination

Lubricant operating temperatures of crankcase oils in wet engines are normally sufficient to control and prevent infestation. Any problems are usually associated with ships which have shut down their lubricant system, allowing temperatures to drop and water to accumulate.

There are many thousands of types of microbes, but only a few are able to grow in lubricating oils at the elevated operational temperatures in use. Provided the plant treatment equipment is functional, the purifier heater temperature is maintained, and the water content kept at a minimum, the lubricating oil system will be both environmentally and nutrient deficient. Such conditions will prevent microbes from establishing themselves. Should heavy contamination of the system occur; however, this self-regulating mechanism will be unable to prevent microbial proliferation (Fig. 7.4). Fortunately, the symptoms of microbial contamination will not occur immediately, allowing engineers to implement remedial physical and/or chemical decontamination programmes.

The visual symptoms of microbial contamination of lubricants are as follows:

- The slimy appearance of the oil, the slime tending to cling to the crankcase doors
- Rust films
- Honey-coloured films on the journals, later associated with corrosion pitting
- Black stains on white-metal bearings, pins, and journals
- Brown or grey/black deposits on metallic parts
- Corrosion of the purifier bowl and newly machined surfaces
- Sludge accumulation in the crankcase and excessive sludge at the purifier discharge
- Paint stripping in the crankcase

Operational symptoms of microbial contamination in lubricants are as follows:

- Additive depletion
- Rancid or sulphitic smells
- Increase in oil acidity or sudden loss of alkalinity

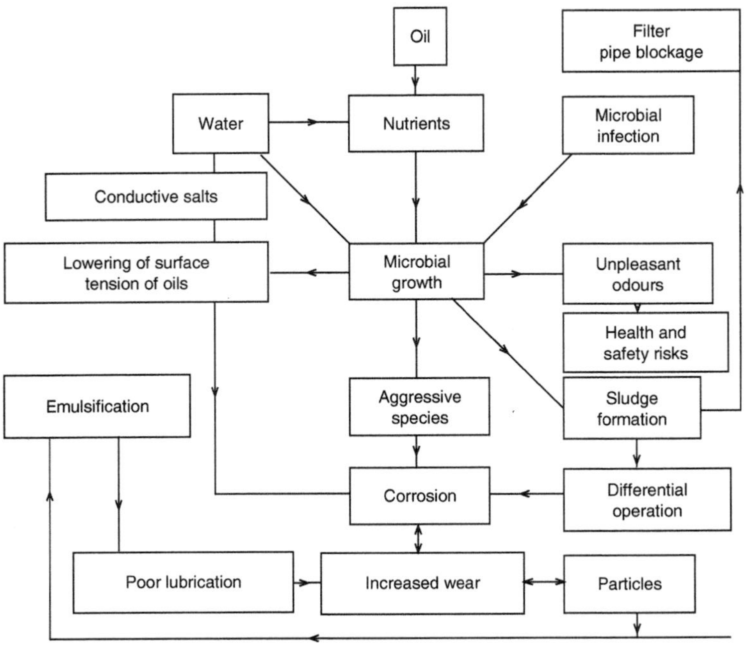

Fig. 7.4 Interactive problems associated with microbial corrosion (Lloyd's Register).

- Stable water content in the oil, which is not resolved by the purifier
- Filter plugging in heavy weather
- Persistent demulsification problems
- Reduction of heat transfer in coolers

Microbial growth occurs in the water associated with the lubricant, and the phenomenon is therefore characteristic of crankcase oils in wet engines, particularly those with water-cooled pistons. Lubricant infection can be identified by slimy film formation on the crankcase doors. There may also be a rancid odour and white-metal parts may be stained black. As the problem progresses, filters choke up, organic acids are formed, and the oils tend to emulsify.

If SRB is present—a problem particularly associated with laid-up tonnage—copious pitting of ferrous and nonferrous metals may occur (Fig. 7.5).

As microorganisms feed upon the additives within the oil, the lubricity of the oil may be impaired, and its viscosity altered, and there is a greater resultant acidity and increased potential for emulsification and corrosion. Hydrogen sulphide may also be produced as a by-product. Serious corrosion problems will result within weeks of the initial contamination if these factors occur at the same time.

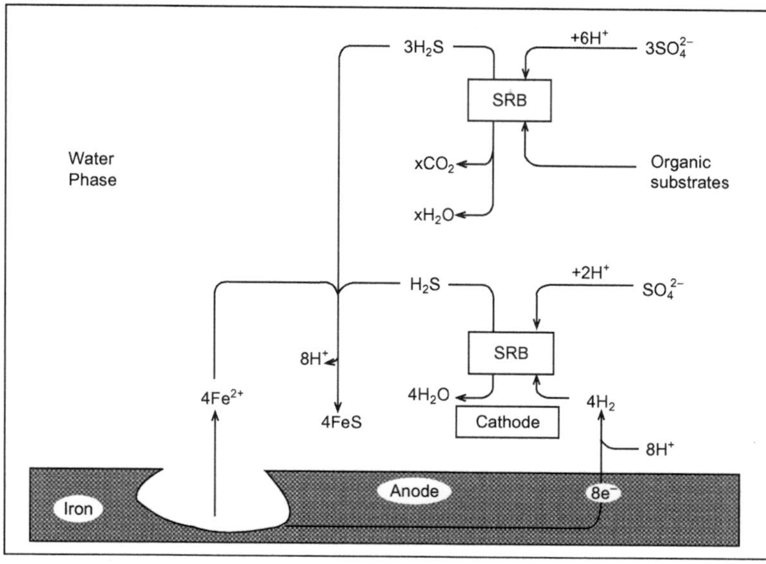

Fig. 7.5 Anaerobic corrosion by SRB.

Sources of contamination within the lubricant are the fuel, cooling water, and sea water. Cooling water in particular has featured as a common contaminant of crankcase oil at engine operating temperatures since the use of chromates as corrosion inhibitors is banned. The use of chromates also acted as an effective antimicrobial biocide.

Prevention is better than cure and it is known that microbial growth is retarded by extreme alkaline conditions. To date, there have been no microbial problems reported with medium-speed engines during operation when using highly alkaline lubricants of BN 12–40. At elevated temperatures, lubricating oils tend to be self-sterilizing. Unfortunately, however, there still remains the possibility of infestation from contaminated bilges and tank tops, which may inadvertently leak into the oil system.

Since the majority of operational microbial problems in lubricating oils arise due to infection from cooling water, sea water from cooler leaks, and overflows of contaminated bilges, the effectiveness and efficiency of plant treatment are important remedial measures.

The following procedures are recommended to avoid microbiological problems:

- Ensure that the water content of crankcase oil does not rise above 0.5% by weight.

- Check that purifier suction is as near to the bottom of the sump as possible.
- Maintain a minimum oil temperature after the purifier heater of 70°C or higher for at least 20 s and/or 80°C for 10 s.
- Circulate the volume of oil in the sump via the purifier at least once every 8–10 h.
- Regularly check that coolant corrosion inhibitor concentrations are at the manufacturer's recommended values.
- Monitor the microbial population of the cooling water and prevent water leaks into the oil system.
- Test for microbial contamination of the oil system and monitor whether the oil after the purifier is sterile.
- Prevent ingress of contaminated bilge water.
- Inspect storage tanks and regularly check for water.

CHAPTER EIGHT

Performance

An important parameter for a marine diesel engine is the rating figure, usually stated as brake horsepower (bhp) or kilowatts per cylinder at given revolutions per minute.

Although engine builders talk of continuous service rating (csr) and maximum continuous rating (mcr), as well as overload ratings, the rating which concerns a shipowner most, is the maximum output guaranteed by the engine builder at which the engine will operate continuously day in and day out. It is most important that an engine is sold for operation at its true maximum rating and that a correctly sized engine be installed in the ship in the first place; an underrated main engine, or more particularly an auxiliary engine, will inevitably be operated at its limits most of the time. It is wrong for a ship to be at the mercy of two or three undersized and thus overrated auxiliary engines, or the main engine that needs to operate at its maximum continuous output to maintain the desired service speed.

Prudent shipowners usually insist that the engines be capable of maintaining the desired service speed fully loaded when developing not more than 85% (or some other percentage) of their rated bhp. However, such a stipulation can leave the full-rated power undefined and therefore does not necessarily ensure a satisfactory moderate continuous rating, hence the appearance of continuous service rating and maximum continuous rating. The former is the moderate in-service figure; the latter is the engine builder's set point of mean pressures and revolutions which can be carried by the engines continuously.

Normally a ship will run sea trials to meet the contract trials speed (at a sufficient margin above the required service speed), and the continuous service rating should be applied when the vessel is in service. It is not unknown for ship owners to then stipulate that the upper power level of the engines in service should be somewhere between 85% and 100% of the service speed output, which could be as much as 20% less than the engine maker's guaranteed maximum continuous rating (Fig. 8.1).

Pounder's Marine Diesel Engines and Gas Turbines
https://doi.org/10.1016/B978-0-08-102748-6.00008-6

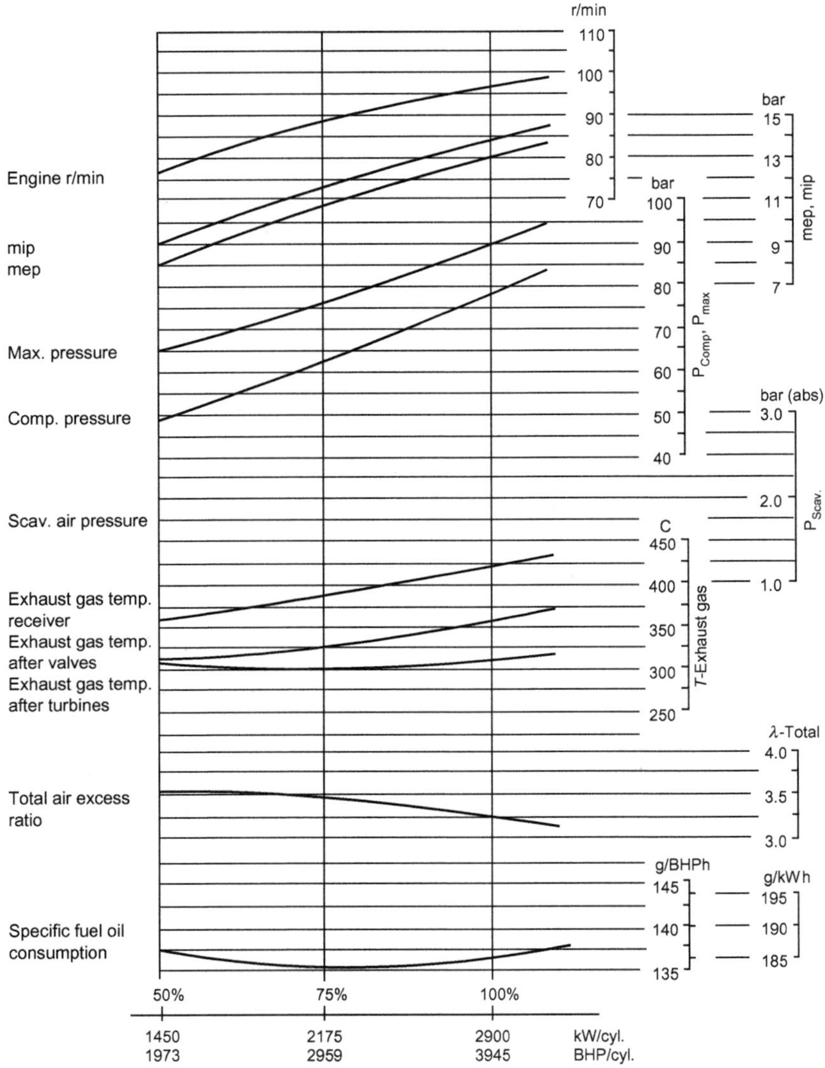

Fig. 8.1 Typical performance curves for a two-stroke engine.

Maximum rating

The practical maximum output of a diesel engine may be said to have been reached when one or more of the following factors operate:

1. The maximum percentage of fuel possible is being burned effectively in the cylinder volume available (combustion must be completed fully at the earliest possible moment during the working stroke) (Fig. 8.2).

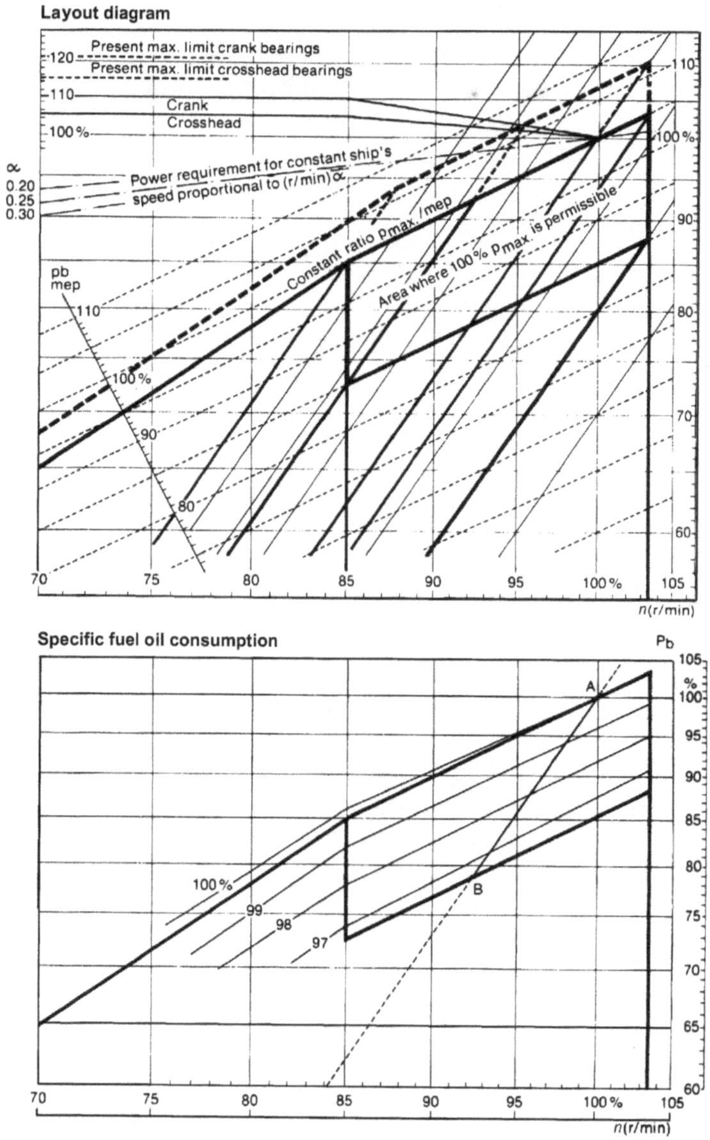

Fig. 8.2 Layout diagrams showing maximum and economy ratings and corresponding fuel consumptions.

2. The stresses in the component parts of the engine generally, for the mechanical and thermal conditions prevailing, have attained the highest safe level for continuous working.
3. The piston speed and thus revolutions per minute cannot safely be increased.

For a given cylinder volume, it is possible for one design of engine effectively to burn considerably more fuel than one of the other designs. This may be the result of more effective scavenging and higher-pressure turbocharging, by a more suitable combustion chamber space and design and a more satisfactory method of fuel injection. Similarly, the endurance limit of the materials of cylinders, pistons, and other parts may be much higher for one engine than for another; this may be achieved by the adoption of more suitable materials, by better design of shapes, thicknesses, etc., more satisfactory cooling and so on. A good example of the latter is the bore cooling arrangements, which is now commonly adopted for piston crowns, cylinder liner collars, and cylinder covers in the way of the combustion chamber.

The piston speed is limited by the acceleration stresses in the materials, the speed of combustion, and the scavenging efficiency: that is, the ability of the cylinder to become completely free of its exhaust gases in the short time of one-part cycle. Within limits, so far as combustion is concerned, it is possible sometimes to increase the speed of an engine if the mean pressure is reduced. This may be of importance for auxiliary engines built to match a given alternator speed.

For each type of engine, therefore, there is a top limit beyond which the engine should not be run continuously. It is not easy to determine this maximum continuous rating; in fact, it can only be satisfactorily established by exhaustive tests for each size and type of engine, depending on the state of development of the engine at the time.

If a cylinder is overloaded by attempting to burn too much fuel, combustion may continue to the end of the working stroke and perhaps also until after exhaust has begun. Besides suffering an efficiency loss, the engine will become overheated and piston seizures or cracking of engine parts may result; or, at least, sticking piston rings, as well as dirty and sticking fuel valves, will be experienced.

Exhaust temperatures

The temperature of the engine exhaust gases can be a limiting factor for the maximum output of an engine. An exhaust temperature graph plotted with mean indicated pressures as abscissae and exhaust temperatures as ordinates will generally indicate when the economical combustion limit and sometimes when the safe working limit of an engine, has been attained. The economical limit is reached shortly after the exhaust temperature begins to curve upwards from what was, previously, almost a straight line.

Very often the safe continuous working load is also reached at the same time, as the designer naturally strives to make all the parts of an engine equally suitable for withstanding the respective thermal and mechanical stresses to which they are subjected.

When comparing different engine types, however, exhaust temperature cannot be taken as proportionate to mean indicated pressure. Sometimes it is said and generally thought that engine power is limited by exhaust temperature. What is really meant is that torque is so limited and exhaust temperature is a function of torque and not of power. The exhaust temperature is influenced by the lead and dimensions of the exhaust piping. The more easily the exhaust gases can flow away, the lower their temperature, and vice versa.

Derating

An option available to reduce the specific fuel consumption of diesel engines is derated or so-called ‘economy’ ratings. This means the operation of an engine at its normal maximum cylinder pressure for the design continuous service rating, but at lower mean effective pressure and shaft speed.

By altering the fuel injection timing to adjust the mean pressure/maximum pressure relationship, the result is worthwhile saving in fuel consumption. The horsepower required for a particular speed by a given ship is calculated by the naval architect, and, once the chosen engine is coupled to a fixed-pitch propeller, the relationship between engine horsepower, propeller revolutions, and ship speed is set according to the fixed propeller curve. A move from one point on the curve to another is simply a matter of giving more or less fuel to the engine.

Derating is the setting of engine performance to maximum cylinder pressures at lower than normal shaft speeds, at a point lower down the propeller curve. For an existing ship and without changing the propeller, this will result in lower ship speed, but in practice when it is applied to new buildings, the derated engine horsepower is that which will drive the ship at a given speed with the propeller optimized to absorb this horsepower at a lower than normal shaft speed.

Savings in specific fuel consumption by fitting a derated engine can be significant. However, should it be required at some later date to operate the engine at its full output potential (normally about 15%–20% above the derated value), the ship would need a new propeller to suit both higher revolutions per minute and greater absorbed horsepower. The injection timing would also have to be reset.

Multiple engine operating profiles

The advent of electronically controlled engines has made it quite simple to manage the derating process by electronic rather than mechanical means. The operational profile of the engine can be programmed to behave differently at the various loads which give the flexibility to optimize the engine for the shipowner/operator. These adjustments are normally done by the engine manufacturer before the engine is delivered to their customers. These different engine settings are commonly referred to as Maps.

An electronically controlled engine has the possibility to have several Maps prepared by the engine manufacturer. A change of Map can be limited to be available only to the engine manufacturer, but it is also possible to open this up to the ship operator. A change may be controlled by different means, which are programmed by the engine manufacturer. The Map settings could be optimized for fuel consumption at, for example, low load and high load operations.

In 2016, Norway proposed to the IMO's MEPC that, in order to reduce greenhouse gas emissions from ships, it should be made possible for shipowners to make use of various Maps for engines installed on their ships. The reason behind the request was the large number of offshore support and service vessels owned and operated by Norwegian owners. These ships have varying operating profiles being required to travel between jobs similar to a cargo ship and then spend long periods working using dynamic position systems that can put very high loads on the engines. The ships also spend periods in port idling. They generally have multiple engines and are diesel-electric ships.

The Norwegian proposal described an engine with two Maps; one optimized for high load and the other for low load. Switching between different Maps depending on ship operation could be a good strategy in order to improve fuel combustion and to reduce the ship's fuel consumption, and thereby reduce the overall emissions. The proposal recognized that using different Maps would influence NOx emissions.

Unrestricted changes in Maps could be regarded as an irrational emission control strategy as defined in regulation 2.13 of MARPOL Annex VI, and is not allowed according to regulation 13.9 of MARPOL Annex VI: '*Irrational emission control strategy means any strategy or measure that, when the ship is operated under normal conditions of use, reduces the effectiveness of an emission control system to a level below that expected on the applicable emission test procedures*'.

If the Maps switches are automatically dependent on load, that would be regarded as a defeat device as defined in regulation 2.6 of MARPOL Annex VI, which is not allowed according to regulation 13.9 of MARPOL Annex VI: '*Defeat device means a device which measures, senses, or responds to operating variables (e.g., engine speed, temperature, intake pressure or any other parameter) for the purpose of activating, modulating, delaying, or deactivating the operation of any component or the function of the emission control system such that the effectiveness of the emission control system is reduced under conditions encountered during normal operation unless the use of such a device is substantially included in the applied emission certification test procedures*'.

There has been support for, and arguments against, the Norwegian proposal not least because of its conflict with the NOx Code. The proposal has been discussed at every MEPC and PPR subcommittee since its introduction. Whether it will be permitted in due course remains to be seen but the restriction does negate one of the main advantages of electronic engine control.

Mean effective pressures

The term brake mean effective pressure (bmep) is widely quoted by engine builders and is useful for industrial and marine auxiliary diesel engines that are not fitted with a mechanical indicator gear. However, the term has no useful meaning for shipboard propulsion engines. It is artificial and superfluous as it is derived from measurements taken by a dynamometer (or brake), which are then used in the calculation of mechanical efficiency. Aboard ship, where formerly the indicator and now electronic pressure transducers producing PV diagrams, are the means of recording cylinder pressures, mean indicated pressure (mip) is the term used, particularly in the calculation of indicated horsepower.

Many ships now have permanently mounted torsionmeters. By using the indicator to calculate mean indicated pressure and thus indicated horsepower, and the torsionmeter to calculate shaft horsepower (shp) from torque readings and shaft revolutions, the performance of the engine both mechanically and thermally in the cylinders can be readily determined.

Instruments such as pressure transducers, indicators, tachometers, and pressure gauges (many of which are of the electronic digital or analogue type of high reliability) allow the ship's engineer to accurately assess the performance of the engine at any time.

The values of bhp, mean indicated pressure and revolutions per minute are, of course, capable of mutual variation within reasonable limits, the horsepower developed per cylinder being the product of mean indicated (or effective) pressure, the revolutions per minute, and the cylinder constant (based on bore and stroke). The actual maximum values for horsepower and revolutions to be used in practice are those quoted by the engine builder for the given continuous service rating.

Propeller slip

The slip of the propeller is normally recorded aboard ship as a useful pointer to overall results. While it may be correct to state that the amount of apparent slip is no indication of propulsive efficiency in new ship design, particularly as a good design may have a relatively high propeller slip, the daily variation in slip (based on ship distance travelled compared with the product of propeller pitch and revolutions turned by the engine over a given period of time) can be symptomatic of the changes in the relationship of propulsive power and ship speed; and slip, therefore, as an entity, is a useful parameter. The effects on ship speed 'over the ground' by ocean currents are sometimes considerable.

For example, the following current may be as much as 2.5%, and heavy weather ahead may have an effect of more than twice this amount.

Propeller law

An engine builder is at liberty to make the engine mean pressure and revolutions what he will, within the practical and experimental limits of the engine design. It is only after the maximum horsepower and revolutions are decided and the engine has been coupled to a propeller that the propeller law operates in its effect upon horsepower, mean pressure, and revolutions.

shp varies as V^3
shp varies as R^3
T varies as R^2
P varies as R^2
where
shp = aggregate shaft horsepower of engine, metric, or imperial
V = speed of ship in knots

$R =$ revolutions per minute
$T =$ torque, in kg m or lb. ft. $= Pr$
$P =$ brake mean pressure, in kg ft. cm^2 or lb. ft. in^2
$r =$ radius of crank, in m or ft.

If propeller slip is assumed to be constant:

$$\text{shp} = KV^3$$

where $K =$ constant from shp and R for a set of conditions.

But R is proportional to V, for constant slip,

$$\therefore \quad \text{shp} = K_1 R^3$$

where $K_1 =$ constant from shp and R for a set of conditions.

But

$$\text{shp} = \frac{p \times A \times c \times r \times 2\pi \times R}{33,000}$$
$$= K_1 R^3 \text{ (imperial)}$$

when $A =$ aggregate area of pistons (cm^2 or in^2), $c = 0.5$ for two-stroke, 0.25 for four-stroke engines, or

$$T = PAcr = \frac{33,000}{2\pi \times R} \times K_1 R^3 = K_2 R_2 \text{ (imperial)}$$

or

$$T = PAcr = \frac{4500}{2\pi \times R} \times K_1 R^3 = K_2 R_2 \text{ (metric)}$$
$$\text{i.e. } T = K_2 R^2$$

where $K_2 =$ constant, determinable from T and R for a set of conditions.

$$PAcr = T \text{ or } \frac{T}{Acr} = \frac{K_2}{Acr} \times R^2$$
$$\text{i.e. } P = K_3 R^2$$

where $K_3 =$ constant determinable from P and R for a set of conditions.

The propeller law index is not always 3, nor is it always constant over the full range of speeds for a ship. It could be as much as 4 for short high-speed vessels, but 3 is normally satisfactory for all ordinary calculations. The index for R, when related to the mean pressure P, is one number less than that of the index for V.

Propeller law is most useful for engine builders at the testbeds where engine loads can be applied with the dynamometer according to the load and revolutions calculated from the law, thus matching conditions to be found on board the ship when actually driving a propeller.

Fuel coefficient

An easy yardstick to apply when measuring machinery performance is the fuel coefficient:

$$C = \frac{D^{2/3} \times V^3}{F}$$

where

$C =$ fuel coefficient
$D =$ displacement of ship in tons
$V =$ speed in knots
$F =$ fuel burnt per 24 h in tons

This method of comparison is applicable only if ships are similar, are run at approximately corresponding speeds, operate under the same conditions, and burn the same quality of fuel. The ship's displacement in relation to draught is obtained from a scale provided by the shipbuilders.

Admiralty constant

$$C = \frac{D^{2/3} \times V^3}{\text{shp}}$$

where $C =$ admiralty constant, dependent on ship form, hull finish, and other factors.

If C is known for a ship, the approximate shp can be calculated for given ship conditions of speed and displacement.

Apparent propeller slip

$$\text{Apparent slip } (\%) = \left(\frac{P \times R - 101.33 \times V}{P \times R}\right) \times 100$$

where

P = propeller pitch in ft.

R = speed of ship in knots

101.33 is one knot in ft./min

The true propeller slip is the slip relative to the wake stream, which is something very different. The engineer, however, is normally interested in the apparent slip.

Propeller performance

Many variables affect the performance of a ship's machinery at sea, so the only practical basis for a contract to build to a specification and acceptance by the owner is a sea trial where everything is under the builder's control. The margin between the trial trip power and sea service requirements of speed and loading must ensure that the machinery is of ample capacity. One important variable on the ship's performance is that of the propeller efficiency.

Propellers are designed for the best combinations of blade area, diameter, pitch, number of blades, etc., and are matched to a given horsepower and speed of propulsion engine; and in fact, each propeller is specifically designed for the particular ship. The search for increased efficiency for vessels has identified that frequently, the assumptions made about a vessel's mode of operation at the design stages does not match actual operation. Significant efficiency improvements can often be made once the ship is in service by considering a redesign of the propeller and changing where appropriate. In once instance, such a change on a ferry produced a 17% reduction in fuel consumption. This is at the high end of the scale but savings into lower double figures have been made in other cases. It is important that the engine should be able to provide heavy torque when required, which implies an ample number of cylinders with the ability to carry high mean pressures. However, when a propeller reaches its limit of thrust capacity under headwinds, an increase in revolutions can be to no avail.

In tank tests with models for powering experiments, the following particulars are given.

Quasi-propulsive coefficient (Q)

$$Q = \frac{\text{model resistance} \times \text{speed}}{2\pi \times \text{torque} \times \text{rev/min}} = \frac{\text{work got out per min}}{\text{work put in per min}}$$

Total shaft horsepower at propeller (EHP)

$$\text{EHP} = \frac{\text{ehp} \times p}{\text{Q}}$$

where

EHP = effective horsepower for the model as determined by tank testing;

p = increase for appendages and air resistance equal to 10%–12% of the naked model EHP, for smooth water conditions.

The shp at the propeller for smooth sea trials is about 10% more than that in tank tests. The additional power, compared with sea trials, for sea service is about 11%–12% more for the South Atlantic and 20%–25% more for the North Atlantic. This is due to the normal weather conditions in these areas. The size of the ship affects these allowances: a small ship needs a greater margin. By way of example, 15% margin over trial conditions equals 26.5% over tank tests.

The shp measured by torsionmeter abaft the thrust block exceeds the EHP by the power lost in friction at the stern tube and plummer blocks and can be as much as 5%–6%.

The bhp exceeds the torsionmeter measured shp by the frictional power lost at the thrust block. Bhp can only be calculated on board ship by multiplying the recorded indicated horsepower by the mechanical efficiency stated by the engine builder.

Required bhp for engine = EHP + sea margin + hp. lost at a stern tube and plummer blocks + hp. lost at thrust block.

Typical values for the quasi-propulsive coefficient (Q) are as follows: tanker 0.67–0.72, slow cargo vessel 0.72–0.75, fast cargo liner 0.70–0.73, ferry 0.58–0.62, and passenger ship 0.65–0.70.

Power buildup

Figs 8.3 and 8.4 are typical diagrams showing the propulsion power data for a twin-screw vessel. In Fig. 8.3 curve A is the EHP at the trial draught; B is the EHP corrected to the contract draught; C is the shp at trial draught on the Firth of Clyde; D is the shp corrected to the contract draught; E is the power service curve from voyage results; F shows the relation between the speed of the ship and engine revolutions on trials, and G is the service result. The full-rated power of the propelling engines is 18,000 bhp; the continuous service rating is 15,000 bhp.

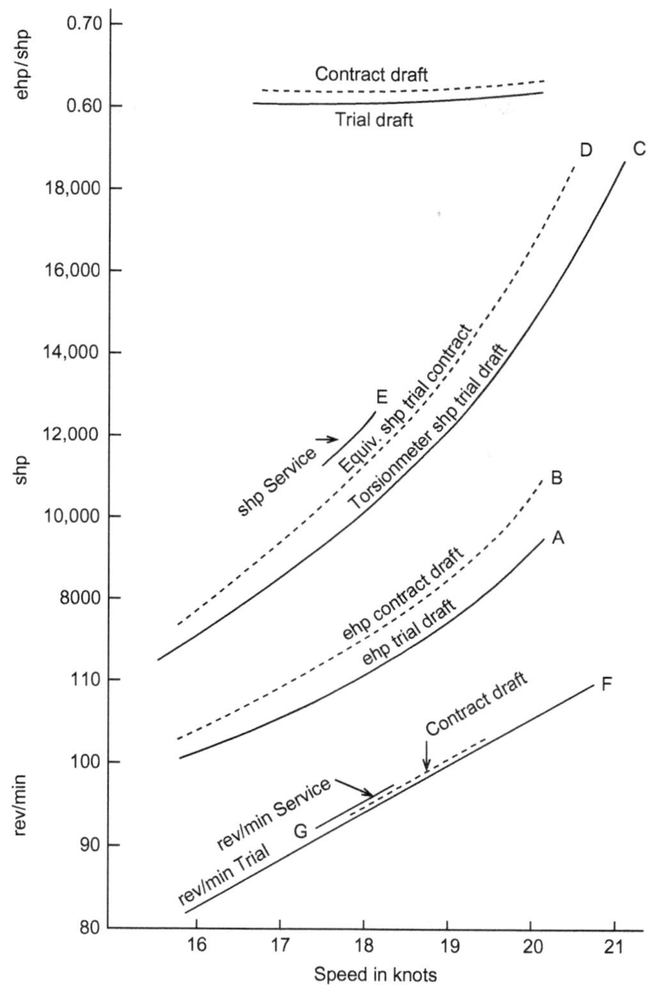

Fig. 8.3 Propulsion data.

In Fig. 8.3, curves A–D show shp as ordinates and the speed of the ship as abscissae. In Fig. 8.4 powers are shown as ordinates, revolutions as abscissae.

Fig. 8.5 shows the relationship between revolutions, power, and brake mean pressure for the conditions summarized in Figs 8.3 and 8.4. A fair line drawn through the observed points for the whole range shows the shp to increase approximately as the cube of the revolutions and the square of the bmep. For the range 95–109 rev/min, the index increases to 3.5 for the power and 2.5 for the bmep. Between 120 and 109 rev/min, a more closely drawn curve shows the index rises to 3.8 and the bmep to 2.8. In

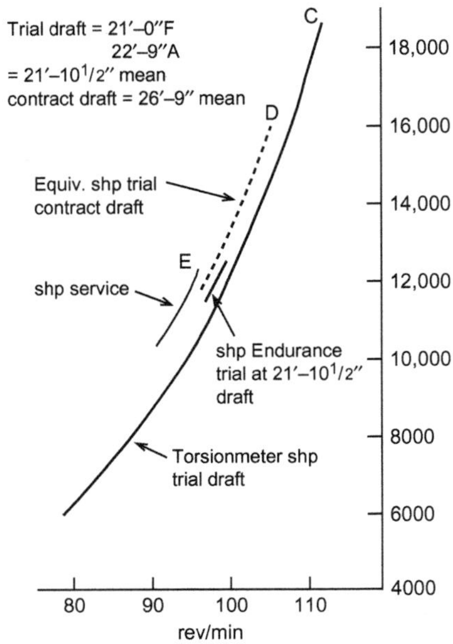

Fig. 8.4 Propulsion data.

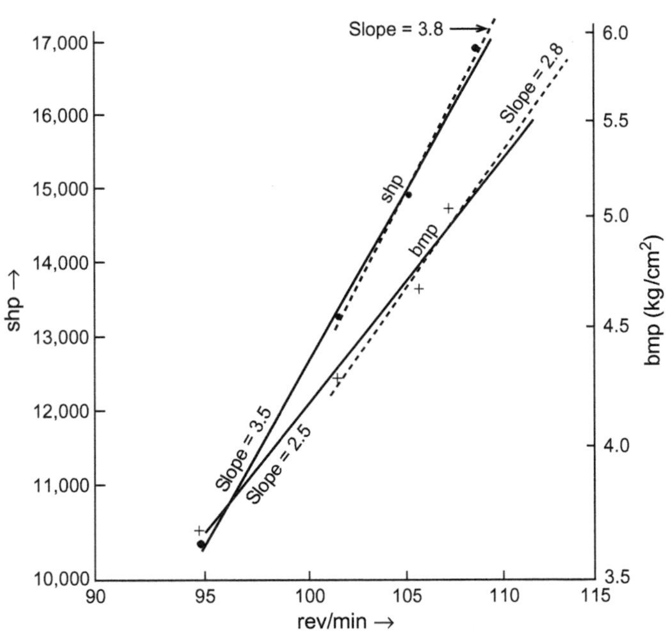

Fig. 8.5 Engine trials: power, revolutions, and mean pressure.

Fig. 8.5 ordinates and abscissae are plotted to a logarithmic base, thus reducing the power/revolution and the pressure/revolution curves to straight lines, for simplicity.

Trailing and locking of propeller

In Fig. 8.6, the normal speed/power curves for a twin-screw motor vessel on the measured mile and in service are shown.

The effect upon the speed and power of the ship when one of the propellers is trailed, by 'free-wheeling', is indicated in the diagram. The effect of one of the propellers being locked is also shown.

Fig. 8.7 shows the speed/power curves for a four-screw motorship under the following conditions:

1. When all propellers are working
2. When the vessel is propelled only by the two centre screws, the outer screws being locked
3. When the ship is being propelled only by the two wing screws, the two inner screws being locked.

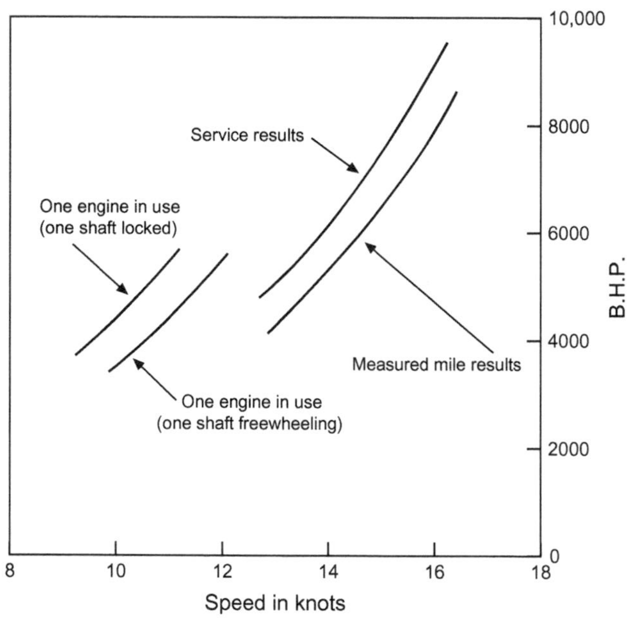

Fig. 8.6 Speed/power curves of a twin-screw vessel.

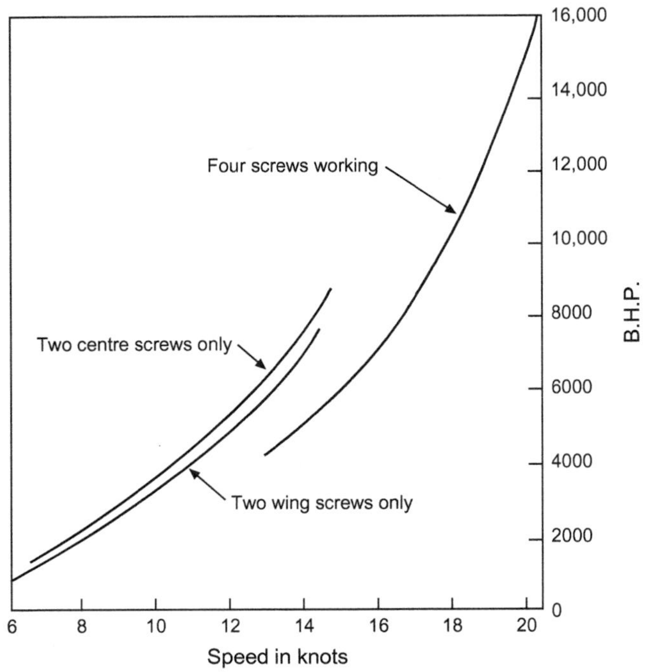

Fig. 8.7 Speed/power curves of a quadruple-screw ship.

Astern running

Fig. 8.8 summarizes a series of tests made on the trials of a twin-screw passenger vessel, 716 ft. long, 83 ft. 6 in. beam, trial draught 21 ft. forward, 26 ft. aft, 26,000 tons displacement.

As plotted in Fig. 8.8, tests I–VI show distances and times, the speed of approach being as stated in column 2 in Table 8.1. The dotted curves show reductions in speed and times.

The dotted curves A–F, respectively, correspond to curves I–VI. In test I, after the ship had travelled, over the ground, a distance of two nautical miles (1 nm = 6080 ft.), the test was terminated and the next test has begun.

In Table 8.2 a typical assortment of observed facts related to engine stopping and astern running is given. Where two or three sets of readings are given, these are for different vessels and/or different engine sizes.

Trials made with a cargo liner showed that the ship was brought to rest from 20 knots in 65 s. Another cargo liner, travelling at 16 knots, was

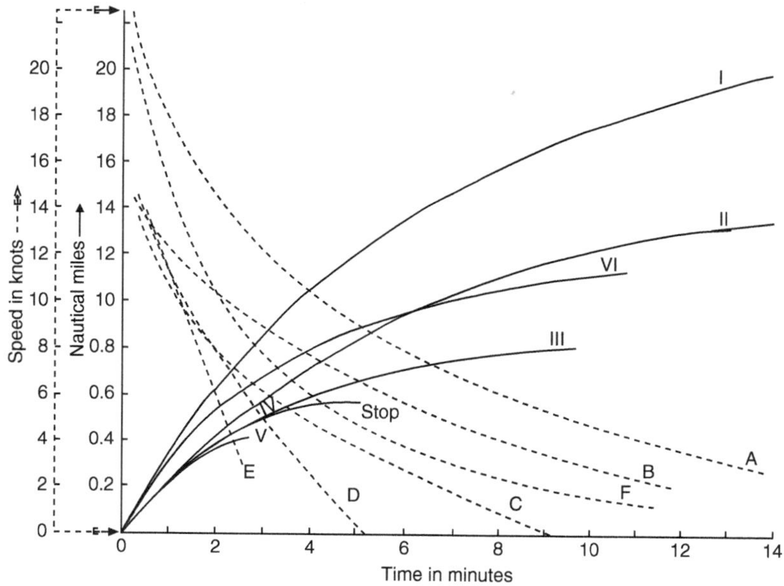

Fig. 8.8 Ship stopping trials.

Table 8.1 Ship stopping trials.

1	2	3	4		5	6
Test no.	**Ahead speed of approach knots (rev/min)**	**Propellers**	**Propellers stopped (min) P.**	**S.**	**Ship stopped (min)**	**Distance travelled (nautical miles)**
I	23.0 (119)	Trailing; unlocked	16.4	14.9	–	2.0
II	14.5 (75)	Trailing; unlocked	12.5	12.1	17.0	1.4
III	13.5 (75)	Trailing; locked	1.5	1.5	13.0	0.8
IV	15.0 (75)	Ahead running checked; no additional astern power	1.3	1.3	5.2	0.5
V	14.8 (75)	Engine stopped; astern as quickly as possible	0.7	0.8	3.1	0.4
VI	22.4 (116)	Trailing; unlocked	1.5	1.6	15.0	1.1

Table 8.2 Engine reversing and ship stopping.

Ship	Engine type	Ahead (rev/min)	Time for engine stopping (s)	Engine moving astern (s)	Astern running rev/min	Astern running s	Ship stopped min	Ship stopped s
Large passenger	D.A. 2C. (twin)	65	–	30	–	–	4	2
Small fast passenger	S.A. 2C. tr. (twin)	217	53	63.5	160	72.5	2	15
		195	35	45	160	60	1	54
		220	45	59	170	81	2	5
Passenger	Diesel electric (twin)	92	110	–	80	225	5	0
Cargo	S.A. 2C. (twin)	112	127.5	136	100	141	–	
Cargo	D.A. 2C. (single)	90	24	26	90	115	2	26
		88	12	14	82	35	3	25
		116	35	40	95	50	–	
Cargo	S.A. 2C. (single)	95	31	33	75	45	–	
		110	12	21	110	53	3	21
		116	10	55	110	110	4	6

brought to a stop in a similar period. The engine, running full power ahead, was brought to 80 rev/min astern in 32 s and had settled down steadily at full astern revolutions in 50 s.

A shipbuilder will think of ship speed in terms of the trial performance in fair weather, but to the shipowner, ship speed is inevitably related to a scheduled performance on a particular trade route. Sea trials are invariably run with the deadweight limited to fuel, freshwater, and ballast. Because of the difference between loaded and trials draught, the hull resistance may be 25%–30% greater for the same speed. This has a consequential effect upon the relation between engine torque and power, and in the reaction on propeller efficiency.

Adverse weather, marine growth, and machinery deterioration necessitate a further power allowance if the service speed is to be maintained. The mean wear and tear of the engine may result in a reduction of output by 10%–15%, or a loss of speed by up to one knot may be experienced.

When selecting a propulsion engine for a given ship, a suitable power allowance for all factors such as weather, fouling, wear, and tear, as well as the need to maintain the service speed at around 85% of the maximum continuous rating, should all be taken into consideration (C.T.W.).

CHAPTER NINE

Engine and plant selection

Choosing a propulsion engine or engines and the most suitable plant configuration for a given newbuilding or retrofit project is not a simple decision. It dictates careful study of the machinery options available and the operating profile of the ship.

In the past, the shipowner or the designer had the straight choice of a direct-coupled low-speed two-stroke engine driving a fixed-pitch propeller or a geared medium-speed, four-stroke engine driving either a fixed or controllable pitch (CP) propeller. Today, ships are entering service with direct-coupled (and sometimes geared) two-stroke engines driving fixed or CP propellers, geared four-stroke engines or high-/medium-speed diesel–electric propulsion plants. Diverse diesel–mechanical and diesel–electric configurations can be considered (Figs. 9.1 and 9.2).

Low-speed engines are dominant in the mainstream deep-sea tanker, bulk carrier, and container ship sectors while medium-speed engines are favoured for smaller cargo ships, ferries, cruise liners, RoRo freight carriers,

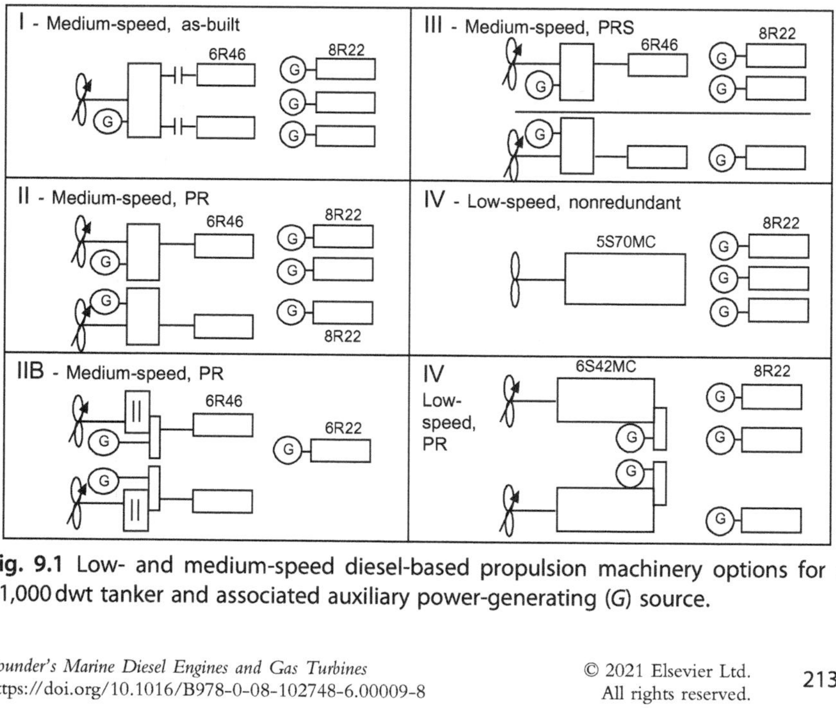

Fig. 9.1 Low- and medium-speed diesel-based propulsion machinery options for a 91,000 dwt tanker and associated auxiliary power-generating (*G*) source.

Pounder's Marine Diesel Engines and Gas Turbines
https://doi.org/10.1016/B978-0-08-102748-6.00009-8

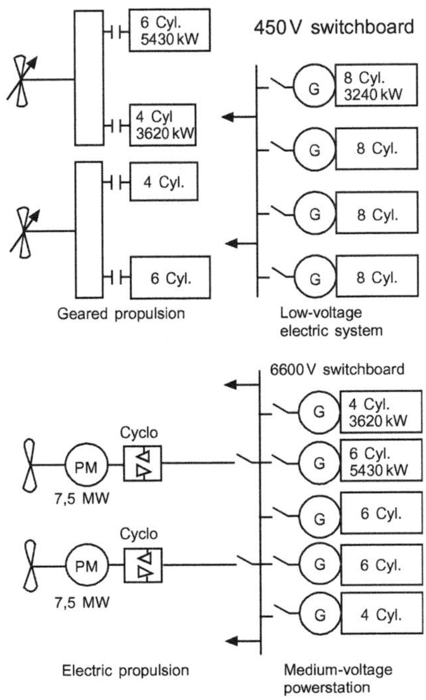

Fig. 9.2 Schematic representations of geared medium-speed diesel propulsion plant with gensets (top) and diesel–electric plant serving propulsion and auxiliary power demands (bottom).

and diverse specialist tonnage such as icebreakers, offshore support, and research vessels. The overlap territory has nevertheless become more blurred in recent years: very small bore low-speed engines can target short sea and even river/coastal vessels while new generations of high-powered large bore medium-speed designs contest the traditional deep-sea arenas of low-speed engines.

Many shipowners may remain loyal to a particular type or make and model of engine for various reasons, such as reliable past-operating experience, crew familiarity, spares inventories, and good service support. But new or refined engine designs continue to proliferate as engine-builders seek to maintain or increase market share or to target a new sector. In some cases, an owner who has not invested in newbuildings for several years may have to consider a number of models and plant configurations which are unfamiliar.

Much decision-making on the engine focuses on cost considerations—not just the initial cost but the type of fuel which can be reliably burned, maintenance costs, desirable manning levels, and availability/price of spares.

The tendency now is to assess the total life cycle costs rather than the purchase price of the main engine. Operating costs over, say, 20 years may vary significantly between different types and makes of engine, and the selected plant configuration in which the engine has to function. Key factors influencing the choice of the engine may be summarized as:

- Capability to burn fuels that meet existing and known future regulations and which will be available in the intended area of operation.
- Compliance of engine with relevant regulations, e.g. NOx Code.
- The maintenance workload: the number of cylinders, valves, liners, rings, and bearings requiring periodic attention in relation to the number of the crew carried (bearing in mind that lower manning levels and less experienced personnel are now more common).
- Suitability for unattended operation by exploiting automated controls and monitoring systems.
- Propulsive efficiency: the ability of the engine or propeller shaft to be turned at a low enough speed to drive the largest diameter (and hence the most efficient) propeller.
- Size and weight of the propulsion machinery.
- Potential to convert engine in the future for use with alternative fuels.
- Cost of the engine

The size of the machinery space is largely governed by the size of the main engine which may undermine the cargo-carrying capacity of the ship. The available headroom is also important in some ships, notably ferries with vehicle decks, and insufficient headroom and surrounding free space may make it difficult or impossible for some engines to be installed or overhauled. EEDI regulations favour smaller engines as this allows for a greater cargo capacity thus influencing the calculation of the EEDI rating.

Diesel–mechanical drives

The direct drive of a fixed-pitch propeller by a low-speed two-stroke engine remains the most popular propulsion mode for deep-sea cargo ships. At one time a slight loss of propulsive efficiency was accepted for the sake of simplicity but the introduction of long stroke and, later, super- and ultra-long stroke cross head engines have reduced such losses. For a large ship a direct-coupled speed of, say, 110 rev/min is not necessarily the most suitable since a larger diameter propeller turning at speeds as low as 60 rev/min is more efficient than one of a smaller diameter absorbing the same horsepower at 110 rev/min. The longer stroke engines now available develop their rated outputs at speeds ranging from as low as 55 rev/min (very large bore models) up to around 250 rev/min for the smallest bore models. It is now possible to specify a direct-drive engine/propeller combination that will yield close to the optimum propulsive efficiency for a given ship design.

Large bore low-speed engines develop high specific outputs, allowing the power level required by many ship types to be delivered from a small number of cylinders. Operators prefer an engine with the fewest possible cylinders, as long as problems with vibration and balance are not suffered. Fewer cylinders obviously influence the size of the engine and the machinery space, the maintenance workload, and the amount of spares which need to be held in stock.

In most deep-sea ships, the height restriction on machinery is less of a problem than length, a larger bore engine with fewer cylinders, therefore, underwriting a shorter engine room and more space for cargo. Larger bore engines also generally return a better specific fuel consumption than smaller engines and offer a greater tolerance to heavy fuels of poor quality.

A direct-coupled propulsion engine cannot operate unaided since it requires service pumps for cooling and lubrication, and fuel/lube oil handling and treatment systems. These ancillaries need electrical power which is usually provided by generators driven by medium- or high-speed diesel engines. Many genset engine-builders can now offer designs capable of burning the same heavy fuel grade as the main engine as well as marine diesel oil or blended fuel (heavy fuel and distillate fuel mixed in various proportions, usually 70:30) either bunkered as an intermediate fuel or blended onboard. 'Unifuel' installations—featuring main and auxiliary engines arranged to burn the same bunkers—are now common.

Auxiliary power generation

The cost of auxiliary power generation can weigh heavily in the choice of the main machinery. Developments have sought to maximize the exploitation of waste heat recovery to supplement electricity supplies at sea, to facilitate the use of alternators driven by the main engine via speed-increasing gearing or mounted directly in the shaftline, and to power other machinery from the main engine.

Gear-based constant frequency generator drives allow a shaft alternator to be driven by a low-speed engine in a fixed-pitch propeller installation, with full alternator output available between 70% and 104% of a propeller speed. A variety of space-saving arrangements are possible with the alternator located alongside or at either end of the main engine equipped with compact integral power take-off (PTO) gear. Alternatively, a frequency converter system can be specified to serve an alternator with a variable main engine shaft speed input in a fixed pitch or CP propeller installation.

The economic attraction of the main engine-driven generator for electrical power supplies at sea is that it exploits the high thermal efficiency, low-specific fuel consumption, and low-grade fuel-burning capability of the ship's diesel prime mover. Other advantages are that the auxiliary diesel gensets can be shut down, yielding benefits from reduced running hours in terms of lower fuel and lubricating oil consumptions, maintenance demands, and spares costs.

System options for electricity generation have been extended by the arrival of power turbines which, fed with exhaust gas surplus to the needs of modern high-efficiency turbochargers, can be arranged to drive alternators in conjunction with the main engine or independently.

These small gas turbines are also in service in integrated systems linking steam turbo-alternators, shaft alternators, and diesel gensets; the various power sources—applied singly or in combination—promise optimum economic electricity production for any ship-operating mode. Some surplus electrical output can also be tapped to support the propulsive effort via a shaft alternator switched to function as a propulsion motor.

Such a plant is exploited in a class of large low-speed engine-powered container ships with significant reefer capacity whose overall electrical load profile is substantial and variable. Crucial to its effectiveness is a computer-controlled energy management system which co-ordinates the respective contributions of the various power sources to achieve the most economical mode for a given load demand.

Integrated energy-saving generating plants have been developed over the years for application mainly to large container ships. The systems typically exploit waste heat (from the low-speed main engine exhaust gas, scavenge air, and cooling water) to serve a steam turbo-alternator, air-conditioning plant, heaters, and distillers. System refinements were stimulated by the diminishing amount of energy available from the exhaust gas of low-speed engines, in terms of both temperature and volumes, with the progressive rise in thermal efficiencies. The ability of the conventional waste heat boiler/turbo-alternator set to meet electrical demands at sea was compromised, any shortfall having to be plugged by supplementary oil firing of an auxiliary boiler or by running a diesel genset and/or shaft alternator. The new integrated systems, some also incorporating power gas turbines, maximize the exploitation of the waste heat available in ships whose operating profiles and revenues can justify the added expense and complexity.

Wider application of 'cold ironing'—shutting down the diesel gensets in port and tapping shoreside electrical power supplies—will be stimulated by concerns over emissions in highly populated areas, a trend initiated in US west coast container terminals.

Most recently, energy storage systems have been developed which allow for energy to be stored in battery packs. The energy can come from any of several diverse sources. It can be obtained by running auxiliary engines at optimum speed and load rather than uneconomic parameters and storing any unneeded energy; from shore power supplies; from solar or wind-powered systems installed on the ship; or even from small fuel cell systems.

When required the stored energy can be used to provide PTI to the main engine—a motor-generator installed on the shaft can allow PTO when the main engine is running and PTI when it is not. It can also supply power to other load users around the ship from the bridge, accommodation, and even machinery. A suitable battery system can store sufficient energy to allow for the main engine to be started even if starting air is not available.

Geared drives

The most common form of the indirect drive of a propeller features one or more medium-speed four-stroke engines connected through clutches and couplings to a reduction gearbox to drive either a fixed-pitch or CP propeller (Figs. 9.3 and 9.4). The CP propeller eliminates the need for a direct-reversing engine while the gearing allows a suitable propeller speed to be selected. There is inevitably a loss of efficiency in the transmission but in most cases, this would be cancelled out by the improvement in propulsive efficiency when making a comparison of direct-coupled and indirect drive engines of the same horsepower. The additional cost of the transmission can also be offset by the lower cost of the four-stroke engine since two-stroke designs, larger and heavier, cost more. The following generic merits are cited for geared multimedium or high-speed engine installations:

- Ships with more than one main engine benefit from enhanced availability through redundancy: in the event of one engine breaking down another can maintain navigation. The number of engines engaged can also be varied to secure the most economical mode for a given speed or deployment profile. Thus, when a ship is running light, partially loaded, or slow steaming, one engine can be deployed at its normal

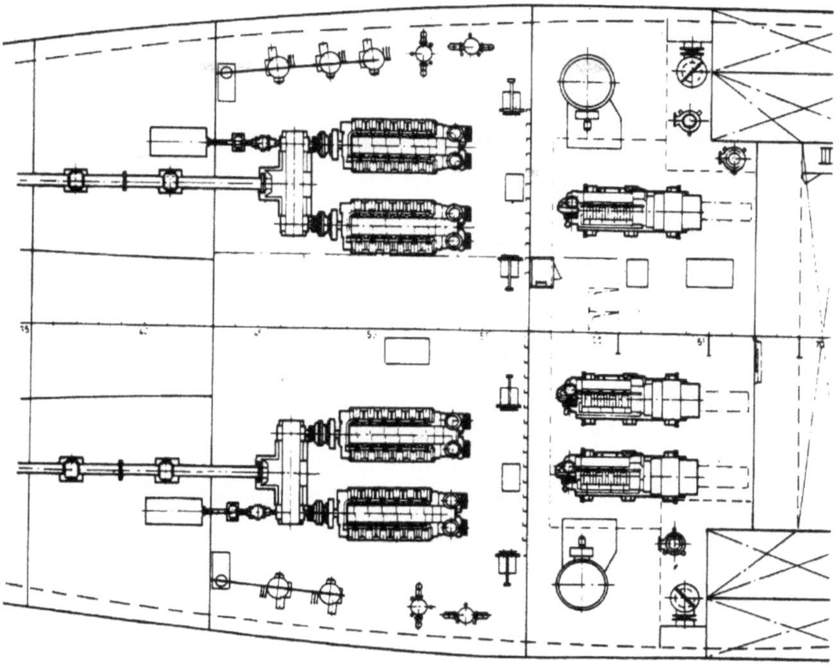

Fig. 9.3 A typical multiengine geared twin-screw installation.

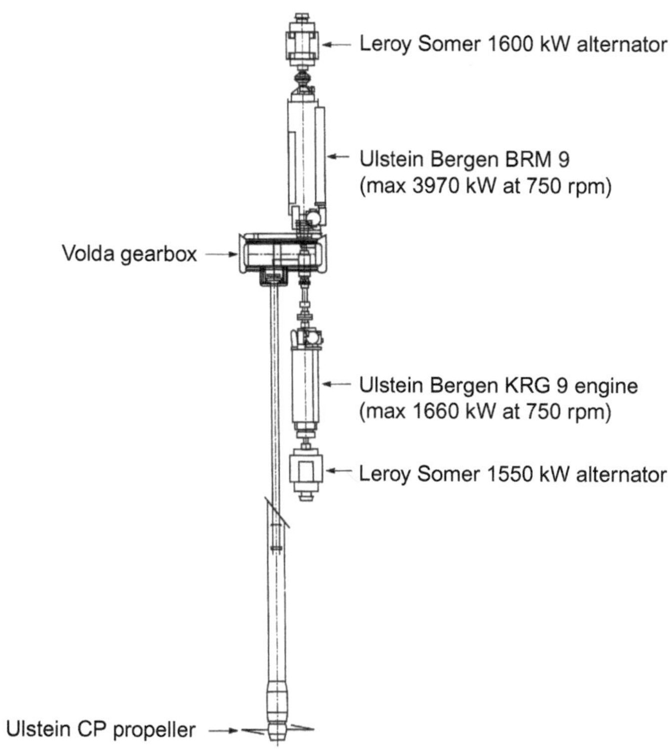

Fig. 9.4 One of the medium-speed geared propulsion/auxiliary power lines of a Norwegian coastal ferry.

(high-efficiency) rating and some or all of the others shut down. In contrast, in similar operational circumstances, a single direct-coupled engine might have to be run for long periods at the reduced output with lower efficiency.

- The ability to vary the number of engines deployed allows an engine to be serviced at sea, easing maintenance planning. This flexibility is particularly valued in an era when port turnround times are minimized. Engines can also be overhauled in port without the worry of authorities demanding an unscheduled shift of berth or sudden departure.
- By modifying the number of engines per ship and the cylinder numbers per engine to suit individual power requirements, the propulsion plant for a fleet can be standardized on a single-engine model, with consequent savings in spares costs and inventories, and benefits in crew familiarity. The concept can also be extended to the auxiliary power plant through 'uniform machinery' installations embracing main and genset engines of the same model.
- Compact machinery spaces with low headrooms can be created, characteristics particularly valued for RoRo ferries.

Designers of marine gearing, clutches, and couplings have to satisfy varying and sometimes conflicting demands for operational flexibility, reliability, low noise, and compactness from transmission systems.

Advances in design, materials, and controls have contributed to innovative solutions for versatile single- and multiengine propulsion installations featuring PTOs for alternator drives and power take-ins (PTIs) to boost the propulsive effort. In many propulsion installations, a gearbox is expected to: (a) determine the propeller speed and direction of rotation, and provide a reversing capability; (b) provide a geometric coupling that can connect and separate the flow of power between the engine and propeller shaft or waterjet drive; and (c) absorb the thrust from the propeller.

Impressive flexibility of operating modes can be arranged from geared multiengine propulsion plants which are, in practice, overwhelmingly based on four-stroke machinery (a number of geared two-stroke engine installations have been specified, however, for special purpose tonnage such as offshore shuttle tankers).

Father-and-son layouts

Operational flexibility is enhanced by adopting a so-called 'father-and-son' (or 'mother-and-daughter') configuration: a partnership of similar four-stroke engine models, but with different cylinder numbers, coupled to a common reduction gearbox to drive a propeller shaft. Father-and-son pairs have been specified to power large twin-screw cruise vessels and ferries. An example is provided by the 1995-built P&O liner *Oriana* whose 40,000 kW propulsion plant is based on two 9-cylinder and two 6-cylinder MAN B&W L58/64 medium-speed engines. Each father-and-son pair drives a highly skewed Lips CP propeller via a Renk Tacke low-noise gearbox which also serves a 4200 kW shaft alternator/propulsion motor.

Propulsion can be affected by either the father-and-son engines together; the father engines alone or the son engines alone; and with or without the shaft alternators operating as propulsion motors (fed with electrical power by the diesel gensets).

Diesel–electric drive

An increasingly popular form of indirect drive is diesel–electric propulsion featuring power stations based on multiple medium-speed main gensets. New generations of AC/AC drive systems exploiting cycloconverter, synchroconverter, and pulse-width modulated converter technology have widened the potential of electric propulsion after years of confinement to specialist niches, such as icebreakers, research vessels, and cable layers. Converters allow a propulsion motor to be driven over its whole speed range from zero to nominal speed in forward and reverse modes.

Diesel–electric propulsion is now dominant in cruise ships and new-generation liquefied natural gas (LNG) carriers, and references are established in North Sea shuttle tankers, short-sea and deep-sea chemical carriers, and Baltic RoRo passenger/freight ferries. The mode is also increasingly popular for offshore supply and support tonnage.

A number of specialist groups contest the high-powered electric drive market, including ABB Marine and Siemens, which highlight system capability to cope with sudden load changes, deliver smooth and accurate ship-speed control, and foster low noise and vibration. Applications have been stimulated by continuing advances in power generation systems, AC drive technology, and power electronics.

Two key AC technologies emerged to supersede traditional DC electric drives: the cycloconverter system and the Load Commutated Inverter (LCI) or Synchrodrive solution. Both are widely used in electric propulsion but, although exploiting the same basic Graetz bridge prevalent in DC drive systems, they have different electrical characteristics. Development efforts have also focused on a further AC system variant, the voltage source pulse-width modulation (PWM) drive.

A variable-speed AC drive system comprises a propulsion motor (Fig. 9.5) and a frequency converter, with the motor speed controlled by the converter altering the input frequency to the motor.

Both Cycloconverter and PWM drives are based on advanced AC variable-speed technology that matches or exceeds the performance characteristics of conventional DC drives. Each features motor speed control with maximum torque available from zero to full speed in either direction, facilitating operation with simple and robust fixed pitch (rather than CP) propellers. Smooth control underwrites operation at very low speeds, and vector control yields a rapid response to enhance plant and ship safety. Integrated full electric propulsion (IFEP) is now standard for cruise liners and specified for a widening range of other ship types—including naval vessels—in which the economic and operational benefits of the central power station concept can be exploited.

After several years during which AC systems were favoured, the last 10 years have seen a resurgence of DC systems. In 2010, ABB announced

Fig. 9.5 One of two 15,000 kW STN Atlas propulsion motors installed on the cruise liner *Costa Victoria*.

its On-board DC Grid with the first vessel to employ it—a PSV named *Dina Star*—being delivered in 2013.

The On-board DC Grid was described by ABB as a step forward in optimized propulsion and is an extension of multiple DC links that already exist in propulsion and thruster drives. On-board DC Grid-enabled combination of the advantage of AC components with a new DC distribution. Just as variable speed drive allows the electric propulsion motors to be run at their optimum working point, On-board DC Grid allows the diesel engines to run at variable speed for top fuel efficiency at each load level. And, the On-board DC Grid enables full flexibility in combining energy sources, including renewables.

Advantages claimed for the concept included a 27% reduction in fuel consumption, less maintenance of generators, improved dynamic response, and lower footprint. It would also require fewer components and cabling and allow for other energy sources to be added more easily.

Factors that led ABB's designers to adopt the alternative method of DC power distribution were the fact that the power for largest power consumers—the propulsion motors—must be processed as a DC input to the variable frequency inverter that performs the speed control of the motor. Actually, distributing at a DC level allows the losses in a switchboard and transformer to be eliminated from this power flow.

In addition, when diesel-electric generators run at constant rpm fuel efficiency is compromised. By distributing electricity in DC, it is no longer required to maintain a fixed electrical frequency. Consequently, the engines can run at variable speed and therefore at the best possible operating point to achieve the best fuel efficiency along with the full power range.

Each power source and consumer on the On-board DC Grid is an AC or DC 'island' and the only connection between them is the DC bus. This yields two advantages:

- Each power source and consumer can be controlled and optimized independently.
- Complex interactions that can arise between units that share an AC connection will never occur. Consumers fed by the On-board DC grid are designed not to interact even under fault conditions.

In the On-board DC Grid solution, energy storage may be included to improve the system's dynamic performance. Diesel engines are slow to handle large, quick load changes. By using batteries or super capacitors to provide power for a short time, the ship's control capabilities can be improved. This will benefit especially vessels with dynamic positioning. Energy storage can also be used to absorb rapid power fluctuations seen by diesel engines thereby improving their fuel efficiency.

Siemens has a similar system which it markets as BlueDrive PlusC. Both systems allow for simpler automated power management on the vessel.

Electric propulsion requires motors to drive the propellers and gensets to supply power. It seems somewhat illogical to use electric generators, switchgear, and motors between prime movers and propeller when direct coupling or geared transmission to the shaft may be all that is necessary. There are obviously sound reasons for some installations to justify the complication and extra cost of electric propulsion, Lloyd's Register citing:

Flexibility of layout: The advantage of electric transmission is that the prime movers and their associated generators are not constrained to have any particular relationship with the load—a cable run is a very versatile medium. In a propulsion system, it is, therefore, possible to install the main diesel gensets and their support services in locations remote from the propeller shaft (Fig. 9.6). Diesel gensets have even been installed in containers

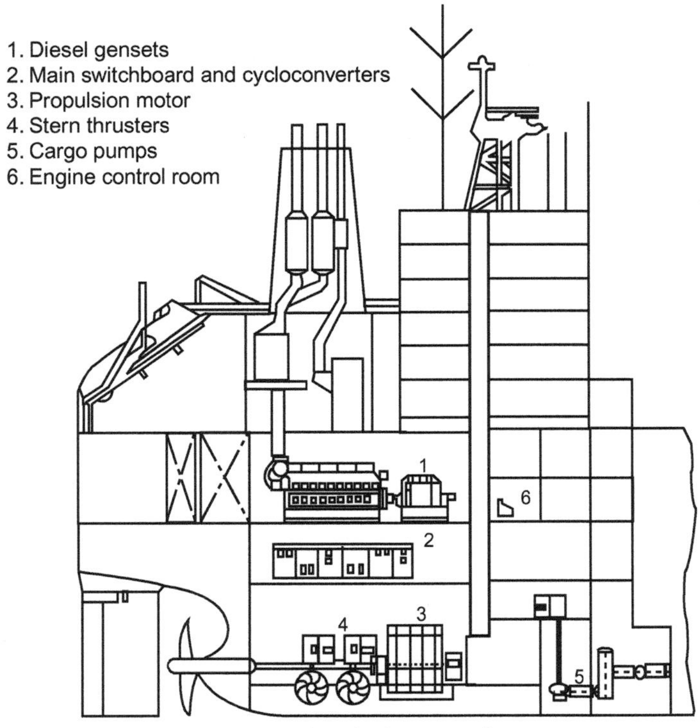

Fig. 9.6 Machinery arrangement in a diesel–electric tanker.

on deck to provide propulsive power, and other vessels equipped with a 10,000 kW generator mounted in a block at the stern above the RoRo deck.

Another example of the flexibility facilitated by an electric propulsion system is a semi-submersible rig installation whose gensets are mounted on the main deck and propulsion motors in the pontoons. In cruise ships, ferries, and tankers the layout flexibility allows the naval architect to create compact machinery installations releasing extra revenue-earning space for passenger accommodation/amenities and cargo.

Opportunities to design and build tankers more cost-effectively are also offered. The optimized location of the main machinery elements allows the ship's overall length to be reduced and steel costs correspondingly lowered for a given cargo capacity; alternatively, the length of the cargo tank section can be extended within given hull dimensions.

Diesel–electric solutions facilitate the modularization and delivery of factory-tested turnkey packages to the building yard, with the complete main gensets mounted on common bedplates ready for coupling to their support systems. The compactness of the machinery outfit fosters shorter runs for the cabling and ancillary systems, and the engine casing and exhaust gas piping are also shortened.

It is difficult for a diesel–electric plant to match the fuel economy of a single direct-drive low-speed main engine which is allowed to operate at its optimum load for the long transoceanic leg of a tanker's voyage. But some types of tanker (oil products and chemical tonnage, as well as North Sea shuttle carriers) are deployed in varied service profiles and spend considerable time at part loads: for example, in restricted waters, during transit in ballast and manoeuvring.

A diesel–electric tanker can exploit the abundance of electrical power to drive a low-noise cargo pumping outfit, perhaps reducing time in port and (particularly in the case of a dynamically positioned shuttle carrier) to serve powerful bow and stern thrusters during sensitive manoeuvring at the loading buoy (Fig. 9.7).

Load diversity: Certain types of tonnage have a requirement for substantial amounts of power for ship services when the demands of the propulsion system are low: for example, tankers and any other ship with a significant cargo discharging load. A large auxiliary power-generating plant for cruise ships and passenger ferries is dictated by the hotel services load (air conditioning, heating, and lighting) and the heavy demands of transverse thrusters during manoeuvring. The overall electrical demand may be 30%–40% of the installed propulsion power and considerable standby capacity is also called

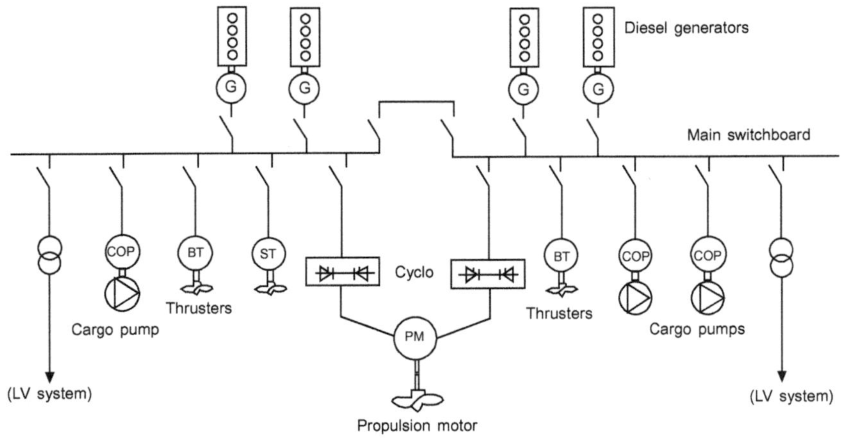

Fig. 9.7 Diesel–electric 'power station' meeting the load demands of propulsion, cargo pumping, thrusters, and hotel services in a tanker.

for to secure system redundancy for safety. These factors have helped to promote the popularity of the medium-speed diesel–electric 'power station' concept for meeting all propulsion, manoeuvring, and hotel energy demands in large passenger ships.

Economical part-load running: This is best achieved when there is a central power station feeding propulsion and ship services. A typical medium-speed diesel–electric installation features four main gensets (although plants with up to nine sets are in service) and, with parallel operation of all the sets, it is easy to match the available generating capacity to the load demand. In a four-engine plant, for example, increasing the number of sets in operation from two that are fully loaded to three partially loaded will result in three sets operating at 67% load: not ideal but not a serious operating condition.

It is not necessary to operate gensets at part load to provide the spare capacity for the sudden loss of a set: propulsion load reduction may be available instantaneously and, in most ships, a short-term reduction in propulsive power does not represent a hazard. The propulsion plant controls continuously monitor-generating capability, any generator overload immediately resulting in the controls adjusting the input to the propulsion motors. During manoeuvring, propulsion plant's requirements are below the system's capacity and the failure of one generator is unlikely to present a hazardous situation.

Ease of control: The widespread use of CP propellers means that control facilities once readily associated with electric drives are no longer able to command the same premium. It is worth noting, however, that electric drives are capable of meeting the most exacting demands with regard to dynamic performance, exceeding by a very wide margin anything that is required of a propulsion system.

Low noise: An electric motor can provide a drive with very low-vibration characteristics, a quality valued for warships, research vessels, and cruise ships where, for different reasons, a low-noise signature is required. In the case of warships and research vessels, it is noise into the water which is the critical factor; in cruise ships, it is structure-borne noise and vibration to passenger spaces that are desirably minimized. The electric transmission allows the prime movers to be mounted in such a way that the vibration transmitted to the seatings, and hence to the water or passenger accommodation, is minimized. The use of double-resilient mounting may be necessary for very low-noise installations.

Environmental protection and ship safety: Controls to curb noxious exhaust gas emissions—tightening nationally, regionally, and globally—also favour the specification of electric transmission since the constant speed running of diesel prime movers optimized to the load is conducive to lower NOx emission levels. An increasing focus on higher ship safety through redundancy of propulsion plant elements is another positive factor. A twin-propeller installation is not the only method of achieving redundancy. Improving the redundancy of a single-screw ship can also be secured by separating two or more propulsion motors in different compartments and coupling them to a reduction gear located in a separate compartment. The input shafts are led through a watertight bulkhead.

Podded propulsors

Significant economic and technical benefits in ship design, construction, and operation are promised by electric podded propulsors, whose appeal has extended since the early 1990s from icebreakers and cruise liners to offshore vessels, ferries, and tankers.

A podded propulsor (or pod) incorporates its electric drive motor in a hydrodynamically optimized submerged housing which can be fully rotated with the propeller(s) to secure 360 degrees azimuthing and thrusting capability (Fig. 9.8).

The motor is directly coupled to the fixed pitch propeller(s) mounted at either or both ends of the pod. Pusher and tractor or tandem versions can be specified, with input powers from 400 to 30,000 kW supplied by a shipboard

Fig. 9.8 Three 14,000 kW Azipod propulsors power Royal Caribbean Cruises' Voyager-class liners.

generating plant. The merits cited for pods over traditional diesel–electric installations with shaftlines highlight:

- Space within the hull otherwise reserved for conventional propulsion motors can be released and exploited for other purposes.
- More creative freedom for ship designers since the propulsors and the prime movers require no direct physical connection.
- Steering capability is significantly better than with any conventional rudder system; stern thruster(s) can be eliminated, along with rudder(s) and shaftline(s).
- Excellent reversing capability and steering during astern navigation, and enhanced crash stop performance.
- Low-noise and vibration characteristics associated with electric drive systems are enhanced by the motor's underwater location; and hull excitations induced by the propeller are very low, thanks to operation in an excellent wake field.
- Propulsor unit deliveries can be late in the shipbuilding process, reducing 'dead time' investment costs; savings in overall weight and ship construction hours are also promised.

Only a small number of makers are present in the sector especially in the larger sizes where ABB is effectively the only player now active. More than 100 cruise ships feature ABB Azipods and the technology has also been installed on gas carriers and tankers. Because of its dominance, the brand name Azipod is frequently used wrongly as a generic term for any podded propulsor.

Combined systems

A combination of geared diesel and diesel–electric drive is exploited in some offshore support vessels whose deployment profile embraces two main roles: (1) to transport materials from shore bases to rigs and platforms and (2) to standby at the offshore structures for cargo and anchor handling, rescue, and other support operations. In the first role, the propulsive power required is that necessary to maintain the free-running service speed. In the second role, little power is required for main propulsion, but sufficient electrical power is vital to serve winches, thrusters, and cargo-handling gear.

The hybrid solution for these dual roles is to arrange for the main propellers to be driven by medium-speed engines through a reduction gear which is also configured to drive powerful shaft alternators. By using one shaft alternator as a motor and the other to generate electrical power, it is possible to secure a diesel–electric main propulsion system of low horsepower and at the same time provide sufficient power for the thrusters. Such a twin-screw plant could simultaneously have one main engine driving the alternator and the propeller, while the other main engine is shut down and its associated propeller driven by the shaft alternator in propulsion motor mode.

A further option is to fit dedicated propulsion motors as well as the shaft alternators, yielding greater plant flexibility, and fuel-saving potential but at the expense of extra cost and complexity. For the above power plants to operate effectively CP main and thruster, propellers are essential.

Combined medium- or high-speed diesel engine and gas turbine (CODAG) systems feature in some large fast ferry propulsion, while some large diesel–electric cruise liners feature medium-speed engines and gas turbines driving generators in a CODLAG or CODEG arrangement.

Gas turbines enjoyed a brief spell of popularity in cruise ships and ferries around the turn of the century but within a few years were out of favour because of high fuel costs. In all 17 cruise ships and around 30 other commercial vessels were fitted with GE Marine gas turbines in various configurations with diesel engines. In 2016 GE Marine predicted a resurgence in interest and also potentially for container ships and other vessel types. This has not yet materialized. Gas turbines do however remain popular for military applications.

CHAPTER TEN

Turbocharging

Turbocharging

A naturally aspirated engine draws air of the same density as the ambient atmosphere. Since this air density determines the maximum weight of fuel that can be effectively burned per working stroke in the cylinder, it also determines the maximum power that can be developed by the engine. Increasing the density of the charge air by applying a suitable compressor between the air intake and the cylinder increases the weight of air induced per working stroke, thereby allowing a greater weight of fuel to be burned with a consequent rise in specific power output.

This boost in charge air density is accomplished in most modern diesel engine types by exhaust gas turbocharging in which a turbine wheel driven by exhaust gases from the engine is rigidly coupled to a centrifugal-type air compressor. This self-governing process does not require an external governor.

The power expended in driving the compressor has an important influence on the operating efficiency of the engine. It is relatively uneconomical to drive the compressor direct from the engine by the chain or gear drive because some of the additional power is thereby absorbed and there is an increase in specific fuel consumption for the extra power obtained. About 35% of the total heat energy in the fuel is wasted to the exhaust gases, so by using the energy in these gases to drive the compressor, an increase in power is obtained in proportion to the increase in the charge air density.

A turbocharger comprises a gas turbine driven by the engine exhaust gases mounted on the same spindle as a blower, with the power generated in the turbine equal to that required by the compressor.

There are a number of advantages of pressure charging by means of an exhaust gas turbocharger system:

- A substantial increase in engine power output for any stated size and piston speed, or conversely a substantial reduction in engine dimensions and weight for any stated horsepower

Pounder's Marine Diesel Engines and Gas Turbines
https://doi.org/10.1016/B978-0-08-102748-6.00010-4

- An appreciable reduction in the specific fuel consumption rate at all engine loads
- A reduction in initial engine cost
- Increased reliability and reduced maintenance costs, resulting from less exacting conditions in the cylinders
- Cleaner emissions (see p. 30)
- Enhanced engine operating flexibility (Fig. 10.1)

Larger two-stroke engines may be equipped with up to four turbochargers, each unit serving between three and five cylinders.

Turbocharging had been applied to four-stroke diesel engines in the 1930s, the 20–30 year gap before general adoption in two-stroke engines in the 1950s and 1960s reflecting the fundamental differences between the cycles. In the four-stroke cycle, the gas exchange process works by filling and emptying with little valve overlap. Early turbochargers thus had adequate efficiency to deliver the necessary quantities of charge air required by contemporary four-stroke engines.

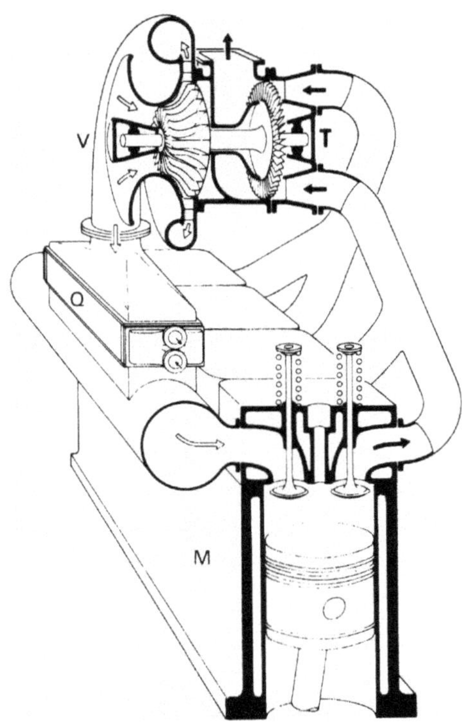

Fig. 10.1 Basic arrangement of exhaust turbocharging. M, diesel engine; Q, charge air cooler; T, turbocharger exhaust turbine; and V, turbocharger compressor.

In contrast, the two-stroke cycle is a gas flow process. The working processes of intake, compression, combustion, expansion, and scavenging must all be completed in one engine revolution. In the two-stroke cycle, the gases must be blown through by the scavenging air during the overlap between scavenge and exhaust openings instead of clearing the combustion gases by piston displacement, as in a four-stroke cycle. There needs to be a clear pressure drop between the scavenge air and exhaust manifolds to secure the necessary flow through the cylinder for good scavenging. Two-stroke engines, therefore, call for greater efficiency from the turbochargers than four-stroke engines to deliver the required larger air quantity for scavenging, while having comparatively less exhaust energy to drive the turbine.

It took some years of development before turbochargers were available with sufficient efficiency for large two-stroke engines, and for engine designers/builders to evolve techniques for matching engine and turbocharger.

Four-stroke engines

Exhaust gas turbocharged four-stroke marine engines can deliver three times or more power than naturally aspirated engines of the same speed and dimensions. Even higher power output ratios are achieved on engine types featuring two-stage turbocharging, where turbochargers are arranged in series. At one time almost all four-stroke engines operated on the pulse system, though constant pressure turbocharging has since become the norm as it provides greater fuel economy while considerably simplifying the arrangement of exhaust piping.

In matching the turbo blower to the engine, a free air quantity in excess of the swept volume is required to allow for the increased density of the charge air and to provide sufficient air for through-scavenging of the cylinders after combustion. For example, an engine with a full-load brake means effective pressure (bmep) of 10.4 bar would need about 100% of excess free air, about 60% of which is retained in the cylinders, with the remaining 40% being used for scavenging. Modern engines carry bmeps of up to 28 bar in some cases, requiring greater proportions of excess air, which is made possible by the latest designs of the turbocharger with pressure ratios as high as 6:1.

In a typical case, to ensure adequate scavenging and cooling of the cylinders, a valve overlap of approximately 140° is normal with the air inlet valve opening at 80° before the top dead centre (TDC) and closing at

40° after bottom dead centre (BDC). The exhaust valve opens at 50° before BDC and closes at 60° after TDC.

For a low-rated engine, with a bmep of 10.4 bar compared with 5.5 bar for a naturally aspirated engine, a boost pressure of some 0.5 bar is required, corresponding to a compressor pressure ratio of 1.5:1. As average pressure ratios, today are between 2.5 and 4, considerable boost pressures are being carried on modern high bmep medium-speed four-stroke engines.

Optimum values of power output and specific fuel consumption can be achieved only by the utilization of the high-energy engine exhaust pulses. The engine exhaust system should be so designed that it is impossible for gases from one cylinder to contaminate the charge air in another cylinder, either by blowing back through the exhaust valve or by interfering with the discharge of gases from the cylinder. During the period of valve overlap, it follows that the exhaust pressure must be less than the air charging pressure to ensure effective scavenging of the cylinder to remove residual gases and cooling purposes. It has been found in practice that if the period between the discharges of successive cylinders into a common manifold is less than about 240°, then interference will take place between the scavenging of one cylinder and the exhaust of the next. This means that engines with more than three cylinders must have more than one turbocharger or, as is more common, separate exhaust gas passages leading to the turbine nozzles.

The exhaust manifold system should be as small as possible in terms of pipe length and bores. The shorter the pipe length, the less likelihood there is of pulse reflections occurring during the scavenging period. The smaller the pipe bore, the greater the preservation of exhaust pulse energy, though too small a bore may increase the frictional losses of the high-velocity exhaust gas to more than offset the increased pulse energy. Sharp bends or sudden changes in the pipe cross-sectional area should be avoided wherever possible.

Two-stroke engines

Turbocharging and scavenging arrangements for two-stroke engines are now fairly standardized. All such engines are today uniflow scavenged, with air entering the cylinder through ports around the full circumference of the cylinder liner at the bottom of the piston stroke and exhausting via a single poppet-type valve in the centre of the cylinder cover. Up to four turbochargers operating on the constant pressure system may be installed, depending on the engine size, all mounted high on the engine side, outboard, and beneath the exhaust manifold.

This fosters an efficient configuration, with the scavenge air cooler and its associated water separator unit located immediately below the turbocharger and adjacent to the scavenge air space on the piston underside. (In some special cases, however, with smaller engines having a single turbocharger, the turbocharger and air cooler can be arranged at the end of the engine.)

Compared with four-stroke engines, the application of pressure charging to two-stroke engines is more complicated because, until a certain level of speed and power is reached, the turbo blower is not self-supporting. At low engine loads, there is insufficient energy in the exhaust gases to drive the turbo blower at the speed required for the necessary air-mass flow. In addition, the small piston movement during the through-scavenge period does nothing to assist the flow of air, as in the four-stroke engine. Accordingly, starting is made very difficult, and off-load running can be very inefficient; below certain loads, it may even be impossible.

A solution was found by having mechanical scavenge pumps driven from the engine arranged to operate in series with the turbo blowers. However, electrically driven auxiliary blowers are standard on modern engines to supplement the scavenge air delivery when the engine is operating below around 30% load.

Two-stroke engine turbocharging is achieved by two distinct methods, respectively, termed the 'constant pressure' and 'pulse' systems. It is the constant pressure system that is now used by all low-speed two-stroke engines.

For constant pressure operation, all cylinders exhaust into a common receiver, which tends to dampen out all the gas pulses to maintain an almost constant pressure. The advantage of this system is that it eliminates complicated multiple exhaust pipe arrangements and leads to higher turbine efficiencies and hence lower specific fuel consumptions. An additional advantage is that the lack of restriction, within reasonable limits, on exhaust pipe length permits greater flexibility in positioning the turbo blower relative to the engine. Typical positions are at either or both ends of the engine, at one side above the air manifold or on a flat adjacent to the engine.

The main disadvantage of the constant pressure system is the poor performance at part-load conditions, and, owing to the relatively large exhaust manifold, the system is insensitive to changes in engine operating conditions. The resultant delay in turbo blower acceleration, or deceleration, results in poor combustion during transition periods.

For operation under the pulse system, the acceptable minimum firing order separation for cylinders exhausting to a common manifold is about 120°.

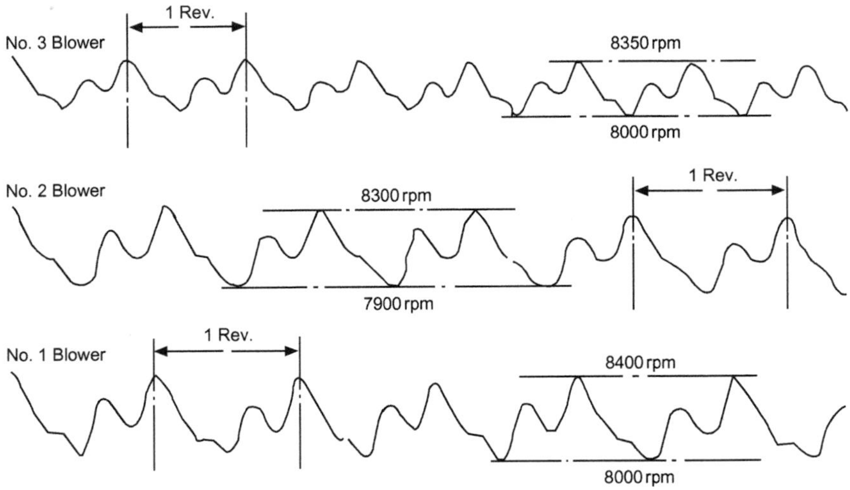

Fig. 10.2 Cyclic variations in turbocharger revolutions.

The sudden drop in manifold pressure, which follows each successive exhaust pulse, results in a greater pressure differential across the cylinder during the scavenging period than that obtained with the constant pressure system. This is a factor which makes for better scavenging.

Fig. 10.2 shows the variation in the rotational speed of a turbo blower during each working cycle of the engine; that is, one revolution. This diagram clearly shows the reality of the impulses given to the turbine wheel. The fluctuation in speed is about 5%. Each blower is coupled to two cylinders, with a crank spacing of 135° and 225°. It is now standard practice to fit charge air coolers to turbocharged two-stroke engines, the coolers being located between the turbochargers and the cylinders (Fig. 10.3).

Charge air cooling

The increased weight or density of air introduced into the cylinder by pressure charging enables a greater weight of fuel to be burned, and this in turn brings about an increase in power output. The increase in air density is, however, fractionally offset by the increase of air temperature resulting from adiabatic compression in the turbo blower, the amount of which is dependent on compressor efficiency.

This reduction of air density due to increased temperature implies a loss of potential power for a stated amount of pressure charging. For example, at

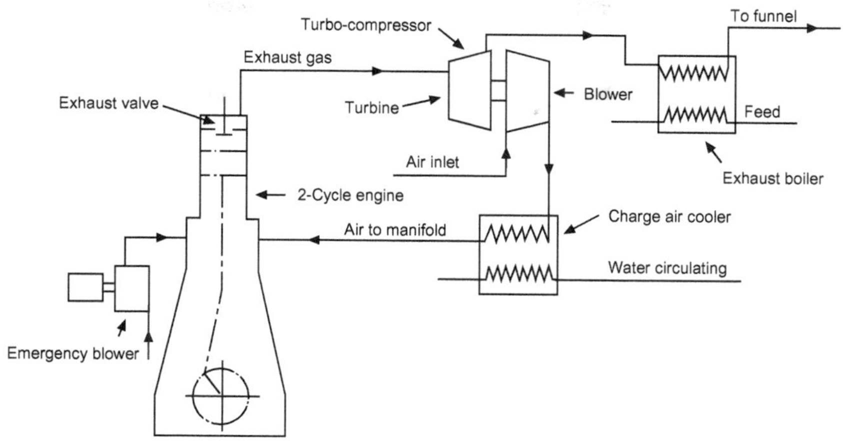

Fig. 10.3 Exhaust gas turbocharging system with charge-air cooling.

a charge air pressure of, say, 0.35 bar, the temperature rise is of the order of 33°C—equivalent to a 10% reduction in charge air density. As the amount of pressure charging is increased, the effect of turbo blower temperature rise becomes more pronounced. Thus, for a charge air pressure of 0.7 bar, the temperature rise is 60°C, which is equivalent to a reduction of 17% in the charge air density.

Much of this potential loss can be recovered by the use of charge air coolers. For moderate amounts of pressure charging, cooling of the charge air is not worthwhile, but especially for two-stroke engines, it is an advantage to fit charge air coolers, which are standard on all makes of two-stroke and most medium-speed four-stroke engines.

Charge air cooling has a double effect on engine performance. By increasing the charge air density, it thereby increases the weight of the air-flowing into the cylinders, and by lowering the air temperature it reduces the maximum cylinder pressure, the exhaust temperature, and the engine thermal loading. The increased power is obtained without loss and, in fact, with an improvement in fuel economy. It is important that charge air coolers should be designed for the low-pressure drop on the air side; otherwise, to obtain the required air pressure the turbo blower speed must be increased.

The most common type of cooler is the water-cooled design with finned tubes in a casing carrying seawater over which the air passes. To ensure satisfactory effectiveness and a minimum pressure drop on the charge air side and on the waterside, the coolers are designed for airspeeds of around 11 m/s and water speeds in the tubes of 0.75 m/s. Charge air cooler effectiveness is defined as the ratio of charge air temperature drop to available temperature drop between air inlet temperature and cooling water inlet temperature. This ratio is approximately 0.8.

Scavenging

It is essential that each cylinder should be adequately scavenged of gas before a fresh charge of air is compressed; otherwise, the fresh charge is contaminated by residual exhaust gases from the previous cycle. Furthermore, the cycle temperature will be unnecessarily high if the air charge is heated by mixing with residual gases and by contact with hot cylinders and pistons.

In the exhaust turbocharged engine, the necessary scavenging is obtained by providing a satisfactory pressure difference between the air manifold and the exhaust manifold. The airflow through the cylinder during the overlap period has a valuable cooling effect; it helps to increase the volumetric efficiency and to ensure a low cycle temperature. The relatively cooler exhaust also allows a higher engine output to be obtained before the exhaust temperature imposes a limitation on the satisfactory operation of the turbine blades.

In two-stroke engines, the exhaust/scavenge overlap is necessarily limited by the engine design characteristics. In Fig. 10.4, a comparison of the exhaust and scavenge events for poppet valve engines and opposed-piston engines is given. In the poppet valve engine, the camshaft lost motion coupling enables the exhaust preopening angle to be 52° ahead and astern. In the

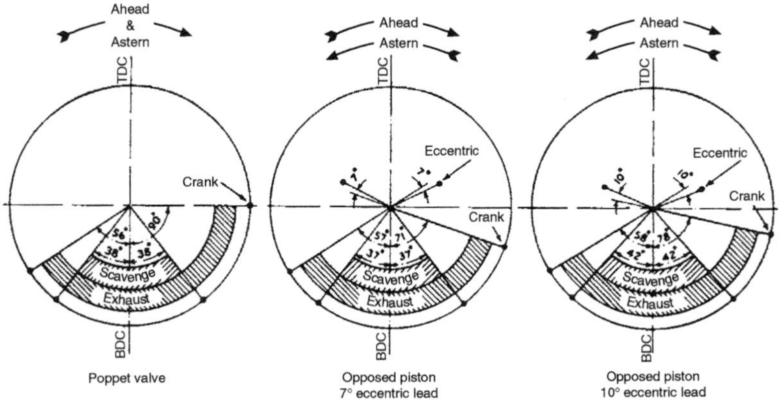

Fig. 10.4 Single-acting two-stroke engine timing.

opposed-piston engine, the exhaust preopening angle is only 34° ahead and 20° astern. Against this, however, the rate of port opening in opposed-piston engines is quicker than that in poppet valve engines. It should be noted, however, that opposed-piston low-speed engines are no longer in production, and poppet valves are used in the majority of new designs.

In the four-stroke engine, the substantial increase in power per cylinder obtained from turbocharging is achieved without the increase of cylinder temperature. In the two-stroke engine, the augmented cylinder loading—if any—is without significance.

A strange fact is that the engine exhaust gas raises the blower air to a pressure level greater than the mean pressure of the exhaust gas itself. This is because of the utilization of the kinetic energy of the exhaust gas leaving the cylinder and the energy of the heat drop as the gas passes through the turbine. In Fig. 10.5, the blower pressure gradient A exceeds the turbine pressure gradient G by the amount of the scavenge gradient. The design of the engine exhaust pipe system can have an important influence on the performance of turbo blowers.

The test results of turbocharged engines of both two- and four-stroke types will show that there is an increase in the temperature of the exhaust gas between the cylinder exhaust branch and the turbine inlet branch, the

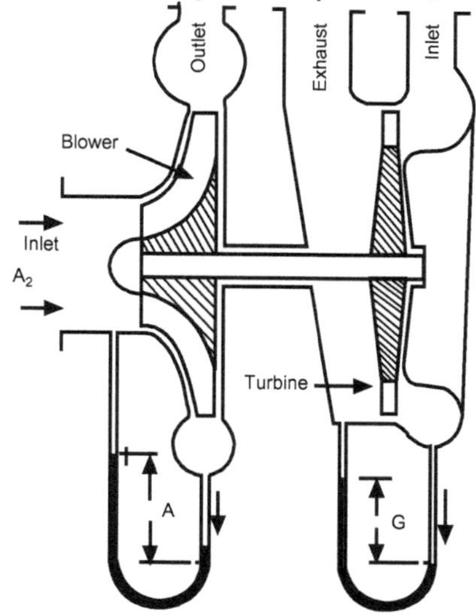

Fig. 10.5 Scavenge gradient.

rise being sometimes as much as 95°C. The reason for this apparent anomaly is that the kinetic energy of the hot gas leaving the cylinder is converted, in part, into additional heat energy as it adiabatically compresses the column of gas ahead of it until, at the turbine inlet, the temperature exceeds that at the cylinder branch. At the turbine some of the heat energy is converted into horsepower, lowering the gas temperature somewhat, with the gas passing out of the turbine to the atmosphere or to a waste heat recovery boiler for further conversion to energy. Though the last thing to emerge from the exhaust parts or valves is a slug of cold scavenging air which can have a cooling effect on the recording thermometer, adiabatic compression can still be accepted as the chief cause of the temperature rise.

Matching of turbo blowers

The correct matching of a turbo blower to an engine is extremely important. With correct matching, the engine operating point should be close to optimum efficiency, as shown by the blower characteristic curve (Fig. 10.6). In this diagram, the ordinates indicate pressure ratio; the abscissae show blower capacity.

During the initial shop trials of a new engine, with the turbocharger recommended by the manufacturer, test data are recorded and analyzed. If the turbo blower is correctly matched, nothing more needs to be done. Should the matching be incorrect, however, the turbo blower will supply charge air at either too low or too high a pressure, or surging may occur at the blower. Mismatching can usually be corrected by a change of turbine capacity and/or blower diffuser.

Compressors can be designed that maintain high efficiency at a constant pressure ratio over a wide range of air mass flows by providing alternate forms of diffuser for any one design of the impeller. This range can be further extended by the use of different impeller designs, each with its own set of diffusers, within a given frame size of turbo blower.

Turbocharger surging

Surging is a common hazard which can be initiated by many events that result in a shifting of the turbocharger operating point towards a surge line. Among the possible reasons for surging is insufficient engine room ventilation; clogging of the turbocharger intake silencer, compressor, scavenge air cooler, turbine gas inlet protection grid, turbine nozzle ring, and exhaust

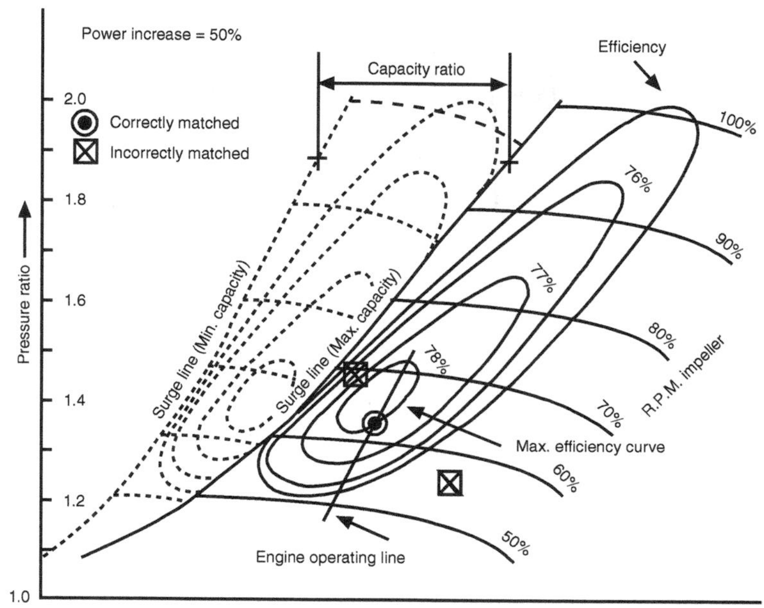

Fig. 10.6 Turbo blower characteristics.

gas economizer and silencer; deposits on the turbocharger turbine blades; damage to scavenge air flaps; damage or wear of fuel injection valves; excessive lubricating oil feed rate; and wear of turbocharger components such as nozzle ring, turbine blades, cover ring, or shroud ring. Careful attention to engine maintenance and turbocharger cleaning is thus the best prevention against surging.

Too low an air-mass flow at a given speed, or pressure ratio, will cause the blower to surge, while too high a mass flow causes the blower to choke, resulting in the loss of pressure ratio and efficiency at a given speed. The blower impeller, as it rotates, accelerates the airflow through the impeller, and the air leaves the blower with a velocity that is convertible into pressure at the diffuser. If, for any reason, the rate of airflow decreases, then its velocity at the blower discharge will also decrease; thus there will come a time when the air pressure that has been generated in the turbo blower will fall below the delivery pressure. There will then occur a sudden breakdown of air delivery, followed immediately by a backward wave of air through the blower, which will continue until the delivery resistance has decreased sufficiently for air discharge to be resumed. 'Surging' is the periodic breakdown of air delivery.

In the lower speed ranges surging is manifested variously as humming, snorting, and howling. If its incidence is limited to spells of short duration, it may be harmless and bearable. In the higher speed ranges, however, prolonged surging may cause damage to the blower, as well as being most annoying to engine room personnel. Close attention to the surging limit is always necessary for the design and arrangement of blowers.

If one cylinder of a two-stroke engine should stop firing, or is cut out for mechanical reasons, when the engine is running above say 40%–50% of the engine load, it is possible that the turbo blower affected may begin to surge. This is easily recognized from the repeated changes in the pitch of the blower noise. In these circumstances, the engine revolutions should be reduced until the surging stops or until firing can be resumed in all cylinders.

Turbocharging systems

Four basic turbocharging arrangements are now used on marine diesel engines (Fig. 10.7):

Constant pressure method—All the exhaust pipes of the cylinders of an engine end in a large common gas manifold to reduce the pressure pulses to a minimum with a given loading. The turbine can be built with all the gas being admitted through one inlet, and therefore a high degree of efficiency is reached. For efficient operation, the pressure generated by the compressor should always be slightly higher than that of the exhaust gas after the cylinder.

Impulse turbocharging method—The exhaust gas flows in pulsating form into the pipes leading to the turbine from where it flows out in a

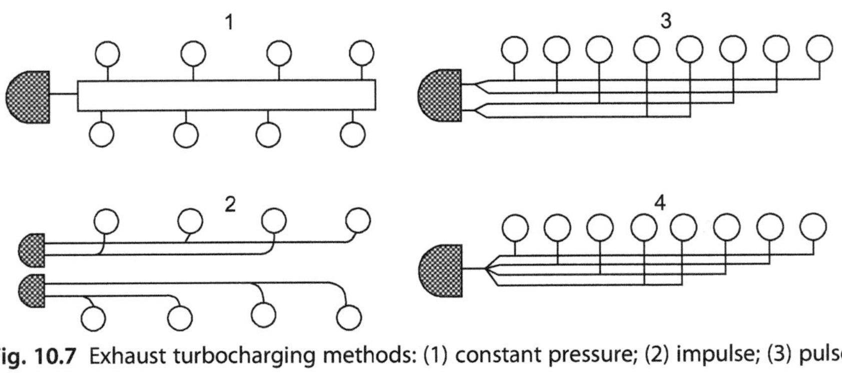

Fig. 10.7 Exhaust turbocharging methods: (1) constant pressure; (2) impulse; (3) pulse converter; and (4) multi-pulse system.

continuous stream. Each of the gas pulses from the separate cylinders is fed to a corresponding nozzle ring segment of the expansion turbine. By overlapping the opening times of inlet and exhaust valves after the pulse decay, efficient scavenging of the cylinders is possible.

Pulse converter method—Interactive interference limits the ways in which the normal impulse system can be connected to groups of cylinder exhaust pipes. However, with a pulse converter, such cylinder groups can be connected to a common ejector. This prevents return flows and has the effect of smoothing out the separate impulses. It also improves turbine admission, increases efficiency, and does not mechanically load the blading as much as normal impulse turbocharging.

Multi-pulse method—This is a further development of the pulse converter. In this case, a number of exhaust pipes feed into a common pulse converter together with a number of nozzles and a mixing pipe. With this form of construction, the pressure waves are fed through with practically no reflection because the turbine nozzle area is larger. The multi-pulse method makes a noticeable performance increase possible compared with the normal impulse turbocharging method.

Turbocharger construction

An example of turbocharger construction is provided by ABB Turbo Systems' (Brown Boveri) VTR design, which remained unchanged for many years (Fig. 10.8).

The diesel engine exhaust gases enter through the water-cooled gas inlet casing (50), expand in the nozzle ring (30), and supply energy to the shaft (20) by flowing through the blading (21). The gases exhaust to the open air through the gas outlet casing (60), which is also water cooled, and the exhaust piping. The charge air enters the compressor through an inlet stub (82) or the silencer filter (80). It is then compressed in the inducer and the impeller (25), flows through the diffuser (28), and is fed to the engine via the pressure stubs on the compressor casing (74).

Air and gas spaces are separated by the heat insulating bulkhead (70). In order to prevent exhaust gases from flowing into the balance channel (2) and the turbine side reservoir, barrier air is fed from the compressor to the turbine rotor labyrinth seal via channel X.

The rotor (20) has easily accessible bearings (32, 38) at both ends, which are supported in the casing with vibration damping spring elements. Either roller or plain bearings are used, but for the most common construction

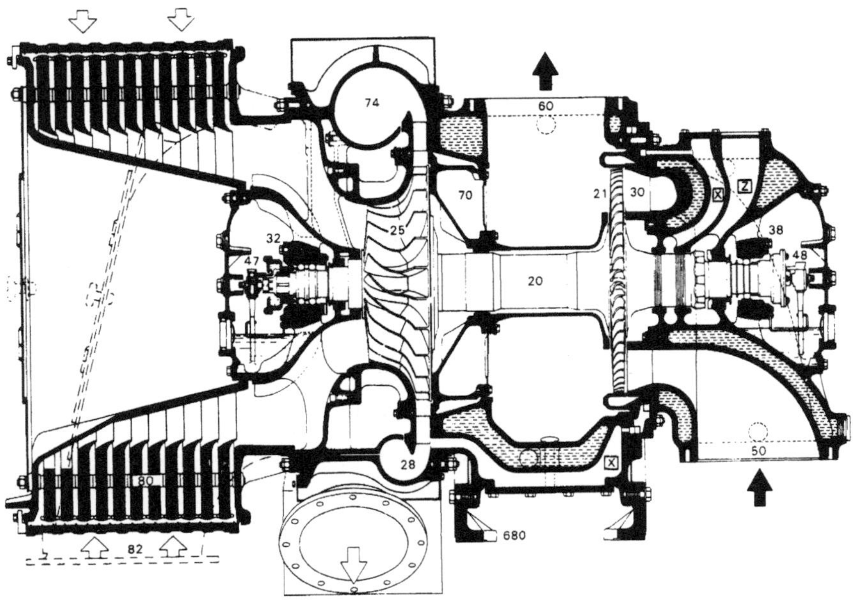

Fig. 10.8 Cross-section of Brown Boveri-type VTR 501 turbocharger.

using roller bearings a closed-loop lubrication system with an oil pump directly driven from the rotor is used (47, 48). These pumps are fitted with oil centrifuges to separate out the dirt in the lubricating oil. The bearing covers are each fitted with an oil filter, an oil drain opening, and an oil gauge glass. On models with plain bearings, where the quantity of oil required is large, these are fed from the main engine lubricating oil system.

A key feature of this VTR design was the modular construction to match a wide range of diesel engine types. The separate modules of the turbocharger, as shown in Fig. 10.8, are the silencer filter (80), the air inlet casing (82), the compressor housing (74), the gas outlet casing (60), the outlet casing feet (680), and the gas inlet casing (50). The fixing screws are placed so that the radial position of all the separate casings can be arranged in any position relative to the other.

Compressor wheels for MAN turbochargers are almost exclusively milled from an aluminium-forged blank. Cast aluminium alloys were once used as well but superseded as mechanical strength requirements increased. Titanium alloys may also be exploited for applications imposing high air outlet temperatures at high-pressure ratios. Radial turbine wheels are purchased as casting blanks of a nickel-base alloy. 'Built' rotors are used only for axial turbine wheels (the cast or forged individual blades are retained in the disc by a fir-tree foot). For smaller diameter ranges, progress in casting technology has yielded axial turbine wheel castings incorporating the blades (integral wheels).

Turbocharger performance and developments

A typical pressure–volume curve for a turbocharger is shown in Fig. 10.9, with reference to the operating characteristics of a two-stroke diesel engine. The operating curve runs almost parallel to the surge line. Every value of the engine output corresponds to a point on the curve, and this point, in turn, corresponds to a particular turbocharger speed which is derived automatically. There is, consequently, no need for any system of turbocharger speed control.

If, for example, at a specific turbocharger speed, the charge air pressure is lower than normal, the conclusion may be drawn that the compressor is contaminated. By spraying a certain amount of water into the compressor, the

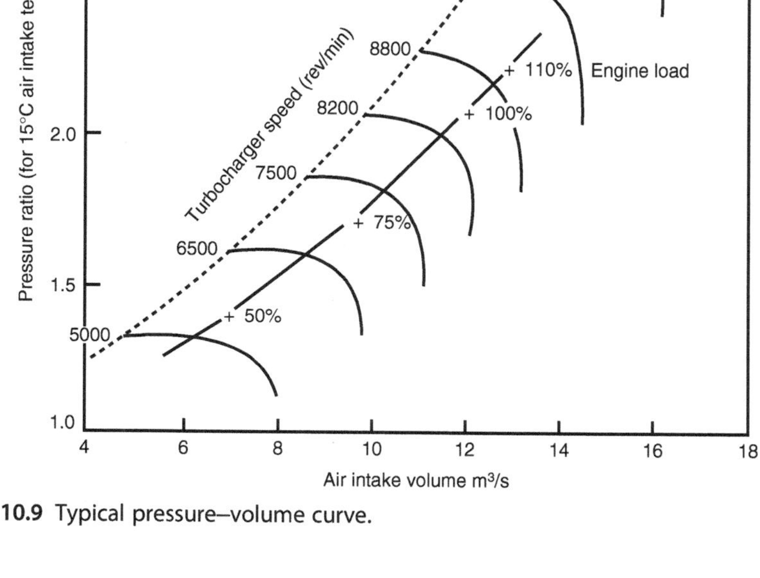

Fig. 10.9 Typical pressure–volume curve.

deposit can be removed, provided it has not already become too hard. A special duct for water injection is arranged on ABB's VTR turbochargers.

The operating curve and the behaviour of the turbocharger may vary, depending on whether constant pressure or pulse turbocharging is employed and also on how any volumetric (i.e. mechanically driven) compressors and turbochargers are connected together. By observing changes in behaviour, it is usually possible to deduce causes and prescribe the measures necessary to rectify the problems.

Turbocharger design and charging system refinements continue to underwrite advances in diesel engine fuel economy and specific output. Turbocharger developments also contribute to improved engine reliability (lower gas and component temperatures) and operational flexibility (e.g. better part-load performance). Rising engine mean effective pressure ratings call for an increase in charge air pressures, with turbocharger efficiencies remaining at least at the same level (Fig. 10.10).

Two-stroke engines typically operate at bmeps of up to 19.5 bar, with turbocharger efficiencies of just over 68% and pressure ratios of up to 4. Raising the bmep to 21 bar calls for an increase in turbocharger efficiency to 70% and a pressure ratio rise to 4.2. A higher bmep requires a higher pressure ratio to keep the air excess ratio at the same level; at the same time, the turbocharger efficiency must also be increased to maintain the air mass flow and cylinder mass purity at desired values.

A valuable development in recent years in matching engines and turbochargers was the 'low port' concept, which took advantage of the availability

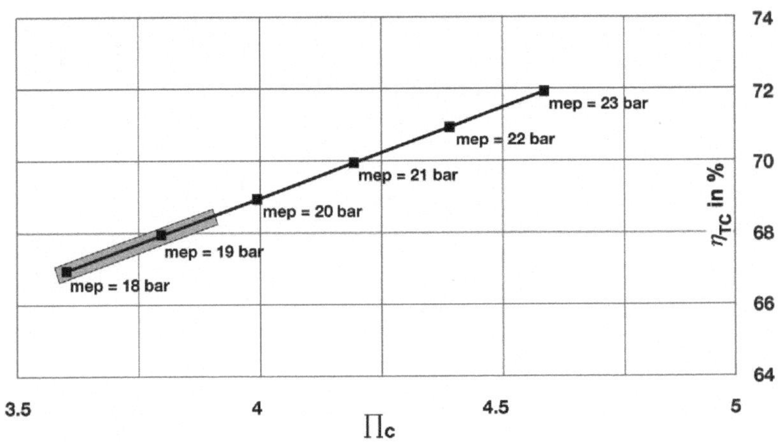

Fig. 10.10 Trends in engine bmep with respect to turbocharger overall efficiency and pressure ratio.

of new turbochargers with higher efficiencies (around 68% instead of the previous 65%). As the required scavenge air mass flow could be delivered using less exhaust gas energy, the height of the scavenge airports could be reduced and the opening of the exhaust valve delayed—lengthening the effective working expansion stroke and thus reducing specific fuel consumption. An important consideration was that this benefit could be obtained without raising key component temperatures. At the same time, the exhaust temperature was not reduced too low, so retaining the potential for waste heat recovery from an economizer.

Suitable turbocharger layouts are achieved through a well-established procedure. An appropriate turbocharger specification laid down during the engine design process is based on the calculation of certain parameters, notably, scavenge air pressure and full-load and part-load turbocharger efficiencies. The desired safety margin against surging—usually 15%—is also specified. In subsequent engine shop tests, the scavenge air pressure is adjusted, based on checking compliance with defined requirements such as the level of turbocharger efficiency, efficiency curve shape, and related tolerances. If necessary, the turbine nozzle ring is changed.

Tests are also made to check the engine's stability against turbocharging surging, which can be achieved in one of the three ways: the exhaust back pressure can be increased at full engine power; fuel can be cut off to one engine cylinder close to the turbocharger while the engine is running at part load, and the dynamometer load can be suddenly reduced by around 25% from the full load.

While satisfying increasing demands for higher efficiencies and charge air pressures, turbocharger designers must also address the requirement for compact configurations, reliability, and minimal maintenance: qualities fostered by designs with fewer components. All servicing and overhaul work should ideally match the intervals allowed for the major engine components. In addition, safe engine operation dictates a maximum width of the compressor map in order to avoid compressor surging (liable to occur due to the changes in operating conditions influenced by hull fouling, which has the same effect as speed reduced at constant torque).

Demand for higher charge air pressures necessitates higher circumferential velocities of the compressor wheel and raises the sound level; this in turn has promoted the development of more effective intake silencers for turbochargers. Turbocharger development also seeks to raise the specific flow rate for a given frame size to achieve as compact a unit as possible for cost efficiency, space saving, and ease of installation.

Turbocharger designers have pursued higher pressure ratios, overall thermodynamic efficiencies, and specific volumetric flow rates, as well as lower noise levels, from more simple and compact modular configurations with uncooled casings, inboard plain bearings lubricated directly from the engine oil circuit, and significantly fewer parts than earlier generations. Refinements have sought easier servicing and enhanced reliability and durability.

Certain trends in engine parameters can be identified that will dictate matching developments in turbocharger technology, Wärtsilä highlighting higher specific power output to achieve less weight per kilowatt, higher engine efficiency, lower exhaust emissions, improved engine reliability, and longer times between overhauls and lower manufacturing costs. These will call for further turbocharger development addressing higher scavenge air pressures, higher turbocharger efficiency, more compact dimensions, and competitive prices.

Bmep ratings increased from around 17 bar for two-stroke engines delivered in 1990 to 19 bar in 2000. Ratings of 22 bar are not unreasonable to expect by 2010. Such a level dictates an increased pressure ratio and efficiency from turbochargers because it is important that certain boundary conditions are maintained to sustain engine reliability, curb smoke levels during part-load running, and retain current specific fuel consumption rates. The wider use of variable turbine geometry (VTG), two-stage turbocharging, and turbochargers with auxiliary drives can also be anticipated.

Turbocharging and emissions/Miller cycle

High turbocharger efficiency contributes to reduced carbon dioxide emissions by improving engine efficiency. The same turbocharger, combined with a 'smart' turbocharging system that is able to guarantee an optimum air/fuel ratio under all conditions, would contribute even more to the control of soot emissions.

ABB Turbo Systems explains that the role that turbocharging can play in reducing NOx emissions is not easily recognized. This is because NOx is produced during combustion at very high temperatures—something which can hardly be influenced by changes in the turbocharging system since the flame temperature depends on local conditions in the cylinder and not on the global mean air/fuel ratio. Nevertheless, an improvement can be achieved through a joint development programme involving both the turbocharging system and the engine, aimed at reducing the temperatures of the working cycle in the cylinders.

The first idea was turbocooling in which the charge air is cooled in a process that makes use of a special turbocharger. If the precompressed air is further compressed in a second-stage compressor, then cooled and expanded through a turbine, very low temperatures can be obtained at the cylinder inlet. ABB Turbo Systems reports that the first evaluations revealed that the available turbocharger efficiencies for this process were not high enough for reasonable engine efficiencies.

The Miller cycle promises much better results. The idea is similar to that on which turbo-cooling is based. The charge air is compressed to a pressure higher than that needed for the engine cycle, but the filling of the cylinders is reduced by suitable timing of the inlet valve. Thus, the expansion of the air and the consequent cooling take place in the cylinders and not in a turbine. The Miller cycle was initially used to increase the power density of some engines. Reducing the temperature of the charge allows the power of a given engine to be increased without making any major changes to the cylinder unit. When the temperature is lower at the beginning of the cycle, the air density is increased without a change in pressure (the mechanical limit of the engine is shifted to a higher power). At the same time, the thermal load limit shifts due to the lower mean temperatures of the cycle.

Promising results were obtained on an engine in which the Miller cycle was used to reduce the cycle temperatures at constant power for a reduction in NOx formation during combustion: a 10% reduction at full load was achieved, while fuel consumption was improved by around 1%. This was mainly due to the fact that with the Miller cycle—at the same cylinder pressure level—the heat losses are reduced due to the air/fuel ratio being slightly higher and the temperatures lower.

The significantly higher boost pressure for the same engine output with the same air/fuel ratio has mitigated against the widespread use of the Miller cycle. If the maximum achievable boost pressure by the turbocharger is too low in relation to the desired mean effective pressure for the engine, the Miller cycle results in a significant derating. Even when the achievable boost pressure is high enough, the fuel consumption benefits of the Miller cycle are partially neutralized when the efficiency of the compressor and turbine of the turbocharger decreases too rapidly at high compression ratios. The practical implementation of the Miller cycle requires a turbocharger or turbocharging system (see the next section) capable of achieving very high compressor pressure ratios and offering excellent efficiency at high-pressure ratios.

Two-stage turbocharging

Lower NOx and carbon dioxide emissions are among the key drivers of marine diesel engine development. The former can be efficiently reduced by cooling down the combustion process using a Miller cycle, whereby inlet valve closure (IVC) timings are normally affected before BDC. (The Miller cycle was originally intended to increase the indicated mean effective pressure of petrol engines and based on shortening the compression stroke and lowering the charge temperature in the cylinder.)

In the Miller cycle, the combustion air is compressed to a much higher pressure than is needed to fill the cylinder for the desired air/fuel ratio. Closure of the inlet valve is timed so that just the right amount of air is sucked into the cylinders. The charge air thus expands, resulting in a lower temperature at the beginning of the cycle.

High degrees of early IVC timings, however, demand high boost pressures for which a powerful solution is a two-stage turbocharging system capable of delivering pressure ratios of up to 10. Engine efficiency is improved by such a system, which further results in lower carbon dioxide emissions. Higher engine efficiency is yielded by higher turbocharger efficiencies from two-stage turbocharging along with a more optimum division between compression and expansion strokes from the Miller cycle. Introducing a Miller cycle in combination with the two-stage turbocharging thus efficiently reduces both NOx and CO_2 emissions. (It is one of the few measures that can be applied to an internal combustion engine to simultaneously cut NOx emissions and fuel consumption.)

A major difference with the Miller cycle compared with a normal diesel cycle is the decreased compression due to the additional expansion of the cylinder load after IVC before BDC. The total compression work is thereby reduced compared with a standard cycle. Since the total expansion work remains the same, there is a positive effect on engine efficiency, which results in lower overall CO_2 emissions. This is also ensured by an increased air receiver pressure guaranteeing that the pressure inside the cylinder remains the same.

Another major difference is seen in the overall temperature level inside the cylinder. The additional expansion of the cylinder load after IVC before BDC ensures that a considerably lower temperature is attained in the cylinder. This difference is maintained and even increased during the high-pressure cycle. Lower NOx emissions, lower exhaust/component temperatures, and lower cooling losses (i.e. better efficiency) are among the positive effects of lower temperatures inside the cylinder.

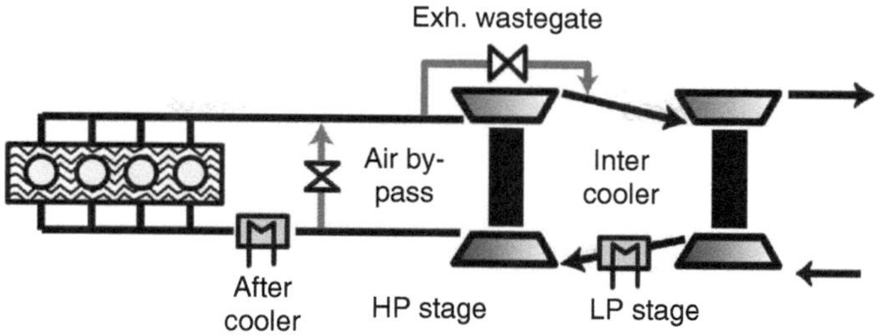

Fig. 10.11 Schematic diagram of the two-stage turbocharging system based on ABB TPS turbochargers used by Wärtsilä in conjunction with Miller cycle early inlet valve closure timings.

Intercooling between the low- and high-pressure turbocharger stages is a key factor because, by allowing an approximation of isothermal compression, it improves the air compression process and consequently reduces the required compression energy. The efficiency gain depends on the intercooling temperature and on the overall pressure ratio. Intercooling is not the only reason for the higher efficiency with two-stage turbocharging. Another is found on the gas side: the expansion energy available for the two turbines is higher than that with a single stage because the losses in the high-pressure stage increase the energy at the low stage inlet. Furthermore, the efficiencies of the compressor and the turbine can be generally higher due to the lower specific loading (Fig. 10.11).

Turbocharger cut out and sequential turbocharging

When slow steaming became a common ship operating strategy in the period after the economic crash of 2008, many ships especially those with large two-stroke engines and multiple turbochargers faced a problem with engines running at loads insufficient to keep the turbochargers operating efficiently.

Various means to overcome this were practiced from blanking off the turbocharger completely through the removal of rotating parts and bearings to fixing of valves in the gas and airlines. MAN developed a Turbocharger Cut Out system with swing gate valves that permit a frequent change between full- and part-load operation without manually having to instal and remove blanking plates. Other engine and turbocharger manufacturers also developed systems or advised on methods that could be safely employed.

MAN Turbocharger Cut Out.

ABB's response at the request of WinGD was the Flexible integrated Turbocharging System for Two-Stroke Engines (FiTS2) first shown in December 2017. The new sequential turbocharging system allows maximum fuel savings for two-stroke engines at the part and low load, and at the same time maintains the flexibility to go to full engine output immediately.

The concept enables two-stroke engine builders to offer more flexibility, increasing the potential for higher efficiency and lower operating costs. NOx emissions from engines fitted with FiTS2 will also remain within IMO Tier II limits. Additional abatement technologies like SCR or EGR can be used to comply with IMO Tier III NOx emissions and will not greatly affect the system while still delivering fuel saving benefits.

To optimize engine efficiency via improved turbocharging in low and part load, the engine runs in lower loads with only one turbocharger in operation, whereas at higher loads (typically above 50%–60% engine load) two turbochargers operate simultaneously. The same principle is applied for very large engines—with FiTS2 they will run with two turbochargers in lower loads and with all three turbochargers for higher load operation. The specially designed cut-off valves for the FiTS2 system are flow-optimized and integrated with the turbocharger casings. The valves can be operated rapidly and automatically under load, without interrupting the operation of the engine up to full load.

Cut out of one turbocharger for lower engine loads leads to higher scavenging air pressure, increasing the compression ratio, and firing pressure optimized by special tuning of FiTS2. The result is enhanced engine efficiency and lower specific fuel consumption while remaining compliant with emission regulations. The sequential turbocharging system enables fuel savings of up to 6 g/kWh (3%–5%) compared with fuel consumption levels typically achievable with conventional turbocharging systems on low-speed diesel engines.

The wide compressor maps of ABB's A100-L and A200-L turbocharger series allow switching under load without surging and there are no requirements for additional by-pass valves or pipework. An additional benefit comes from the possibility to switch off the electrically powered auxiliary blowers at 25% engine load, instead of around 35% load. This enables additional savings via reduced power consumption and lower blower maintenance costs due to significantly fewer operating hours.

Turbocharger cleaning

Fouling on the compressor wheel and diffuser as well as on the turbine nozzle ring and turbine blades degrades turbocharger performance. The result can be increased exhaust temperatures, higher engine component temperatures, and increased fuel consumption, or, in extreme cases, an increased danger of surging. Dirt deposited on the compressor side is the usual dust from the environment along with oil mist from the engine room. On the turbine side, the deposits are normally noncombustible components of the heavy fuel oil and surplus cylinder lubricating oil.

The solution is to clean the turbocharger components at regular intervals, usually carried out while the engine is running but at reduced speed. Dry cleaning of the turbine needs to be executed every 24–48 h using a granular medium, with turbine wet cleaning every 48–500 h using water alone. Wet cleaning of the compressor is performed every 25–100 h.

Burst tests

The rise in bmep ratings of engines in recent years has dictated a considerable increase in the power density of the turbocharger rotors. Burst tests are performed to ensure that even in the worst-case scenario (e.g. a fire in the exhaust system)—in which the compressor or turbine wheel can burst—all parts remain within the turbocharger casings and do not endanger personnel. The turbine casing design, which includes inner and outer burst protection rings, seeks optimum containment.

Turbocharger designers and manufacturers

The number of organizations producing marine equipment is gradually shrinking due to mergers and acquisitions. This is especially true for engine and turbocharger manufacturers. Both make use of licensees for much of their production although some still make all of their own products. The following pages include details of the offering of the main players in the turbocharger market. There are many smaller players beyond those mentioned who are equally capable of producing innovative and reliable products. Some, such as Rolls-Royce subsidiary MTU will produce for their own engines while others supply to a limited range of engine makers only.

ABB Turbocharging

Switzerland-based ABB Switzerland Ltd., Turbocharging (the company name was changed from ABB Turbo Systems in 2020), through its heritage company Brown Boveri and Cie (BBC), has strong links with the pioneering days of turbocharger development by Alfred Büchi and marine engine applications of turbocharging. Büchi applied for his first patent in 1905, but it was not until after World War I (during which the first turbochargers were applied to aero engines) that turbochargers entered seagoing service. The 1925-commissioned liners *Preussen* and *Hansestadt Danzig* were the first ships with exhaust gas turbocharged machinery, the output of the MAN 10-cylinder four-stroke engines being raised from 1250 kW (naturally aspirated) to 1840 kW by BBC turbochargers.

BBC's launch of the VTR. series in 1944 marked the end of an era in which each turbocharger was custom built for the application and made available a range of volume-produced standard turbochargers for serving engines with power outputs from 370 to 14,700 kW. Until then, turbochargers had been applied commercially to four-stroke engines, but it now became possible to turbocharge two-stroke engines equipped with engine-driven scavenging air pumps. In order to eliminate the scavenging pumps and reduce fuel consumption, however, the turbocharging system had to be a pulse system. This feature was successfully introduced in 1952 on a B&W engine powering the Danish tanker *Dorthe Maersk*, the first ship to be equipped with a turbocharged two-stroke engine. The two VTR 630 turbochargers boosted engine output by some 35% to 5520 kW.

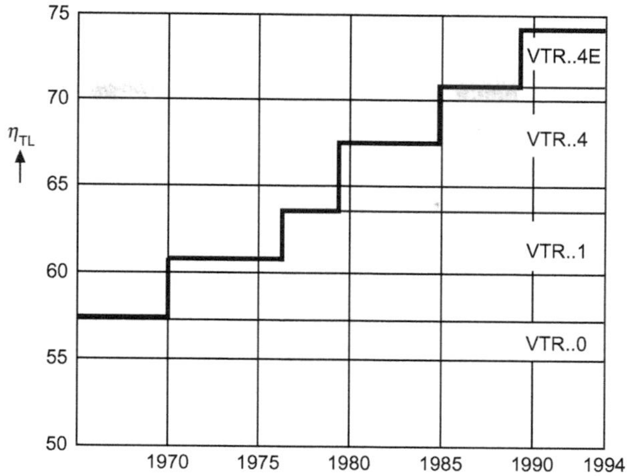

Fig. 10.12 The total efficiency of successive ABB VTR turbocharger generations, which has increased significantly.

The use of VTR-type turbochargers on two- and four-stroke engines proliferated rapidly from the 1950s, the series subsequently benefiting from continual design refinements to raise overall efficiencies and compressor pressure ratios (Fig. 10.12). A turbocharger efficiency factor of 74.7% was achieved by a VTR 714E model in 1989 (contrasting with the 50%–55% efficiency of BBC turbochargers in the early 1950s), and a VTR 304P model attained a pressure ratio of 5:1 in 1991. Apart from performance improvements, R&D has sought enhanced capability to serve engines burning low-grade fuels and operating under severe loading conditions, with reliability and minimal maintenance.

ABB Turbocharging's current programme embraces the following series:

- *VTR series*: designed for operation with low-speed two-stroke and medium-speed four-stroke heavy-duty engines with outputs from around 700 to 18,500 kW per turbocharger. Despite having been eclipsed in terms of technology this range remains popular with shipowners.
- *VTC series*: for engines with outputs from 1500 to 3200 kW per turbocharger; particularly favoured where compactness is required, which is achieved by using internal plain bearings lubricated by the engine lube oil system.
- *RR series*: a lightweight, low-cost turbocharger of robust design offering high efficiency, high gas inlet temperature capability, and good

compressor characteristics; mainly applied to high-speed engines and smaller medium-speed engines with ratings between 500 and 1800 kW per turbocharger.

- *TPS-D/E series*: for modern high-speed diesel engines, small medium-speed diesel and gas engines with power outputs ranging from 500 to 3000 kW.
- *TPS-F series*: a compact turbocharger for small medium-speed diesel engines, large high-speed diesel and gas engines in the 400 to 3300 kW power range; three series, respectively, offer full-load pressure ratios of up to 4.75, 5.0, and 5.2.
- *TPL-A series*: designed for modern four-stroke medium-to-large diesel and gas engines with outputs from 2500 to 12,500 kW.
- *TPL-B series*: designed primarily for large modern two-stroke diesel engines, six frame sizes covering an engine output range from 3000to 28,000 kW per turbocharger.
- *TPL-C series*: designed for advanced medium-speed four-stroke diesel and gas engines in the power range from 3000 to 10,000 kW per turbocharger.
- *A100 generation*: embraces three series (-H, -M, and -L) covering the requirements of high- and medium-speed diesel and gas engines (compressor ratios of up to 5.8) and of low-speed engines (pressure ratio of up to 4.7).
- *A200 generation*: development of A100 series embraces three series (-H, and -L). More compact for the same airflow or offering 30% higher flow rate for similar size turbocharger.

VTR..4 series turbochargers are fitted with single-stage axial turbines and radial compressors. The combination of various turbine blade heights with a large number of guide vane variants (nozzle rings)—as well as different compressor wheel widths and various corresponding diffusers—fosters optimum matching of engines and turbocharger operating characteristics.

External bearings result in lower bearing forces and allow the use of self-lubricating antifriction bearings. Sleeve bearings with external lubrication (engine lubrication) are available as an alternative for larger turbocharger types. Any imbalance of the rotor shaft caused by the contamination of foreign particles is absorbed by damping spring assemblies in the bearings, the lower bearing forces associated with external bearings being an advantage in this context. External bearings are also accessible for servicing and can be removed without dismantling the compressor wheel, and the rotor shafts can be removed without dismantling the gas pipes.

ABB Turbocharging recommends the use of self-lubricating antifriction bearings whose reduced friction promotes higher mechanical efficiency and other advantages when starting up the engine and during manoeuvring. Another benefit cited is the ability of this type of bearing to withstand short-term oil failure. Oil centrifuges in the closed lubrication system of the antifriction bearings separate out any dirt particles and ensure constant lubrication with pure lube oil, even in an inclined position. The turbine oil used has a positive effect on service life. Sleeve bearings with external lubrication are available, but these normally involve substantially higher costs, particularly for large turbochargers. Separate oil filtering systems and a standby tank for emergency lubrication must be provided.

The gas inlet, gas outlet, and compressor casings of the VTR turbocharger are split vertically and bolted together. The three casings and the supports can be rotated with respect to one another in increments of 15° or 30° to give the engine designer freedom in mounting the turbocharger to the engine. A large selection of gas inlet casing variants is available with different numbers and arrangements of inlets. Flexibility in the matching of the turbocharger to various charging systems, engine types (vee or inline), and numbers of cylinders is thereby fostered.

Uncooled casings are available as an alternative to the water-cooled casing on VTR..4 turbochargers for applications where heat recovery is paramount (Fig. 10.13). All the gas ducts are entirely uncooled and not in contact with cooling water at any point, making the greatest possible amount of exhaust heat available for further exploitation (e.g. steam generation for electricity production or ship's services).

Functional reliability is fostered by arranging for the bearing housing on the turbine end to be cooled with a small amount of water, helping to keep the lubricating oil temperature low. The simultaneous cooling of the jacket of the gas outlet casing makes it possible for the temperature of the entire surface of the casing to be kept within the limits stipulated by classification societies for the prevention of fire and protection against accidental contact.

Air to be compressed is drawn in either through air suction branches or filter silencers. An integral part of every silencer is a filter which intercepts coarse dirt particles from the intake air and counteracts compressor contamination. The filter can be cleaned during operation without having to dismantle the silencer. The compressor noise is reduced in line with international regulations by felt-covered, shaped plates.

Addressing the demands of low-speed two-stroke engine designers, ABB Turbo Systems introduced the VTR..4D turbocharger series with a

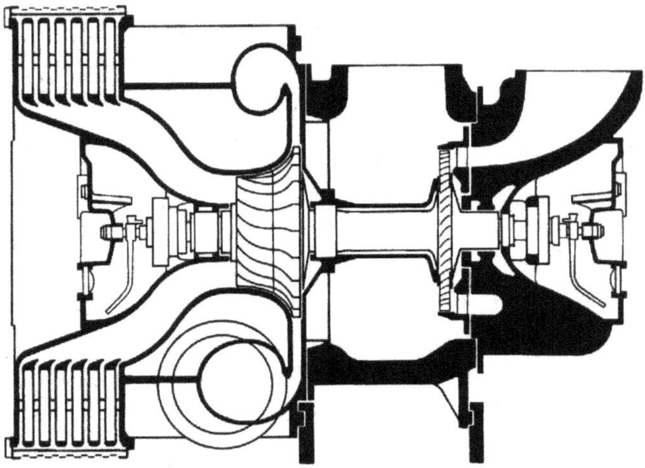

Section through water-cooled VTR turbocharger

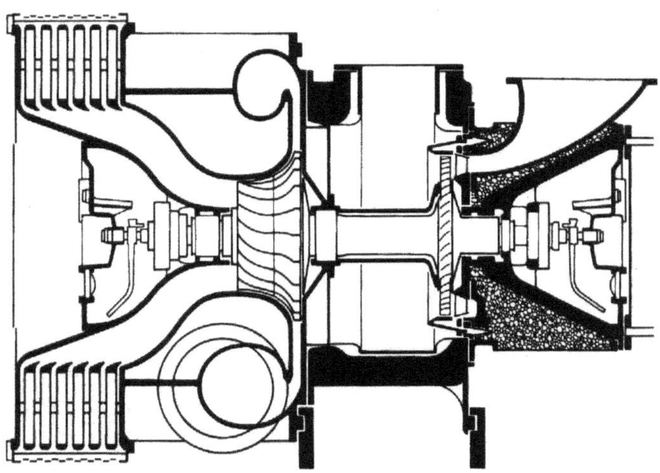

Section through uncooled VTR turbocharger

Fig. 10.13 Water-cooled and uncooled ABB VTR turbochargers.

compressor pressure ratio of up to 4. Three model sizes (VTR454D, VTR564D, and VTR714D) cover engine outputs from 5000 to 70,000 kW. Efficiency levels for contemporary low-speed engines can be met by the established VTR..4 and VTR..4E turbochargers (Figs 10.14 and 10.15). The significantly higher efficiency of the VTR..4E series allows surplus exhaust energy to be exploited in a power gas turbine connected mechanically to the engine crankshaft or a generator (see the section on Turbo Compound Systems). But neither series is able to meet fully the

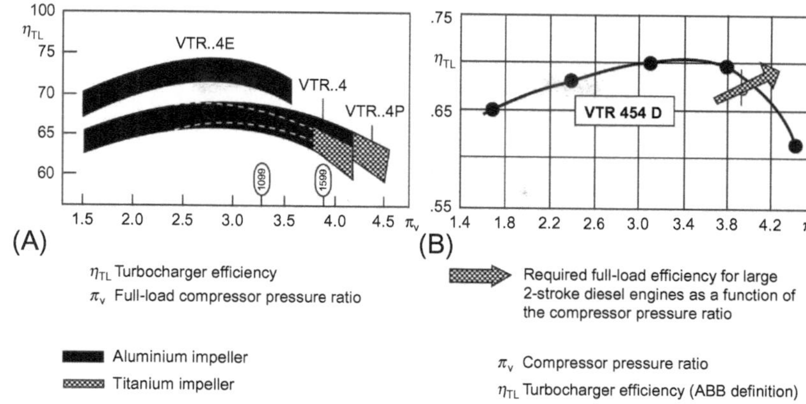

Fig. 10.14 (A) Turbocharger efficiency as a function of the compressor pressure ratio for today's ABB turbochargers (valid for applications on large two-stroke diesel engines). (B) Turbocharger efficiency as a function of the compressor pressure ratio for the VTR 454D along a typical two-stroke diesel engine operating line.

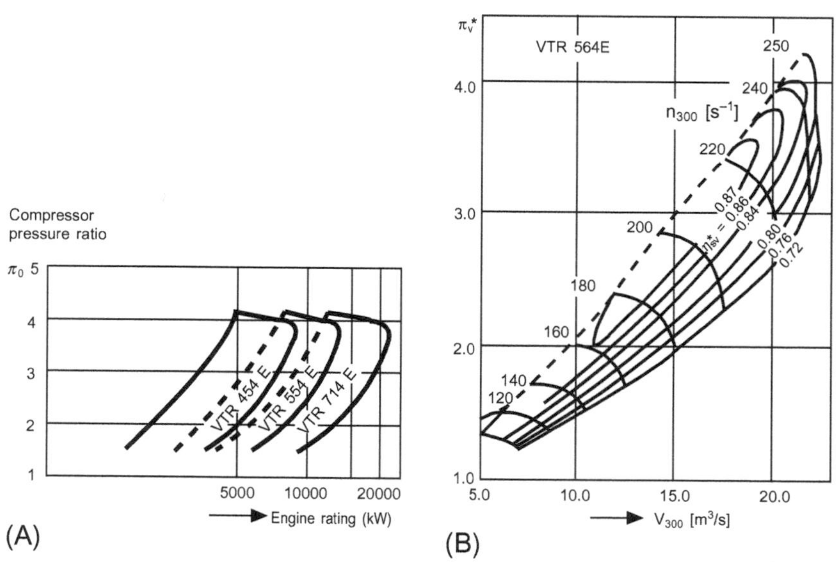

Fig. 10.15 (A) Range of application of VTR..4E turbochargers and (B) the performance map of a VTR 564E model.

demands of low-speed engines running with mean effective pressures of up to 20 bar, requiring compressor pressure ratios of 3.9 at full-load and turbocharger efficiencies of 67% and higher. The VTR..4 does not reach the required efficiency level, and the aluminium alloy compressor wheel of the VTR.4E limits the achievable pressure ratio. The VTR.P turbocharger, developed for four-stroke engines (see next page), does not represent a solution either, although it offers an efficiency improvement at higher pressure ratios compared with the VTR..4.

Development goals for the VTR..4D series were defined as a compressor pressure ratio of 4 while improving efficiency, the highest level of reliability despite an increase in mechanical load and the retention of key elements of the proven VTR design and its external dimensions, and hence facility for full interchangeability.

A turbocharger combining the pressure ratio of the VTR..4/4P and the high efficiency of the VTR..4E was targeted. The thermodynamic potential of existing compressors and turbine designs led to a strategy for developing a new series optimally blending well-proven components and matching the fitting dimensions of the VTR..4E. The high-pressure compressor of the VTR..4P turbocharger was selected for thermodynamic and economic reasons, the designer citing its:

- High-efficiency level for the whole range of competitive volume flow rate.
- The greater margin for further increases in mean effective pressure and overload margin.
- Use of aluminium as the impeller material instead of the significantly more expensive titanium; a full-load pressure ratio of 4.2 is underwritten.

Two compressors of different diameters are available to cover a competitive volume range at the required high-efficiency level within the same overall turbocharger envelope.

The turbine of the VTR..4E turbocharger was chosen for the VTR..4D series since it offers efficiency advances of up to around 3.5 points over the VTR..4/4P turbine. This is due to better recovery of kinetic energy at the turbine wheel outlet, resulting from an optimally adapted outlet diffuser in combination with a newly enlarged exhaust gas casing. The latest antifriction bearing technology was tapped to benefit the VTR..4D series, particularly on the turbine side. Improved mechanical reliability is promised from a fully machined steel cage with surface treatment, an optimized oil feed for better lubrication, optimized smaller rollers for less slip, and a larger cage guiding area.

Like other ABB turbochargers, the VTR..4D is designed to ensure that in the worst-case scenario—bursting of the impeller or turbine—all the parts remain within the casings. Casing material was changed from grey (laminar) cast iron to ductile (nodular) cast iron and the turbine enclosed in a protective shroud. Burst tests on the compressor and turbine confirmed the containment capability.

ABB Turbo Systems' VTR..4P series addresses the needs of engine builders seeking to raise the output of medium-speed designs. Such engines may run with meps of up to 24–25 bar, requiring a turbocharger compressor ratio of 3.8–4.0 and an overall turbocharger efficiency exceeding 60%. Meps of up to 28 bar, required by some four-stroke engine designs, dictate a compressor ratio of around 4.2–4.5 while maintaining or even improving the overall turbocharger efficiency. The VTR..4P series offers ratios of up to 5:1 from a single-stage turbocharging system whose simplicity is favoured over more complex two-stage arrangements.

Advanced flow calculations, stress analyses, CAD/CAM methods, and laboratory tests resulted in a one-piece compressor wheel with backward-curving main and splitter blades discharging through an airfoil diffuser. An aluminium alloy impeller will suit most needs, but titanium can be specified for extreme applications. The turbine disc is of a high-grade nickel alloy, with new configurations applied for the blade roots and damping wires.

The VTR..4P programme was evolved to cover the demands of engines with outputs ranging from around 1000 to 10,000 kW. The high-performance units are designed to be physically interchangeable with VTR..4 models for convenient retrofitting as well as for application to new engines. The turbocharger can be flanged directly onto existing air and exhaust gas ducting to improve engine performance by increasing the boost pressure. The pressure-optimized series complemented the established efficiency-optimized VTR..4E series in the ABB Turbo Systems' portfolio.

TPS series

Advances in small medium-speed and large high-speed diesel engine technology called for more compact turbochargers with increased pressure ratios at higher overall efficiency levels. ABB Turbo Systems addressed the market in the mid-1990s with the TPS series, designed to serve four-stroke engines with outputs from 500 to 3200 kW per turbocharger (Fig. 10.16).

To ensure the full range of pressure ratios required by such engines, the TPS..D/E turbocharger is available with two completely different compressor stages. One stage is designed for pressure ratios of up to 4.2 (TPS..D) and the other for pressure ratios of up to 4.7 (TPS..E). Thanks to mechanically optimized single-piece designs incorporating bore-less compressor wheels, pressure ratios of up to 4.5 can be reached for a wide range of applications using aluminium alloy as the compressor material. Both compressors feature

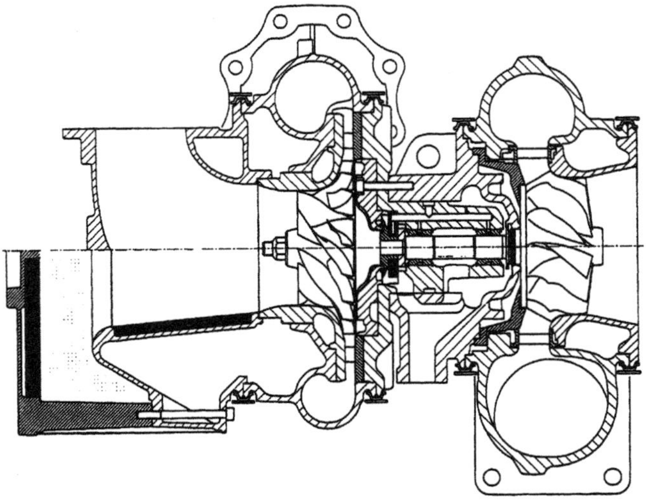

Fig. 10.16 ABB Turbo Systems' TPS turbocharger for small-speed, medium-speed, and large high-speed engines.

a single-piece aluminium alloy wheel with splitter-bladed impeller and backswept blades for high efficiency and a wide compressor map. Peak efficiencies of more than 84% are obtainable.

Experience with the RR-type turbocharger was tapped by the designers in creating the TPS series. The resulting configuration is modular and offers a high level of flexibility with a minimum of subassemblies. A significantly reduced number of parts are used in comparison with the RR type, which itself is a simple design. A modular configuration fosters the easy integration or fitting of options, such as a jet assist system. Special attention was paid during design and development to ease of maintenance and simple mounting of the turbocharger on the engine.

Criteria for turbine development included the ability to operate at gas temperatures of up to 750°C and yield a long lifetime on heavy fuel operation and under extreme load conditions. The requirement for a large volumetric range had to be met with a minimum number of parts. The considerable aerodynamic potential was offered by the turbine of the RR.1 turbocharger, which had also demonstrated excellent reliability and the highest efficiencies. Operation at pressure ratios far above today's levels was possible with the turbine for which a nozzle ring was developed to satisfy the goals for the TPS turbocharger. A new scroll casing and nozzle ring were designed using proven inverse calculation methods that ensure losses are kept very low and turbine blading is not too strongly excited.

Thermodynamic measurements showed that the turbine efficiency of the new design was as high as that of its predecessor. Blade vibration excitations are even lower.

Previously, stress distribution for the turbine wheel was optimized by trial and error: a geometry was defined, and the resulting stresses calculated, the geometry then being varied until the stresses lay at an acceptable level. Using the latest computer-aided tools for optimizing structures, the stress level was reduced by around 30% over the earlier version.

Turbocharger bearings must deliver a good load-carrying capability with respect to both static and dynamic forces, good stability, and minimal mechanical losses from a cost-effective design fostering modest maintenance. The TPS turbocharger bearings are lubricated directly from the engine lube oil system.

An increase in the turbocharger pressure ratio means a rise in the load exerted on the thrust bearing. A turbine wheel with a high back wall was chosen to counteract the main thrust of the compressor as far as possible. Extensive calculations and temperature measurements were performed on the thrust bearing, both in steady-state operation and with the compressor in surge mode, to optimize its dimensions and to achieve the lowest possible bearing losses without compromising on load-carrying capability.

The key design factor for the radial plain bearings is their stability characteristic. Considered in terms of this criterion, ABB Turbo Systems notes that a squeezed film damped multilobe plain bearing would be the best choice. But such a bearing has higher losses than that with freely rotating bushes. The latter, however, exhibits an inadequate stability characteristic. A new bearing design with rotating hydraulically braked multilobe bearings was therefore developed to blend the respective merits of the two types.

By providing a tangential lube oil supply, it is possible to reduce the speed at which the floating bushes rotate by around 20% compared with freely rotating bushes. The designer says that this measure has ensured stability which is effectively as good as that with the squeezed film damped multilobe plain bearings, and mechanical losses can be reduced by more than 20%.

As noted earlier, considerably higher boost pressure is needed to reduce the temperature in the cylinders during the Miller process. Turbochargers offering even higher compression pressure ratios are therefore dictated. For such engines, ABB developed three new TPS models covering an engine power range from 500 to 3300 kW. The first, designated the TPS-F33, was introduced in 2000/2001 and followed 2 years later by the TPS-F32 and in 2004 by the TPS-F31. Based on the TPS-D/E

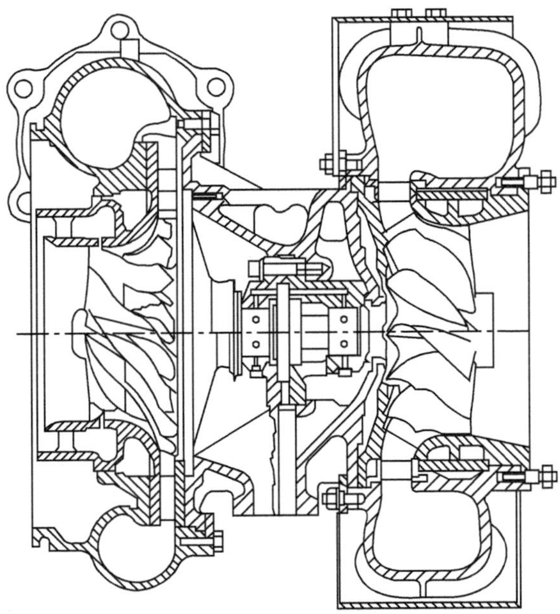

Fig. 10.17 Cross-section of ABB TPS57-F33 turbocharger.

platform—with the same outline dimensions fostering interchangeability and engine upgrades—they achieve full-load pressure ratios of up to 4.75, 5.0, and 5.2, respectively, using an aluminium alloy compressor wheel. The increased performance was sought while retaining high efficiency, reliability, long lifetime, and ease of maintenance (Fig. 10.17). The new compressors developed for the TPS-F series underwrote a 15% increase in the flow rates that could be covered by a given impeller size.

A recent addition to the TPS-F family extended the application range down to small high-speed and medium-speed heavy fuel-burning engines with ratings from 400 to 750kW. This new TPS44-F model is the fifth and the smallest member of the compact series. Three radial compressor stages support operator efforts to reduce emissions and raise fuel economy while gaining higher powers from their engines. The TPS44-F31 and F32 versions were introduced to the market first, offering full-load pressure ratios of up to 4.8; a TPS44-F33 variant delivers an enlarged volume flow.

The TPS-F was the first ABB turbocharger to feature recirculation technology: a bleed slot around the compressor wheel, which, by improving the flow field, increases the surge margin. The effect of the slot is to enlarge the map width without compromising the compressor's high efficiency.

Variable turbine geometry

In the mid-1990s, developments in the diesel and gas engine markets led to a TPS turbocharger version with VTG. One reason was the increasing popularity of single-pipe exhaust systems for diesel engines. When conventional turbochargers are used with these systems, part-load operation tends to be difficult, load response is poor and smoke, and particle emissions can be high. Gas engine performance had also progressed impressively due to increased efficiency and bmeps, high-altitude capability, and controlled air-to-fuel ratios.

It was not possible, however, to simply use conventional turbochargers with these engines. Solutions ranged from installing a wastegate or throttle mechanism to the special matching of the turbocharger, but each had its drawbacks. An 'adjustable' turbocharger was considered the ideal solution for the candidate diesel and gas engines. Apart from eliminating the losses associated with a wastegate, a turbocharger equipped with VTG is more flexible in applications with changing operating or ambient conditions.

Precise control of the air-to-fuel ratio (so-called Lamda regulation) is achieved with an innovative nozzle ring that enables the effective turbine area to be varied without any significant drop in turbine efficiency. The clearances for the movable nozzle blade are reduced almost to zero by springs that push the blades against the opposing casing wall. Numerous VTG turbochargers have been supplied for Bergen gas engines specified for land-based power plants.

ABB offers the VTG nozzle ring as an option for TPS turbochargers (frame sizes 57 through 61), TPL65-A turbochargers, and the new A100 series serving diesel and gas engines but plans wider application of the technology. Precise adjustment of engine operation to actual working requirements is facilitated over a wide range of ambient conditions. The part-load operation, load acceptance, and engine emission levels can all be optimized by controlled rotation of the movable nozzle ring vanes.

TPL series

A completely new generation turbocharger from ABB Turbo Systems—the TPL series—was launched in 1996 to meet the demands of the highest output low- and medium-speed engines, the designer having set the following development goals:

- High compressor pressure ratio for increased engine power output

- High turbocharger efficiency levels for reduced engine fuel consumption and lower exhaust gas temperatures
- High specific flow capacity, resulting in a compact design and low weight
- High reliability, long lifetime, and extended periods between overhauls
- Easy maintenance, even for machines operated under adverse conditions (such as those serving engines burning low-quality heavy fuel oils)

The uncooled design features an overall number of components reduced by a factor of six compared with the VTR series, contributing to easier maintenance and extended times between overhauls (see Fig. 10.18A and B).

A modular configuration provides the TPL series with the flexibility to meet the turbocharging requirements of two- and four-stroke engines. Two-stroke engines with outputs from 3000 to 28,000 kW per turbocharger are served by the TPL-B variant introduced in 1999, while the original TPL-A model addresses demands from four-stroke engines delivering from 2500 to 12,500 kW per turbocharger.

Despite differences in their thermodynamics and design, the TPL-A and TPL-B series share established features, for example, the basic modular TPL concept and the central bearing housing with the complete bearing assembly. All TPL turbochargers consist of a central cartridge unit with the rotor assembly. The casings on the compressor side and the turbine side are connected by flanges, and all casings are uncooled. Complete dismantling of the turbocharger, including the nozzle ring, is possible from the 'cold side', leaving the exhaust pipes in place and untouched.

All TPL turbochargers feature the same bearing technology, with plain bearings selected to secure a long lifetime. The nonrotating, radial bearing bushes are suspended in a squeeze film damper, which increases the stability of the rotor while reducing the dynamic forces and hence the load on the radial bearings. The main thrust bearing comprises a free-floating disc between the rotating shaft and the stationary casing. This bearing concept halves the speed gradient in the axial gap so that the losses and the risk of wear are significantly reduced; thanks to this technology, the thrust bearing is also very tolerant to inclinations of the rotor.

Since two-stroke engines usually have a lubrication system driven by electric pumps rather than by the engine itself, an emergency oil supply has to be provided to ensure that the turbocharger runs out safely in the event of a black out. The TPL-B turbocharger can be optionally equipped with an integrated emergency oil tank mounted on top of the bearing casing. Since this system is based on gravity, it does not require any auxiliary support.

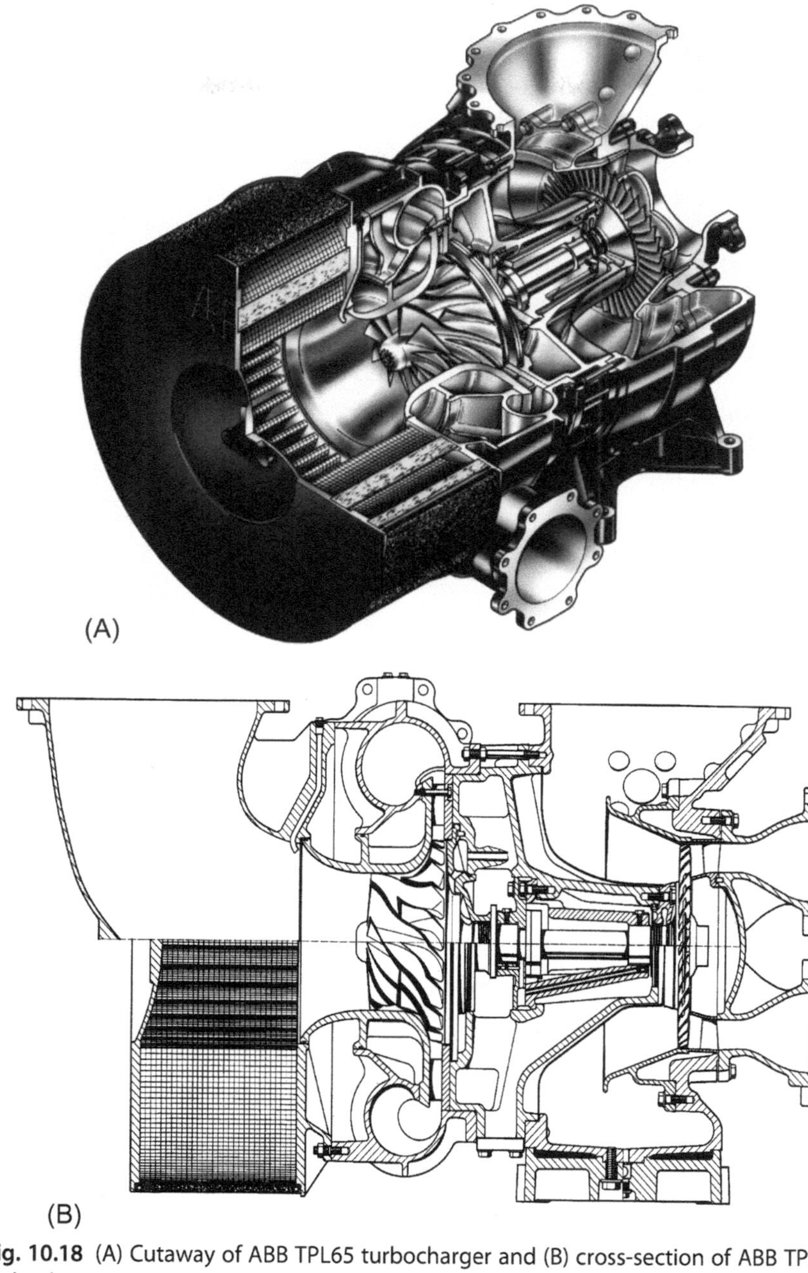

Fig. 10.18 (A) Cutaway of ABB TPL65 turbocharger and (B) cross-section of ABB TPL65 turbocharger.

Thermodynamic improvements from the TPL-B series derive from a new axial turbine family (the TV10), which was especially developed for turbocharging systems for two-stroke engines. Five turbine wheels and over 20 nozzle rings can be specified for each TPL-B model, these options allowing the turbocharger characteristic to be adapted to the requirements of the application within a wide range.

Two new radial compressor stages were developed for the TPL family. The TPL-B compressor features a single-piece aluminium alloy wheel with splitter-bladed impeller design and backswept blades for high efficiency and a wide compressor map. Peak efficiencies exceeding 87% are reportedly obtainable. Enlarged compressor diameters further increased the volume flow, thereby fostering optimized matching of the turbochargers to the engine application.

Market studies led to ABB's decision at the end of 1997 to develop a turbocharger to boost low-speed engines required to propel the increasingly larger container ships entering service, surpassing the output and performance of what was then the world's most powerful design, the TPL85-B. The resulting TPL91-B turbocharger—which extended the volume flow range of the established TPL-B series by over 20%—was designed to yield advanced thermodynamic performance combined with the highest reliability and lowest possible life-cycle costs. It was also optimized for integration and mounting as well as for convenient servicing, important factors because of the heavier individual elements and restricted engine room space for onboard overhauling.

Built at the end of 2000, the first prototype underwent extensive qualification tests on a burner test rig in the following year. Engine shop tests began in 2004 with three TPL91-B turbochargers boosting what was then the world's most powerful electronically controlled two-stroke engine, a 12-cylinder MAN B&W K98ME model.

An engine power range from 20,000 to 28,000 kW per turbocharger is covered by the TPL-91B (Fig. 10.19), whose key performance data and design features include:

- Maximum volume flow rates: 56 m^3/s at a compressor ratio of 3.5 and 59 m^3/s at a ratio of 4.0
- Compressor pressure ratio: 4.2
- Turbocharger efficiency $\geq$70%
- Compact frame and construction with the short rotor and integrated lubrication oil tank

Fig. 10.19 ABB's largest series, the TPL91 turbocharger.

- Minimized specific length, volume, and weight
- Inducer casing bleed system for wide compressor maps (no pulsating noise at part load)
- Design optimized for easy servicing and long times-between-overhauls.

Turbocharger efficiencies are higher than 70% for a wide range of pressure ratios in both full-load and part-load operation, fostering low fuel consumption and low engine-side component temperatures in those modes.

The conceptual design of the TPL-91B mirrors that of all the other models in the TPL-B series, with a central cartridge unit and a rotor assembly, which can be removed for servicing as an alternative to the standard method of dismantling from both ends. A number of innovative features, however, reflect the special area of application.

A two-stroke engine's lubrication system is usually driven by electric pumps rather than by the engine itself, so an emergency oil supply has to be provided to ensure that the turbocharger runs out safely if the system shuts down. TPL-B turbochargers normally have the emergency oil tank mounted on top of the bearing casing. To significantly reduce the outer dimensions of the turbocharger, however, and give engine builders as well as shipyards more freedom in arranging machinery, the TPL-91B has its emergency lube oil system integrated in the bearing casing.

ABB also developed a short rotor concept for the TPL-91B as a contribution to compactness. A new gas outlet configuration and TV11 turbine allowed the rotor to be made even 24 mm shorter than that of the TPL85-B turbocharger, the next largest one in the series. The complete TPL91-B turbocharger is in fact only marginally (around 300 mm) longer than the TPL85-B, despite its 20% higher volume flow rate.

ABB reports that a result of the very high compressor pressure ratios of modern turbochargers is that loading is closer and closer to their physical limits, with an obvious adverse effect on component lifetimes. Usually made of aluminium alloy, the compressor wheel is especially susceptible to ageing. Compressor cooling was introduced in the TPL91-B to lower the temperature in the wheel's outer area and hence slow the ageing process. This design feature allows a 3% increase in turbocharger speed for the same exchange interval of 100,000 h.

To simplify installation and provide better access for servicing, ABB designers also positioned the oil inlet and outlet sideways in the lower part of the bearing casing.

Large bore two-stroke engines are boosted by up to four turbochargers, depending on the number of cylinders. Pulsating compressor noise, typically registered at around 30%–50% engine load, can be an issue due to today's higher power outputs. ABB notes that although it has no critical influence on turbocharger operation, this noise phenomenon is unacceptable to many customers. Noise prevention was therefore a high priority in designing the TPL91-B since the turbocharger was intended for the highest-powered engines. To solve the problem, an inducer casing bleed system already applied successfully as an option in many TPL85-B installations was selected. Engine tests with the system (standard on the TPL91-B) demonstrated that it almost completely eliminated pulsating noise.

ABB's first TPL80-B12 turbocharger with a new CV14 compressor stage was tested in 2007 on a Mitsui-MAN B&W 7S60MC-C low-speed engine. In addition to this larger compressor stage, the turbocharger has an improved turbine with an increased stagger angle and a modified gas inlet casing with a larger gas flow area. The turbocharger design's extended capacity allows engines to be fitted either with smaller models or with fewer units for a given duty. Market introduction of these larger capacity versions of the TPL-B series was planned to start from mid-2007 with the TPL80-B, with the TPL77-B, TPL73-B, and TPL85-B (featuring a CV16 compressor stage) following.

TPL-C series

Building further on the TPL series, ABB developed the TPL-C for advanced medium-speed engines, delivering the first example in early 2004. By mid-2006 over 250 units were in service or on order, typically serving Wärtsilä 32C and 46F, MaK M43C, and HiMSEN diesel engines as well as Bergen gas engines.

TPL-C turbochargers are available in three sizes (TPL67-C, TPL71-C, and TPL76-C) covering an engine power range from 3000 to 10,000 kW per unit. Easier maintenance was sought from the compact, modular design, the lube oil inlet, and outlet located at the bottom of the turbocharger to facilitate servicing, especially in confined spaces. Installation flexibility was also pursued different casing types and a range of options promoting optimized matching to the specific engine application.

Turbocharger efficiencies of up to 70% are achieved, depending on the TPL-C model size and specification. Airflow recirculation for an improved surge margin, wide compressor maps, and reduced compressor noise (particularly at part load) are among the design characteristics.

Increasing adoption of Miller timing (see earlier) on new four-stroke engine designs—contributing to cleaner emissions by reducing the temperature in the cylinders—dictates very high boost pressures for a constant bmep. Such demands are addressed by the TPL-C series with compressor pressure ratios of up to 5.0 (and even 5.2 with optional compressor cooling).

Since the engines for which the TPL-C turbocharger is designed may use a pulse or quasi-constant pressure charging system, two dedicated turbine designs are offered. The turbine blading for the two smaller model sizes (TPL67-C and TPL71-C) for quasi-constant and pulse charging is stabilized by lacing wire. The turbine for the larger TPL76-C model in quasi-constant pressure systems features wide-chord blading.

TPL-C turbochargers can be delivered with an axial or radial gas inlet casing with either single- or multi-entry for optimum matching to the engine's charging system. Three large-diameter radial compressors cover a broad range of volume flow rates, the innovative design features including a bleed slot in the compressor wheel shroud for airflow recirculation. ABB explains that by improving the flow field, this measure increases the compressor surge margin and enlarges the map width without compromising efficiency. Strong secondary flows and separations are avoided, which also helps to reduce the compressor noise.

Compressor cooling, available as an option, extends the field of application for aluminium alloy wheels, offering an economic alternative to titanium impellers when the highest pressure ratios are required. Such cooling is also an option for customers requiring longer exchange intervals for these key components. Good mechanical behaviour with minimal vibration levels is promoted by a sturdy foot and newly designed filter silencer, which further contributes to a reduction in noise levels.

A100 series

ABB's A100 series was launched in 2007 to offer compressor pressure ratios of up to 5.8 for high- and medium-speed four-stroke diesel and gas engines, and up to 4.7 for low-speed two-stroke engines (Fig. 10.20). Such performance represented a significant step in the development of single-stage, high-efficiency, and high-pressure designs with aluminium compressor wheels.

Three A100 series (-H for high-speed engines, -M for medium-speed engines, and -L for low-speed engines) are designed to achieve the highest compressor pressure ratios for four-stroke engines as well as the highest efficiency at high-pressure ratios for two-stroke engines. The turbochargers can be exploited to raise the power density of engines and/or lower NOx emissions.

New compressor stages with further optimized impellers and new diffusers are incorporated. Compressor cooling is standard for four-stroke engine applications to underwrite the highest pressure ratios while retaining compressor replacement intervals comparable to standard turbochargers.

Fig. 10.20 ABB's new generation A100 series turbocharger.

Newly developed turbine stages and gas inlet and outlet casings are also featured.

The first A100 generation model, the A140-M/H for high- and medium-speed engines, was released in 2008 to offer full-load pressure ratios of up to 5.8 in continuous operation. The A135-M/H followed shortly afterwards followed by an A130 and an A145 extending the range to cover flows from below 1 to above 4 m^3/s.

A100-series frame sizes have the same outer dimensions as the TPS turbochargers, and also like the TPS design, have the oil inlet and outlet ducts integrated into the foot. This enables existing TPS-turbocharged engine platforms to adopt A100 solutions without major mounting changes.

A100 turbochargers are of modular construction with a minimized number of components and designed to enable matching to individual engine applications. Different casing materials are available for different turbine inlet temperatures.

Specific design and configuration features allow A100-M models to serve small medium-speed engines arranged for heavy fuel burning and pulse turbocharging. Since exhaust gas temperatures associated with such engines are usually lower than those with high-speed engines, the bearing casing of the A100-M models can be supplied with or without water cooling. Coated nozzle rings and multi-entry turbine inlet casings are among other options.

An ABB-developed cooling technology allowed the continued use of aluminium for the compressor wheels of the A100 series, even at very high-pressure ratios and without compromising the high reliability and long component exchange intervals expected from the material by operators. A change to more expensive titanium components was thereby avoided. Cooling with compressor air was shown by an extensive test programme to be the most efficient as well as the easiest and least costly solution for the engine builder to implement. The concept was already proven in the field, having been offered as an optional feature for the larger ABB TPL-C turbochargers for a number of years.

The A100 turbocharger casings reflect the much higher mechanical demands imposed on them. In designing the casings, ABB worked closely with engine builders to secure the same compactness as the TPS units as well as optimum mounting of the turbocharger on the engine console. The safety of the containment concept—vital in view of the significantly increased power density—was confirmed both numerically and experimentally on the test rig.

A stronger shaft (required because of the higher power transmission) was also a factor in designing the A100 bearing assembly, which was based on TPS turbocharger bearing technology. On the turbine side, the casing centring concept which had proved successful with the TPS-F turbocharger was retained.

Three completely new compressor stages, each with different wheel blading, allow the compressor volume flow range of the TPS-F turbochargers to be covered by the A100-M/H models with significantly higher pressure ratios. New high-pressure diffusers and compressor blading were developed in addition to innovative wheel cooling to facilitate the full-load pressure ratios of around 5.8 with a single-piece aluminium compressor wheel.

A range of compressor stages is available for every turbocharger frame size, allowing optimum matching in every application. The compressor map shows high efficiencies, excellent map widths, and more than adequate overspeed margins. An 80% compressor efficiency is attained on a typical generator line for a full-load pressure ratio of 5.8.

A new generation of mixed-flow turbines was developed for the A100 turbochargers in addition to the existing TPS mixed-flow turbine stage. A characteristic of the new turbine family is the larger operating range, allowing the high-pressure ratio potential of the new compressor stage to be exploited over an even wider range of applications. The turbine design is optimized in each specific volume flow range so that the individual stages exhibit higher turbine efficiencies than the current TPS turbine stages. Further development of the sealing technologies has reduced the bypass flows so that flow losses are also lower. In particular, this has allowed a substantial improvement in turbocharging performance at higher boost pressures.

ABB designed the A100-L programme for advanced two-stroke engine generations planned or under development, taking special account of the market's increasing focus on reducing emissions. The A100-L series was headed in 2008 by the A175-L model for boosting medium bore low-speed two-stroke engines designed to benefit from compressor pressure ratios of up to 4.7 at the highest turbocharger efficiencies. (ABB continues to supply its TPL-B family for engines requiring pressure ratios of up to 4.2.)

New compressor stages with further optimized impeller blading, new diffusers, and new air casings are incorporated. Turbines and gas chambers are also new, along with an integrated emergency lube oil tank originally developed for larger TPL-B turbochargers. VTG is available as an option.

In 2019, the smallest A100-L starts the series, from a capacity of below $10\,m^3/s$, with the largest A190-L capable of exceeding $50\,m^3/s$. The range includes the A160-L, A165-L, A170-L, A175-L, A180-L, A185-L, and A190-L.

A200 series

Some 5 years after the A100 series was launched in 2007, ABB announced the new A200 series at the SMM exhibition in Hamburg in late 2012. Based on the A100 but able to deliver around 30% more airflow, the new series permits for a more compact turbocharger to be fitted for the same or better flow rate. A major selling point of the new series was the fact that a smaller turbocharger generally means lower maintenance costs as spare parts are lower in cost.

At the beginning of 2019, the A200-L comprised four variants: A265-L, A270-L, A275-L, and A280-L. In July 2019, ABB announced two new models the A255-L and A260-L. Scaled down from the earlier A100-L and A200-L designs, their compact design allows installation flexibility in modern, space-optimized engine rooms.

They were developed to assist in improving small-bore two-stroke efficiency so as to meet increasingly stricter EEDI requirements for ships of between 10,000 and 40,000 dwt. They are designed to operate on all fuel types and on engines specified for both IMO Tier II or Tier III emissions levels and can be used on low-pressure dual-fuel engines to optimize combustion, improving engine efficiency and allowing for the reduction of GHG emissions. The high turbocharging efficiency provided by the new A255-L and A260-L turbochargers ensures that a high air-fuel ratio is maintained in high-load operation, and thus reducing the tendency for fast combustion which can adversely affect operation and emissions. The A255-L and A260-L turbochargers also facilitate the utilization of emission abatement technologies for NOx and SOx reduction.

Initially, the A200 series was available only in L versions for low-speed engines but in 2019, an H version was made available for lean-burn high-speed gas engines. As with the L versions, this is a development of the earlier A100 series. The A100-H models are suited to engines operating up to 22 bar bmep, whereas the A200-H takes this to 24 bar bmep.

The A200-H has further pushed the limits to pressure ratios up to 6.5 without compromising efficiencies. The A200-H is forecast to achieve above 69%; an improvement of 3% over the A100-H generation to ensure engine output and efficiencies can be increased above current single-stage limits.

For the A200-H models, aluminium alloy material has been retained for the compressor wheel and the product design is based on a requirement for minimum 40,000 h of life. The A200-H also has a new turbine stage, specifically developed to enable the compressor to be driven to the levels required to achieve 6.5–1 pressure ratios. All turbine side components, including the gas inlet, nozzle ring, turbine wheel, and diffuser, are designed to meet the highest efficiencies and pressure ratios with the necessary component lives being met despite the high rotor speeds required.

With many gas engines now operating in more variable and cyclic applications, there are increased opportunities to move away from more traditional hours-based maintenance regimes to a more technology-based approach, considering all aspects of speed, temperature, stop-start cycles, and ambient fluctuations seen by the turbocharger. The A200-H series will either utilize increased instrumentation or access to engine information, allowing life assessments and maintenance recommendations to be made based on installation-specific operational data.

Online analysis of this collected data will allow analytical assessment of the shaft and rotor components at every inspection interval, typically at 20,000 running hours for natural gas applications. Where confirmed, the installation will then be warranted as fully capable of operation to the next inspection point. At the 40,000-h and successive inspection points, there will then be the option, subject to condition and analytical assessment, for the extension of rotor lifetime, rather than the more rigid conventional approach of replacement of parts.

Through the assessment of the real operation condition of the A200-H, there is a significant potential to reduce turbocharger maintenance cost by extending the rotor lifetime and avoiding one rotor exchange event during the engine lifetime.

Safety and integrity are critical features of the new A200-H and all components are designed to operate in arduous conditions of service. In peak shaving applications, for example, fast engine starts are required so advanced analytical tools have been used to optimize thermal and mechanical stresses in all components, including casings. Combined with high numbers of start-stop cycles and high rotor speeds, containment safety against compressor or turbine burst is also a critical aspect and this has always been a top priority for ABB. Design experience, gained with the latest turbocharger types, has been used to identify and eliminate potential weak points at an early development stage.

The A200-H will be offered in three frame sizes: A230-H, A235-H, and A240-H. Two rotor sizes will be available for each frame, along with matching components, to allow a wide range of volume flows within each frame. Both rotors use a standard bearing housing arrangement and there is a choice of two bearing concepts for the same turbocharger.

An ABB patented bearing module design supports robust plain bearings, but a newly developed ball bearing system is also available. This offers increased mechanical efficiency, particularly at part load where power losses are as low as 15% of the equivalent plain bearing loss. This reduction enables more rapid turbocharger and engine start times to be achieved. Both bearing types operate with standard engine oil, but oil consumption is reduced significantly if the ball bearing system is adopted. Upgrading to the ball bearing system can be also be carried out with module exchange, with no other changes required.

The A200-H uses the same engine mounting arrangements as A100-H series, but a further configuration option is the availability of mounting either with a hydraulic fastening or a clamping nut system.

Power2

Between the A100 and A200 series in 2010, ABB developed a two-stage turbocharging system which it called Power2. As mentioned earlier in this chapter, emissions reduction and efficiency were the drivers in this development and in particular the impending NOx Code restrictions.

ABB's first Power2 systems were employed in shore-based power plants but in 2013, it developed Power2 800-M as the second generation of the technology. With pressures of up to 12 bar and turbocharging efficiencies over 75%, the second-generation Power2 leveraged the full potential of extreme Miller cycles to minimize NOx reduction and fuel consumption.

Power2's high efficiency comes as a result of its sophisticated design, which was always planned as a dedicated two-stage turbocharging solution rather than a marriage of components. Its compressor and turbine stages make a special contribution in this regard.

Two-stage turbocharging implies specific requirements having to be met by the thermodynamic components that differ from those of single-stage systems. Pressure ratios of compressors and turbines in low- and high-pressure turbochargers of two-stage systems are generally different from those of components in single-stage turbochargers. With respect to flow areas, differences in fluid density resulting from specific temperatures and pressures in the high- and low-pressure stages need to be considered. In addition, performance maps of compressors and turbines need to be matched perfectly in order to allow unrestricted engine operation over the entire load range.

The high-pressure turbocharger features two compressor stage variants, which are specified according to the desired overall pressure ratio. The compressor stage known from the first generation yields optimum performance for applications with pressure ratios up to $\pi_C = 10$. For pressure ratios up to $\pi_C = 12$, an entirely new high-pressure compressor stage design was developed. Both stages are optimized, focusing on the compressor map width and efficiency at their specific operating range. Both stages operate at an efficiency level of up to 84% at the design pressure ratio.

Axial turbines are used for both the low- and high-pressure stage. The turbine stage designs take into account the diverging needs of the high- and low-pressure sides. These result from different temperature levels and flow area requirements. The axial turbine design for the high-pressure turbocharger combines the advantages of high specific volume flow, excellent efficiency even at part load, and good acceleration behaviour due to low mass moment of inertia.

The requirements for the low-pressure turbine stage also differ significantly from those for single-stage turbines. The required range of specific volume flows in the case of the low-pressure turbine is shifted towards a lower level than for single-stage turbines. Consequently, this lower requirement has been exploited in the design of the new low-pressure turbine stage to gain higher stage efficiency.

In addition to the compressor and turbine design, the design framework around these components needed to be considered to achieve optimum system performance. Therefore, the design of the flow channels and subsystems was optimized specifically during the development of the turbochargers.

During the development of the new turbocharger generation, service friendliness was considered from the very beginning. The designated goal was to reduce the service time of the complete two-stage system to below the reference value of current single-stage turbochargers.

For the second-generation turbochargers, the proven cartridge concept has been enhanced in order to further reduce service time. The idea behind the extractable cartridge is a turbocharger which has an outer shell, consisting of compressor casing and gas casings, and a cartridge group which contains the entire interior of the turbocharger. In order to exchange the cartridge during service, only the air inlet casing together with the insert wall of the compressor needs to be removed. All other interfaces to the engine, such as oil connections, air outlet flange connection, gas inlet, and gas outlet flange connections, remain untouched. Furthermore, there is no need to remove the insulation of the turbocharger anymore, since only the flange connection of the air inlet needs to be accessed during service.

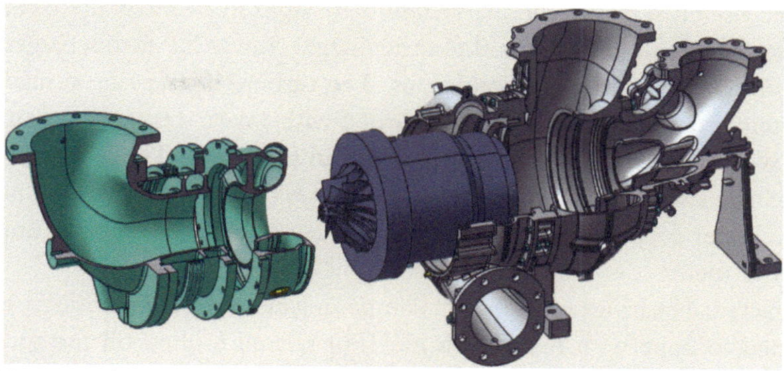

Servicing is made easier by making use of exchangeable cartridges.

ABB MXP series

Although most turbocharger development is directed at those used on main propulsion engines. It should not be forgotten that most ships also have auxiliary engines driving generators for electrical power. These engines must be turbocharged and, for all but the smallest genset, the engines must also comply with NOx Code requirements and the power will count towards that used in the EEDI calculations.

In 2016 at the CIMAC conference in Helsinki, ABB delivered a technical paper describing its first dedicated turbocharger for marine auxiliary engine applications which had been developed in cooperation with the Japanese shipbuilder IHI Corporation.

MXP has been designed for auxiliary engines with power output up to 2 MW, operating under HFO conditions. In this dedicated product, the complexity of parts required in turbochargers with broader usage has been simplified for the specific operational requirements of this market. These include ease of maintenance by the crew, based on condition, as well as optimized load response behaviour, and improved efficiency at part load, which both contribute to fuel savings.

A simplified service approach was also a significant focus in product development as MXP allows for condition-based maintenance, optimizing operational costs. As a 'user-friendly' turbocharger, it enables easy and fast service preparation and simplicity of replacing parts. Self-service maintenance by the ship's crew onboard additionally lowers costs while maximizing application availability and flexibility.

Because most marine auxiliary engines in larger ships are set up to run on HFO which can contaminate the turbocharger, the MXP turbocharger has been designed run in these conditions. The turbine nozzle ring is made of high-temperature resistant steel and treated with a wear-resistant material to cope with the highly erosive particles within the HFO. The design of the nozzle ring also features fewer, but larger, vanes compared to existing products to better cope with fouling and wear effects when the engine is running on HFO fuel.

The radial bearings and adjacent dampening elements have similarly been designed to cope with high levels of HFO related fouling on the turbine rotor and any consequent unbalancing. Its bearing casing is protected from hot exhaust gases by using a cast heat shield while the impellers have been designed to resist wear by minimizing the penetration of HFO residues and abrasive dirt particles into the turbine. The air filter silencer has also been made of anticorrosive sheet metal.

Particular attention has been put to the need for rotor balancing during service which has been minimized and can be avoided altogether once the rotor has been mechanically cleaned; a cleaned and brushed rotor will not require any balancing during an overhaul. Eliminating of rotor balancing is a key factor in simplifying maintenance by the ship's crew.

Fewer parts mean fewer steps in assembly and reassembly. The complete bearing unit incorporates the radial and axial bearing parts, the thruster ring, and the compressor side shaft sealing arrangement. The unit can be preassembled and stored on board as a spare part in case servicing is required. Another feature which facilitates servicing is the fact that all the screws within a particular frame size have the same thread dimension strength class and tightening torque. Maintenance can, therefore, be carried out using standard tools.

The maintenance and service work for an MXP turbocharger is simple and intuitive and is carried out using the 'ABB Ability Self-Maintenance for MXP' app, which has been specially developed for the MXP. This tool contains all operation manuals, animated instructions, and a wear-based evaluation of the turbocharger components and their replacement in the event of service.

The MXP range has been promoted at recent marine equipment exhibitions and is said to be available in three frame sizes with airflow rates between 0.6 and 3.5 m^3/h.

Components of MXP Auxiliary engine turbocharger.

Napier turbochargers

UK-based Napier Turbochargers, which became part of Wabtec, a US-based railway technology provider in February 2013, continues to offer a range of turbochargers for marine engines developing up to 12,000 kW.

Based on the designer's original cartridge construction, Napier 7-series turbochargers when introduced were suitable for engines with outputs from 2000 to 6500 kW per turbocharger. Developed from the company's earlier generation, these NA297, NA307, NA357, and NA397 models achieve pressure ratios of up to 5:1 and an overall efficiency exceeding 70%. Optimization for different applications and fuels is facilitated by alternative compressor options, and an innovative lip seal secures effective turbine end sealing at all operating loads.

Cost and complexity are reduced by the air-cooled operation, while the cartridge construction enhances on-site maintenance and cuts downtime. Service life is extended, and through-life costs reduced by a squeeze film bearing system.

Proven features from the 7 series, including cartridge construction and lip seal turbine sealing, were retained for the short-lived generation 8 series, whose development was stimulated by the need to offer turbochargers capable of yielding pressure ratios in excess of 5:1 and serving engines with outputs of up to 10,500 kW per turbocharger. Exploiting aerodynamic and mechanical advances, the new axial-turbine series comprises NA298, NA358, NA398, and NA498 models. A pressure ratio of 5.5:1 could be achieved in continuous operation with an overall efficiency of up to 70%, while reportedly maintaining a service life similar to the 7-series products.

Enhanced levels of component durability were also claimed, while the air-cooled design contributes to reduced overall weight. Inboard bearings were applied for the first time on a Napier turbocharger of this size, the hydrodynamic units securing stable rotor dynamic performance and underwriting a long life. The bearing arrangement facilitates the use of axial and radial exhaust gas flow into the turbocharger, extending the options for the engine builder.

Two separate compressor wheel designs are offered to provide differing characteristics for optimized efficiency in diverse applications. Other features include a circular exhaust flange (eliminating the need for transition pieces) and a mounting flange incorporating both the lube oil inlet and outlet. Another concept pioneered by Napier—cartridge construction—benefits serviceability of the 7 series and 8 series, this configuration allowing operators to remove and repair the rotor and bearings through a simple procedure without disturbing the exhaust connections. Good access is also provided to key functional areas such as the nozzle ring.

Titanium compressors, offering enhanced aerodynamic potential and mechanical properties, were initially planned to replace the aluminium components, but design refinements enabled the required performance to be attained with aluminium.

Smaller engine requirements are addressed by the Napier 047, 057, and 067 all-radial air-cooled turbochargers in the 7-series, which are suitable for engines developing outputs from 500 kW to around 1700 kW and yield pressure ratios of up to 5. Typical applications include the Wärtsilä 20 engine and other *c*.-200 mm bore genset engines. The 047 model features inboard bearings and is uncooled. In common with other turbochargers in the series, the bearings are designed to run on the engine's lube oil system and to yield an extremely long life, even when operating on contaminated lube oil. The 047 model incorporates 50% fewer parts than its predecessor, the Napier CO45. Cost savings result from using a reduced number of components, and maintenance is simplified, with no special tooling required or clearances to set. Simple clamps are used to hold the casings together and provide total indexability of connecting flange positions.

Earlier axial-flow turbine Napier 457 and 557 models, which serve engines from 4000 to 10,000 kW output, featured long-life outboard plain hydrodynamic bearings fed directly from the engine lube oil system and water-cooled casings.

High efficiency at all pressure ratios was sought from the 5- and 7-series through advanced compressor and turbine aerodynamic designs (Fig. 10.21).

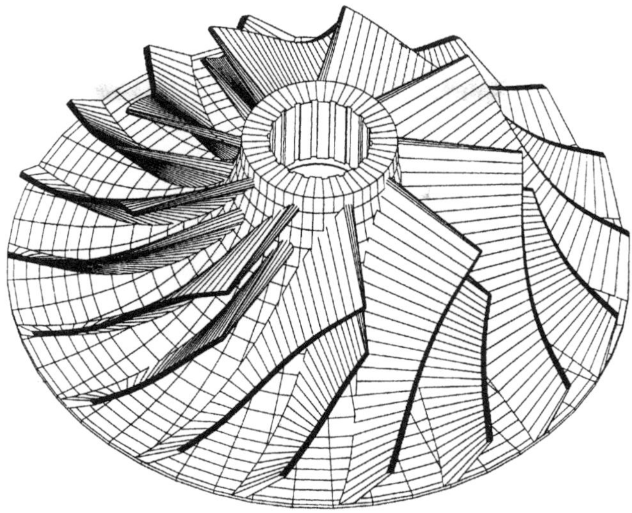

Fig. 10.21 A 3D CAD visualization of the Napier 457 turbocharger impeller, showing the one-piece intervened swept-back design with tip rake.

The compressor stage features a one-piece aluminium alloy wheel manufactured on five-axes milling machines. Vanes with sweepback and rake are specified to secure high performance, and a divergent circular arc (DCA) diffuser design yields maximum pressure recovery from a compact configuration. A high-efficiency fabricated nozzle with profiled blades serves the turbine section of the small and large axial-flow turbochargers.

Napier turbochargers exploit the heavy fuel bearing design to secure a long service life, even when operating with the poorest quality engine lubricating oil: a bearing life exceeding 20,000 h is reportedly common.

Along with most other turbocharger manufacturers, Napier needed to address the demand for higher pressure ratios and volume flows in order to meet the NOx Tier III requirements that came into effect in 2016. The company's response was the new NT1 series that was first announced in 2014 as a replacement for the 8 series. The series is comprised of three models intended for large bore four-stroke engines operating on any fuel type and with outputs ranging from 2 to 8 MW.

The models were labelled NT1-10A, NT1-12A, and NT1-14A covering engine outputs as shown below. At the time of their introduction, Napier claimed that the compression ratio of 6.0:1 for all three turbochargers were the highest on the market. The design process had made use of CFD to achieve the efficiency of the turbochargers which is ensured by a stress-resistant aluminium alloy compressor intended to keep costs down compared to more expensive materials such as titanium.

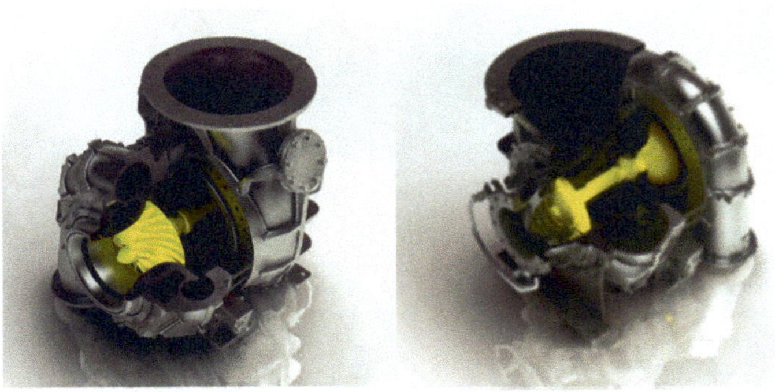

All three models have a wide compressor map thanks to antisurge technology allowing peak efficiency operation in different applications, with increased surge margin eliminating the risk of surge during operation. The turbine has a fabricated nozzle ring designed for high efficiency and resilience to thermal transients. The hydrodynamic journal bearings feature a squeeze film damper for vibration resilience. The casing has waste ports on gas inlet and outlets, rotatable casings, and a number of axial and radial gas inlet options to meet a wide range of engine and engine room requirements.

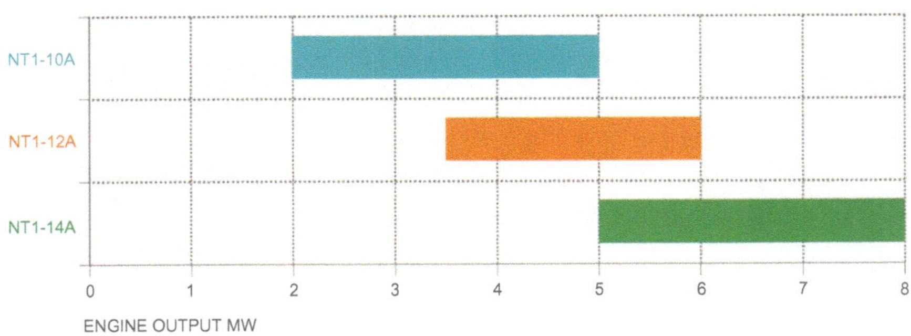

After the launch of the NT1 range, Napier indicated a larger variant, the NT1-16A was in the pipeline and would be suitable for engines up to 12MW. It does not feature in the production portfolio because late in 2016, Napier announced a new design of compressor wheel for the range which allowed a 15% greater airflow at the same pressure ratio.

The additional airflow from the revised LC compressor stage (the original compressor is referred to as L) can allow the NT1-14A to be matched to a 12 MW engine. In addition, fitting an LC compressor to the smallest model in the range would allow it to replace the middle variant. This would reduce costs and save as much as 880 kg in weight as well as being space saving. Both the L and LC compressors are available as options for the NT1 range. It is possible for the new compressor to be fitted as an upgrade although the turbocharger should have been matched to the engine operating profile based on the existing compressor, so performance benefits are not necessarily needed.

Developments and improvements made during the 8 and NT1 series have wherever possible been incorporated into the latest variants in the 7 series range which continues in production. Peak efficiency can now be as high as 84% in some applications.

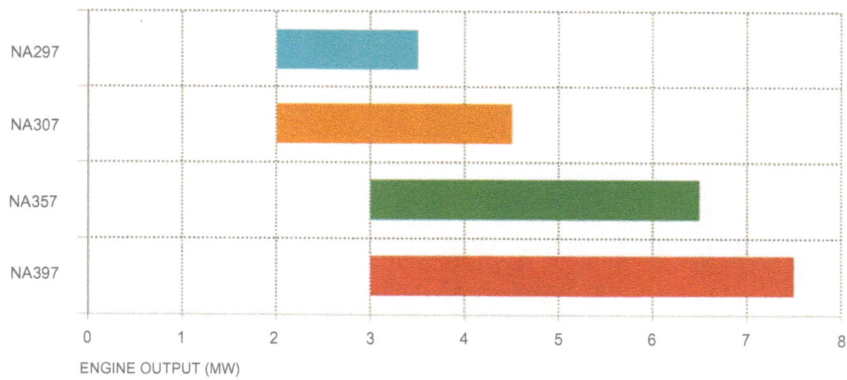

MAN Energy Solutions

A full range of turbochargers for two- and four-stroke engines is offered by MAN of Germany, its PBS Turbo subsidiary in the Czech Republic and overseas licensees. The current programme embraces the older NR/R series, the NR/S series and the NA/S/T9 series, the TCR radial series, and TCA axial series and the latest TCT and TCX generations introduced to cover two-stage turbocharging and to meet more stringent NOx Code Tier III requirements; in addition, the PT/PTG power turbine series is available for turbo compound systems.

NA and NR series

The NA/S and NR/S series turbochargers were progressively introduced to the market in the 1990s, their uncooled casings and inboard plain bearing features subsequently being adopted by other turbocharger manufacturers.

The radial-flow turbine NR/S series was designed for safe control of the charge air pressure ratio of up to 4.5, adequate extension of the application range, a higher efficiency level, improved reliability in all operating ranges, unrestricted heavy fuel capability, and reduced maintenance demands. Seven models—the NR12/S, NR14/S, NR17/S, NR20/S, NR24/S, NR29/S, and NR34/S—formed the programme, which overlapped with the smallest of the established axial-flow turbine NA/S series. The totally water-free series (Fig. 10.22) was a further development of the established NR/R turbocharger, featuring a revised compressor and turbine wheel. MAN's proven system of internal plain bearings was retained, the axial thrust now being absorbed by a separate thrust bearing. The five models in the radial-flow turbine NR/R series serve engine outputs from around 400 to 4400 kW per turbocharger.

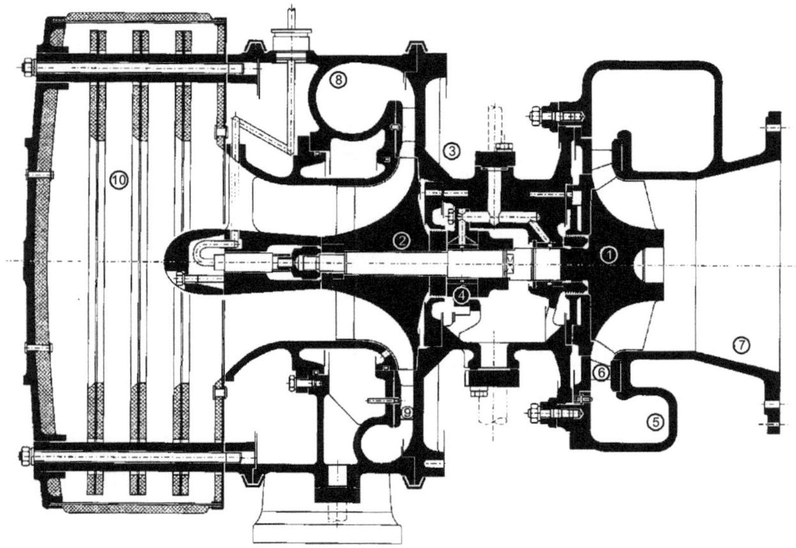

1 Turbine rotor with shaft
2 Compressor wheel
3 Bearing casing
4 Bearing bush
5 Gas inlet casing
6 Nozzle ring
7 Outlet diffuser
8 Compressor casing
9 Diffuser
10 Silencer

Fig. 10.22 MAN NR/S radial-flow turbocharger.

The development targets for the axial-flow turbine NA/S and NA/T9 series turbochargers listed the following performance criteria:

- Medium-speed four-stroke engines: a compressor ratio in excess of 4, and a turbocharger efficiency in excess of 60%
- Low-speed two-stroke engines: a compressor ratio in excess of 3.6, a turbocharger efficiency of at least 64% (without an associated power turbine system) and a turbocharger efficiency of at least 69% (with a power turbine system)
- A pressure ratio of up to 4.5 for the NA/S and 4 for the NA/T9

MAN carried out extensive theoretical and experimental investigations on the key individual turbocharger components: the compressor, turbine, and bearings. Apart from higher efficiencies from the turbine and compressor, the goals included the attainment of high vibratory stability for the bladings and a smooth running performance at low bearing temperatures.

A one-piece radial compressor wheel with continuously backswept blades was selected to secure an uprated pressure ratio and flow rate and improved vibratory stability. The wheel carries 10 main blades and 10 splitter blades. The introduction of splitter blades reportedly yielded a marked increase in efficiency for pressure ratios >3.5, and an adequate safety margin against surging in the part-load range was maintained. Various coatings, such as Keplacoat and Alumite, have been investigated to protect the compressor wheels against corrosive attack undermining the fatigue strength under reversed bending stresses.

A few percentage points improvement in efficiency was also gained by the development of a new turbine blade and by optimizing the outlet diffuser. The investment-cast turbine blade offers a significant cost reduction over a forged blade while retaining the material properties and the fatigue strength under reversed bending stresses. The resistance of the blading to vibrations and 'exceptionally smooth' running at pressure ratios of up to 4.5 were promoted by the incorporation of a floating bush in the turbine bearing system.

The S-type turbochargers are characterized by uncooled turbine casings and bearing housings, dispensing with the previous double-walled heavy casings, and cooling connections. The water-cooled bearing casing was retained, however, for the NA/T9 turbocharger. Tests with an NA57/T9 turbocharger serving a MAN B&W 6L60MC low-speed engine achieved an efficiency of around 67.5% at the 100% load point; the charge air pressure ratio was around 3.6. The maximum efficiency measured at near-70% engine load was around 69%. A high-efficiency level over the remaining load range was also logged.

Axial-turbine NA series and radial-turbine NR series turbochargers are still manufactured by MAN and its licensees to serve appropriate engines that remain in production. The NA series now embraces six models covering maximum supercharged engine outputs per turbocharger from 3600 to 24,500 kW. The seven-model NR series covers an engine power range from 670 to 5400 kW per turbocharger.

TCA series

In 1999/2000 MAN decided to develop a completely new turbocharger generation to replace the NA series, resulting in the launch in 2002 of the TCA series with radial-flow compressor wheel and axial-flow turbine blading. The designers pursued the following goals: high specific flow rates, high efficiencies, low noise emissions, ease of maintenance and reduced servicing times, ease of engine mounting, and high reliability with long service life.

The six-model TCA series addresses two- and four-stroke engines requiring from around 3000 to 30,000 kW output per turbocharger. All elements of the design were optimized with regard to flow control and stress reduction using 3D CFD and FEM calculations.

The TCA series was headed into production by the TCA77 model (Fig. 10.23) with additional frame sizes from the TCA33 to the TCA99 planned to follow progressively and cover the requirements of the highest-powered two- and four-stroke engines. The largest projected model was subsequently deemed unnecessary, thanks to the performance of the TCA88-25 model, which can serve two-stroke engines with 27,300 kW output per turbocharger (Fig. 10.24).

Fewer turbochargers to boost a given low-speed engine are thus enabled. An eight-cylinder MAN B&W K98MC/ME or MC-C/ME-C engine, for example, can be specified with two TCA88-25 turbochargers instead of the

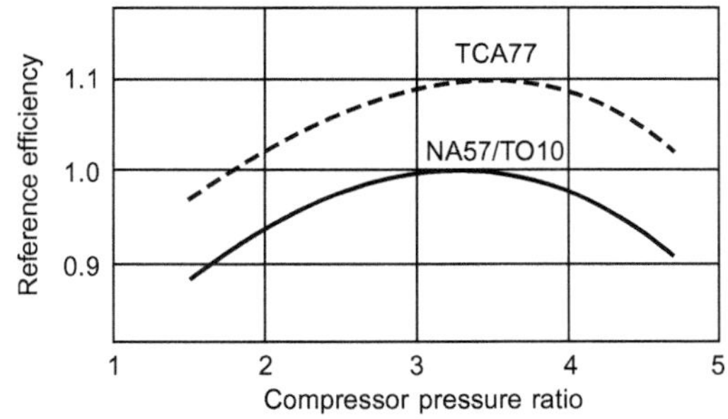

Fig. 10.23 Efficiency map of the prototype MAN TCA77 turbocharger plotted against the maximum efficiency of the earlier NA57 design.

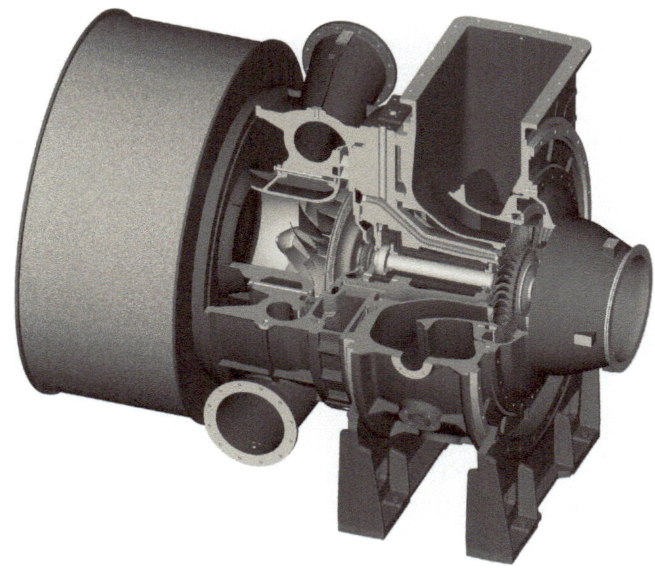

Fig. 10.24 MAN's axial-flow TCA series turbocharger.

previous outfit of three, and 12-cylinder engines of the same type can be boosted by three such turbochargers instead of the previous four.

Regular improvements since the initial introduction have seen pressure ratios increased and the engine outputs also lifted to 30,000 kW for the two-stroke versions.

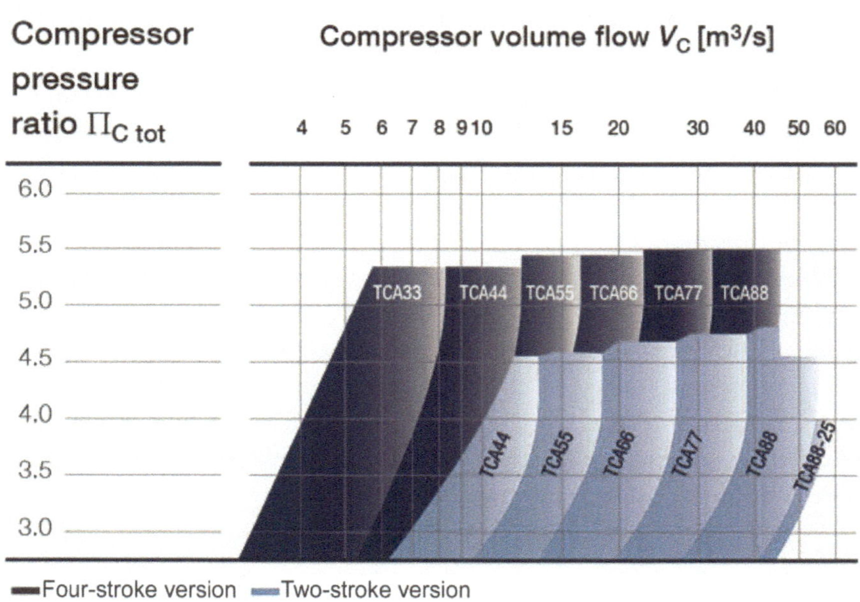

MAN TCA series

Model	Maximum engine output (kW)			Mass (kg)
	2-stroke engines (le = 9* kg/kWh)	4-stroke engines (le = 6.5* kg/kWh)	Max speed	
TCA 33		5400	27,800	1140
TCA 44	7000	7900	27,700	1970
TCA55	9600	10,400	20,000	3290
TCA66	14,000	14,800	16,900	5510
TCA77	20,000	21,000	14,200	9250
TCA88	30,000	30,000	12,000	15,790

Increased efficiency was secured by newly developed wide-chord turbine blades, which are very stiff and highly wear resistant, and supported in the turbine disc with fir-tree feet. A damping wire thus became unnecessary, easing maintenance work. Efficiency was also raised by a new optimized turbine outlet diffuser and turbine inlet casing, while a new turbine nozzle ring extended durability.

Running on two internal radial bearings with floating bushes and a separate thrust bearing, the advanced rotor design achieves higher safety margins. Bearing points with small diameters were realized by optimizing both the rotor dynamics and the shaft/hub connection of the compressor wheel and turbine shaft. The resulting low circumferential speeds of the bearing bushes and minimization of friction losses contribute to a higher total efficiency. Very quiet running and minimal bearing wear are cited for the floating journal bearing bushes, which optimize damping behaviour, and the bearing body has an expected service life of up to 50,000 h.

Mechanical losses are minimized by a reduced shaft diameter and high-performance thrust bearings. Easy handling and short maintenance times were sought from the support design; neither the compressor casing nor the turbine rotor, for example, has to be disassembled when removing the thrust bearing.

Very high centrifugal forces result from a compressor wheel operating at a circumferential speed of well over 500 m/s. A standard aluminium alloy design and a specially adapted machining process ensure a long service life for a component subjected to high-pressure conditions. An optimized geometry is said to secure high efficiency at a safe surge margin over the whole operating range, and noise is considerably reduced. A new type of force and form-locking compressor wheel attachment on the turbine shaft enables assembly and disassembly of the wheel without the need for an unwieldy hydraulic tool. The new MAN-patented shaft/hub connection substantially reduces replacement times and eases maintenance.

Efficiency is enhanced by the newly developed compressor volute and compressor diffuser vanes, while optional internal recirculation (IRC) increases the surge margin, and optional jet assistance provides fast rotor acceleration.

TCA nozzle rings are manufactured from high-resistance materials to secure a long service life. Optimum matching of the turbocharger to the specific engine is smoothed by individually adapting nozzle rings. A new variable turbine area (VTA) option enables the flow cross-section to be adapted to the corresponding load condition of the engine, thereby reducing fuel consumption and emissions, especially in part-load operation.

Optimized mounting to the engines is facilitated by enabling all casings to be installed and turned in 15° steps, and newly developed one- and two-socket compressor spirals and diffusers foster optimum matching of the turbocharger to the engine.

Modular design allows the turbocharger to cover as broad a range of applications as possible. Air is sucked in either via an intake air silencer, a 90° intake casing or an intake manifold; exhaust gas flows into the turbine via either an axial- or radial-flow casing. The design features the separate casing-base concept proven in the NA series. The compressor volute is available with either one or two opposing thrust pieces: the single-piece variant has advantages for two- and four-stroke inline cylinder engines, while the two-piece volute offers advantages when attaching a single turbocharger to V-type four-stroke engines.

Both the compressor wheel and the turbine can be adjusted to engine requirements by choosing from a range of meridians and blading configurations. Diffusers and nozzle rings that are very finely stepped in their mass-flow areas allow the turbochargers to be fine-tuned to the engine. A nozzle ring capable of adjustment during operation is available optionally for maximum variability.

A minimal number of external connections were sought. On the engine side and in the plant system itself the only additional connections—apart from the air and exhaust gas connections—are an oil inlet and drain pipe, together with a breather connection for the bracket. No cooling water, or lock-air connections, additional oil overflow piping, or external lube oil supply are required. A special emergency operation facility is integrated into the turbocharger to allow continued operation under emergency conditions. All the hot casing components are equipped with a new temperature- and noise-reducing casing with a sheet metal cover, significantly reducing the risk of engine room personnel coming into contact with excessively hot surfaces.

Uncooled casings are designed in accordance with the 'pipeless engine' principle, all supply pipes being fully integrated within the casing. Only one oil supply and discharge pipe and a venting pipe have to be connected externally. A robust structure and the material thickness of the casing walls means that additional burst protection is not required. Safety is further enhanced by the rigid connection of the turbine outlet, bearing and compressor casing using tie rods.

In order to limit bearing losses, the shaft diameter of the rotating assembly was reduced, paying special attention to rotor dynamics so that, despite the reduction, it was possible to increase rotor stability compared with the NA series turbochargers. The rotating assembly is mounted in the bearing casing on two floating-sleeve radial bearings. Since the floating-sleeve design significantly reduces the rate of shear in the bearing gaps, in comparison with the fixed-sleeve type, both bearing losses and bearing wear are reduced.

The thrust bearing is equipped with a floating centre plate which acts in the same way as the floating-sleeve radial bearings. The thrust bearing is located outside the radial bearing on the compressor side, facilitating removal without having to dismantle the radial bearing and rotating assembly. MAN notes that the fact that the thrust bearing is subject to far greater stresses than radial bearings (because of the axial thrust of the turbine and compressor) makes this a significant feature for ease of maintenance.

Bearing points are supplied with oil by a ring channel arranged in the inside of the bearing casing. Oil feed, either from the right or left side of the turbocharger, lube oil distribution to all the bearing points, and the supply from the emergency lube oil service tank are via this ring channel.

No bearing case is water cooled. The heat input from the compressor and turbine is dissipated in the lube oil flung off the shaft of the rotating assembly. The oil mist thus generated can drop down the walls of the generously dimensioned interior of the bearing casing, thereby evenly absorbing the heat to be dissipated. The bearing casing has its own air vent, ensuring that the leakage air, which the compressor inevitably forces into the casing through the shaft seal of the rotating assembly, does not increase crankcase pressure in the engine but dissipates it directly.

All airflow components were optimized with regard to flow configuration and stress reduction using 3D CFD and FEM analyses. The result was a turbine with wide-chord blades arranged in a fir-tree root in the turbine disc. A characteristic feature of such blades is their very high chord-to-height ratio, and thus creating a compact, very stiff, and hard-wearing blade. The turbine blades can be of varying angles and lengths. With the aid of advanced design tools, it is now possible to dispense with lacing wire to dampen exhaust-generated vibrations, even in four-stroke engine applications. MAN explains that apart from improving the blade profile, this has significantly boosted efficiency.

As in the NA series, the turbine shaft is friction-welded to the turbine disc. The disc (which in the case of the TCA77 turbocharger must absorb a centrifugal force per blade equivalent to the weight of a fully loaded truck and trailer) is made from a forged steel alloy. The nozzle rings, with a new blade profile design, and a carefully matched turbine outlet diffuser, also contribute to efficiency gains. A key element of the development work was designing the patented turbine outlet diffuser, which effectively converts kinetic energy remaining downstream of the turbine wheel into pressure. Simultaneously, the outlet diffuser acts as integrated burst protection within the turbine (see later).

The compressor wheel is operated at circumferential speeds well in excess of 500 m/s, generating considerable centrifugal forces. To withstand these forces, the wheel is made of a high-strength aluminium alloy. For applications where the components will come into contact with corrosive media, a special corrosion-resistant coating can be applied to the compressor wheel.

A new design of compressor volute and new nozzle ring designs contribute to optimized turbocharger matching and high efficiency, while a new compressor mounting system was developed to ease removal. The compressor wheel is deformed by the powerful centrifugal forces acting on it; this tends to cause it to expand across its diameter, inevitably provoking a shortening in the axial direction and causing the bore to widen. The new design ensures good centring of the wheel both on the inlet side and to the labyrinth ring in the plant system itself. A tie bolt compensates for the axial shortening, and a tightening system allows a small torque wrench to produce the required tie bolt tensioning. The excellent centring makes it possible to remove and replace the compressor wheel without the need to rebalance the rotating assembly.

Noise emission reduction was an important goal, achieved by a compact design (the noise emitting surface area is greatly minimized) along with newly developed insulation for the turbine outlet and compressor casing.

Noise emissions in a turbocharger occur primarily at the compressor wheel. Any measures aimed at reducing noise, therefore, concentrate on the cold area. Noise is generated by a combination of pressure fluctuation in way of the compressor wheel inlet and the interaction between the wheel and the diffuser. Inlet side noise emission was reduced on the TCA series by optimizing the fluid mechanics of the compressor wheel, seeking the best compromise between high compressor efficiency and noise emission. In addition, the silencer was designed with radially arranged curve plate elements, dampening noise emissions to a high degree while keeping intake pressure losses low.

In order to reduce noise radiation in the engine's charge air manifold, the number of blades in the diffuser was matched to that in the compressor; furthermore, the compressor volute was provided with a noise-reducing casing. All of these measures together significantly reduced the level of noise emissions compared with the NA series, and comfortably achieved the goal of attaining emissions below the 105 dB(A) at 1-m distance level demanded by seafarer associations and classification societies.

A new radial silencer design is reportedly capable of reducing sound emissions down to well under that level. The shape of the radially arranged damping segments fosters optimum flow conditions and contributes to total turbocharger efficiency. A filter mat over the circumference of the silencer is configured for easy removal and cleaning when required.

In the event of the worst possible form of turbocharger damage—the rupture of a compressor wheel or one or more of the turbine blades—steps must be taken to ensure that no fragments can escape from the turbocharger casings and cause injury. All the antiburst measures used on the TCA series are based on the idea that the forces occurring cannot be absorbed by solid, rigid elements. The strategy instead is to employ flexible elements to transform the kinetic energy generated by the individual ruptured components into friction heat, thereby preventing the casing components from failing as a result of local overload.

The compressor volute, bearing casing, and turbine gas outlet casing, for example, are held together by 24 very strong tie bolts. The compressor volute is of very rigid design, while the insert piece directly surrounding the compressor wheel is made flexible by means of long tie bolts. When the compressor wheel bursts, the compressor vanes are crushed against the insert piece. The remaining components of the hub body become jammed between the insert piece and the bearing casing. The insert piece is dragged towards the silencer, using up a major part of the kinetic energy in the component fragments. The fragments then become lodged between the compressor volute and the bearing casing; the remaining kinetic energy in the fragments is completely dissipated.

On the turbine side, a controlled failure point was targeted at the connection between the turbine blade and the disc. Burst protection is, therefore, easier on the turbine side than on the compressor side since, in the worst-case scenario, one or more turbine blades can fracture when, for example, one of the engine's exhaust valves fractures. The components surrounding the turbine impeller are designed to be able to absorb the kinetic energy of several broken blades.

TCR series

New standards for radial-flow turbochargers were sought by MAN in designing the TCR series introduced in 2004 and which is now offered in seven sizes—from TCR10 to TCR22—to cover four-stroke engine outputs from 600 to 6850 kW per turbocharger with pressure ratios of up to 5.4. The three largest models (TCR18, TCR20, and TCR22) also address two-stroke engines with outputs from 2700 to 7000 kW.

Part- or full-load optimization is facilitated by a range of different rotor/nozzle ring flow area combinations. New CFD-optimized profiled rotor blades, nozzle rings, and inlet and outlet casings contribute to higher efficiency.

Experience from the NR series turbocharger's radial and axial bearing designs was tapped to benefit the TCR, the floating radial bearing bushes being coupled with new solutions such as the nonrotating radial bearings. The new 'semi-floating' bearing design with squeeze oil damper fosters reliable operation and low wear. The most evident modification is the axial bearing, which is now arranged between the radial bearings to create a space-saving design. All bearings are housed in a bearing body.

New compressor wheel geometry underwrites improved efficiencies; depending on the application, part- and full-load optimization is facilitated by flow area combinations. An internal recirculation (IRC) system can provide a larger compressor characteristic map with increased surge margin capability. IRC increases the stable operating range of the compressor and enables the choice of a favourable engine operating line inside the compressor map. In this way, it is possible to achieve higher compressor efficiencies at high load; this leads to both lower fuel consumption and a reduction in mechanical and thermal stress in the compressor wheel compared with a compressor having the same pressure ratio but without IRC.

Milled from a high-resistance aluminium alloy, the compressor wheel can withstand intake conditions within a wide temperature spectrum. A new type of wheel attachment and retainer simplifies maintenance and extends the service life of a component subject to great stresses. An optional jet assist facility supports rotor acceleration.

Nozzle rings are manufactured from an extremely resistant material to foster a long service life. Optimized matching of the turbocharger to a specific engine is pursued through individually selected flow areas. A variable nozzle ring, optionally available, is recommended by MAN for engines with frequently changing loads. Such a ring enables adaptation of the flow cross-section to the given load conditions to be optimized (see section Variable Turbine Area technology).

As with the TCA series turbochargers, a new cast aluminium silencer lowers sound emissions to <105 dB(A), and the flow-optimized design of the radially arranged silencer segments contributes to turbocharger efficiency. The filter mat on the silencer circumference is easily cleaned.

Again following TCA turbocharger practice, the uncooled casing of the TCR series was designed in accordance with the 'pipeless engine' principle, with all supply pipes integrated in the casing. The compressor and turbine casings feature measuring connections for monitoring the pressure and temperature; a speed pick-up is also included in the scope of supply. Manufactured from SiMo steel or ductile cast iron, the casings are designed and tested to be containment proof and are heat insulated to comply with SOLAS requirements. Claw connections facilitate swift disassembly and allow a casing to be rotated into any required position.

A wide inspection hatch incorporated in the turbine outlet casing offers good access to the turbine wheel for checking and cleaning, and integrated turbine and compressor washing devices are optionally available.

The DIN flange of the flow-optimized turbine outlet casing allows it to be easily fitted to the exhaust gas pipe; alternatively, a simple and space-saving 90° elbow can be supplied. Easier maintenance was addressed by the designers by minimizing the number of screws and overall components and developing a set of special tools for servicing.

Variable turbine area technology

Both TCA and TCR series turbochargers are available optionally with VTA technology, whereby a nozzle ring equipped with adjustable vanes replaces the fixed vane ring used in the standard turbochargers (Fig. 10.25).

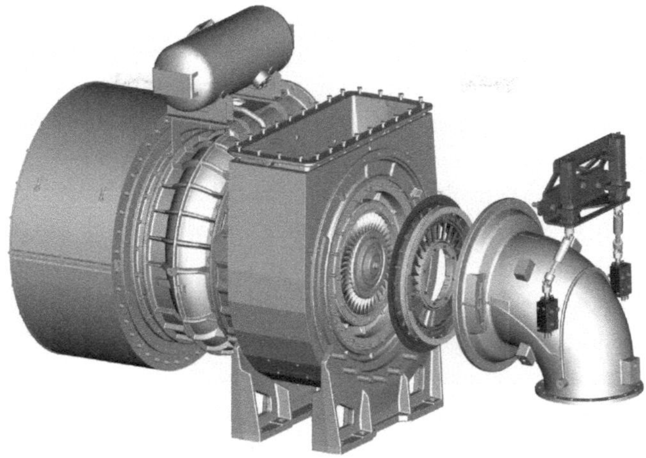

Fig. 10.25 MAN's VTA-equipped turbochargers feature a nozzle ring with adjustable blades, illustrated here on an axial-flow TCA model.

Adjusting the vane pitch regulates the pressure of the exhaust gases impinging on the turbine, and thus enabling the compressor output to be varied and optimized at all points on the engine's performance map. The quantity of charge air can be more precisely matched to the quantity of fuel injected, fostering reduced specific fuel consumption and emissions along with improved dynamic behaviour of the engine–turbocharger system.

In order to minimize thermal hysteresis and improve adjustment accuracy, each vane has a lever directly connected to a control ring, which is actuated by an electric positional motor with integrated reduction gear. Control of the vane position is fully electronic, with feedback or open-loop control featuring mapped vane adjustment. A comprehensive range of control signals can be used, including charge air pressure after the compressor and exhaust gas temperature before and after the turbocharger. Control packages can thus be tailored precisely to the specific application, whether that involves mechanically controlled engines or engines with electronic management.

VTA adjustable vanes are manufactured from heat- and erosion-resistant steel alloy. Careful selection of fits and materials is made to ensure system operation under all conditions without sticking, particularly in applications on engines burning heavy fuel oil.

Shop tests and shipboard trials with VTA-equipped turbochargers have reportedly demonstrated the expected improvements at part load in terms of fuel economy, reduced emissions of soot and unburnt hydrocarbons, and enhanced engine response under load changes. A lower specific fuel consumption, improved torque and engine acceleration, lower combustion chamber temperatures, and savings in electrical power to drive the auxiliary blowers are among the benefits cited from the higher scavenging pressures in part-load operation. Retrofits to existing turbochargers can be facilitated by MAN supplying complete packages for such projects, including the VTA nozzle ring, the actuator, and the control system.

VTA technology is approved for TCA turbochargers from the TCA55 to the TCA88. Almost 400 engines with VTA are now in use in marine or power generation and as a consequence of performance data collected, MAN has now opted to fit VTA turbochargers as standard on its MAN 51/60 engines.

TCT series

This series of turbocharger is specifically aimed at two-stroke engines and was first announced in 2014 with the expectation that the first models would be in production in 2016. In fact it would be 3 more years before production began. Development of the TCT was a decision taken to ensure compliance with impending NOx regulations and especially the Tier III phase due to come into effect in 2016.

MAN described the motivation for the TCT series in 2014 saying that low fuel consumption and low emissions (particularly NOx) are the twin development objectives of large new-generation two- and four-stroke engines. They are achieved by using the Miller process, which involves the introduction of a special timing system for the inlet valve in a four-stroke engine and the exhaust valve in a two-stroke engine. However, the ways in which this process is implemented differently.

Fundamentally, the temperature at the start of the combustion phase is lowered by reducing the compression ratio in the cylinder during the compression stroke, and thus cutting the directly related production of NOx. In order to create similar circumstances under the conditions of final compression pressure and cylinder charging (the combustion air ratio), part of the compression must, therefore, be shifted to the turbocharging phase in order to obtain similar conditions in relation to the final compression pressure. Higher turbocharger compressor pressure ratios are thus necessary.

In four-stroke engines, the gas exchange phase can also be used to make a considerable positive contribution to work, if the scavenging gradient between the inlet and exhaust is sufficiently positive. This is achieved by a higher degree of turbocharging efficiency, as provided by two-stage turbocharging with intercooling—as opposed to single-stage turbocharging—in parallel to the potential for maximum charging pressures. In the new TCX range, MAN had developed in parallel with the TCT a turbocharger specifically for two-stage turbocharging of four-stroke diesel and gas engines with a total pressure ratio > 10.

Conversely, a positive scavenging gradient cannot be used for a contribution to work in two-stroke engines, but only for scavenging. The scavenging gradient required is moderate. Unlike four-stroke engines, a very high scavenging pressure (a very high-pressure ratio) also appears to obstruct optimal scavenging. This means that high degrees of turbocharging efficiency at high compressor pressure ratios, which can be achieved by single-stage turbocharging, are required by two-stroke applications. In its TCT range, MAN has developed a turbocharger which has been specifically matched to the requirements of two-stroke engines, without making any compromises. The turbocharger efficiency has been increased by 5% by using newly developed compressor and turbine wheel geometry, consequently increasing the waste heat recovery potential by 30% by means of a Turbo Compound System. At the same time, the new TCT range promised to deliver a 10% greater airflow at a turbocharging pressure 25% higher, while being 30% smaller and 40% lighter than the existing TCA range.

Maintenance was also a design factor with increased time between overhauls and a smaller component count and ease of assembly and disassembly being key points. The introduction of the TCT series in 2019 means that MAN's two-stage turbocharging solution can now achieve efficiency levels of up to 80%.

MAN is currently offering just the TCT40 and TCT60 types but expects the other TCT sizes to come on stream by 2020 and 2021. The TCT is not only aimed at low-speed engines but is also eminently suitable for two-stage turbocharging on four-stroke engines. MAN TCT turbochargers are suitable for conventional fuel and dual-fuel engines in both marine and power applications ranging from 6 MW up to 24 MW output per turbocharger.

MAN TCT series	Supercharged engine output
Turbine type	**[kW]**
TCT30	7500
TCT40	9460
TCT50	12,000
TCT60	15,120
TCT70	19,040
TCT80	24,030

TCX series

MAN has a long history in two-stage turbocharging and first evaluated the technology back in the 1970s. In late-2000, MAN focused on the technology to improve its four-stroke portfolio, resulting in the development of MAN ECOCHARGE, a combination of low- and high-pressure stages with intercooling that significantly improves engine performance.

In 2016, the company marked a milestone with the 1000th two-stage turbocharger installation. Most of these were in nonmarine installations but like most engine and turbocharger makers, MAN had been developing suitable products to meet NOx Tier III requirements.

A year previously, MAN introduced the new TCX range to provide the high-pressure stage of a two-stage turbocharging set up. The low-pressure stage would feature a TCR turbocharger. The company named the two-stage project ECOCHARGE. The first pilot used MAN TCR20 and MAN TCX17 units—combined as a single, two-stage turbocharging system on a MAN 12V35/44 gas engine on a test bed at its Augsburg, Germany works. The company said that the ECOCHARGE system had been developed for both gas- and HFO-burning engines.

The combination of a MAN TCR20—for the low-pressure stage—and an MAN TCX17—for the high-pressure stage—offers significant potential for today's demand for more economical and environmentally friendly engine operation. This is because two-stage turbocharging supplies engines with improved scavenging air pressures ranging from 5 to >10 bar, as well as significantly improved turbocharging efficiencies.

MAN ECOCHARGE offers major benefits for engine builders. The increase in turbocharging efficiencies, in comparison to single-stage turbochargers, is mainly related to the intercooler—positioned between the low-pressure-stage and high-pressure-stage turbochargers—that significantly reduces the energy required to compress the intake air to high pressure. The resulting, higher efficiencies have an instantaneous impact on the engine by advantageously increasing the air pressure over the cylinder during the scavenging process. In addition, greater turbocharging efficiency fosters the reduction of NOx emissions through the Miller cycle while the improved scavenging efficiencies provided by the ECOCHARGE system make the engine more fuel efficient.

The higher power-density generated by the ECOCHARGE technology presents the opportunity to choose between significantly boosting an engine's power output or reducing engine size. The ECOCHARGE system's increased efficiencies and higher cylinder rating facilitate the use of a smaller engine with the same power output of a larger unit using traditional, single-stage turbocharging.

Today the TCK series comprises eight frame sizes; TCX09, TCX11, TCX13, TCX15, TCX17, TCX19, TCX21, and TCX23 suited to engines with a maximum output of 1700 kW for the smallest model through to 23,900 kW for the TCX23. The TCX has a turbine of mixed-flow design and can operate at high air-side pressures and in engine exhaust gas temperatures of up to 650°C. The pressure ratio of the two-stage combination is up to 10.5:1.

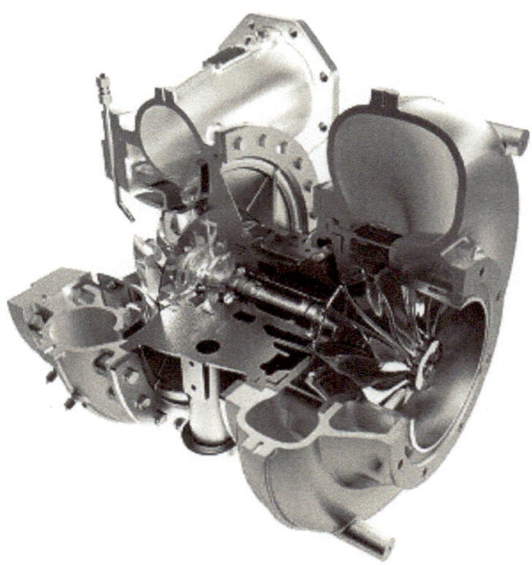

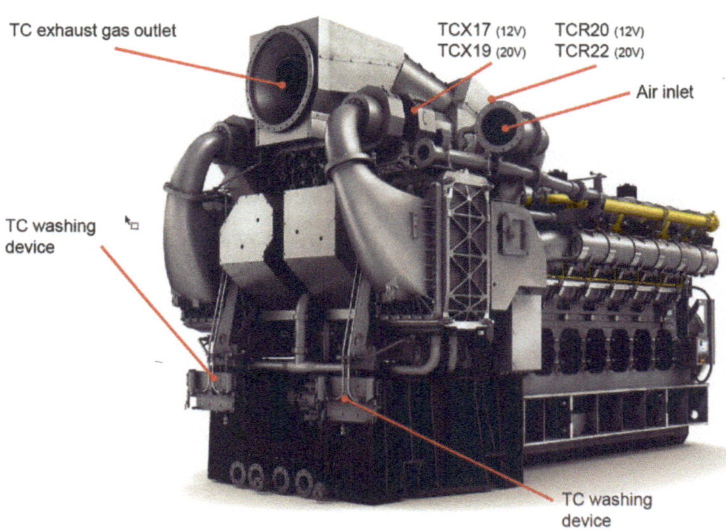

Mitsubishi heavy industries

Successive generations of MET axial turbochargers introduced since 1965 by Mitsubishi have featured nonwater-cooled gas inlet and outlet casings. The diagram below shows the evolution of the MET turbocharger to the point where today the latest generation is the MET-MB series. Efficiency demands and the growth in container ship size were the main influencers of development.

The biggest driver in development as with all makers is the need for better pressure ratios to improve efficiency. From 1990, when the reduction in sfoc was the main customer requirement, demand for higher pressure ratios and greater efficiency was addressed by the Japanese designer with the development of the MET-SD series (Fig. 10.26), an evolution of the MET-SC programme. The main features of the earlier series were retained, but the new models exploited a more efficient compressor wheel incorporating 11 full and 11 splitter blades arranged alternately.

The MET-SD turbocharger shadowed the efficiency of the MET-SC equivalent in the low-speed range but yielded 4%–5% more in the high-speed range when the pressure ratio exceeds 3.5. The air, gas, and lube oil pipe installations were unchanged, as were the mounting arrangements.

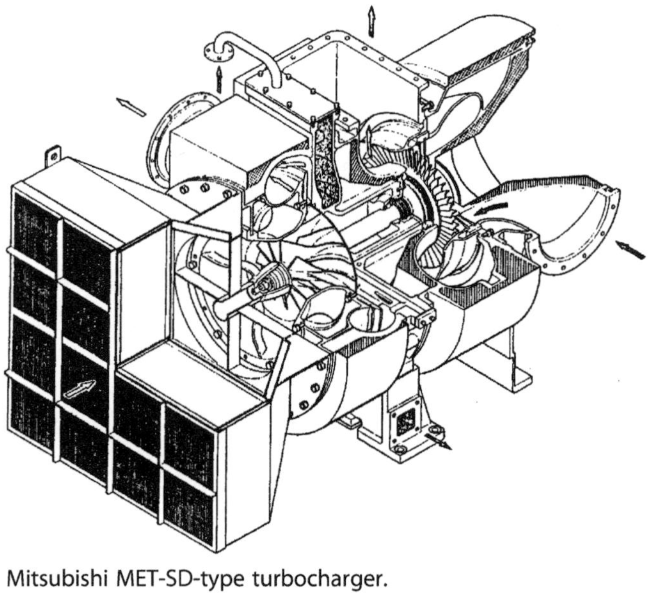

Fig. 10.26 Mitsubishi MET-SD-type turbocharger.

The five models in the MET-SD programme individually covered the turbocharging demands of engines with outputs ranging from 1800 to 18,000 kW at a pressure ratio of 3.5. The airflow ratings from the smallest (MET33SD) to the largest model (MET83SD) at that pressure ratio vary from 3.5 to 37.5 m^3/s.

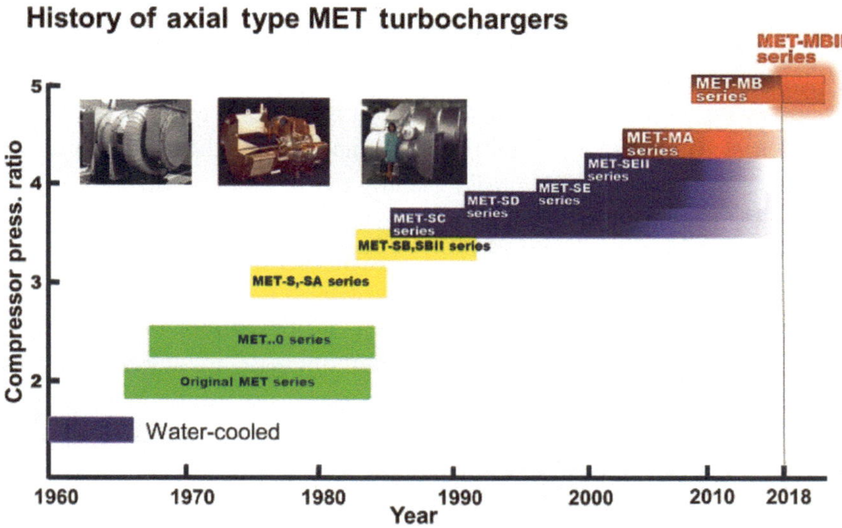

The MET turbocharger rotor is supported by an internal bearing arrangement based on a long-life plain bearing lubricated from the main engine system. The large exhaust gas inlet casing is of two-piece construction comprising a detachable inner casing fitted with the turbine nozzle assembly and a fixed outer casing containing the rotor shaft fitted with the turbine wheel. The inner casing can be readily unbolted and pulled from the outer casing, complete with the nozzle assembly, and thus exposing the turbine wheel and nozzle assembly for inspection or servicing. If necessary, the turbine wheel and rotor can be removed from the outer casing.

Mitsubishi targeted pressure ratios of 4 and beyond with the subsequent MET-SE/SEII series, which shares the same main characteristics of earlier MET generations:

- Completely noncooled (no sulphuric acid corrosion of the casing)
- Inboard bearing arrangement (ease of release)
- Sliding bearings (long life).

In addition, as with previous models, an inner/outer double-walled structure was adopted for the gas inlet casing for ease of release on the turbine side for maintenance checks, and a lubricant header tank is incorporated in the structure for effective lubrication of the turbocharger rotor following the emergency shutdown of the engine. MET-SE turbochargers also exploit wide chord-type blades for improved performance in the high-pressure ratio region.

The seven models in the MET-SE/SEII programme—embracing 33SE to 90SE types—deliver airflow rates from 3.1 to 53 m^3/s at a pressure ratio of 3.6, satisfying engines with outputs ranging from 1300 to 24,100 kW per turbocharger (Fig. 10.27). Total turbine efficiency is enhanced by a compressor impeller featuring 3D profile blading to maximize aerodynamic performance, and turbine wheel blading that requires no damping wire.

Introduced in 2001 to boost two-stroke engines powering large container ships, the MET90SE design had an airflow range some 30% higher than the 83SE (formerly the highest capacity model) which was only 10% larger. The higher capacity from a modest increase in dimensions was achieved by proportionally enlarging the rotor shaft and developing a new compressor impeller wheel and turbine blades. Mitsubishi notes that expanding the capacity of such wheels and blades without compromising reliability and performance is a key challenge for turbocharger designers.

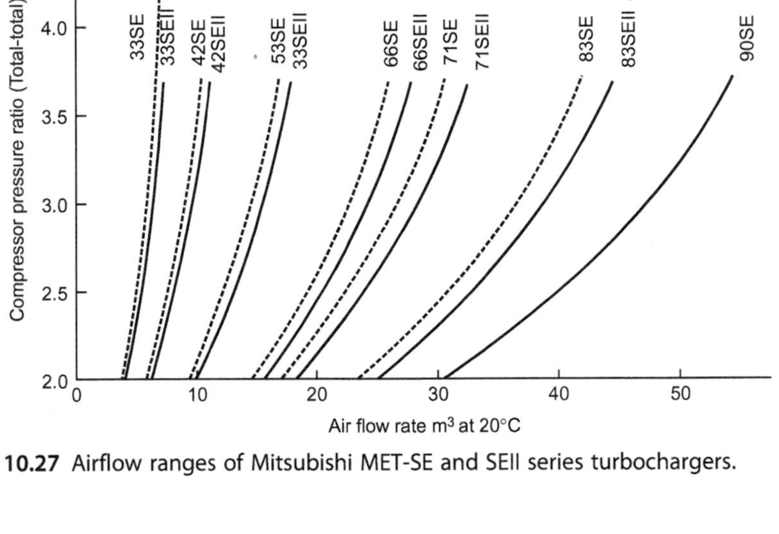

Fig. 10.27 Airflow ranges of Mitsubishi MET-SE and SEII series turbochargers.

Different compressor impeller materials—such as titanium and cast steel—have been investigated as an alternative to the forged aluminium traditionally used in MET turbochargers. A pressure ratio of 4 with an aluminium impeller was nevertheless achieved, thanks to an impeller air cooling system and a redesigned compressor blade. By reducing the impeller backward curvature angle from 35° to 25°, the pressure ratio could be raised without the need to increase the impeller peripheral speed.

Mitsubishi successfully developed a cast stainless steel compressor impeller in preference to an aluminium alloy component for its MET-SH turbocharger. Raising the compressor pressure ratio to the higher levels increasingly demanded by the market dictates attention to compatibility between the impeller strength and design and the aerodynamic design. Mitsubishi cites, respectively, the problems of higher impeller peripheral speed, elevated impeller blade temperature and increased impeller blade surface load, and the problem of optimizing the aerodynamic characteristics of the higher speed airflow in the impeller.

Limitations in resolving all these problems are presented by an impeller of aluminium alloy, a material which Mitsubishi has traditionally used for MET turbochargers because of its high specific strength and excellent machinability. Running the aluminium alloy impeller at temperatures above the ageing treatment level (190–200°C), however, threatens a reduction in material strength and should be avoided. Depending on the suction air temperature, the impeller blade could reach this temperature level when operating at a pressure ratio of around 3.7.

An aluminium alloy forging also experiences progressive creep deformation at temperatures of around 160°C and above, calling for consideration in design to preventing the stress from becoming excessive at impeller blade areas subject to high temperature.

The stainless steel impeller casting for the SH series turbocharger delivered improved high-temperature strength and made it possible to raise the maximum working pressure ratio to 4 and higher, without reducing the impeller peripheral speed. Improved material strength allowed the impeller blade to be smaller in thickness than its aluminium alloy counterpart, despite the higher pressure ratio and increased peripheral speed. This feature in turn made it possible to increase the number of blades while ensuring the necessary throat area, hence improving the impeller's aerodynamic characteristics. The stainless steel impeller of the MET42SH turbocharger, the first production model in the MET-SH series, features 15 solid-formed full and splitter blades.

A higher density impeller casting called for special consideration of the rotor stability. The rotor shaft of the SH series turbocharger is larger in diameter than that of the SD series, with the compressor side journal bearing located outside the thrust bearing to increase the distance between the two journal bearings. Tests showed that, despite the higher mass of the stainless steel impeller, the third critical speed for the rotor is around 50% higher than the maximum allowable speed, underwriting rotor security.

A turbocharger total efficiency of 68% was logged by the MET42SH version fitted with impeller blading having a large backward angle and operating at a pressure ratio of 4.3 on a diesel genset.

In 2004, Mitsubishi's new MET-MA turbocharger series resulted from further optimization of components such as blades, gas inlet, and outlet casings to yield greater performance while retaining the proven basic construction of the previous series. An improved aerodynamic performance for increased turbocharging efficiency was delivered. The turbine capacity is approximately the same as that of the former MET-SEII series, with priority placed on higher efficiency.

MET-MA turbochargers have approximately the same outer dimensions as the MET-SE and SEII series and, apart from the gas outlet, the same interfaces for air, gas, and lube oil. Easier maintenance was sought from the detachable gas inlet inner casing concept. Compressor design follows that of the ME-SEII series.

Eight frame sizes extend from the MET33MA to the MET90MA models and cover the power requirements per turbocharger of engines with ratings from 2000 to 27,700kW at a pressure ratio of 3.8. The airflow range of the MA series spans 5.4–66.3kg/s. An MET60MA model—suitable for engine outputs of 10,000kW per turbocharger—bridged the gap in size between the MET53 and MET66 models. The MET66SE model formerly used with engines of this rating can be replaced by the MET60MA, and thus reducing the space requirement without sacrificing turbocharger performance. The MET83MA turbocharger is specified for engines requiring a pressure ratio of around 4, and an efficiency of least 74% has been confirmed in the pressure ratio range of 3–3.5; this is 2–3 percentage points higher than that for the preceding MET83SEII model.

The gas inlet casing arrangement used for MET-SB series turbochargers developed in 1980 and subsequently applied through to the SEII series features a detachable inner casing. This allows maintenance to be carried out on the turbine internals without disconnecting the engine exhaust pipe from the turbocharger. A smaller casing and optimized gas flow were introduced in the MET-MA series. While gas flow was previously routed within the gas inlet inner casing, the new design places the flow between the inner and outer casings. In addition, partition plates are deployed so as to prevent gas swirling and to facilitate a smoother flow.

Velocity distribution studies indicate that flow through the new gas inlet casing is more uniform in the MET-MA turbocharger compared with the previous series, which is beneficial for both greater turbine efficiency and reduced turbine blade vibration. The new inlet gas casing has the extra advantage of enabling easier inspection and cleaning of the nozzle ring since both inlet and outlet sides of the ring are exposed when the inner casing is removed.

While the new turbine blades are of the same number and height dimensions as before, the blade profile was redesigned for greater turbine efficiency as well as for added strength. Resonance with the number of nozzle vanes induces substantial vibratory stress on the turbine blades. Such resonance is particularly noticeable when the natural frequency is the third mode or higher, and the vibration mode configuration is complex. Furthermore, coupled vibration with the turbine disc must be considered. Mitsubishi notes that methods for measuring and evaluating vibratory stress are thus important in ensuring the reliability of newly developed turbine blades.

An asymmetric nozzle ring was developed to reduce turbine blade vibratory stress in resonance with the number of nozzle vanes for application in cases where the nozzle has a low reaction (i.e. where the nozzle area is small). The nozzle blades, which are normally spaced at the uniform pitch, feature a differing pitch at approximately half the periphery. This results in the timing of the excitation induced by the nozzle wake being interrupted at half-revolution intervals as the turbine blades rotate, and thus enabling reduced turbine blade vibratory stress during resonance. An asymmetric nozzle results in a reduction of around 40% in the maximum vibratory stress.

Asymmetric nozzle rings adopted as the standard for MET-SEII and MA turbochargers with small area nozzle specifications are said to offer equivalent safety to medium and large area nozzles; they also have no impact on turbine performance. The development of the MA series saw two sizes of compressor wheels being offered.

In common with other major manufacturers, Mitsubishi responded to demand new products to meet increasingly strict NOx requirements and introduced a new range in 2008. The MET-MB series was based on the MA which continued into production alongside the MB series for several more years before being phased out in 2017. The pressure ratio of the new MB series was increased to 6.0 compared to 4.0 of the MA series.

The MB series was available with a choice of impeller types. The V-type has a straight leading edge on the splitter vanes whereas the U-type has a curved leading edge. The new series allowed a higher maximum speed and a generally more robust design.

The MB series was further changed in 2018 when the MET-MBII series was announced beginning production in 2019. Turbochargers of this series are smaller and lighter than the MET-MB series, with a 16% greater airflow capacity. A new compressor wheel with the optimum number of blades and blade angle distribution was developed for the MET-MBII series, aiming to increase the capacity while maintaining high performance. The turbine blade throat distribution was also optimized to develop a new turbine with higher exhaust gas pressure recovery performance in the gas outlet casing.

Meanwhile, the casing, a component excluding the silencer, is the same as the casing used in the conventional MET-MB series. There are 10 frame sizes in the MBII series as there were in the MB but this is one more than the older MA series. The increase in airflow and corresponding efficiency of the MBII range could make it possible to drop down two frame sizes for the same output.

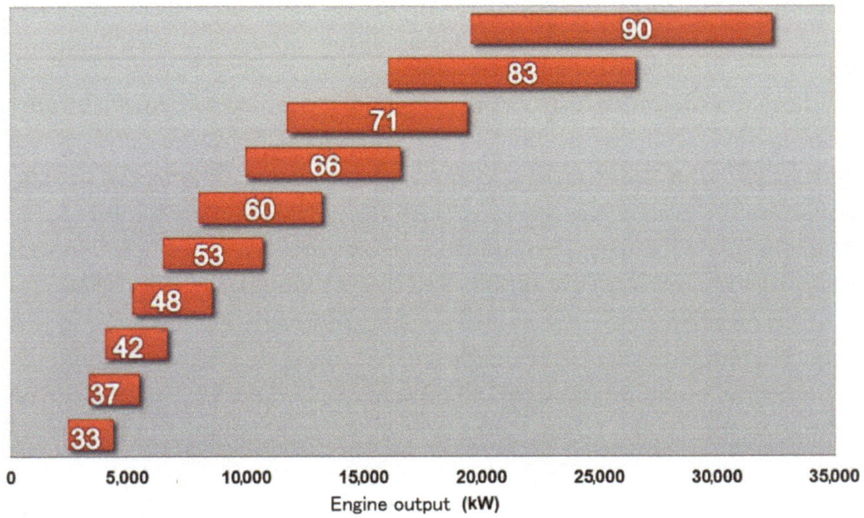

MET-MBII series engine matching.

To meet the slow-steaming requirements of some ship operators, Mitsubishi developed its own version of variable turbocharging. The result is the MHI-VTI (variable turbine inlet) introduced in 2012 and designed to maximize turbocharger efficiency at low-load conditions by increasing the scavenging air pressure so that the pressure is maintained even during the partial load operation. Using the VTI system, the area of the nozzle throat can be narrowed to increase boost and improve fuel efficiency by 2–3 g/kWh. Compared with a turbocharger cut-out arrangement, VTI offers a simple, reliable, and flexible solution for increasing the scavenge air pressure of two-stroke engines during partial-load operation, improving the engine's fuel efficiency.

The VTI mechanism switches the nozzle between two conditions, based on the addition of a ring to divide the nozzle into inner and outer sections. By closing a valve on the inlet casing, flow to the inner ring is shut off, effectively reducing the nozzle area. A simple construction, however, is retained and the MET turbocharger construction also allows turbine blades and nozzle to be inspected and cleaned by removing only the inner section of the turbine inlet casing, reducing the time and effort required for inspection and maintenance, and therefore cutting in-service costs. VTI was available as a retrofit option for MET-MA and MB engines using a gas inlet casing incorporating a valve and nozzle designed for the VTI.

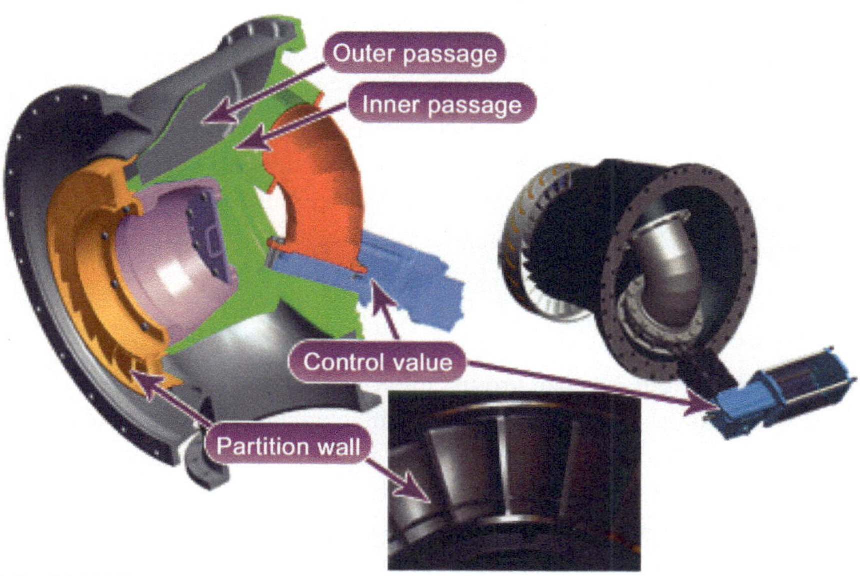

Mitsubishi VTI structure.

For ease of maintenance on the VTI versions of MA and MB turbochargers, a ring was added to the existing nozzle to divide the nozzle into the inner and outer sides. Exhaust gas flows from the engine to the casing and the VTI casing is divided into two parts allowing the turbine blades and nozzle to be inspected and cleaned by removing the inner casing only, facilitating maintenance work. Since the VTI Turbochargers have the same structure as MET Turbochargers, maintenance work can also be performed easily by simply removing the inner casing, reducing the time required to open and inspect the turbocharger.

As well as the axial turbochargers covered above, Mitsubishi has also been active in the production of radial models intended for four-stroke engines and some small-bore two-strokes. Three series of radial engines were in production through to 2018. The initial MET-SR series was replaced in 2002 by the MET-SRII which continues in production and complemented by the MET-SRC series introduced in 2004. The SR series goes back to the 1980s and even existed in a variable turbine geometry variant. Long before the idea re-emerged in the 21st century as a solution for NOx reduction.

The current range of SRC turbochargers includes five frame sizes (MET18SRC, MET22SRC, MET26SRC, MET30SRC, and MET37SRC) covering engines with power outputs from 400 to 4500 kW.

In 2018, Mitsubishi announced a new range of radial turbochargers. The MET-ER series offers improved performance and more compact size compared to the MET-SR series (MET-SR-MET-SRC), which reached a cumulative production total of 15,000 units in 2018. Comparing systems with 1000 kW output, the new MET16ER is around 40% smaller than the existing MET18SRC because of the greater airflow capacity. It also offers improved maintainability with around 30% fewer components.

KBB turbochargers

A subsidiary of the Ogepar group, the German specialist KBB Turbochargers (formerly Kompressorenbau Bannewitz) offers radial- and axial-flow turbine turbochargers to serve four- and two-stroke engines.

The sixth-generation HPR series embraces four models (HPR 3000, 4000, 5000, and 6000) with radial turbines covering an engine power range from 500 to 3000 kW with a maximum pressure ratio of 5 and overall efficiencies ranging from 63% to 68%. Series production began in 2003, typical references now embracing Caterpillar MaK, Hyundai HiMSEN, Wärtsilä, Yanmar, and SKL engines.

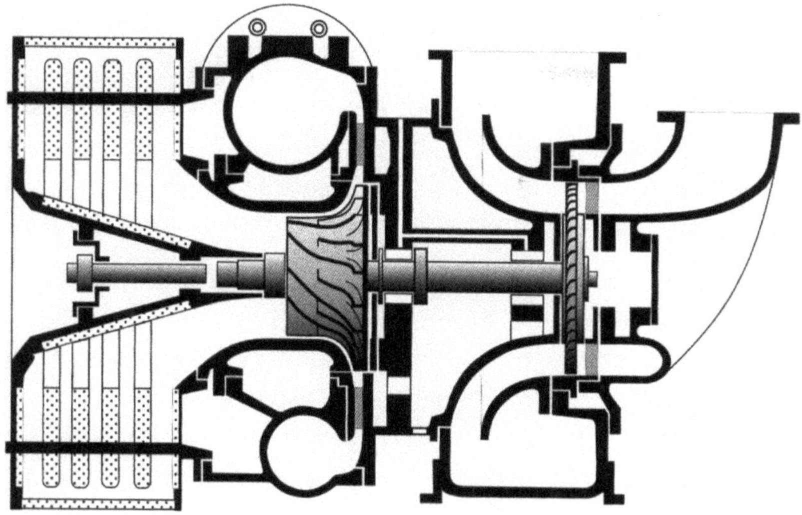

Fig. 10.28 KBB Turbochargers' axial-turbine M-series design.

The axial-turbine M series (Fig. 10.28) which once featured four models (M 40, 50, 60, and 70) serving engines with outputs from 900 to 8000 kW with a maximum pressure ratio of 4 and overall efficiencies ranging from 65% to 70% is now represented only by the M40. This model is designed for engines from 900 to 1800 kW.

A new generation turbocharger development by KBB was launched in 2009, the six-model ST27 radial-flow series addressing high- and medium-speed diesel and gas engines with outputs from 300 to 4800 kW with a maximum pressure ratio of 5.5. Continued development of the ST27 resulted in improvements to meet new NOx rules. A new design of compressor in 2016 allowed the pressure ratio to be increased to 6.0 and models with this modification are titled ST27-EP. The ST27-EP series is KBB's latest generation of turbochargers for the single-stage, high-pressure turbocharging of medium-speed engines between 500 and 6000 kW per unit. The range of both the ST27 and ST27-EP is now limited to five models.

Two-stage turbocharging features in the KBB portfolio in the form of the K2B series. First offered in 2013, K2B turbochargers provide the possibility of turbocharging diesel, heavy fuel oil, and gas engines with an engine output between 0.5 and 5 MW. The maximum pressure ratio of a turbocharging group comprising a low-pressure stage and a high-pressure stage may be 10:1.

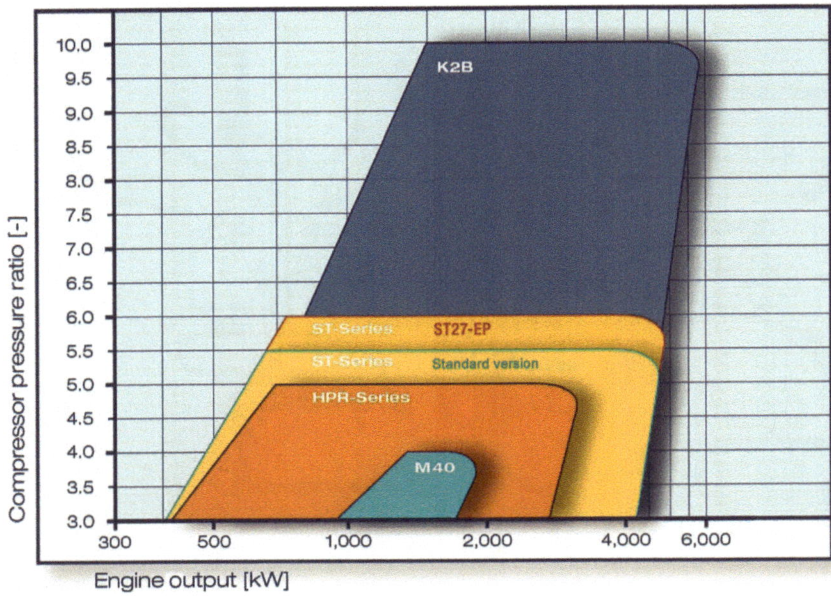

Turbocharger retrofits

The benefits of turbocharger retrofits are promoted by manufacturers, arguing that new generation models—with significantly higher overall efficiencies than their predecessors—can underwrite a worthwhile return on investment. Retrofitting older engines with new high-efficiency turbochargers can be economically justified when one or more of the following conditions prevail, according to ABB Turbo Systems:

- The engine suffers from high exhaust temperatures, air manifold pressure does not rise to its normal level, and full load can no longer be attained.
- Need to reduce fuel costs and/or exhaust emissions.
- Need to increase engine operating safety for an extended period.
- Engine performance requires optimizing for slow steaming or another new operating profile.
- Unsatisfactory lifetimes are experienced with key turbocharger elements due to erosion or corrosion; older generation turbocharger spares from some manufacturers are difficult to obtain and/or are costly, and long delivery time is quoted for replacement turbochargers.

The performance benefits from a retrofit (Fig. 10.29) is reportedly realized in lower exhaust temperatures (down by around 50–80°C) and lower fuel consumption (reduced by 3%–5%). The uncooled gas casings of new turbochargers also foster a longer lifetime for critical parts subject to wear from erosion and corrosion. Furthermore, ABB Turbo Systems notes that the ball and roller antifriction bearings normally fitted to its VTR..4-type turbochargers deliver better starting characteristics, improved part-load performance, and faster acceleration. Enhanced reliability is also promised by modern turbochargers.

Turbo compound systems

The high efficiency of some modern large turbochargers is comfortably in excess of that required to pressure charge engines, allowing some of the exhaust gas to be diverted to drive a power turbine. A number of these turbo compound installations have entered service, mainly in conjunction with MAN B&W and Wärtsilä (Sulzer) low-speed engines, the investment particularly worthwhile in an era of high fuel prices.

The power gas turbine is arranged either to supplement directly the propulsive effort of the engine via power take-in (PTI) gearing to the crankshaft or to drive a generator via power take-out/power take-in (PTO/PTI) gearing. Sulzer's efficiency booster system—based on a concept of ABB Turbo Systems—promised fuel savings of up to 5.5 g/kWh, raising the maximum engine efficiency to 54% with virtually no effect on the exhaust gas energy available for conventional waste heat recovery (Fig. 10.30).

Apart from yielding worthwhile fuel savings throughout the engine load range, the efficiency booster system can be an attractive alternative to engine derating for an equivalent brake specific fuel consumption. Compared with a derated engine, the system may allow the saving of one cylinder (depending on the type of engine and its power/speed rating), with consequent benefits in both a lower first cost and a shorter engine room.

The efficiency booster is described as a straightforward on/off the system, which can be exploited above 40%–50% engine load without undermining engine reliability. Exhaust energy is recovered simply by diverting a proportion of the exhaust gas from the exhaust manifold to a standard ABB turbocharger-type power turbine arranged in parallel with the engine turbochargers. The power turbine—forming a compact module with its integrated epicyclic gear—is mechanically coupled to the Wärtsilä (Sulzer) RTA low-speed engine's integral PTO gear. The recovered energy is thus directly fed to the engine crankshaft, allowing the engine to run at a correspondingly reduced load to deliver the same total power to the propeller shaft—but with

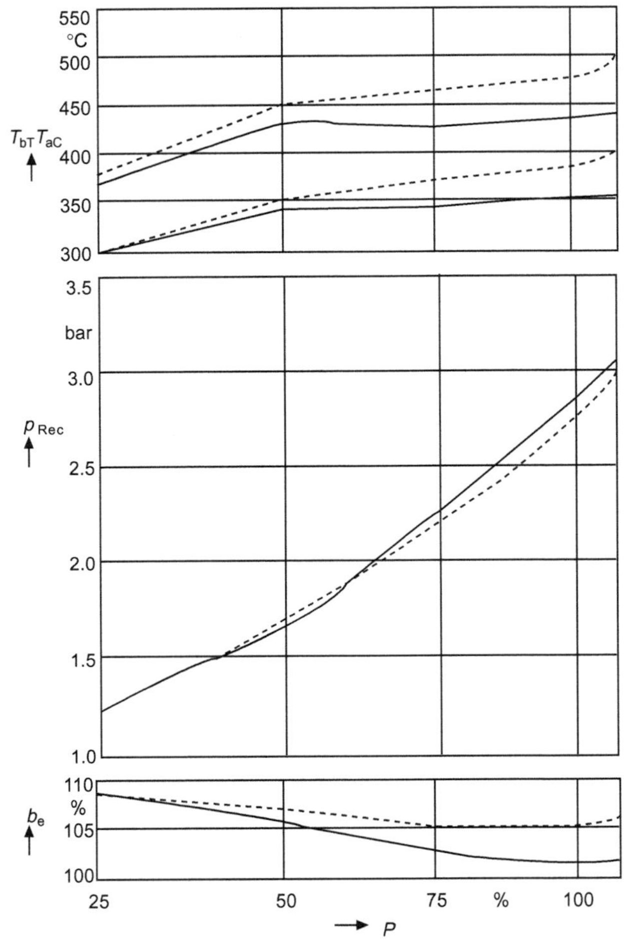

Fig. 10.29 Turbocharger retrofit benefits exhaust-gas temperatures and specific fuel consumption of an ABB VTR..1 turbocharger exhibiting low efficiency (dashed lines) compared with a replacement VTR..4 turbocharger operating with markedly higher efficiency (continuous lines).

correspondingly lower fuel consumption. The result depends on the engine's contract-maximum continuous rating since fuel savings increase with bmep.

The simple mechanical arrangement, based on standard components, underwrites a low first cost and offers a high transmission efficiency. It also allows the efficiency booster to be shut off at engine part load when manoeuvring.

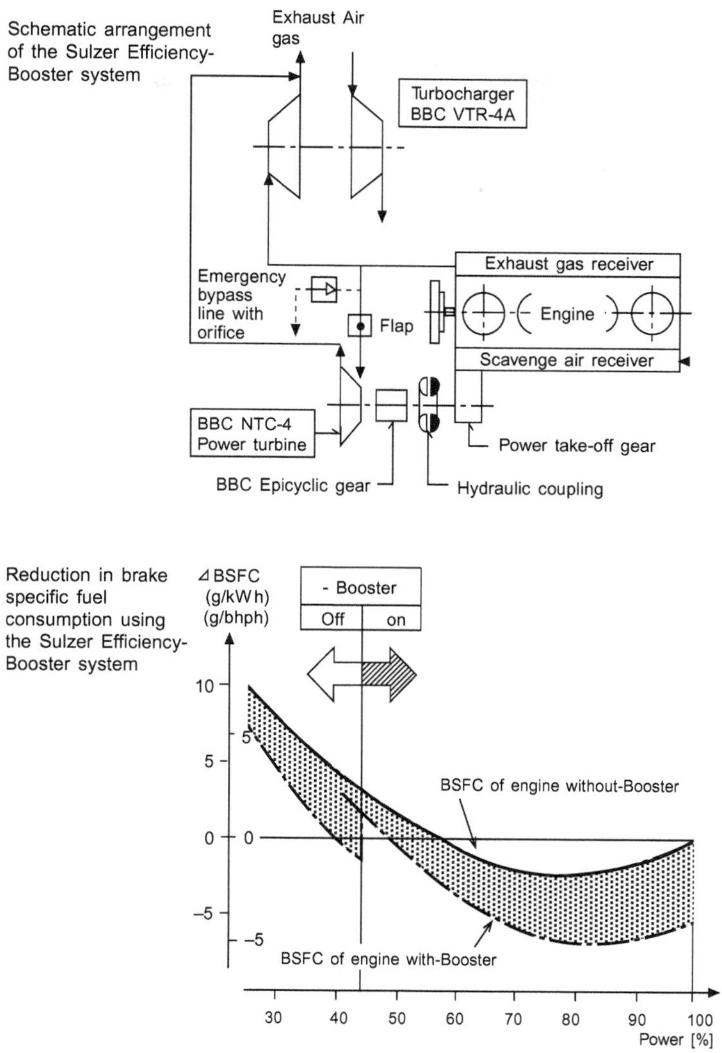

Fig. 10.30 Sulzer efficiency booster system layout and performance.

Similar benefits are reported for MAN's turbo compound system/power take-in (TCS/PTI) arrangement, which can be fitted to its MC low-speed engines with bore sizes from 600 to 900mm (Fig. 10.31). The system is not applicable to smaller models because of their size and the lower efficiency of the smaller turbochargers used. MAN and ABB power turbines are used, respectively, in association with the MAN and ABB turbochargers serving the engine. At engine loads above 50% of the optimized power, the power turbine is connected to the crankshaft via a hydraulic clutch and flexible coupling.

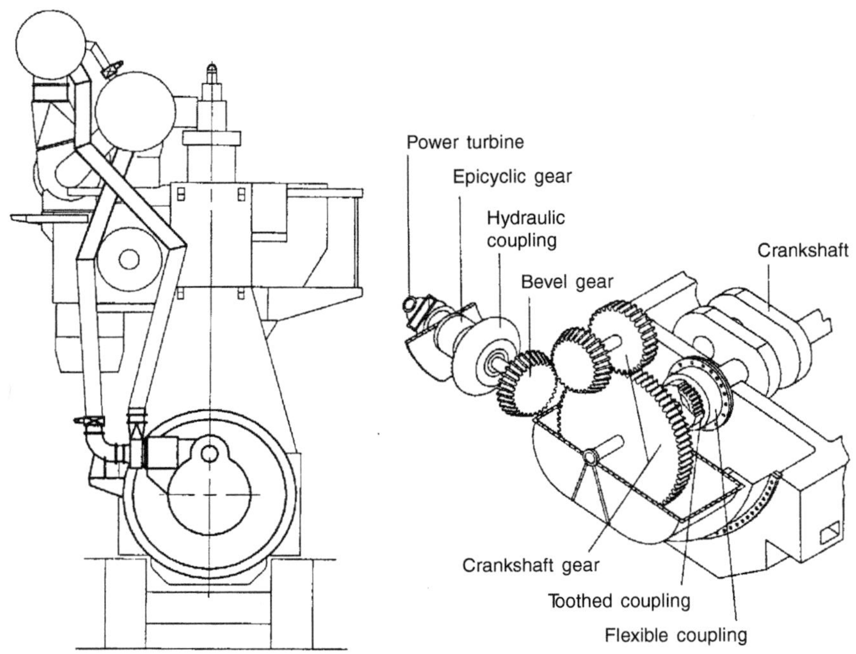

Fig. 10.31 Arrangement of MAN TCS/PTI system on the main engine and the power turbine drive to the crankshaft.

The amounts of exhaust gas actually available for the TCS depends on the turbocharger efficiency and turbocharger matching chosen for the individual case. TCS power falls off at lower loads owing to the smaller gas amounts: at 50% of the optimized engine power, the output of the TCS/PTI unit is around 25%, returning negligible savings in the specific fuel consumption of the engine. The maximum power turbine output obtainable is 4.2% of the engine's optimized power, the yield depending on the turbocharger selected.

MAN's PTO/PTI system combines a PTO serving a generator and a TCS/PTI unit (power turbine coupled to the crankshaft gear). The combined benefits of both systems are achieved without incurring the full first cost associated with each as some of the most costly elements (e.g. the crankshaft gear) are common.

Engines equipped with high-efficiency turbochargers can operate without auxiliary blowers at lower loads (down to around 25%) than engines with standard turbochargers (around 35%).

Based on its MET-MA turbocharger's turbine design, Mitsubishi has developed the MPT-A series power turbine for application in advanced waste heat recovery turbo compound systems.

Turbocharger generator/motors

Advances in high-speed generator technology as well as in high-capacity electronic switching devices have fostered the practical development of compact, high-output generators, and frequency converters suitable for use in combination with turbochargers. Such so-called hybrid turbochargers—in which a high-speed generator is attached directly to the turbocharger rotor shaft—offer these advantages over conventional turbo compound systems based on a power gas turbine, according to Mitsubishi:

- Elimination of piping and valves associated with feeding exhaust gas to a power turbine
- Control of turbocharger performance by raising and lowering generator output
- Use of the generator as a motor as a substitute for the auxiliary blower of two-stroke engines; motor mode can also be applied to accelerate the turbocharger rotor and improve turbocharger performance, particularly at the low load range of the engine
- High efficiency due to energy conversion by the turbocharger turbine.

Working in conjunction with a specialist generator manufacturer, Mitsubishi has developed a suitable high-speed/high-capacity generator for use with hybrid turbochargers. The four-pole permanent magnet synchronous machine has a rated speed of 18,700 rev/min and a maximum permissible speed of 22,300 rev/min, with a maximum output rating of 252 kW.

Mitsubishi summarizes that hybrid turbochargers incorporating a generator/motor enable the combustion air to be adjusted optimally in response to engine operating conditions by raising and lowering the amount of output power generated, or—in motor mode—the amount of electrical power used to assist the turbocharger (representing a possible new mode of controlling diesel engines).

Working with Calnetix Technologies of California, Mitsubishi's current iteration of The Hybrid Turbocharger has a high-speed permanent magnet electric generator, integrated into it, enabling electrical power to be generated from the engine exhaust. The power generated can be sufficient to allow one of the ship's auxiliary engines to be taken offline.

The Calnetix Magnaforce generator is housed within the turbocharger air inlet silencer, resulting in the total package length being only 313 mm greater than for a standard MET-MB turbocharger. When the engine is running at full load, the generator can deliver electrical power up to the equivalent of 5% of the main engine power output. Furthermore, when the vessel is slow steaming, the generator can act as a motor to boost the turbocharger's speed, enhancing engine performance and reducing fuel consumption.

CHAPTER ELEVEN

Fuel injection

The emphasis in this chapter is on four-stroke medium- and high-speed engines with camshaft-actuated individual jerk pumps for each cylinder or common rail (CR) fuel injection systems and does not apply, except in a general way, to low-speed two-stroke engines whose fuel injection systems are discussed in later chapters under individual designs.

The systems and equipment described below have generally stood the test of time and countless hours in operation in thousands of ships. With the new fuels being developed to meet the IMO 2020sulphur cap, some of the accepted properties of fuels and their effect on equipment may change. Although the sulphur levels in fuels have usually been the main influence on acidity, in some of the new low-sulphur fuels acidity levels may actually be higher due to the presence of animal or vegetable fatty acids in the fuels.

Injection and combustion

The essence of a diesel engine is the introduction of finely atomized fuel into the air compressed in the cylinder during the piston's inward stroke. It is, of course, the heat generated by this compression, which is normally nearly adiabatic, that is crucial in achieving ignition. Although the pressure in the cylinder at this point is likely to be anything up to 230 bar, the fuel pressure at the atomizer will be of the order of 1300–1800 bar.

High injection pressures at full load are beneficial in terms of fuel economy, emissions, and the ability to accept inferior fuel. Most modern medium-speed engines attain 1200–1800 bar in the high-pressure (HP) injection pipe, although some recent engine designs achieve as much as 2300 bar when pumping heavy fuel. For reasons of available technology, the earliest diesel engines had to use compressed air to achieve atomization of the fuel as it entered the cylinder (air blast injection), and while airless (or solid) injection delivered a significant reduction in parasitic loads, it also presented considerable problems in the need for high precision manufacture, and the containment of very high and complex stresses.

Pounder's Marine Diesel Engines and Gas Turbines
https://doi.org/10.1016/B978-0-08-102748-6.00011-6

The very high standard of reliability and lifetime now attained by modern fuel injection systems is the result of considerable investment in R&D and production technology by designers and equipment manufacturers. Ultra-high precision machining facilities for flat and curved surfaces are dictated for injection systems operating at pressures of up to 2000 bar and beyond, calling for minimal manufacturing tolerances down to 0.001 mm and the highest cleanliness in all processes. A particularly challenging component is the injector needle, which, in some systems, requires microscopic helical grooves to be machined into the pin shaft to hydraulically absorb lateral forces with the aid of the fuel.

In the early days of airless injection, many ingenious varieties of combustion chamber were used, sometimes mainly to reduce noise or smoke, or to ease starting, but often in part to reduce or to use the modest injection and combustion pressures. A growing emphasis on economy and specific output, coupled with materials development and advances in calculation methods allowing greater loads to be carried safely, has left the direct injection principle dominant in modern medium- and high-speed engine practice.

In direct injection systems, the fuel is delivered directly into a single combustion chamber formed in the cylinder space (Fig. 11.1), atomization being achieved as the fuel issues from small drillings in the nozzle tip.

For complete combustion of the fuel to take place, every droplet of fuel must be exposed to the correct proportion of air to achieve complete oxidation, or to an excess of air. In the direct injection engine, the fuel/air mixing is achieved by the energy in the fuel spray propelling the droplets into the hot, dense air. Additional mixing may be achieved by the orderly movement of the air in the combustion chamber, which is called 'air swirl'. Naturally aspirated engines usually have a degree of swirl and an injection pressure of around 800 bar. Highly turbocharged engines with four-valve heads have virtually no swirl, but typically have an injection pressure of 1200–1800 bar to provide the mixing energy.

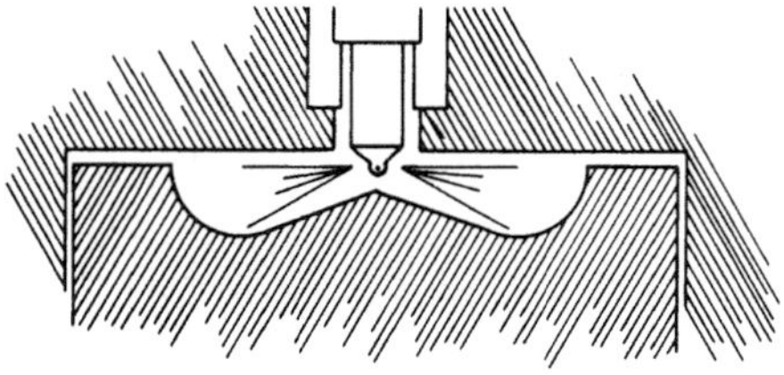

Fig. 11.1 Cross-section of the direct injection combustion chamber.

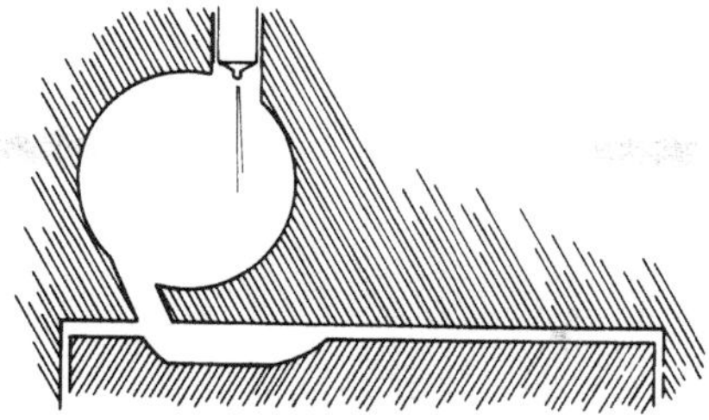

Fig. 11.2 Indirect injection: the Ricardo prechamber.

Where indirect injection is exploited, some high-speed engines retain a prechamber in the cylinder head into which fuel is injected as a relatively coarse spray at low pressure, sometimes using a single hole. Combustion is initiated in the prechamber, the burning gases issuing through the throat of the chamber to act on the piston (Fig. 11.2).

Fuel/air mixing is achieved by a very high air velocity in the chamber, the air movement scouring the walls of the chamber and promoting good heat transfer. Thus, the wall can be very hot—requiring heat-resistant materials—but it can also absorb too much heat from the air in the initial compression strokes during starting and prevent ignition. It is these heat losses that lead to poor starting and inferior economy. Further forms of assistance, such as glow plugs, have therefore sometimes been necessary to achieve starting when ambient pressures are low. The throttling loss entailed by the restricting throat also imposes an additional fuel consumption penalty.

One engine designer, SEMT-Pielstick, achieved an ingenious combination of the two systems by dividing the prechamber between cylinder head and piston crown. At the top dead centre, a stud on the piston enters the prechamber to provide a restricted outlet. On the expansion stroke, the restriction is automatically removed and fuel economy comparable with normal direct injection engines is attainable (Fig. 11.3).

Direct injection, too, has variants, which reflect the fact that, despite considerable expenditure on research into its mechanism, the detail of how combustion develops after ignition is achieved is still largely empirical. The essentials are as follows:

1. At least some of the fuel injected is atomized sufficiently finely to initiate combustion. The ignition cannot take place until a droplet of fuel has reached the temperature for spontaneous ignition. Since heat is taken up as a function of surface area (proportional to the square of the diameter) and the quantity of heat needed to achieve a temperature rise is a

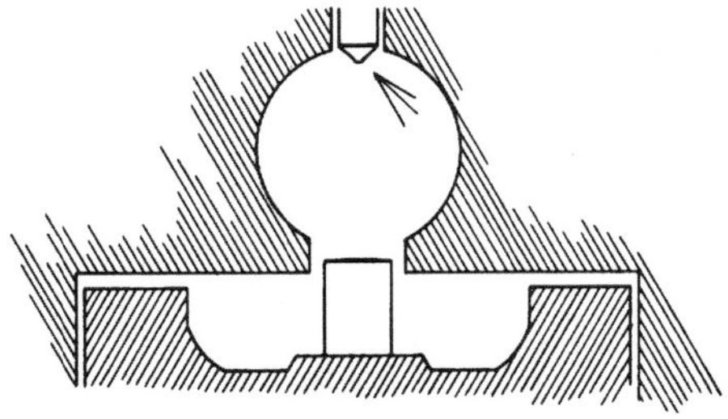

Fig. 11.3 The variable geometry combustion chamber of SEMT-Pielstick.

function of volume (varying as the cube of the diameter), only a small number of fine droplets are needed to initiate combustion. High-speed photography of combustion does indeed show that ignition takes place in a random manner near the injector tip and usually outside the main core of the spray.

2. The fuel should mix with the air in order to burn. Since most of the air in a roughly cylindrical space is, for geometrical reasons, near the periphery, most of the fuel must penetrate there, and this is easier with large droplets; hence, also, the use of wide core angles and multiple spray holes.
3. Under no circumstances must fuel reach the liner walls, or it will contaminate the lubricating oil. An advantage of combustion spaces formed in piston crowns is that the walls of the chamber form a safe target at which spray may be directed. This type of combustion chamber in the piston has the further advantage that, during the piston descent, the air above the piston periphery is drawn into the combustion process in a progressive manner.
4. The injection period should be reasonably short and must end sharply. Dribble and secondary injection are frequent causes of smoke and also of lubricating oil becoming diluted with fuel or loaded with insoluble residues. Dribble is the condition where fuel continues to emerge from the nozzle at pressures too low to atomize properly; it is caused by bad seating faces or slow closure.

The secondary injection is what happens when the pressure wave caused by the end of the main injection is reflected to the pump and then again to the injector, reaching it with sufficient pressure to reopen the injector at a relatively late stage in combustion. Any unburned or partly burned fuel may find its way onto the cylinder walls and be drawn down by the piston rings to the sump.

Injector

Working backwards from the desired result to the means to achieve it, the injector has to snap open when the timed HP wave from the pump travelling along the HP pipe has reached the injector needle valve. Needle lift is limited by the gap between its upper shoulder and the main body of the holder. Needle lift is opposed by a spring, set to keep the needle seated until the 'blow-off pressure' or 'release pressure' of the injector is reached by the fuel as the pressure wave arrives from the pump. This pressure is chosen by the engine builder to ensure that there is no tendency for the needle to reopen as the closing pressure waves are reflected back and forth along the HP pipe from the pump. The setting also has some effect on injection delay and the quality of injection; it is usually chosen to be between 200 and 300 bar.

The needle valve is invariably provided with an outer diameter on which the fuel pressure acts to overcome the spring pressure and cause the initial lift. This brings into play the central diameter of the needle so that it snaps open to the lift permitted.

The needle and the seat cones are usually ground to a different angle so that contact is made at the larger diameter (Fig. 11.4). When the needle is only slightly open, the greater restriction to flow is at the outer rather than the inner diameter of the seat. This ensures that as the needle lifts, there is a

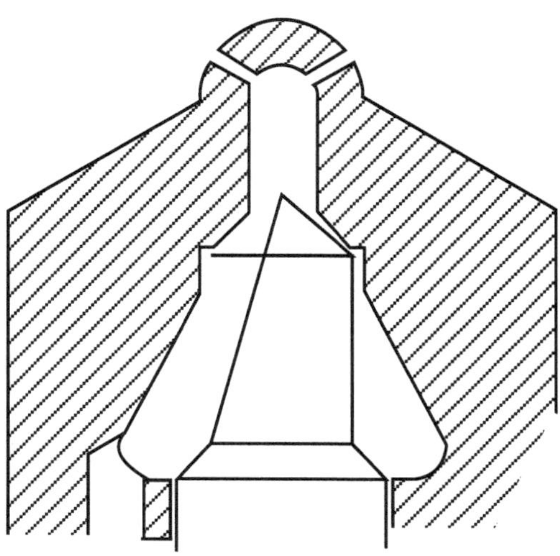

Fig. 11.4 Detail of needle and seat showing differential angle (exaggerated).

sudden change of pressurized area and the needle force against the spring changes correspondingly, giving a very rapid lift to fully open; and conversely when closing. (This is why it is bad practice to lap the needle to its seat.)

Control of the temperature of the injector, particularly of the sensitive region round the needle seat and the sac, is very important. This is especially so when heavy fuels are used, both because the fuels themselves have to be heated and because they tend to burn with a more luminous flame that increases the heat input to all the metal surfaces. If the tip temperature rises too high, the fuel remaining in the sac after injection boils and spits from the spray holes: it is not properly burned and forms a carbon deposit around each spray hole. Such carbon formation, which builds up fairly rapidly around the boundary of the sprays, causes a rapid deterioration in the quality of combustion.

With smaller engines of up to about 300-mm bore, it is usually sufficient to rely on the passage and recirculation of the pumped fuel itself to achieve the necessary cooling. The injector clearances may, in fact, be eased slightly to promote such circulation. This back leakage, normally about 1% of the pumped fuel, is led away to waste, or via a monitoring tank to the fuel tank. For larger engines, it is usually necessary to provide separate circuits specifically for coolant flow, either of treated water, light fuel, or lubricating oil.

All of these considerations lead to a relatively bulky injector and thus to tricky compromises with the available space within the cylinder head on single piston engines, where space is also needed to permit the largest possible valve areas, cages, and cooling. Nonetheless, there are performance advantages in highly rated engines in reducing the inertia of the components directly attached to the needle of the injector; and this leads fuel injection equipment manufacturers to try to find space for the injector spring at the bottom rather than at the top of the injector, as in Fig. 11.5.

Fuel line

Fig. 11.6 shows a typical fuel line pressure diagram and also the needle lift diagram for an engine having a camshaft speed of 300 rev/min, both at high load (Fig. 11.6A) and at low load (Fig. 11.6B). Fig. 11.6A shows a very brief partial reopening of the nozzle, or secondary injection, after the main opening period. In Fig. 11.6, the breaks in each trace are made to indicate intervals of 3 cam degrees.

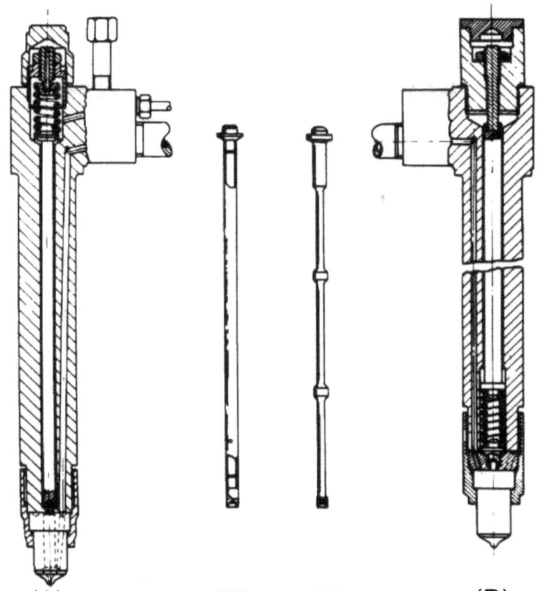

(A) (B) (C) (D)

Fig. 11.5 Comparison of traditional and low inertia injectors: (A) conventional injector, (B) and (C) lightweight spindles, and (D) low-inertia injector (Delphi Bryce).

The fuel line has to preserve as much as possible of the rate of pressure rise, the maximum pressure, and the brevity of injection that the fuel pump has been designed to achieve. To do this, the fuel line must have the lowest practical volume and the maximum possible stiffness against dilation. There is also a limit to the minimum bore because of pressure loss considerations.

The only real degree of freedom left to the designer is to make the fuel line as short as possible. Even so, a given molecule of fuel may experience two or three injection pulses on its way from the pump chamber to the engine cylinder in a medium-speed engine running at full load, rising to 30 or more at idling.

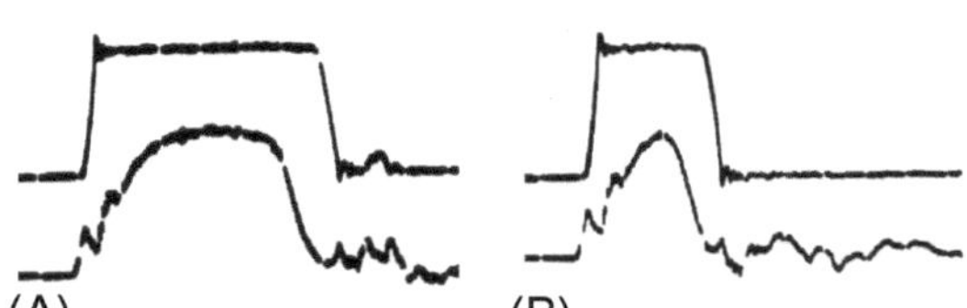

(A) (B)

Fig. 11.6 Fuel line pressure (lower line) and needle lift diagrams. (A) At high load. (B) At low load (Delphi Bryce).

If the fuel pump is mounted at a greater distance from the injector, the elastic deformation of the longer pipe and compression of the larger volume of fuel will dissipate more and more of the pressure developed at the fuel pump and prolong the period of injector needle lift compared with the pump delivery. In all medium-speed engines, this compromises on injection characteristics, and its adverse effect on fuel economy (and heat flux in combustion chamber components) is considered unacceptable, and each cylinder has an individual fuel pump with an HP fuel line as short as possible.

The smaller the engine, the more manufacturers (in fact all manufacturers of automotive engine sizes) opt for the production convenience of a camshaft pump (i.e. the pump elements serving all or several cylinders are mounted in one housing with their own camshaft) and accept any compromise in performance. In such engines, a given molecule of fuel may endure 20 or more injection pulses while migrating along the HP pipe at full load. In such cases, the elastic characteristics of the HP pipe (and the trapped fuel) come to equal the pumping element as a major influence on the qualities and duration of injection.

Naturally, all injection pipes must have identical lengths to equalize cylinder behaviour. On a 150-kW auxiliary engine of approximately 150-mm bore running at 1200 rev/min, the effect of a 600-mm difference in pipe length, due to running pipes each of different length directly from the pump to the cylinder it served, was 8° in effective injection timing, and consequently about the 25-bar difference in firing pressure.

Camshaft or block pumps are, of course, much cheaper, and for a higher speed engine, one maker (Deutz) adopted the innovative solution of grouping up to three or four pump elements in one housing, each housing being mounted close to the cylinders concerned.

The HP fuel lines are made of a very high-quality precision-drawn seamless tube almost totally free from internal or external blemishes. They must, moreover, be free of installation stress or they risk fracture at the end connections, and they must be adequately clipped, preferably with a damping sleeve, if there is any risk of vibration.

Fuel line failures due to pressure are not as frequent as they once were, thanks to improvements in the production of a seamless tube. Nevertheless, current regulations insist that in marine use fuel lines are sleeved or enclosed to ensure that in the event of a fracture, escaping fuel is channelled to a tank where an alarm may be fitted. This is to avoid the fire hazard of fuel finding a surface hot enough to ignite it, or of accumulating somewhere where it could be otherwise accidentally ignited. Most designers now also try to ensure that leaking fuel cannot reach the lubricating oil circuit via the rocker gear return drainage or by other routes.

Several manufacturers shorten and segregate the HP line by leading it to the injector through, rather than over, the cylinder head. Some others incorporate the pump and the injector in a single component, avoiding the HP pipe altogether (see the next section).

Pump

The final item for consideration is the fuel injection pump, which is the heart of the fuel injection system. Essentially, the pumping element is a robust sleeve or barrel, which envelops a close-fitting plunger. Each one is produced to a very fine surface finish of 1–2 μm in and to give a clearance of approximately 7–10 μm depending on the plunger diameter.

In its simplest form (Fig. 11.7), the barrel has a supply port at one side and a spill port at the other. (In early practice these ports were combined.) The plunger is actuated by the fuel cam through a roller follower. Before, and as the plunger starts to rise (Fig. 11.7A), the chamber is free to fill with fuel and to spill back into the gallery in the pump housing outside until the rising top edge of the plunger closes the supply and spill ports (Fig. 11.7B). Thereafter, the fuel is pressurized and displaced through the delivery valve towards the injectors. The initial travel ensures that the plunger is rising fast when displacement commences so that the pressure rises sharply to the desired injection pressure.

When the plunger has risen far enough, a relieved area on it uncovers the spill port, the pressure collapses, and injection ceases (Fig. 11.7C). The relief on the plunger has a helical top (control) edge so that the rotation of the plunger by means of the control rod varies the lift of the plunger during

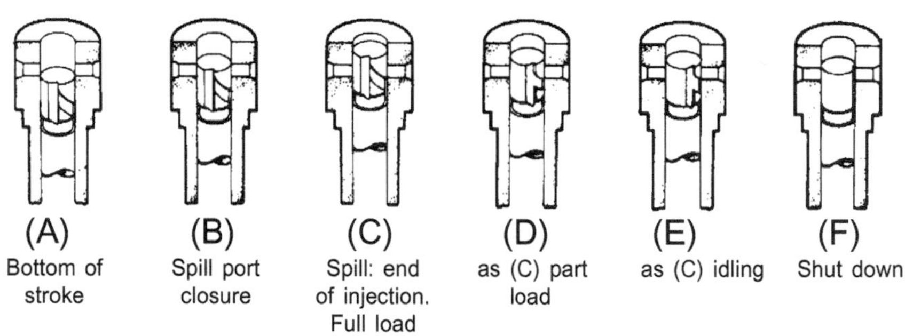

Fig. 11.7 Principle of operation of the fuel injection pump (Delphi Bryce).

which the spill port is closed, and therefore the fuel quantity injected and the load carried by the engine, as shown in Fig. 11.7C–E. At minimum setting, the helical edge joins the top of the plunger, and if the plunger is rotated so that this point, or the groove beyond it, coincides with the spill port, the latter is never closed. In that case, no fuel is pumped and the cylinder cuts out (Fig. 11.7F). The area of the plunger between the top and the control edge is termed 'the land'.

There are, of course, many complications of this simple principle. The first is that a delivery valve, essentially a nonreturn valve, is needed at the pump outlet to keep the fuel line full, as shown in Fig. 11.8. The delivery valve usually has an unloading collar to allow for the elasticity of the pipe and fuel and to help eliminate the pressure pulsations, which cause secondary injection. Some types of delivery valve, however, sense and unload the line pressure directly.

The second cause of complication is that when the spill port opens and a pressure of anything up to 1600 bar is released, cavitation and/or erosion is likely to occur, affecting both the housing directly opposite the port, and the plunger land, which is still exposed to the port at the instant of release. Modern designs seek to eliminate these effects by using, respectively, a sacrificial (or in some cases a specially hardened) plug in the housing wall, and by special shaping of the ports.

Another complication is the filling process. When the pump spills, further plunger lift displaces fuel through the spill port. As it falls on the descending flank of the cam, the opposite effect applies until the falling control edge closes the spill port. Further fall, under the action of the plunger return spring, has to expand the fuel trapped above the plunger until the top

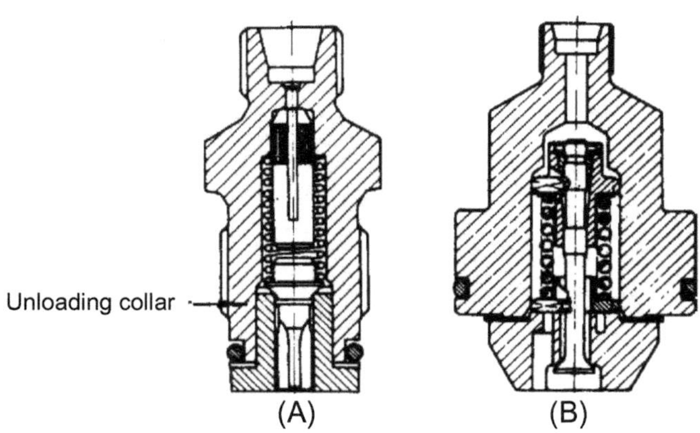

Fig. 11.8 Delivery valves (A) volume unloading and (B) direct pressure unloading (Delphi Bryce).

edge uncovers the supply port again. It can do this only by expanding trapped air and fuel vapour, and by further vaporization, so that a distinct vacuum exists, and this is exploited to refill the pump.

There is also the effect of dilation due to pressure, the need for lubrication, and the need to prevent fuel from migrating into spaces where it could mingle with crankcase oils.

Early pumps could get by without undue complications: the fuel itself provided adequate plunger lubrication, and fuel which had leaked away in order to do this was simply rejected to a tank which was emptied to waste.

Today, grooves are provided on the plunger or in the barrel wall. The first groove collects the leaking fuel. Where there are two grooves, the second will spread lubrication, but there is sometimes an additional groove between them to collect any further migration of either fuel or lubricant. Fig. 11.9 shows a section of a modern heavy-duty fuel pump incorporating most of the features described earlier.

The pump is actuated by a follower that derives its motion from the cam. Block pumps incorporate all of these components in common housing for several cylinders, while individual pumps used on some larger engines incorporate the roller. This not only makes servicing more straightforward but also ensures unified design and manufacturing responsibility.

The roller is a very highly stressed component, as is the cam, and close attention to the case hardening of the cam and the roller transverse profile is essential to control Hertzian (contact) stresses. The roller is sometimes barrelled very slightly, and it must bear on the cam track parallel to the roller axis under all conditions. On the flanks of the cam, this is a function of slight freedom for the follower assembly to rotate, while on the base circle and the top dwell of the cam, it depends on precise geometry in the assembly.

The roller pin requires positive lubrication, and the cam track must at least receive a definite spray. Lubrication channels have to be provided from the engine system through the pump housing to the roller surface, its bearings, and the lower part of the plunger.

Timing

The desired timing takes into account all the delays and dynamic effects between the cam and the moment of ignition. The desired timing is that which ensures that combustion starts and continues while the cylinder pressure generated can press on the piston and crank with the greatest mechanical advantage and is complete before the exhaust valve opens at 110–130°

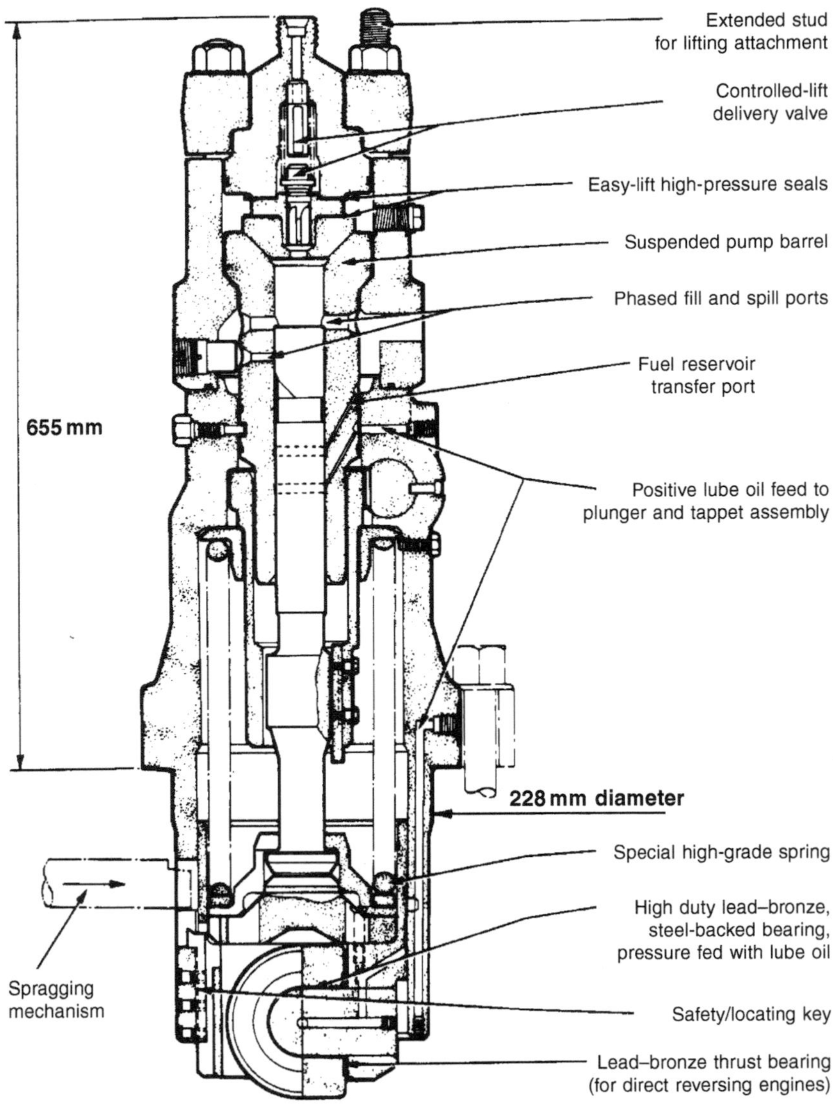

Fig. 11.9 Cross-section of Bryce FFOAR-type flange-mounted unit pump suitable for four-stroke engines with crankshaft speeds of up to 450 rev/min (Delphi Bryce).

after top dead centre. The engine builder always specifies this from his/her development work, but it usually means that the only settable criterion, the moment of spill port closure (SPC), is about 20–25° BTDC.

In earlier practice, this was done by barring (or manually turning over) the engine with the delivery valve on the appropriate cylinder removed, and with the fuel circulating pump on (or more accurately with a separate small gravity head) and the plunger rotated to a working position. When the fuel ceased to flow, the spill port had closed. Sometimes this criterion was used to mark a line on a window in the pump housing through which the plunger guide could be seen, and to coincide with a reference line on the guide.

The accuracy of this method is, however, disappointing. There are differences between static and dynamic timings, which vary from pump to pump, and many makers now rely on the accuracy of manufacture and specify a jig setting based on the geometry of the pump and the SPC point to set the cam rise or follower rise at the required flywheel (crank) position.

Some makers allow users to make further adjustments of the timing according to the cylinder pressure readings, but this places a premium on the accuracy, not only of the calibration of instruments but also of the method itself. It should be borne in mind that neither the instruments nor the method is as accurate as the pump manufacture by an order of magnitude.

If the cam is fixed, it is usually possible, within stated limits, to adjust the height of the tappet in relation to the cam. Raising the cam has the effect that the plunger cuts off the supply spill port (and starts injection) sooner for a given cam position, and vice versa (Fig. 11.10). The limit is set at one extreme by the need for there to be 'top clearance' at the top of the follower

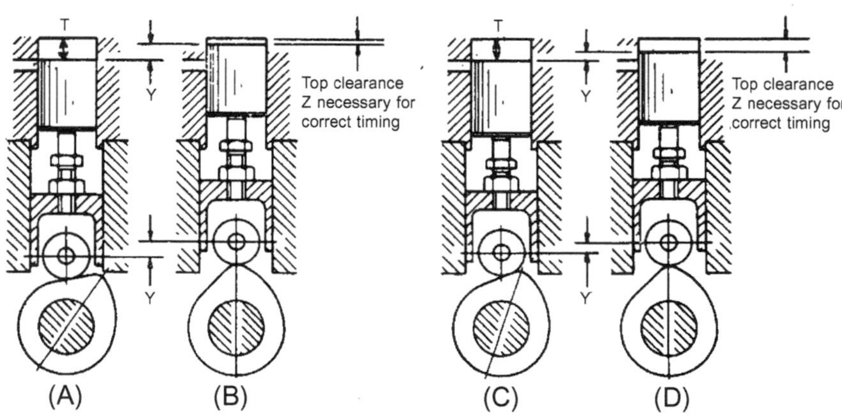

Fig. 11.10 Adjusting timing by top clearance. Configuration of the cam, tappet and pump at the commencement of injection (A) and (C), and maximum cam and follower lift (B) and (D). In each case, the pump is set correctly for the required injection point. Part (A) shows the case for an over-retarded cam. Part (C) shows the case for an over-advanced cam within permitted limits of top clearances. In either case Y (the residual lift) + Z = the timing dimension of the pump, T (GEC Ltd).

lift (i.e. of the plunger stroke) between the top of the plunger and the underside of the delivery valve in order to avoid damage. Too small a clearance also means that injection is commencing too near the base of the cam, and therefore at too low a velocity to give the required injection characteristics. On the other hand, too large a clearance (to compensate for an over-advanced cam by lowering the plunger) risks interfering with the proper spill function, which terminates injection.

The cam with its follower may be designed to give a constant velocity characteristic throughout the injection stroke. Recent designs with a high initial velocity aim to give a constant pressure characteristic, but many manufacturers find that the performance advantage or stress advantage over the constant velocity design is too small to warrant the greater production cost.

Uniformity

Bearing in mind the random way in which ignition occurs during each injection, and perhaps also because the filling process of the pump chamber is largely dependent on suction, it is not surprising that successive combustion cycles in the same cylinder at the same steady load are not identical. A drawcard taken slowly to show a succession of cycles will show this visibly, and a maximum pressure indicator will show by feel, or otherwise, that maximum pressures vary from cycle to cycle, perhaps by 5 bar or more. (Any governor corrections will affect, too, but progressively over several cycles.) These variations are a fact of life, and not worth worrying about unless they are 20 bar or more. It does mean that if indicator cards are being taken as a check on cylinder performance, several cycles per cylinder should be taken to ensure that a properly representative diagram is obtained.

Differences will also occur between cylinders for reasons of component tolerance (unless compensated adequately by the setting method), and because of the dynamic behaviour of the camshaft, which deflects in bending due to cam loading and torsionally due to the interaction of the cam loadings of adjacent cylinders. These will characteristically fall into a pattern for a given engine range and type.

While engine builders should (and usually do) allow for variations in cylinder behaviour to occur, it is obviously good practice to minimize these to achieve maximum economy with optimum stresses. However, it would be wise to bear in mind that it is inherently very difficult to make pressure and temperature measurements of even the average behaviour of combustion gases in the cylinders.

To achieve, without laboratory instrumentation techniques, an accuracy which comes within an order of magnitude of the standard of accuracy to which fuel injection components are manufactured and set is impossible. This is particularly true of exhaust temperature as a measure of injection quantity, notwithstanding that this practice has been generally accepted by many years of use. Fuel injection settings on engines should not, therefore, be adjusted without good reason, and then normally only on a clear indication that a change (not indicative of a fault) has occurred.

Fuel quantity imbalance is far from being the only cause of a variation in the pattern of exhaust temperatures, nor the timing of firing pressure. Consider, for instance, variations in compression ratio, charge air mass, scavenge ratio, not to mention calibration errors and the accuracy of the instruments used.

Fuel injection pumps (and injectors) have their own idiosyncrasies. The relationship between output and control rod position (the calibration) is not absolutely identical in different examples of pumps made to the same detail design. The output of a family of similar pumps can be made sensibly identical at only one output, namely, at the chosen balance point. At this output, a stop, or cap, on the control rod is adjusted to fit a setting gauge (or, in older designs, a pointer to a scale), so that it can be reproduced when setting the pumps to the governor linkage on the engine.

Obviously, the calibration tolerances will widen at outputs away from the balance point, which is usually chosen to be near the full-load fuelling rate. As the tolerances will be physically widest at idling and will also constitute a relatively larger proportion of the small idling quantity (perhaps 35%), the engine may not fire equally well on all cylinders at idling, but this sort of difference is not sufficient to cause any cylinder to cut out unless another fault is present.

When a pump is overhauled, it is advisable that its calibration should be checked, and the balance point reset. This ideally entails the use of a suitable heavy-duty test rig, but in the largest sizes of the pump, it is sometimes more practical to rely on dimensional settings (D.A.W.).

Developments and trends

Fuel injection systems have a significant influence on the combustion process and hence a key role to play in improving engine fuel consumption and reducing noxious exhaust emissions. The following characteristics of an injection system are desirable in achieving these goals:

- Injection pressures during the whole process should be above 1000–1200 bar for a good spray formation and air–fuel mixture; a tendency in practice to 1600–1800 bar and higher is noted.
- The total nozzle area should be as small as possible in relation to cylinder diameter for good combustion, particularly at part load.
- Total injection duration should be 20° of crank angle or less for achieving a minimum burning time in order to exploit retarded combustion for reduced NOx emissions without loss in efficiency. A high compression ratio is desirable.
- High pressures at the beginning of injection promote reduced ignition delay, while increased mass flow can result in overcompensation and increased pressure gradients. Consequently, rate shaping is necessary in some cases, particularly with high-speed engines.
- High-aromaticity fuels cause increased ignition delay in some cases. Preinjection with high injection pressures is necessary and can achieve nonsensitivity to fuel quality.
- Electronically controlled adjustment of injection timing should be applied for optimized NOx emissions at all loads, speeds, and other parameters.
- The load from the torque of the injection equipment on the camshaft and/or the gear train should be as low as possible in order to prevent unwanted additional stresses and noise.
- For safety reasons, even a total breakdown of electrical and other power supplies should not result in the engine stopping.

To understand fully the complex hydraulic events during the fuel injection process, the Swiss-based specialist Duap says that it is essential to appreciate the function of all the elements (pump, pressure pipe, fuel valve, and nozzle) forming the injection system.

It is commonly believed by nonspecialists that the plunger within the pump pushes the fuel upwards like a pillar, thus affecting the immediate injection of fuel into the combustion chamber. The reality, however, is that the plunger moves with such a high speed that the fuel likely to be conducted is highly compressed locally (elasticity of fuel). The compressed fuel now generates a pressure wave, which runs through the pipe and valve, causing the nozzle to open and inject. The pressure wave of the fuel is forced through the system at a speed of around 1300 m/s. In the same way that sound in the air is reflected by houses and hills, the wave is reflected between the fuel valve and the pump. It may therefore easily run back and forth between those components several times before the nozzle is actually forced to open and inject.

Duap warns that this process underlines the importance of maintaining the injection system in an excellent condition and ensuring that the fuel is properly treated and free of dirt. If, for example, the nozzle spray holes are partially blocked by extraneous elements or carbon particles, the pressure wave may not be sufficiently reduced within the system. This eventually results in the destruction of the fuel pump cam or other vital parts of the injection system upon the next stroke.

The task of the injection system is to feed fuel consecutively to each cylinder within a very short period of time (0.004–0.010 s, depending on the engine revolutions per minute). It is also essential that the same amount of fuel is delivered with each stroke: deviations in the quantity supplied to different cylinders will adversely affect the performance of the engine and may result in crankshaft damage due to resonance.

All of the key elements of the system must be manufactured to very small tolerances; the clearance between plunger and barrel, for example, is not larger than 4–16 μm (depending on the plunger diameter). Such a high-precision clearance also dictates an adequate surface quality (within 0.2–0.5 μm), a finish which can be achieved only by careful grinding and lapping.

Before the fuel is pushed into the pressure pipe (linking the pump and fuel valve), it has to pass the pressure valve, which has several tasks. The first task is to separate the pump hydraulically from the pressure pipe after the fuel has passed the valve; the second is to smooth the pressure wave running back and forth within the pipe: this calming down is necessary to secure the proper closing of the fuel valve without having additional and uncontrolled injection, and the third is to maintain a certain pressure within the pressure pipe for the next injection stroke (rest pressure). The fulfilment of these tasks can be guaranteed only by using pressure valves manufactured to very tight tolerances (bearing in mind that the valves play an important role in ensuring that identical amounts of fuel are delivered to each engine cylinder).

Another highly underestimated component of the fuel injection system is the pressure pipe. Despite its small size and apparent simplicity, Duap notes that the pipe has to endure pressure waves of up to 1800 bar; even the smallest mark may thus lead to fracture.

The last link in the injection system chain is the fuel valve (nozzle holder assembly and nozzle), which is designed to inject and atomize the fuel into the combustion chamber when the pressure wave has reached a predetermined strength.

Unit injector versus pump/pipe/injector

A unit injector is more likely to be considered for engines today than in the past, but the choice between unit injector (Fig. 11.11) and pump/pipe/injector (Fig. 11.12) systems can be complex. The UK-based specialist Delphi Bryce (formerly Lucas Bryce and now part of the Woodward Governor group) cites the following merits and demerits of the two configurations:

1. *Unit injector positive features*: hydraulic volumetric efficiency (smaller plunger), longer nozzle life, fewer components, no high-pressure pipe, and fast end of injection. More complex cylinder head designs and higher cylinder head loads are negative features.
2. *Pump/pipe/injector system positive features*: less space required by the injector, facilitates the use of residual fuels and water injection, simple pushrod

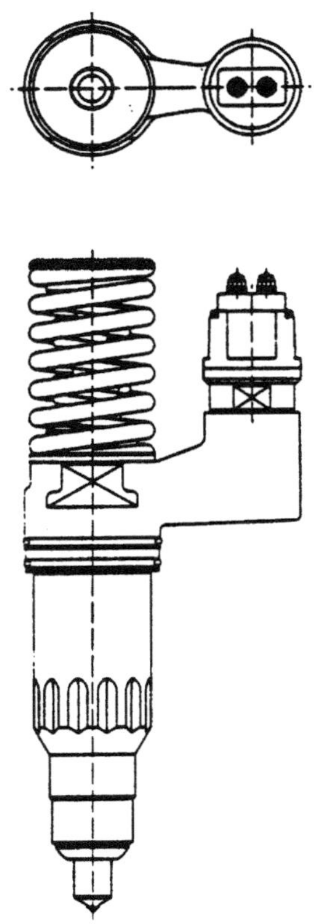

Fig. 11.11 Outline of the electronic unit injector (Delphi Bryce).

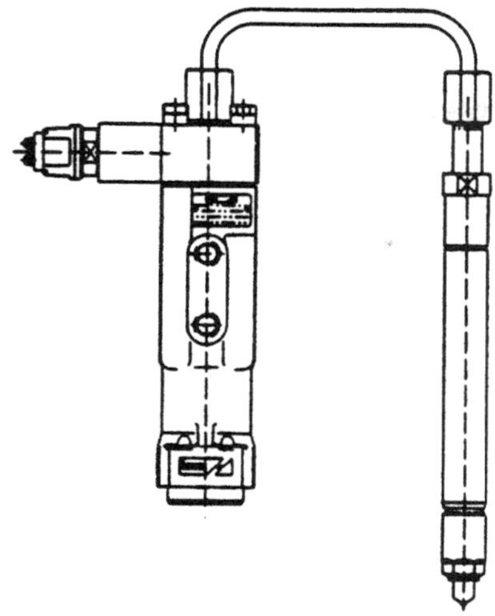

Fig. 11.12 Pump/pipe/injector system (Delphi Bryce).

mechanism, generally easier servicing, and higher low load operating pressure. Large trapped fuel volume and parasitic power loss are negative features.

Unit injectors

The initial concern with the application of unit injectors is usually cylinder head installation and the drive. Increased space demands and the need to provide fuel passages (and cooling passages in the case of residual fuel operation) can conflict with cylinder head air and exhaust passages and cooling requirements. Such issues can be resolved through close cooperation at the design stage.

A unit injector has a low trapped fuel volume, which yields several major benefits:

- Pumping efficiency is high, allowing smaller plungers to be used and hence fostering reduced camshaft loading, which can be important as a means of minimizing camshaft diameters.
- Very responsive to drive input, thus reducing hydraulic delays. The injection rate diagram more closely follows, in both time and shape, the cam profile, making it easier to optimize injection system performance requirements.
- Fast end of injection with limited HP wave effects; this eliminates the need for a delivery valve and, experience has shown, leads to enhanced durability, particularly of the nozzle.

The fuel system drive must be sufficiently stiff to counteract the high operating loads, or the hydraulic performance benefits can be lost. A typical unit

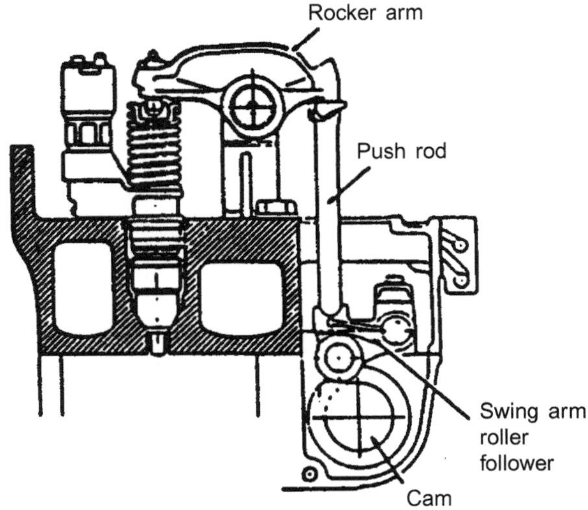

Fig. 11.13 Typical unit injector drive (Delphi Bryce).

injector drive is illustrated in Fig. 11.13. The responsiveness of unit injectors can lead to greater pressure turndown at reduced speed and loads greater than that exhibited by pump/pipe/injector systems. This can be addressed by cam design selection. The cam design can be tailored to achieve maximum plunger velocity when delivering low fuel quantities; Fig. 11.14 shows the effect of using a falling rate cam on the injection pressure.

Unit injectors offer many advantages, particularly as operating pressures increase towards 2000 bar, but they have installation drawbacks. In practice, unit injectors are limited to engines of under 300-mm bore. In larger engines, Delphi Bryce suggests that they are bulky and difficult to handle, and their withdrawal height from the cylinder head can pose a problem in some applications.

Pump/pipe/injector systems

Engine designs with bore sizes larger than 300 mm can benefit from the advantages of pump/pipe/injector systems. The primary requirement for installation is to minimize the trapped HP fuel volume, which is usually achieved by having a high camshaft and the HP connection to the injector as low as possible through the cylinder head. The resulting low trapped fuel volume maximizes hydraulic efficiency and responsiveness.

Pressure wave effects in the HP passages will still be sufficient to require a delivery valve to eliminate secondary injection and cavitation erosion.

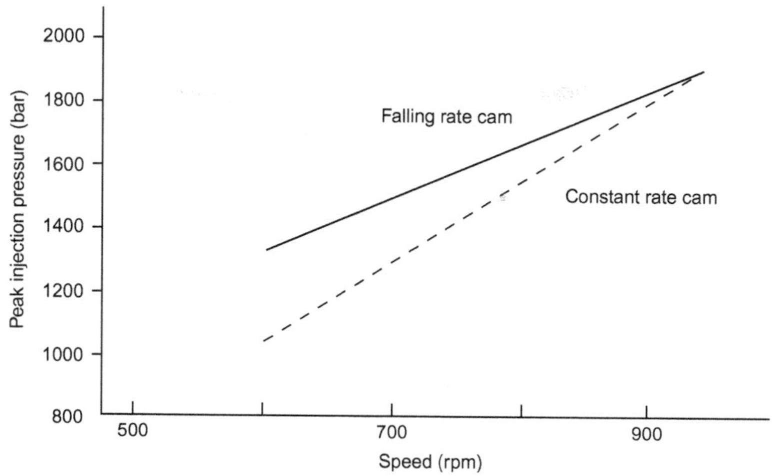

Fig. 11.14 Effect of cam design on the injection pressure (Delphi Bryce).

Pressure unloading delivery valves have been used for many years and, with experience, have become reliable. The need to create the fastest possible pressure collapse at the end of injection and a trapped residual pressure and to ensure the durability of the valve components, led to a valve design with a series of tailoring features (Fig. 11.15). Valve opening pressures, an unloading feature on the pro-flow valve and a control orifice can all be used to tune hydraulic performance.

Pump/pipe/injector systems have hydraulic disadvantages but offer flexibility. Elements and valves can be changed relatively easily during servicing

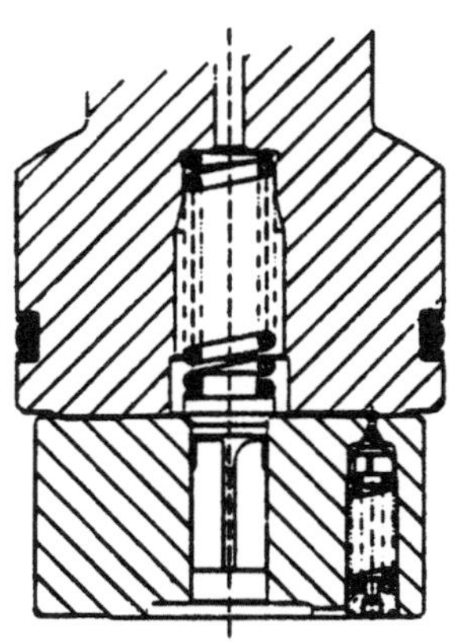

Fig. 11.15 Typical pressure unloading delivery valve design (Delphi Bryce).

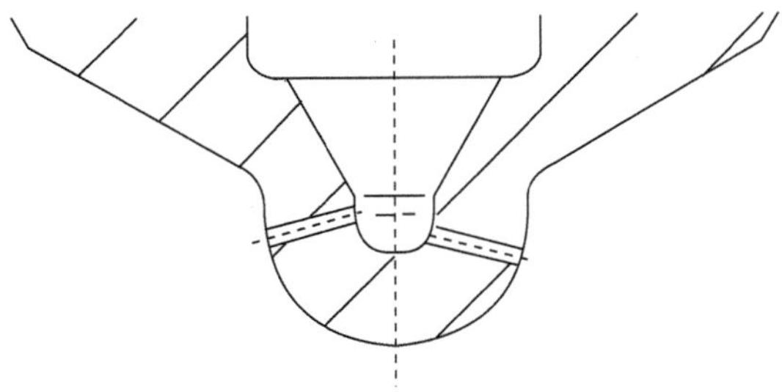

Fig. 11.16 Typical nozzle 'sac end' design; low volume for reduced emissions (Delphi Bryce).

and to offer uprating opportunities. Pump/pipe/injector systems also facilitate water injection and HP gas injection, which are of increasing interest to engine designers.

The nozzle is the interface between the injection system and the combustion process: good nozzle performance is essential for good combustion. The detailed design of the nozzle seat, sac, and spray hole entry is vital in minimizing pressure-flow losses and maximizing energy in the spray (Fig. 11.16). This ensures minimum droplet size and an even spray distribution and penetration. Abrasive flow machining techniques are typically exploited to secure rounded edges and polished surfaces to improve discharge coefficients.

Major durability and reliability concerns for injection equipment have been cavitation erosion at HP spill and element seizure. Cavitation erosion is being addressed by improved delivery valve designs and pressure decay techniques at the barrel port/fuel gallery. To promote seizure resistance at the increased plunger velocities required to generate higher pressures, it is necessary to improve element geometries and control surface textures. The use of surface coatings is also becoming widespread.

Programs are available to design fuel cams and analyse unit injector rocker arrangements. The results can be used to predict injection pressure, period, and rate. Accurate system modelling is a considerable aid in reducing application development time and cost; a wide range of variables can be analysed with confidence.

Electronic fuel injection

It was inevitable that electronic fuel injection technology, successfully introduced for high-speed diesel engines, would eventually be applied to benefit medium-speed designs.

Lucas Bryce (which became Delphi Bryce, now part of the Woodward Governor group) first demonstrated electronic fuel injection in the late 1970s, showing that improved fuel consumption and emissions could be achieved. The market was not ready; however, the technology was new and there was then no environmental legislation to provide the incentive to improve the performance of engine emissions. The introduction of electronic fuel injection in the heavy truck market to meet stringent emission controls gave that impetus and the investment to advance the technology.

The Lucas Bryce system comprised a single-cylinder, electronically controlled plunger pump, which supplies fuel to the injector via an HP pipe, and is driven directly from the engine camshaft; alternatively, an electronically controlled unit injector may be used (the choice depends on the engine design). Fuel control is achieved by solenoid valves operated from the electronic control unit, which is called the smart drive unit (SDU) by Delphi Bryce. Lucas Electronics also developed an engine governor, called a control and protection card (CPC); in some cases, however, the governor is supplied by specialist manufacturers or the engine builder.

With larger volume fuel flows to handle, Lucas Bryce's main effort was to achieve a suitable hydraulic performance while using the solenoid/valve capsule developed for the truck engine market by its then sister company Lucas EUI Systems. This approach avoided the potentially prohibitive cost required for a special production line for a low volume solenoid assembly.

Fuel is introduced via a port into space above the pump plunger. As the plunger rises beyond the port, fuel is displaced through the open solenoid and past the secondary valve seat. The solenoid is energized, the valve closes, and pressure builds up above the secondary valve. The pressure drops across the secondary valve orifice, and then closes it rapidly. With the spill passages closed, fuel is forced through the delivery valve to the injector. At the computed time for the end of injection, the solenoid is de-energized and pressure collapses above the secondary valve, causing it to open and spill fuel. The injector nozzle then closes to finish injection.

As well as limiting harmful engine emissions, Delphi Bryce claimed its electronic fuel injection system offered enhanced reliability, diagnostic capability, optimized timing, and fuel control for all load and speed conditions, including transient operation.

Electronically controlled fuel injection equipment is now playing a key role in meeting exhaust emissions legislation. In isolation, however, electronic systems do not provide a total solution. Other important complementary developments are higher injection rates, higher injection pressures, balanced systems, initial rate control, optimized spray pattern, flow-controlled nozzles, and low-sac volume nozzles.

The move to electronic control of fuelling brings advantages in the accuracy of delivery and timing and underwrites sophisticated scheduling of those parameters over the full range of engine operating conditions. Developments are expected in many areas, Woodward Governor (Delphi Bryce) citing:

- Pilot injection/rate shaping and control
- Skip cylinder firing and reduced cylinder operation under low loads
- Individual cylinder balancing by end-of-line test or adaptive software compensating for in-service wear
- Dual-fuel engine management systems
- Variable fuel quality compensation
- Combustion diagnostics
- Advanced closed-loop controls with 'smart sensors'
- Water injection controls
- Engine health monitoring/failure prediction for preventative maintenance
- Data collection for engine life history monitoring

Electronic fuel control offers the opportunity to cut out fuel to individual cylinders, thereby keeping cylinders equally warm and preventing other problems which can occur on unfired cylinders such as lubricating oil 'souping' in the exhaust passages. Cylinder cut-out is also applied on some engines to secure more stable idling.

Generally, Woodward Governor (Delphi Bryce) explains that if a transducer can be fitted to the engine to measure a feature which can be beneficial to enhancing engine control, then it can be incorporated into an electronic control/engine management system. An obvious example is cylinder pressure, whereby fuel delivery can be continually adjusted within agreed maximum limits throughout the operating period to the next overhaul to compensate for cylinder wear and injection system wear.

Common rail injection systems

In mechanical fuel injection systems, the injection pressure is a function of engine speed and load. At low load, therefore, the pressure drops, and the result is very large fuel droplets, which survive as such until they contact the combustion space surfaces. CR fuel injection technology offers the possibility to maintain the same high injection pressure (and hence small droplets) at all loads down to idling (Fig. 11.17). Highly efficient and clean

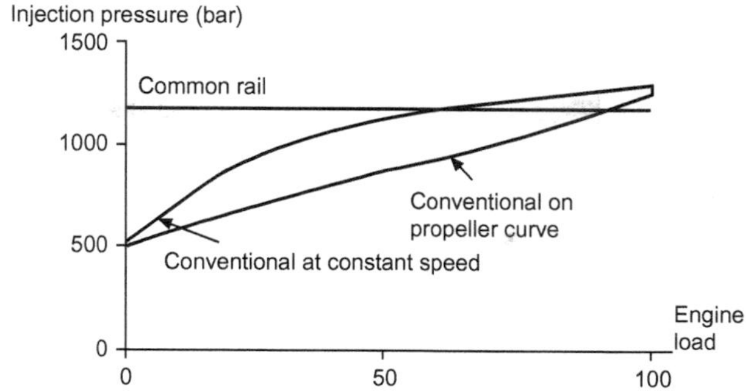

Fig. 11.17 In contrast to conventional systems, CR fuel injection allows a high injection pressure and small fuel droplets to be maintained down to idling (Wärtsilä).

combustion is thus fostered across the engine operating range, delivering economic, and environmental benefits.

Optimum injection pressure and timing can be selected for a given operating mode, irrespective of engine speed because the generation of fuel pressure and the injection of the fuel is not interconnected. In addition, pilot and postinjection patterns can be exploited to meet differing demands: for example, invisible exhaust at the lowest loads and NOx reductions at medium loads without undermining fuel economy.

Successful automotive diesel applications of CR systems, in car and truck power units, were largely driven by stricter emission regulations. These called for flexible fuel injection systems offering injection rate shaping, free adjustment of injection pressure, the variable start of injection, and pre- and postinjection features. Tougher emission curbs on shipping stimulated the transfer of CR technology to marine engines, although major developments were dictated to enable system components to handle heavy fuel oils, working under conditions of high temperature and high viscosity, and with fuels containing abrasive particles and other aggressive elements.

CR systems existed in the early 1900s in mechanically activated form and were applied for years on the Doxford opposed-piston two-stroke marine engines. The comparatively recent proliferation of the concept in four- and two-stroke engine applications resulted from advances in fast and reliable rail valves and electronic controls. Developments in materials and manufacturing technology also fostered the creation of CR systems capable of handling pressures of 1500 bar and higher.

MTU Friedrichshafen, the German high-speed engine designer, was a pioneer in adopting the technology in the modern era and worked on

CR systems in conjunction with specialist fuel injection partner L'Orange from 1990. Its first-generation system was introduced on the 165-mm-bore Series 4000 engine in 1996. Development progressed to a second-generation system (on the 265-mm-bore Series 8000 engine) in 2000, with a third-generation system benefiting the 130-mm-bore Series 2000CR design in 2004.

Finland-based Wärtsilä developed a CR system suitable for application across its heavy fuel-burning medium-speed engine programme, the first seagoing W46 installation entering cruise ship service in 2001. MAN Diesel of Germany followed up with a system for its 320-mm and 480-mm-bore medium-speed designs, and other European, Japanese, Korean, and US designers—including Caterpillar Marine Power Systems (MaK)—have introduced or plan to offer CR-equipped engines.

Whatever the detailed differences in system architectures and control mechanisms, CR fuel injection can be expected to mature and eventually dominate the four-stroke engine market in the future, whether for propulsion or genset drives. Some existing engines can also benefit from retrofitted systems.

The following advantages are cited for CR systems over conventional mechanically and electronically controlled jerk pumps:

- Any injection pressure can be used, at any engine load or speed operation point, with the injection rate control close to the nozzle; this makes it possible to improve performance over the full engine load range.
- Injection timing can be varied during engine running to optimize engine performance without the timing range being limited by cam working lengths; this is beneficial for emissions optimization as well.
- Camshaft peak torque is smaller; this allows a smaller camshaft to be used and makes possible greater engine power density and smaller engine outlines.
- The CR pumping principle gives higher efficiency (no helix spill) and the potential to reduce mechanical drive losses, underwriting lower fuel consumption.
- The design is simpler, allowing the elimination of the one pump-per-cylinder principle, the fuel rack control shaft, and the mechanical governor; this results in cost reductions for the overall fuel injection system.

Among the potential operational benefits are lower fuel consumption, lower NOx emissions, no visible smoke at any load (and the possibility to start the engine without visible smoke), load cycling without smoke, and lower maintenance costs.

MTU common rail system for high-speed engines

Development work on CR fuel injection systems by MTU from 1990 benefited the German designer's Series 4000 high-speed model which was introduced in 1996 as the first production engine in its class to feature such a system (Fig. 11.18). The group's Stuttgart-based subsidiary L'Orange was responsible for developing the system components (Figs 11.19 and 11.20).

Research and development progressed from a pilot project on single-cylinder engines and an eight-cylinder demonstrator engine. The positive results encouraged MTU to develop the Series 4000 engine from the outset with a CR injection system so that its benefits could be realized not just in terms of injection but in the overall design. The system embraces the following key elements:

- The injectors in the cylinder heads
- The HP pump, driven by a gear system
- The fuel rail (fuel supply line running along the engine length)
- The fuel lines between the fuel rail and the injectors

One fuel rail is provided for the injectors and their respective solenoid valves on each bank of cylinders. The fuel pressure is generated by an HP pump driven via a gear train at the end of the crankcase.

Fig. 11.18 L'Orange CR fuel injection system installed on an MTU Series 4000 high-speed engine, showing the engine-driven HP pumps feeding the CR and injector pipes.

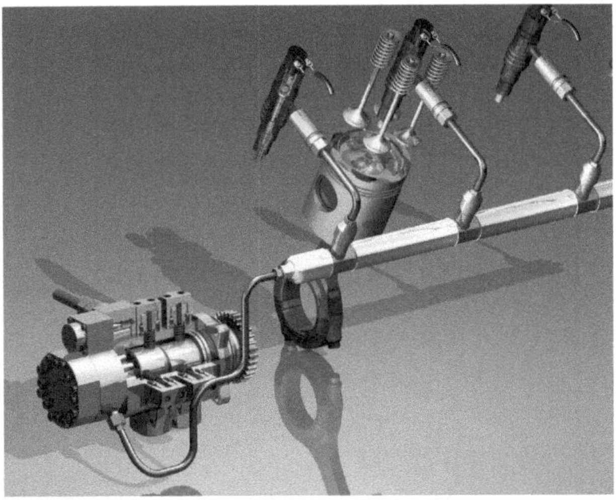

Fig. 11.19 The main elements of the L'Orange CR fuel injection system are a low-pressure fuel supply line to the HP pump, an HP accumulator (CR), the fuel injectors, and a double-walled HP fuel line.

The electronic control system controls the amount of fuel delivered to the injectors by means of the solenoid valves, and the injection pressure is also adjusted to the optimum level according to the engine power demand. Separate injection pumps for each cylinder are eliminated and hence the need for a complicated drive system for traditional pumps running off the camshafts (which, due to high mechanical stresses, calls for reinforced

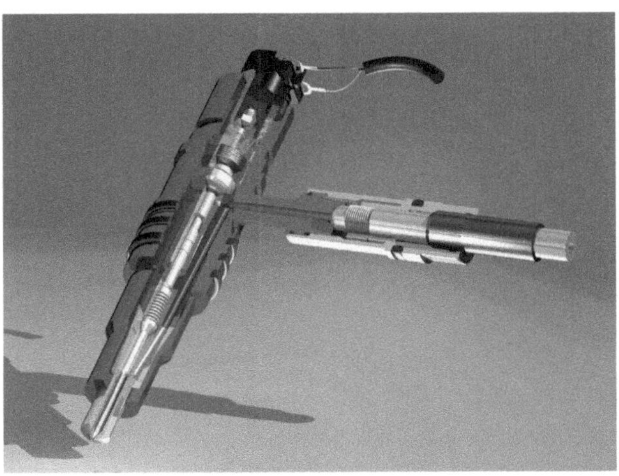

Fig. 11.20 An L'Orange LEAD CR injector.

gearing systems or even a second system of gears on larger engines). Fewer components naturally foster greater reliability.

Another benefit cited for the CR system is its flexible capability across the engine power band and the fact that it is able to deliver the same injection pressure at all engine speeds from rated speed down to idling. In the case of the Series 4000 engine, that pressure is around 1200 bar, corresponding to 1600–1800 bar on a conventional system.

Success is reflected in the fuel consumption (below 195 g/kWh) and emission characteristics of the engine. Future fuel injection systems must be distinguished by extremely flexible quantity control, the commencement of injection timing and injection rate shaping: capabilities achieved only with the aid of electronic control systems.

Conventional injection systems with mechanical actuation include in-line pumps, unit pumps with short HP fuel lines, and unit injectors. A cam controls the injection pressure and timing while the fuel volume is determined by the fuel rack position. MTU asserts that for future engines with high injection pressures, however, the in-line pump system can be ignored because it would be hydraulically too 'soft' due to the long HP lines.

A comparison has been made between the unit pump and unit injector systems, assuming the unit injector drive adopts the typical camshaft/push rod/rocker arm principle. Using simulation calculations, the relative behaviour of the two systems was investigated for a specified mean injection pressure of 1150 bar in the injector sac. MTU explains that this time-averaged sac pressure is a determining factor in fuel mixture preparation, whereas the frequently used maximum injection pressure is less meaningful. The pressure in a unit pump has been found to be lower than that in a unit injector but, because of the dynamic pressure increase in the HP fuel line, the same mean injection pressure of 1150 bar is achieved with less stress in the unit pump.

With the unit injector, the maximum sac pressure was 1670 bar (some 60 bar higher than that in the unit pump). To generate 1150 bar, the unit injector needed 3.5 kW, some 6% more power. During the ignition delay period, 12.5% of the cycle-related amount of fuel was injected by the unit pump compared with 9.8% by the unit injector. The former is, therefore, overall the stiffer system. Translating the pressure differential at the nozzle orifice and the volume flow into the mechanical energy absorbed, the result was a higher efficiency of 28% for the unit pump than 26% for the unit injector.

MTU reports that from the hydraulic aspect, the unit pump offers benefits in that there is no transfer of mechanical forces from the push rod drive to the cylinder head, and less space is required for the fuel injector (yielding better design possibilities for inlet and exhaust systems). With conventional systems, the volume of fuel injected is controlled by the fuel rack; and matching the individual cylinders dictates appropriate engineering effort. The effort is increased significantly if injection timing is effected mechanically.

The engineering complexity involved in enabling fuel injection and timing to be freely selected can be reduced considerably by using a solenoid valve to effect time-orientated control of fuel quantity. To produce minimum fuel injection quantity, extremely short shift periods must be possible to ensure good engine speed control. Activation of the individual solenoid valves and other prime functions, such as engine speed control and fuel injection limitation, is executed by a microprocessor-controlled engine control unit (ECU). Optional adjustment of individual cylinder fuel injection calibration and injection timing is thus possible with the injection period being newly specified and realized for each injection phase. Individual cylinder cut-out control is only a question of the software incorporated in the ECU.

With cam-controlled injection systems, the injection pressure is dependent on the pump speed and the amount of fuel injected. For engines with high mean effective pressures in the lower speed and low load ranges, this characteristic is detrimental to the atomization process as the injection pressure drops rapidly.

Adjusting the injection timing also influences the in-system pressure build-up. For example, if the timing is advanced, the solenoid valve closes earlier, fuel compression starts at a lower speed, and thus leads to lower injection pressures, which, in turn, undermines mixture preparation.

To achieve higher injection pressures, extremely steep cam configurations are required. As a result, high torque peaks are induced into the camshaft, which dictates a compensating degree of engineering effort on the dimensioning of the camshaft and gear train and may even call for a vibration damper.

MTU argues that while the solenoid valve-controlled system has a number of advantages, it retains the disadvantages of the conventional systems; and in the search for a new flexible injection system, it represents only half a step. A full step is achieved only when the pressure generation, fuel quantity, and injection timing functions can be varied independently by exploiting the CR injection system.

In the CR system configuration for a V16-cylinder engine (Fig. 11.21), the HP pump delivers fuel to the rail, which is common to all cylinders. Each injector is actuated in sequence by the ECU as a function of the crankshaft angle. The injector opens when energized and closes when de-energized. The amount of fuel injected per cycle is determined by the time differential and the in-system pressure. The actual in-system pressure is transmitted to the control unit via a pressure sensor, and the rail pressure is regulated by the ECU via the actuator in the fuel supply to the HP pump.

In such a system the injector incorporates several functions. The nozzle needle is relieved by a solenoid valve and thus opened by the fuel pressure. The amount of fuel injected during the ignition delay period is regulated by the nozzle opening speed. After the control valve is de-energized, an additional hydraulic valve is activated to secure the rapid closure of the nozzle needle and, therefore, a minimum smoke index. This servo-assisted injector allows the opening and closing characteristics to be adjusted individually and effected very precisely. MTU claims that it is capable of extremely high reaction speeds for controlling minimum fuel quantities during idle operation or pilot injection.

Compared with a conventional injection system, the pumping force is considerably lower: pressure generation is accomplished by a multi-cylinder, radial piston pump driven by an eccentric cam. Pressure control is realized by restricting supply flow. Locating the HP pump on the crankcase presents no problems, while the deletion of the fuel injection control cams from the camshaft allows that component to be dimensioned accordingly.

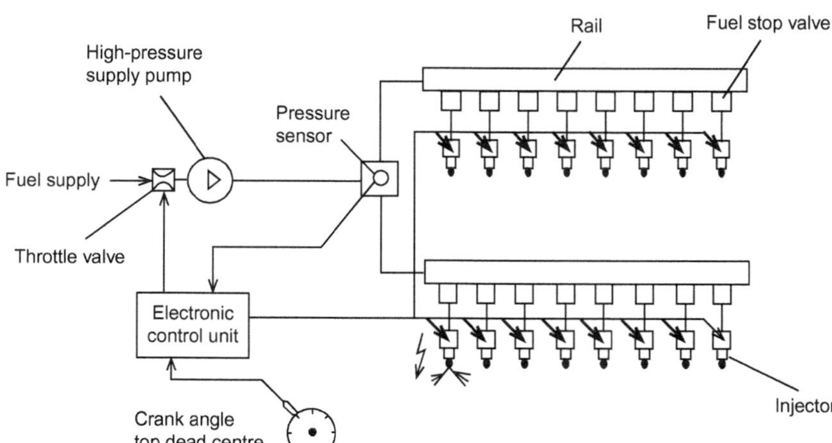

Fig. 11.21 CR injection system arranged for a V16-cylinder high-speed engine (MTU).

The rail system is required to supply all injectors with fuel at an identical pressure, and its design is based on criteria such as minimum fuel injection quantity deviations between cylinders, faster pressure build-up after the start, and minimum pressure loss from the rail to the nozzle sac. Simulation exercises have shown that pressure fluctuations can be very low.

The ECU determines the engine speed and calculates the amount of fuel to be injected, based on the difference between actual and preset engine speeds. The individual injectors are energized as a function of crankshaft angle and firing order.

Control of the CR in-system pressure is also carried out by the ECU. The advantages inherent in the solenoid valve-controlled conventional injection system (regarding speed control due to the improved actuator dynamic) can also be applied to the same extent to the CR system: at the moment of each injection, the governor-generated data can be immediately processed. MTU says that pressure regulation is highly effective. Load shedding (required e.g. when a waterjet propulsion unit emerges from the sea during rough conditions) can be affected from full to zero loads in just 10 ms.

While this leads to a rapid rise in engine speed, the governor reduces fuel injection to zero, which, in turn, causes a pressure increase in the rail system as the individual injectors become inactive. The pressure regulator reacts and restricts the HP pumps to a maximum of 1330 bar (a pressure excess of 130 bar over the specified value of 1200 bar). In addition, the ECU assumes the control and monitoring functions that are standard for MTU engines, including sequential turbocharging control.

With CR injection, the complete system is permanently subjected to extremely HPs. In the event of failure of a single injector, the engine is protected in that the injector is cut off from the fuel supply by the fuel stop valve: no fuel can enter the combustion chamber. The engine can, however, still be operated at reduced (get-you-home) power.

A mechanical pressure relief valve is actuated if, due to a pressure regulating system malfunction, the pressure rises to an unacceptable level. Leaks in the system are identified by the pressure monitoring system.

MTU reports that during tests the unit pump system displayed the best characteristics due to the high specified injection pressures. Greater flexibility, improved dynamics, reduced design effort, and as a result, lower costs are offered by the unit pump system with solenoid valve control. MTU asserts that only a CR injection system, however, makes it possible to achieve injection rate shaping over the complete operating range and thus to underwrite the specifications for future diesel engines.

The HP pump for the Series 4000 fuel system was changed from a four- or eight-cylinder radial piston pump to a four-cylinder inline unit to increase its performance and achieve a higher delivery rate for the V20-cylinder engine model.

MTU series 8000 engine CR system

MTU subsequently applied an electronic CR system to the group's largest-ever high-speed engine, the 265-mm bore Series 8000, introduced in 2000 (see Chapter 26), whose V20-cylinder version develops up to 9000 kW at 1150 rev/min. As on the Series 4000 engine, the system enables infinite variation of injection timing, volume, and pressure over the entire engine performance curve. The supply of pressurized fuel to the injectors, however, is achieved by different means. On the Series 4000 engine, there is a rigid fuel accumulator (CR) for each row of cylinders, while the Series 8000 engine has a separate accumulator for each cylinder located in the vicinity of the injector and with a capacity of 0.4 L.

This arrangement facilitates a smaller overall engine width and optimum progression of combustion and reduces the interplay between the injectors and the pressure fluctuations in the fuel pipes. Only two HP pumps are required to generate substantial pressure and the necessary flow volume for the entire engine.

The L'Orange system for the Series 8000 engine is based on two 3-cylinder HP fuel pumps producing a maximum rail pressure of 1800 bar. These pumps are integrated in the gear drive at the front side of the engine and driven by the crankshaft with a gear ratio of 1.65 at speeds of up to 1800 rev/min. The HP part of the injection system, with the functions of HP generation and accumulation as well as fuel metering, comprises the following:

- Two HP fuel pumps with integrated pump accumulators.
- Up to 20 accumulators, each equipped with a flow-limiting valve.
- High-pressure fuel lines.
- Up to 20 electromagnetically controlled injectors.

Double-walled HP fuel lines distribute the fuel from the pumps along with the engine to each cylinder. An accumulator serving each cylinder acts as a storage space to maintain an almost constant fuel pressure during injection, despite the relatively small cross-section in the fuel supply lines. The compressibility of the fuel ensures an almost constant injection pressure during injection. The intervals between injections are used to replace the injected fuel mass in the accumulators and build up the fuel pressure again. The accumulator incorporates a so-called fuel-stop valve, which interrupts the fuel supply to the relevant injector in cases of excessive fuel flow rates (e.g. caused by leakages or continuous injection). The engine management system will then adjust the power output of the engine accordingly.

Rounded-off spray hole intake geometries on the electronically controlled injector ensure an optimally sprayed injection of fuel into the combustion chamber. The start of injection and the quantity injected are accurately controlled to the microsecond by the engine control system in line with the engine operating condition.

The system fuel pressure is available at the nozzle seat as well as in the control unit above the plunger for needle lift control (the so-called connecting rod). Compared with the usual two-cycle systems, this offers the advantage of a faster operating response. The different-sized pressurized surfaces of connecting rod and needle result in a hydraulic closing force. When the solenoid is energized by an electrical signal, the armature is lifted. This opens an accurately machined cross-section of the outlet throttle in the control unit, causing a rapid drop in the hydraulic pressure on the connecting rod. The nozzle needle lifts off the seat and starts the injection process, which is stopped when the solenoid is de-energized. The closing flank of the injection characteristics can be shaped very rapidly through an intermediate valve after the control unit; it opens additional supply cross-sections during the closing of the solenoid.

Fuel is injected in the precise quantity and at exactly the moment required for the conditions under which the engine is operating. The pump's intake flow throttle ensures that only the precise amount of fuel required is pressurized, resulting in very high pump efficiency and consequently low fuel consumption.

The advantages of the CR injection system are fully exploited in conjunction with the engine control system. The flexibility of the injection system with respect to injection pressure and timing allows the option to access and optimize the engine operating parameters throughout the performance map (e.g. addressing fuel consumption and emissions).

For each operating condition of the engine, the engine control system analyses the sensor information and computes them into a setting for system fuel pressure, injection start, and timing. The computed data take into account engine speed, power demand, and transient operating conditions such as engine warm-up after cold starting. The scope also embraces surrounding conditions and possible faults in the drive system, with appropriate countermeasures automatically being initiated. In addition, the engine control system monitors the limits to certain operating parameters (e.g. maximum pressure in the fuel system) to avoid component damages.

Wärtsilä CR system for medium-speed engines

A CR fuel system developed by Wärtsilä in conjunction with Delphi Bryce (now part of the Woodward Governor group) for application to the group's heavy fuel-burning medium-speed engines is modular, with common components designed to serve engines with cylinder numbers from 4 to 20. Development was stimulated by increasing market pressure, especially from passenger tonnage, for no visible smoke emissions under any circumstances.

The Wärtsilä system (Figs 11.22 and 11.23) is based on a pressurizing pump and a fuel accumulator for every two cylinders, the series of accumulators interconnected by small-bore piping. Connections from every accumulator feed the injectors of two cylinders (in engines with an odd number of cylinders, an extra accumulator is installed with a connection to one cylinder).

Functionally, the main reason for choosing such a system is that by splitting up the fuel volumes in several accumulators, the risk of pressure waves in the CR is avoided. From a safety viewpoint, an advantage is that HP fuel exists only in the same area as in a conventional engine (in the hot box). Third merit cited is economic, the system using a camshaft that is present anyway to drive the valves.

The computer-controlled fuel injection system opens up possibilities to optimize engine performance, thanks to the freedom to choose key parameters: injection timing, injection pressure independent of engine speed, maximum injection quantity dependent on charge air pressure and rotational speed, optimized injection during starting, and optimized injection timing at load pickup.

Each CR fuel pump is driven by a two-lobe cam, whose pumping profile can be considerably slower than the conventional jerk pump type that it replaces. In a CR system there is no longer the requirement to pump fuel to the injector directly within the typical range of 6–10 μs; this fosters a reduction in mechanical drive noise and drives energy consumed. The interconnected accumulator arrangement reduces the pulsation migration ability along the rail. In addition, the rail components are rigidly fixed to the engine block, reducing the risks of piping leakage. Each accumulator is connected

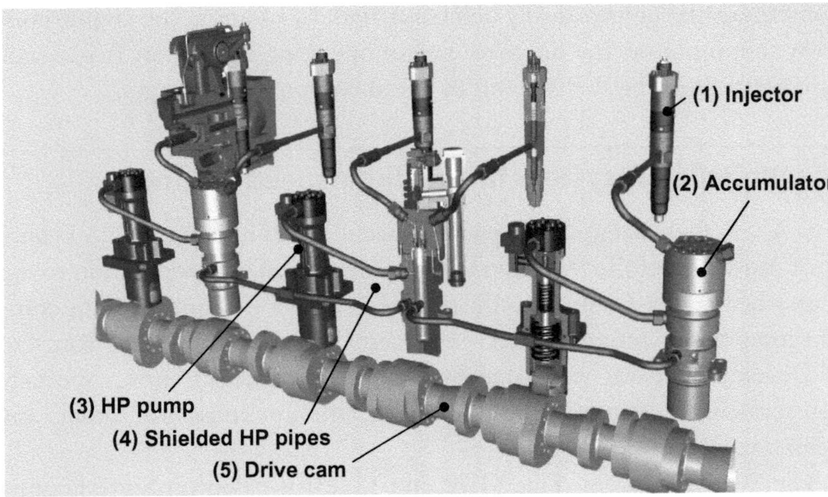

Fig. 11.22 CR fuel system for Wärtsilä medium-speed engines.

directly to two injectors, and each pipeline contains a hydraulic flow fuse for safety reasons; each accumulator also contains a pressure sensor.

Wärtsilä's engine control system (WECS) regulates the CR system, monitoring several system parameters; it also controls fuel pressure in the rail and injection quantities and timings according to specified maps. The solenoids are not responsible for all control; pressurized engine oil is used to

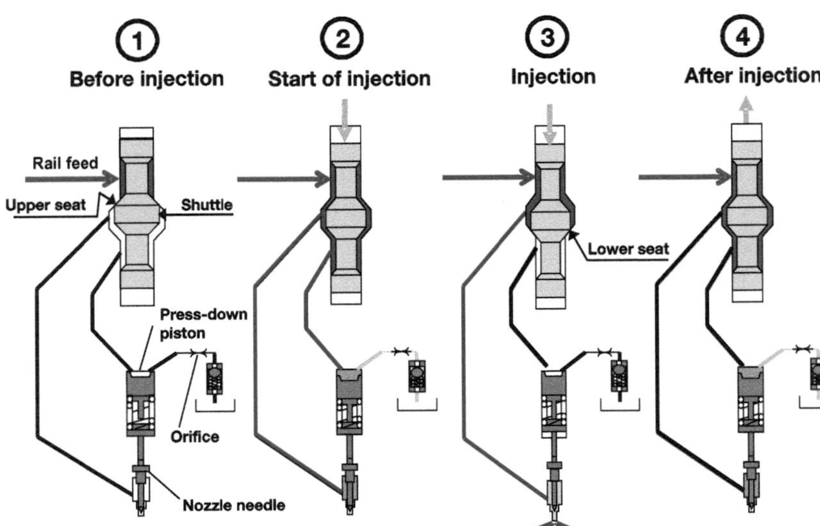

Fig. 11.23 Schematic diagram showing the function of the Wärtsilä CR fuel injector.

achieve the necessary forces in the injectors and stop solenoids (this oil system is termed 'control oil').

The CR pump is constructed with a tappet and barrel element integrated into one housing. The use of heavy fuel oil makes it preferable to divide the pump internally; this places the pumping element in a fuel-only environment, which eliminates the risk of lacquer accumulation from the fuel and chemical interaction with the engine oil. The pump can be driven by two- or three-cam lobes. The two-lobe alternative is preferred both for the roller/roller pin oil film duty cycle and because it creates fewer pumping pulses to settle into the rail prior to injections.

The volumetric efficiency of the pump is typically above 80%. This level is generally higher than that of jerk pumps with a plunger spill helix, where the end of the plunger stroke marks the rejection of fuel into the return circuit. Smaller plunger sizes are thus allowed, which in turn significantly reduces the maximum camshaft drive torque for an equivalent pumping pressure target.

Charging of the CR is controlled by regulating the volume of fuel allowed to enter the pumping chamber; this is affected by a flow control valve installed in either the barrel or the pump housing. The flow control valve itself is rotated by a rotary solenoid, which has a position feedback sensor for fault detection prior to engine start or during engine operation. Another detection device is incorporated in the barrel. A thermocouple sensor is installed close to the pumping chamber to detect delivery valve leakage; if the delivery valve were to leak, fuel from the rail would be repumped, causing a temperature increase in the pumping chamber.

Each accumulator, mounted rigidly on the engine block, receives fuel from the HP pump, stores it under HP, and distributes it to the injectors. The accumulator itself is an empty space. Flow fuses located at outlets from the accumulator to the injector limit the quantity of fuel that can be transferred from the accumulator to the injector if the latter malfunctions. The accumulator is also fitted with a device that makes it easy to find a leaking pipe if the leak alarm is activated.

At least one of the accumulators is also equipped with a start-up and safety valve (SSV), which opens the flow path from the accumulator and is controlled by the WECS. The SSV is opened when the engine is in the prewarming phase or when the engine is to be stopped. Incorporated in the SSV is a mechanical safety valve to protect the system against excessive rail pressure, this valve operating completely independently of the main valve. Connected to the SSV is an empty space in the fuel outlet channel, whose function is to dampen the pressure waves when the SSV is opened.

The fuel injector was designed for efficiency and safe operation, with four main elements: nozzle, shuttle valve, solenoid valve, and piston acting on the nozzle needle. The injector is fitted with a conventional nozzle that is linked to the CR fuel source via the shuttle valve. Two key functions are performed by the shuttle valve: it controls injections and depressurizes the injector needle. In its closed (static) position, the valve isolates the injector passages and nozzle from the pressure source, so that pressure is not applied to the nozzle seat between injections. The shuttle is opened by applying a hydraulic load using the control oil source. The control oil is switched on and off using a fast-acting solenoid valve (a reliable automotive component produced in high volumes).

Wärtsilä explains that such an approach fosters full flexibility of the start of injection and fuelling quantity. The piston above the nozzle, together with any pressure on it, acts to hold the nozzle closed; pressure is normally applied only to the piston at the start and end of injections. The piston arrangement has the following three functions:

1. The pressure on the piston at the start of injection slightly delays opening of the nozzle, allowing time for the shuttle to complete its travel—even for the smallest injections. This promotes fuelling consistency because the need for partial shuttle travel at the lowest outputs is avoided.
2. At the end of injection, the pressure on the piston helps to close the nozzle. The fuel pressure at the nozzle seat at the end of injection is much higher than that in conventional systems, and thus each injection is terminated very rapidly.
3. *A safety benefit*: if the shuttle becomes stuck in a partially open position, the pressure is continuously applied to the piston, this force keeping the nozzle closed in this fault condition.

These design features resulted in an injector with a high mean operating pressure, capable of working up to 2000 bar. Injections start crisply and terminate with the nozzle being closed while injecting a nearly full pressure. The injector was also designed for robust safety in the event of a fault condition:

- Rail pressure is not applied to the nozzle seat between injections; this ensures that the fuel is cut off if the nozzle needle is prevented from fully closing.
- The piston acting on the nozzle needle prevents injection if the shuttle seizes or is prevented from fully closing.

Engine operation on heavy fuel oil is addressed by the following features:

- During the engine's prestart warm-up phase, the pump flow control valves are shaken and opened separately every 5 s. Heated HFO flows through each pump in turn and then onward to the rail and back to the tank via the bypass SSV valve system.
- The fuel injectors do not have HFO present in the solenoid armature area; a servo oil system is used.
- In the fuel injectors, the rail pressure does not act directly on the nozzle needle, reducing the risk of needle sticking; and the rail pressure is released from the nozzle needle between injections to improve safety.

The Wärtsilä CR system was designed for retrofitting to the group's existing engines by simply removing the original conventional injection pumps and replacing them with CR fuel pumps and accumulators that fit exactly into current engine blocks.

Acknowledgements

Duap, Fiedler Motoren Consulting, Delphi Bryce (Woodward Governor), MTU, L'Orange, and Wärtsilä Corporation.

See also chapters on Caterpillar Marine Power Systems (MaK), MAN Energy Solutions, MTU, and Wärtsilä four-stroke engines. Two-stroke engine fuel injection systems, including electronically controlled configurations, are discussed in Chapter 13 and the MAN Energy Solutions, Japan Engine, and WinGD low-speed engine chapters.

CHAPTER TWELVE

Waste heat recovery

The average modern marine diesel engine is nominally around 50% thermally efficient. While that is a great advance on engines of the 1970s and 1980s when oil crises and high bunker prices caused ship operators to demand more efficient engines, it means that there is more to be done in the search for increasing efficiency. The IMO's increasingly stringent EEDI requirements will require more efficiency from oil and gas-fuelled engines or ways in which some of the lost energy content can be better utilized.

One way this might be achieved is through the use of a waste heat recovery system (WHRS). Some of the lost energy is already recovered in many ships by way of the exhaust gas boiler or economizer. This is a relatively simple heat exchanger which allows for a supply of hot water for the ship's domestic purposes or steam for powering pumps or used for heating oil fuel or cargoes.

It works by the hot exhaust gases being passed through tubes surrounded by water that becomes heated to quite high temperatures. In some ships with large engines, the amount of steam produced is sufficient to drive a turbine but in others, the heated water can be sent to an oil fired boiler where it is further heated to produce steam for use in for example the steam-powered cargo pumps of a tanker or for cargo heating.

Electricity from waste heat

In the early 21st century, some of the leading engine makers began investigating the use of more sophisticated means of recovering the waste heat beyond using an economizer. In specialist shore-based situations, something like 90% of wasted energy can be recovered from both diesels and steam turbines, and it is usually used to provide space and water heating in buildings such as hospitals and factories. Achieving that sort of efficiency is almost impossible in a ship, but some savings are possible depending upon ship and engine type and operating profile of the ship.

Pounder's Marine Diesel Engines and Gas Turbines
https://doi.org/10.1016/B978-0-08-102748-6.00012-8

MAN Energy Solutions was a pioneer in this and developed the thermo efficiency system (TES) concept combining a power turbine and a turbo generator running off the exhaust gas. The primary source of waste heat of the main engine is exhaust gas heat dissipation, which accounts for about half of the total waste heat, i.e. about 25% of the total fuel energy. In a standard engine version, the exhaust gas temperature after the turbocharger is relatively low, and just high enough to produce the necessary steam for a ship's heating purposes by means of the exhaust gas-fired boiler. However, the main engine with changed timing and exhaust gas bypass—which redistributes the exhaust gas heat from high amount/low temperature to the low amount/high temperature—increases the effect of utilizing the exhaust gas heat, but at the same time may slightly reduce the efficiency of the main engine itself.

The MAN TES consists of an exhaust gas-fired boiler system, a steam turbine (often called a turbo generator), an exhaust gas turbine (often called a power turbine), and a common generator for electric power production. The turbines and the generator are placed on a common bedplate. The exhaust gas turbine is driven by a part of the exhaust gas flow, which bypasses the turbochargers. The exhaust gas turbine produces extra output power for electric power production, which depends on the bypassed exhaust gas flow amount.

When a part of the exhaust gas flow is bypassed the turbocharger, the total amount of air and gas will be reduced, and the exhaust gas temperature after the turbocharger and bypass will increase. This will increase the obtainable steam production from the exhaust gas-fired boiler.

The exhaust gas bypass valve will be closed for engine loads lower than about 50% SMCR, which means that the exhaust gas temperature will be reduced when operating below 50% SMCR. The power output from the exhaust gas turbine is transmitted to the steam turbine via a reduction gear with an overspeed clutch, which is needed in order to protect the exhaust gas turbine from overspeeding in case the generator drops out.

The total electric power output of the TES—which reduces the ship's fuel costs—is only a gain provided that it can replace the power output of other electric power producers on board the ship. Otherwise, a shaft power motor connected to the main engine shaft could be an option, but this extra system is rather expensive.

In general (without a shaft power motor installed), when producing too much electric power, the (high pressure) superheated steam to the steam turbine is controlled by a speed control governor through a single throttle valve, which means that the surplus steam is dumped via a dumping condenser. When the generator is operating in parallel with the auxiliary diesel generators, the governor operates in the normal way to give correct load sharing.

Working with UK engineering specialist Peter Brotherhood, MAN developed two versions of TES. The total output of the TES system with a dual-pressure boiler and steam turbine was estimated as high as 11.2% of the engine's shaft power. A simplified single-pressure version TES would be less efficient but would still allow generation of up to 7500 kW of electricity from a 12K98ME-C. Confirmation tests were successfully completed at Mitsui in Japan in 2005 on a 10K98MC engine. The results verified calculations and expectations related to heat load, emission data, and performance figures.

A WHRS may be a good means of recovering some of the lost energy but it is not suitable for all ship types. Firstly, such a system works best with very large engines although Alfa Laval, which manufactures boilers and economizers has developed a system able to recover heat energy from auxiliary engine exhausts. This can be used to meet steam needs in port and for vessels whose long-stroke main engines produce cooler exhaust, it can also provide a needed boost in steam production en route.

Whether from the main engine or main and auxiliaries, it will generally be shipped above Panamax size that are best suited. Arguably it is container vessels with large numbers of reefer containers that can benefit most. For these ships, it is normally necessary to have one or more auxiliary engines running to provide electric power for the reefer connections. A WHRS could provide part or all of the electric power needed whenever the main engine is running.

Most of the WHRS installed have been on container ships but with many ship operators now exploring the potential of energy storage systems and hybrid propulsion systems, there are new possibilities. The electricity produced by a WHRS could be stored in a battery system and used for a variety of purposes including propulsion in ports by way of a shaft motor. This would enable the ship to be manoeuvred without the production of any exhaust as the main engine could be stopped.

MAN Energy Solutions has installed a number of WHRS and explains that the main engine control and the WHRS control must be able to function as an integrated part of the total control system of the ship, which means that integration between the systems must be in place.

The development of the WHRS and main engine controls is based on the ship owner's demand for full control and optimum fuel consumption in all operational modes. Focus is very much on fuel economy because of the continuously increasing fuel oil prices. A central part of this is the ship's power management system, which controls the different power sources on-board, so that minimum fuel consumption can be pursued for the different sailing conditions.

It is therefore very important to understand the complexity of the power systems (propulsion and electric power sources) and their interfacing when planning the control functions and interfaces for the main engine and WHRS control.

The engine control and the WHRS control have to be strongly integrated because the WHRS steam turbine and power turbine use the energy in the exhaust gas to recover energy. The dynamics of the main engine will influence the behaviour of the steam turbine and power turbine.

The connection between the main engine and WHRS is seen in Fig. 12.1, showing the exhaust gas flow between the engine, power turbine, and exhaust boiler. Two exhaust connection on engine exhaust receiver is necessary for WHRS arrangements including power turbine.

The basis for this layout of the connections between the engine and WHRS components is two sided as the control must first be able to protect the engine against unacceptable conditions and second ensure maximum available power for the WHRS. This is done by controlling the exhaust gas bypass amount, keeping it within acceptable limits.

This is ensured by controlling the exhaust gas bypass (EGB) control valve via the engine control system and the power turbine control valve. The sum of exhaust bypasses must always be within the allowed bypass amount selected for the particular engine with WHRS tuning.

In 2014, MAN installed its TCS-PTG (turbo compound system with power turbine and generator) on a pair of the container ship. This system operates by diverting some of the exhaust gas flow to a power turbine generator inserted into the exhaust system parallel to the turbochargers. It drives an electrical generator via a reduction gearbox and receives up to 13% of exhaust gas flow, diverted from the main engine exhaust gas receiver. High-efficiency turbochargers are a prerequisite for the system to be operated effectively. The ships on which the systems are installed are each powered by a MAN B&W6S80MEC9.2 low-speed engine, rated at 27,060 kW. Each TCS-PTG20 unit can supply 920 kW to the board grid.

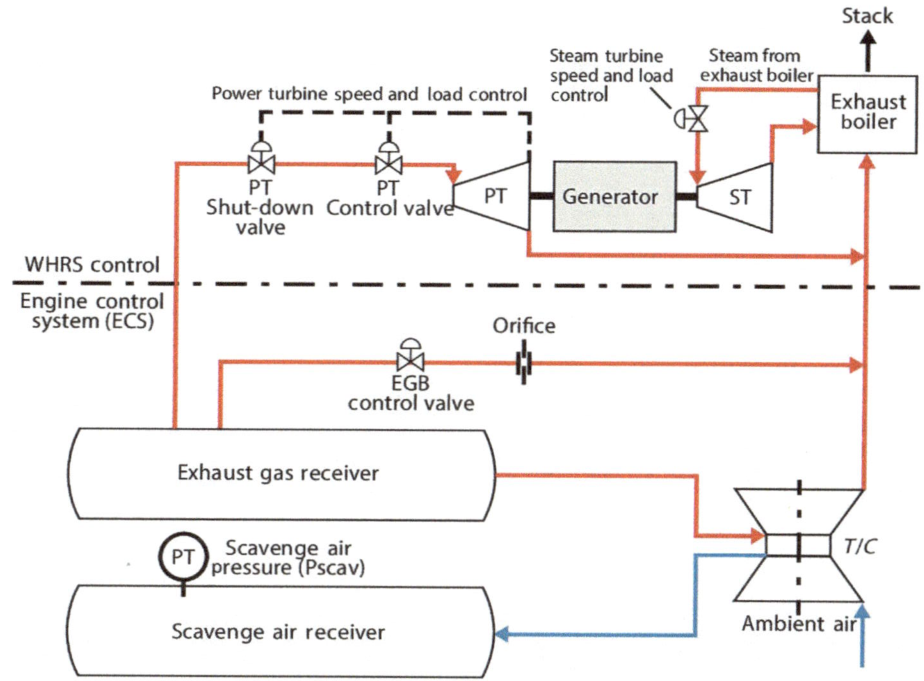

Fig. 12.1 Exhaust gas flow between the engine, power turbine, and exhaust boiler of MAN Waste heat recovery system.

In 2010 Wärtsilä announced that it was linking up with Turboden of Italy to develop, market, and distribute another method of waste heat recovery based on the organic Rankine cycle. The system was planned to appear in 2011 but has not been used for marine applications.

The Rankine cycle is a thermodynamic cycle that converts heat into work energy. Water is the most commonly used working fluid in Rankine cycle systems, which generate the majority of the electricity used throughout the world. In the organic Rankine cycle, the working fluid is not water but any of various organic fluids such as n-pentane or toluene. Because these liquids have a lower boiling point than water the system can operate at lower temperatures and make use of lower grade heat.

While the ORC is already a well-established form of energy production with an increasing application base, it is new to the marine market. The ORC technology provides many advantages, including high thermodynamic cycle efficiency, the ability to operate at lower heat source temperatures, simple start-up procedures, automatic and continuous operation without operator intervention, simple maintenance procedures, and long life cycles. More importantly, because less heat is required, the potential for smaller vessels to be able to employ WHRS is significantly enhanced.

Although Wärtsilä did not pursue the development of an ORC WHRS, the concept has not been forgotten with Mitsubishi and Calnetix Technologies jointly involved in the development of the Hydrocurrent Organic Rankine Cycle (ORC) Module 125EJW. An example was installed in 2016 on the AP Moller vessel *Arnold Maersk*. That system has an output of 125 kW. It received type-approval from both ClassNK and Lloyd's Register.

CHAPTER THIRTEEN

Low-speed two-stroke engines—Introduction

Low-speed two-stroke engine designers have invested heavily to sustain their dominance of the mainstream deep-sea propulsion sector formed by tankers, bulk carriers, container ships, and LNG carriers as well as several other ship types. The long-established supremacy reflects the perceived overall operational economy, simplicity, and reliability of single, direct-coupled cross head engine plants. The term low-speed fits well with the very large engines which typically operate between 80 and 120 rpm, but it also extends to the smaller bore sizes where speeds increase up to 170 rpm or occasionally higher.

Other factors are the continual evolution of engine programmes by the designer/licensors in response to or the anticipation of changing market requirements, and the extensive network of engine-building licensees in key shipbuilding regions, notably Korea, Japan, and China. Many of the standard ship designs of the leading yards, particularly in Asia, are based on low-speed engines.

The necessary investment in R&D, production, and overseas infrastructure dictated to stay competitive, however, took its toll over the decades. Only three low-speed engines designer/licensors—MAN B&W Diesel, Mitsubishi, and Sulzer—survived into the 1990s to contest the international market. MAN B&W Diesel was superseded in 2007 by MAN Diesel, which retained MAN B&W as a brand name for its low-speed engines, and again in 2018 when the company began operating as MAN Energy Solutions but still retaining the MAN B&W brand.

The Sulzer diesel engine business became part of the Wärtsilä Corporation in 1997 and the Sulzer brand is no longer used in the Wärtsilä portfolio. The Swiss company, Wärtsilä Switzerland Ltd., responsible for the low-speed, two-stroke engine within Wärtsilä, was merged with China State Shipbuilding Corporation (CSSC), one of the main licensees in early 2015 and renamed Winterthur Gas & Diesel Ltd. (WinGD). In 2016, Wärtsilä Corporation transferred its remaining shares of WinGD to CSSC making WinGD 100% owned by CSSC. The engine brand was changed from Wärtsilä to WinGD.

Pounder's Marine Diesel Engines and Gas Turbines
https://doi.org/10.1016/B978-0-08-102748-6.00013-X

There is now only one other manufacturer of low-speed two-stroke engines, the Japan Engine Corporation. This company was formed in 2017 by the merger of Mitsubishi Heavy Industries Marine Machinery and its main licensee Kobe Diesel. As well as being an engine designer and manufacturer in its own right, Mitsubishi was also previously a licensee of Wärtsilä and had co-operated on design projects with that company.

Historically MAN Energy Solutions has enjoyed the lion's share of the two-stroke market with an average of over 70% of all sales, WinGD has a share of around 20% with the balance going to Japan Engine Corporation. MAN engines are produced by eight licensees in China, four in Japan, and three in South Korea.

The roll call of past contenders in two-stroke engine design and construction includes names either long forgotten or living on only in other engineering sectors: AEG-Hesselman, Deutsche Werft, Fullagar, Krupp, McIntosh and Seymour, Neptune, Nobel, North British, Polar, Richardsons Westgarth, Still, Tosi, Vickers, Werkspoor, and Worthington. The last casualties were Doxford, Götaverken, and Stork, some of whose distinctive engines remain at sea in diminishing numbers.

Low-speed engine manufacturers and licensee production.

	MAN	WinGD	J-Eng
CSSC Marine Power Co., Ltd., Jiangsu, China	X	X	
CSSC-MES Diesel Co., Ltd., Shanghai, China	X	X	
Dalian Marine Diesel Co., Ltd., Dalian, China	X	X	
Hefei Rong An Power Machinery Co, Hefei, China		X	
Hudong Heavy Machinery Co., Shanghai, China	X	X	
Jiangsu Antai Power Machinery Co., Ltd., Jingjiang, China		X	
JingJiang Dingkun Machinery Co., Jingjiang, China	X		
Qingdao Haixi Marine Diesel Engine Co., Qingdao, China	X		
Yichang Marine Diesel Engine Co., Ltd., Yichang, China	X		
Yuchai Marine Power Co., Ltd., Zhihai City, China	X	X	
Hitachi Zosen Corporation, Osaka, Japan	X	X	
IHI Power Systems (Diesel United), Tokyo, Japan		X	
Japan Engine Corporation, Kobe, Japan		X	X
Kawasaki Heavy Industries Ltd., Kobe, Japan	X		
Mitsui E&S Machinery Co., Ltd., Tokyo, Japan	X		
Mitsui sublicensee: Makita Corporation, Kagawa, Japan	X		
HSD Engine Co., Ltd. (Doosan), Changwon, S. Korea	X	X	
Hyundai Heavy Industries Co. Ltd. Ulsan, S. Korea	X	X	
STX Heavy Industries Co. Ltd. Changwon, S. Korea	X		

The pioneering designs in the low-speed sector displayed individual flair within generic classifications, which offered two- or four-stroke, single- or double-acting, and single- or opposed-piston configurations. The still concept even combined the diesel principle with a steam engine: heat in the exhaust gases and cooling water were used to raise steam, which was then supplied to the underside of the working piston.

Evolution decreed that the surviving trio of low-speed cross head engine designers should exploit a common basic configuration: two-stroke engines with constant pressure turbocharging and uniflow scavenging via a single hydraulically operated exhaust valve in the cylinder head.

All three companies now offer predominantly electronically controlled engines although traditional camshaft-controlled designs are still available. By 2016, more than 90% of all new two-stroke engines were electronically controlled models. Operators who retain a preference for the mechanically controlled models will probably find a licensee prepared to produce one, but it should be expected that their demise is not long away.

Current programmes embrace mini-to-large bore models with short-, long-, and superlong-stroke variations to match the propulsive power demands and characteristics of the most deep-sea (and even some coastal/short sea) cargo tonnage. Installations can be near optimized for a given duty from a permutation involving the engine bore size, the number of cylinders, selected output rating, and running speed.

Bore sizes range from 260 to 980 mm (MAN did include a 1080 mm bore engine in its catalogue around 2005 but none were sold), stroke/bore ratios up to 4.4:1, in-line cylinder numbers from 4 to 14, and rated speeds from around 55 to 250 rev/min. Specific fuel consumptions as low as 154 g/kWh are quoted for the larger bore models whose economy can be enhanced by optional turbo compound systems in which power gas turbines exploit exhaust energy surplus to the requirements of modern high-efficiency turbochargers.

The first decade of the century has seen the addition of intermediate bore sizes to enhance coverage of the power/speed spectrum and further optimize engine selection. Both MAN and Wärtsilä (Sulzer) also extended their upper power limits in the mid-1990s with the introduction of superlarge bore models—respectively, of 980 and 960 mm bore sizes—dedicated to the single-screw propulsion of new generations of large container ships with service speeds of 25 knots or more.

The 12-cylinder version of the MAN B&W K98ME design, delivering 74,760 kW, highlighted the advance in specific output achieved since the

1970s when the equivalent 12-cylinder B&W K98GF model yielded just under 36,800 kW. Large bore models tailored to the demands of new generation VLCC and ULCC propulsion were also introduced.

Successively larger and faster generations of post-Panamax container ships drove the development of specific output and upper power limits by low-speed engine designers. To ensure that container ship designs with capacities exceeding 10,000 TEU could continue to be specified with single engines, both MAN Diesel and Wärtsilä extended their respective MC/ME and RTA/RT-flex programmes. Wärtsilä raised the specific rating of the RTA96C design by 4% to 5720 kW/cylinder at 102 rev/min and introduced an in-line 14-cylinder model delivering up to 80,080 kW. (The previous power ceiling had been 65,880 kW from a 12-cylinder model.) The debut 14RT-flex96C engine (80,080 kW at 102 rev/min) was completed by Wärtsilä licensee Hyundai Heavy Industries in 2007 to power the first of a series of 8600 TEU container ships.

MAN Diesel responded to the challenge with a 14-cylinder variant of the K98MC and MC-C series, offering outputs up to 80,080 kW at 94 or 104 rev/min, which have since been superseded by the electronically controlled 14K98ME model with ratings up to 87,220 kW at 97 rev/min on a mean effective pressure of 19.2 bar. The first such engine (a Mark 7 version rated for 84,280 kW at 104 rpm)—the most powerful diesel engines ever commissioned—was due for delivery in 2008 to power a container ship. However, construction of the series was delayed and the choice of engine was altered to a 12-cylinder version of lower power.

In 2003, MAN Diesel opened another route to higher powers: a 1080-mm bore version of the MC engine (now the ME-C) with a rating of 6950 kW/cylinder at 94 rev/min, the 14-cylinder K108ME-C model thus offering 97,300 kW at 94 rev/min on a mean effective pressure of 18.2 bar. No K108ME-C engine—the most powerful design in the world—was ever sold, the need obviated by further uprating of the K98ME/ME-C models, whose Mark 9 version would be capable of delivering the same output as its standard larger bore derivative.

The economic crash of 2008 virtually ended the need for fast container ships and while existing ships adopted a strategy of slow steaming to both reduce fuel consumption and remove shipping capacity, newbuild projects usually specified a slightly reduced service speed. The advent of the EEDI regulations aimed at improving the efficiency of new ships also heralded a demand for more compact and efficient engines in all ship types.

Engine designers responded with a new generation of super and ultra-long stroke engines with a much greater emphasis on the whole power and propulsion configuration through to propeller. The last MAN K98ME engine installed in a newbuild ship was in 2015 by which time 644 ships had been built with a version of the engine since the first installation in 2000.

MAN's replacement for the K98ME engine was the G95ME-C, an electronic engine with a 30 mm smaller bore but with an ultra-long stroke of 3460 mm compared to the 2400 mm of the K98ME-C. Where the first 12-cylinder K98ME-C's produced 74,760 kW, a 2019 12-cylinder version of the G95ME-C has an output of 82,440 kW.

Between 2008 and 2019, the size of the largest container ships in service has increased from around 14,000 to 23,756 TEU. Some designs now have twin-engine and twin-propeller propulsion ships but the largest vessel in service in 2019—*MSC Gulsun*—is fitted with an 11-cylinder G95ME-C9.5 engine producing 75,570 kW at 80 rpm and driving a 10.4-m diameter propeller for a service speed of 23.2 knots.

There has been no less development activity by all designers at the other end of the power spectrum, with significant smaller bore engine developments (350–500 mm) targeting market opportunities from small ship propulsion. Electronic control, initially confined to larger bore designs, has also been extended to smaller bore models, such as MAN Diesel's new ME-B programme, which exploits a stroke/bore ratio of 4.4.

Parallel development by the designer/licensors seeks to refine existing models and lay the groundwork for the creation of new generations of low-speed engines. The emphasis in the past has been on optimizing fuel economy and raising specific outputs, but reliability, durability, and overall economy are now priorities in R&D programmes, operators valuing longer component lifetimes, extended periods between overhauls and easier servicing.

Lower production costs through more simple manufacture and easier installation procedures are also targeted, reflecting the concerns of engine builder/licensees and shipyards. More compact and lighter weight engines are appreciated by naval architects seeking to maximize cargo space and deadweight capacity within the given overall ship dimensions.

A number of features have further improved the cylinder condition and extended the time between overhauls through refinements in piston design and piston ring configurations. A piston cleaning ring incorporated in the top of the cylinder liner now controls ash and carbon deposits on the piston topland, preventing contact between the liner and these deposits, which would otherwise remove part of the cylinder lube oil from the liner wall.

Computer software has smoothed the design, development, and testing of engine refinements, and new concepts, but the low-speed engine groups also exploit full-scale advanced hardware to evaluate innovations in components and systems.

Sulzer (since absorbed first into Wärtsilä and more recently into WinGD) began operating its first technology demonstrator in 1990, an advanced two-stroke development and test engine designated the 4RTX54 whose operating parameters well exceeded those of any production engine. Until then, this group had used computer-based predictions to try to calculate the next development stage. Extrapolations were applied, sometimes with less than desirable results. The 4RTX54 engine, installed at the Swiss designer's Winterthur headquarters, allowed practical tests with new parameters, components, and systems to be carried out instead of just theory and calculations. Operating data gathered in the field could be assessed alongside results derived from the test engine.

The four-cylinder 540-mm bore/2150-mm stroke engine had a stroke/bore ratio approaching 4:1 and could operate with mean effective pressures of up to 20 bar, maximum cylinder pressures up to 180 bar, and mean piston speeds up to 8.5 m/s. For operating without a camshaft—reportedly the first large two-stroke engine to do so—the RTX54 was equipped with combined mechanical, hydraulic, and electronic (mechatronics) systems for fuel injection, exhaust valve lift, cylinder lubrication, and starting, as well as controllable cooling water flow. The systems underwrote full flexibility in engine settings during test runs.

Sulzer's main objectives from the technology demonstrator engine were to explore the potential of thermal efficiency and power concentration, to increase the lifetime and improve the reliability of components, to investigate the merits of microprocessor technology, and to explore improvements in propulsion efficiency. A number of concepts first tested and confirmed on the 4RTX54 engine were subsequently applied to production designs. The upgraded RTA-2U series and RTA84T, RTA84C, and RTA96C engines, for example, benefit from a triple-fuel injection valve system in place of two valves. This configuration fosters a more uniform temperature distribution around the main combustion chamber components and lower overall temperatures despite higher loads. The significantly lower exhaust valve and valve seat temperatures are also yielded.

An enhanced piston ring package for the RTA-2U series was also proven under severe running conditions on the 4RTX54 engine. Four rings are now used instead of five, the plasma-coated top ring being thicker than the others and featuring a preprofiled running face.

Significant fuel savings in the part-load range were realized from the RTA84T 'Tanker' engine, which further exploits load-dependent cylinder liner cooling and cylinder lubrication systems refined on the 4RTX54. The 4RTX54 was replaced as a research and testing tool in 1995 by the prototype 4RTA58T engine adapted to serve as Wärtsilä's next two-stroke technology demonstrator. This RTX-3 research engine, installed in the Diesel Technology Centre in Winterthur, subsequently played a key role in developing the electronically controlled common rail RT-flex design in Wärtsilä's programme.

A commitment to sustained development was underlined by investment in a new RTX-4 research and laboratory engine, a four-cylinder 600-mm-bore facility, which is based on an RT-flex60 design but evolved from a clean sheet via CAD models and simulations. This engine started running in April 2008.

Initially developing 10,160 kW at a nominal speed of 114 rev/min, the RTX-4 was designed to allow its power output to be increased in refining technologies for future market requirements.

The importance of curbing exhaust gas emissions to progressively lower limits was reflected in the RTX-4 engine's after-treatment systems support, which includes a selective catalytic reduction (SCR) unit capable of reducing NOx emissions discharge to the atmosphere by more than 90% from the level at the engine outlet. In addition, the installation was prepared for investigating particulate filters and scrubbers. Scavenge air humidification and exhaust gas recirculation (EGR) systems for NOx emissions reduction can also be examined.

In 2010, Wärtsilä initiated a major project to adapt its low-pressure gas engine technology for use in its two-stroke engine portfolio. Subsequently, a new test engine, the RTX-5, based on a commercially available six-cylinder RT-flex50 engine, was installed in the Trieste engine laboratory in Italy in March 2011. The resulting tests with the low-pressure low-speed dual-fuel engine demonstrated that low-speed engine performance can fully comply with IMO Tier III NOx limits when operating on gas.

In 2012, Wärtsilä announced the X-Generation of engines featuring a high stroke-to-bore ratio, thus achieving a lower engine speed and reduced brake specific fuel consumption. They also incorporated Wärtsilä's second-generation electronically controlled, common-rail fuel injection system to give fuel savings across the entire load range of the engine.

Another important feature of the Wärtsilä X-Generation engines is that a single electronic control system, the Wärtsilä UNIC, controlled the fuel injection, the exhaust valve operation, and the cylinder lubrication. The first engines were delivered in 2014.

In November 2013, Wärtsilä conducted full-scale testing on the gas of the RTX-5 engine and announced a full new range of engines based on its established models. The first engine utilizing this technology was the RT-flex50DF which received its first sale in December 2013. Wärtsilä said its entire portfolio of two-stroke engines would be available as low-pressure dual-fuel versions with the RT-flex50DF in production in late 2014 and other sizes available for delivery during 2015 and 2016.

Before the first RT-flex50DF engine was delivered, Wärtsilä and China State Shipbuilding Corporation (CSSC) agreed to establish a joint venture to eventually take over Wärtsilä's two-stroke engine business. The joint venture was duly established in 2015 as Winterthur Gas & Diesel Ltd. In 2016, Wärtsilä Corporation transferred its remaining shares of WinGD to CSSC making WinGD 100% owned by CSSC. WinGD has continued the development of both the X-generation and the RT-Flex engine series and has developed dual-fuel versions as well as improved models.

The widest flexibility in operating modes and the highest degree of reliability are cited by MAN Energy Solutions as prime R&D goals underwriting future MAN B&W two-stroke engine generations, along with ease of maintenance, production cost reductions, low-specific fuel consumption and high plant efficiency over a wide load spectrum, high tolerance towards varied heavy fuel qualities, easy installation, continual adjustments to the engine programme in line with the evolving power and speed requirements of the market, compliance with emission controls, and integrated intelligent electronic systems.

Continuing refinement of the MAN B&W MC and ME low-speed engine programmes and the development of intelligent engines are supported by an R&D centre adjacent to the group's Teglholmen office and factory in Copenhagen. At its heart is the 4T50MX research engine, an advanced testing facility. Although the four-cylinder 500-mm bore/2200-mm stroke engine was based on the MC series, it was designed to operate at substantially higher ratings and firing pressures than any production two-stroke engine then available.

The engine has been used to test and demonstrate numerous improvements and has been modified at various times. In 2010, it was rebuilt as a 4T50ME-GI engine to operate on natural gas. This was the first application of dual-fuel to low-speed two-stroke engines.

Dual-fuel operation requires the injection of both pilot fuel-oil and gas fuel into the engine's combustion chamber via different types of valves arranged in the cylinder head. The ME-GI engine head is fitted with two valves for gas injection and two for pilot fuel. The pilot-oil valve is a standard ME fuel-oil valve.

The fuel gas supply system of the 4T50ME-GI-X test engine was provided by Daewoo Shipbuilding & Marine Engineering CO., Ltd. and is based on a high-pressure cryogenic pump system. The system consists of a cryogenic storage tank, a feed pump, suction drum, high-pressure cryogenic pump, pulsation damper, vaporizer, and gas flow/pressure/temperature control system.

MAN B&W ME-C and ME-GI engines are broadly similar and share the same efficiency, output, and dimensions. In comparison, the ME-GI engine's key components are its modified exhaust receiver, modified cylinder cover with gas-injection valves and gas-control block, an expanding top gallery platform, high-pressure fuel-supply pipes, and mounted gas-control units.

Commercial sales of the ME-GI type got underway in 2012 and the first engine was delivered in 2014 and went into service in 2015. This first order was for an 8L70ME-C8-GI engine, sized for a 3100 TEU containership built in the United States. By February 2015 sales had exceeded the 100 marks.

The ME-GI concept has since been developed further and now exists in the basic dual-fuel GI, ethane burning ME-GIE, a liquid gas methanol ME-LGIM, and an LPG burning ME=LGIP variant.

In 2019 MAN Energy Solutions MAN Energy Solutions Two-Stroke Business Unit began the development of a low-pressure gas engine as a complement to its existing, successful dual-fuel ME-GI engine. The project is being conducted in conjunction with Hyundai Heavy Industries and has been initiated in response to customer demand from the LNG carrier segment.

The decision to pursue the development of a low-pressure gas engine came apparently in response from shipowners in the gas carrier market. Some, however, believe that it may be intended as competition for the low-pressure Otto engines produced by WinGD which have been commercially successful partially due to lower costs of fuel gas supply system installations.

MAN Energy Solutions said it plans to develop an LP gas engine with the intention of offering LP dual-fuel LNG engines in late 2021 or early 2022. The initial focus will be on developing 70 bore engines for the gas carrier market. In order to meet the 2022 commercial target, MAN Energy Solutions has invested in expanding its research capacity in Copenhagen, establishing an LP technology development team alongside the existing HP technology development team.

Japan Engine Corporation has the smallest share—below 10%—of the low-speed two-stroke market but it has a loyal following for its engines which have a long pedigree stretching back to 1955 when developed by Mitsubishi. The engines were turbocharged from the outset employing pulse turbocharging in 1973. In 1975, Mitsubishi developed two-stage turbocharging anticipating a trend that was to come to the fore across all engine makers some 40 years later as a means of meeting the NOx Code rules.

In 2006, the first electronically controlled engines were introduced as part of the UEC-LSE series. In 2015, the UEC50LSH-Eco, was released as efficiency and NOx control rules became more stringent. The first 6UEC50LSH-Eco-C2 was completed by Kobe Diesel. To meet Tier III NOx rules, the Mitsubishi Low-Pressure EGR (LP-EGR) system was applied to a 6UEC45LSE-Eco-B2 engine in 2015.

Intelligent engines

Both MAN Diesel and Sulzer (before being acquired by Wärtsilä) demonstrated a 'camshaftless' operation with their research engines, applying electronically controlled fuel injection and exhaust valve actuation systems. Since then engine R&D has been directed at working towards a future generation of highly reliable 'intelligent engines': those which monitor their own condition and adjust parameters for optimum performance in all operating regimes, including fuel-optimized and emissions-optimized modes. An 'intelligent engine-management system' will effectively close the feedback loop by built-in expert knowledge.

Engine performance data can be constantly monitored and compared with defined values in the expert system; if deviations are detected, corrective action is automatically taken to restore the situation to normal. A further step would incorporate not only engine optimizing functions but also management responsibilities such as maintenance planning and spare parts control.

Already inshore installations, engines are often remotely monitored and operated by the engine provider using data taken directly from the engine and transferred in real time to a control centre. Repair and maintenance of the engines may also be contracted to the engine provider with remote assistance being provided when physical intervention is necessary. Ship operators have so far been reluctant to follow this direction although several do now send operating data to the engine companies who can issue advice on maintenance strategies.

All this has become possible only with electronic engines because to meet the operational flexibility target, it is necessary to be able to change the timing of the fuel injection and exhaust valve systems while the engine is running. To achieve this objective with cam-driven units would involve a substantial mechanical complexity, which would undermine engine reliability.

The concept is illustrated in Fig. 13.1 whose upper part shows the operating modes, which may be selected from the bridge control system or by the intelligent engine's own control system. The centre part shows the brain of the system: the electronic control system which analyses the general engine condition and controls the operation of the engine systems shown in the

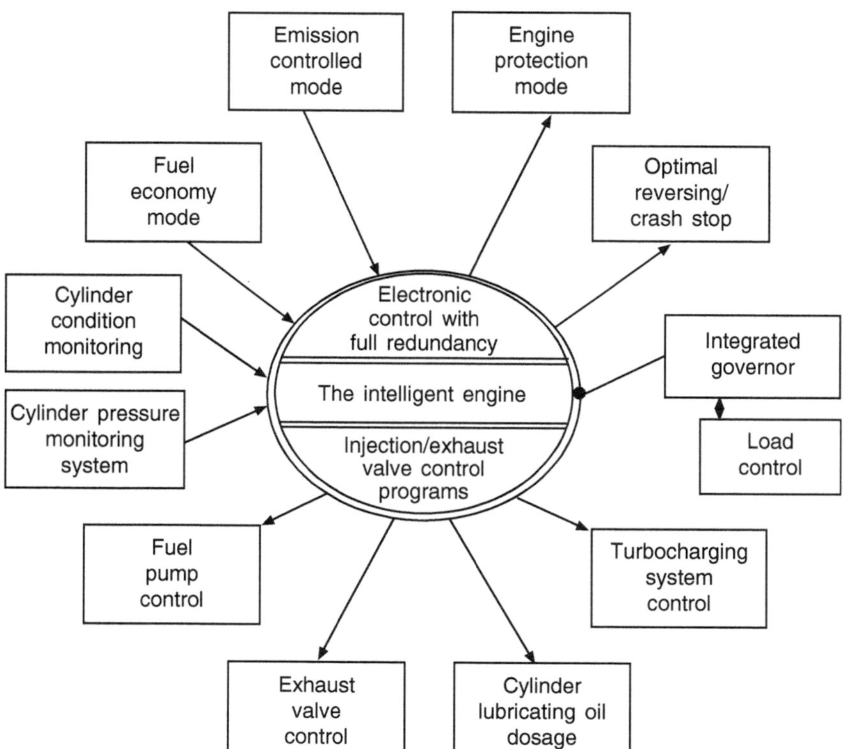

Fig. 13.1 Schematic diagram of the intelligent engine (MAN Diesel). *From MAN Energy Solutions.*

lower part of the diagram (the fuel injection, exhaust valve, cylinder lube oil, and turbocharging systems).

To meet the reliability target, it is necessary to have a system which can actively protect the engine from damage due to overload, lack of maintenance, and maladjustments. A condition monitoring system must be used to evaluate the general condition of the engine, thus maintaining its performance and keeping its operating parameters within prescribed limits. The condition monitoring and evaluation system is an online system with an automatic sampling of all 'normal' engine performance data, supplemented by cylinder pressure measurements. The system will report and actively intervene when performance parameters show unsatisfactory deviations. The cylinder pressure data delivered by the measuring system are used for various calculations:

- The mean indicated pressure is determined as a check on cylinder load distribution as well as total engine output.
- The compression pressure is determined as an indicator of excessive leakage caused by, for example, a burnt exhaust valve or collapsed piston rings (the former condition is usually accompanied by an increased exhaust gas temperature in the cylinder in question).
- The cylinder wall temperature is monitored as an additional indicator of the piston ring condition.
- The firing pressure is determined for control of injection timing and mechanical loads.
- The rates of pressure rise ($\mathrm{d}p/\mathrm{d}t$) and heat release are determined for combustion quality evaluation as a warning in the event of 'bad fuels' and to indicate any risk of piston ring problems in the event of high $\mathrm{d}p/\mathrm{d}t$ values.

The cylinder condition monitoring system is intended to detect faults such as blow-by past the piston rings, cylinder liner scuffing, and abnormal combustion. The detection of severe anomalies by the integrated systems triggers a changeover to a special operating mode for the engine—the 'engine protection mode'. The control system will contain data for optimum operation in a number of different modes such as 'fuel economy mode', 'emission controlled mode', 'reversing/crash stop mode', and various engine protection modes. The load limiter system (load diagram compliance system) aims to prevent any overloading of the engine in conditions such as heavy weather, fouled hull, shallow water, too heavy propeller layout, or excessive shaft alternator output. This function will appear as a natural part of future governor specifications.

The fuel injection system is operated without a conventional camshaft using high-pressure hydraulic oil from an engine-driven pump as a power source and an electronically controlled servo system to drive the injection pump plunger. The general concept of the intelligent fuel injection (InFI) system and the intelligent valve actuation (InVA) system for operating the exhaust valves is shown in Figs. 13.2 and 13.3. Both systems, when operated in the electronic mode, receive the electronic signals to the control units. In the event of failure of the electronic control system, the engine is controlled by a mechanical input supplied by a diminutive camshaft giving full redundancy.

Unlike a conventional, cam-driven pump, the InFI pump has a variable stroke and will pressurize only the amount of fuel to be injected at the relevant load. In the electronic mode (i.e. operating without a camshaft), the system can perform as a single injection system as well as a preinjection

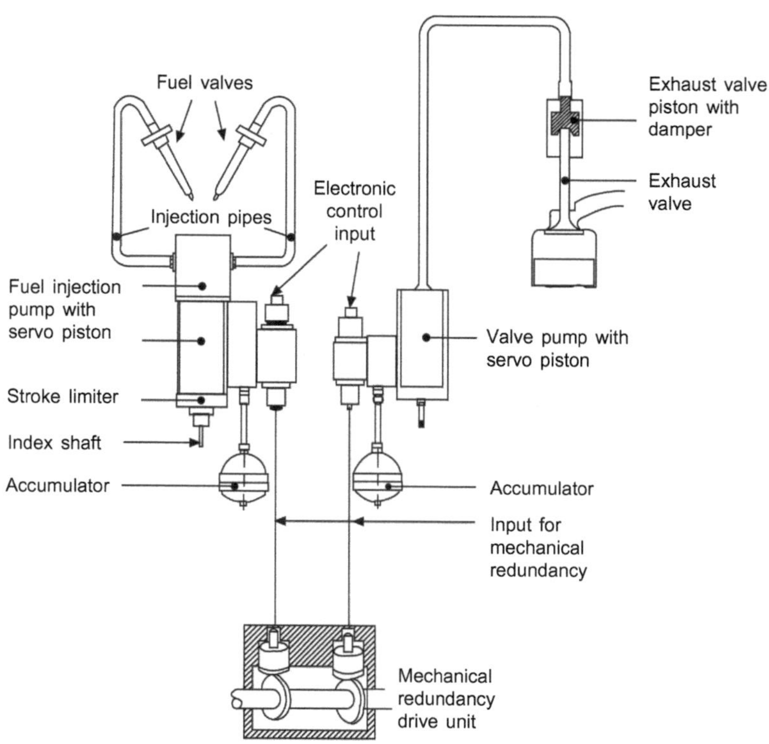

Fig. 13.2 Electronically controlled hydraulic systems for fuel injection and exhaust valve operation on MAN Diesel's 4T50MX research engine.

Fig. 13.3 Cross-section of MAN B&W S70ME-C electronically controlled engine.

system with a high degree of freedom to modulate the process in terms of injection rate, timing, duration, pressure, single/double injection, cam profile, and so on.

Several optimized injection patterns can be stored in the computer and chosen by the control system in order to operate the engine with optimum injection characteristics at several loads: from dead slow to overload as well as for starting, astern running, and crash stop. The changeover from one to another of the stored injection characteristics is affected from one injection to the next. The system is able to adjust the injection amount and injection timing for each cylinder individually in order to achieve the same load (mean indicated pressure) and the same firing pressure (P_{max}) in all cylinders, or, in protection mode, to reduce the load and P_{max} on a given single cylinder if the need arises.

The exhaust valve system (InVA) is driven on the same principles as the fuel injection system, exploiting the same high-pressure hydraulic oil supply and a similar facility for mechanical redundancy. The need for controlling exhaust valve operation is basically limited to timing the opening and closing of the valve. The control system is thus simpler than that for fuel injection.

Cylinder lubrication is controllable from the condition evaluation system so that the lubricating oil amount can be adjusted to match the engine load. Dosage is increased in line with load changes and if the need is indicated by the cylinder condition monitoring system (e.g. in the event of liner scuffing and ring blow-by). Such systems are already available for existing engines.

The turbocharging system control will incorporate control of the scavenge air pressure if a turbocharger with variable turbine nozzle geometry is used, and control of bypass valves, turbo-compound system valves and turbocharger cut-off valves if such valves are incorporated in the system. Valves for any SCR exhaust gas cleaning system installed will also be controlled.

Operating modes may be selected from the bridge control system or by the system's own control system. The former case applies to the fuel economy modes and the emission-controlled modes (some of which may incorporate the use of an SCR system). The optimum reversing/crash stop modes are selected by the system itself when the bridge control system requests the engine to carry out the corresponding operation. Engine protection mode, in contrast, will be selected by the condition monitoring and evaluation system independently of actual operating modes (when this is not considered to threaten ship safety).

Although the majority of new engines are now electronically controlled, gathering of data necessary to build a sufficiently large bank of experience to predict problems and maintenance issues is not yet something that has been achieved as comprehensively as might have been hoped. There are various reasons for this, a reluctance by shipowners to share information, the cost of communications, and lack of bandwidth for streaming data and changes in ownership of vessels after relatively short periods. Furthermore, the number of engines for each type is quite low.

In 2018, MAN Energy Solutions service division, PrimeServ developed a new tool called PrimeServ Assist which offers subscribing ship operators real-time assistance in anticipating problems and faults on connected engines. The information from the sensors on the engine and associated machinery is transmitted via the ship's communication systems to a PrimeServ shore facility and constantly monitored electronically. Information and advice from the shore centre can be transmitted back to the ship or to the owner's superintendents using apps for mobile telephones or tablets.

CHAPTER FOURTEEN

MAN B&W low-speed engines

MAN Energy Solutions' roots are closely entwined with the early days of the diesel engine through both its German and Danish branches, respectively, MAN and Burmeister & Wain (B&W). Both groups evolved distinctive low-speed two-stroke crosshead engine designs before MAN acquired control of B&W in 1980. MAN subsequently discontinued the development of its own loop-scavenged engines at Augsburg and centred all low-speed design and R&D operations in Copenhagen, pursuing the refinement of the MAN B&W uniflow-scavenged programme in the shape of the MC series. The company's low-speed engines continue to be branded as MAN B&W.

The prototype, an L35MC model, entered service in 1982 and the first large bore example, a six-cylinder L60MC engine, was started on the Christianshavn testbed in September 1983. The full L-MC programme was introduced in 1982 with bore sizes of 350, 500, 600, 700, 800, and 900 mm. Refinement of the MC design and the introduction of new bore sizes (260, 420, 460, and 980 mm) and stroke options continued through the 1980s/1990s in line with service experience and market trends. The portfolio eventually embraced 11 bore sizes from 260 to 1080 mm and K (short stroke), L (long stroke), and S (superlong stroke) variations, with stroke–bore ratios ranging from 2.44 to 4.2:1 and rated speeds from 56 to 250 rev/min. Although it appeared in the portfolio, no 1080 mm bore marine engine was ever built.

Output demands from around 1100 to 97,300 kW were covered by 4–14-cylinder in-line models in the MC programme, whose individual rating envelopes are shown in Fig. 14.1. Additions to the portfolio and key parameter refinements were made over the years progressing to Mark 7 versions with mean effective pressures of 19 bar or higher. Performance curves are typified by those for an S60MC engine (Fig. 14.2).

Higher rated Mark 9 versions of 800- and 900-mm-bore S and K engines were released in 2006, an mep of 20 bar, a maximum firing pressure of 160 bar, and a mean piston speed of 9 m/s underwriting power increases of up to 25%. Smaller bore MC-C models had already benefited from an mep rise to 20 bar introduced in 2004, delivering 5% power increases.

Pounder's Marine Diesel Engines and Gas Turbines
https://doi.org/10.1016/B978-0-08-102748-6.00014-1

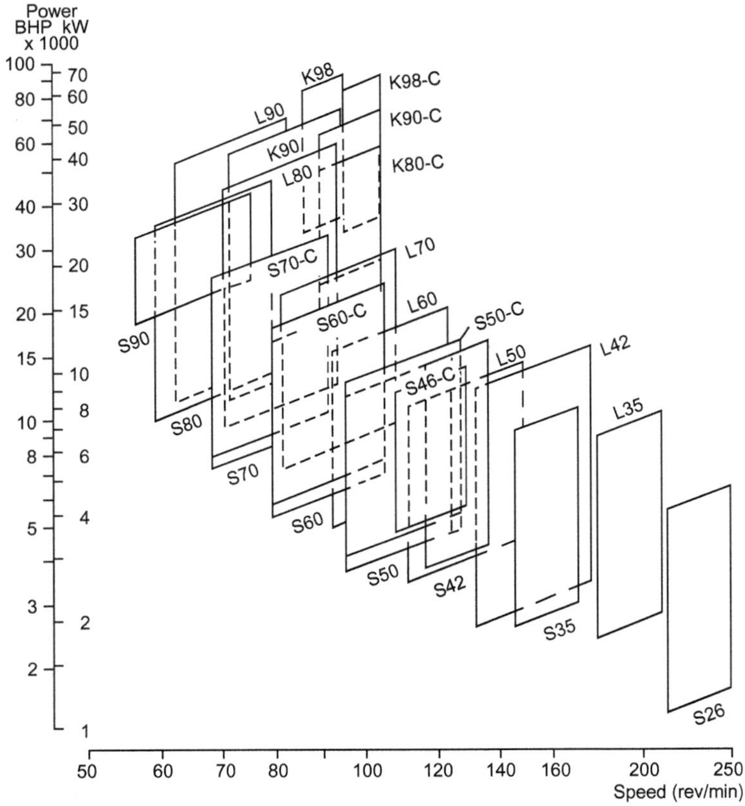

Fig. 14.1 MAN Diesel's MAN B&W MC engine programme.

Electronically controlled ME versions of the 500–1080-mm bore models (detailed below) were progressively introduced from 2001, offering the same outputs as their MC engine equivalents. ME-B models with bore sizes from 350 to 600 mm extended the electronic engine programme from 2006, with stroke–bore ratios up to 4.42:1 and mean effective pressures of 21 bar.

The layout diagram in Fig. 14.3 (in this case for an S60MC engine) shows the area within which there is full freedom to select the combination of engine power (kW) and speed (rev/min) that is the optimum for the projected ship and the expected operating profile. The engine speed (horizontal axis) and engine power (vertical axis) are shown in percentage scales. The scales are logarithmic, which means that, in this diagram, exponential curves such as propeller curves (third power), constant mean effective pressure curves (first power), and constant ship speed curves (0.15–0.30 power) are straight lines. The constant ship speed lines, shown at the uppermost part

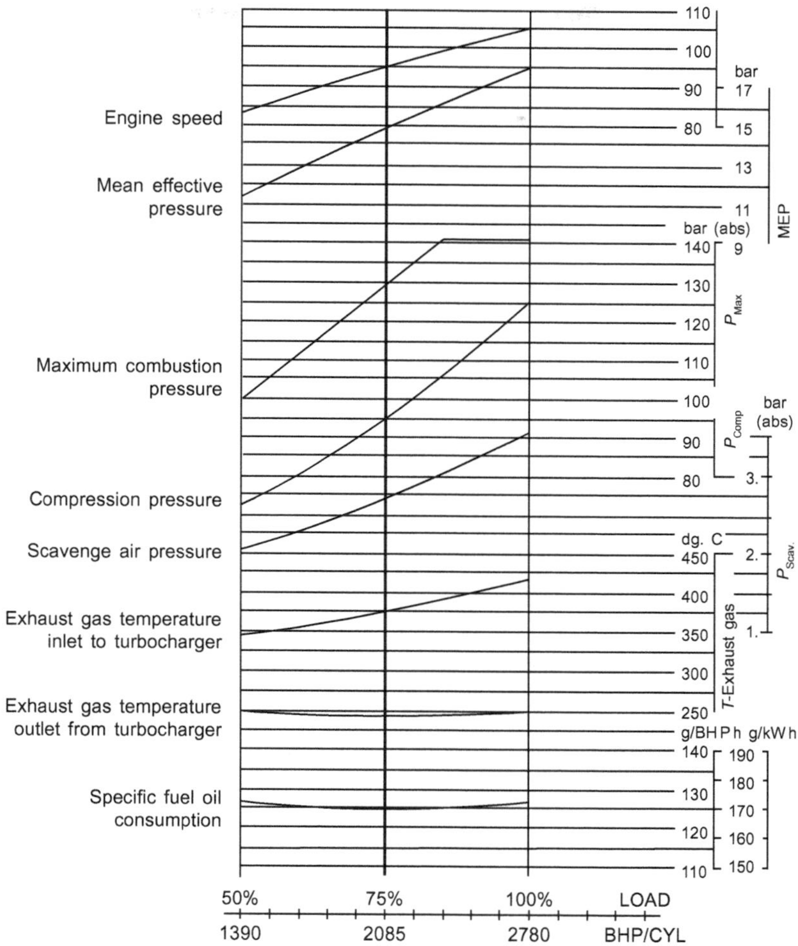

Fig. 14.2 Performance curves for S60MC engine.

of the diagram, indicate the power required at various propeller speeds in order to keep the same ship speed, provided that, for each ship speed, the optimum propeller diameter is used, taking into consideration the total propulsion efficiency.

An engine's layout diagram is limited by two constant mean effective pressure lines L1–L3 and L2–L4, and two constant engine speed lines L1–L2 and L3–L4. The L1 point refers to the engine's nominal maximum continuous rating (mcr). Based on the propulsion and the engine running points, the layout diagram of a relevant main engine can be drawn in.

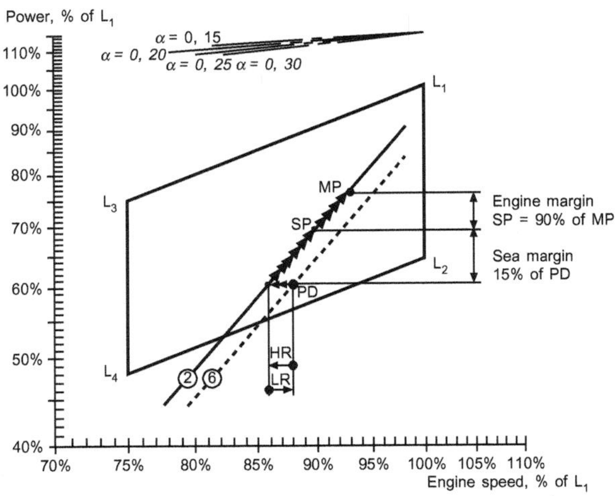

Line 2	Propulsion curve, fouled hull and heavy weather (heavy running)
Line 6	Propulsion curve, clean hull and calm weather (light running)
MP	Specified MCR for propulsion
SP	Continuous service rating for propulsion
PD	Propeller design point
HR	Heavy running
LR	Light running

Fig. 14.3 Ship propulsion running points and engine layout for the S60MC model. Line 2, propulsion curve, fouled hull and heavy weather (heavy running); Line 6, propulsion curve, clean hull, and calm weather (light running); MP, specified MCR for propulsion; SP, continuous service rating for propulsion; PD, propeller design point; HR, heavy running; LR, light running.

The specified mcr point (M) must be inside or on the limitation lines of the layout diagram; if not, the propeller speed has to be changed or another main engine model must be chosen. It is only in special cases that point M may be located to the right of line L1–L2.

The specified mcr is the maximum rating required by the yard or owner for the continuous operation of the chosen engine. Point M can be any point within the layout diagram. Once the specified mcr point has been selected and provided that the shaftline and auxiliary equipment are dimensioned accordingly, that point is now the maximum rating at which an overload of 10% is permissible for 1 h in 12 h. The continuous service rating (S) is the power at which the engine is normally assumed to operate; point S is identical to the service propulsion point (SP) unless a main engine-driven shaft generator is installed.

All MC engines can be delivered to comply with the IMO speed-dependent NOx emission limits for the exhaust gas, measured according to the ISO 8178 test cycles E2/E3 for heavy-duty diesel engines. NOx emissions from a given engine will vary according to the engine load and the optimizing power. Specific fuel consumption (sfc) and NOx emission levels are interrelated parameters, and an engine offered with both a guaranteed sfc and the IMO NOx limitation will be subject to a tolerance of 5% on the fuel consumption.

MC engine design features

All MC series engine models are based on the same design principles and aim for simplicity and reliability, the key elements comprising the following:

Bedplate. The rigid bedplate for the large-bore engines is built up of longitudinal side girders and welded crossgirders with cast steel bearing supports. For the smaller bore engine types, the bedplate is of cast iron. It is designed for long, elastic holding-down bolts arranged in a single row and tightened with hydraulic tools. The main bearings are lined with white metal, and the thrust bearing is incorporated in the aft end of the bedplate. The aft-most crossgirder is therefore designed with ample stiffness to transmit the variable thrust from the thrust collar to the engine seating.

Framebox. This is a single welded unit for the large-bore models and a cast-iron unit for the smaller types, the design contributing to a very rigid engine structure. The framebox is equipped on the exhaust side with a relief valve and on the camshaft side with a large door for each cylinder providing access to the crankcase components.

Cylinder frame. The cast-iron cylinder frames on the top of the framebox make another significant contribution to the rigidity of the overall engine structure. The frames include the scavenge boxes, which are dimensioned to ensure that scavenge air is admitted uniformly to the cylinders. Stay bolts are tightened hydraulically to connect the bedplate, the framebox, and cylinder frames, and form a very rigid unit.

Crankshaft. The conventional semi-built, shrink-fitted type crankshaft is provided with a thrust collar. The sprocket rim for the camshaft chain drive is fitted on the outer circumference of the thrust collar in order to reduce the overall length of the engine—except for the high cylinder numbers of the 800-mm bore model and upwards in the programme for which the chain drive is located between two cylinders. An axial vibration damper is integrated on the free end of the crankshaft.

Connecting rod. In order to limit the height of the engines, a relatively short connecting rod, comprising few principal parts, is specified. The large area of the lower half of the crosshead bearing allows the use of white metal or tin–aluminium on the small-bore engine models. Floating guide shoes mean that most of the alignment work formerly required with crosshead engine pistons can be eliminated. The crankpin bearing for all engine models features thin shells lined with white metal.

Cylinder liner. A simple symmetrical design fosters low lubricating oil consumption and low wear rates. The liner is bore cooled on the larger engine models and available in two different configurations—with or without insulation of the cooling water jet pipes—so as to match the cooling intensity closely to the different engine ratings. The joint between liner and cover is located relatively low. This arrangement means that a larger part of the heat-exposed combustion chamber is contained in the steel cylinder cover rather than in the cast iron cylinder liner. Smaller bore engines feature a simple slim-type liner without cooling bores. For both types of liner, adequate temperature control of the liner surfaces safeguards against cold corrosion caused by the condensation of sulphuric acid (originating from the sulphur content of heavy fuel) and, at the same time, ensures stable lubrication conditions by preventing excessive temperatures (Fig. 14.4).

Cylinder cover. This is a solid steel component provided with bored passages for cooling water, a central bore for the exhaust valve and bores for fuel valves, a safety valve, a starting valve, and an indicator valve.

Piston. An oil-cooled piston crown of heat-resistant chrome–molybdenum steel is rigidly bolted to the piston rod to allow distortion-free transmission of the firing pressure. The piston has four ring grooves, which are hard chrome plated on both upper and lower surfaces of the grooves. A cast iron piston skirt (with bronze sliding bands on the large-bore engines) is bolted to the underside of the piston crown (Figs 14.5 and 14.6).

Piston rod. The rod is surface treated to minimize friction in the stuffing box and to allow a higher sealing ring contact pressure. The piston rod stuffing box, providing effective sealing between the clean crankcase and the 'combustion area', has a proven record of very low amounts of drainage oil.

Camshaft. The fuel injection pumps and the hydraulic exhaust valve actuators are driven by the camshaft. Cams are shrink-fitted to the shaft and can be individually adjusted by the high-pressure oil method. Like its predecessors, the MC engine uses a chain drive to operate the camshaft and thus secures high

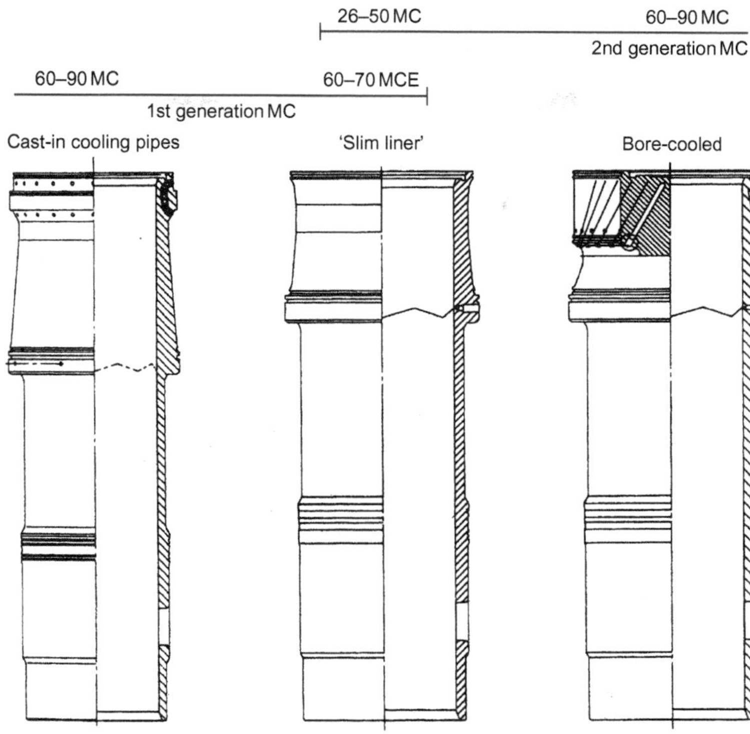

Fig. 14.4 Cylinder liner designs.

reliability since a chain is virtually immune to foreign particles. It also enables the camshaft to be positioned higher, shortening the hydraulic connections to the fuel valves and the exhaust valves, and, in turn, minimizing timing errors due to elasticity and pressure fluctuations in the pipe system.

Exhaust valve. Hydraulic oil supplied from the actuator opens the exhaust valve, and the closing force is delivered by a 'pneumatic spring', which leaves the valve spindle free to rotate. The closing of the valve is damped by an oil cushion on top of the spindle. The rotation force is provided by exhaust gas acting on vanes fitted to the valve stem. Extended service life from the valve is underwritten by Nimonic valve spindles and hardened steel bottom pieces, specified as standard on the large-bore engines. The bottom piece features patented 'chamber-in-seat' geometry.

Fuel pump. Larger engine models incorporate pumps with variable injection timing for optimizing fuel economy at part load; the start of fuel

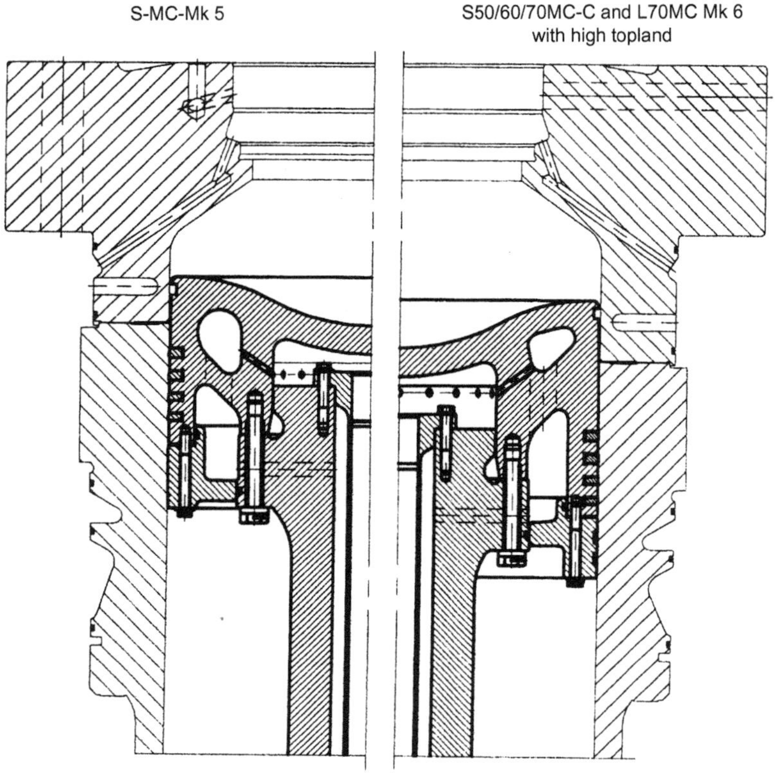

Fig. 14.5 Piston/ring pack assembly MC vs MC-C engines.

injection is controlled by altering the pump barrel position via a toothed rack and a servo unit. The individual adjustment can be made on each cylinder. In addition, collective adjustment of the maximum pressure level of the engine can be carried out to compensate for varying fuel qualities, wear, and other factors. Both adjustments can be effected while the engine is running. The pump is provided with a puncture valve, which prevents fuel injection during normal stopping and shutdown.

Fuel oil system. The engine is served by a closed pressurized fuel oil system, with the fuel preheated to a maximum of 150°C to ensure a suitable injection viscosity. The fuel injection valves are uncooled. The fuel system is kept warm by the circulation of heated fuel oil, thus allowing pier-to-pier operation on heavy fuel oil.

Reversing mechanism. The engine is reversed by a simple and reliable mechanism, which incorporates an angularly displaceable roller in the fuel pump drive of each cylinder. The link connecting the roller guide and

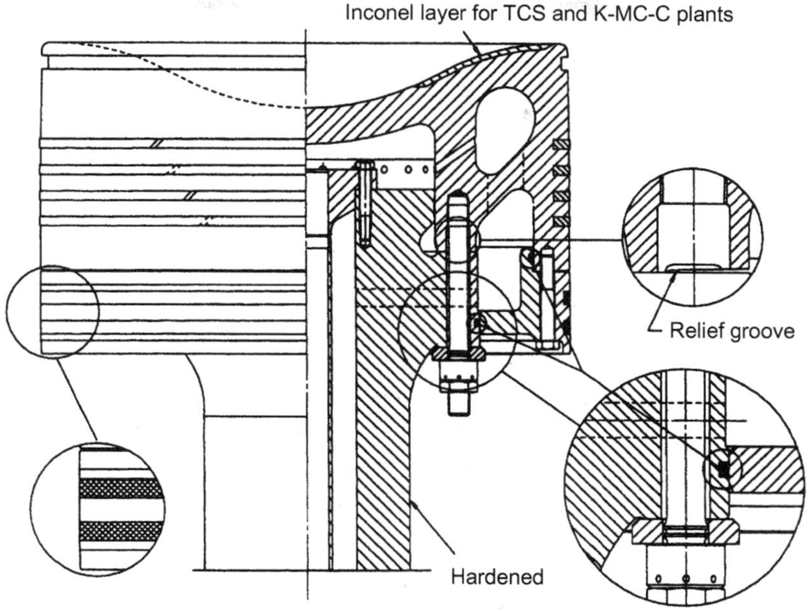

Fig. 14.6 Piston assembly.

the roller is self-locking in the ahead and astern positions. The link is activated by compressed air, which has proved to be a very reliable method since each cylinder is reversed individually. The engine remains manoeuvrable even if one cylinder fails: in such a case the relevant fuel pump is set to the zero index position.

Shaft generators. All MC engines can be arranged to drive shaft generators. The power take-off/Renk constant frequency (PTO/RCF) system is MAN Diesel's standard configuration for PTOs from 420-mm bore models and upwards coupled to a fixed-pitch propeller. The system is mounted on brackets along the bedplate on the exhaust side of the engine. The generator and its drive are isolated from torsional and axial vibrations in the engine by an elastic coupling and a tooth coupling mounted on a flange on the free end of the crankshaft. The drive comprises a three-wheel gear train. To ensure a constant frequency, the planetary gear incorporates a hydraulic speed control arrangement, which varies the gearing ratio. The frequency is kept constant down to 70% of the main engine's specified mcr speed, corresponding to

30% of the engine's specified mcr power, making the system suitable for fixed pitch propeller installations.

The PTO/gear constant ratio (GCR) system is MAN Diesel's standard solution for PTO in plants featuring controllable pitch propellers. The system comprises a compact unit with a step-up gear coupled directly to the generator, which is located above the elastic coupling.

Power turbines. Larger MC engines (from 600-mm bore upwards) can be specified with a turbo compound system (TCS), which exploits exhaust energy surplus to the requirements of a high-efficiency turbocharger to drive a power gas turbine (see Chapter 12). For engine loads above 50% of the optimized power, the gas turbine is mechanically or hydraulically connected to the crankshaft. Power is fed back to the main engine, thus reducing the total fuel consumption. The standard system, designated TCS/power take-in (PTI), is delivered as a complete unit built on the engine. A combined PTO/PTI unit embracing a PTO with the associated generator and a TCS/PTI unit with a power turbine coupled to the crankshaft gear is also offered.

Programme expansion

New models have entered the MC programme since its launch in response to propulsion market trends at both ends of the power spectrum. A notable addition in 1986—the S26MC series—took the low-speed engine deep into small-ship propulsion territory, offering an output per cylinder of 365 kW at 250 rev/min. The nominal power of the 260-mm-bore/980-mm-stroke design has since been raised to 400 kW at 250 rev/min and the 4- to 12-cylinder programme covers an output range from 1100 to 4800 kW. With a stroke–bore ratio of 3.77:1, this rating corresponds to a mean piston speed of 8.2 m/s and a mean effective pressure of 18.5 bar. A firing pressure of around 170 bar yields an sfc of 179 g/kWh at the mcr. Reliable operation on poor quality heavy fuel oil with a viscosity of up to 700 cSt/50°C is promised.

The key components of the S26MC engine (Fig. 14.7) are based on those well proven in the established larger bore models but with modifications to suit the intended small-ship market. Design amendments since its introduction include an improved axial vibration damper and refinements to the crankshaft. The cylinder cover and fuel injection system have also been

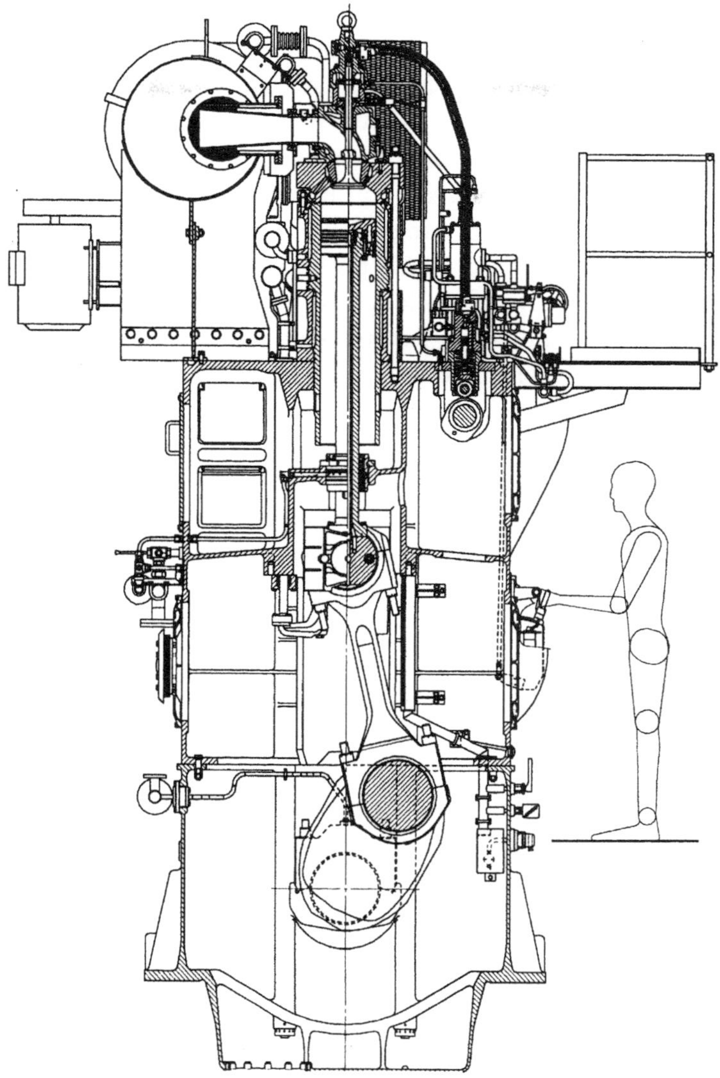

Fig. 14.7 S26MC engine cross section.

upgraded, the auxiliary blower arrangement is simplified, and the exhaust valve provided with a new closing system with built-in damping.

Similar refinements have benefited the L35MC engine, which pioneered the MC programme. Upgrading in 1992 sought improved performance and reduced maintenance costs, the new maximum power rating of 650 kW/cylinder at 210 rev/min, mean effective pressure of 18.4 bar, and mean piston speed of 7.35 m/s calling for increased diameters of the main and crankpin journals, and the reinforcement of the main bearing housing and bedplate. The crosshead bearing, thrust bearing, and piston and piston rod were also reinforced, and the cylinder frame, camshaft arrangement, and auxiliary blower arrangement were simplified.

A longer stroke S-version of the 350-mm bore design was introduced in 1993, this S35MC model delivering 700 kW/cylinder at 170 rev/min (since increased to 740 kW/cylinder at 173 rev/min). A number of components were modified in line with the longer stroke (1400 mm compared with 1050 mm), but the main design features were unchanged.

The strong and rigid main structure of the S35MC is essentially the same as the L35MC, but the bedplate and framebox are wider and higher due to the longer stroke of the semi-built crankshaft. The connecting and piston rods are necessarily longer, but the cylinder frame is identical. The main bearings are wider in order to keep the bearing load inside the area ensuring good service results. Most of the components in the S35MC combustion chamber are also similar to those of the L-version. The exhaust valve design was unchanged but an improved sealing system was introduced to further reduce the wear of the valve stem. The cylinder cover is slightly higher to allow for the increased combustion space.

Multilevel cylinder lubrication was adopted to ensure sufficient lubrication of the liner. A modified piston pack comprises two high rings in grooves 1 and 2, and two rings of normal height in grooves 3 and 4. All of the rings are made from improved material. The fuel pump plunger diameter was increased to satisfy the larger injection volume per stroke, and improved sealing was introduced to prevent any fuel from leaking into the main lube oil, which is common to the crankcase and camshaft. The camshaft diameter was increased to accept the larger torque from the fuel pumps.

The S35MC and the 1994-launched S42MC series were conceived to improve the propulsive efficiency of plants for small-to-medium-sized ships by lowering the engine speed, these 350-mm bore and 420-mm bore models, respectively, distinguished by stroke–bore ratios of 4:1 and 4.2:1. The later introduction of the S46MC-C model (Table 14.1) (also exploiting a 4.2:1 stroke–bore ratio, then the highest of any production engine in the world) addressed the needs of smaller tankers and bulk carriers, which can also be served by the S42MC and L or S50MC engines. The 460-mm bore model extended the number of ideal combinations of power, speed, and the number of cylinders for a given project (Table 14.1 and Fig. 14.8).

Table 14.1 S46MC-C engine.

Bore	460 mm
Stroke	1932 mm
Stroke–bore ratio	4.2:1
Speed	129 rev/min
Mean effective pressure	19 bar
Output/cylinder	1310 kW
Mean piston speed	8.3 m/s
Cylinders	4–8
Specific fuel consumption (mcr)	174 g/kW h

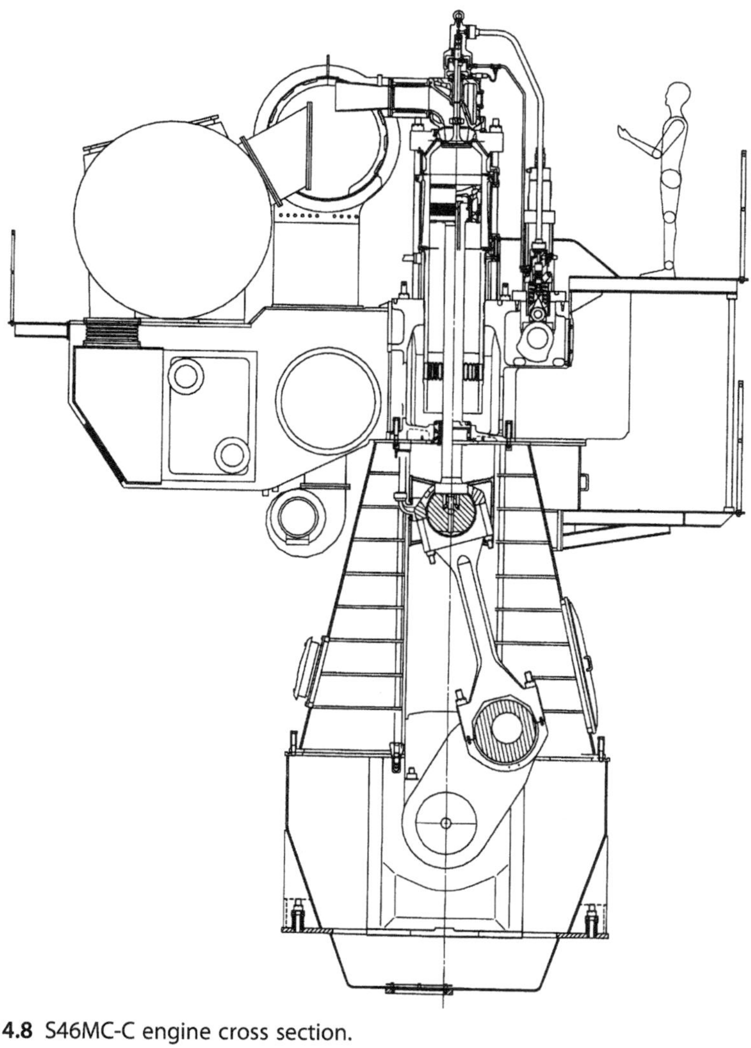

Fig. 14.8 S46MC-C engine cross section.

Choice in the mid-bore range (500, 600, and 700 mm models) was widened from 1996 by uprated versions supplementing the existing S50MC, S60MC, and S70MC engines. These new S50MC-C, S60MC-C, and S70MC-C models, which share the same design features as the S46MC-C, are more compact (hence the C-designation which is also used for the subsequent compact version of the electronically controlled ME engines) and offer higher outputs than their established equivalents (an L70MC engine is shown in Fig. 14.9). Stroke–bore ratios were raised from 3.82 to 4:1, and the increased power ratings correspond to a rise in the mean effective pressure to 19 bar.

Supporting the higher ratings are modified turbocharging and scavenge air systems as well as modifications of the combustion chamber configuration and bearings. Installation space savings were achieved by reducing the overall length of the C-engines (by around 1000 mm in the case of the six-cylinder S50MC-C engine) and the overhauling height requirement compared with the original designs. The masses are also lower (by 25 t or 13% for the 6S50MC-C), which yields benefits in reduced vibration excitations. The MC-C engines can be 100% balanced.

VLCC project planners were given a wider choice with the introduction of a longer stroke 900-mm bore model, the S90MC-T design, tailored to the propulsion of large tankers (Table 14.2). The layout flexibility enables operators to select maximum continuous speeds between 64 and 75 rev/min in seeking optimum propeller efficiency. The design parameters (Table 14.2) addressed the key factors influencing the selection of the VLCC propulsion plant, defined by MAN B&W as the projected ship speed, the propeller diameter that can be accommodated and engine compactness. An sfc of 159 g/kWh was quoted for a derated S90MC-T engine served by high-efficiency turbocharger(s).

The dimensions of the S90MC-T engine were required not to exceed those of the 800-mm bore S80MC series, which had established numerous references in the 'new generation' VLCC market. MAN Diesel suggested that VLCCs with speed requirements of more than 15 knots could take advantage of the S90MC-T engine, either as a full-powered six-cylinder model or as an economy-rated seven-cylinder model. The S80MC engine, in seven-cylinder form, was considered an attractive solution for VLCCs required to operate with speeds of up to 15 knots.

Refinements also sought to maintain the competitiveness of the S80MC design. Engine length, and hence engine room length, represents dead

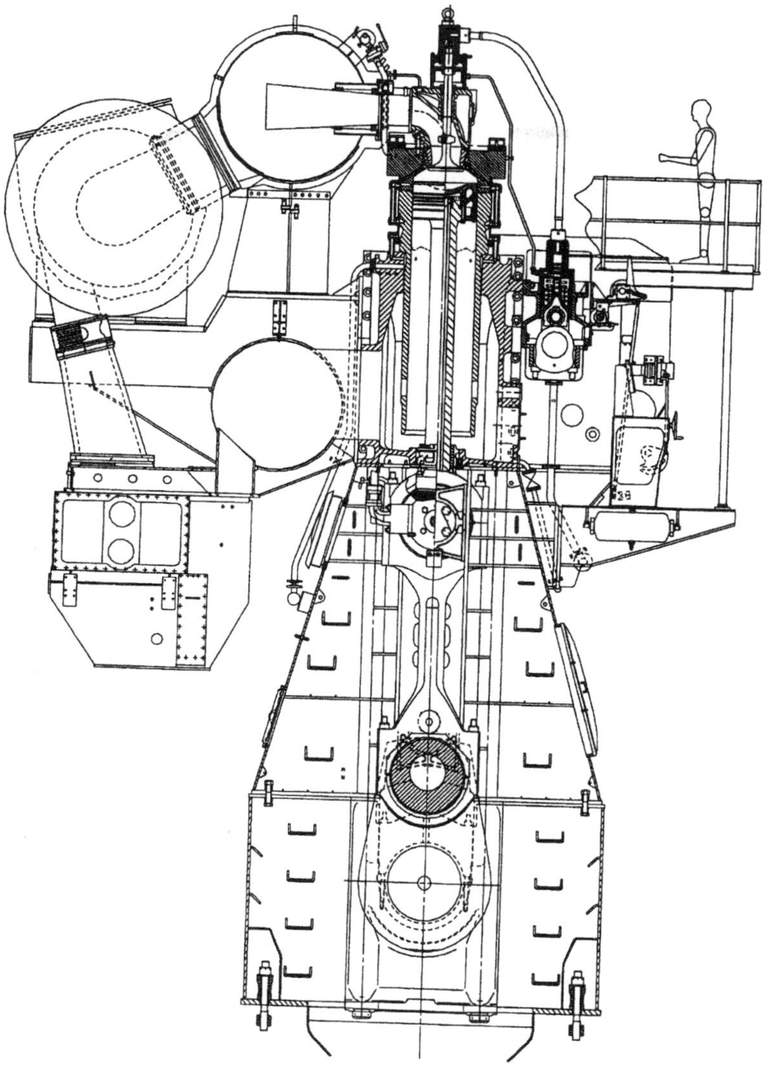

Fig. 14.9 L70MC Mk 5 engine cross section.

Table 14.2 S90MC-T engine.

Bore	900 mm
Stroke	3188 mm
Stroke–bore ratio	3.54:1
Nominal speed	75 rev/min
Mean piston speed	7.97 m/s
Mean effective pressure	18 bar
Output/cylinder	4560 kW
Cylinders	5, 6, 7
Specific fuel consumption (mcr)	166 g/kWh

volume and is normally to be minimized. The S80MC, therefore, benefited from a remodelling of its bedplate and chain drive/thrust bearing to shorten the length by 700 mm. The engine could then be pushed deeper aft into the hull, promoting increased cargo capacity from a given overall ship length or reduced ship length for the same capacity. The introduction of an optimized fuel system in conjunction with improved timing (already implemented on smaller bore MC engines) raised thermal efficiency equivalent to a reduction in sfc approaching 3 g/kW h. The sfc can be further improved at part load by specifying a high-efficiency turbocharger and an adjustable exhaust gas bypass, an option which can be applied in conjunction with a TCS power turbine system.

The S-MC-C series was expanded by a 900-mm bore design, the first example of which entered service in mid-2000 (Table 14.3); the first units of another new type, the S80MC-C, were ordered in the same year. The 7S80MC-C and 6S90MC-C models have proved popular for VLCCs and ULCCs requiring higher service speeds than before, their outputs, respectively, 7% and 15% higher than the 7S80MC Mark 6 engine (Fig. 14.10).

Demands for slightly higher outputs from the L60MC and L70MC engines were addressed by adopting S-MC-C design principles for these two types, with a mean effective pressure of 19 bar and a mean piston speed of approximately 8.5 m/s. The output of the L60MC-C engine is 2230 kW/cylinder at 123 rev/min, while the L70MC-C engine delivers 3110 kW at 108 rev/min. Among the targeted applications of the L-MC-C series were feeder container ships.

Table 14.3 MAN B&W 80- and 90-bore Mark 9 engines.

	S80ME-C	K80ME-C	K90ME	K90ME-C
Bore (mm)	800	800	900	900
Stroke (mm)	3450	2600	2870	2600
Speed (rev/min)	78	104	94	104
Mean effective pressure (bar)	20	20	20	20
P_{max} (bar)	160	160	160	160
Mean piston speed (m/s)	9	9	9	9
Cylinders	6–9	6–12	6–12	6–12
Output/cylinder (kW)	4510	4530	5720	5730
Power range (MW)	27.1–40.6	27.2–54.4	34.3–68.6	34.4–68.8

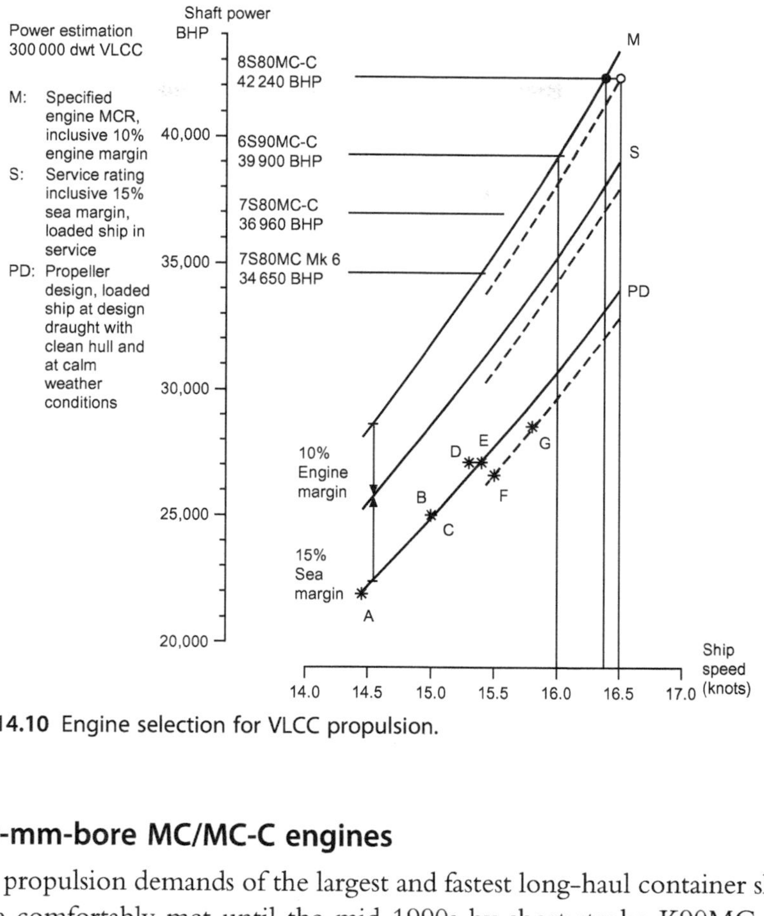

Fig. 14.10 Engine selection for VLCC propulsion.

980-mm-bore MC/MC-C engines

The propulsion demands of the largest and fastest long-haul container ships were comfortably met until the mid-1990s by short-stroke K90MC and K90MC-C engines (Fig. 14.11) with an upper output limit of 54,840kW. In 1994, market interest in a new generation of 6000 TEU-plus ships with service speeds of 25 knots or more stimulated the introduction of the K98MC-C series, which, in 12-cylinder form, could deliver 68,520kW (Fig. 14.12). This 980-mm bore/2400-mm stroke model shared key operating parameters with the 900-mm bore MC/MC-C engines, notably a mean effective pressure of 18.2bar and a mean piston speed of 8.32m/s. A propeller speed requirement of 104rev/min was selected as the design basis, resulting in a stroke–bore ratio of 2.45 and an output per cylinder of 5710kW. (A cylinder bore of 980mm was not new to the marine engine industry: in 1966 B&W introduced a K98 design, whose 12-cylinder version developed 33,530kW; a number of these K98FF and K98GF engines remain in service.)

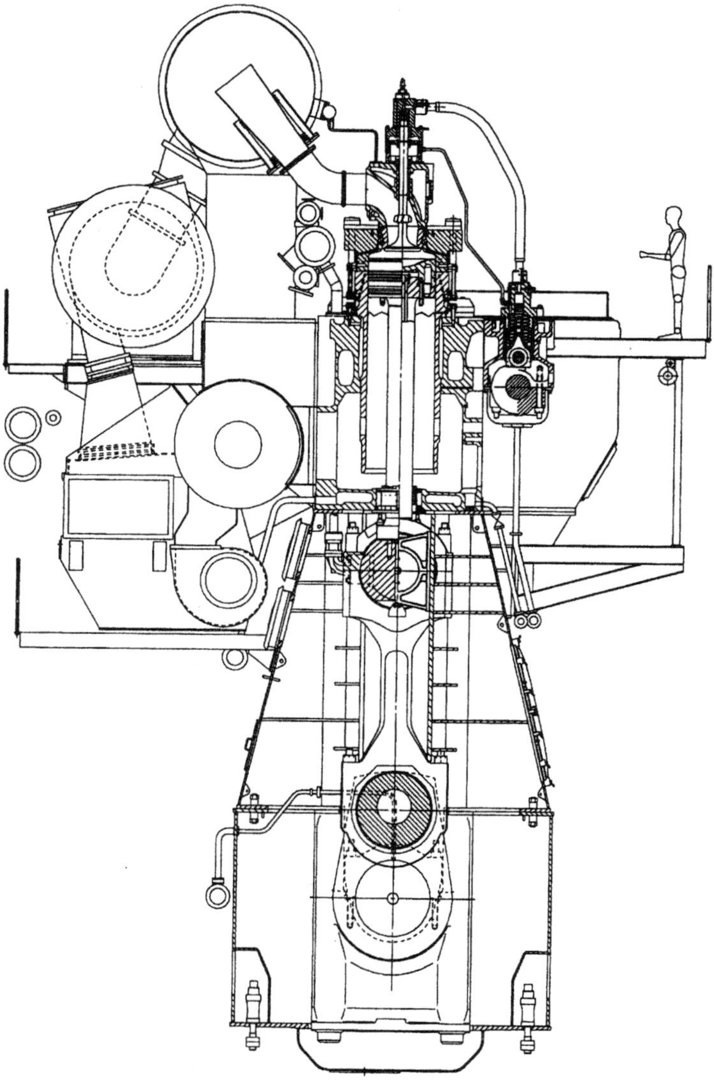

Fig. 14.11 K90MC-C engine cross section.

A longer stroke K98MC version was introduced in 1997 to complement the K98MC-C engine, with the same mcr but at a lower speed of 94 rev/min, and the number of cylinder options for both series was expanded from 9 to 12 to 6–12. The first K98MC engine, a seven-cylinder model, was tested in mid-1999 (Fig. 14.13). The most powerful diesel engines ever commissioned

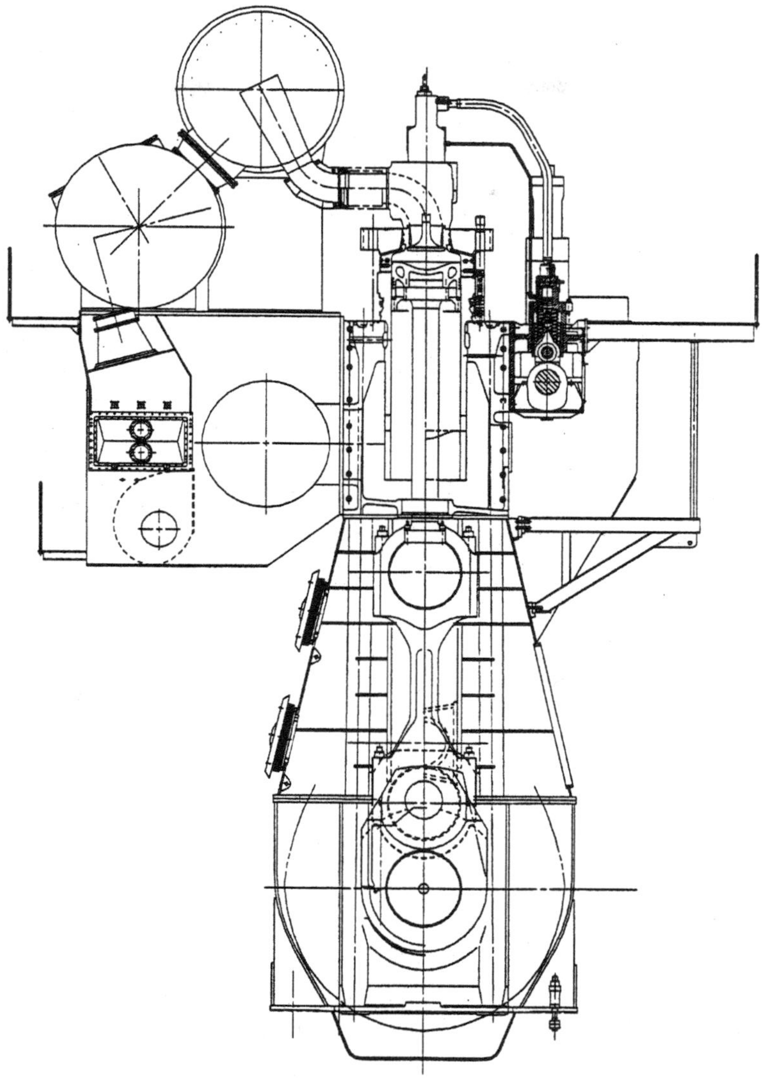

Fig. 14.12 K98MC-C engine cross section.

at the time—12-cylinder K98MC models with a maximum rating of 68,640 kW at 94 rev/min—entered service in 7500 TEU container ships in 2001/2002. The programme was extended by a 14-cylinder K98MC/MC-C model taking the output limit up to 80,080 kW.

Another route to higher powers for container ships was provided by MAN Diesel in 2003 with the release of a 1080 mm bore version of the

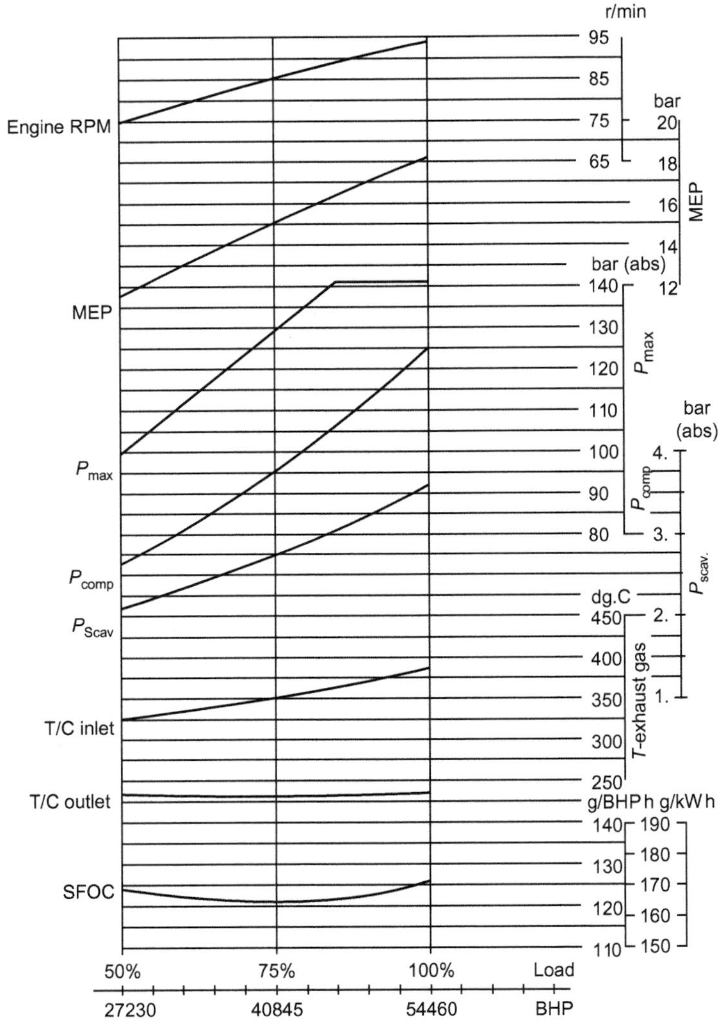

Fig. 14.13 Performance curves for 7K98MC engine at ISO reference conditions.

K-MC/MC-C series developing 6950 kW/cylinder at 94 rev/min. The 14-cylinder model of this K108 design, the world's most powerful diesel engine, could thus offer 97,300 kW. By then an electronically controlled ME engine, the K108 was dropped from the official programme in 2008, having been displaced by uprated (Mark 7) K98ME and MC designs, which, in 14-cylinder form, can deliver up to 87,220 kW. The first 14K98ME-C7 engine was due for completion in 2009.

Large-bore engines

The structural design of the K98MC, K98MC-C, S90MC-C, and S80MC-C engines is primarily based on that of the compact S-MC-C medium bore models. The main differences between the established K90MC/MC-C engines and the new design are as follows.

Bedplate

All the new large-bore engines are designed with thin-shell main bearings of white metal. The rigidity of the bearing housing was increased substantially to reduce stress levels and deformations. Furthermore, the change from the traditional stay bolts to the twin stay bolt design, with bolts screwed into the top of the cast bearing part, means that the geometry of the main bearing structure is simplified and improved casting quality is achieved. Finally, to improve the fatigue strength, a material with upgraded mechanical properties was specified. The use of twin stay bolts, fitted in threads in the top of the bedplate, has almost eliminated deformation of the main bearing housing during bolt tightening. The match between main bearing cap and saddle, secured in the final machining, is thus maintained in the operational condition, yielding a highly beneficial effect on main bearing performance.

A triangular plate design engine framebox was adopted for later K98MC/MC-C engines when this was introduced on the S90MC-C and S80MC-C engines. The new design provides continuous support of the guide bar, thus ensuring uniform deformation of the bar and a more even contact pattern between guide shoe and guide bar that enhances guide shoe performance. In addition, the continuous support contributes to a significant reduction of the stress level in the areas in question. The holes in the supporting plate make it possible to inspect all longitudinal welding seams from the back, and thus ensure that the specified quality is secured.

Main bearing

Tin–aluminium (Sn40Al) has proved very reliable as the bearing material for the smaller engines, the alloy having a much higher fatigue strength at elevated temperatures. The material may be applied to large-bore engines to maintain the reliability of the main bearings. Dry barring on the testbed in some cases caused the light seizure of Sn40Al main bearing shells on S-MC-C engines, but the problem was eliminated either by prelubrication with grease and high-viscosity oil or by PTFE coating the running surface of the shells.

Engine alignment

Traditionally, bedplate alignment, especially on large tankers, has been performed on the basis of a precalculated vertical position of the bearings, as well as of the engine as such, and possibly also involving an inclination of the entire main engine. On completing the precalculated alignment procedure, it has been normal practice to check the alignment by measuring the crankshaft deflections. Such checks are normally carried out either in drydock or with the ship afloat alongside at the yard in a very light ballast condition.

Owing to repeated cases of bearing damage, presumably caused by the lack of static loads in the normal operating conditions (ballast and design draught), MAN Diesel introduced modified alignment procedures for bedplate and shafting (crankshaft and propulsion shafting) as well as modifying the vertical offset of the main bearing saddles. In the modified bedplate alignment procedure, the importance of the so-called sag of the bedplate is emphasized in order to counteract the hog caused by hull deflections as a result of the loading down of the ship, and partly by deformations due to the heating up of the engine and certain tanks.

Combustion chamber

A reconfigured combustion chamber (Fig. 14.14) was developed for the new large-bore engines (800-mm bore and above), the key features being:

- *Piston crown with high topland.* In order to protect the piston rings from the thermal load from combustion, the height of the piston topland was increased (Fig. 14.15). The resulting increased buffer volume between the piston crown and the cylinder wall improves conditions for the rings and allows longer times-between-overhauls (TBOs). The high topland was first introduced in the mid-1990s, the positive service experience leading to its use for all-new engine types.
- *Piston crown with Oros shape.* With increasing engine ratings, the major development challenge with respect to the combustion chamber components is to control the heat load on them. The short-stroke large-bore engines have a rather flat combustion chamber because of the relatively smaller compression volume; this makes it more difficult to distribute the

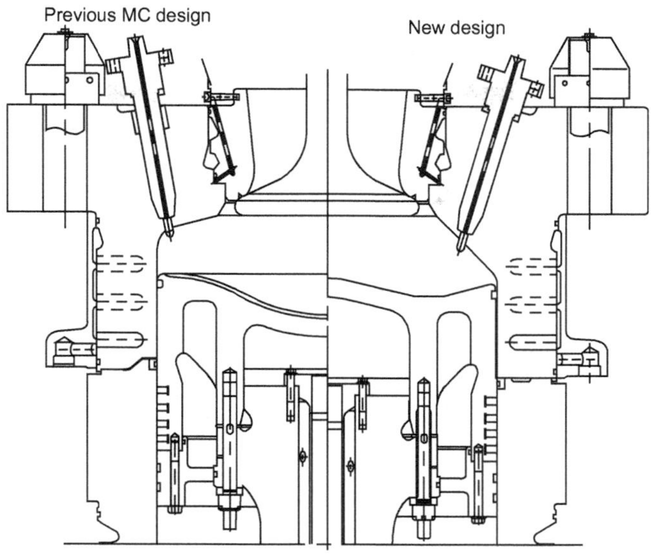

Fig. 14.14 New and previous designs of MC engine combustion chamber.

injected fuel oil without getting closer to the combustion chamber components. Furthermore, the higher rating means an increased amount of fuel injected per stroke. All this makes it more difficult to control the heat load on the components in short-stroke engines compared with long-stroke engines of the same bore size.

The heat load on the cylinder liner has been reduced by lowering the mating surface between cylinder liner and cylinder cover as much as possible (to just above the uppermost piston ring at the top dead centre). This means that the greater part of the heat load is absorbed by the cylinder cover, which is made of steel and thus more resistant to high heat loads. In addition, the cylinder cover is water cooled, making it relatively easy to control the temperature level.

The piston is cooled by system oil, which means a lower cooling efficiency compared with the cooling of the cylinder cover. Oil cooling of the piston, however, offers a number of advantages. The optimum way of reducing the temperature level on the piston is to reduce the heat load on it, this being secured by redesigning the shape of the combustion chamber, including the piston, to provide more space around the fuel valve nozzles. The new piston shape was termed Oros (Greek for 'small mountain').

The result of the increased distance from the fuel valve nozzles to the piston surface was simulated by computational fluid dynamics (CFD)

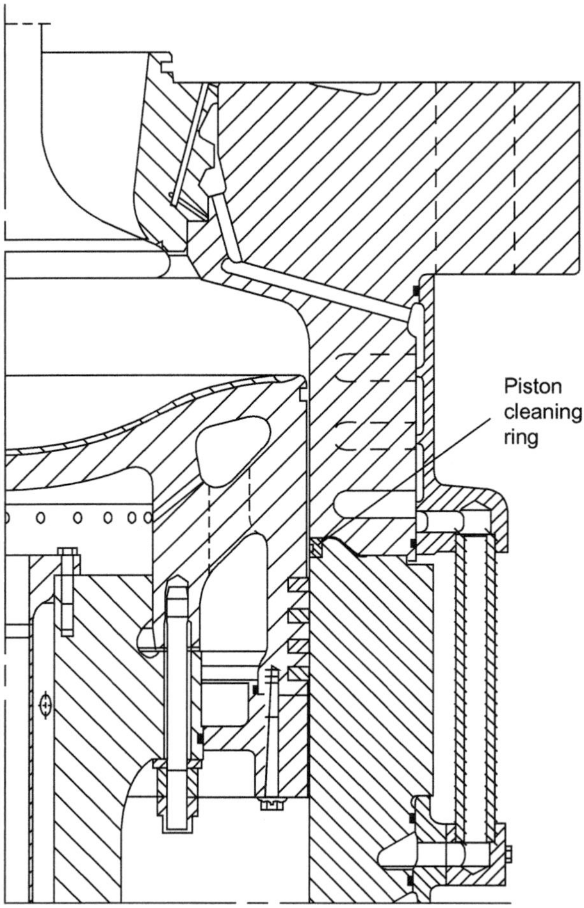

Fig. 14.15 Piston with high topland and PC ring in the cylinder liner.

analyses, and the optimum shapes of the piston crown and cylinder cover determined from these simulations. Tests on several engine types verified the simulations. A significant reduction in temperature was obtained after development tests on K90MC engines with various layouts of fuel oil spray pattern.

Temperature measurements on the piston crown and exhaust valve are shown in Fig. 14.16. The reduction in maximum piston temperature was approximately 90°C, this result being attained without impairing the temperature level on the oil side of the piston or the temperature on the exhaust valve.

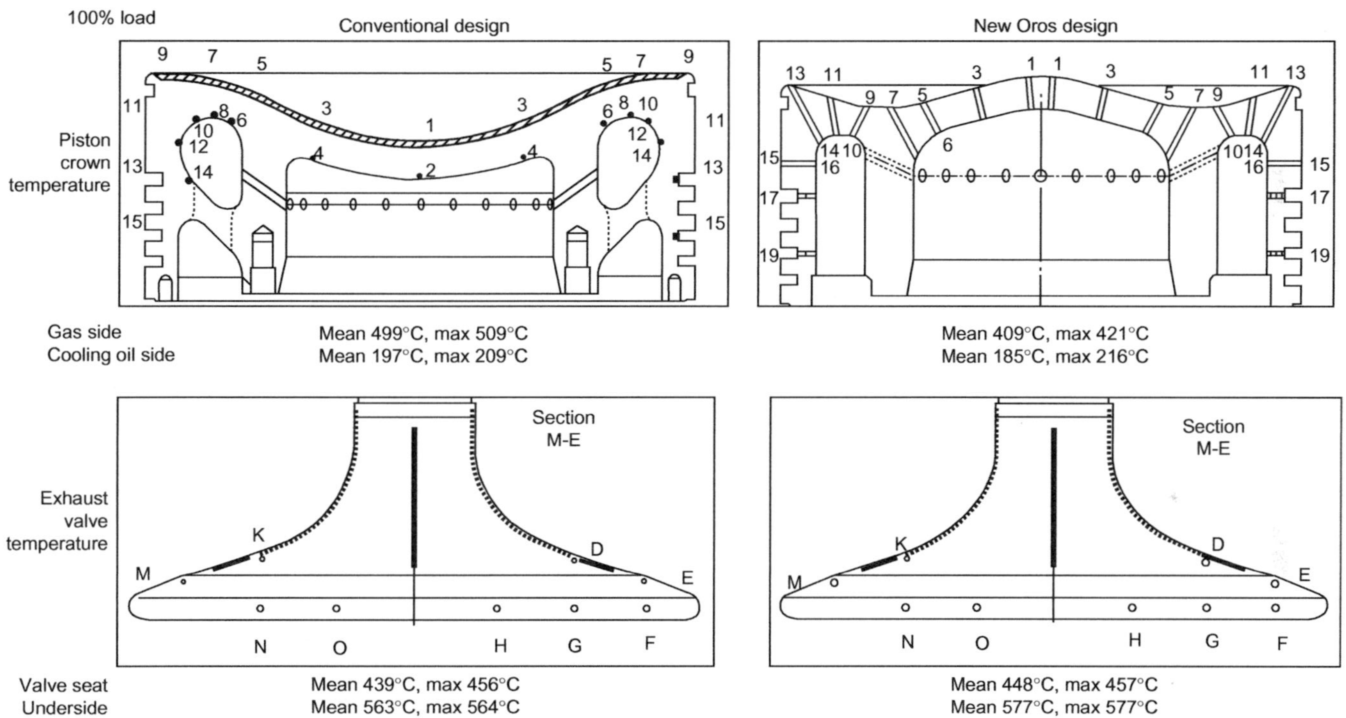

Fig. 14.16 New Oros-shaped combustion chamber geometry (*right*) has contributed to reduced temperature measurements.

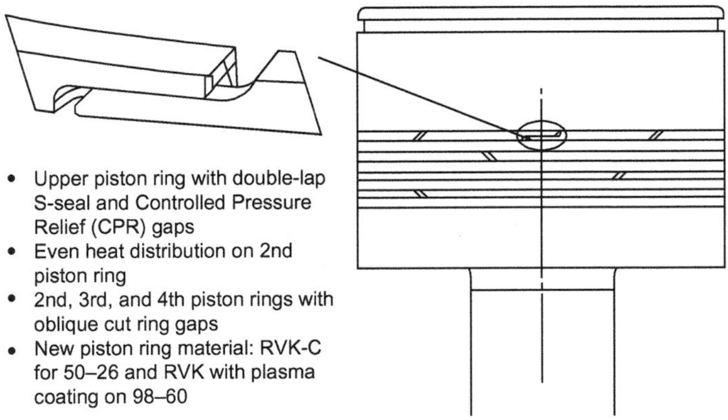

Fig. 14.17 Configuration of CPR top piston ring.

- *Piston ring pack.* The CPR top ring with relief grooves is now standard on all MC engines and has proved very effective in protecting the cylinder liner surface as well as the lower piston rings against excessive heat load. The CPR ring has a double lap joint, and an optimum pressure drop across the top piston ring is ensured by relief grooves (Fig. 14.17). With increasing mean indicated pressures, the traditional angle-cut ring gap may result in a higher thermal load on the cylinder liner; this load is significantly reduced by the CPR ring as no gas will pass through its double lap joint. The relief grooves ensure an almost even distribution of the thermal load from the combustion gases over the circumference of the liner, resulting in a reduced load on the liner as well as on the second piston ring.

Measurements confirmed that the peak temperature on the No. 2 piston ring was reduced from 300°C in association with an oblique cut top ring to 150°C with the CPR top ring. No. 2 ring retains its spring force, and TBOs are considerably extended. Furthermore, the pressure drop across the top piston ring has been optimized with respect to wear on the liner, piston rings, and ring grooves. Thanks to the double lap joint, the pressure drop will be almost constant irrespective of the wear on the liner and rings. This contrasts with the traditional angle-cut ring, with which the cylinder condition slowly deteriorates as the liner wears. MAN Diesel asserts that with the CPR rings, a continually good cylinder condition and low wear rate can be expected over the whole lifetime of the liner.

Al coating of the sliding surface of piston rings was introduced to ensure safe running-in. The aluminium–bronze alloy coating, a type of bearing material, has proved effective in protecting ring and liner surfaces during the running-in period. Al-coated rings make it possible to load up a completely new engine on the testbed within only 5 h, fostering time and fuel savings; running-in can also be performed with reduced cylinder lubrication. The lifetime of the coating is 1000–2000 h, depending on such factors as the cylinder oil feed rate and surface roughness of the liner. The technical explanation is that no scuffing occurs with the combination of aluminium–bronze/grey cast iron as the wear components in the cylinder. Al-coated rings allow the normal increase in cylinder oil feed rate after changing rings and/or liners to be dispensed with.

- *Piston ring groove.* In order to extend the interval before reconditioning of the piston ring grooves is required, the chromium layer has been increased to 0.5 mm and induction hardening of the grooves before chrome plating introduced. The chromium layer is thus better supported by the base material and the risk of cracking of the brittle chrome reduced, fostering a longer service life.
- *Piston cleaning (PC) ring.* Incorporated in the top of the cylinder liner, the PC ring has a slightly smaller inner diameter than the liner and hence scrapes off ash and carbon deposits built up on the piston topland (Fig. 14.18). Without such a ring, contact between the topland and the liner wall could wipe off the injected cylinder oil, preventing the lubricant from performing its optimized role. In some cases, deposit formation on the topland could cause bore polishing of the liner wall, contributing to the deterioration of the cylinder condition. Introducing the PC ring eliminates contact between deposits on the topland and the liner, promoting an enhanced cylinder condition and lube oil performance.

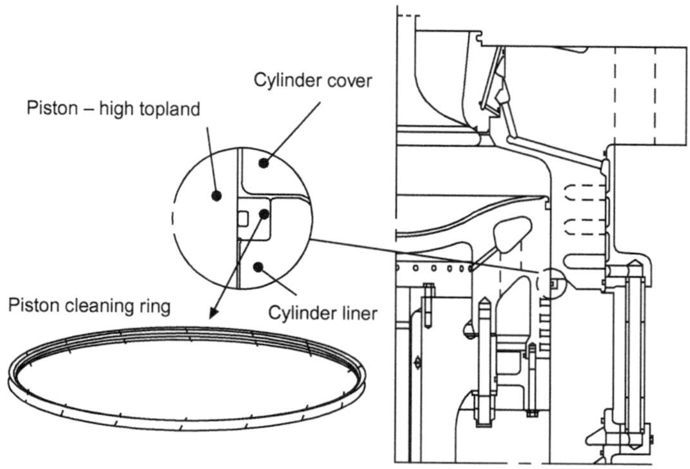

Fig. 14.18 Location of PC ring in the cylinder liner.

Cylinder liner

The liners of the large-bore engines are bore cooled; the cooling intensity is adjusted to maintain an optimum temperature level and to ensure optimum tribological conditions for the cylinder lube oil. For years MAN Diesel used a wave-cut liner surface, which was modified a few years ago to a semi-honed surface to facilitate the running-in of the highly loaded engines. The original wave-cut had a depth of approximately 0.02 mm. Experience showed that a deeper cut is advantageous in increasing the lifetime of the oil pockets, which, together with the Al-coated piston rings, leads to very safe running-in. An improved semi-honed wave-cut liner surface was therefore introduced.

Fuel valve

The fuel valve design was changed some years ago from a conventional type to the mini-sac type. The aim was to reduce the sac volume in the fuel nozzle and curb dripping, thus improving combustion; introducing the mini-sac valves reduced the sac volume to approximately one-third of the original. To improve combustion even further, however, a new slide-type valve was introduced for all large-bore engines, completely eliminating the sac volume. A significant improvement in combustion is accompanied by reduced NOx, smoke, and particulate emissions. The reduced particulates also improve the cylinder condition, and the wear rates of cylinder liner, piston rings, and ring grooves are also generally lower with slide-type fuel valves.

Slide-type valves were introduced on the K98MC engine from the beginning, their positive effect confirmed on the testbed. With fuel nozzles optimized with respect to heat distribution and sfc, the NOx emission values associated with this engine are described as very satisfactory.

Electronic high-pressure cylinder lubricator

Cutting cylinder lube oil consumption—which represents a significant annual cost, especially for large-bore engines—has been an important R&D target, resulting in the development of a computer-controlled high-pressure cylinder lubricator. The development of the electronic Alpha Lubricator started in 1997 and the prototype was installed on a 7S35MC engine in the following year; refinements were carried out on MAN Diesel's 4T50MX research engine.

System flexibility makes it possible to choose any number of engine revolutions (four, five, six, etc.) between injections of a specific amount of oil into the cylinder. For example, lubrication can be affected every fifth revolution in the compression stroke and every tenth in the expansion stroke if that turns out to be optimal. The aim is to inject the cylinder oil exactly where and when it is needed: in the piston ring pack as it passes the lube oil quills. Thanks to the high pressure, it is possible to establish an injection period that starts just when the uppermost piston ring is passing the quills and ends exactly when the lowermost ring is passing. Furthermore, injection is in the tangential direction, ensuring the optimal distribution of the oil in the complete ring pack and ring grooves.

The Alpha Lubricator features a small piston for each lubrication quill in the cylinder liner. Power for injecting the oil is derived from the system pressure, supplied by a pump station. A conventional common rail system is used on the driving side and a high-pressure positive displacement system on the injection side. Equal amounts of oil are supplied to each quill and the highest possible safety margin against clogging of individual quills is secured. The basic oil feed rate can be set by a screw, which limits the stroke of the main lubricator piston.

All MC/MC-C/ME/ME-C engines (new and in service) can benefit from the system; large-bore engines are provided with two Alpha Lubricators per cylinder and small-bore engines with one. The actual savings from the reduced cylinder oil feed rates achieved depend on the engine size, but MAN Diesel has suggested 20%-plus cost reductions as possible. Further reductions are promised by applying the 'sulphur handle' to dose oil in proportion to the amount of sulphur entering the cylinder in the fuel; this Alpha Adaptive Cylinder oil control (ACC) is now standard on Alpha Lubricators.

Water mist catcher

Water droplets entering the cylinder can have a negative effect on the cylinder condition by disturbing the lube oil film on the cylinder liner surface; all of the two-stroke engines are therefore equipped with a water mist catcher. With increasing pressures and air amounts, however, it has become more difficult to obtain adequate efficiency.

Seeking the optimum method of separating water droplets from the scavenge air, MAN Diesel carried out advanced CFD calculations to find the trajectories of the droplets as they leave the cooler. An analysis of the influence of different guide vanes on the airflow made it possible, by calculation, to predict that all droplets larger than 0.1 mm would be separated before the air entered the traditional water mist catcher element.

A new airflow reversing chamber was designed with a precatcher guide plate diverting the flow in front of the water mist catcher element and securing 80% water separation before the normal element. In addition, inspection covers were fitted before and after the air cooler element and improved drains (to prevent the re-entry of water) applied.

Electronically controlled ME engines

In an MC engine, the camshaft mechanically controls fuel injection and exhaust valve operation, a linkage that delivers very limited timing flexibility. Electronically controlled MAN B&W ME-series engines, introduced to the market from 2001, dispense with the camshaft and exploit hydraulic–mechanical systems supported by electronic hardware and software for activating fuel injection and the exhaust valves. ME engines became available in bore sizes from 500 to 1080 mm offering the same outputs as their MC engine counterparts (Fig. 14.19).

Electronically controlled fuel injection and exhaust valve actuation (Fig. 14.20) allow individual and continuous adjustment of the timing for each cylinder, securing the following key benefits.

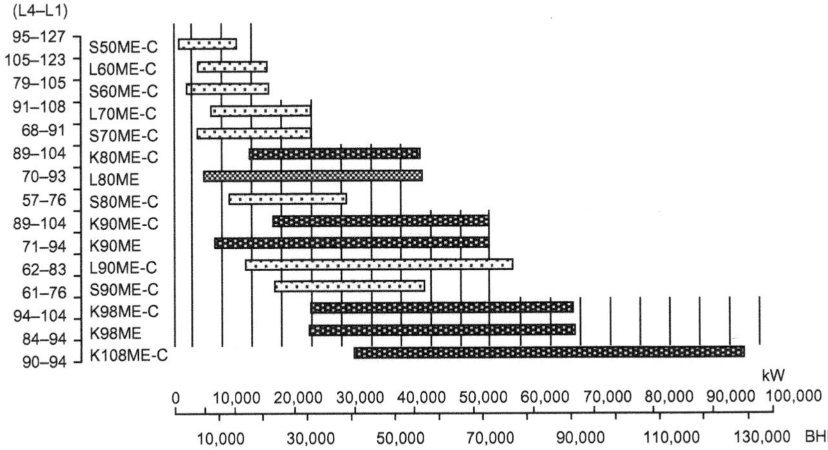

Fig. 14.19 The original electronically controlled ME engine programme (the K108ME-C was withdrawn in 2008 and a 650-mm bore S65ME-C model added in 2004).

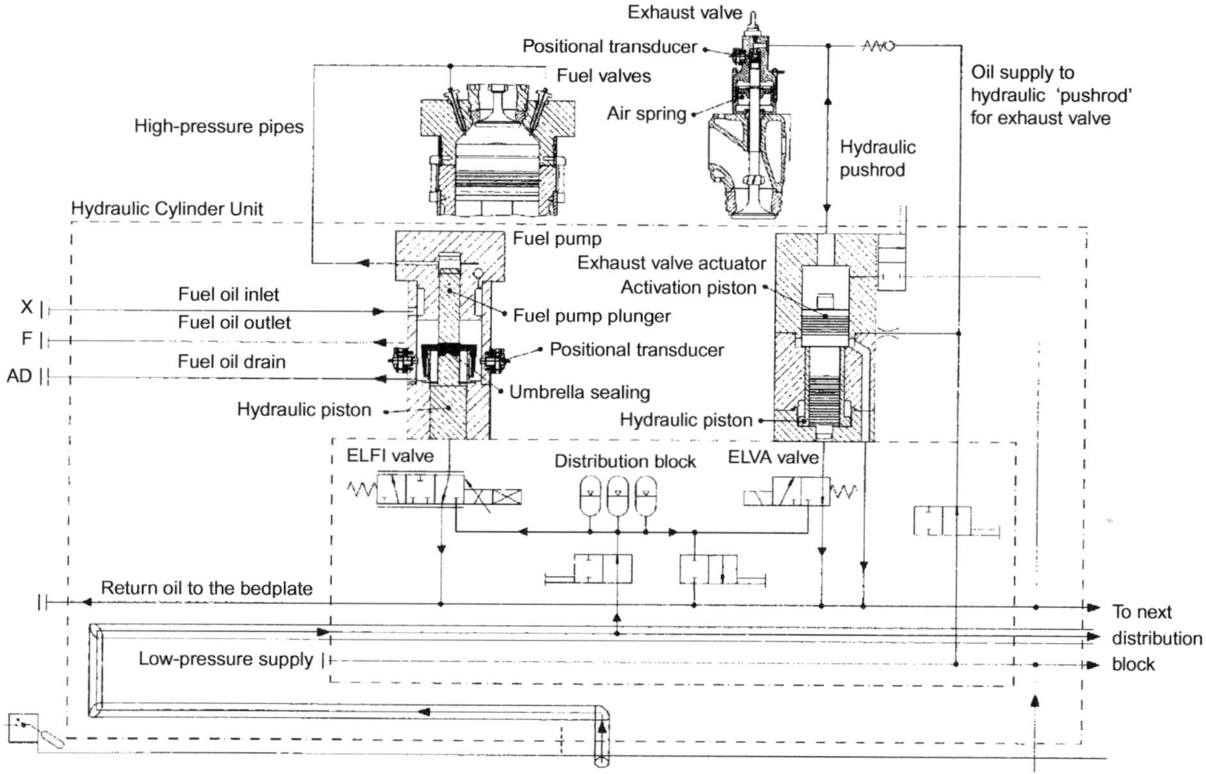

Fig. 14.20 Mechanical–hydraulic system for activating the fuel pump and exhaust valve of an ME engine cylinder.

Reduced fuel consumption

- Fuel injection characteristics can be optimized at many different load conditions, while a conventional engine is optimized for the guarantee load, typically at 90%–100% mcr.
- Constant P_{max} in the upper load range can be achieved by a combination of fuel injection timing and variation of the compression ratio (the latter by varying the closing of the exhaust valve). As a result, the maximum pressure can be kept constant over a wider load range without overloading the engine, leading to significant reductions in sfc at part load.
- Online monitoring of the cylinder process ensures that the load distribution among the cylinders and the individual cylinder's firing pressure can be maintained at 'as new' standard over the lifetime of the engine.

Operational safety and flexibility

- The engine's crash stop and reverse running performance are improved because the timing of the exhaust valves and fuel injection can be optimized for these manoeuvres as well.
- 'Engine braking' may be obtained, reducing the ship's stopping distance.
- Swifter acceleration of the engine, since the scavenge air pressure can be increased faster than normal by opening the exhaust valve earlier during acceleration.
- Dead slow running is improved significantly: the minimum revolutions per minute are substantially lower than those for a conventional engine; dead slow running is much more regular, and combustion is improved thanks to the electronic control of fuel injection.
- Electronic monitoring of the engine (based on MAN Diesel's CoCoS-EDS system) identifies running conditions that could lead to performance problems. Damage due to poor ignition quality fuel can be prevented by injection control (preinjection).
- The engine control system incorporates MAN Diesel's online overload protection system (OPS) feature, which ensures the engine complies with the load diagram and is not overloaded (often the case in shallow waters and with 'heavy propeller' operation).
- Maintenance costs will be lower (and maintenance easier) as a result of the protection against general overloading as well as overloading of

individual cylinders, and also because of the 'as new' running conditions for the engine, further enhanced by the ability of the diagnosis system to give early warning of faults and thus enable proper countermeasures to be taken in good time.

Exhaust gas emissions flexibility

- The engine can be changed over to various 'low emission' modes, its NOx exhaust emissions reduced below the IMO limits if dictated by local regulations.
- By appropriate selection of operating modes, ships may sail with lower exhaust gas emissions in special areas, where this may be required (or be more economical due to variable harbour fee schemes) without having negative effects on the sfc outside those areas.

The following components of the conventional MC engine are eliminated in the ME engine: chain drive for camshaft, camshaft with fuel cams, exhaust cams and indicator cams, fuel pump actuating gear, including roller guides and reversing mechanism, conventional fuel injection pumps, exhaust valve actuating gear and roller guides, engine-driven starting air distributor, the electronic governor with the actuator, regulating shaft, mechanical engine-driven cylinder lubricators, and engine side control console.

These elements are replaced on the ME engine by an electrohydraulic platform comprising a hydraulic power supply (HPS), a hydraulic cylinder unit (HCU) with electronic fuel injection (ELFI) and electronic exhaust valve actuation (ELVA), an electronic Alpha Cylinder Lubricator, an electronically controlled starting valve, a local control panel, a control system with governor, and a condition monitoring system.

ME engine systems

Valuable experience was gained by MAN Diesel from its 4T50MX research engine at Copenhagen, operated from 1993 to 1997 with a first-generation intelligent engine (IE) system. Second-generation IE systems fitted to the engine in 1997 aimed for simplified design, production, and installation of the key electronically controlled fuel injection and exhaust valve actuation systems. Subsequent R&D focused on transforming the electronic elements into a modular system, whereby some of the individual modules could also be applied to conventional engines. This called for the development of a new computer unit and large software packages, both of which had to comply with the demands of classification societies for marine applications.

The second-generation IE system is based on an engine-driven high-pressure servo oil system, which provides the power for the hydraulically operated fuel injection and exhaust valve actuation units on each cylinder. Before the engine is started, the hydraulic power system (or servo oil system) is pressurized by a small electrically driven high-pressure pump. Fine-filtered main system lube oil is used as the actuating medium supplied by engine-driven multi-piston pumps at around 200 bar (Fig. 14.21).

Fuel injection system

A common-rail servo oil system applies the cool, clean, and pressurized lube oil to power the fuel injection pump of each cylinder. Each cylinder unit is provided with a servo oil accumulator to ensure sufficiently swift delivery of oil in accordance with the requirements of the injection system, and to avoid heavy pressure oscillations in the associated servo oil pipe system. The movement of the pump plunger is controlled by a fast-acting proportional control valve (a so-called NC valve), which, in turn, is controlled by an electric linear motor that receives its control input from a cylinder control unit. The fuel injection pump features well-proven technology, and the fuel valves are of a standard design.

Second- and third-generation fuel injection pumps are much simpler than the first-generation design, and their components are smaller and easier to manufacture (Fig. 14.22). A major feature of the third-generation pump is its ability to operate on heavy fuel oil; the pump plunger is equipped with a modified umbrella design to prevent heavy fuel from entering the lube oil system. The driving piston and injection plunger are simple and kept in contact by the fuel pressure acting on the plunger and the hydraulic oil pressure acting on the driving piston. Both the beginning and the end of the plunger stroke are controlled solely by the fast-acting NC valve, which is computer controlled.

Optimum combustion (and thus optimum thermal efficiency) calls for an optimized fuel injection pattern, which in a conventional engine is generated by the fuel injection cam shape. Large two-stroke engines are designed for a specified maximum firing pressure and the fuel injection timing is controlled so as to reach that pressure with the given fuel injection system (cams, pumps, injection nozzles).

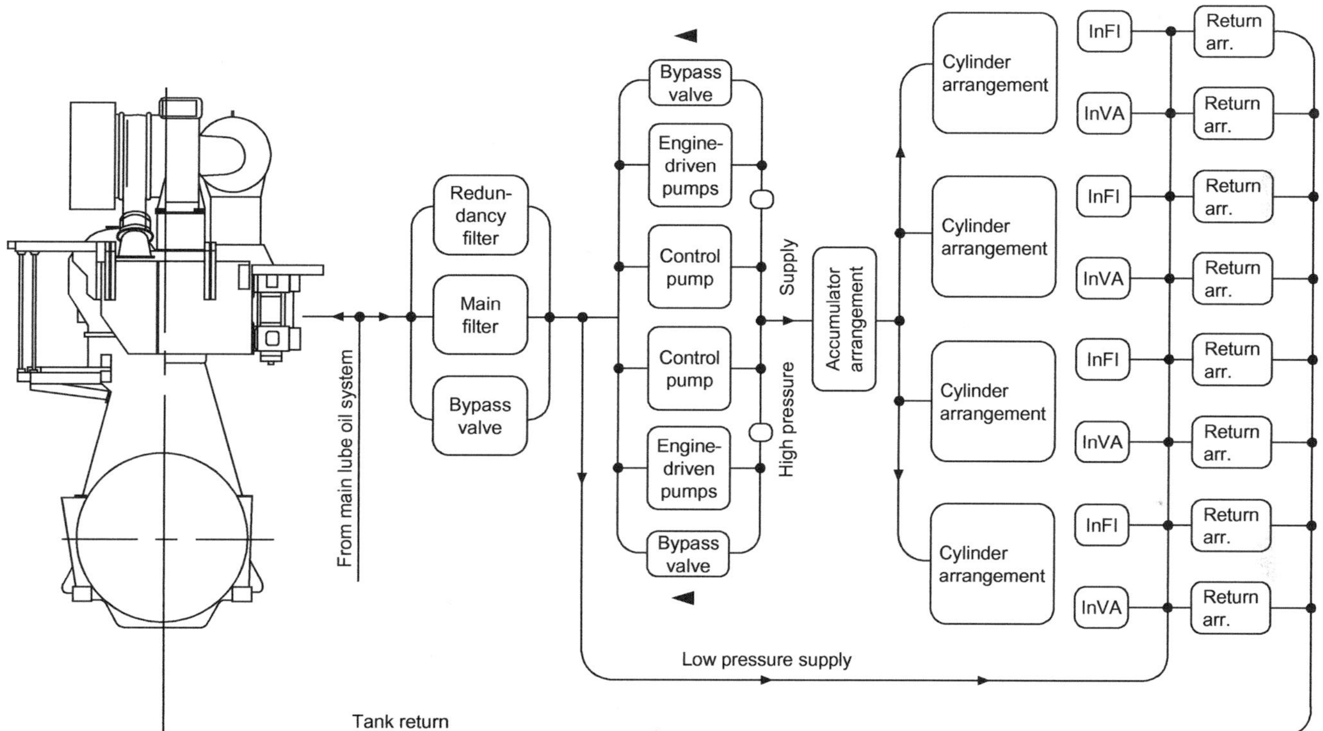

Fig. 14.21 Servo oil system flow diagram for ME engine.

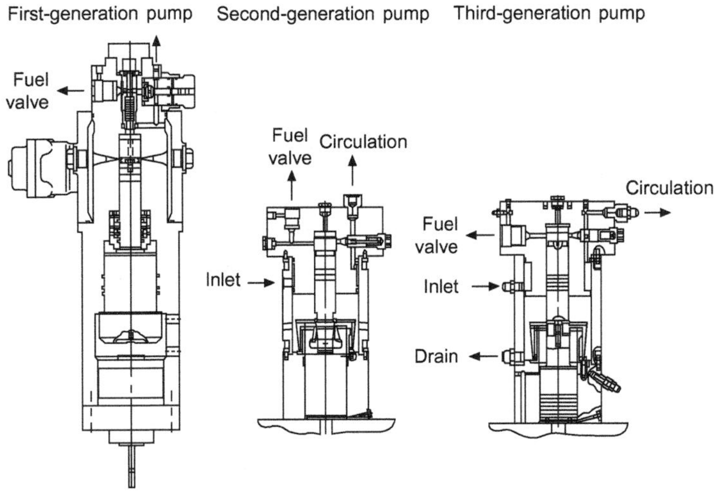

Fig. 14.22 Development of fuel injection pump for ME engine.

For modern engines, the optimum injection duration is around 18–20° crank angle at full load and the maximum firing pressure is reached in the second half of that period. To secure the best thermal efficiency, fuel injected after the maximum firing pressure is reached must be injected (and burned) as quickly as possible in order to obtain the highest expansion ratio for that part of the heat released. From this, it can be deduced that the optimum 'rate shaping' of the fuel injection is one showing an increasing injection rate towards the end of injection, thus supplying the remaining fuel as quickly as possible. The camshaft of the conventional engine is designed accordingly, as is the fuel injection system of the ME engine. In contrast to the camshaft-based injection system, however, the ME system can be optimized at a large number of load conditions.

MAN Diesel claims that the fuel injection system for the ME engines can execute any sensible injection pattern needed to operate the engine. It can perform as a single-injection system as well as a preinjection system with a high degree of freedom to modulate the injection in terms of injection rate, timing, duration, pressure, or single/double injection. In practice, a number of injection patterns are stored in the computer and selected by the control system for operating the engine with optimum injection characteristics from dead slow to overload, as well as during astern running and crash stop. Changeover from one to another of the stored injection characteristics may be executed from one injection cycle to the next.

Exhaust valve actuation system

The exhaust valve is driven by the same servo oil system as that for the fuel injection system, using cool pressurized lube oil as the working medium. The necessary functionality of the exhaust valve is less complex than fuel injection, however, calling only for the control of the timing of its opening and closing. This is arranged by a simple fast-acting on/off control valve.

Well-proven technology from the established MC engine series was retained. The actuator for the exhaust valve system is of a simple, two-stage design. The first-stage actuator piston is equipped with a collar for damping in both directions of movement. The second-stage actuator piston has no damper of its own and is in direct contact with a gear oil piston transforming the hydraulic system oil pressure into oil pressure in the oil push rod. The gear oil piston includes a damper collar that becomes active at the end of the opening sequence when the exhaust valve movement will be stopped by the standard air spring.

Control system

Redundant computers connected in a network provide the control functions of the camshaft (timing and rate shaping). The engine control system—an integrated element of the IE—comprises two engine control units (ECU), a cylinder control unit (CCU) for each cylinder, a local control terminal, and an interface for an external application control system. Both the ECU and the CCU were developed as dedicated controllers, optimized for the specific needs of the IE (Fig. 14.23).

The ECU controls functions related to the overall condition of the engine. It is connected to the plant control system, the safety system, and the supervision and alarm system and is directly connected to sensors and actuators. The ECU's function is to control the action of the following components and systems:

- The engine speed in accordance with a reference value from the application control system (an integrated governor control)
- Engine protection (overload protection as well as faults)
- Optimization of combustion to suit the running condition
- Start, stop, and reversing sequencing of the engine
- Hydraulic (servo) oil supply (lube oil)
- Auxiliary blowers and turbocharging.

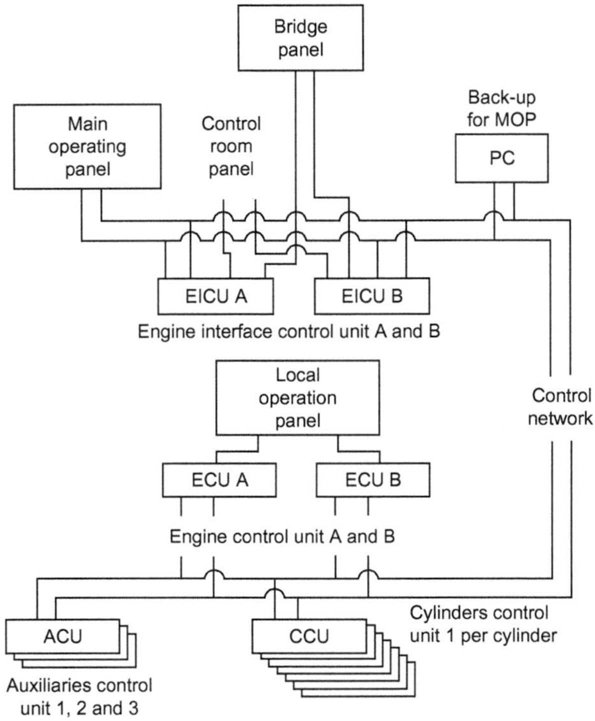

Fig. 14.23 Control system configuration for ME engines.

The cylinder control unit is connected to all of the functional components to be controlled on each cylinder, its purpose to control the activation of fuel injection, the exhaust valve, the starting valve, and the cylinder lubricator for a specific cylinder. Since each cylinder is equipped with its own controller (the CCU), the worst consequence of a CCU failure is a temporary loss of power from that particular cylinder (e.g. similar to a sticking fuel pump on a conventional engine). The engine controller (ECU) has a second ECU as a hot standby, which, in the event of a failure, immediately takes over and continues the operation without any change in performance (except for the decreased tolerance for further faults until the repair has been completed).

In the event of a failure in a controller, the system will identify the faulty unit, which can be simply replaced with a spare. As soon as the spare is connected, it will automatically be configured to the functions it is to replace, and resume operation. As both the ECU and the CCU are implemented in the same type of hardware, only a few identical spares are needed. If failures occur in connected equipment—sensors, actuators, wires—the system will locate the area of the failure and, through built-in guidance and test facilities, assist the engine operating staff in the final identification of the failed component.

Cylinder pressure measuring system

Reliable measurement of the cylinder pressure is essential for ensuring sustained 'as new' engine performance. A conventional mechanical indicator in the hands of skilled engineroom staff can provide reasonable data on this parameter, but the process is quite time consuming, and the cylinder pressure data derived are not available for analysis on a computer. Some valuable information is therefore less likely to be used in a further analysis of the engine condition.

A computerized measuring system with a high-quality pressure pick-up connected to the indicator bore—pressure measuring system (PMI) offline—was developed by MAN Diesel for application to its MC engines. For the ME, however, online measurements of the cylinder pressure are necessary, or at least highly desirable. In this case, the indicator cock cannot be used because the indicator bore would clog up after a few days of normal operation.

The strain-pin type of pressure sensor was applied instead. Here, the pressure sensing element is a rod located in a bottom hole in the cylinder cover, in close contact with the bottom of the hole, and close to the combustion chamber surface of the cylinder cover. The sensor thus measures the deformation of the cover caused by the cylinder pressure without being in contact with the aggressive combustion products. The position of the sensor also makes it easier to prevent electrical noise from interfering with the cylinder pressure signal.

The pressure transducer of the above-mentioned offline system is used for taking simultaneous measurements for calibrating the online system. A calibration curve is determined for each cylinder by feeding the two signals into the computer in the calibration mode. The fact that the same high-quality pressure transducer is used to calibrate all cylinders means that the cylinder-to-cylinder balance is not at all influenced by differences between the individual pressure sensors.

Both online and offline systems provide the user with valuable assistance in keeping the engine performance at 'as new' standard, extending the TBOs and reducing the workload of the crew. The systems automatically identify the cylinder being measured without any interaction from the person carrying out the measurement (because the system contains data for the engine's firing order). Furthermore, compensation for the crankshaft twisting is automatic, exploiting proprietary data for the engine design. If there is no such compensation, the mean indicated pressure will be measured wrongly, and when the figure is applied to adjust the fuel pumps, the cylinders will not have the same true uniform load after the adjustment, although it may seem so. (Crankshaft twisting may lead to errors in mean indicated pressure of some 5%, if not compensated for.)

The computer executes the tedious task of evaluating the 'indicator card' data, which are now in computer files, and the cylinder pressure data can be transferred directly to MAN Diesel's CoCoS-EDS engine diagnostic system for inclusion in the general engine performance monitoring. The result presented to engineroom staff is far more comprehensive, comprising a list of necessary adjustments. These recommendations take into account that the condition of the nonadjusted cylinders changes when the adjustments are carried out; it is not necessary therefore to check the cylinder pressure after the adjustment.

Starting air system

The starting air valves on the ME engine are opened pneumatically by electronically controlled on/off solenoid valves, which replace the mechanically activated starting air distributor of the MC engine. Greater freedom and more precise control are yielded, while the 'slow turning' function is maintained.

ME engines in service

Seeking to confirm the efficiency and reliability of the ME systems in regular commercial service, the Hitachi-MAN B&W 6L60MC main engine of the chemical/product carrier *Bow Cecil*—delivered in October 1998—was prepared for the systems during its production. The mechanical–hydraulic system components were installed on the upper platform of the engine in parallel with the conventional camshaft. The camshaft system was used during the sea trial and the early operating period of the tanker until the electronic components were installed in September 2000. The engine was subsequently commissioned as the first ME engine in service in November 2000.

System flexibility is accessed by different running modes, selected either automatically to suit specific operating conditions or manually by the operator to meet specific goals such as 'low fuel consumption' or 'limited exhaust gas emissions' (Fig. 14.24). The mode selection screen of the human–machine interface in the control room allows the operator to switch from fuel economy and emission control modes, as well as to switch between governor control modes such as constant speed and constant torque.

The first engine built from the start as an electronically controlled ME design—a 7S50ME-C model—was completed by MAN B&W Alpha Diesel in Denmark in early 2003 and entered service in November that year. Subsequent delivery references have expanded to embrace 500-, 600-, 650-, 700-, 800-, 900-, and 980-mm-bore ME or ME-C models for diverse ship types.

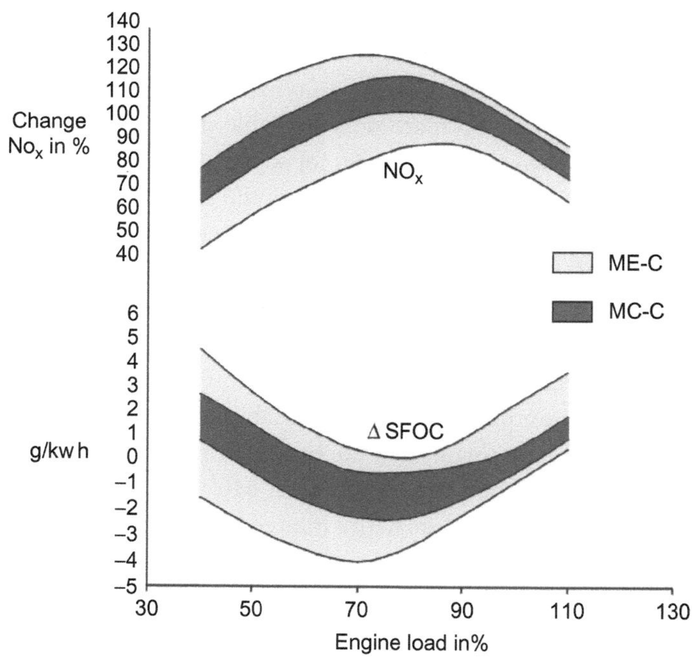

Fig. 14.24 Flexibility of the electronically controlled ME concept compared with a traditional camshaft-controlled MC-C engine in terms of NOx emissions and sfc.

A 650-mm-bore design with a stroke–bore ratio of 4.2 was added to the ME programme in 2004 to meet the propulsion requirements of Cape-size bulk carriers, Suezmax tankers, and 2000–3000 TEU container ships. A power band from 14,350 to 22,960 kW at 95 rev/min is covered by five- to eight-cylinder models in this S65ME-C range. The first example, a seven-cylinder model, was completed in 2006 for powering an ice-class Suezmax tanker.

Mark 9 and 10 large-bore engines

Introduced in the 1980s, MAN B&W 900-mm-bore K90MC and MC-C engines targeted the largest container ships of the time, which had capacities of up to 4000 TEU and service speeds of up to 24 knots. The 800-mm-bore K80MC-C model addressed propulsion projects for ships up to 3000 TEU with similar speeds. Numerous references were gained for the K80 and K90 engines in medium-to-large container ships over many years, but the largest vessels are now typically specified with the 980-mm-bore K98 engines in either mechanical MC or electronic ME versions. In VLCC propulsion, as well as the very large bulk carrier market, the S80MC was commonly selected for tonnage with limited speed requirements, while faster and larger ships were typically specified with S80MC-C or S90MC-C engines (both are now available in ME form).

With full order books at major yards for some time ahead in 2005/2006, MAN Diesel took the opportunity to sharpen the competitiveness of the S/K80ME-C and K90ME/ME-C engines through a comprehensive updating and uprating to Mark 9 status (Table 14.3 and Fig. 14.25).

Options for owners specifying propulsion plant for large tanker, bulk carrier, and container-ship tonnage were extended by the modified 800-mm-bore engine designs. The new K80ME-C and S80ME-C Mark 9 models were higher output variants of the established shorter stroke engines of this bore size, which remained in the production programme. Both K80 and S80 series could be specified in ME-C or MC-C versions.

VLCCs and large bulk carriers with a limited need for speed were targeted by the S80ME-C engine, available in 6–9-cylinder models covering a power range from around 27,000 kW to just over 40,000 kW at 78 rev/min. With an output band from 27,000 to 54,400 kW at 104 rev/min covered by 6–12-cylinder models, the K80ME-C series was tailored to the requirements of medium and large container-ship propulsion. Improved power-to-weight

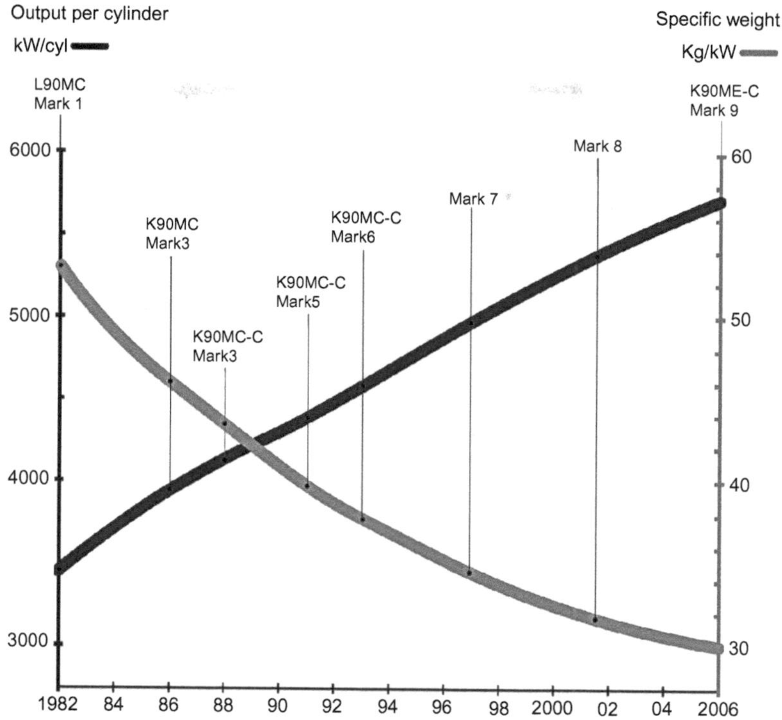

Fig. 14.25 Trends in output per cylinder and specific weight for MAN B&W 90-bore engines from Mark 1 to Mark 9 status.

ratios were achieved, reportedly without compromising reliability from a combination of new and proven developments and design features (Figs 14.26 and 14.27) as follows:

- *An Oros combustion chamber*: reduced surface temperatures, elimination of piston top coating, and less piston top stress
- *Slide fuel valve*: less deposits, improved combustion process, less visible smoke, and reduced emissions
- *Piston ring pack configuration*: CPR piston rings, PC rings, high topland piston crown, increased height of first piston ring, and copper band on the piston skirt
- *W-seat*: increased TBOs for the exhaust valve seat, reduced number of grindings, and fewer dent marks
- An Alpha Lubricator

A higher mean effective pressure of 20 bar, maximum firing pressure of 160 bar, and a mean piston speed of 9 m/s contributed to power increases of up to 25% for the 800 mm and 900-mm-bore S and K designs, without

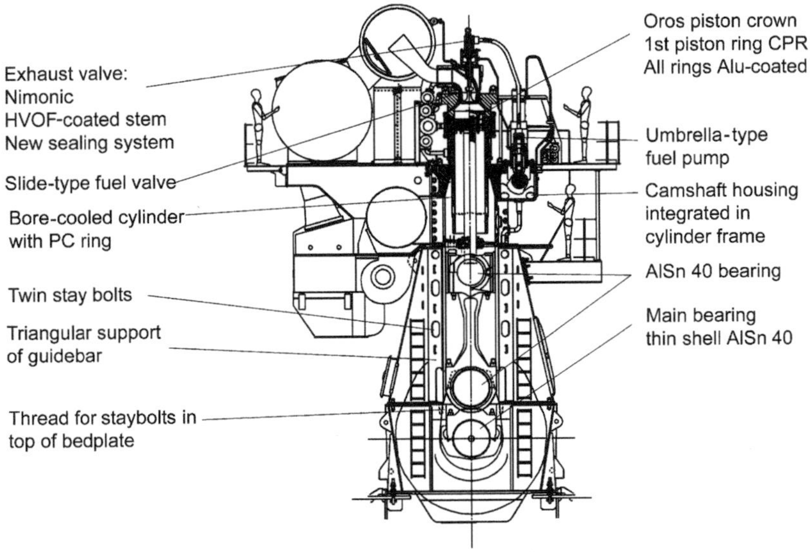

Fig. 14.26 Design features of MAN B&W MC and ME engines with increased ratings.

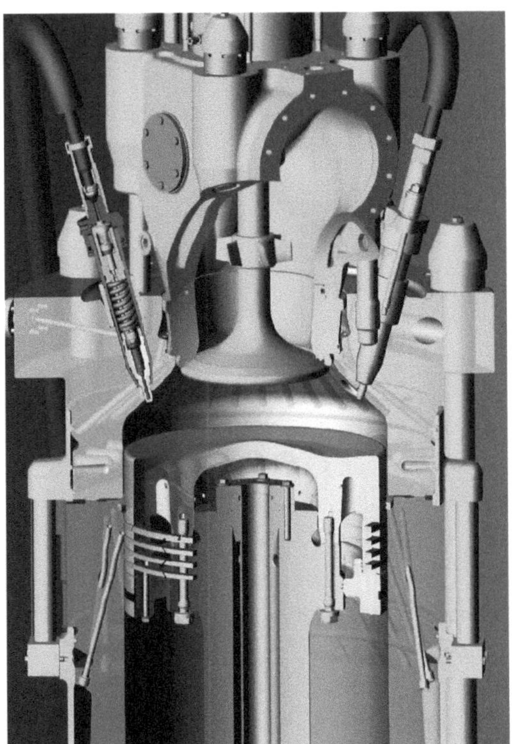

Fig. 14.27 Modified 80-bore engines feature an Oros combustion chamber, W-seat for the exhaust valve, an improved piston ring pack configuration, and a slide fuel valve.

undermining sfc. A lower production cost and weight per kilowatt were also reported, along with enhanced reliability and extended service intervals fostered by design features adopted from the new S65ME-C engine, which also support the higher ratings and enhanced compactness.

The combination of mep and P_{max} ensured an unchanged sfc despite a rise in power output of up to 25% from the modified K-type engines (albeit compared with the engines originally designed in the late 1980s). The power increase was such that the modified models were able to replace contemporary engines of a different design having a larger cylinder bore as well as new engines of a different design with a slightly larger bore.

It was decided to launch the updated engines only in relevant cylinder numbers: the K80/90 designs in 6–12-cylinder versions and the S80ME-C design in six to nine-cylinder versions. A higher specific output enables engines with fewer cylinders to be specified for a given power requirement, thereby reducing maintenance costs.

A number of well-proven design features were adopted from other engines, one of which—differentiated cylinder distances—yields a large weight saving and reduces the engine length. Typically, it is possible to exploit a shorter cylinder distance in the forward cylinder units because the torque to be transferred through the crank throws is smaller than the aft end of the engine. For a 12-cylinder K90ME/ME-C Mark 9 engine, this means that the six foremost cylinder units can have a cylinder distance of 1492 mm compared with 1602 mm for the six aftmost units. The resulting weight saving on such an engine is more than 25 t and the reduction in overall length of more than 0.5 m.

MAN Diesel reports that the benefits for an eight-cylinder version of this engine are even more impressive because the shorter cylinder distance can be applied on all cylinders: the weight saving is more than 34 t and the overall engine length is reduced by almost 1 m. A similar result is gained for the K80ME-C engine.

In order to reduce production costs, no machining of the bearing side girder walls is applied in the bedplate. The main bearings are of the proven thin shell design using white metal as the material. Further reducing production costs, the building-in of the bearing housing is modified in that the 25/65 mating face between the cap and bearing support in the existing design is changed to a horizontal assembly. This simplification dictates a pair of high friction plates between the cap and bearing support to enable the assembly to handle the shear forces.

For the updated engines with 9–12 cylinders, the designers specified a 360° thrust bearing yielding a weight saving over the traditional 240° bearing while achieving a stronger bedplate structure in the aft end. The diameter of the thrust cam was also reduced, as was the thrust bearing length.

The thrust bearing housing forms a vessel and contains an oil bath providing additional lubrication for the bearing segments. Other benefits cited include only one type of segment and optimized use of the segments according to the magnitude of the thrust at the fore and aft ends by using 12 segments fore (360°) and eight segments aft (240°). A further advantage of this design is that the thrust acts on the centreline of the crankshaft whereas with the traditional design the force is somewhat offset from the centreline. For engines with fewer than 9–12 cylinders, the 240° thrust bearing is retained to meet the lower thrust force to be transferred.

A flexible thrust cam—introduced in the S65ME-C engine—features machined grooves that optimize the load in the thrust bearing, yielding reduced oil film pressure, hence lowering the maximum bearing pressure, increased film thickness, and improved load distribution on the thrust segments, thereby reducing the area and thus the thrust cam dimensions. Efforts were also made to reduce production costs and weight in the structure of the thrust bearing section. Calculations showed that the vertical stiffeners could be removed without reducing the strength of the structure.

Like other newer engines among the ME/ME-C/MC-C types, the updated designs feature a triangular plate framebox (Fig. 14.28) with twin stay bolts. Essentially, the design is like the S-MC-C framebox with ribs. The main difference is the substitution of ribs with vertical plates, forming a continuing triangular profile from top to bottom. In this way, an even

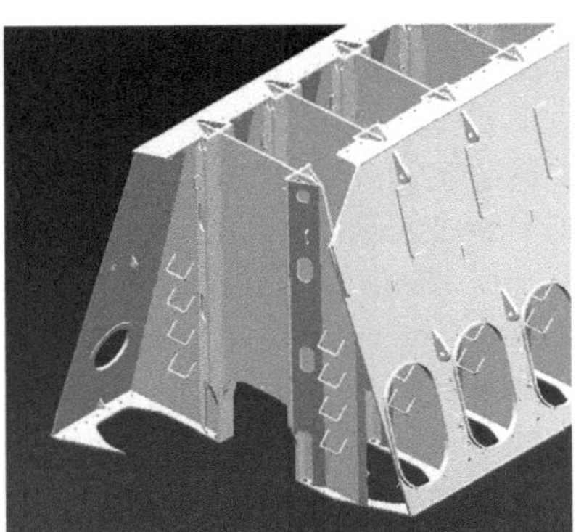

Fig. 14.28 A triangular plate framebox provides increased strength.

stiffness is derived along the guide bar, which improves conditions for the sliding surface of the guide shoes. Contact simulations verified this improvement and service experience has reportedly shown an excellent bearing condition.

In general, production costs (especially welding costs) are reduced. With the lower stress level in the structure, weld joints with lower quality requirements could be used, ideally suited to robot welding techniques. After laboratory and service testing, it was decided to omit the specification of postweld-heat treatment (PW-HT) for the frameboxes of triangular plate guide bar design, allowing engine builders to benefit from the cost saving.

The cylinder frames of the updated engines are of cast design as standard, with a welded structure as an option. The decision reflected the preferences of MAN B&W engine licences in recent years, whose foundry capacity had generally increased after earlier shortages. Another factor is that machining of the top plate of the welded cylinder frame demands substantial plano-miller time.

A number of inherent advantages are nevertheless acknowledged for the welded cylinder frame, such as increased rigidity and reduced weight. In addition, the welded design makes it possible to integrate the scavenge air receiver in the frame; the chain case is integrated as well, securing a total weight saving of 30%. Furthermore, dimensions are smaller, thereby lowering the total engine height.

Large-bore MAN B&W engines have exploited the Oros combustion chamber since the late 1990s (and it is now also specified for medium-bore designs). In terms of cylinder condition, the design has reportedly proven very successful, enabling overhauling intervals for the combustion chamber components to be extended. The Oros chamber is formed by the piston and cylinder cover geometry, which concentrates the hot combustion gases around the fuel nozzles, and the high topland piston, which allows a low position of the mating surface between cylinder liner and cover (thereby reducing liner wall temperatures and giving controlled cold corrosion, which is favourable for refreshing the liner surface) (Figs 14.29 and 14.30).

Piston ring packs of the latest design also benefitted the updated engines (Fig. 14.31), the change in design introduced to increase resistance against scuffing. Hard coating on the first and fourth rings was introduced to K90 and K98 engines, and the position of the CL grooves was also changed to secure a more even heat distribution. The CL groove depth was reduced by 20%.

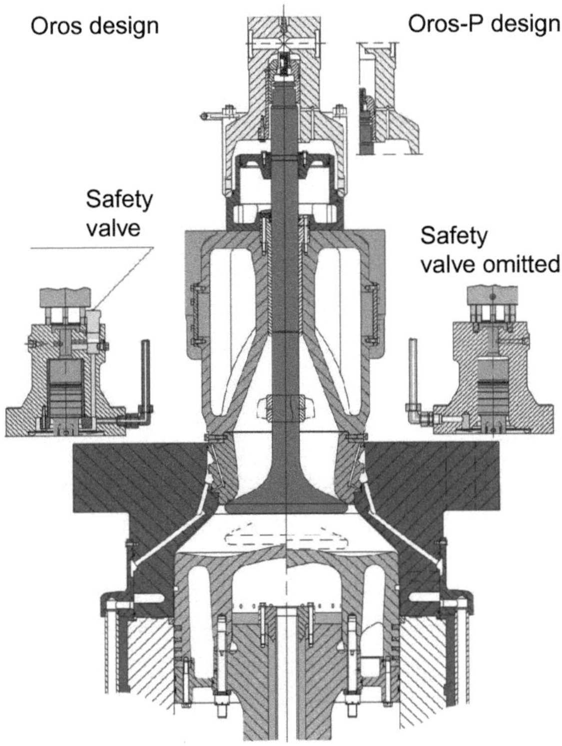

Fig. 14.29 The Oros P relief bore design (right) is specified for the updated K-type engines, omitting the safety valve on the exhaust valve actuator.

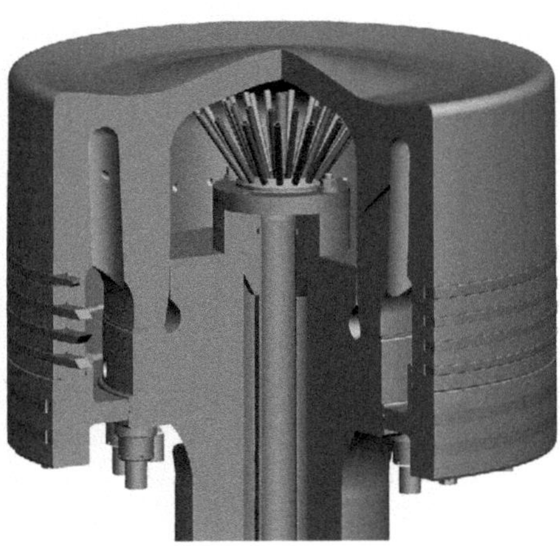

Fig. 14.30 A spray cooling insert secures a comfortably low piston temperature for the Mark 9-rated large-bore engines.

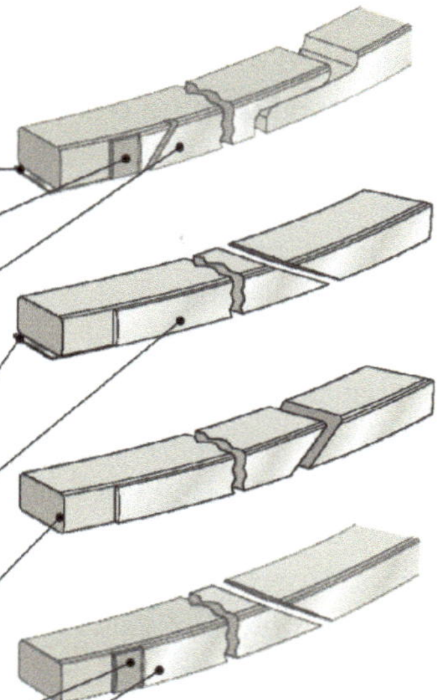

Fig. 14.31 Hard-coated/semi Alu-coat ring package.

However, this ring pack needed further development when the 9.2 Tier II engines with maximum pressures above 180 bar became the standard. To cope with this development, MAN designed the gas-tight three-ring pack.

Gas-tight three-ring pack on a 6G70ME-C9.2.

PC rings are incorporated in the top of the liner to avoid the build-up of carbon deposits on the topland, which would otherwise harm the cylinder condition by scraping the lube oil off the walls. Overall, the combustion chamber design is said to foster low wear rates at low cylinder oil feed rates, long TBOs, and good scuffing resistance.

With a view to omitting the safety valve on the exhaust valve actuator, the Oros P version of the Oros design was chosen for the updated K engines (Fig. 14.29). The reason was that the Oros piston leaves less room for fitting the exhaust valve spindle than with the traditional piston design. To prevent contact between piston and spindle, therefore, a mechanical stop is built into the exhaust valve air spring, and thus limiting the maximum spindle lift.

The exhaust valve design previously used for engines without an Oros combustion chamber featured relief bores in the exhaust valve oil cylinder, which would be uncovered in the event of excessive spindle lift. With the mechanical lifter, however, excessive lift and pressure relief using such bores were no longer possible; for that reason, a safety valve was built into the exhaust valve actuator.

Using the Oros P version, designed with a slightly lower piston dome, creates enough space to allow a slightly larger exhaust valve spindle lift. This makes it possible to exploit the well-proven relief bore design of the exhaust valve oil cylinder, and the safety valve on the actuator is no longer necessary. MAN Diesel asserts that such a solution represents a cost saving for engine licensees and a simplification for operators, yet with performance identical to the original Oros design.

A spray cooling insert is now provided to obtain a comfortably low piston temperature (Fig. 14.30). Improved cooling is confirmed by piston crown temperature measurements taken on the testbed and on engines in service. Comparative measurements with and without a spray cooling insert revealed a temperature reduction at the central part of the crown, both on the combustion chamber side and on the cooling side; the insert thus increases both of the margins against hot corrosion and underside coke formation.

An exhaust valve unit of a modified type uses a low-grade cast iron housing with water cooling around the spindle guide and bore-cooled twin stud bosses. The upper part of the housing forms the bottom of the air spring and the lower part forms part of the cooling chamber in the cylinder cover. Apart from eliminating cold corrosion of the exhaust valve duct, the lower cooling intensity slightly increases the exhaust gas energy, thus improving the potential for waste heat recovery.

A W-seat of low height is used and a high spindle guide lubricated by the controlled oil level system was introduced. The system implies a minimum oil level inside the air cylinder, secured by a high outlet (overflow) to the safety valve. This ensures a level high enough to have oil on the top of the sealing ring, the oil passing through three bores in the flange. Furthermore, the safety valve adjustment corresponds to this high oil level. No sealing air arrangement or special type of sealing ring is required by this design, which means that the old type ring can be used regardless of the type of valve spindle.

By 2014, the scavenge air receiver and the exhaust gas receiver had undergone a series of design improvements. Modifications have been carried out on the auxiliary blower inlet, scavenge air cooler housing, turbocharger foundation, scavenge air receiver, and exhaust gas receiver. The improved rigidity has reduced the vibration level on the turbocharger, scavenge air cooler housing, and exhaust gas receiver. The safety factor against cracks in the related parts has been increased while at the same time achieving a simplification such as a reduced number of exhaust receiver bottles on the largest engines.

Modifications to the ME electronic control system reflected experience gained from engines in service. Improvements have benefited the hydraulic servo oil supply, for which there are now two standards: engine driven and electrically driven.

Engine-driven oil supply. The well-proven design with a chain-driven gearbox, which drives three to five high-pressure pumps, was selected for the updated engines. A chain is particularly preferred on large-bore engines because of its reliability, easy installation, and high tolerance against relative movements of the crankshaft. Furthermore, the chain is used for driving the second-order moment compensator on six-cylinder engines specified with such attachments.

Compared with the initial ME/ME-C large-bore engine assembly, however, the new design has been simplified. It comprises small accumulator blocks (one for each pump) and a safety block positioned on the aft-most HCU. The large double-walled supply pipe with forged connection pipes was replaced by standardized high-pressure pipes. Easing installation and avoiding vibrations from the pumps, high-pressure hoses were installed between the pumps and HP pipes.

Electrically driven oil supply. If this arrangement is preferred, the complete gearbox and chain drive are replaced by electrically powered HP pumps, which are similar to the mechanical pumps. The chain drive section of the framebox is omitted, and thus making the engine structurally simpler than an engine with a mechanical servo oil supply.

Ultra-long stroke engines

After 2008, when shipping was suffering from the economic crash, overcapacity in many segments, and with the IMO discussing the impending EEDI regulations along with other efficiency measures, shipowners began looking for further efficiencies from engines and ship designs. MAN's response was to develop an ultra-long stroke variant of its low-speed engine design. This engine would have a lower rpm and be designed to work with larger diameter propellers.

MAN announced the new G-type engine in 2010 claiming that it could reduce fuel consumption by 7% when installed as part of an integrated propulsion package and matched to an appropriate propeller. The first bore size decided upon was the 800 mm model with the engine being designated G80ME-C (see Table 14.4 for comparison with long-stroke model).

MAN said that tankers and bulk carriers have traditionally used MAN B&W S-type engines with their long stroke and low engine speed as prime-movers, while larger container vessels have tended to use the shorter-stroke K-type with its higher engine speed. However, because container vessel speeds were being reduced by slow steaming strategies, some owners newbuildings had been specified with S80ME-C9 and S90ME-C8 engines because of the opportunity these engines offer to employ larger propeller diameters.

As new generations of efficiently designed ships of all types are expected to be compatible with propellers with larger diameters than current designs facilitating higher efficiencies and adaptation of the aft-hull design to accommodate the increase in propeller size. MAN considered that such new designs could offer potential fuel-consumption savings of some 4%–7%, and a similar reduction in CO_2 emissions. Simultaneously, the engine itself

Table 14.4 Comparison of S-type and G-type 800 mm bore engines.

Engine Data	**S80ME-C9**	**G80ME-C9**
Power kW/cyl	4510	4450
Engine speed rpm	78	68
Stroke mm	3450	3720
MEP bar	20.0	21.0
Mean piston speed m/s	8.97	8.43
Length mm (7-cylinder)	12,034	12,500
Dry mass tonnes (7-cyl)	910	960
SFOC at L1 (g/kWh)	168	167

could achieve a higher thermal efficiency using the latest engine process parameters and design features. Typically, an overall saving of 4.6% was expected—3.6% as a result of the higher propulsion efficiency and 1% engine savings.

Other bore sizes would eventually adopt the longer stroke although somewhat confusingly, the S90 and G90 engines of 2019 have the same stroke length even if some other characteristics are different. The first G engines to pass type-approval testing were G80ME-C models in March 2013, throughout the year, G60ME-C, G50ME-B, and G70ME-C models followed in that order. The first S90ME-C had passed its TAT in 2012.

A new model was added at the top end of the large-bore series in July 2013. The G95ME-C9 with a bore of 950mm and a stroke of 3460mm, provides 6870kW/cylinder at 80rpm and 21 bar MEP (in L1) and was introduced as a supplement to the successful S90ME-C9/10 engine types, allowing the engine to be further derated thanks to the larger cylinder bore and/or fewer cylinders to be installed. It was aimed at large container ships which had now reached sizes in excess of 18,000 teu.

In 2016, MAN introduced three more new models into its large-bore portfolio.

The engines are weight optimized, compared to their Mk 9 counterparts, and form the new design platform for a new Mk 10 engine portfolio. The key to the new platform is the development of the TCEV (top controlled exhaust valve) and FBIV (fuel booster injection valve) components.

The first three announced models were the S60ME-C10, the S70ME-C10, and the G90ME-C10 delivering 2490, 3430, and 6240kW per cylinder, respectively. Design targets in developing the new models were reducing the complexity of the hydraulic system and increase system performance; the new TCEV and FBIV technologies have been developed within this scope. The design changes allow a weight reduction of up to 10% per kW depending upon model and also permit a higher P_{max} to reduce fuel consumption.

The Mk 10 platform is based on a much more mass-optimized design platform that results in lighter engines with reduced overall length, width, and height compared to its Mk 9 counterpart. The purpose of the TCEV is to integrate the exhaust actuator, the hydraulic push rod, and the HCU block into the exhaust valve. By doing so, the dynamic behaviour is improved with no long hydraulic push rod.

The FBIV and the TCEV technologies are well suited for integration on the cylinder cover of the engine as the control of the valves for fuel injection is separated from the control of the exhaust valve through using the well-known electronic fuel injection valve (ELFI) and the new proportional exhaust valve actuator (PEVA). Integration of FBIV and TCEV leads to a considerable weight reduction, because the baseplate, HCU, pressure booster, high-pressure fuel oil pipes, and exhaust actuator are eliminated. In combination, these two technologies also offer improved hydraulic dynamics and flexibility.

As well as introducing the new models in 2016, MAN declared the G-type to be its fastest-selling engine design of all time with more than 1500 being sold in 6 years.

ME-B engines

Electronically controlled fuel injection was extended to smaller bore MAN B&W two-stroke engines in mid-2006 with the launch of the ME-B series, initially comprising 350- and 400-mm-bore models but later extended to embrace 460-, 500-, and 600-mm-bore versions (Figs 14.32–14.34). By 2012, the series was further expanded with the launch of the S30ME-B9 and the ultra-long-stroke G40ME-B9, G45ME-B9, and G50ME-B9 engines, with the result that the ME-B programme then comprised a total output range from 3200 to 19,040 kW (Table 14.5).

Development was driven by a perceived market requirement for the lowest possible propeller speed in relation to bore size, leading to a stroke–bore ratio of over 4.4:1 for the S35, S40, and S50 designs (similar to that of MAN Diesel's 4T50ME-X research engine). The 46 and 60-bore models have slightly lower ratios of 4.2:1 and 4:1, respectively. Type approval tests for the first production engine, a six-cylinder S40ME-B version, were completed in December 2007, since when models of all bore sizes have flowed into service.

The ME-B has proved to be a popular engine with around 2000 having been specified for vessels in service and under construction. However, meeting the increasingly strict environmental regulations coming into play from 2016 onwards has seen most shipowners gradually switch to fully electronic ME-C versions with the result that in the 2019 portfolio only the S30 ME-B is included.

It was decided at an early stage of development to retain a slim camshaft for controlling the exhaust valves, thereby also reducing the required capacity of the electronic control system. While this small camshaft operates the

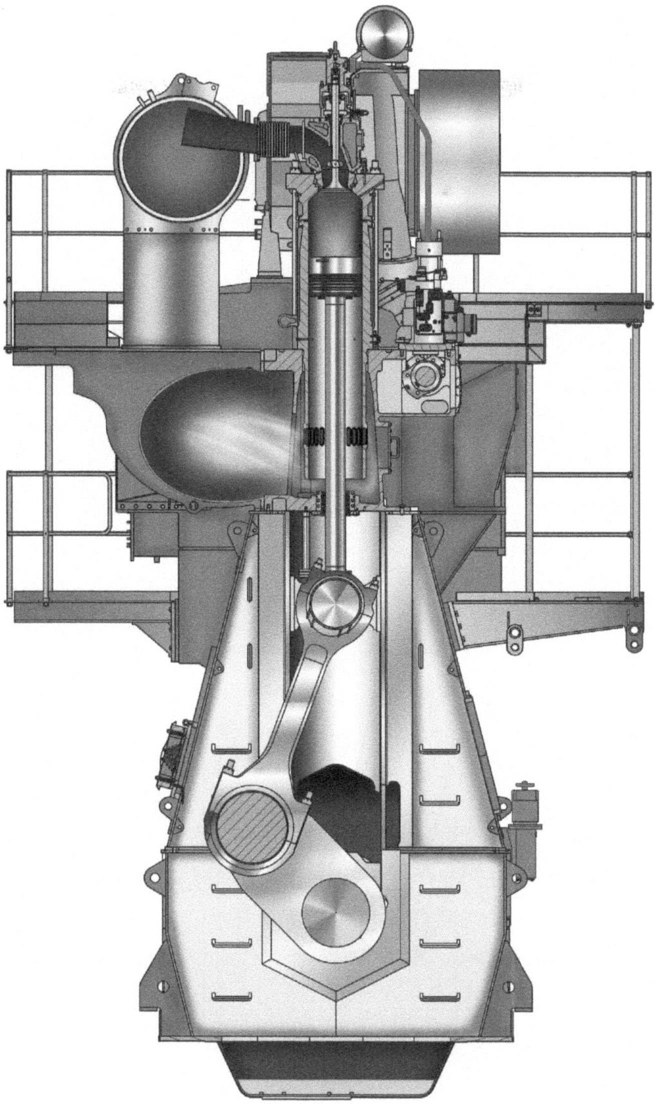

Fig. 14.32 Cross section of S50ME-B engine.

valves mechanically in the traditional manner, an electronically controlled fuel injection system adapted from ME practice enables the ME-B engines to meet IMO Tier II emission requirements.

Fuel injection is performed by one fuel booster per cylinder, similar to the established system for larger bore ME engines. The pressure boosters are mounted on HCU—two boosters per unit—with hydraulic oil supplied to

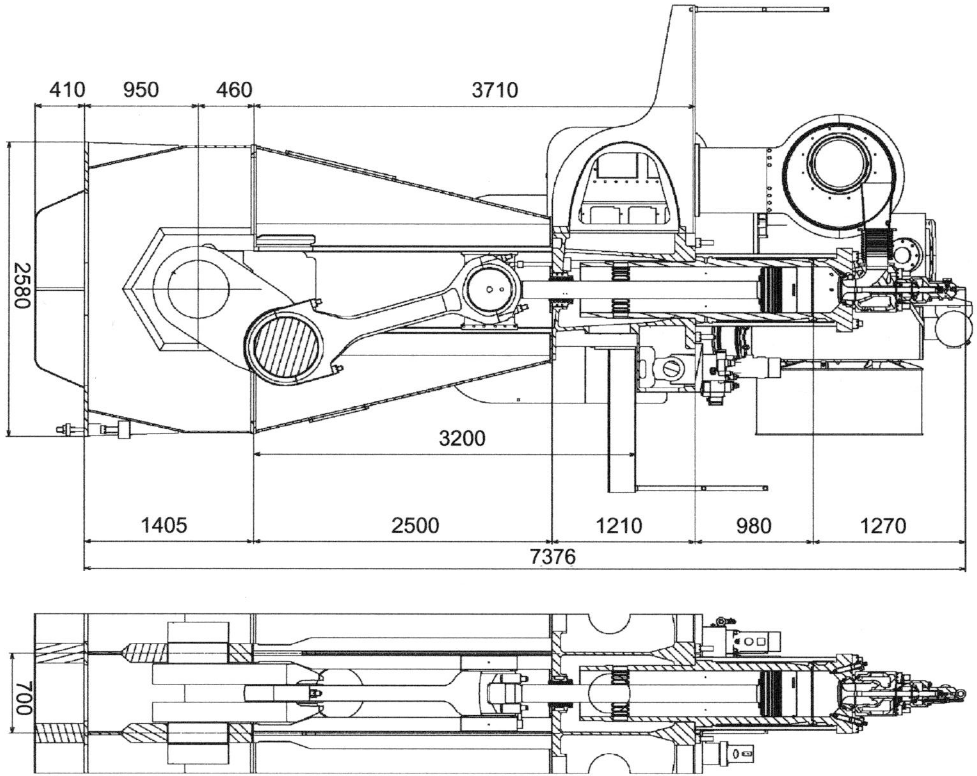

Fig. 14.33 Cross sections of S40ME-B engine.

Fig. 14.34 ME-B engine combustion chamber.

Table 14.5 MAN B&W ME-B engine programme.

	S35ME-B	S40ME-B	S46ME-B	S50ME-B9[a]	S60ME-B
Bore (mm)	350	400	460	500	600
Stroke (mm)	1550	1770	1932	2214	2400
Mean effective pressure (bar)	21	21	20	21	20
Speed (rev/min)	167	146	129	117	105
Mean piston speed (m/s)	8.6	8.6	8.3	8.6	8.4
Output/cylinder (kW)	870	1135	1380	1780	2380
Cylinders	5–8	5–8	5–8	5–9	5–8
Power range (MW)	4.35–6.96	5.67–9.08	6.9–11	8.9–16	11.9–19

[a]Engine is available in Mark 7, 8, and 9 variants with different ratings.

the HCUs via a single oil pipe enclosed in the camshaft housing. In addition to two pressure boosters, the HCU is equipped with two ELFI valves and two Alpha Lubricators. The accumulators used in the HCUs of the established ME engine design are replaced by one buffer of hydraulic oil serving each HCU, which in turn serves the injection needs of two cylinders.

Compression of the oil with respect to its bulk modulus accounts for the accumulator effect. The hydraulic system pressure is 300 bar (increased from the 250 bar of the standard ME system). The higher pressure makes it possible to omit the accumulators normally installed on the HCU; instead, a hydraulic volume for each of the two cylinders is placed below the HCU inside the camshaft housing. Two electrically driven pumps installed at the front end of the engine provide hydraulic power for the injection system. In the event of failure of one pump, more than 50% engine power will remain available, enabling 80% ship speed to be maintained.

The ME-B fuel injection system offers the same possibility for rate shaping as the established ME engines. The injection is controlled by a proportional valve allowing continuous change of the injection pressure; typically, a gradual pressure increase during injection is desirable. Specific fuel consumption, as well as emissions, is influenced by the injection profile. One profile is often favourable for sfc but at the cost of high NOx emissions, while the opposite effects result from another profile. The freedom to choose the fuel injection profile is thus a valuable tool in minimizing sfc while keeping emissions within given limits. Fuel valves similar to those of S35MC and S42MC engines are used for the S35/40ME-B models, the fuel nozzles being of the proven slide type already introduced as standard on all small-bore engines.

The engine control system uses the same multi-purpose controllers (MPCs) as the traditional ME control system. One MPC can control injection on two cylinders, but one of the MPCs also controls the hydraulic pressure and therefore only controls injection on one cylinder. A six-cylinder ME engine requires 13 MPCs, while a corresponding small-bore ME-B model may need only four MPCs.

Most of the design concepts of the large-bore Mark 9 engines are shared by the ME-B series, including a piston with high topland, PC ring, twin stay bolts, thin shell bearings, low-friction guide shoes, and thrust cams. The structural parts of the S35/40ME-B engines were designed to secure the rigidity and strength required for the higher output. The *bedplate* is of welded design, but the normal cast part for the main bearing girders is made from rolled steel plates. The aim was to achieve homogeneity of the material used for that bearing area with no risk of casting imperfections arising during final machining.

A new *main bearing* concept was under development at the launch of the engines to decrease the production cost of the bedplate. In the new design, the bedplate and bearing caps are finish machined as separate parts, with the caps positioned by means of the bearing shell and kept in position with the force from hydraulically tightened studs and high-friction coated shims placed between bearing cap and bedplate. The proven main bearing design used on all MAN B&W small-bore engine designs, however, was to be retained until the new development was completed.

A *framebox* of the proven triangular guide plane design with twin stay bolts—now standard on all updated MAN B&W engine types—was specified to provide excellent support for the guide shoe forces.

Two possibilities are available for the *cylinder frame*—nodular cast iron (KF) or a welded design with integrated scavenge air receiver—but it was decided to use the cast solution because of the material's high strength and high E-modulus in countering the high ignition force. Furthermore, compared with C3Cu material, the weight of a six-cylinder S35ME-B cylinder frame can be reduced by 3 t, corresponding to a 12% cost reduction for the frame.

A semi-built *crankshaft*, in one piece, is made of S34CrNi or S42Cr1 material. Even though the stroke–bore ratio was raised for the new engines, the cylinder distance is only slightly increased. Comprehensive FEM calculations were performed to ensure that the geometry (including the journal diameters) of the shaft was optimized for maintaining the rigidity, shrink-fit, and stresses at the same level as for the MC-C engines.

Based on the design introduced for the MC-C engine, the *thrust bearing* is very compact. Addressing the increased propeller thrust resulting from the higher engine power, a flexible thrust cam secures a more even force distribution on the pads. The overall dimension of the parts can thus be smaller than that with the old design.

A *connecting rod* design based on that applied throughout the small-bore engine programme (initially introduced for the L35MC model) is used. To reduce the production cost and oscillating forces, however, the new design is a combination of those used in the MC-C and 35MC engines. The *crosshead pin* design is taken from the S50MC-C engine, with the bearing dimensions based on long experience with the S35MC engine (exploiting a hardened running surface for the pin). The guide shoe is of the new low-friction design.

Bearings are of the same thin-shell design as for the previous small-bore engine generation, the material on all large bearings being Sn40Al. The relative loads on the large bearings are in all cases below the design targets.

The increased power of the ME-B engines dictated careful investigation of the *combustion chamber* to compensate for the higher ignition pressure and thermal load, and also to raise the reliability of the components and further extend the TBOs (Fig. 14.34).

A slim *cylinder liner* used for other small-bore MC-C/ME engines was also applicable to the ME-B 35 and 40-bore types, but the material was upgraded to Tarkalloy A to counteract the higher ignition pressure. A PC ring was also introduced to prevent bore polish.

The bore-cooled *piston* has a semi-high topland and a crown shaped in conjunction with the combustion chamber to cope with the increased power. Extensive FEM calculations were carried out in evolving the crown geometry to enable the high firing pressure and mean pressure to be withstood. A *piston ring pack* similar to those used for the established small-bore engines features a No. 1 ring with high CPR and Nos. 2–4 rings with angle cut. All of the rings have an Al-coat on the running surface for safe running-in. If a prolonged TBO is requested, a special ring pack with hard coating on the running surfaces of the No. 1 ring can be specified as an option.

As with the larger bore ME engines, an Alpha Lubricator is standard for the ME-B models, bringing the benefits of an ACC lubrication mode to smaller bore designs: a low total cylinder lube oil consumption while maintaining an excellent cylinder condition.

Exhaust valves are actuated by a light camshaft (small diameter and small exhaust cam) driven by a chain located at the aft end of the engine; the chain size is smaller than the MC-type chains (Fig. 14.35). Pursuing common spare parts for all four engines, the exhaust valves used for the S35ME-B and S40ME-B models are the same as those for the S35MC and S42MC models, respectively. The DuraSpindle design has a W-seat bottom piece.

Appropriate *turbocharger* models from MAN Diesel, ABB, and Mitsubishi can be specified for the ME-B engines. The turbocharger can be aft mounted, as for the existing engines, or exhaust side mounted. The latest type of water mist catcher and water drain arrangement were adopted. MAN Diesel's TCA turbochargers with variable nozzle ring technology are optionally available, compensating (in terms of part-load fuel consumption) for the absence of variable exhaust gas valve timing.

An *auxiliary blower* of the new integrated type, introduced on the Mark 8 version of the S50MC-C engine, yields a cost reduction of around 40% over

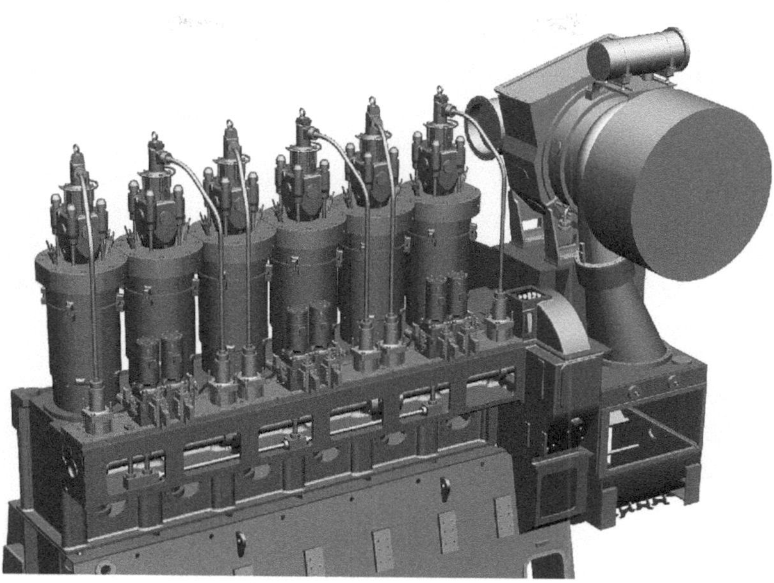

Fig. 14.35 An ME-B engine showing the camshaft-activated exhaust valve system and HCUs serving two cylinders.

the existing design. Maintenance is also lowered as the belt drive is omitted for the integrated blowers.

ME-GI gas and alternative fuel engines

Although dual-fuel four-stroke medium speed engines for marine use debuted around 2001, it was considered that applying the technology to two-stroke crosshead engines was more difficult. When Wärtsilä first installed dual-fuel engines in LNG carriers, it was to make use of boil off gas from the cargo in a more efficient way than the conventional steam turbines did. At that time, MAN considered that the high value of the cargo was best preserved by re-liquefying the BOG and returning it to the cargo with the ship being propelled by conventional low-speed diesel engines. Several LNG carriers were built to this concept.

As more attention focussed on meeting the Tier II and Tier III NOx requirements and the reduction in SOx in emission control areas, MAN began exploring the possibility of dual-fuel engines for marine use having already employed the idea in shore-power production. Unveiled in 2011, the first fruit of this research was the ME-GI engine. This was the culmination of many years' work that began in the 1990s with the company's prototype MC-GI dual-fuel engine that entered service at a power plant in

Chiba, near Tokyo, Japan in 1994. This engine had been approved by some classification societies for marine use but there was little interest from the shipowning community.

After 20,000 h of operation in this peak-shaving stationary application with a Mitsui-built 40 MW 12K80MC-GI engine in Japan, Hyundai became the first MAN Energy Solution two-stroke licensee to demonstrate the latest ME-GI concept in Korea in late-2012, with Mitsui following in April 2013. At the time, MAN Energy Solution predicted a broad, potential market for its ME-GI engine.

The ME-GI was significantly different from dual-fuel four-stroke engines in that it is a high-pressure diesel and does not employ the Otto cycle as the other engines do. Technically, there is only a little difference between fuel and gas-burning engines, but the GI engine provides optimal fuel flexibility.

Relatively few components need to be modified or added to the engine itself, these being the oil and gas injection system. The ME-GI can run on fuel oil alone or on 95% LNG (5% fuel oil is needed as a pilot fuel) or any combination in between (Fig. 14.36).

The gas supply line is designed with ventilated double-wall piping including HC sensors for safety shutdown. For control of the gas engine,

Fig. 14.36 Fuel injection system for ME-GI engines.

the GI control and safety system added to the proven ME control system. Apart from these systems on the engine, the engine itself and its auxiliaries will comprise some new units. The most important ones, apart from the gas supply system, are:

- Ventilation system for venting the space between the inner and outer pipe wall of the double-wall piping
- Sealing oil system, delivering sealing oil to the gas valves separating control oil and gas. This system is fully integrated on the engine, and the shipyard no longer needs to consider this installation.
- Inert gas system that enables purging of the gas system on the engine with inert gas.
- Control and safety system, comprising a hydrocarbon analyser for checking the hydrocarbon content of the air in the double-wall gas pipes.

The GI control and safety system are designed to fail to a safe condition. All failures detected during gas fuel running, including failures of the control system itself, will result in a gas fuel stop and change-over to HFO fuel operation. Blowout and gas-freeing purging of the high-pressure gas pipes and the complete gas supply system follows. The change-over to fuel oil mode is done without any power loss on the engine.

The high-pressure gas supply flows through the main 'chain' pipe that connects each cylinder's gas valve block and accumulator. This 'chain' pipe design performs two important tasks: separating each cylinder unit from the rest in terms of gas dynamics, utilizing the well-proven design philosophy of the ME engine's fuel oil system and acting as flexible connections between the stiff main pipe system and the engine structure, safeguarding against extra stresses in the chain pipes caused by the inevitable differences in thermal expansion of the gas pipe system and the engine structure.

The buffer tank, containing about 20 times the injection amount per stroke at the maximum continuous rating (MCR), also performs two important tasks: it supplies the gas amount for injection at a slight, but predetermined, pressure drop, and it forms an important part of the safety system.

Because the gas supply piping is of the common rail design, the gas injection valve must be controlled by an auxiliary control oil system. This, in principle, consists of the ME hydraulic control oil system and an ELGI and ELWI (electronic gas injection) valve system, supplying high-pressure control oil to the gas injection valve, thereby controlling the timing and opening of the gas valve.

ME-GI injection system dual-fuel operation requires the injection of both pilot fuel and gas fuel into the combustion chamber. Different types of valves are used for this purpose. Two are fitted for gas injection and two for pilot fuel. The auxiliary medium required for both fuel and gas operation is as follows: high-pressure gas supply, fuel oil supply (pilot oil), and control oil supply for actuation of gas injection valves.

The gas injection valve complies with the traditional design principles of the compact design. Gas is admitted to the gas injection valve through bores in the cylinder cover. To prevent gas leakage between the cylinder cover/gas injection valve and the valve housing/spindle guide, sealing rings made of temperature and gas resistant material have been installed. Any gas leakage through the gas sealing rings will be led through bores in the gas injection valve to the space between the inner and the outer shield pipe of the double-wall gas piping system, where the leakage will be detected by HC sensors.

The gas acts continuously on the valve spindle at a maximum pressure of about 300 bar. To prevent gas from entering the control oil actuation system via the clearance around the spindle, the spindle is sealed with sealing oil at a pressure higher than the gas pressure (25–50 bar higher).

The pilot oil valve is a standard ME fuel oil valve without any changes, except for the nozzle. HFO, MGO, or MDO can be used as pilot oil. The oil valve design allows operation solely on fuel oil up to MCR. The gas engine can be run on fuel oil at 100% load at any time, without stopping the engine. For prolonged operation on fuel oil, it is recommended to change the nozzles and gain an increase in efficiency of around 1% when running at full engine load.

Gas fuels correspond to low-sulphur fuels, and for this type of fuel, MAN recommends a cylinder lube oil of TBN40 to be used. A heavy fuel oil with a high sulphur content requires a cylinder lube oil of TBN 70. Compared with HFO operation, gas gives a cleaner exhaust. Having very low or no sulphur, SOx are negligible in the exhaust gas. Particulates are reduced considerably as they are emissions of NOx and CO_2.

If running on oil fuels, all typical NOx reduction techniques can be used as necessary to meet regulations except water emulsification. In the ultimate event, an SCR catalyst can cut NOx emissions by up to 98% but an EGR system is also an option.

If the EGR system is combined with gas operation, the engine can easily fulfil Tier III. The NOx level when operating on gas is 20%–30% lower than HFO operation, and only around 30% of the exhaust gas needs to be bypassed across the EGR, this will lead to higher efficiency on gas compared with HFO operation in Tier III zones. Cleaning of the EGR scrubber water is another issue that becomes a lot easier when operating the engine on gas because the exhaust from gas contains limited particulate matter and no SOx.

A standard ME-GI engine is designed to run on LNG which is mainly composed of methane. A variant, the ME-GIE was developed later for running on ethane which, as a fuel, can be derived from cargoes of ships carrying ethane or ethylene.

A year after announcing the ME-GI engines, MAN began the development of a new model designed to run on other low flashpoint fuels than LNG. The fuels contemplated would be fuels that were liquid rather than gaseous. The result is the MAN B&W ME-LGI engine which can run on methanol, LPG, and other liquid fuels as well as fuel oil. Although the operation principle of the ME-LGI engine is similar to the ME-GI concepts, differences in fuel properties mean that the ME-LGI components and auxiliary systems will be different from the ME-GI.

The ME-LGI engine can be delivered in different versions, depending on the low flashpoint liquid fuel type used. The first type developed was the ME-LGIM designed for running on methanol followed by the ME-LGIP which is fuelled by LPG. Fuels for the ME-LGI engines are categorized by their vapour pressure at 60°C with methanol and ethanol being below one bar and LPG and DME (dimethyl ether) above one bar.

Fuel injection is accomplished by a booster fuel injection valve (BFIV), using 300 bar hydraulic power to raise the fuel pressure to injection pressure. The BFIV is an integration of MAN's fuel oil booster design and its old slide injector design, which have been in use since the early 2000s. Both designs are well proven, and the combined solution has the advantage that it minimizes the total weight and removes the HP pipes in-between. By using this design, the total inertia increases the response time of the valve, and tests in service have demonstrated better-controlled injection profiles.

SOLAS currently requires all fuels other than LNG (which is covered by the IGF Code) to have a flashpoint of at least 60°C. This is under review but until an international regulation is adopted, the use of low flashpoint fuels must be approved by a ship's flag state.

Methanol has a flashpoint of 11°C so MAN uses a double-wall design of all its methanol components, and all leakages are monitored and collected in the double barrier, so there are no problems related to its use. It is generally accepted that methanol is far easier to handle than LNG. To be able to use methanol fuel on the ME-LGI, the cylinder covers are equipped with the fuel booster injector valves designed specifically for methanol operation.

For a 50-bore engine, this means that each cylinder cover is equipped with two additional methanol booster injectors. A liquid gas injection (LGI) block is fitted on the cylinder cover.

Cylinder cover with fuel booster injectors and LGI control block.

This block contains a control valve for methanol fuel injection, a sealing booster activation valve, a forced suction activation valve, an LGI purge valve, and methanol fuel inlet/outlet valves. All pipes for hydraulic oil and fuel are double walled. Furthermore, the double-walled pipes for methanol fuel are vented with ventilation air.

The methanol booster injector valve must be cooled and running surfaces must be lubricated. For this purpose, a combined sealing and cooling oil system delivering a 50 bar system oil pressure has been integrated on the engine, and the system both lubricates all running surfaces and controls that the temperature in the booster valve is lower than maximum 60°C.

The sealing oil pressure is generated internally in the booster valve in order to avoid contamination of the hydraulic oil operating the valve. The sealing oil has other advantages. It avoids methanol from going into the umbrella system and further down into the drain oil system. The cooling oil and sealing oil systems are fully integrated into the engine design, including equipment for continuous monitoring of methanol contamination in the oil system. If methanol is detected in the system, the engine will switch to fuel oil mode, and the methanol will be purged from the engine.

At the same time, the cooling oil pump supply side will be switched to clean system oil, and the oil circuit will be flushed with clean oil. Then, the clean oil will be collected together with the contaminated oil in the cooling oil tank, and the system will only be able to continue operation when no methanol is detected in the tank.

To ensure the correct temperature of the BFIV, the oil is cooled in a heat exchanger, which is connected to the low-temperature cooling system. When the fuel is injected, the combustion condition is monitored with pressure mean indicator (PMI) sensors located in each of the cylinder covers. The injection pressure is approximately 500 bar.

Three combustion conditions are monitored such as the compression pressures, combustion pressures, and expansion pressures. The 8-bar pressurized methanol is delivered to the engine via double-walled pipes, ventilated with dry air taken from the starting air system. To suck in air, a ventilation system is fitted at the outlet. All methanol fuel equipment is made in a double-walled design, and any methanol leakage will develop into methanol vapour.

This is monitored by HC sensors located close to the outlet of the double-walled piping system. In case of a too high methanol vapour content in the ventilation, the safety system will shut down the operation on methanol and return to operate on sole fuel oil. This switch is done smoothly and without any power loss.

By late 2019, MAN had made sales for all of the engine types designed for running on gas and alternative fuels. While a large number of these have yet to be completed, the number of ships involved exceeds 130. This is no small achievement as the bunkering infrastructure is still at an early stage and only ships suited to specific operational areas where fuels are available to have been specified with these engines.

MAN's new competitor—WinGD—is also experiencing goods sales figures with its Otto cycle low-pressure dual-fuel two strokes and it would appear that MAN would like to compete here as well. In October 2019, MAN announced that in cooperation with its licensee Hyundai Heavy Industries, it was developing a new engine to be known as the ME-GA. It is a low-pressure engine with five- and six-cylinder models in the 700 mm-bore size, intended as the first offerings.

MAN cited interest from the LNG carrier sector as the main motivator. Its ME-GI engines have mainly been supplied for use in twin-engine LNG carriers so the new model would tap this market because it would not require the costly fuel gas supply arrangements, such as the high-pressure compressor for boosting the pressure of boil-off gas to the 300 bar injection pressure needed for the ME-GI engine.

Table 14.6 MAN B&W ME-GA engines.

Type	**Dual-fuel two-stroke**
Bore	700 mm
Stroke	3256 mm
Stroke/bore ratio	4.65
Power per cylinder L1 rating	2830 kW@78 rpm
Power/cyl L2 rating	2260 kW@78 rpm
Power/cyl L3 rating	2400 kW@66 rpm
Power/cyl L4 rating	1920 kW @66 rpm
Mean effective pressure L1 rating	17.4 bar
Engine versions cylinder numbers	5 and 6
Power range(SMCR) 5 cyl model	9600–14,150 kW
Power range(SMCR) 6 cyl model	11,520–16,980 kW

The first G70ME-C10.5-GA will produce up to 2830 kW per cylinder and be NOx Tier III compliant when in gas mode without additional exhaust treatment. Output at the L1 rating point would be 14,150 and 16,980 kW, respectively, at a crankshaft speed of 78 rpm in five- and six-cylinder configurations. However, the layout diagram in each case provides a high degree of flexibility and choice as to contractually specified maximum continuous power and speed combinations. At the L4 SMCR, for instance, the per-cylinder output is 1920 kW at 66 rpm. The design will have the same outline and footprint as that of the G70ME-C10.5 engine, which can deliver up to 3100 kW per cylinder at 78 rpm (Table 14.6).

CHAPTER FIFTEEN

Japan Engine Corporation low-speed engines

The dominance enjoyed for many years by MAN Diesel and Wärtsilä in the low-speed two-stroke cross head engine sector was challenged by Mitsubishi, the only other contender in the international arena.

By 2018 MAN had changed its name to MAN Energy Solutions, Wärtsilä had withdrawn from the two-stroke sector having disposed of its interests to Winterthur Gas & Diesel (WinGD) a wholly owned subsidiary of China State Shipbuilding Corporation and Mitsubishi had merged in 2017 with its former licensee Kobe Diesel to form Japan Engine Corporation or J-Eng as it styles itself.

The company has developed and built a range of UE engines since 1955. Most of its low-speed engines, unlike MAN B&W and Wärtsilä engines, were built by the group itself rather than by overseas licensees. The complete UEC programme was handled by its Kobe factory while engines of up to 600-mm bore were built by its Yokohama works and the Japanese licensees Akasaka and Kobe Diesel. UEC engine technology was extended to new licensees in China and Vietnam in 2005.

An experimental UEC model built in 1952 yielded a power output of 50% higher than its predecessor, the MS engine, which had been developed from the early 1930s. The first commercial UEC engine, a nine-cylinder UEC 75/150 model developing 8830kW, was introduced in 1955. The UEC engine now shares a common design philosophy with MAN B&W MC and Wärtsilä's RTA series, exploiting constant pressure turbocharging and uniflow scavenging through a central hydraulically opened/pneumatically closed exhaust valve in the cylinder cover. (The UEC designation refers to Uniflow-scavenged, Exhaust gas turbocharged, and Cross head-type engine.)

The original UEC engines—progressing from A to E type—featured impulse turbocharging, three mechanically actuated exhaust valves per cylinder, and a single centre-injection fuel valve. This arrangement fostered high performance with a smooth exhaust gas outlet and good combustion, but exhaust valve lifetime was sometimes short.

Pounder's Marine Diesel Engines and Gas Turbines
https://doi.org/10.1016/B978-0-08-102748-6.00015-3

A two-stage turbocharging system (unique in contemporary production low-speed engine practice) was introduced for the UEC-E type in 1977 to boost specific output and achieve a mean effective pressure of 15 bar. Low- and high-pressure turbochargers were linked in series (Fig. 15.1). To match this increase in maximum cylinder pressures, the engine frame, bearings, and combustion chamber components were made more rigid than those of the UEC-D-type predecessor while still retaining the monobloc cast iron construction (three elements held by tie bolts). A new piston crown of specially reinforced molybdenum cast steel was specified with a combination of radial

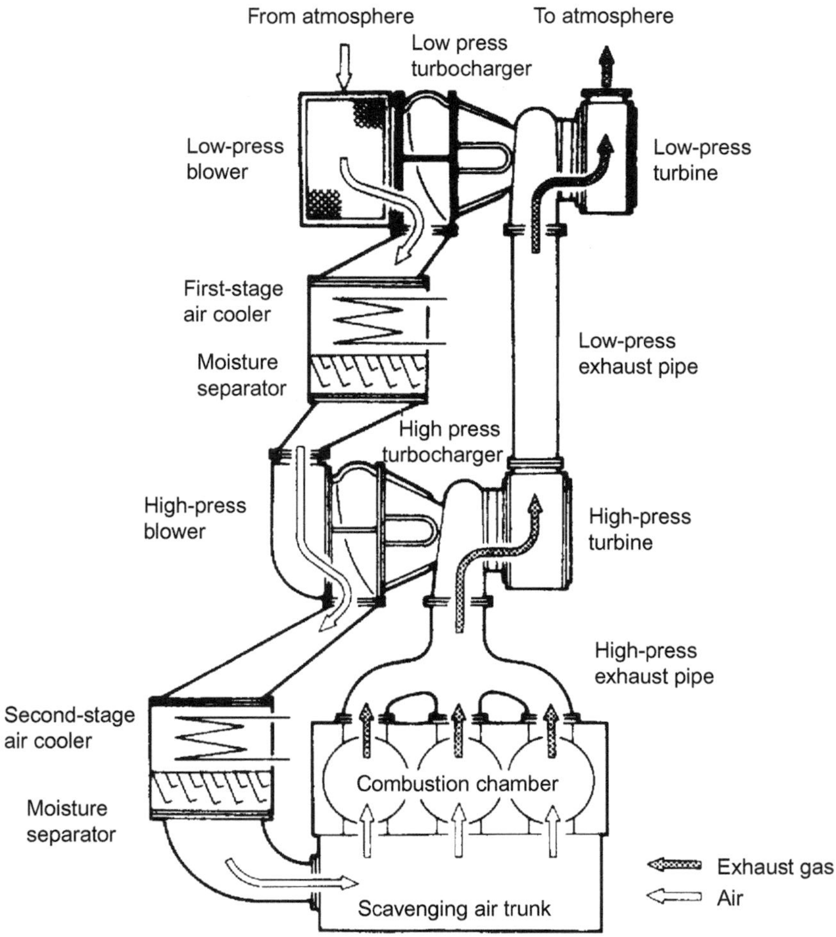

Fig. 15.1 Two-stage turbocharging system for UEC-E engines.

and circular ribs behind the frame plate to withstand the rise in maximum pressure (increased from 90 to 110 bar).

The UEC-E engine was comparatively short-lived in the programme, however, the steep rise in bunker prices in the late 1970s dictated a stronger focus on fuel economy rather than sheer output. Mitsubishi, therefore, introduced the long-stroke UEC-H-type model in 1979, abandoning impulse turbocharging in favour of constant pressure turbocharging based on the group's own MET-S turbocharger. The H-type also broke with previous UEC engine practice in using one large-diameter exhaust valve in the centre of the cylinder cover and twin side-mounted fuel injection valves (Fig. 15.2). A significant reduction in fuel consumption was yielded. The H-type is detailed later in this chapter. The HA-type, which followed in 1982, adopted the higher efficiency non-water-cooled MET-SB turbocharger and an improved combustion system, further enhancing fuel economy. Output was also raised by 11%–14%.

Advances in fuel economy were pursued with the development of longer stroke/lower speed L-type engines from 1983 with successive refinements embodied in the LA-type (1985), the LS-type (1986), and the LSII-type (1987). Stroke-bore ratios were increased to achieve lower direct-coupled propeller speeds and hence higher propulsive efficiency. Efforts to improve specific fuel consumption were reflected in rises in the maximum cylinder pressure.

For most of the 1980s, Mitsubishi was content to concentrate on the small-to-medium size ship propulsion sector with just four bores (370, 450, 520, and 600 mm). Four- to eight-cylinder models covered outputs of up to around 15,000 kW. The UEC-LA models, with a stroke–bore ratio

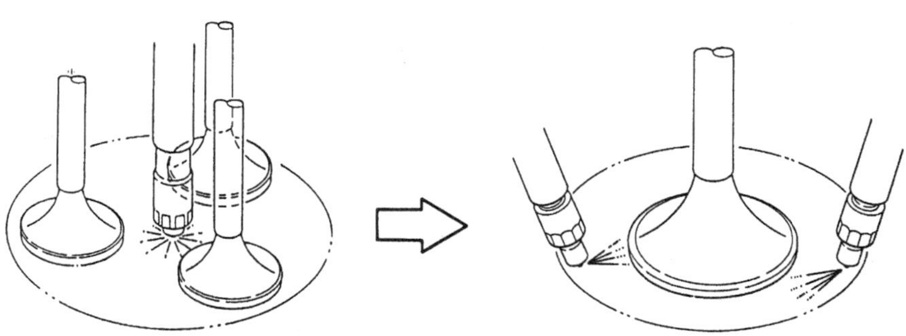

Fig. 15.2 The UEC-A to E series engines featured three exhaust valves and a centre injection fuel valve (*left*). Later series (H to LS) is equipped with one exhaust valve and two side-injection fuel valves (*right*).

of 3.17:1, were supplemented in 1985 by LS options for the 520 and 600 mm models (3.55 and 3.66:1 ratio, respectively).

A return to the large bore arena, long surrendered to MAN Diesel and Sulzer, was signalled in 1987 by the launch of the UEC 75LSII series with a maximum rating of 2940 kW/cylinder. The 750-mm bore/2800-mm stroke design retained the main features of the LS engines but exploited an even higher ratio (3.73:1) and lower running speeds (63–84 rev/min). Mitsubishi stepped up its challenge in the VLCC and large bulk carrier propulsion markets in 1990 with the introduction of an 850-mm bore/3150-mm stroke version. This UEC 85LSII series is now available in five- to nine-cylinder versions, offering a maximum rating of 3860 kW/cylinder and hence an upper output of 34,740 kW.

The propulsion market created by a new generation of large Panamax and post-Panamax container ships requiring service speeds of up to 25 knots was addressed by Mitsubishi in 1992 with its UEC 85LSC series, a shorter stroke (2360 mm) faster running derivative of the UEC 85LSII (Fig. 15.19). The design is the highest-rated engine in the group's programme, with an output of 3900 kW/cylinder at 102 rev/min. The series covers a power band of up to 46,800 kW from 5- to 12-cylinder models.

In 1991, a determination to secure small-ship propulsion business led to the introduction of a 330-mm bore/1050-mm stroke LSII design (Fig. 15.3). The series succeeded the 370-mm bore LA models, which pioneered the modern 'mini-bore' two-stroke engine from the late 1970s for small general cargo tonnage, tankers, colliers, feeder container ships, and even large fishing vessels. The LS and more advanced UEC 33LSII engine now develop 566 kW/cylinder at 215 rev/min and covers an output band from 1230 to 4530 kW at 162–215 rev/min with four- to eight-cylinder models.

Mitsubishi enhanced its medium-power programme in 1992 by upgrading the 500-mm and 600-mm bore models to LSII status. Later power upratings of 3%–5% for these UEC 50LSII/60LSII models, without major design changes, were introduced to meet the demands of larger ships and higher service speeds. The specific outputs, respectively, rose to 1445 and 2045 kW per cylinder. With a stroke–bore ratio of 3.9:1, the 50LSII series embraces four-to-eight-cylinder versions covering an output range of up to 11,560 kW at 127 rev/min.

The small bore programme was strengthened in 1995 with 370 and 430 mm UEC LSII designs to attract business from Handy-sized bulk carriers, feeder container ships, small tankers, and gas carriers. The engines

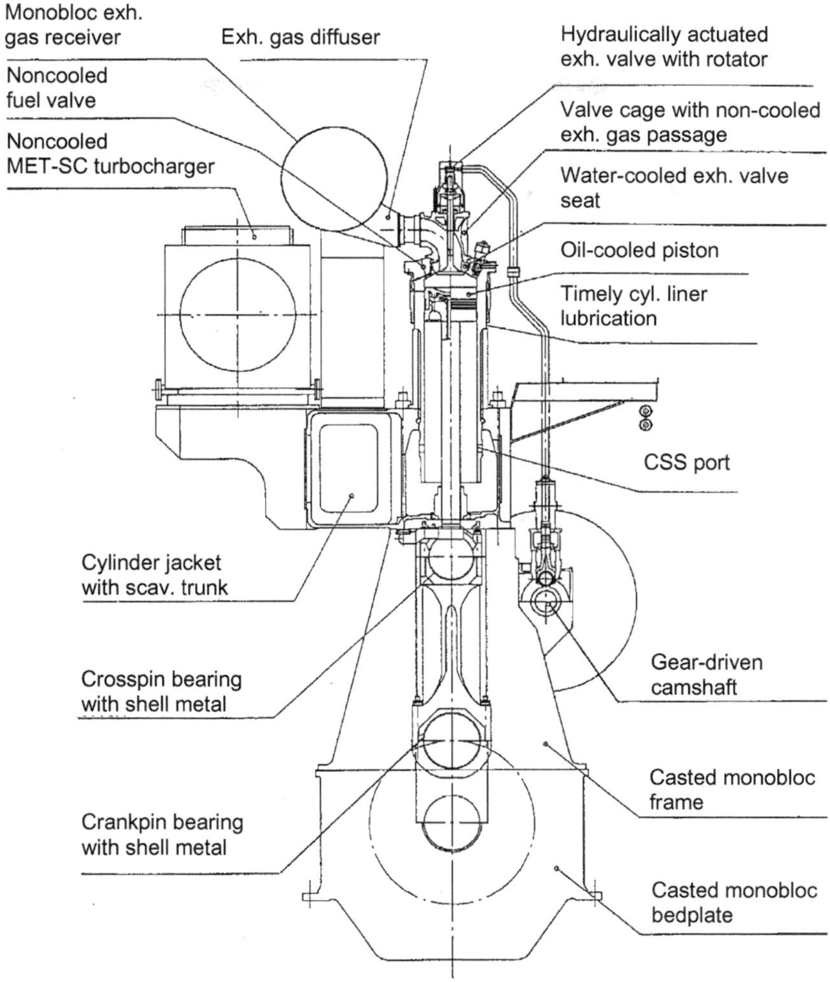

Fig. 15.3 Construction of UEC 33LSII engine.

plugged a gap between the established 330 and 500-mm bore LSII series. The UEC 37LSII engine offers an output of 772 kW/cylinder at 186 rev/min from five-to-eight-cylinder versions, while the UEC 43LSII engine delivers 1050 kW/cylinder at 160 rev/min from four-to-eight-cylinder models. The first UEC 37LSII engine, a six-cylinder model, completed testing in early 2000, and the debut order for a 43LSII engine, a seven-cylinder model, was logged in 2002 (Table 15.1).

Development of the LSII design from 1998 resulted in the LSE series, headed by a 520-mm bore model (UEC 52LSE) targeting feeder container

Table 15.1 UEC37LSII engine data.

Bore	370 mm
Stroke	1290 mm
Stroke–bore ratio	3.49
Output	772 kW/cyl
Cylinders	5–8
Power range	2095–6180 kW
Speed	186 rev/min
Mean piston speed	8 m/s
Mean effective pressure	18 bar
Maximum cylinder pressure	147 bar
Specific fuel consumption	175 g/kWh

ships and Panamax bulk carriers. The series was extended by a 680-mm bore model (UEC 68LSE) for propelling tonnage including Aframax and Suezmax tankers, Cape-size bulk carriers, and container ships. The LSE models share a more compact configuration and a 25% reduction in the overall number of components compared with the LSII engines. Benefits were also gained in production costs, reliability, and maintainability. The mean effective pressure rating and mean piston speed were, respectively, raised by 5% and 6% to 19 bar and around 8.5 m/s. The first 450 mm-bore LSE engine, a six-cylinder model, was completed in summer 2008. A successor to the UEC 52LA engine, this 45LSE has proven particularly popular for powering Handy-size bulk carriers built in Japan (Table 15.2).

An extension of the same 'total economy' LSE concept later resulted in a 600-mm bore version and in 2013 an 800-mm bore variant was added to the

Table 15.2 UEC-LSE engine data.

	52LSE	**60LSE**	**68LSE**
Bore (mm)	520	600	680
Stroke (mm)	2000	2400	2690
Stroke–bore ratio	3.85	4	3.96
Output (kW/cyl)	1705	2255	2940
Cylinders	4–12	5–8	5–8
Power range (kW)	3700–20,460	7650–18,040	10 050–23,520
Speed (rev/min)	127	105	95
Mean piston speed (m/s)	8.47	8.4	8.52
Mean effective pressure (bar)	19	19	19
Specific fuel consumption (g/kWh)	160–167	160–166	159–165

See also Table 15.3.

range. A new range—the LSH series was added in 2014 that aimed at slow steaming vessels. The range currently comprises five models—two with a 420 mm bore and three with a 500 mm bore.

In 2018 Japan Engine Corporation as the company is now known, announced preliminary details of a new LSJ series designed to meet the requirements of NOx, SOx, and EEDI rules. In a departure from previous ranges, the LSJ engine is designed to run only on MGO rather than a range of fuels from HFO to MGO. The idea behind this is that by using only the cleanest distillate fuel, the engine will have no difficulty in complying with the 2020 global sulphur cap and by making use of either EGR or SCR depending on bore size, also be able to meet Tier III NOx emission levels.

A proposed dual-fuel engine planned before the change of structure appears to have been shelved. The current UEC programme embraces models with bore sizes of 330, 370, 430, 450, 500, 520, 600, 750, and 800 mm in LS, LSII, LSE, LSH, and LSJ versions, covering an output band from 1655 to 35,520 kW.

UEC design details

Structural and design details are exemplified by the UEC 75LSII and 85LSII engines (Fig. 15.4). The bedplate and column are of simple welded monobloc construction, and the partition for each cylinder consists of a single welded plate. Side forces are supported by horizontal steel ribs arranged behind the guide bar. The pistons are of the well-proven central support type and are fitted with four rings (Fig. 15.5). Hard chromium plating applied to the running surfaces of the rings improves wear resistance under high pressure and corrosion resistance in a low-temperature corrosive environment. Chromium plating the ring grooves has the same purpose, as well as improving the durability of the rings by ensuring normal movement in the grooves.

The lower part of the cylinder liners features controlled swirl scavenging (CSS) ports, which promote improved scavenging efficiency and hence favourably influence the fuel consumption and thermal load. The angle of the lower scavenging port is directed towards the centre of the cylinder to discharge combustion gas on the piston top, and the angle of the upper port is directed sideways to discharge gas around the inner surface of the liner. The CSS configuration (Fig. 15.6) reportedly raises scavenging efficiency by 5%–7% over the conventional port arrangement. Two-row bore liner cooling is incorporated, and cylinder lubrication is timed so as to reduce oil consumption. Corrosive liner wear, caused by low temperatures, is inhibited by controlled positioning of the cooling bores.

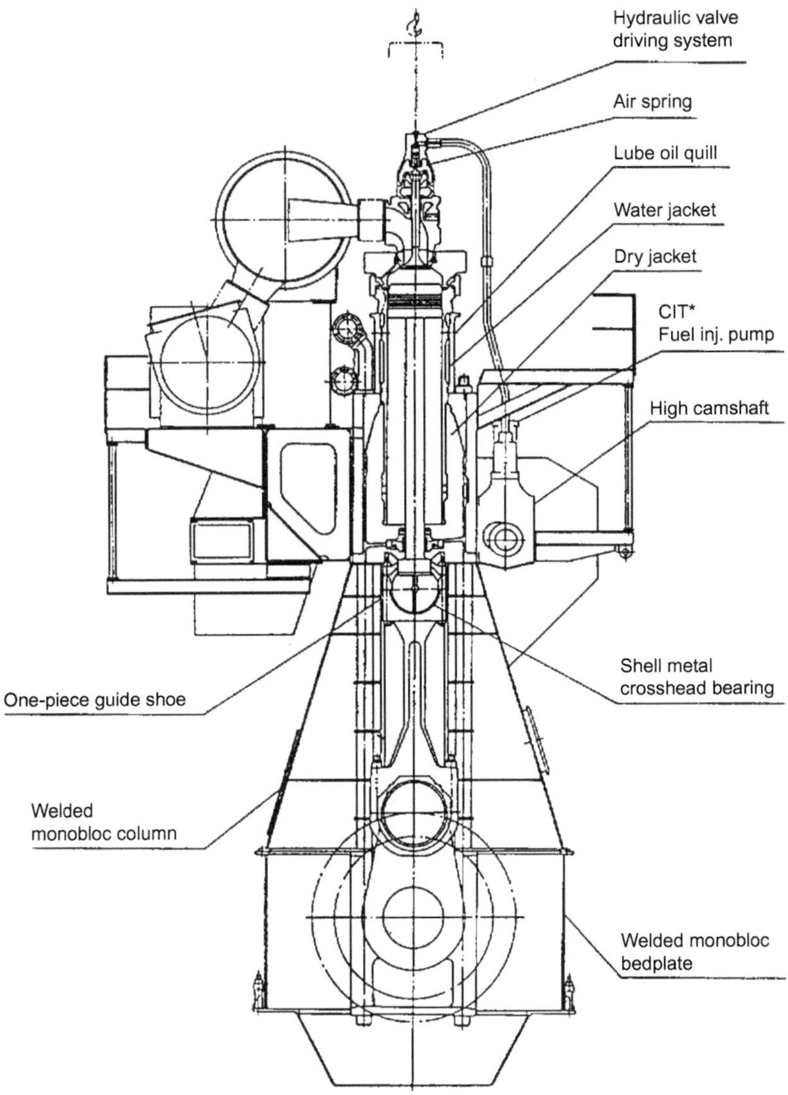

*CIT: Controllable injection timing

Fig. 15.4 Construction of UEC 75LSII.

The single exhaust valve of each cylinder has a hydraulic rotation system, which rotates the valve to improve running control conditions. The rotation wings are fitted on the actuator piston on the exhaust valve side, an arrangement that is said to offer less obstruction to exhaust gas passage than with the traditional rotator

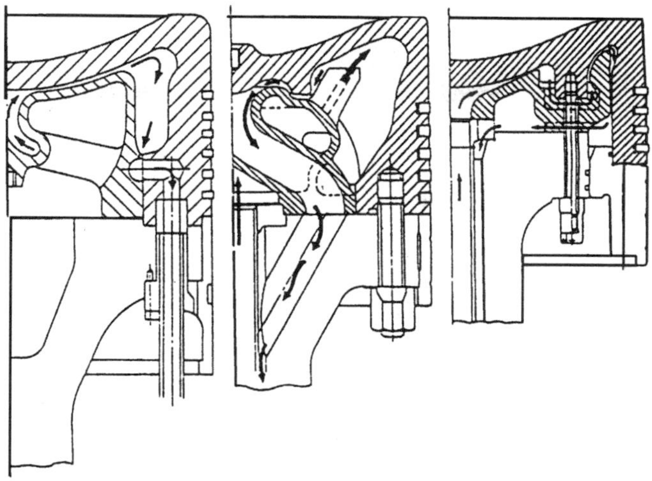

Fig. 15.5 Improvements in piston design from the original UE engine component (*left*) to the HA model (*centre*) and the current L and LS types (*right*).

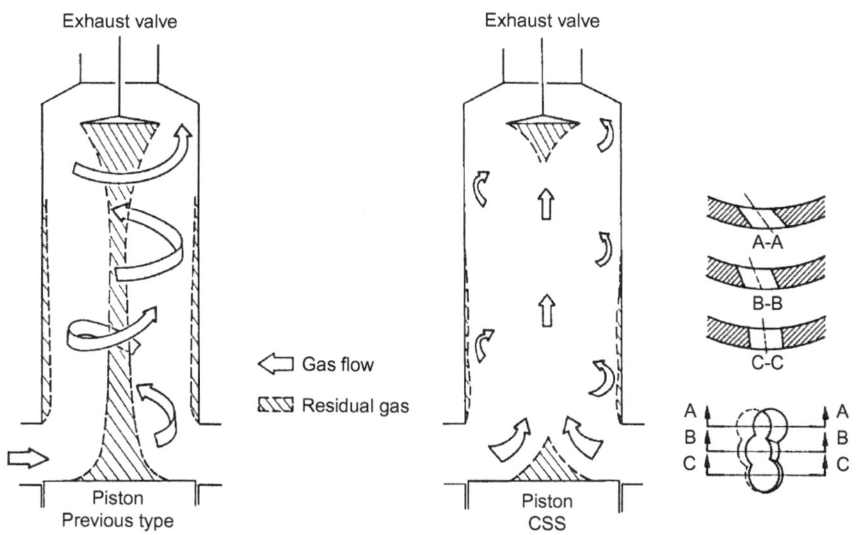

Fig. 15.6 Comparison of the gas flow in the cylinder with original UE engine scavenging and with controlled swirl scavenging (CSS).

wing mounted on the valve spindle. A water-cooled valve seat is specified.

The Bosch fuel pumps have control racks to adjust the quantity of fuel supplied. The controllable injection timing (CIT) pumps also ensure that engine performance is improved at the service load (Fig. 15.7). Two

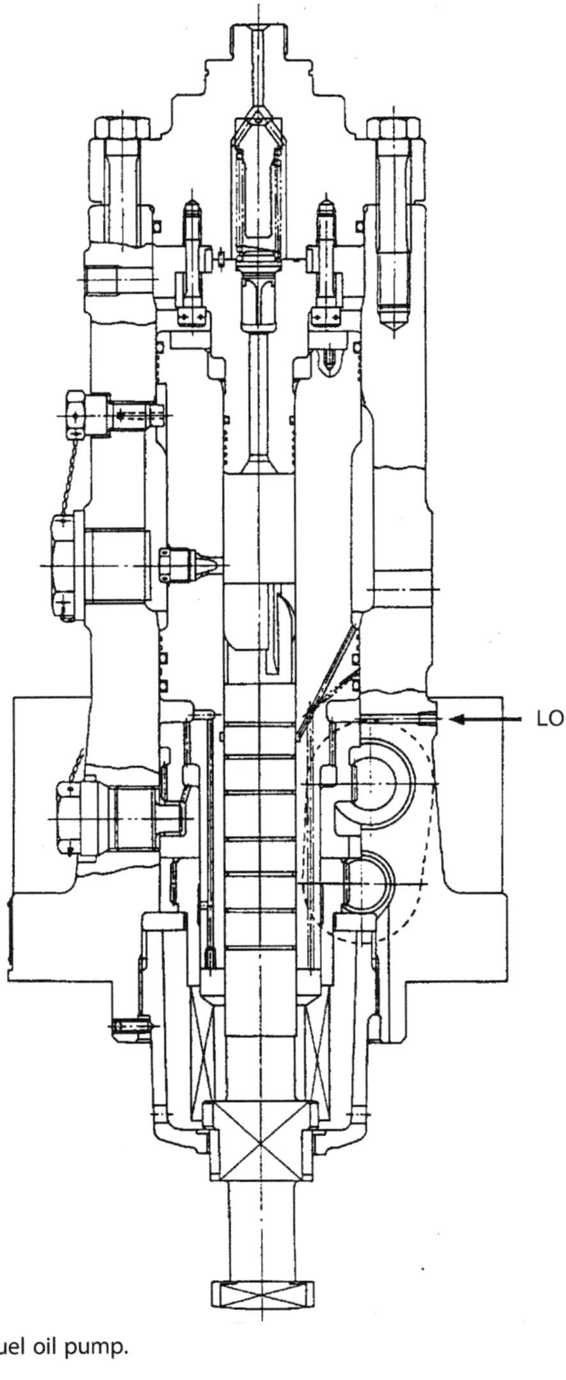

Fig. 15.7 CIT fuel oil pump.

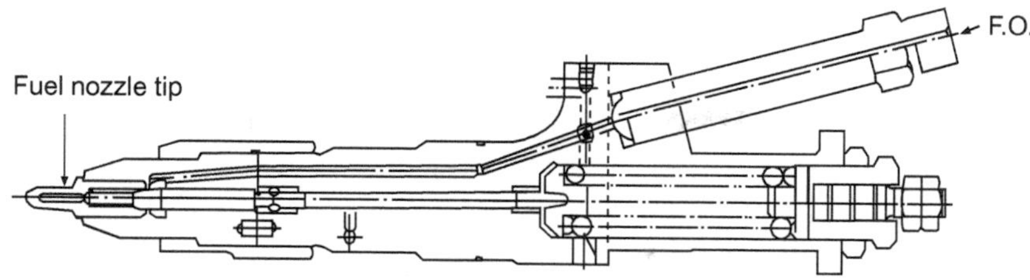

Fig. 15.8 Lightweight (20 kg) noncooled fuel injection valve for UEC LSII engine.

side-injection non-water-cooled fuel injection valves (Fig. 15.8) are arranged on each cylinder cover; a steam passage is provided for warming the valve body when the engine is stopped.

An optional automatic flexible rating (AFR) control system can be specified to reduce fuel consumption during part-load operation by coordinating the engine output with the operational conditions of the ship. The system (Fig. 15.9) controls the fuel delivery timing of the injection pump by using the CIT system to produce maximum cylinder pressure and also controls the turbine nozzle area of the variable geometry (VG) turbocharger to secure an optimum scavenging air pressure. The CIT fuel injection pumps, offered as standard with LSII engines, are operated by a governor when the AFR control system is not installed.

A mean time-between-overhaul of 4 years was targeted for the LSII engines, Mitsubishi, citing contributions from the following key elements:

- Piston rings and cylinder liners, yielding respective wear rates of 0.2 mm/1000 h and 0.02 mm/1000 h or less.
- Reliable hydraulically rotated exhaust valves with stellite-coated stems and seats.
- Easily maintained fuel injection valves.
- High-performance piston gland packing (stuffing box drainage of >10 L/day/cylinder, and a low wear rate scraper ring).

Mitsubishi's R&D priorities mirror those of its two rivals in the low-speed engine arena, with emphases on further upratings of designs, enhancing reliability and durability, and reducing noxious emissions. Its resources include the group's main research centre at Nagasaki. Mitsubishi has investigated all aspects of exhaust gas de-NOxing technology including the direct injection of water into the combustion chamber. Its stratified injection system based on a fuel valve designed to inject fuel–water–fuel sequentially has achieved NOx reductions of around 50% without significantly increasing fuel consumption and is offered as an option for UEC–LSE engines.

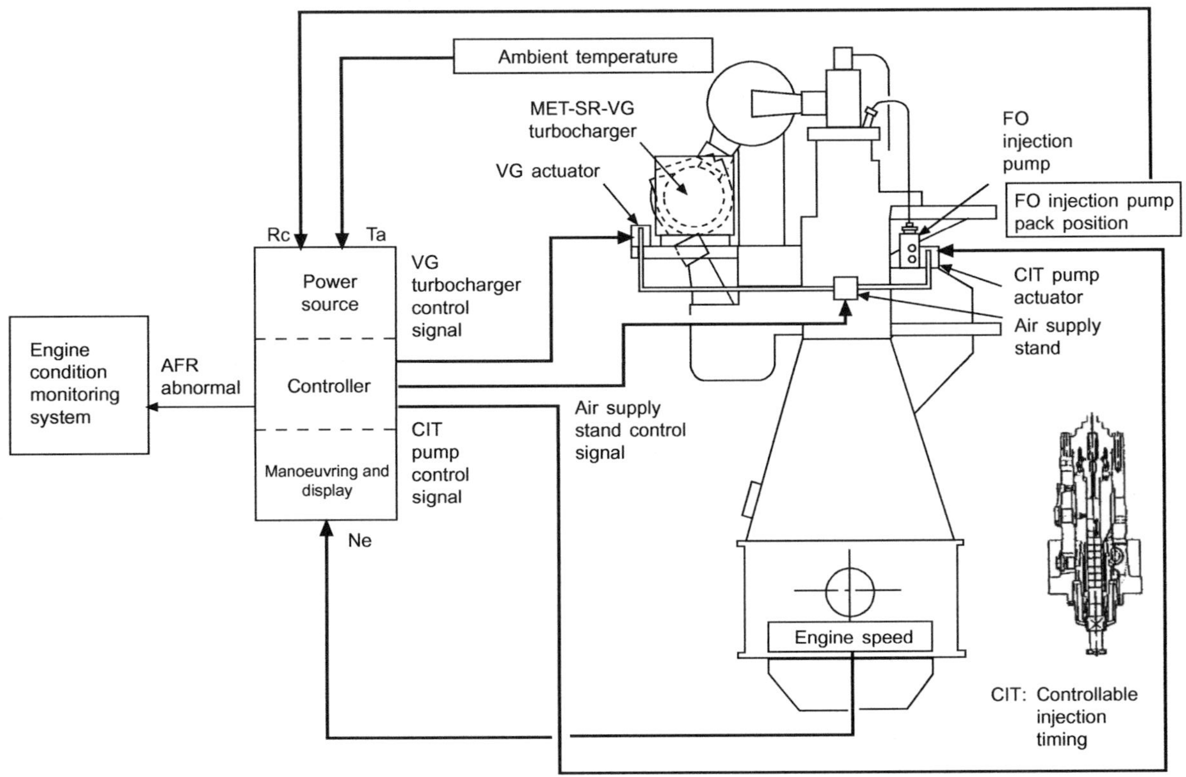

Fig. 15.9 A UEC engine served by an AFR control system.

New 350- and 400-mm bore models

A joint venture by MHI with the Wärtsilä Corporation to design and develop a new low-speed engine was announced at end-2002. The Japanese group has been a prolific licensed builder of Sulzer engines since 1925, a marque then under the Wärtsilä umbrella. The resulting 500-mm bore/2050-mm stroke design was realized in two versions: the conventional camshaft-controlled Sulzer RTA50C and the electronically controlled RT-flex50C.

A strategic alliance forged between the groups in September 2005 resulted in the launch of a new range of 350 and 400 mm-bore designs in 2008, in both mechanically and electronically controlled versions. The engines cover the power ratings and running speeds associated with propelling diverse small-to-medium size cargo tonnage. Tapping experience from earlier and contemporary generations of small bore two-stroke engines, the designers sought compliance with IMO Tier II emission regulations, low fuel consumption and cylinder oil feed rates, high reliability, and long times-between-overhauls. Electrical power requirements, weight and dimensions were other development factors as well as competitive manufacturing costs.

Mitsubishi's programme was thereby extended with the UEC 35LSE and 40LSE designs (see Table 15.3), which exploited a traditional camshaft for mechanical actuation of the fuel injection pumps and exhaust valve operation.

Table 15.3 Small-bore UEC–LSE engines.

	35LSE	40LSE	45LSE
Bore (mm)	350	400	450
Stroke (mm)	1550	1770	1840
Output/cycle (mcr, kW)	870	1135	1245
Speed (rev/min)	167	146	130
Mean effective pressure (bar)	21	21	19.6
Mean piston speed (m/s)	8.6	8.6	7.97
Cylinders	5–8	5–8	5–8
Power range (kW)	3475–6960	4550–9080	6225–9960

35LSE and 40LSE engines jointly developed with Wärtsilä.

UEC eco-engine

Mitsubishi followed MAN Diesel and Wärtsilä in dispensing with the traditional large camshaft and developing a UEC eco-engine engine with electronic control of fuel injection, exhaust valve actuation, starting air valve, and cylinder lubrication.

Studies on electronic control solutions for two-stroke engines were started by MHI in 1988, the fundamental systems subsequently verified on single-cylinder 330-mm and 450-mm bore research engines at the group's Nagasaki R&D centre over almost 15 years. Based on the test results, the eco-engine project was initiated in early 2000. A seven-cylinder UEC 33LSII stationary engine at MHI's Kobe shipyard retrofitted to eco status in the late 2001 proved system functionality and reliability in daily service under various operating modes for over 2 years.

The first commercial production engine, an eight-cylinder 600-mm-bore UEC 60LSII-eco model developing 16,360 kW at 105 rev/min, completed tests in 2005 before delivery for powering the large pure car/truck carrier *Lyra Leader*. Design work extending the concept to other bore sizes has been completed, but no further eco-engines had entered service by end-2008. The current programme offers eco-engine versions of the UEC 85LSII, 60LSII, 50LSII, 68LSE, 60LSE, 52LSE, and 50LSE models (Fig. 15.10).

In an eco-engine, the fuel injection pump and exhaust valve driving gear are actuated by hydraulic oil at a pressure of 320 bar, the pressurized oil stored in an accumulator block at each cylinder. Pressure compensation during actuation is effected by the volume in the accumulator chamber, making a pressurized gas enclosed-type accumulator unnecessary.

Hydraulic power for fuel injection and exhaust valve actuation is controlled by an on–off type solenoid valve unit and engine control system. The timing of injection and exhaust valve opening/closing are also controlled electronically to achieve the best solution for any operating mode and to match changing conditions. The fuel injection pump is similar in structure to the conventional mechanical unit but simplified. Two sets of on–off type solenoid valves are mounted on the pump to control the injection pattern with respect to the operating load and improve the trade-off between thermal efficiency and NOx emissions. Stricter emission regulations in the future could be met by MHI's stratified fuel–water injection system (see Chapter 3).

Fig. 15.10 Mitsubishi's first production eco-engine on test in 2005, an eight-cylinder 60LSII model subsequently installed in a pure car/truck carrier.

A feedback control function is applied in controlling the fuel injection volume to compensate for the equivalent thermal load and individual cylinder control. The fuel pump stroke is monitored by twin gap sensors at each cycle, underwriting system reliability (Figs 15.11 and 15.12).

Exhaust valve opening and timing are controlled by the electro-hydraulic system using the on–off solenoid valve unit, the timings optimized to the operating load. For precise timing control, the feedback control function is applied by monitoring the valve lift. The actuating mechanism is otherwise similar to the conventional system, reportedly inheriting its reliability and maintenance procedure. Solenoid valves are key elements of the eco-engine, the application demanding a very quick response, high flow rate, and long life cycle. MHI started developing suitable valves in 1999, the most important goal being reliability. Endurance testing corresponding to around 6 years of actual operation confirmed satisfactory performance.

A conventional starting air control valve is eliminated, and solenoid valves and a control air pipe are added. Electronically controlled starting valves are said to achieve a better performance and flexibility in starting and crash-astern procedures.

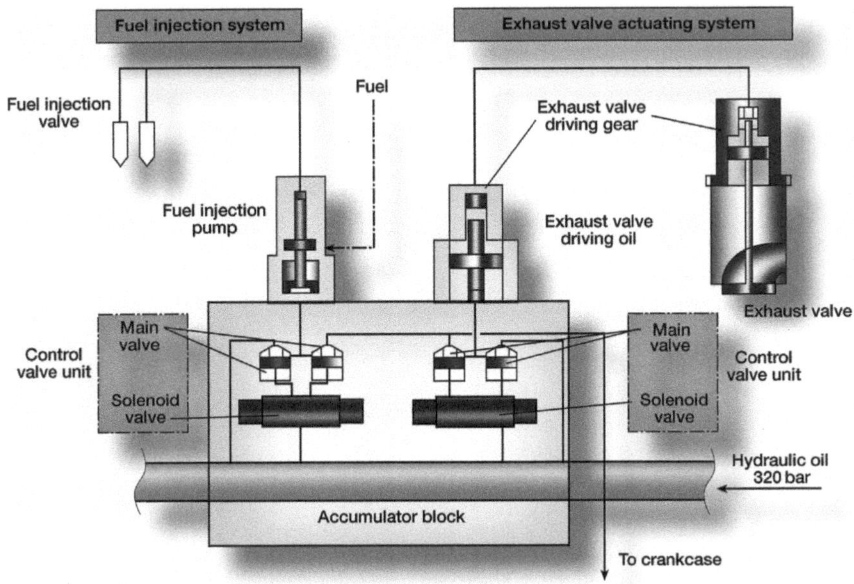

Fig. 15.11 Schematic diagram of eco-engine fuel injection and exhaust valve actuation and control systems.

Hydraulic power for the systems is provided by an installation based on engine-driven pumps, which supply pressurized oil mainly during engine operation, and electrically driven pumps, which maintain system pressure when the engine is stopped. The engine-driven axial piston pumps are coupled through a gear drive to the crankshaft.

An engine control system is prepared for the 'eco main controller' (EMC) installed in the control room and interfaced with the remote control system. A local control box and 'eco cylinder controller' (ECC) are mounted on the engine and connected by a duplicated network line. For redundancy, the EMC system comprises two duplicated units operating in parallel and performing the same task. If the active EMC fails, the standby unit takes over control without interruption. Among its functions, the controller carries out the following tasks: speed governing; start/stop sequences; timing control of fuel injection, exhaust valve and starting air system actuation; control of the hydraulic oil supply system; alternative operation and control modes; network functioning and monitoring malfunctions in the entire control system.

Each ECC, mounted on the individual cylinders, executes the orders for fuel injection, exhaust valve actuation, and starting air system timing. The

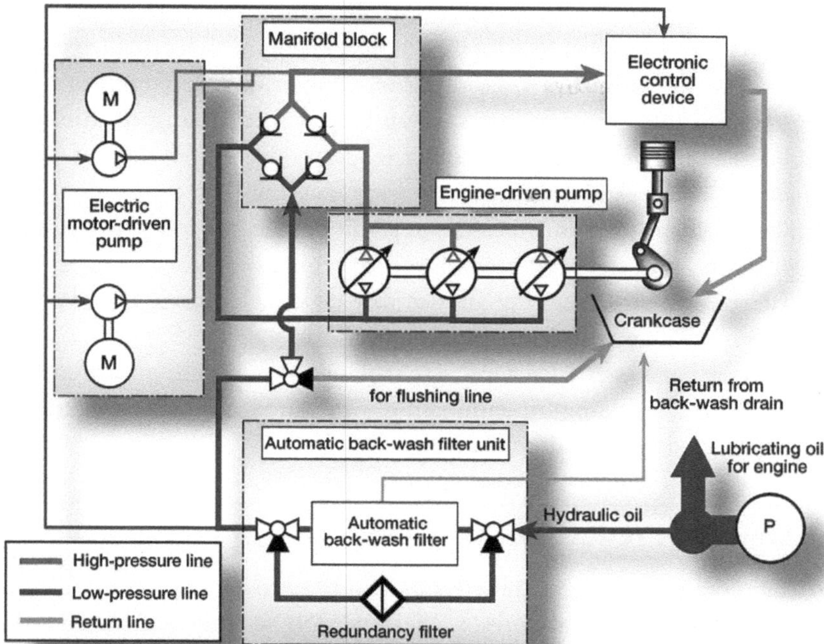

Fig. 15.12 Configuration of hydraulic oil power supply system for the eco-engine.

capability to control two cylinders provides redundancy: if one ECC fails, one serving another cylinder can take over automatically.

An excellent experience is reported from the pioneering *Lyra Leader* installation, with better performance across the load range and lower operating costs than camshaft-controlled engines. The electronically controlled cylinder lubrication (ECL) system can be specified for the conventional UE engines to reduce lube oil consumption and improve piston ring and cylinder liner running conditions.

MHI summarizes the benefits of the eco-engine systems as follows:

- *Environmental friendliness*. Smokeless operation can be achieved by the appropriate fuel injection pressure at any load, and reductions in NOx emissions can be secured by tuning the fuel injection timing and pattern at any load.
- *Higher fuel economy*. Fuel injection and exhaust valve actuation timing can be optimized flexibly by electronic control according to the engine operating load, atmospheric conditions, and fuel oil properties; lower specific fuel consumption ratings can thus be achieved, particularly at part load.

- *Easier control (better manoeuvrability).* Stable, continuous low load operation (even at extremely low loads) with good engine performance is fostered by the improved combustion conditions arising from an appropriate fuel injection pressure and optimized timing of injection and exhaust valve actuation at lower loads.
- *Higher reliability.* A fuel injection pressure favourable to the combustion condition at any load enhances the reliability of the hot components such as piston crown, rings, cylinder liner, and exhaust valve.
- *Flexible operation.* Operating modes can be easily changed, and fine-tuning is possible during engine running.
- *Reduced maintenance.* Electronic control significantly simplifies engine assembly and servicing by eliminating some large mechanical elements and their associated maintenance.

Emissions reduction

Two operating modes are selectable for the eco-engine: low emission and economy modes. If the former is selected, NOx emissions can reportedly be reduced by 10%–15% without any penalty in fuel consumption; if the economy mode is chosen, the fuel consumption can be lowered by 1%–2% while retaining the same level of NOx emissions.

New emission control technologies have been pursued by MHI for all UEC engines to counter IMO third-stage NOx regulations, including two water injection systems: a stratified fuel–water injection system (already installed in stationary power plants) and an independent water injection system.

With water injection valves arranged separately from the fuel injection valves, the independent system was tested on the bench before installation and trials on a 6UEC 50LSE engine in the shop. Water is injected with appropriate timing controlled by solenoid valves according to the crank angle; since the electronic control system is similar to that of the eco-engine, the independent water injection system is easy to integrate with that design.

Unlike a stratified fuel–water or emulsion injection system, it is not necessary with the independent water injection system to increase the fuel pump capacity or the size of the camshaft driving gears and cams; additionally, a larger volume of water can be injected to achieve a greater reduction in NOx emissions. NOx reduction is relative to the amount of water injected, MHI reporting that a 69% cut was achieved in the engine test, with a water injection amount of 169% to fuel quantity at 75% load (the maximum weighting factors in E3 mode in the IMO NOx calculation formula). Furthermore, a 68% NOx reduction was obtained at 100% load, with a 134% water injection to fuel ratio.

Although the targeted 80% NOx reduction could not be achieved, the results from an independent system—with higher than 50% reductions—had never been secured by MHI with stratified fuel–water injection or emulsified fuel. A remaining challenge was to reduce the amount of water injection required (e.g. by optimizing timing and the atomizer specification) and hence lower the volume of water to be generated or carried onboard for this purpose.

UEC LSH series

In early 2014, Mitsubishi introduced the UEC-LSH series of super-long stroke engines purposely developed for ships operating a slow-steaming strategy. Initially, it was planned to include a 600 mm bore engine, but the range presently consists only of 420 and 500 mm bore versions. The first production engine was delivered in 2015 as a UEC50LSH-Eco-C2 model.

Target vessel types include Supramax and Handysize bulkers, medium-range (MR) tankers, vehicle carriers, and small containerships. The engines have a bore to stroke ratio of 4.60, giving the 500 mm bore engine a 2300 mm stroke and the smaller 420 mm model a stroke of 1930 mm.

Both bore sizes come in either 5-, 6-, 7-, or 8-cylinder variants. The 500 mm bore engine is available as either a NOx Tier II engine or two models of Tier III engines—one with a lower sfoc figure. In its highest output configuration, the engine will deliver 1780 kW/cylinder at 108 rpm reducing to 985 kW/cylinder at 81 rpm. At full load, sfoc is quoted at 164.0 g/kWh reducing to 158.0 g/kWh at the P4 point. There is a 1.6 g/kWh penalty in sfoc when the engine is fitted with EGR for meeting NOx Tier III. The UEC50LSH-Eco represents an adaptation of the electronic control of earlier UEC series and in that electronically controlled fuel oil injection system is combined with camshaft-controlled exhaust valve actuation. Hydraulic oil for the Ecosystem is supplied by an electric pump.

The turbocharger is arranged on the exhaust side and a variable turbine inlet (VTI) version of the proprietary MET turbo is available for slow-speed operation. The LSH generation has been conceived for continuous low-load operation down to 20% load. The engines also feature the in-house developed advanced electronically controlled lubrication (A-ECL) system which can reduce the minimum dosage down to 0.5 g/kWh, depending on engine conditions. The A-ECL system can further reduce the cylinder oil feed rate, compared with a mechanical lubricating system, particularly under partial load operation, by controlling cylinder oil consumption according to the mean effective pressure.

UEC LSJ series

Developed from the LSH series, the LSJ is J-Eng's first new engine since the merger of Mitsubishi and Kobe Diesel. It shares the same dimensions, bore and stroke of the LSH engine but is an interesting departure aimed at meeting the NOx Tier III levels and as a solution for the 2020 sulphur cap. It is specifically adapted to run only on MGO which means that it will easily allow compliance with the 2020 sulphur rules but only with what might be considered the most expensive fuel choice.

It is also anachronistic in its way of meeting NOx emission levels as it employs a combination of direct water injection and EGR. The latter is necessary to meet Tier III NOx rules. It is an electronically controlled camshaftless engine and, because it runs on low viscosity MGO alone, the fuel line needs no heating for highly viscous fuels such as HFO. The water injection system requires additional equipment in the way of pumps and controllers.

J-Eng expects the first test engine to be ready in late 2019 with commercialization taking place in 2020 or 2021. As well as the 420 and 500 mm bore models based on the LSH series, a smaller 350 mm bore engine is planned for the LSJ range.

Cylinder lubrication

MHI addressed the concerns of operators on cylinder lubrication costs by offering the swirl injection principle (SIP) lubricating system from Danish specialist Hans Jensen Lubricators for UEC engines. Spray injection creates a thin and uniform oil film distribution on the liner surface, which is effective in reducing waste oil splashed up and down by the piston ring movement. Additionally, the system can reportedly achieve a reduction in ring and liner wear rates, improve the liner condition, and reduce particulate matter.

The cylinder lubricator pumps oil to the SIP valves mounted in the cylinder liner wall. The valves are equipped with a nozzle for spraying tiny droplets of oil tangentially to the liner wall, covering a large area above the next injection valve. Spraying from all valves ensures coverage of the entire circumference of the liner. Injection takes place when the exhaust valve closes and well before the piston passes the injection quills in the upward movement.

The centrifugal power of the scavenging air swirl ensures that the droplets settle horizontally well distributed on the liner wall. This allows the piston rings to effect the vertical distribution of the injected cylinder oil when they pass in the continued upward movement. The designers claim that an optimum distribution of oil is thus achieved with a minimum oil quantity injected. The spring in the SIP valve works at a pressure of approximately 40 bar. Apart from securing an adequate pressure for cylinder oil to become a suitable spray when leaving the nozzle, it also ensures that the oil in the tube between cylinder lubricator and SIP valve stays under pressure (Fig. 15.13).

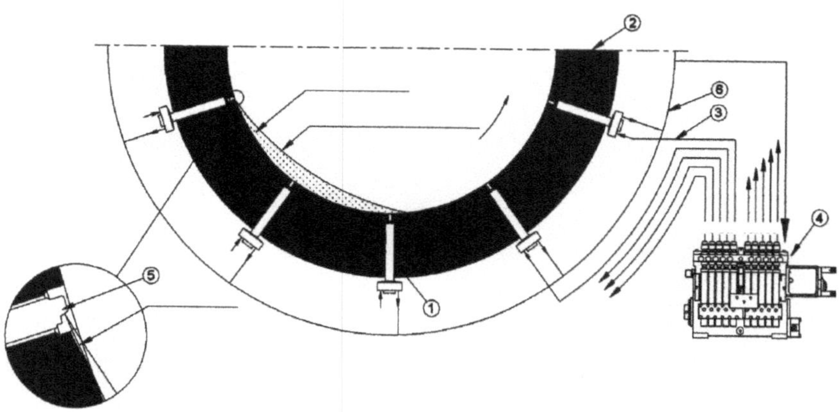

Fig. 15.13 Hans Jensen Lubricators' SIP cylinder lubrication system benefits UEC engines 1 = SIP valve, 2 = cylinder liner, 3 = pressure pipe, 4 = lubricator, 5 = nozzle, and 6 = return pipe.

CHAPTER SIXTEEN

WinGD (Wärtsilä/Sulzer) low-speed engines

Winterthur Gas and Diesel (WinGD) may be one of the newest names in marine diesel engine design and manufacture but the name hints at a heritage that stretches back to the very beginning of motorized ships. The company has its headquarters and R&D facilities in Winterthur, Switzerland where the Sulzer company had been based since the 19th century. Sulzer developed a marine engine in 1905 but the first application in an ocean-going ship the *Monte Penedo* was not until 1912, the same year in which the Danish ship *Selandia* powered by two diesel engines was built by Burmeister and Wain in Copenhagen.

WinGD is the successor to Wärtsilä having been founded in 2015 as a joint venture between the Finland-based company which had acquired New Sulzer Diesel in 1997 and China State Shipbuilding Corporation (CSSC). The following year, Wärtsilä transferred its stake to CSSC and withdrew from the low-speed two-stroke market. It does however continue to service the engines through its global network of service centres.

New Sulzer Diesel itself was founded as a spin-off from the parent Sulzer company in 1990 some 4 years after production of engines had ceased at the Winterthur plant in 1986. Wärtsilä committing to the sustained development of the successful RTA low-speed two-stroke engines inherited from the Swiss designer alongside its own range of four-stroke medium-speed engines. Launched at the end of 1981, the uniflow scavenged, constant pressure turbocharged RTA series with a single poppet-type exhaust valve (Figs 16.1 and 16.2) represented a break with Sulzer's traditional loop-scavenged valveless two-stroke concept.

The original RTA series embraced six models with bore sizes of 380, 480, 580, 680, 760, and 840 mm, and a stroke–bore ratio of 2.86 (compared with the 2.1 ratio of the loop-scavenged RL series). These RTA38, RTA48, RTA58, RTA68, RTA76, and RTA84 models—collectively termed the RTA-8 series—were supplemented in 1984 by the longer stroke RTA-2 series comprising 520 and 620-mm bore models and the RTA84M model. The RTA-2 series was extended again in 1986 by a 720-mm bore model.

Pounder's Marine Diesel Engines and Gas Turbines
https://doi.org/10.1016/B978-0-08-102748-6.00016-5

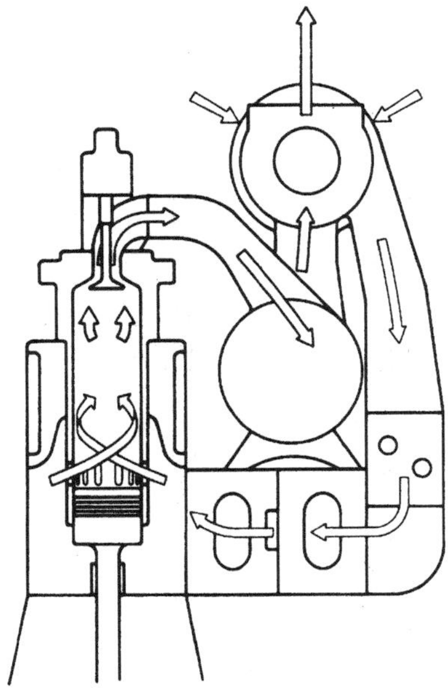

Fig. 16.1 Uniflow scavenging system for RTA engine series.

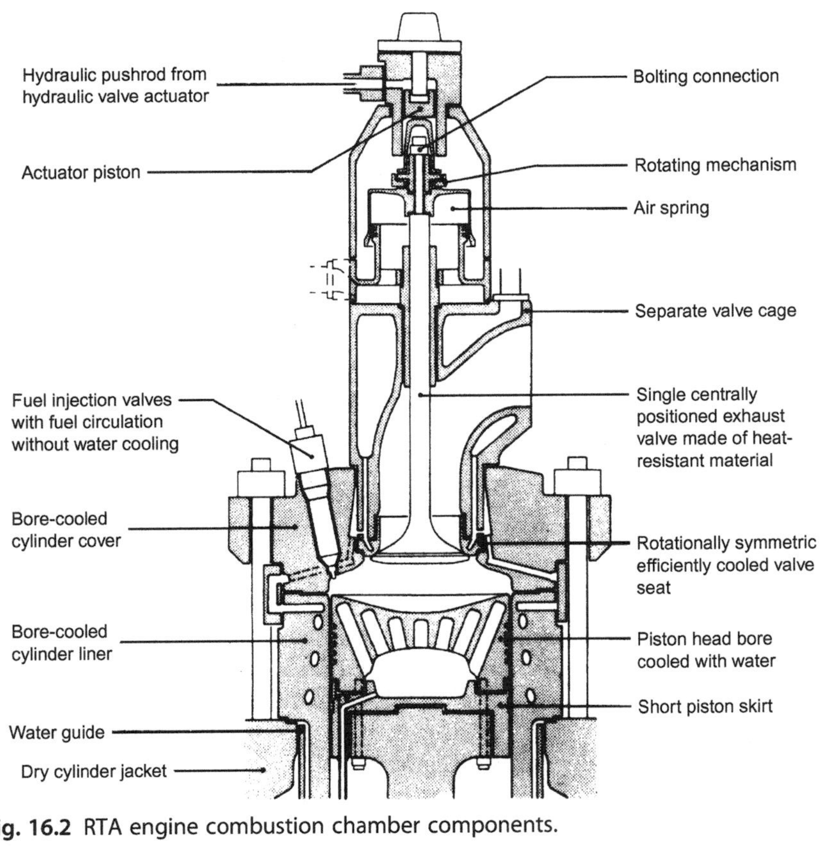

Fig. 16.2 RTA engine combustion chamber components.

An uprated 840-mm bore design, the RTA84C, was introduced in 1988 to offer higher outputs in the appropriate speed range for propelling large cellular container ships.

A higher stroke–bore ratio (3.47) for the RTA-2 series secured lower engine rotational speeds and hence higher propulsion efficiencies. The ratings of the RTA-2 engines were increased in 1987 in line with the power level already offered by the RTA72 model. Various design improvements, together with higher ratings, were introduced for the RTA-2 series in 1988 at the same time that the RTA84C engine was launched. Further upgrading of the RTA-2 series—to RTA-2U status—came in 1992, with a 9% rise in specific power output.

The RTA range was subsequently extended by the introduction in 1991 of the RTA84T 'Tanker' engine design, which, with a stroke–bore ratio of 3.75 and a speed range down to 54 rev/min, was tailored for the propulsion of VLCCs and large bulk carriers. The first production engine entered service in 1994. Even longer strokes (4.17 stroke/bore ratio) were adopted in 1995/1996 for RTA-T versions of the 480-, 580-, and 680-mm bore models, whose key parameters addressed the power and speed demands of bulk carriers up to Cape-size and tankers up to Suezmax size. These RTA48T, RTA58T, and RTA68T designs have more compact dimensions than earlier models, giving ship designers more freedom to create short engine rooms, while reduced component sizes and weights facilitate easier inspection and overhaul. Shipyard-friendly features were also introduced to a smooth installation.

In 1994 Sulzer anticipated demand for even higher single-engine outputs from large and fast container ships by announcing the RTA96C engine, a short stroke (2.6 strokes/bore ratio) 960-mm bore design.

RTA engines of all bore sizes have benefited from continual design refinements to increase power ratings and/or enhance durability and reliability, based on service experience. In early 1999 the RTA programme was streamlined by phasing out some older types (such as the 380-mm bore model), which had fallen out of demand, a measure also dictated by the need to ensure that all engines could comply with the IMO Annex VI Tier I NOx emission regulations.

In mid-1999 the RTA60C model was introduced as the first Sulzer engine designed from the bedplate up to embody RT-flex electronically controlled common-rail fuel injection and exhaust valve actuation systems. Starting with the 580- and 600-mm bore designs, RT-flex series versions progressively became available as options throughout the RTA programme.

The RT-Flex series of engines were subject to constant improvement but not all engines were modified at the same time. As with the earlier RTA engines, the engines of a given bore size were optimized either to container ships or to slower-moving tankers and bulkers by way of a longer stroke for the tanker version. These were identified by the suffix C or T following the bore size in the engine nomenclature. So, an RT-Flex82C was designed for containership installation whereas the RT-Flex82T was for tankers or bulkers. The RT-Flex50 did not have the distinction.

In addition to the C or T suffix, improved versions were distinguished by an additional letter either B, D, or E. C was omitted to avoid confusion with the container or tanker distinction. There was also a separate ER notation to indicate extended range but this was used only on the RT-Flex58T-D ER engine.

In 2007 Wärtsilä initiated the development of its Generation X engines based on existing engines. Wärtsilä said at the time that "The "X" in Generation X stands for RT-flex technology, extra low speed, extremely reliable, and efficient. The first Generation X engines—the X35 and X40 became available in 2012.

The next development—and the last under the Wärtsilä brand—was the development in 2013 of the first low-speed dual-fuel engine, the RT-Flex50DF. In announcing the development in November 2013, Wärtsilä indicated its entire portfolio of low-speed engines would eventually be made available in DF form including the new Generation X series.

Since its foundation, WinGD has followed the same naming system for the engines including Generation X which were inherited after the business transfer. Predictably the range of older engine types still available has thinned leaving just three bore sizes and six engine variants for the RT-Flex range. The Generation X series now just called X has seven bore sizes and 17 engines. Only one bore size in the X series does not have a dual-fuel variant. As well as the 17 production engines, four new short-stroke engines were announced in late 2019 but not in production.

RTA series service and production

There are still a very small number of Sulzer engines in service that precede the RTA engines and significant numbers of early RTA engines with service records extending to 35 years. The RT-Flex models began outselling the RTA engines around 2005 but the RTA enjoyed further development and consistent popularity for many more years extending beyond the introduction of the Generation X engines to as recently as 2014.

With more than 1000 RTA engines of different types still in service and likely to be so for at least 10 years more, it is justified to include the following full history of this important engine type, especially as it was both a precursor and a contemporary of the RT-Flex which was based upon it.

Engine selection

Selecting a suitable main engine model to meet optimally the power demands of a given project dictates attention to the anticipated load range and the influence that the operating conditions are likely to have throughout the entire life of the ship. Every RTA engine had a layout field within which the power/speed ratio (=rating) could be selected. It is limited by envelopes defining the area where 100% firing pressure (i.e. nominal maximum pressure) is available for the selection of the contract maximum continuous rating (CMCR).

Contrary to the 'layout field', the 'load range' is the admissible area of operation once the CMCR has been determined. Various parameters have to be considered in order to define the required CMCR: for example, propulsive power, propeller efficiency, operational flexibility, power and speed margins, the possibility of a main engine-driven generator, and the ship's trading patterns.

The layout field (Fig. 16.3) is the area of power and engine speed within which the CMCR of an engine can be positioned individually to give the desired combination of propulsive power and rotational speed. Engines within this layout field will be tuned for maximum firing pressure and best fuel efficiency. The engine speed is given on the horizontal axis and the engine power on the vertical axis of the layout field; both are expressed as a percentage of the respective engine's nominal R1 parameters. Percentage values are used so that the same diagram can be applied to all engine models. The scales are logarithmic so that exponential curves, such as propeller characteristics (cubic power) and mean effective pressure (mep) curves, are straight lines. The layout field serves to determine the specific fuel oil consumption, exhaust gas flow and temperature, fuel injection parameters, turbocharger, and scavenge air cooler specifications for a given engine.

The rating points R1, R2, R3, and R4 are the corner points of the engine layout field. R1 represents the nominal maximum continuous rating

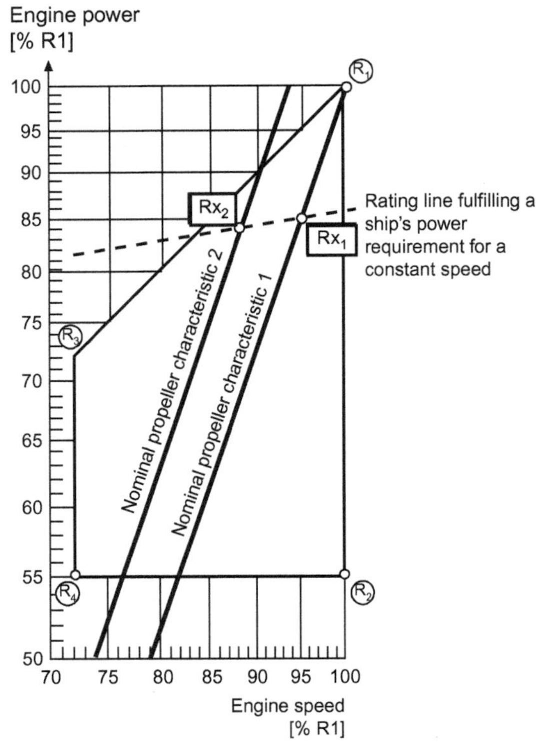

Fig. 16.3 Layout field applicable to all Wärtsilä RTA models. The contracted MCR (R*x*) may be freely positioned within the layout field.

(MCR). This is the maximum power/speed combination available for a particular engine. A 10% overload of this figure is permissible for 1 h during sea trials in the presence of authorized representatives of the engine builder. The point R2 defines 100% speed and 55% power. The point R3 is at 72% power and speed. The line R2–R4 is a line of 55% power between 72% and 100% speed. Points such as R*x* are power/speed ratios for the selection of CMCRs required for individual applications. Rating points R*x* can be selected within the entire layout field.

Sulzer smoothed the path for selecting the optimum model for a given propulsion project with its PC computer-based EnSel engine selection program, which presented a list of all the models that fulfil the power and speed requirements, along with their main data. The input data call for the user to specify the power and speed required, and whether or not an efficiency-booster power turbine system is wanted. The output data then list the engines with the appropriate layout fields and detail their MCR power, speed and specific fuel consumption, main dimensions and weight, and other relevant information.

RTA design features

The RTA design benefited from principles proven in earlier generations of Sulzer R-type engines. The key elements were as follows:

- A sturdy engine structure designed for low stresses and small deflections comprises a bedplate, columns, and cylinder block pretensioned by vertical tie rods. The single-wall bedplate has an integrated thrust block and incorporates standardized large surface main bearing shells. The robust A-shaped columns are assembled with stiffening plates or are of monobloc design. The single cast iron cylinder jackets are bolted together to form a rigid cylinder block (multi-cylinder jacket units for smaller bore engines).
- Lamellar cast iron, bore-cooled cylinder liners with back-pressure timed, load-dependent cylinder lubrication.
- Solid, forged bore-cooled cylinder covers with one large central exhaust valve arranged in a bolted-on valve cage; the valve is made from heat- and corrosion-resistant material and its seat ring is bore cooled.
- Semi-built crankshaft divided into two parts for larger bore engines with a large number of cylinders.
- Running gear comprising a connecting rod, cross head pin with very large surface cross head bearing shells (with high-pressure lubrication) and double-guided slippers, piston rod, and bore-cooled piston crown using oil cooling. All have short piston skirts.

All combustion chamber components were bore cooled, a traditional feature of Sulzer engines fostering optimum surface temperatures and preventing high-temperature corrosion due to high temperatures on one side and sulphuric acid corrosion due to too low temperatures on the other side. At the same time, rigidity and mechanical strength were provided by the cooler material behind the cooling bores (Fig. 16.4).

Comfortable working conditions for the exhaust valve were promoted by hydraulic operation with controlled valve landing speed; air spring; full rotational symmetry of the valve seat, yielding well-balanced thermal and mechanical stresses and deformations of the valve and valve seat, as well as uniform seating; extremely low and even temperatures in valve seat areas due to efficient bore cooling; valve rotation by simple vane impeller; valve

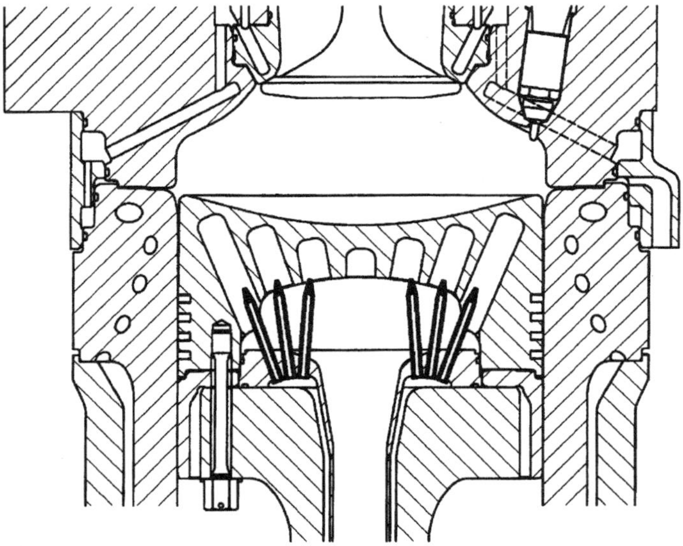

(A)

(B)

Fig. 16.4 The combustion chamber of the RTA series engines is fully bore cooled. Cooling oil spray nozzles on top of the piston rod direct oil into the bores of the piston crown.

actuation free from lateral forces, with axial symmetry and simple guide bushes sealed by pressurized air.

The low exhaust valve seating face temperature secured an ample safety margin to avoid corrosive attack from vanadium/sodium compounds under all conditions. Efficient valve cooling was given by intimate contact with the bore-cooled seat, together with the appropriate excess air ratio in the cylinder. The specific design features of the valve assembly were also said to deter the buildup of seat deposits, seat distortion, misalignment, and other factors which might accelerate seat damage.

- Camshaft gear drive housed in a special double column or integrated into a monobloc column, placed at the driving end or in the centre of the engine for larger bore models with a large number of cylinders.
- Balancer gear mounted on larger bore engines, when required, to counter second-order couples for four-, five-, and six-cylinder models, and combined first- and second-order couples for four-cylinder models.
- A compact integral axial detuner incorporated, if required, in the free end of the engine bedplate.
- The fuel injection pump and exhaust valve actuator were combined in common units for each of the two cylinders. The camshaft-driven injection pump with double valve-controlled variable injection timing (VIT) delivers fuel to multiple uncooled injectors. The camshaft-driven actuators impart hydraulic drive to the single central exhaust valve working against an air spring.
- Constant pressure turbocharging based on high-efficiency uncooled turbochargers; auxiliary blowers support uniflow scavenging during the low load operation. In-service cleaning of the charge air coolers is possible. A standard optional three-stage charge air cooler unit could be specified for heat recovery.
- A standard pneumatic engine control system is based on a remote manoeuvring stand in the control room and an emergency manoeuvring stand on the engine itself. The optional SBC-7.1 electropneumatic bridge control system was matched to the engine and arranged for engine control and manoeuvring from the wheelhouse or bridge wings.
- Standard optional power takeoff drives arranged either at the side or at the free end for connecting an alternator to the main engine.
- Sulzer's efficiency-booster system could be specified for the RTA84C, RTA84T, and RTA84M engines, as well as for the RTA72U, RTA62U, and RTA52U models, to exploit surplus exhaust gas energy in a power recovery turbine.
- RTA engines could satisfy the speed-dependent IMO limits on NOx emissions from exhaust gases. Tuning is facilitated by the electronic variable fuel injection timing (VIT) system.

RTA design developments

The basic RTA engine design was refined over the years and the range expanded to address changing market requirements. The improvements yielded wider power/speed fields, increased power outputs, reduced fuel consumption, lower wear rates, and longer times-between-overhauls (TBOs). Advances in key performance parameters are illustrated in Fig. 16.5. The operating economy has also benefited from the reduced auxiliary power demand of current RTA engines—imposed by electric pumps for cooling water, lubricating oil, and fuel supply—which is some 40% less than that of the RL engines.

RTA-2U series

Sulzer's introduction of upgraded RTA-2 engine designs called for some modifications to match the higher power outputs and maximum combustion pressures involved (summarized in Fig. 16.6). The shrinkfit of the crankshaft was strengthened to suit the increased torque values; the main bearing was adapted to accommodate the higher loads and ensure optimum oil flow, and the cylinder cover material was changed to take advantage of better fatigue strength at higher loads. Some of the design ideas tested on the 4RTX54 research engine and already applied to the RTA84T engine were also adopted for the RTA-2U series.

Three fuel injection valves per cylinder were specified for the RTA62U and RTA72U models (though not, because of its smaller bore size, for the RTA52U model, which retained two valves per cylinder). The reported benefits of the triple-valve configuration are a more uniform temperature distribution around the principal combustion space components (cylinder cover, liner, and piston crown) at the increased maximum combustion pressures, along with even lower temperatures despite the higher loads. Three fuel valves also foster significantly lower exhaust valve and valve seat temperatures.

Other spin-offs from the research engine included a modified cylinder liner bore-cooling geometry whose tangential outlets of the bores aim for optimum distribution of wall temperatures and thermal strains at higher specific loads. The geometry of the oil cooling arrangements of the piston crown was also modified to maintain an optimum temperature distribution. The good piston running behaviour was maintained by retaining established features of the RTA design: multilevel cylinder lubrication, die-casting

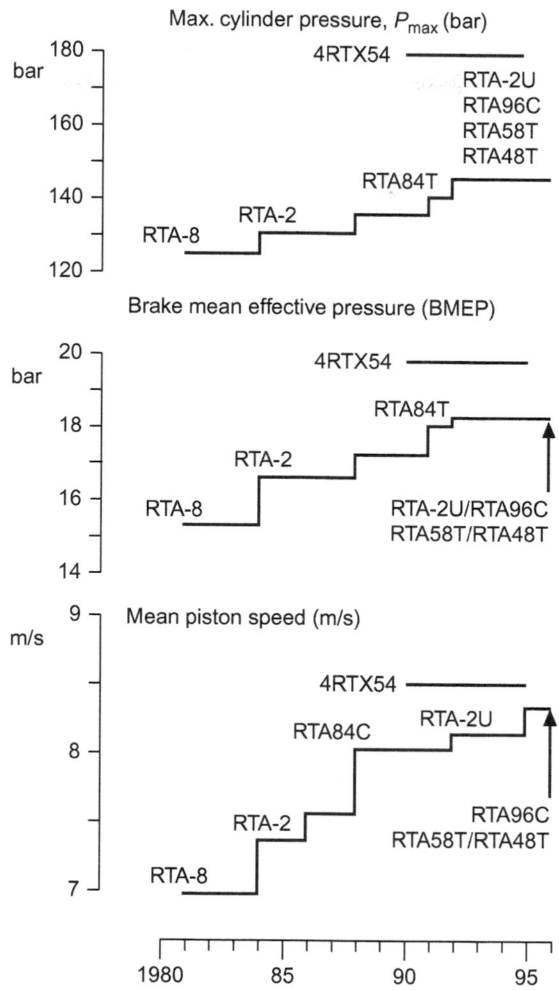

Fig. 16.5 Advances in key parameters of RTA series engines (the RTX54 was a Sulzer research engine replaced in 2008 by the RTX-4 research engine).

technology for cylinder liners and temperature-optimized cylinder liners. Advances in materials technology in terms of wear resistance have permitted engines to run at higher liner surface temperatures. This, in turn, allows a safe margin to be maintained above the increased dew point temperature and thus avoiding corrosive wear.

Some refinements were introduced, however, to match the new running conditions. Four-piston rings are specified for the RTA-2U engines instead of the five previously used. The top ring is now thicker than the other rings

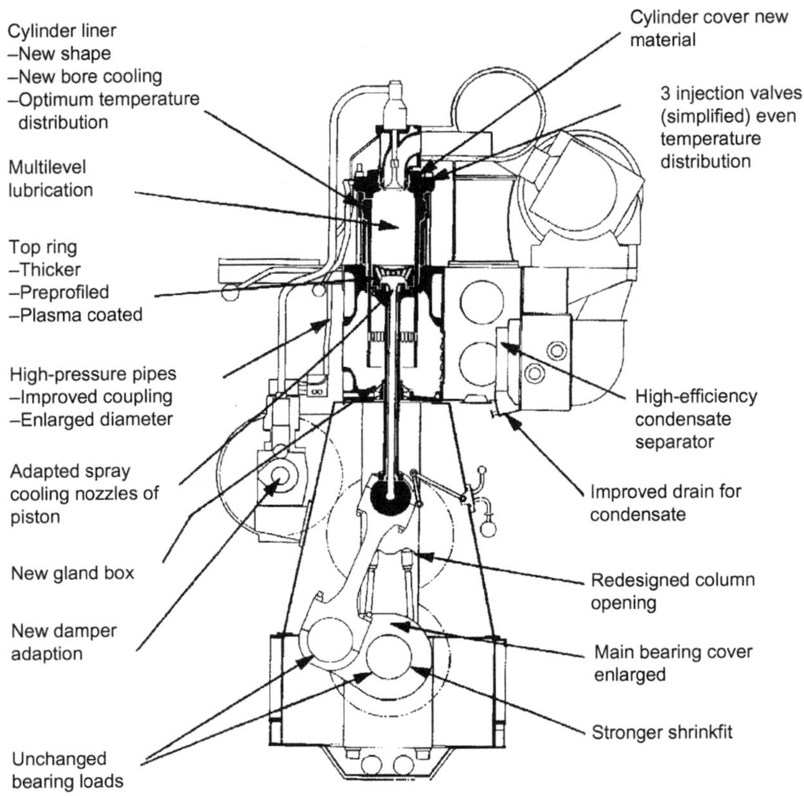

Fig. 16.6 Main improvements introduced in RTA-2U series engines.

and has a preprofiled running face, as on the RTA84C container ship engine. The top ring is also plasma coated. The plasma-coated, preprofiled thicker rings have demonstrated excellent wear results. The radial wear rates were measured at <0.04 mm/1000 h in trials of up to and exceeding 13,000 h of operation.

An important contribution to low wear rates of liners and pistons results from improving the separation and draining of waterborne in the cooled scavenge air before it enters the cylinders, particularly at higher engine loads. The RTA-2U engines were specified with a more efficient condensate water separator in the scavenge airflow after the cooler, along with a more effective drain.

RTA-T series

A number of design modifications to the RTA series were introduced for the ultra-long stroke RTA48T, RTA58T, and RTA68T engines to achieve more compact, lower weight models offering reduced production, installation, and maintenance costs (Figs 16.7 and 16.8). A TBO for the main components of 15,000 h was sought. Cutting the manufacturing cost, despite the greater stroke–bore ratio (4.17) than previous RTA series engines, was addressed by a number of measures: reducing the size and weight of components, simplifying the designs of components and subassemblies and

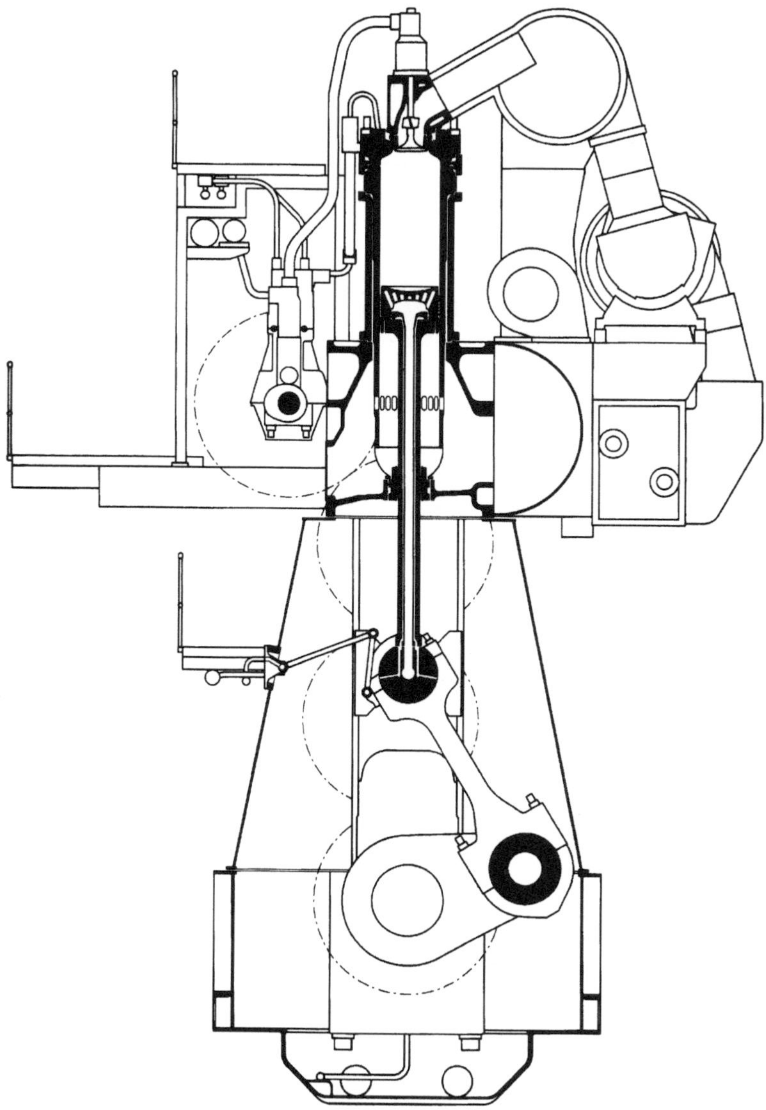

Fig. 16.7 Cross section of RTA58T engine. Note the high camshaft level allowing the use of shorter high-pressure fuel injection pipes for better injection control.

Fig. 16.8 Sulzer 4RTA58 engine on test.

making them easier to produce, reducing the number of parts, and designing to save assembly time.

An example of the design changes made to reduce the sizes and weights of components is provided by the cylinder cover, along with its exhaust valve, housing and valve actuator. An overall weight saving of 30% was achieved on each new cover, largely due to the smaller dimensions. Reducing the distance between cylinder centres by around 9% allowed components such as the bedplate, monobloc columns, and cylinder block to be reduced in size, resulting in weight savings of 13%–14%

(Figs 16.9–16.11). The cylinder block is lower in overall height and thus lighter than in equivalent RTA-2U engines. The freedom for ship designers to create short engine rooms was enhanced by a degree of flexibility in the fore and aft location of the turbocharger and scavenge air cooler module.

The hydraulic jack bolts on the main bearings were eliminated and replaced by simple holding-down bolts that fix the bearing cap directly to the bedplate (Fig. 16.12). In addition, it was possible to simplify the column structure in the region where the jack bolts had needed support and thereby

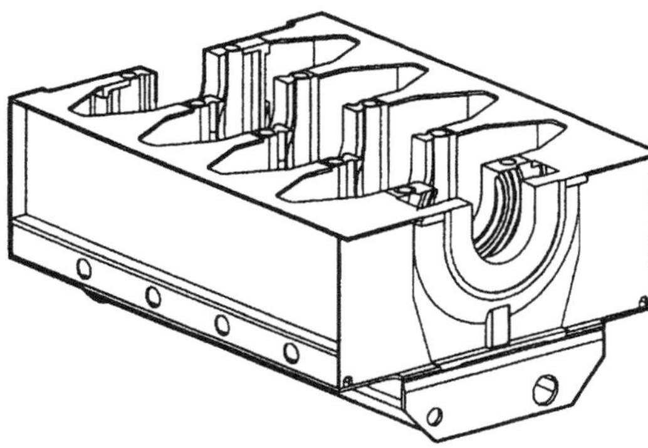

Fig. 16.9 Bedplate of RTA58T engine.

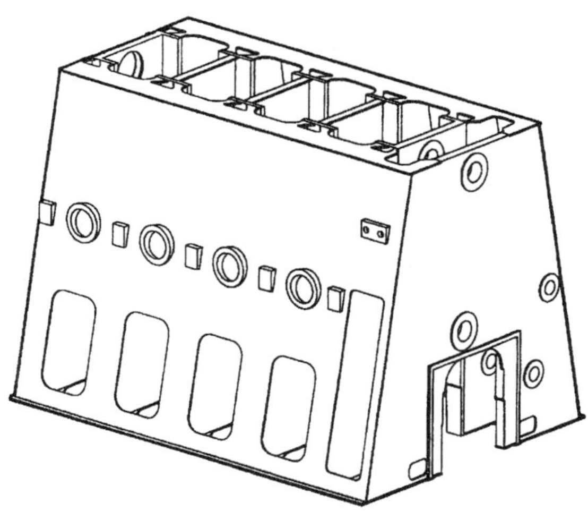

Fig. 16.10 The monobloc columns of the RTA48T and RTA58T engines were designed for simpler welding and machining.

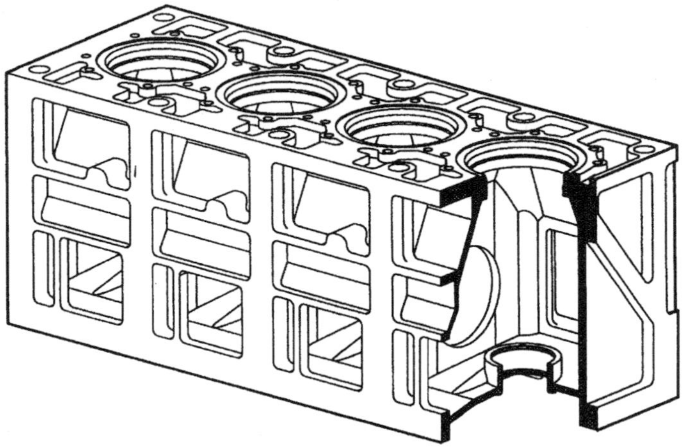

Fig. 16.11 The monobloc dry cylinder blocks of the RTA48T and RTA58T engines were designed for reduced machining, compactness and lower weight.

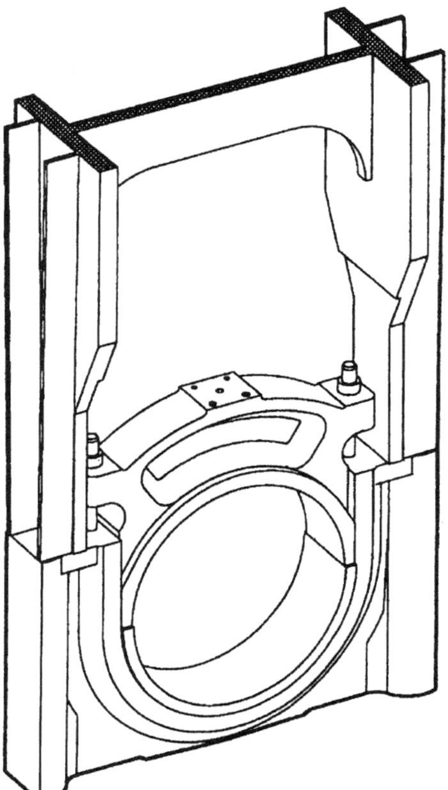

Fig. 16.12 The main bearings of the RTA48T and RTA58T engines use holding-down studs instead of the previous hydraulic jack bolts.

omit machining operations. The scavenge air receiver was simplified by using a 'half-pipe' design, which is welded to the module incorporating the turbocharger and scavenge air cooler.

The result was an easier manufacturing process and significant weight savings compared with equivalent RTA-2U series engines. The single-piece column structure can be manufactured without the need for machining inclined surfaces: all machined surfaces on the columns are either vertical or horizontal by design. Eliminating the jack bolts on the main bearings also allowed the design of the columns to be made more tolerant of welding quality.

The exhaust valve actuator was redesigned for the RTA-T engines. The bush of the actuating piston was eliminated and its function integrated directly into the housing, thus reducing the number of components to be produced and the assembly time. Smaller dimensions and lighter weight (by 27%) were also achieved. A substantial cost saving can be realized if an electrically driven balancer at the forward end of the engine is specified, allowing the previously necessary coupling between balancer and camshaft to be omitted.

Valve-controlled fuel injection pumps are located at the height of the cylinder blocks. The fuel pump blocks can be directly bolted to the cylinder blocks without using any shims. The intermediate gear wheels of the camshaft drive can be aligned more quickly than before by means of a built-in device to move the wheels vertically and horizontally into position.

Engines can be built up from modules, starting with the bedplate and crankshaft, and working up with the columns and cylinder blocks. The modules can be preassembled at convenient, separate locations in the workshop. All pipe connections are designed to support this type of modular assembly and achieve the shortest possible assembly time in the works or, if necessary, in the ship itself.

Piston-running technology was based on experience gained and refined from the RTA-2U and RTA84C engines. The cylinder liner is made of cast iron with sufficient hard-phase content and a smooth machined running surface. Bore cooling and three fuel injection valves underwrite favourable temperature levels and distributions. The top piston ring is preprofiled for easy running-in and plasma coated for low wear rates over extended periods. The multilevel cylinder lubrication system fosters optimum distribution of the lubricating oil and adapts to the longer stroke, and condensate water is taken out of the scavenge air efficiently to avoid contamination of the oil film.

RTA60C engine

A new type added to the programme in mid-1999, the RTA60C engine was designed to serve faster cargo tonnage such as medium-sized container ships, car carriers, Ro-Ro ships, and reefer vessels (Fig. 16.13). A desire for compactness and economical production costs moulded the design, resulting in

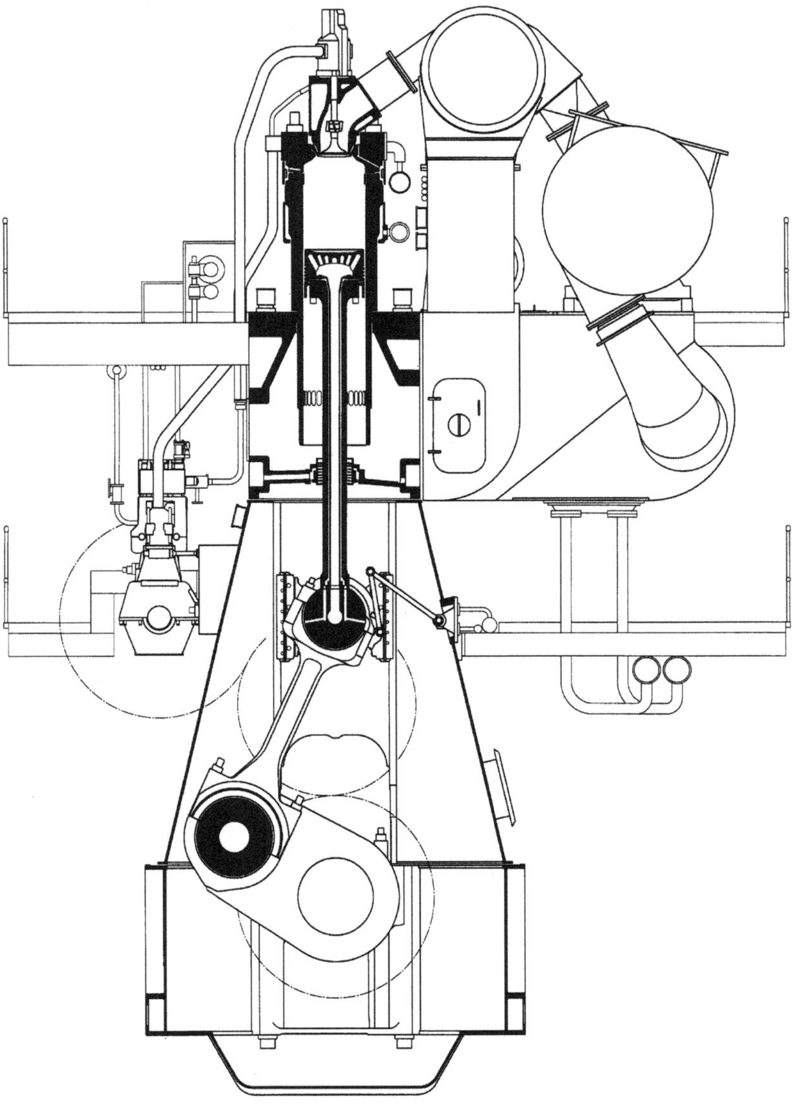

Fig. 16.13 Cross section of RTA60C engine, the first RTA model designed from the start to accept electronic RT-flex systems (camshaft-controlled version illustrated).

an engine with a shorter length and lighter weight (5%–10% lower) for a given power than others in its class.

A higher rotational speed to suit the intended ship types allowed a reduced piston stroke, which, together with the short connecting rod, fostered a significant reduction in engine height: the engine is 8.52 m tall above the shaft centreline and requires a hook height of only 10.4 m for withdrawing pistons (or less by using special tools). The length was minimized by applying measures from the RTA48T and RTA58T engines, a cylinder distance of 1040 mm is achieved. The six-cylinder RTA60C model has an overall length of 7.62 m, including the flywheel and weighs 330 t.

The need to keep engine length to a minimum called for particularly good bearing design; thin-walled white-metal shells are used for cross head, main, and bottom end bearings. The main bearing housing was specifically designed to accommodate and support the thin-walled bearing shell, with four elastic holding-down bolts for each main bearing cap. Two pairs of studs are said to give the most even distribution of holding-down load and also allow the tie rods to be located close to the bearing for efficient transfer of firing pressure loads.

Cylinder covers are secured by eight elastic holding-down studs arranged in four pairs, promoting compactness and contributing to a shorter engine length, and also simplifying manufacture. A support ring between the cylinder block and the collar of the cylinder liner carries both liner and cylinder cover; it also passes cooling water to the cooling bores and to the cover. A simple water jacket provides the necessary cooling for the portion of the liner length immediately below the bore-cooled top flange of the liner.

The cylinder jacket is a single-piece iron casting, its height determined by the space required for the scavenge air receiver. Access to the piston underside is possible from the receiver side of the engine to allow maintenance of the piston rod gland and piston ring inspection. On the fuel side, one door per cylinder can be opened for inspection and to support in-engine work from outside. The tilting-pad thrust bearing is integrated into the bedplate, the pads arranged to ensure a safe and uniform load distribution. The thrust-bearing girder consists of only two steel cast pieces, omitting welding seams in critical corners; the girder is clearly stiffer than in previous designs.

The piston comprises a forged steel crown with a very short skirt; the four-piston rings are all 16-mm thick and of the same geometry. Piston running behaviour benefits from an anti-polishing ring (APR) incorporated at the top of the liner, preventing deposits on the piston top land from damaging the liner running surface and its lubrication film. Keeping the top land clean also ensures a good spread of lubricant over the liner surface while using a lower cylinder oil feed rate. (More details of the APR are provided in the section TriboPack for Extended TBO.)

An exhaust system featuring tangential gas inlet and outlet in the manifold allows a smooth flow of gases from the exhaust valve to the turbocharger inlet with an energy-saving swirl along the manifold. The combined turbocharger and scavenge air cooler module is designed to accommodate one or two turbochargers and allows for different sizes of coolers. The flexibility to locate the turbocharger at the aft (driving end) of the engine is particularly suited to modern all-aft ship designs.

The key operating parameters of the RTA60C (see table) are only slightly higher than those of the B versions of the RTA48T, RTA58T, and RTA68T models, from whose service experience the new engine benefited.

RTA60C engine data

Bore	600 mm
Stroke	2250 mm
Stroke–bore ratio	3.75:1
Output, R1 (mcr)	2360 kW/cyl
Speed range (R1–R3)	91–114 rev/min
Mean effective pressure at R1	19.5 bar
Mean piston speed at R1	8.55 m/s
Maximum cylinder pressure	155 bar
Number of cylinders	5–8
Power range	8250–18,880 kW
Specific fuel consumption	164–170 g/kWh

The RTA60C engine was designed from the outset to smooth its acceptance of Sulzer RT-flex fully electronically controlled fuel injection and exhaust valve systems, which eliminate the camshaft and individual fuel pumps (see the next section). The debut orders for RTA60C engines—in December 2000—specified RT-flex versions.

RTA50C and RT-flex 50C engines

At the end of 2002 Wärtsilä Corporation announced a joint venture with Mitsubishi (now Japan Engine Corporation) to design and develop a new small-bore low-speed engine. The Japanese group, which designs and builds its own UEC two-stroke engines (Chapter 15), had been a prolific licensed builder of Sulzer engines since 1925.

The new 500-mm bore/2050-mm stroke design was planned for release in two versions: the conventional 'mechanical' RTA50C and the electronically controlled RT-flex 50C. Each type offers an MCR of 1620 kW/cylinder at 124 rev/min, five-to-eight-cylinder models covering a power range from 5650 to 12,960 kW at 99 to 124 rev/min. Output and speed ratings target Handymax and Panamax bulk carriers, large product tankers, feeder container ships, and medium-sized reefer vessels. Both versions incorporate TriboPack measures for enhancing piston-running behaviour, underwriting low cylinder wear rates for 3 years between overhauls, and minimizing cylinder lube oil consumption (see p. 423). High efficiency, compactness, and environmental friendliness were set as design goals for a joint working group of engineers from Wärtsilä and Mitsubishi. The first examples of the RT-flex50 engine were tested and type-approved in mid-2005, the design subsequently proving popular in the market.

RTA50C/RT-flex 50C engine data

Bore	500 mm
Stroke	2050 mm
Stroke–bore ratio	4:1
Output (mcr)	1620 kW/cyl
Mean effective pressure at R1	19.5 bar
P_{max}	155 bar
Mean piston speed at R1	8.5 m/s
Cylinders	5–8
Speed range (R1–R3)	124–99 rev/min
Power range	5650–12,960 kW
Specific fuel consumption, full load (R1–R2)	165–171 g/kWh

RTA84T engine refinements

The piston of the RTA84T engine is similar to that of the earlier RTA84M engine but simplified in design. The skirt is bolted directly to the crown, and the increased rigidity of the piston rod can accept higher loads. The crown is oil cooled by the 'jet/shaker' cooling principle, with the oil spray nozzles matched to secure optimum piston temperatures. The piston rod gland design (Fig. 16.14) features bronze rings and a modified scraper/gas sealing package on top compared with previous RTA-series engines. This gland principle confirmed the expected appropriate oil leakage rates in the neutral

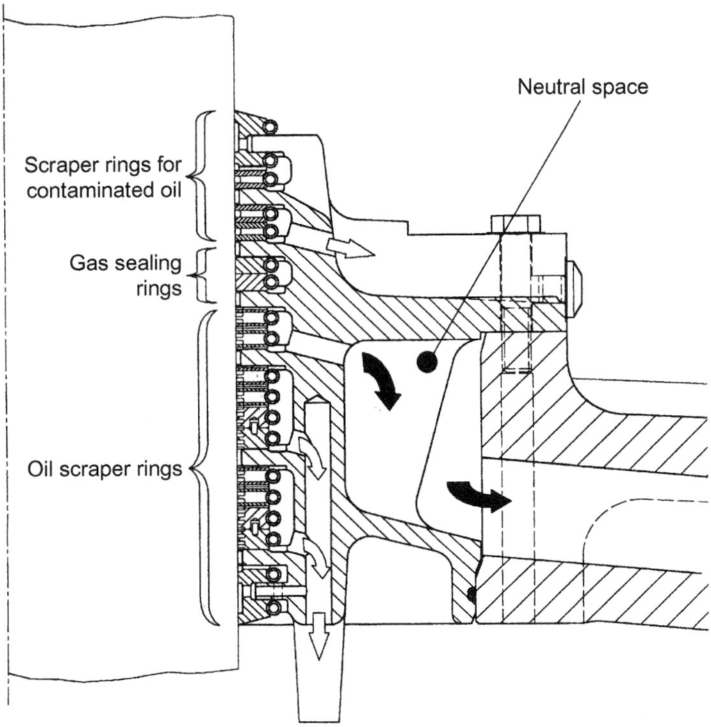

Fig. 16.14 Piston rod gland of RTA84T engine; all rings are made of bronze.

space. The well-proven RTA exhaust valve design was adopted for the RTA84T, embracing an efficiently bore-cooled valve seat, Nimonic valve, hydraulic actuation, and a rotation device.

A deeper cylinder cover, with its lower joint between cover and liner, reduces the portion of the liner exposed to gas pressure and high temperature. The bore cooling principle was adapted to control the temperatures and mechanical and thermal stresses in the components adjacent to the combustion space. The cylinder cover is bolted down with eight studs. Compact uncooled fuel injection valves make it possible to place three nozzles symmetrically in the cover despite the very short cylinder pitch. The nozzle tips are sufficiently long for the cap nut to be shielded by the cylinder cover and hence not exposed to the combustion space (Fig. 16.15).

The camshaft was raised to be close to the cylinder covers and is driven via gear wheels. The hydraulic reversing mechanism of the RTA series was retained unchanged. The decision to move the camshaft up towards the

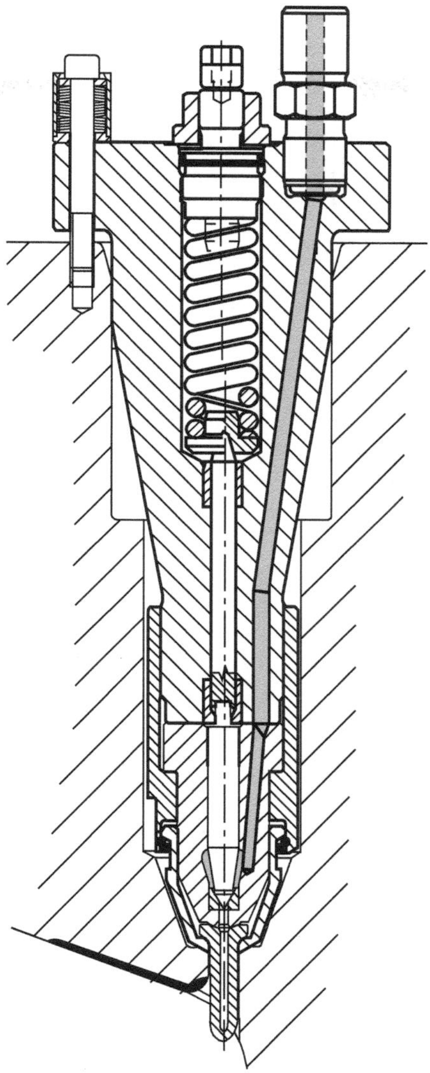

Fig. 16.15 RTA84T fuel injection valve. The nozzle is not exposed to the combustion space, thereby avoiding material burning off.

engine top was taken after evaluating such factors as engine cost, fuel injection system performance, hydraulic valve actuation, and torsional vibration. In particular, with an elongated stroke–bore ratio of 3.75, the advantage of shorter hydraulic connections for injection and valve actuating pipes was significant in avoiding high-pressure losses. The cost of an additional gearwheel (Fig. 16.16) is compensated by the slimmer camshaft, the fuel pump

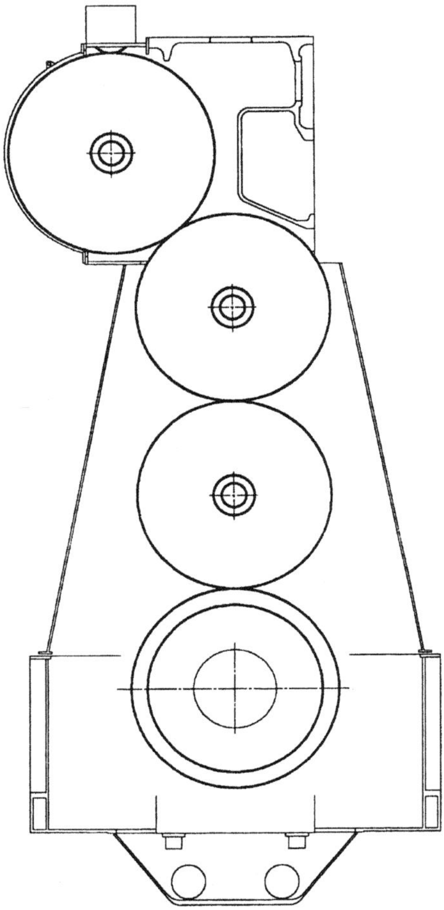

Fig. 16.16 Camshaft drive gear train for RTA84T engine.

arrangement, and the shorter high-pressure injection pipes. Higher injection pressures, and hence more efficient and cleaner combustion, are thus facilitated.

The camshaft-driven fuel pump is of the valve-controlled design used in the longer stroke RTA-2 engines and located at a high level, bolted directly to the side of the cylinder block (Fig. 16.17). The double valve-controlled pump principle, traditional in Sulzer low-speed engines, offers high flexibility and stability over time in controlling the injection timing. Timing is controlled by separate suction and spill valves regulated through eccentrics on hydraulically actuated lay shafts. The pump housing incorporates the

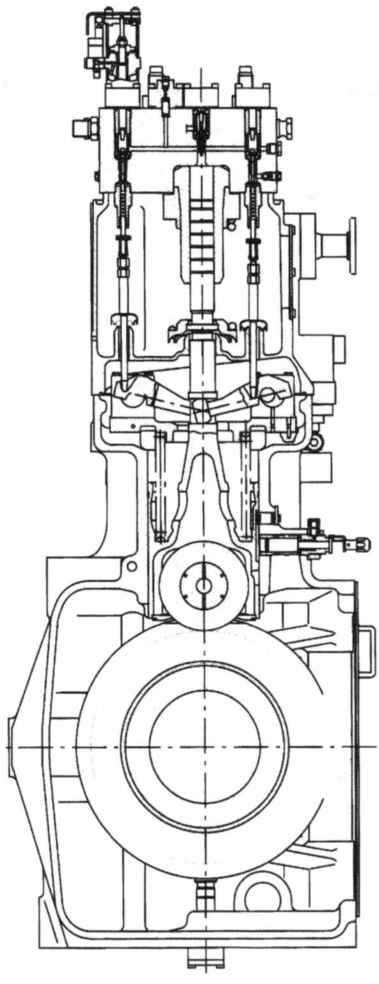

Fig. 16.17 RTA84T fuel injection pump with double control valves.

compact RTA-series arrangement, with a fuel injection pump and exhaust valve actuator modules provided to serve one pair of cylinders.

In comparison with a helix type, the valve-controlled fuel injection pump has a plunger with a significantly greater sealing length. The higher volumetric efficiency reduces the torque in the camshaft. Wärtsilä says additionally that injection from a valve-controlled pump is far more stable at very low loads and rotational shaft speeds down to 15% of the rated speed are achieved. Valve control also offers benefits in reduced deterioration of timing over the years owing to less wear and to freedom from cavitation.

A combination of variable exhaust valve closing (VEC) and VIT devices provides an improved degree of setting flexibility. In the upper load range, the specific fuel consumption is optimized through the electronically controlled VIT system, maintaining the maximum cylinder pressure by injection timing advance. The fuel quality setting (FQS) function is incorporated in the VIT system. Engine efficiency is additionally improved in the lower load range (between 80% and 65% load) via the VEC system (Fig. 16.18). The compression ratio is thereby effectively increased by the early closing of the exhaust valve.

The working process of the engine can thus be 'shaped' for optimum performance over the whole load range, facilitating its adaptation to a particular ship sailing mode. The resulting benefits promised are the lowest possible fuel consumption in the most common load range, and higher exhaust gas temperatures within that load range compared with other solutions,

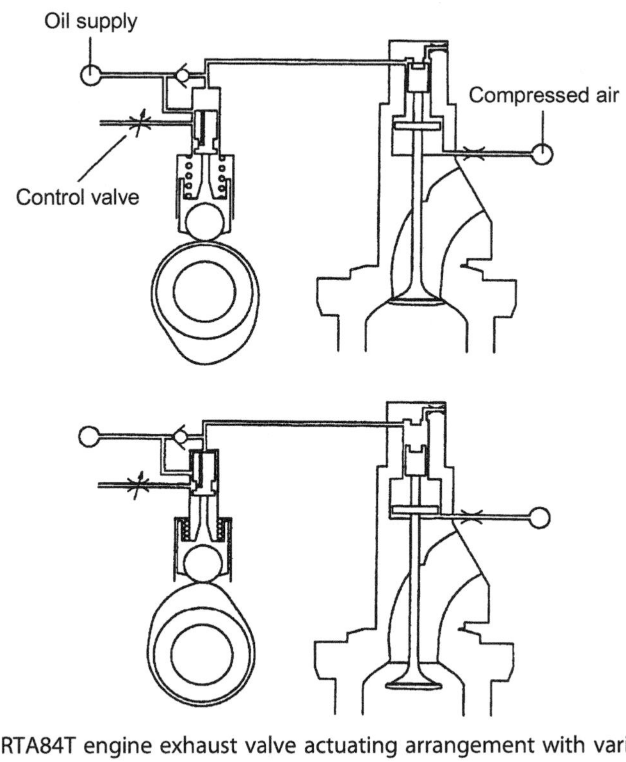

Fig. 16.18 RTA84T engine exhaust valve actuating arrangement with variable closing (VEC) system.

such as variable turbocharger geometry. (Higher exhaust gas temperatures are valued for waste heat recovery.) The VEC, VIT, and FQS devices are electronically regulated from the engine control system. The influence of the VIT/VEC combination on the RTA84T engine's performance characteristics is shown in Fig. 16.19.

A load-dependent cylinder liner cooling system (Fig. 16.20) helps to avoid cold corrosion of the liner surface over the whole load range and

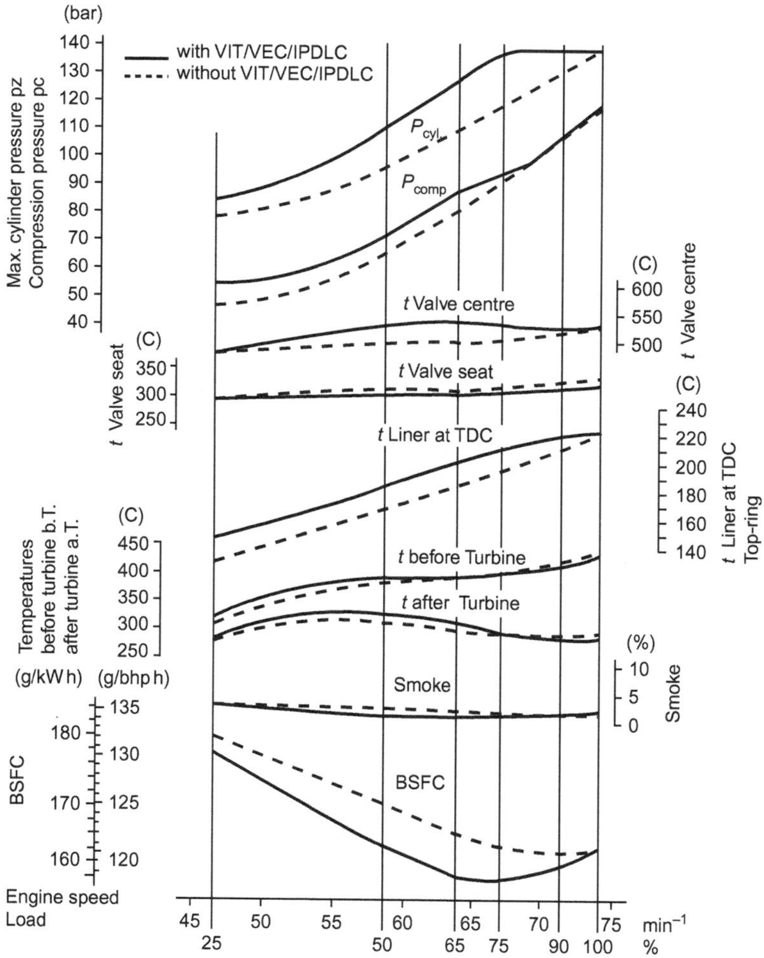

Fig. 16.19 Influence of the VIT/VEC combination (solid line) compared with results without VIT/VEC (dashed line) on the performance characteristics of a 7RTA84T engine. The engine has an RI rating of 27,160 kW at 74 rev/min and exploits an exhaust gas power turbine.

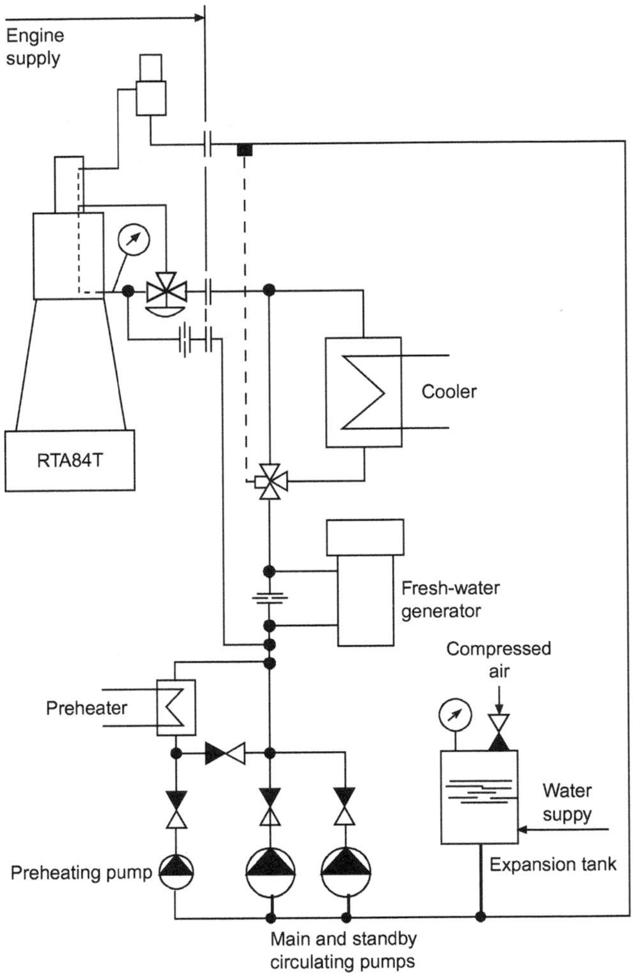

Fig. 16.20 Schematic layout of the RTA84T cylinder jacket cooling water system with a load-dependent controlled bypass past the jacket to the cylinder cover.

hence to reduce wear rates further. The liner temperature is maintained above the prevailing dew point throughout the load range: the liner is efficiently cooled at full load, while over-cooling at part load is avoided. This is achieved by splitting the cooling water supply to the engine into two lines. Only a restricted amount of water is led through the cylinder liner jacket at part load while the main water flow is led directly to the cylinder cover. The water flow is electronically controlled as a function of the engine load.

Cylinder liner lubrication is affected by Wärtsilä's multilevel accumulator system. This allows the cylinder lubricating oil to be distributed in very small amounts at each engine stroke, thereby creating an optimum oil film distribution. The main improvement for the RTA84T was the introduction of an electronically controlled flexible oil dosage promoting low wear rates and low lubricating oil consumption. The oil feed rate is controlled according to engine load and is adjustable as a function of engine condition. A simplified dual-line distributor eliminates the mechanical drive shaft.

Piston-running experience with RTA84T engines in service demonstrated that engines with very long strokes need higher liner wall temperatures than had previously been required by the RTA-2 series (with stroke–bore ratios of around 3.5) to overcome corrosive wear problems when operating on high sulphur fuels. This experience was applied to benefit later RTA84T production models as well as the design of the smaller RTA-T engines. A number of improvements introduced for the latter series have also been applied to the RTA84T, the refinements including design simplifications to make the engine more manufacturing friendly. A significant saving in weight also resulted, the seven-cylinder model being trimmed by some 6.5% to 960 t.

RTA84T-B and RTA84T-D versions

Following the introduction of the RTA48T and RTA58T models in 1995, the same design concepts were applied to the RTA84T 'Tanker' engine—originally introduced in May 1991—resulting in a Version B in 1996. Easier manufacturing and enhanced service behaviour were realized, with no change in power output. In July 1998, a lower specific fuel consumption (down by 2 g/kWh) was gained by applying 'low port' cylinder liners (scavenging air inlet ports with a reduced height) in combination with higher efficiency turbochargers. There is no penalty in either higher component temperatures or too low exhaust gas temperatures, and the low ports give a longer effective expansion stroke in the cycle (Fig. 16.21).

The RTA84T-B engine was uprated at the end of 1998 from 3880 kW/cylinder at 74 rev/min to 4100 kW/cylinder at 76 rev/min to create the Version D through tuning and turbocharging matching measures. The power available from a seven-cylinder engine was thereby increased from 27,160 to 28,700 kW, better addressing the propulsive demands of VLCCs with higher service speeds. Nine-cylinder RTA84T-D models, each developing

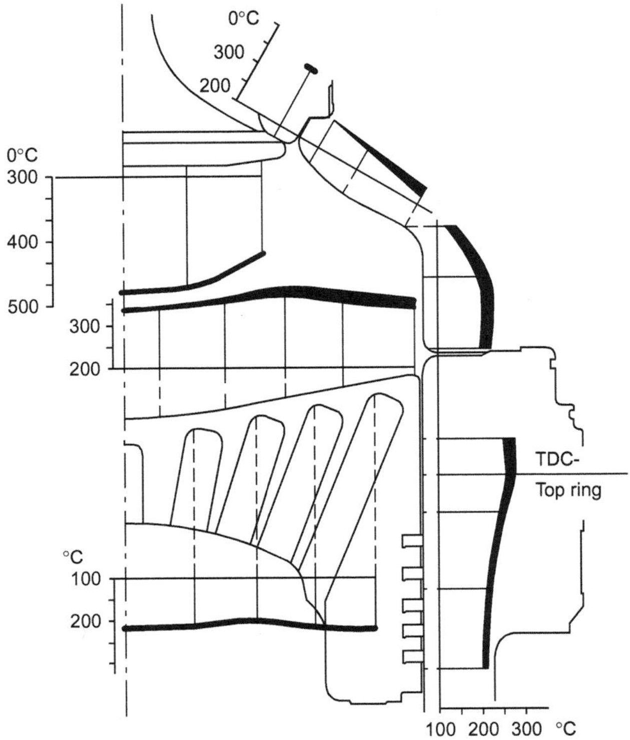

Fig. 16.21 Surface temperatures measured on the combustion chamber components of the RTA84T-B engine at full-load R1 rating. The thickness of the lines represents the circumferential variation in temperature.

36,900 kW, were specified to power a series of 442,500 dwt tankers with a service speed of 16.5 knots.

Structural, running gear, combustion chamber, fuel injection, turbocharging, and scavenge air system design details are similar to those of the RTA84C and RTA96C engines described later. The large stroke-to-bore ratio of the RTA84T, however, allows a relatively deeper combustion chamber with more freedom in the layout of the fuel spray pattern. The semibuilt crankshaft had to cater for the longest stroke ever applied in a Sulzer engine; to limit the crankshaft weight for production, assembly and transport, the main journals and crankpins are bored. The design of the crank has a good transverse width at the upper part of the web, allowing the latter to be slim longitudinally.

The favourable torsional vibration characteristics allow six-cylinder engines to use a viscous damper in many cases instead of a Geislinger damper.

RTA84T-B and RTA84T-D engine data

	RTA84T-D	RTA84T-B
Bore (mm)	840	840
Stroke (mm)	3150	3150
Stroke–bore ratio	3.75	3.75
Power (mcr) (kW/cyl)	4100	3880
Cylinders	5–9	5–9
Speed range (rev/min)	61–76	59–74
Mean effective pressure (bar)	18.5	18
Mean piston speed (m/s)	8	7.77
Maximum cylinder pressure (bar)	144	140
Specific fuel consumption (g/kWh)	160–168	160–168

RTA68T-B engine

A further development of the RTA-T programme resulted in the 680-mm bore RTA68T-B engine, the first example of which entered service in November 2000. The design benefited from all the reliability improvements incorporated in the earlier RTA-T engines, with parameters similar to those of the smaller bore RTA48T-B and RTA58T-B models. TriboPack design measures (see later) contributed to piston-running behaviour, and a new design of piston rod gland featured oil scraper rings with grey cast iron lips: the system oil is fully recirculated and a dry neutral space effected.

RTA68T-B engine data

Bore	680 mm
Stroke	2720 mm
Stroke–bore ratio	4
Power (mcr)	2940 kW/cyl
Cylinders	5–8
Speed range	75–94 rev/min
Mean effective pressure	19 bar
Mean piston speed	8.5 m/s
Maximum cylinder pressure	150 bar
Specific fuel consumption	161–169 g/kWh

RTA84C and RTA96C engines

Introduced in September 1988 for propelling the coming generation of larger and faster container ships, the RTA84C engine was developed from the RTA84, already established in that market. In 1993 its power output was increased by 6%, the cylinder cover modified and the number of fuel nozzles increased from two to three; these measures contributed to a reduction in the thermal load of the combustion chamber. The performance of the series, shown in the following table, equated to a maximum output of 48,600 kW at 102 rev/min from the 12-cylinder model.

RTA84C engine data

Bore	840 mm
Stroke	2400 mm
Stroke–bore ratio	2.86
Power (mcr)	4050 kW/cyl
Cylinders	6–12
Speed range	82–102 rev/min
Mean effective pressure	17.9 bar
Mean piston speed	8.2 m/s
Maximum cylinder pressure	140 bar
Specific fuel consumption	166–171 g/kWh

Increasing sizes of post-Panamax container ships dictated more powerful engines and stimulated the introduction of the RTA96C series in December 1994. The 960-mm bore design was fully based on the RTA84C (Figs 16.22 and 16.23), which it supplemented and later replaced in the programme. The RTA84C series was further superseded by the launch at the end-2005 of the 82-bore design.

The selection of a stroke (2500 mm) some 100 mm longer than that of the RTA84C engine for the RTA96C design (Fig. 16.24) was influenced by demands for the highest reliability. By adopting a longer stroke, the absolute depth of the combustion chamber could be proportionally increased to give more room for securing the best combustion and fuel injection parameters, and better control of temperatures in the combustion chamber components. In addition, the slightly longer stroke fosters a simplified crankshaft design with enhanced reliability, since the shrunk-in main journals do not cut the journal fillets at the inner sides of the crank webs.

A TBO of 3 years from the RTA96C engine's key components was sought, a goal underwritten by a cast iron (preferably die cast) cylinder liner with the necessary amount of wear-resistant hard-phase particles and a smoothly machined and honed surface for quick and trouble-free running-in; bore cooling of all the main combustion chamber components; three fuel injection valves symmetrically distributed in the cylinder cover to

Fig. 16.22 A 12-cylinder RTA84C engine on the test.

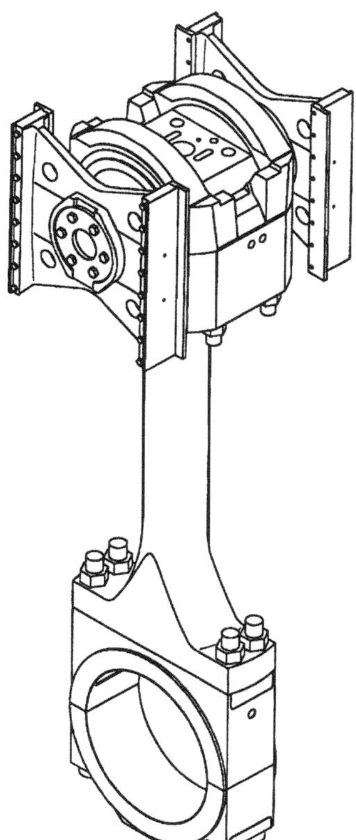

Fig. 16.23 Cross head and connecting rod assembly for RTA84C engine.

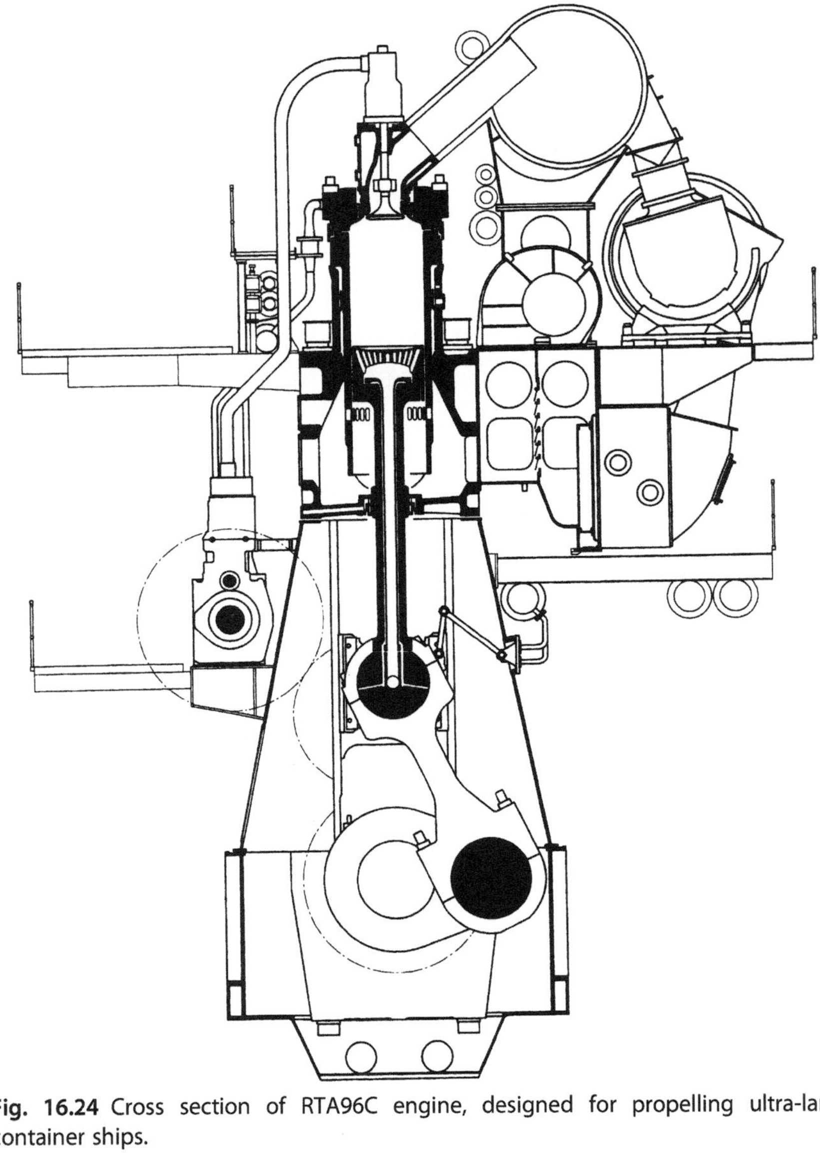

Fig. 16.24 Cross section of RTA96C engine, designed for propelling ultra-large container ships.

contribute to the evenness of temperature distribution; cylinder oil lubrication of the liner surface via two levels of quills to achieve effective and economical distribution; and a top piston ring preprofiled and plasma coated to secure the lowest wear rate to reach the 3-year TBO goal with sufficient margin. Diametral cylinder liner wear rates of around 0.03 mm/1000 h are reported for the RTA96C design, in service. All-new engines further benefit from TriboPack measures to enhance piston-running behaviour (see p. 423).

The semi-built crankshaft comprises combined crankpin/web elements forged from a solid ingot, with the journal pins then shrunk into the crank web. The main bearings have white-metal shells, and the main bearing caps are held down by a pair of jack bolts in the RTA84C and by a pair of elastic holding-down studs in the RTA96C. The cross head bearing is designed to the same principles as for all other RTA engines; it also features a full-width lower half bearing. The cross head bearings have thin-walled shells of white metal to yield a high load-bearing capacity. Wärtsilä low-speed engines retain a separate elevated-pressure lube oil supply to the cross head; this provides hydrostatic lubrication, which lifts the cross head pin off the shell during every revolution to ensure that sufficient oil film thickness is maintained under the gas load: crucial to long-term bearing security.

The combustion chamber was recognized as the most important design area because of its key influence on engine reliability and the high power concentration. The component design was based on established practice and benefited from work carried out for the medium-bore RTA48T and RTA58T engines. Wärtsilä suggests that bore-cooling technology provides an escape from the rule that larger components (resulting from a larger bore) when subjected to thermal loading will also have higher thermal strains. With bore cooling, the thermal strains in the cylinder cover, liner, and piston crown of the RTA96C can be kept fully within the values of earlier generations of RTA engines, as can the mechanical stresses in those components.

The solid forged steel, bore-cooled cylinder cover is secured by eight elastic studs, and the central exhaust valve of Nimonic 80A material is housed in a bolted-on valve cage. Anti-corrosion cladding is applied to the cylinder covers downstream of the injection nozzles to protect the covers from hot corrosive or erosive attack. The pistons feature a forged steel crown with a short skirt; the crown is cooled by combined jet-shaker oil cooling achieving moderate temperatures on the crown and a fairly even temperature distribution across the crown surface. No coatings are necessary.

High structural rigidity with low stresses and high stiffness is important for low-speed engines. The RTA84C and RTA96C designs exploit a well-proven structure with a 'gondola'-type bedplate surmounted by very rigid A-shaped double-walled columns and cylinder blocks, all secured by pretensioned vertical tie rods. Both bedplate and columns are welded fabrications designed for minimum machining. The cylinder jacket is assembled from individual cast iron cylinder blocks bolted together to form a rigid whole. The fuel pumps are carried on supports on one side of the column and the scavenge air receiver on the other side of the cylinder jacket.

Access to the piston under-side is normally from the fuel pump side but is also possible from the receiver side of the engine, to facilitate the maintenance of the piston rod gland and also for inspecting piston rings. The tilting-pad thrust bearing is integrated into the bedplate. The use of gear wheels for the camshaft drive allows the thrust bearing to be very short and stiff, and to be carried in a closed rigid housing.

The three uncooled fuel injection valves in each cylinder cover have nozzle tips sufficiently long for the cap nut to be shielded by the cylinder cover and not exposed to the combustion space. The camshaft-driven fuel injection pumps are of the double-valve controlled type, traditional in Sulzer low-speed engines. Injection timing is controlled by separate suction and spill valves regulated through eccentrics on hydraulically actuated lay shafts. Flexibility in timing is possible through the VIT system for improved part-load consumption, while the FQS lever can adjust timing according to the fuel oil quality. The valve-controlled fuel injection pump, in comparison with a helix type, has a plunger with a significantly greater sealing length. The higher volumetric efficiency reduces the torque in the camshaft; in addition, injection from a valve-controlled pump is far more stable at very low loads, and rotational shaft speeds down to 15% of the rated speed are achieved. Valve control also has the benefits of less deterioration of timing over the years owing to reduced wear and freedom from cavitation.

The camshaft is assembled from a number of segments, one for each fuel pump housing. The segments are connected through SKF sleeve couplings, each segment having an integral hydraulic reversing servomotor located within the pump housing. The camshaft drive is a traditional Sulzer arrangement, affected in this case by three gearwheels housed in a double column located at the driving end or in the centre of the engine. The main gearwheel on the crankshaft is in one piece and flange mounted.

Scavenge air is delivered by a constant pressure turbocharging system based on one or more turbochargers, depending on the number of engine cylinders. For starting and during slow running, scavenge air delivery is augmented by electrically driven auxiliary blowers. The scavenge air receiver incorporates nonreturn flaps, an air cooler, and the auxiliary blowers; the turbochargers are mounted on the receiver, which also carries the fixed foot for the exhaust manifold. Immediately after the cooler, the scavenge air passes through a water separator comprising a row of vanes that divert the airflow and collect the water. Ample drainage is provided to completely remove the condensed water collected at the bottom of the air cooler

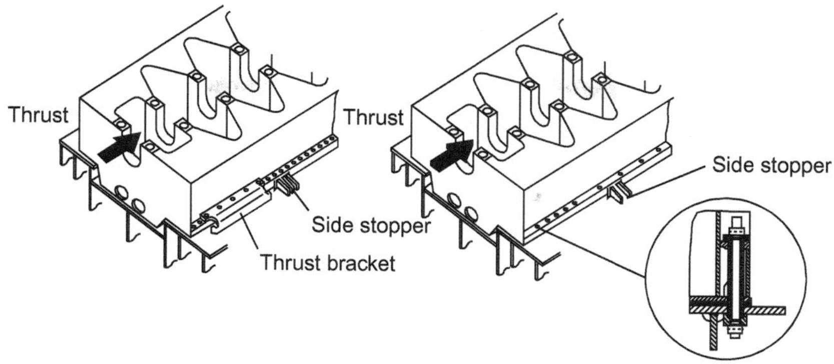

Fig. 16.25 Arrangements for transmitting propeller thrust to the RTA84T, RTA84C, and RTA96C engine seatings; the inset shows the thrust sleeve for the thrust bolts.

and separator. Effective separation of condensed water from the stream of scavenge air is thus accomplished, a necessity for satisfactory piston-running behaviour.

RTA engines have simple seating arrangements with a modest number of holding-down bolts and side stoppers (14 side stoppers are needed for a 12-cylinder RTA96C engine). No end stoppers or thrust brackets are needed as thrust transmission is provided by fitted bolts or thrust sleeves applied to a number of the holding-down bolts (Fig. 16.25). The holes in the tank top for the thrust sleeves can be made by drilling or even flame cutting. After alignment of the bedplate, epoxy resin chocking material is poured around the sleeves. The engine is equipped with an integrated axial detuner at the free end of the crankshaft, and a detuner monitoring system developed by Wärtsilä is standard equipment.

RTA96C uprating and programme expansion

The first RTA96C engine, an 11-cylinder model, started tests in March 1997 and was followed into service by 8-, 9-, and 10-cylinder models and numerous 12-cylinder versions. The design was originally offered with an output of 5490 kW/cylinder and up to 12 cylinders, but the rating was increased by 4% in the year 2000 to 5720 kW/cylinder at 102 rev/min (see the following table). At the same time, the programme was extended to embrace an in-line 14-cylinder model delivering 80,080 kW (the first time such a low-speed engine configuration had featured in a production programme) and suitable for powering single-screw post-Panamax container ships with the capacities of 10,000 TEU-plus and service speeds of up to 25 knots (Fig. 16.26).

The first-ever 14-cylinder low-speed engines (RT-flex96C models) were ordered in early 2005 to provide a series of 8600 TEU container ships with a service speed of 27 knots. The original RT-flex96C crankshaft had sufficient torque capacity for 14 cylinders, but the material was upgraded for this application to enable an increased shrink fit for a greater design margin. The thrust bearing structure in RT-flex96C engines with a mid-gear drive was also revised to reduce deformations and stresses even with the increased thrust in a 14-cylinder model. A common-rail fuel injection system is beneficial to the 14-cylinder engine, which has a mid-gear drive for the fuel and servo oil supply pumps. Two identical rail units are thus installed, each serving seven cylinders.

A 14RTA96C engine measures 27.31-m long overall to the flywheel flange × 10.93-m high over the shaft centreline (or 13.54-m overall) and weighs 2300 t dry (compared with 2050 t for a 12RTA96C). The individual elements, such as the two crankshaft sections, are close to the maximum lifting capacities of contemporary engine builders, which impose a limit on engines with more than 14 cylinders. The camshaft for engines with more than 14 cylinders would need to be split into three parts, and at least two camshaft drives would become necessary. Such engines, if needed, would, therefore, have to adopt the RT-flex common-rail systems for fuel injection and exhaust valve actuation, which eliminate the camshaft (see p. 431).

With a length <15% longer than the established 12-cylinder engine, the longitudinal and torsional rigidity of the 14RTA96C model is considered adequate for the expected ship structures. The torsional vibration characteristics are also considered acceptable for the envisaged firing order.

RTA96C engine uprating data

	Original rating	Revised rating
Bore (mm)	960	960
Stroke (mm)	2500	2500
Stroke–bore ratio	2.6	2.6
Power (mcr) (kW/cyl)	5490	5720 (6030)

RTA96C engine uprating data—cont'd

	Original rating	Revised rating
Speed (rev/min)	100	102
Cylinders	6–12	6–12, 14
Mean effective pressure (bar)	18.2	18.6 (19.6)
Mean piston speed (m/s)	8.3	8.5
Maximum cylinder pressure (bar)	142	145
Maximum output, 12-cycle (kW)	65,880	68,640
Maximum output, 14-cyl (kW)	–	80,080 (84,420)
Specific fuel consumption, full load (g/kWh)	163–171	163–171

Note. A subsequent power uprating raised the output of the RTA96C/RT-flex96C designs to 6030kW/cylinder at 102rev/min on a mean effective pressure of 19.6bar, enabling a 14-cylinder model to offer 84,420kW. The 13- and 14-cylinder models in the programme are only available in electronically controlled RT-flex versions.

RTA96C service experience

Some cases of main bearing damage were restricted to a specific series of container ships and involved the local breaking-out of white metal in the lower shells and some local fretting. The fretting problem was solved by applying a so-called key-slot solution with the aim of increasing the radial forces pressing the bearing shell into the girder and by some other small design modifications. With the exception of one particular main bearing, no connection could be found with the bearing load and shaft orbit that could explain the breaking-out problem. As a countermeasure, the lower shell of the main bearing next to the thrust bearing is now machined together with its respective bearing cap to ensure better bearing geometry. The more precise alignment of the bearing cap and shell, along with a key between the cap and the shell to secure tight fitting of the shell in the bearing girder when tightening down, gradually yielded results.

Some reports of cracks in A-frame columns were first received in mid-1999 after some 5000–12,000 running hours. A project team was established to investigate the matter and all other engines in service were inspected. Although the cracks occurred in seven of the eight engines, they were in only some of the double-walled A-frame columns manufactured to the original design and were of various degrees of severity. The cracks were initiated at the inboard ends of two pairs of horizontal ribs on the columns and propagated vertically and in the direction of the cross head guide rails. The root of the problem was a combination of a lack of welding quality and somewhat higher than anticipated local stress level.

All cracked A-frame columns were repaired by welding. At the same time, the inboard ends of all horizontal ribs in the engines concerned were cut back to limit the stress concentration and thereby eliminate the possibility of cracking. Subsequent measurements showed that the modified shape of the ribs reduces the stress concentrations in the columns down to around half, and thus is not as sensitive to imperfect welding. The same shape of the horizontal rib is now applied on all RTA96C engines, whether repaired or built as new; the shape was adapted from the modified rib design introduced in December 1997 but slightly revised for ease of manufacture.

Some leaks were experienced in the cylinder pressure relief valve located in the cylinder cover. These were caused by corrosion, which arose because the shortened valve was too cold through being too far from the combustion space. An improved valve design replaced those in existing engines, while later engines were fitted with valves of a traditional design which operate at higher temperatures and hence avoid the original problem.

A number of scavenge air receivers suffered cracks in the form of rupturing of the upper and lower longitudinal welded seams; these were found to have been caused by resonant vibration in the dividing wall. As a remedy, additional stiffening was applied to the dividing walls of the affected receivers (and adopted as the standard for new engines) to change the resonant frequency and thereby reduce vibration amplitudes.

Stiffer turbocharger support was developed for new engines after vibration had caused problems in the supports of one of the turbochargers of the first 10-cylinder RTA96C engine. Excessive transverse movement of the exhaust manifolds on the first such models was also recorded. This was remedied by stiffening the manifold support in the transverse direction, and thus halving the manifold movements. Excessive movements of the manifolds also led to the breakage of cooling water discharge pipes; the adoption of flexible pipes of a proprietary make as standard solved the problem.

A number of broken cylinder cover studs have been noted, mostly broken at the first thread at the top but some fractured at the bottom. The lower fractures were initiated by corrosion because of improper sealing; the upper fractures resulted from materials and thread qualities beyond the specification. The design was changed to a wasted stud and a special nut to reduce stress levels and ensure a good safety margin.

TriboPack for extended TBO

The TBO of low-speed marine diesel engines is largely determined by the piston running behaviour and its effect on the wear of piston rings and cylinder liners. Addressing this, Sulzer introduced a package of design measures in 1999, which are now standard on all new RTA engines and retrofittable to existing engines. The TriboPack technology enables the TBO of cylinder components, including piston ring renewal, to be extended to at least 3 years and also allows a further reduction in the cylinder lubricating oil feed rate. The design measures (Fig. 16.26) are as follows:

- Multilevel cylinder lubrication
- Fully and deep-honed cylinder liner with sufficient hard phase
- The careful turning of the liner running surface and deep honing of the liner over the full length of the running surface
- Mid-stroke liner insulation and, where necessary, insulating tubes in the cooling bores in the upper part of the liner
- Preprofiled piston rings in all piston grooves
- Chromium–ceramic coating on the top piston ring
- Running-in coating (RC) piston rings in all lower piston grooves
- Antipolishing ring at the top of the cylinder liner
- Increased thickness of the chromium layer in the piston ring grooves.

A key element of TriboPack is the cylinder liner manufactured in cast iron, which needs a controlled hard-phase content and the best grain structure in the running surface for both good strength and running behaviour. Careful machining followed by full deep honing to remove all damaged hard phase from the liner surface reportedly delivers an ideal running surface for the piston rings, together with an optimum surface microstructure. Wärtsilä

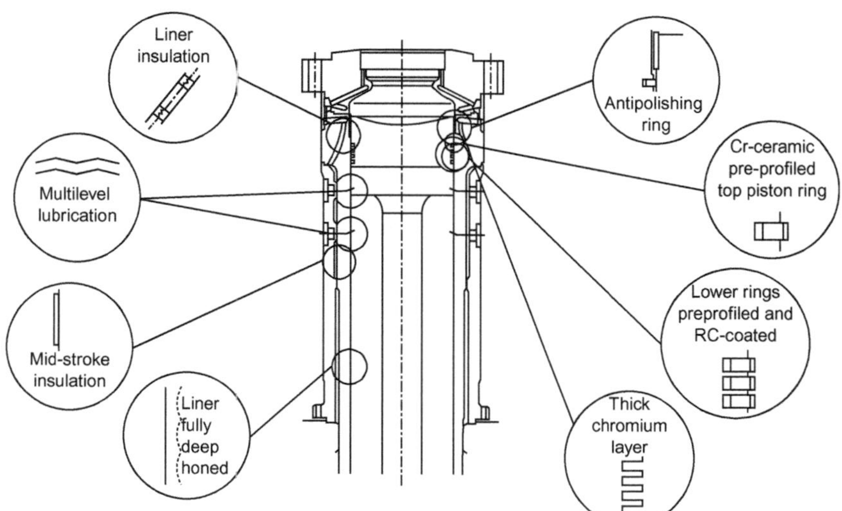

Fig. 16.26 TriboPack design measures for improving piston-running behaviour.

says that deep honing of the full liner running surface is a prerequisite for maximizing the benefits of TriboPack. Its experience has shown that plateau honing of a wave-cut liner is not adequate because, once the plateau is worn down, the rings run on liner metal whose hard phase structure was damaged during machining. This damaged hard phase must be removed by deep honing.

Pistons have four rings, all having the same thickness. The *chrome–ceramic top ring*, proven in Wärtsilä four-stroke engine practice, has a cast iron base material. The running face is profiled and coated with a layer of chromium as a matrix into which a ceramic material is trapped. High operational safety and low liner and ring wear have been demonstrated, with much better resistance to scuffing than any other ring material. Good performance is conditional, however, on using the chrome–ceramic rings in conjunction with a deep-honed liner. The other piston rings have a running-in and anti-scuffing coating, which fosters a safe and swift running-in of the engine when the liners are deep honed.

The APR prevents the buildup of deposits on the top land of the piston, which can damage the lube oil film on the liner and cause bore polishing. Deposit buildup can be heavy in some engines, especially those running on very low sulphur content fuel oil (<1% sulphur) combined with an excessive cylinder lube oil feed rate. If such deposits are allowed to accumulate, they inevitably touch the liner running surface over a large part of the piston stroke. The lube oil film can then be wiped off, allowing metal-to-metal contact between the piston rings and the liner; in the worst case, there can be scuffing.

Applied as standard for some years on Wärtsilä four-stroke engines, the thin alloy steel APR is located in a recess at the top of the liner and has an internal diameter less than the cylinder bore to reduce the clearance to the piston top land. It does not need to be specifically fixed, as the thermal expansion of the hot ring keeps it tightly in place. The steel material was selected to ensure and maintain a high safety margin against thermal yielding. Excessive deposits are scraped off the piston top land at every stroke while they are still soft, thus preventing hard contact between the deposit and the liner wall surface. The oil film on the liner wall remains undisturbed and can fulfil its function. The APR also stops the upward transportation of new lube oil by the layer of deposits to the top of the liner where it is burned instead of being used for lubrication. The ring is thus effective in allowing the lube oil feed rate to be kept down to recommended values.

Load-dependent cylinder lubrication is provided by a *multilevel accumulator system*, the lubricating pumps driven by frequency-controlled electric motors. On the cylinder liner, oil distributors bring oil to the different oil accumulators. For ease of access, the quills are positioned in dry spaces instead of in the way of cooling water spaces.

It is also important that the liner wall temperature is adapted to keep the liner surface above the dew point temperature over the whole of the piston stroke to avoid cold corrosion and maintain good piston-running conditions. The upper part of the liner is bore cooled with cooling water passing through tangential drillings in the liner collar. The mid-stroke region of the liner is cooled by a water jacket, and only the lower part is uncooled. There is often a tendency for liner temperatures to be too low, thus leading to corrosive wear from the sulphuric acid formed during combustion. Wärtsilä applies two insulating techniques to secure better temperature distributions. For some years, PTFE insulating tubes have been fitted in the cooling bores of the liner. As part of TriboPack, the liner is now also insulated in the mid-stroke region by a Teflon band on the waterside. The insulating tubes are adapted according to the engine rating to ensure that the temperature of the liner running surface is kept above the dew point temperature of water over the full length of the stroke and over a wide load range.

Mid-stroke insulation and, where necessary, insulating tubes are therefore used to optimize liner temperatures over the piston stroke. An insulation bandage in the form of Teflon bands with an outer stainless steel shell is arranged around the outside of the liner to raise liner wall temperatures in the mid-stroke region. Mid-stroke insulation is known to be particularly useful for sustained engine operation at low power outputs, while the TriboPack gives an additional safety margin in abnormal operating conditions (e.g. against excessive carbon deposits built up on the piston crown).

While trying to avoid corrosive wear by optimizing liner wall temperatures, it is necessary to keep as much water as possible out of the cylinders. Highly efficient vane-type water separators fitted after the scavenge air cooler and effective water drain arrangements are thus vital for good piston running behaviour. Load-dependent cylinder lubrication is provided by the multilevel accumulator system, which ensures the timely quantity of lube oil for good piston running. The lube oil feed rate is controlled according to the engine load and can also be adjusted according to the engine condition.

Large bore liner wear problems

TriboPack measures resulted in cylinder liner wear rates generally <0.05 mm/1000 h while using lube oil feed rates of 1–1.4 g/kWh, and the wear rates of both top and other piston rings were satisfactory for a 3-year TBO. The lower wear rates allowed the guide rates for cylinder lubrication to be reduced from 1.4 to 1.1 g/kWh for all sizes of RTA and RT-flex engines built to the TriboPack standard. The excellent results did not extend to the largest bore RTA96C and RT-flex96C models, however, which suffered what Wärtsilä described as 'unsatisfactory numbers' of liner and piston ring failures through sudden severe wear.

Further R&D resulted in other design measures, which demonstrated that piston-running behaviour could be improved in these large-bore engines as well. One measure was to introduce an underslung scavenge air receiver design as standard for the RTA96C and RT-flex96C models. With its high-efficiency separator and more efficient drain system, this arrangement considerably improves the efficiency of condensate water removal from the scavenge airflow between the air cooler and the engine cylinders. The receiver design has reportedly proved successful, with virtually no water being carried over to the cylinders, even under high humidity tropical conditions.

In the underslung air receiver, the scavenge air coolers are arranged horizontally beneath the turbochargers. Scavenge air is then swung round 180° in its flow to the cylinders, in the process passing through the vertically arranged water separators in the underslung trunks from the air coolers to the scavenge air receiver. There are ample drainage provisions to completely remove the condensed water collected at the bottom of the separator.

In the cylinder itself, the new design measures for the largest bore engines include lower liner wall temperatures, chromium–ceramic preprofiled rings in all piston grooves, plateau honing of the liner, the omission of oil distribution grooves on the liner surface, bronze rubbing bands on a shorter piston skirt, and improved geometry of the APR. Features such as multilevel cylinder lubrication and chromium plating of piston ring grooves are retained. Attention was given to the stability of the lube oil film on the liner running surface. The lube oil distribution grooves were omitted because it was found that they disturbed the oil film more than it benefits from the oil distribution.

It had become clear that polishing of the liner surface by deposits on the top land of the piston crown must not occur if liner scuffing is to be avoided. An improved APR geometry introduced as standard features a smaller clearance between the ring and the piston crown, and two edges to clean the piston top land during both up and down strokes of the piston.

Another source of disturbance for the oil film is the rubbing of the piston skirt on the liner's running surface. A new shortened skirt with a ground surface was therefore introduced, equipped as standard with two bronze rubbing bands. Such features contribute to reduced friction and hence lower risk of oil film disturbance. A reduced liner wall temperature is possible with chromium–ceramic piston rings, with a positive effect found on the lube oil film stability in RTA96C and RT-flex96C engines. Wärtsilä reports that extended shipboard tests with cylinder liners having lower wall temperatures returned excellent results. Liner wear of 0.05 mm/1000 h confirmed that wear is not increased with the above-mentioned ring pack at the lower temperatures.

Pulse jet cylinder lubrication

Piston-running behaviour was further enhanced from 2006 with the availability of Wärtsilä's new pulse jet lubricating system (Figs 16.27 and 16.28),

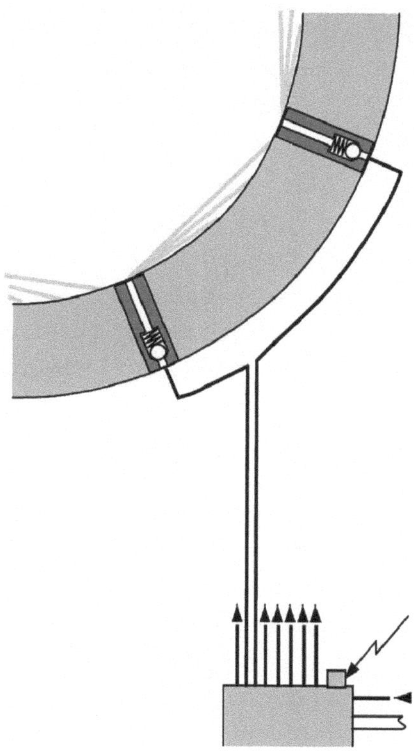

Fig. 16.27 Schematic diagram of Wärtsilä's pulse jet lubricating system, whereby pressurized cylinder lube oil is delivered to multiple jets around the liner.

Fig. 16.28 Arrangement of multiple lube oil jets in the cylinder liner, indicating the pattern of distribution with Wärtsilä's pulse jet lubricating system.

designed to ensure improved cylinder oil distribution on the liner's running surface through precise feed timing and dosage. The system, which could be specified for new RTA and RT-flex engines or retrofitted to engines in service, allows much lower lube oil feed rates to be adopted. Cylinder oil is sprayed onto the liner surface from a single row of quills arranged around the liner, each quill having a number of nozzle holes.

Eight quills (simple and reliable nonreturn valves) are specified for RTA96C and RT-flex96C engines, for example, each quill having five oil jets. A total of 40 lubricating points is thus provided on the liner surface, the jets individually directed to separate points on the surface. There is no atomization and no loss of lube oil to the scavenge air. Cylinder oil is delivered to the quills by a lubricator pump powered by the engine's pressurized system oil. A lubricating module is provided for each cylinder with the associated dosage pump and monitoring electronics.

Metered quantities of oil are delivered under pressure into the piston ring package, from where it is evenly distributed around the liner circumference. The feed rate and timing are electronically controlled via a solenoid valve at the lubricator pump. Full flexibility is provided in the setting of the lubricator timing point, and volumetric metering ensures constant spray patterns across the engine load range. The dosage is precisely regulated even for low feed rates.

Piston rod glands

TBOs of cross head engines are partly defined by the piston rod glands, in the sense that their removal for the exchange of elements is often connected to a withdrawal of the piston and piston rod assembly. The gland elements and piston rods, therefore, need to have a long-life expectation (TBO of 3 years or more). At the same time, they have to assure sealing of the crankcase from the piston underside, limit contamination of crankcase system oil by combustion residues, and keep the oil consumption at a reasonable level for maintaining oil quality.

Recent improvements have introduced additional gas-tight top scraper rings, stronger springs for the other scraper rings, enlarged drain channels for the scraped-off oil, and the exclusive application of bronze scraper rings on fully hardened rods. The drain quantities from the neutral space were reduced by a factor of three. In addition, the scraped-off oil is reusable without any treatment and therefore can be directly fed back internally in the gland box to the crankcase. System oil consumption figures were significantly reduced. The design of the gland box housing was modified, allowing it to be dismantled either upwards during piston overhaul or downwards without pulling the piston.

Complete retrofit packages available for all RTA engines in service comprise the newly designed upper scraper, new middle sealing and new lower scraper groups, and some modifications on the gland box housing. The upper scraper group consists of a two-piece housing with newly designed oil scraper rings made of bronze, newly designed gas-tight sealing rings, and modified tension springs. The oil scraper rings consist of four segments conforming better to the piston rod. Two new seal rings in three parts and adapted tension springs were introduced for the middle seal group. For the lower scraper group, all rings are made of bronze, since Teflon has an inferior performance when there is an increased amount of hard particles in the oil residues coming from the piston underside. Here, the new scraper rings comprise three slotted segments for adaptability to the piston rod; they are provided with grooves at the top to promote draining of the scraped oil.

Wärtsilä stresses that the actual surface condition and shape of the piston rod is of paramount importance. Ideally, the new glands should be used with hardened piston rods. Existing rods can be retained, however, providing their surface condition and geometry is acceptable, before introducing any new stuffing box elements. If the rod is worn down, roughened, or otherwise surface damaged, it can be ground to standard diameters of 2 or 4 mm undersize and then surface hardened.

Exhaust valve behaviour

The exhaust valve is subjected to hot gases, and the temperature resistance of its seat and body is therefore crucial. Nimonic valves combined with proper seat cooling have yielded excellent service behaviour and long lifetimes. When the RTA96C engine was introduced, and its shallow combustion space created difficult conditions for combustion chamber components, some exhaust valves were additionally coated with Inconel alloy. After limited running times, however, there was some cracking of the coating originating from the centre hole, with loosened material, making removal by grinding necessary. Noncoated valves, on the contrary, showed excellent performance, remaining free of cracks after over 14,000 running hours, without any loss of material. Today's standard, therefore, is the noncoated valve.

Piston crowns

Burning-off of material on piston crowns is very dangerous as hole formation leads to direct contact of the combustion flame with the piston cooling oil system and dire consequences. The use of the combined shaker and jet cooling system in RTA engines assures piston crown temperatures below 400 °C and thus eliminates such burning.

Condition-based maintenance

A new tool for managing the condition-based maintenance of two-stroke engines combines the expert knowledge and wide experience of Wärtsilä Corporation engineers. The CBM Management tool is offered in three different solutions, depending on the extent that customers want to outsource this activity.

A wide range of products for diesel engine monitoring, trend analysis, diagnosis, and management already support condition-based maintenance. CBM Management takes these a step further, with the aim of optimizing the balance between long TBOs, low spare part costs, less time off-hire, and high reliability from the engine. Wärtsilä appreciated that although customers use the same engine types in their fleet, they experience quite different maintenance costs and varying engine reliability. This arises from various factors that influence the condition of the engine, such as the quality of fuel, lube oil and spare parts, and the quality of routine inspections and overhauls.

Analysis of cases of extensive damage to engine components had shown that long before the damage occurred it would have been possible to foresee that something was going wrong. With the appropriate expert knowledge and systematic analysis of available information, it is often possible to predict problems and to prevent expensive maintenance costs. Wartsila's solution was to collect the expert knowledge of engineers from its technical service organization worldwide and store it in a common database.

Some data are specific to an individual engine: shop trial results, sea trial results, electronically stored information from general service reports made by Wärtsilä and performance data collected onboard. All of these were combined with the database of expert knowledge, enabling the resulting expert system to automatically generate advice on the condition of the specific engine. The CBM Management tool comprises a data collector (hardware and software) and an expert system (hardware and software).

The *data collector* is a hand-held computer whose main purpose is to enable the ship's engineers to collect data in a standardized way. It is automatically loaded with engine performance data from the control system through a standard interface but manually fed with maintenance and inspection data. For systematic and efficient manual input, the data collector contains a structured inspection guide for the engine, supported by interactive templates for measurement records and performance sheets. This guide exists for the following 12 function groups: piston performance, piston rod gland condition, scavenge airflow, combustion performance, fuel injection pumps, camshaft, engine structure, gears and wheels, bearings, crankshaft, engine control system, and vibration dampers.

The data collector takes into account different conditions of important components such as piston rings, cylinder liners, and bearings. A picture gallery helps the ship's engineers to understand the terminology used. The data collector reduces inspection time and partly replaces the necessary paperwork through the automatically created templates.

The *expert system* is the heart of the tool and runs on a PC using data from the data collector. There are three steps from collected data to real expert advice: calculate trends, assess engine condition by comparing the data with a weighted set of reference cases based on the expert knowledge of Wärtsilä engineers and assess the condition of the engine components and parts, and offer advice to the ship's engineers on achieving optimized TBOs. The system supports the operator with graphs, pictures, trends, and text recommendations. Communication is possible between ships, the operator's headquarters and Wärtsilä's CBM centre in all directions.

Three different solutions are offered to suit the individual requirements of owners and operators:

1. *CBM virtual expert.* The operator performs data collection and analysis, and Wärtsilä's expert knowledge is given as far as possible without further involvement. The onboard engineers carry out data collection, the expert system being used on the ship as well as at the operator's HQ. Such a solution typically suits the needs of owners and operators who have their own technical experts.
2. *CBM report agreement.* The operator collects the data and Wärtsilä undertakes the analysis. The owner or operator uses only the data collector, while the data are analysed at Wärtsilä's two-stroke CBM centre in Switzerland, which sends the customer a regular report. This type of agreement typically suits owners and operators without their own technical office.
3. *CBM inspection agreement.* Wärtsilä carries out both data collection and analysis. Wärtsilä service engineers visit the ship in port, collect the data, and send them to the CBM centre, from where the customer receives a report. Such a solution suits owners and operators without a technical office and who want to outsource inspections.

RT-flex electronic engines

Wärtsilä's RT-flex system, offered as an option for most models in the RTA programme, resulted from a project originated by Sulzer in the 1980s to develop an electronically controlled low-speed engine without the constraints imposed by the mechanical drive of the fuel injection pumps and exhaust valve actuation. Traditional jerk-type fuel injection systems combine pressure generation, timing, and metering in the injection pump with only limited flexibility to influence the variables. In contrast, Wärtsilä's common-rail system separates the functions and gives far more flexibility for optimizing the combustion process with injection and valve timing.

RT-flex engines are essentially the same as their standard RTA equivalents but dispense with the camshaft and its gear drive, jerk-type fuel injection pumps, exhaust valve actuator pumps, and reversing servomotors. Instead, they are equipped with common-rail systems for fuel injection and exhaust valve actuation, and full electronic control of these functions (Figs 16.29 and 16.30).

The following benefits for operators are cited for the RT-flex system:

- Smokeless operation at all running speeds.
- Lower steady running speeds (in the range of 10%–12% nominal engine speed) obtained smokelessly through sequential shut-off of injectors while continuing to run on all cylinders. Very steady running at 7 rev/min has been demonstrated.
- Reduced running costs through lower part-load fuel consumption and longer TBOs.
- Reduced maintenance requirements, with the simpler setting of the engine; the 'as new' running settings are automatically maintained.
- Reduced maintenance costs through precise volumetric fuel injection control, leading to extendable TBOs. The common-rail fuel system and its volumetric control yields excellent balance in engine power developed between cylinders and between cycles, with precise injection timing and equalized thermal loads.

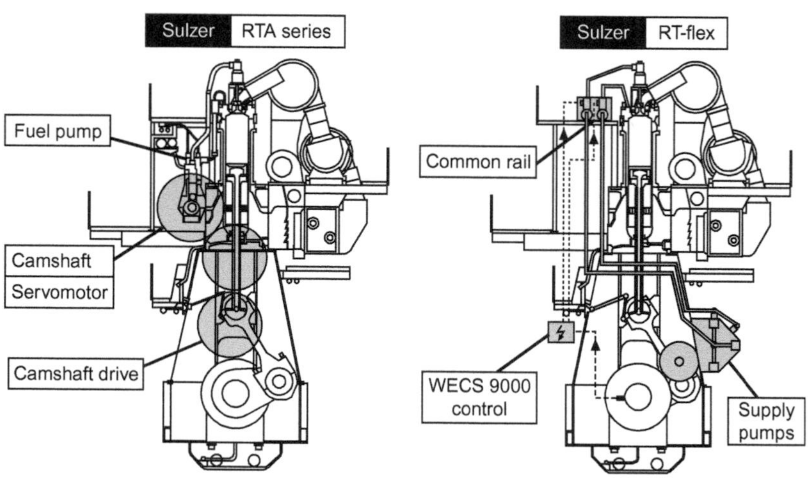

Fig. 16.29 The fuel pumps, valve actuator pumps, camshafts, and drive train of the standard RTA engine are replaced by a compact set of supply pumps and common-rail fuel system on the RT-flex engine.

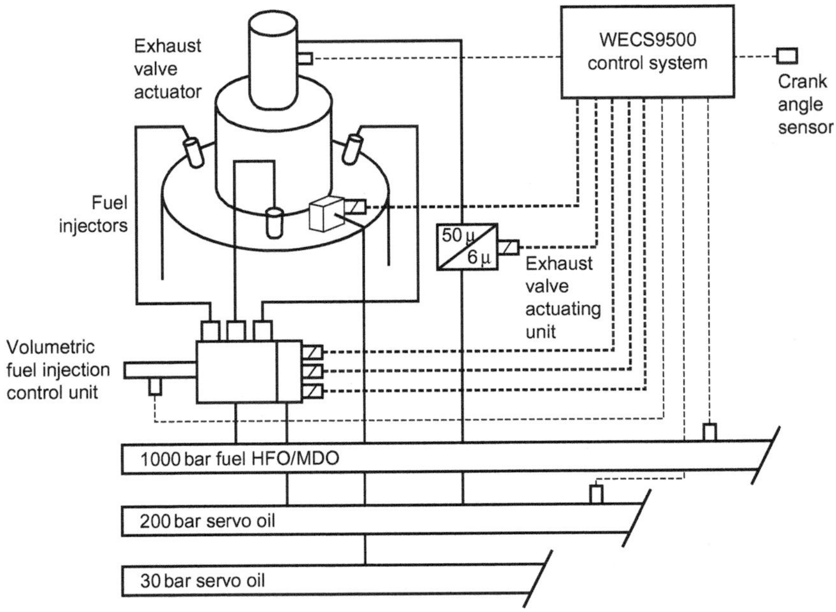

Fig. 16.30 RT-flex electronically controlled common-rail systems.

- Reliability underwritten by long-term testing of common-rail system hardware and the use of fuel supply pumps based on proven Sulzer four-stroke engine fuel injection pumps.
- Higher availability resulting from integrated monitoring functions and from built-in redundancy: full power can be developed with one fuel pump and one servo oil pump out of action. High-pressure fuel and servo oil delivery pipes, and electronic systems, are also duplicated.
- A reduced overall engine weight: approximately 2 t per cylinder lower in the case of a 580-mm-bore RT-flex engine compared with its conventional RTA counterpart.

The common rail for fuel injection is a manifold running the length of the engine at just below the cylinder cover level; the rail and other related pipework are arranged on the top engine platform with ready accessibility from above (Fig. 16.31). The common rail is fed with heated fuel oil at high pressure (nominally 1000 bar) ready for injection into the engine cylinders. The fuel supply unit embraces a number of high-pressure pumps mechanically driven from the crankshaft and running on multi-lobe cams, which increase their supply capacity and hence reduce the number needed. A four-pump set is sufficient for a six-cylinder RT-flex 58T-B engine. The

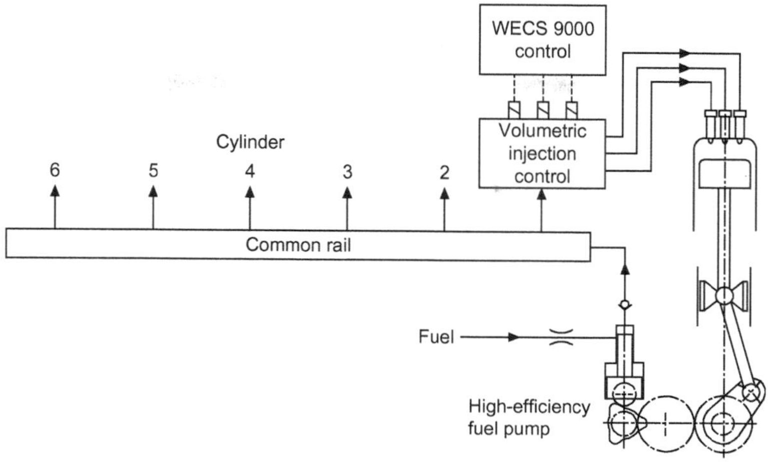

Fig. 16.31 Schematic layout of Wärtsilä RT-flex common-rail fuel system.

pump design, based on fuel injection pumps used in Sulzer four-stroke engines, has suction control to regulate the fuel delivery volume according to engine requirements (Fig. 16.32).

Heated fuel oil is delivered from the common rail through a separate injection control unit for each engine cylinder to the standard fuel injection valves, which are hydraulically operated in the usual way by the high-

Fig. 16.32 Three of the six fuel oil pumps for a Wärtsilä 7RT-flex 60C engine.

pressure fuel. The control units, exploiting quick-acting solenoid rail valves, regulate the timing of fuel injection, control the volume of fuel injected, and set the shape of the injection pattern. The three fuel injection valves in each cylinder cover are separately controlled so that, although they normally act in unison, they can also be programmed to operate separately as necessary. The key features of the common-rail system are defined as follows:

- Precise volumetric control of fuel injection, with integrated flow-out security
- Variable injection rate shaping and free selection of injection pressure
- Stable pressure levels in common rail and supply pipes
- Possibility for independent control and shutting-off of individual fuel injection valves
- Ideally suited for heavy fuel (up to 730 cSt at 50°C) through a clear separation of the fuel oil from the hydraulic pilot valves
- Proven standard fuel injection valves
- Proven high-efficiency common-rail fuel pumps

The fuel injection pressure can be freely selected up to more than 1000 bar over the whole load range. In combination with different injection patterns, this provides the opportunity to optimize the engine in several ways: for example, for low emission levels or improved fuel efficiency at less than optimum loads.

The injection system can be adapted to different patterns such as preinjection, with a small part of the fuel charge injected before the main charge; triple injection, with the fuel charge, injected in three separate short sprays in succession; and sequential injection, with individual actuation of the fuel injection nozzles so that injection timing is different for each of the three nozzles in a cylinder. Different shapes of cylinder pressure profile during the engine cycle can thus be created, which, with a free selection of the rail pressure, allows the optimum pattern to be selected in each case for the loads and performance optimization target of the engine.

Selective shut-off of single injectors is valuable for low manoeuvring speeds or 'slow steaming' as this facility fosters better injection and atomization of the small quantities of fuel needed. In such modes, the common-rail system is controlled to use the three injection valves in sequence. Regulated by an electronic governor, the RT-flex engine demonstrated very steadily running at the lowest speed of 7 rev/min.

Exhaust valves are operated in much the same way as in conventional RTA engines by a hydraulic 'pushrod', but with the actuating energy coming from a servo oil rail at 200 bar pressure. The servo oil is supplied by hydraulic pumps mechanically driven from the same gear train as the fuel supply pumps. An electronically controlled actuator unit for each engine cylinder—operated by hydraulic pressure from the servo oil rail—gives full flexibility for valve opening and closing timing. Two redundant sensors inform the WECS-9500 control system (see below) of the current position of the exhaust valve.

Lube oil from the engine is used as servo oil to keep the system simple and compatible. Before entering the servo oil circuit, the oil is directed through an additional 6-μm filter with an automatic self-cleaning device to ensure reliability and a long lifetime of the actuator units and solenoid valves.

All functions of the RT-flex system are controlled and monitored through the integrated Wärtsilä WECS-9500 electronic control system. This modular system has separate microprocessor control units for each cylinder, with overall control and supervision by duplicated microprocessor control units, which provide the usual interface for the electronic governor and remote control and alarm systems.

The full-load efficiency of RT-flex engines is the same as their conventional RTA engine equivalents, but improvements in part-load fuel economy are gained. This results from the freedom allowed in selecting the optimum fuel injection pressure and timing, and exhaust valve timing, at all engine loads or speeds, while maintaining efficient combustion at all times, even during dead slow running. Similar freedom in exhaust valve timing allows the RT-flex system to keep the combustion air excess high by earlier closing as the load/speed is reduced. Such a facility is not only beneficial for fuel consumption but also limits component temperatures, which normally increase at low load. Lower turbocharger efficiencies at part load normally result in low excess combustion air with fixed valve timing.

Another contribution of the RT-flex system to fuel economy cited by Wärtsilä is the capability to easily adapt the injection timing to various fuel properties influencing poor combustion behaviour. VIT overload had been a traditional feature of Sulzer low-speed engines for many years, using a mechanical arrangement primarily to keep the cylinder pressure high for the upper load range. This is much easier to arrange in an electronically controlled engine.

Environmental performance: A very wide flexibility in optimizing fuel injection and exhaust valve processes enables RT-flex engines to comfortably meet IMO limits on NOx emissions. The most visible benefit cited is smokeless operation at all ship speeds, underwritten by superior combustion.

The common-rail system allows the fuel injection pressure to be maintained at an optimum level irrespective of the engine speed. In addition, at very low speeds, individual fuel injectors are selectively shut off and the exhaust valve timing adapted to help keep smoke emissions below the visible limit. In contrast, engines with traditional jerk-type injection pumps have increasing smoke emissions as engine speed is reduced because the fuel injection pressure and volume decrease with speed and power, and they have no means of cutting off individual injection valves and changing exhaust valve timing (Fig. 16.33).

As all settings and adjustments within the combustion and scavenging processes are made electronically, future adaptations are possible simply through changes in software, which could be easily retrofitted to existing RT-flex engines. A possibility is to offer different modes for different emissions regimes. In one mode, the engine would be optimized for minimum fuel consumption while complying with the global NOx limit, and then, to satisfy local emissions regulations, the engine could be switched to an alternative mode for even lower NOx emissions while the fuel consumption is allowed to rise.

RT-flex engines in service

The RT-flex system was first applied to a full-size research engine in June 1998 at Wärtsilä's facilities in Switzerland and represented the third generation of electronically controlled Sulzer diesel engines. The first RT-flex engine, a six-cylinder RT-flex 58T-B model, was tested in January

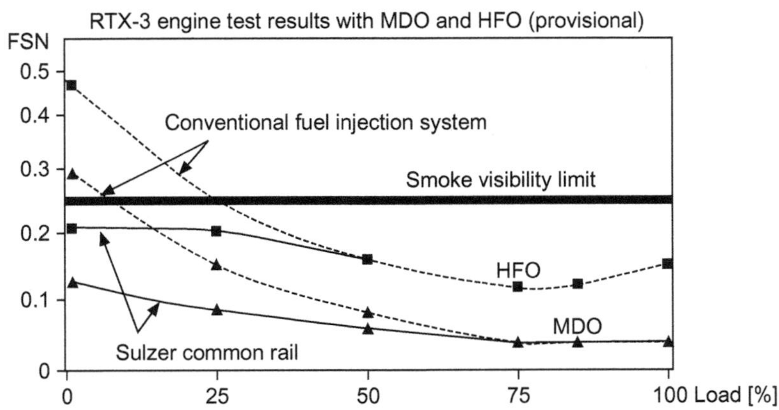

Fig. 16.33 Smoke emissions with conventional fuel injection and with RT-flex common-rail technology for engines burning heavy fuel and marine diesel oi.

Fig. 16.34 A 7RT-flex60C engine on test at Wärtsilä's Trieste factory in Italy.

2001 and later installed as the propulsion plant of a 47,950 dwt bulk carrier, which was handed over in September that year. The engine's slow-running capability was demonstrated by steady operation at speeds down to 12 rev/min. Subsequent orders called for seven-cylinder RT-flex 60C models, the first Wärtsilä low-speed engine designed from the bedplate up with electronically controlled common-rail systems (Fig. 16.34).

The RT-flex option was extended to all bore sizes in the RTA series, the first contract for a 12-cylinder RT-flex 96C engine developing 68,640 kW at 102 rev/min being booked for container ship propulsion in early 2003. Tested in May/June 2004, this debut 12RT-flex96C engine demonstrated an ability to run stably at speeds down to 7 rev/min, fostered by improved combustion which also promotes smokeless operation across the speed range.

82-bore series

Introduced to the market at end-2005, the '82' programme embraces four-engine types—the RT-flex82C, RTA82C, RT-flex82T, and RTA82T—sharing a common platform. The two C-versions target container ship propulsion while the two T-versions address large tankers and bulk carriers. The first two production examples, both eight-cylinder RTA82C models, were tested in early 2008 by the Korean licensee Hyundai

Heavy Industries, which collaborated with Wärtsilä in developing the 820-mm bore designs.

Development was stimulated by a clear need to replace the long-established 840-mm bore RTA84C engine, which had served well in Panamax container ship propulsion for many years but could not meet the power demands of the later tonnage of that class with capacities up to 5000 TEU. Studies led to the choice of 4520kW/cylinder at 102rev/min as the maximum continuous output of the RTA82C and RT-flex82C engines, compared with the 4050kW/cylinder rating of the RTA84C series.

A broadly similar increased propulsion power scenario was also anticipated for the next generation of VLCCs and ULCCs, with capacities larger than 200, 000 and 350,000dwt, respectively. This demand could be addressed by the new RT-flex82T and RTA82T engines with a maximum continuous output of 4520kW/cylinder at 80rev/min compared with the 4200kW/cylinder offered by the established RTA84T-D engine. Similar considerations apply to the propulsion of very large ore carriers of around 300, 000dwt for which the new '82' engines are equally valid.

The possibility of meeting the requirements of two distinctly different market segments—container ships and tankers/bulk carriers—with engines of the same bore size and cylinder output (4520kW/cylinder) opened the way to developing new engine types based on the platform concept. Parameters were thus standardized as far as possible so that many components could be the same for both engine types. All four engines feature an 820-mm bore but two different strokes appropriate for the ship-type targeted. The C-versions have a stroke of 2646mm and the T-versions have a stroke of 3375mm.

During the initial studies, it became clear that although the required power could be readily identified, single running speeds could not be identified as the optimum for the two main targeted propulsion markets for the engines. The solution was found in widening the layout fields to provide a range of speeds at the given maximum continuous rated power output.

The engine layout fields, usually defined by the power/speed ratings R1, R2, R3, and R4, are thus extended to higher speeds defined by the additional rating points R1+ and R2+, respectively, but with 5% greater shaft speed (Fig. 16.35). Any power and speed within this whole engine layout field may be selected as the contracted maximum continuous rating (CMCR) point for an engine.

With the 5% increase in shaft speed at the R1+ point at the same power as the R1 point, the engine is running at a 5% lower mep. This reduced bmep at

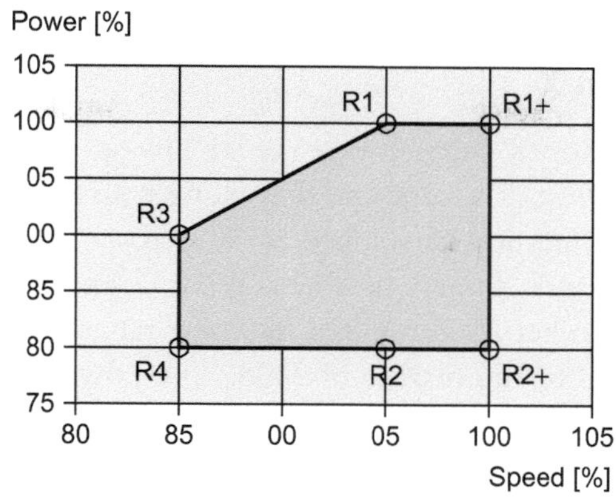

Fig. 16.35 Greater layout flexibility is provided by the '82' bore engines. The usual power/speed layout field for Wärtsilä low-speed engines, defined by the points R1, R2, R3, and R4, is extended to the points R1+ and R2+>

the unchanged maximum combustion pressure (P_{max}) gives this R11 point the benefit of lower specific fuel consumption compared with the R1 point. The higher running speed at the R1+ point offers the possibility of freedom to select a smaller propeller diameter but with the ship sailing at the same fuel consumption (tonnes/day) as at the R1 rating using a larger diameter propeller. An extended layout field also allows the appropriate propeller diameter to be selected in the case of ships with reduced draught, such as container tonnage in the broad size range from 3000 to 5000 TEU.

The C-versions were available in 6–2-cylinder variants covering a power range of 21,720–54,240 kW while the T-versions only went to a maximum of nine cylinders and power output at the top end of 40,680 kW.

By the end of 2012, more than 1000 of the various RT-Flex engines had been sold. The following year, Wärtsilä took the RT-Flex range into the dual-fuel sector with the launch of the RT-flex 50DF. MAN had announced its dual-fuel two-strokes a year previously and was already recording sales. There was however a big difference in that MAN engines were high pressure while Wärtsilä employed the low-pressure Otto cycle in its engines when running in LNG mode and the Diesel cycle when burning oil fuels.

In the Rt-Flex 50DF, when running on LNG gas is not injected at a pressure of 300 bar at the top of the stroke and instead is ported into the engine under a 10 bar pressure at mid-stroke allowing a thorough mix to be achieved by combustion and subsequently burn more evenly. The low-pressure system makes the fuel supply line more economic to produce.

Generation X

The Generation X engines were the last low-speed two-strokes to be developed under Wärtsilä's name and they along with the RT-Flex50DF as the last of its series are now developed by WinGD.

A strategic alliance forged between Wärtsilä and Mitsubishi Heavy Industries in September 2005 resulted in a new range of small two-stroke engines announced to the market in 2008. The power range of the new engines covered 3500–9000 kW. They were to have cylinder bores of 350 and 400 mm and each would be available as a Wärtsilä RT-flex electronically controlled common-rail engine and a Wärtsilä RTA mechanically controlled engine. Mitsubishi Heavy Industries would develop the engines as UEC-LSE series engines.

The new engines extended the low-speed engine range available from Wärtsilä to lower powers for a variety of ship types, such as handysize bulk carriers and product tankers, general cargo vessels, reefers, feeder container ships, and small LPG carriers.

Claimed benefits of the new engines were compliance with NOx Tier II emissions regulations, lower fuel consumption, lower cylinder oil feed rate, greater reliability, and longer times between overhauls.

The Wärtsilä RT-flex35, Wärtsilä RTA35, and Mitsubishi UEC35LSE engine types have cylinder dimensions of 350 mm bore × 1550 mm piston stroke, and a maximum continuous power of 870 kW/cylinder at 167 rpm. The Wärtsilä RT-flex40, Wärtsilä RTA40, and Mitsubishi UEC40LSE have cylinder dimensions of 400 mm bore × 1770 mm piston stroke, and a maximum continuous power of 1135 kW/cylinder at 146 rpm. All types would be available with 5–8 cylinders. The 350-mm bore models thus covering a power range from 3475 to 6960 kW and the 400-mm bore models a band from 4550 to 9080 kW. The RT-flex versions benefited from Wärtsilä's latest common-rail fuel injection technology and full electronic control of injection and exhaust valve operation. Otherwise sharing the same main characteristics and design features as the RT-flex models, the RTA engines exploit a traditional camshaft for mechanical actuation of the fuel injection pumps and exhaust valve pumps. The Mitsubishi version was also mechanically controlled.

	RT-flex35/RTA35	RT-flex40/RTA40
Bore (mm)	350	400
Stroke (mm)	1550	1770
Output/cyl (mcr, kW)	870	1135
Speed range (rev/min)	167–142	146–124
Mean effective pressure (bar)	21	21
Mean piston speed (m/s)	8.6	8.6
Cylinders	5–8	5–8
Power range (kW)	3475–6960	4550–9080

The development of the engines took time and sales began in 2011. Between the 2008 announcement and the first sales, it was decided that instead of labelling the two electronically controlled engines as Flex they would instead be the first of the Generation X engines.

New models were added to the range before the first X35 and X40 models were delivered so that by 2012 Generation X comprised the X35, X40, X62, X72, and X82. The X62 and X72 engines were designed for the wide range of merchant vessels that use mid-size low-speed engines, such as Panamax and Capesize bulk carriers, Aframax and Suezmax tankers, and Feeder and Panamax container vessels.

The X82 was initially to be the B version of the Rt-Flex82 but was rebranded on its release in 2012. At the end of 2012 a larger new engine was added. The X92 was intended for containerships above 8000 teu capacity and is a very low-speed engine with a designed speed of 70–80 rpm. It has a cylinder bore of 920 mm a stroke of 3468 mm giving a bore to stroke ratio of 3.77. With six to 12 cylinder variants available the X92 covered power outputs of 24,420–73,650 kW.

The introduction of the X92 came at a time when environmental regulations were being ramped up with the NOx Tier II already in place, the Tier III requirements just a few years away and the EEDI just having been introduced. The engine was fully compliant with Tier II requirements. It could also be equipped with an SCR catalyst to meet Tier III NOx emission levels and a scrubber to reduce SOx emissions to 0.1%—even with high sulphur fuels. The introduction of the EEDI index also put an emphasis on CO_2 emissions and total vessel efficiency. The internal engine efficiency of the Wärtsilä X92 and the possibility to apply various power take-off (PTO) arrangements for onboard electricity production, were intended to make it easy for shipyards to meet these new requirements.

The last size addition to the Generation X series was the X52 announced in 2013. Described as the most compact engine in its class of very low shaft speed engines. One of the main targets when designing the X52 was to allow for slim hull lines to provide the ship with greater efficiency. Its power output is in the 4080–14,480 kW range, and it comes in 4–8 cylinder configurations.

Further development of some engine types resulted in B versions of the X62 and X72 in 2014. These two engines featured Wärtsilä's extended rating layout whereby the power/speed ranges of the engines are extended to higher speeds at ratings R1 + and R2 + with the same powers as the usual R1 and R2 ratings, respectively. The extended fields offer widened flexibility to select the most efficient propeller speed for lower fuel consumption and propeller type and size.

The final development under Wärtsilä's ownership was the introduction of dual-fuel versions of all engines in the range with the exception of the two smallest sizes and the X82 (although this was listed in the engine programme for 2015). Although all of the Generation X engines were capable of conversion to dual-fuel, anticipating that conversions would be become commonplace to meet future environmental regulations the newly formed WinGD introduced the DF-ready version as an option. The DF-ready engines can be easily converted to dual-fuel, as no major structural components need to be modified. All parts, which are to be replaced at a later conversion, are either typical wear parts or specific X-DF components and systems.

Since coming under the WinGD name, no changes were made until 2019 when in June three new models were announced. These were a D version of the X82 along with a dual-version of the same engine and a dual-fuel version of the X40. Earlier in the year, WinGD had announced that the electronic control systems WECS and UNIC were becoming obsolete as modern control methods required much more capability and ability to be expanded to meet advances in digitalization. As a consequence, the WinGD Integrated Control Electronics (WiCE) was in development and would be rolled from 2020.

WiCE is a scalable platform for machinery control, consisting of several types of modular devices which can be put together to build differently sized arrangements. In the case of main engine control, a distributed system is built and introduces a variety of new concepts and standards, compared to the current systems. Some examples of this are:

- State-of-the-art software and hardware architectures, allowing easy scalability and lifecycle management

- Ethernet ring for data communication and crank angle signal distribution; enabling faster data transmission
- New bus system topology internally, and to external systems, contributing to efficient and lean installation
- Hardware hierarchy with redundant main controllers, gateway units, and one control unit per cylinder, clustering signals, and functions
- A firewall, separating essential engine control functionality from peripherical systems, minimizing the risk of a system breach
- New HW design with powerful logic components and flexible I/O types aimed at future proofing the system
- New development environment and module interface for software development and configuration
- New commissioning and monitoring tools; making the system intuitive for the operator

Given the rather long in-service timespan of the main engine, WiCE was developed as a futureproof system with an extended lifecycle and capacity to fulfil future requirements. The CPU applied has the power that is believed to be sufficient for the years to come. The state-of-the-art hardware and software architectures and interfaces enable relatively straight-forward scalability and testability.

The new X40DF in its standard configuration is designed to run on LNG, offering the first two-stroke dual-fuel engine for smaller vessels. As the engine is already compliant with all existing emission regulations, including IMO Tier III, the entire support system is reduced due to the absence of the need for any exhaust-gas after-treatment system. The X40DF, available in 5–8-cylinder configurations, covering a power range from 2775 to 7480 kW, at 104–146 rpm.

The X82-D engine is 'Dual Fuel ready', with the ability to upgrade to operation with LNG. The cylinder distance has been reduced to 1440 mm from 1505 mm, compared to the previous X82-B version. The brake-specific fuel consumption has also been reduced depending on the rating point and the tuning variant chosen. It will be available in versions from 6 to 9 cylinders, covering a power output from 16,560 kW at 58 rpm to 49,500 kW at 84 rpm.

Key benefits claimed for the X82-D include reduced engine length and weight, due to shorter cylinder distance with adapted bearing, and structural design. The engine offers a wear-resistant ICU MK-III with simpler overhaul concept and the WiCE engine control system with more functionalities supporting prolonged overhaul periods. All cylinder configurations are possible with a one-piece crankshaft design further reducing engine length and with pistons fitted with two-piston rings for reduced friction losses. The X82-D has also been designed with optimized liner design to avoid cold-corrosion occurring within the cylinder.

Fuel consumption is reduced as a result of an increased firing pressure facilitating the significant fuel savings for vessels using the 7X82-D engine compared with its predecessor engine, the 7X82-B.

The X82DF is the latest offering in WinGD's low-pressure X-DF dual-fuel engine portfolio. The engine dimensions are based on that of the X82-D diesel engine design but follow the Otto-cycle combustion principle. It has an integrated gas pressure regulation unit (iGPR), replacing the standalone GVU. Due to its compact design, the iGPR can be arranged on the platform of the engine. The function of the iGPR is controlled directly by the engine control system.

The iGPR consists of several elements: a control valve for the accurate adjustment of the gas pressure for the gas admission; a gas filter and gas flow meter; automatic valves for the venting and inverting of the gas manifolds and an electric control unit with the local panel. All elements are of double-wall design, like the gas manifolds, and fulfil the requirements of marine classification societies.

The prechamber technology and the advanced design of the low-pressure gas admission valves offer a high level of combustion stability and reduced emissions. The X82DF NOX emissions produced are below that of the IMO Tier-III level in gas mode without exhaust gas after-treatment. As a result of the low amount of pilot fuel required (<0.5%) at 75% engine load, low particle and SOx emissions are also achieved.

WinGD's next development was a B version of the X92 engine announced in November 2019. This engine will also feature the WiCE control system. As with all other X series, the X92-B has the capability to cover dual-fuel functionality, in case the engine is converted to dual-fuel at a later date. The tuning options available for the X92-B are Delta tuning, Delta bypass tuning, and low load tuning. Although the engine rating field remains unchanged, due to increased cylinder pressures, the fuel consumption is reduced by up to 2g/kWh depending on the rating point.

To comply with the increased maximum firing pressure the engine structure and powertrain had to be partly modified and reinforced. A new main bearing girder design was developed, reducing the bearing loading significantly, and at the same time reducing the girder weight and consequently manufacturing cost.

Because of the increased maximum cylinder pressure, the loading of the cross head is increased. The X92-B and its dual-fuel counterpart the X92DF will incorporate a new cross head design. The main difference is the increased cross head pin diameter. The new design was introduced without impacting the column design. Furthermore, cross head lubrication pumps are introduced as standard on all X-DF engines. These pumps are installed cascading to the normal bearing lubrication system and increase the oil pressure specific for the cross head bearing. This enhances the hydrodynamic behaviour of the bearing and considerably increases reliability with the new boundary conditions.

In December 2019, four more new engines were announced. Although based on existing X52 and X62 engines, the new models represent a departure being short-stroke variations of the existing engines.

The range, starting with the X52-S2.0 and X62-S2.0 and their dual-fuel counterparts X52DF-S1.0 and X62DF-S1.0, features a short piston stroke that is ideal for vessels with a shallow draught, small propeller diameter or low main deck height.

On the new X62-S2.0 and X62DF-S1.0 engines, the piston stroke has been reduced by 413mm compared to the standard X62 and X62DF—offering a substantial saving in the engine room height needed for installation.

As well as making the engines more compact, the new stroke-to-bore ratio also reduces manufacturing and component cost. Along with design improvements to reduce maintenance and cut operating costs, the new liquid-fuel engines will be fitted with an integrated selective catalytic reduction (iSCR) system. They will be among the first engines to offer the new, compact solution to meeting Tier III NOX limits.

The X62-S2.0 engine has a cylinder bore diameter of 620mm and a piston stroke of 2245mm—compared to a stroke of 2658mm on the standard X62—with a maximum continuous power of 2685kW/cylinder at 108rpm. The engine will be available with five to eight cylinders, covering an overall power range of 7600–21,480kW at 85–108rpm. The X62DF-S1.0 engine has similar dimensions and a maximum continuous power of 2110kW/cylinder, for an overall power range of 6925–16,880kW.

The X52-S2.0 engine has a cylinder bore diameter of 520mm and a piston stroke of 2045mm, with a maximum continuous power of 1910kW/cylinder at 120rpm. The engine will be available with 5–8 cylinders, covering an overall power range of 5425–15,280kW at 95–120rpm. The X52DF-S1.0 engine has a maximum continuous power of 1500kW/cylinder and covers an overall power range of 4950–12,000kW.

WinGD X62DF-S1.0 engine introduced in 2019.

X92DF

In September 2017, French liner operator CMA CGM ordered a series of large 22,000 teu container ships to be built in China. There was speculation at the time that they would have dual-fuel engines and this was confirmed some weeks later when WinGD was awarded the contract to supply a 12X92DF engine for each of the nine ships on order. The first of the vessels—*CMA CGM Jacques Saade* will be completed in 2020.

The engines have a bore of 920 mm and a stroke of 3468 mm and will run at 80 rpm when they will produce 63,840 kW. The height of the engines is 15.92 m and the length 22.87 m. Each engine weighs 2140 t.

In many aspects, the design of the X92DF is a scaled-up version of smaller DF engines. Two gas admission valves (GAVs) are mounted on each cylinder liner. The GAVs are connected to two gas manifolds that run along the engine on both sides. Pneumatic shut-off and vent valves are arranged at both ends of the gas manifolds for a fast release of the gas pressure providing for the required double block and bleed system. Two pilot injectors with their prechambers are arranged in each cylinder cover.

In addition, the three main injectors for the diesel operation are arranged in the same way as with the standard X92 diesel application in each cylinder cover. The scaling-up approach not only reduces the risk of new initial design flaws but also allows to keep the effort for the initial manufacturing of the new, larger components lower.

However, the higher number of cylinders required adaptations of the design and several design iterations during the development. For instance, to avoid disturbing pressure oscillations in the long gas manifolds, the system layout was investigated and optimized by simulating the fluid dynamics in detail. Based on the mechanical finite element method analysis of the gas pipes, additional elastic bellows are applied in the gas manifolds to reduce the stress triggered by thermal expansion. The fast release and venting of the gas pressure from the manifolds is an important safety function, which could be realized with the same fast shut-off and vent valves as on smaller X-DF engines, despite the higher number of cylinders.

Generally, the X-DF engines utilize a lower compression ratio than the diesel X-series engines. Accordingly, the X92DF cylinder liner and cover dimensions were adapted to achieve the desired compression ratio. In addition, due to the higher maximum firing pressure, the design was generally reinforced. Therefore, the X92DF and the derived X92-B engines will build 82 mm higher than the previous X92 pure diesel design.

As the ideal cylinder liner wall temperature (LWT) in gas and in diesel operation are quite different, a new cylinder liner cooling concept was developed. The new concept features active LWT monitoring and control, based on an on-engine cooling water recirculation pump and cooling water flow control valve. In diesel mode, by active cooling water recirculation, the LWT can be increased to avoid sulphuric acid condensation and cold corrosion, while in gas mode a lower LWT is applied.

As the X92DF rail unit incorporates additional piping and systems required for X-DF operation, such as the iCAT, the unit is slightly wider than the one of the previous X92. As on other X-DF engines, the rail unit feet are re-arranged to allow easy access to the gas admission valves, mounted on the cylinder liners. In all X-DF engines, the scavenge air receiver and exhaust manifold are designed to withstand a potential gas explosion. The applied calculated pressure is 10 bar. The necessary reinforcements were made to comply with this requirement.

CHAPTER SEVENTEEN

Legacy low-speed engines

Burmeister & Wain (B&W)

A pioneering Danish designer and builder of low-speed two-stroke cross head engines, Burmeister and Wain (B&W), focused in the post-War era on high-pressure turbocharged single-acting uniflow-scavenged models with one centrally located exhaust valve. The uniflow system fosters effective scavenging of the exhaust gases from the cylinder in an even progressive upwards movement with low flow resistance through to the exhaust manifold via a large diameter poppet-type exhaust valve (Fig. 17.1).

Until 1978, all B&W engines operated on the impulse turbocharging system (Fig. 17.2), but the search for enhanced economy in the wake of contemporary fuel crises dictated a change to the constant pressure system, yielding a saving of around 5% in specific fuel consumption (Fig. 17.3). In the years before its takeover by MAN of Germany in 1980, B&W introduced a number of engine series—successively the K-GF, L-GF, L-GFCA, and L-GB programmes—before launching the MC design portfolio, which is still evolving under MAN Diesel's MAN B&W brand (see Chapter 10).

All of these designs were improved variants of the K-GF series, which itself introduced many notable design innovations such as box-frame construction, intensively cooled cylinder components, and hydraulically actuated exhaust valves.

A major change to the structure of the company followed its purchase by MAN, resulting in Copenhagen becoming the centre for two-stroke engine development of all designs bearing the MAN B&W trademark.

The last surviving ships (November 2019) with engine designs predating the MC type engines were built in 1986. In total, there are around 200–250 ships with these older engines still in existence. The inclusion of the marque in this edition reflects the importance of the brand in the MAN B&W development heritage.

Pounder's Marine Diesel Engines and Gas Turbines
https://doi.org/10.1016/B978-0-08-102748-6.00017-7

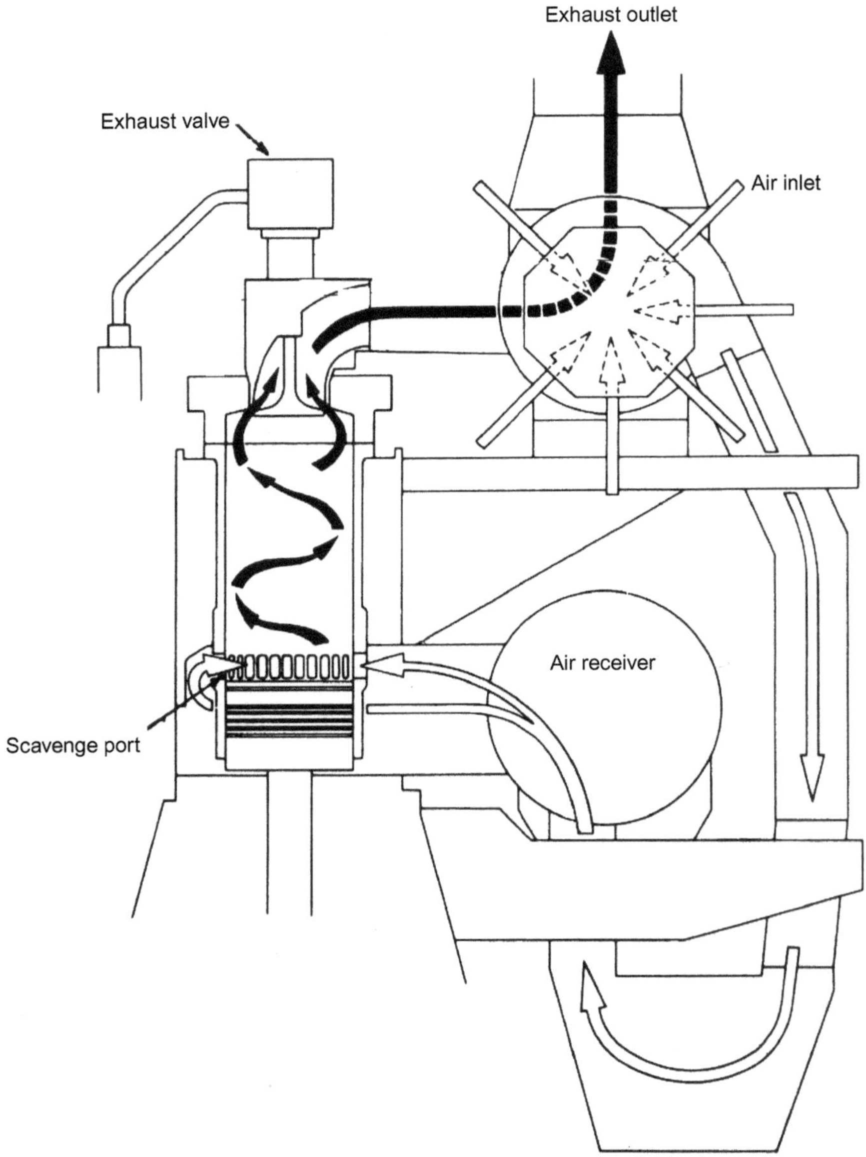

Fig. 17.1 B&W uniflow scavenging.

K-GF type engines

B&W two-stroke engines based on the K-GF series featured many new design features. Notable were the box-type construction of the crankcase and the use of a hydraulically actuated exhaust valve instead of the

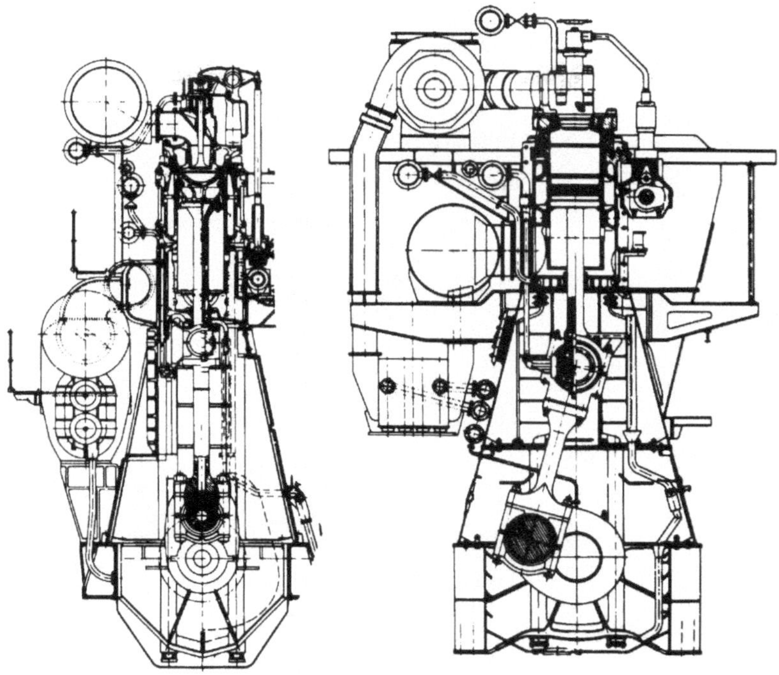

Fig. 17.2 B&W uniflow scavenged engines of 1950 (left) and 1970 (right) with impulse turbocharging.

conventional mechanical rocker arm system used in the earlier K-EF types. This model was rapidly superseded by the L-GF and L-GB models.

L-GF and L-GB engines

The oil crisis in 1973 and the resulting massive increase in fuel prices stimulated engine builders to develop newer engines with reduced specific fuel consumption. B&W's answer was the L-GF series combining constant pressure turbocharging with an increase in piston stroke: an increase in stroke of about 22% results in a lowering of the shaft speed by around 18%, leading to greater propulsive efficiency when using larger diameter propellers and around 7% increase in thermal efficiency.

The design of the L-GF series engine was heavily based on that of the K-GF, and the necessary changes were mainly those relating to the increase in piston stroke and modifications to components to embody thermodynamic improvements (Fig. 17.4). A major component design change was

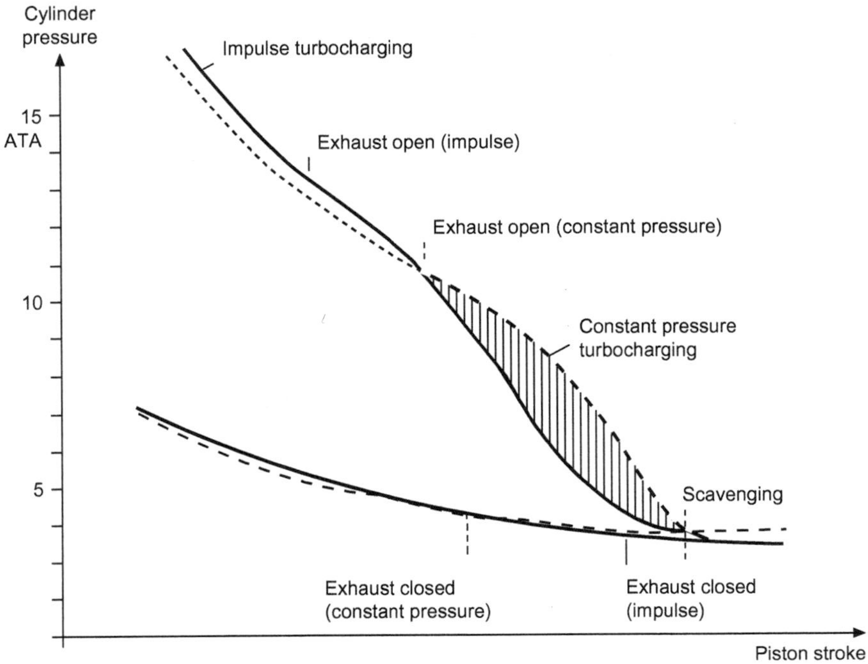

Fig. 17.3 Working diagrams for impulse and constant pressure turbocharging.

the cylinder liner, made longer for the increased stroke, and featuring cooling bore drillings forming generatrices on a hyperboloid to ensure efficient cooling of the high liner collar without cross borings with high stress concentration factors.

The cylinder covers are of the solid plate type used for the K-GF, while the pistons are also of the original oil-cooled type but with somewhat improved cooling caused by the stronger 'cocktail shaker' effect resulting from the greater quantity of oil in the elongated piston rod. The designs of the crankshaft, cross head, bearings, and exhaust valve actuating gear are, in the main, similar to the components introduced with the K-GF series.

Constant pressure turbocharging

The increased stroke and thus reduced revolutions per minute of L-GF engines compared with K-GF engines was not aimed at developing more power but rather an improvement in ship propulsive efficiency. The uniflow scavenging system has the advantage of a good separation between air and gas during the scavenging process, and the rotating flow of air along the cylinder contributes to the high scavenging efficiency and clean air charge.

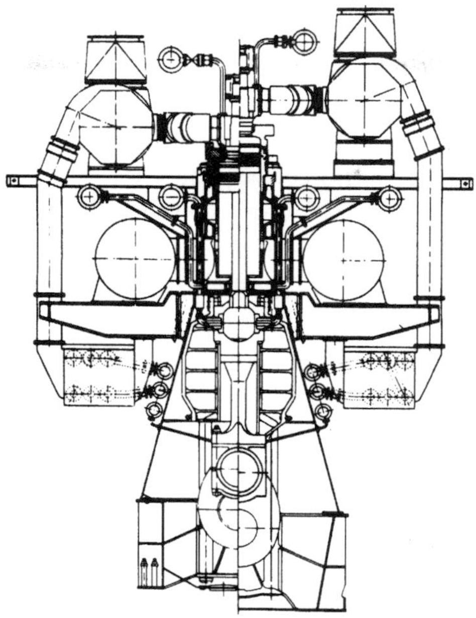

Fig. 17.4 Short- and long-stroke 67GF engines (left and right).

As an engine's mean indicated pressure increases the amount of exhaust gas energy supplied during the scavenging period, relative to the impulse energy during the blow-down period, constant pressure turbocharging is advantageous also for uniflow scavenged engines with unsymmetrical exhaust valve timing. Theoretical calculations for improved fuel economy showed a possible gain of 5%–7% in specific fuel consumption by using constant pressure turbocharging.

The most obvious change with constant pressure turbocharging is that the exhaust pipes from each valve body led to a common large exhaust gas receiver instead of to the turbochargers as in the impulse system. When the cylinders exhaust into a large receiver, the outflow of gas is quicker because the large gas impulse at the commencement of the exhaust period is levelled out in the gas receiver, and the outflow of gas will not be retarded, as in the case of impulse turbocharging where a pressure peak is built up in the narrow exhaust pipe before the turbocharger.

The opening of the exhaust valve can be delayed about 15 degrees, thereby lengthening the expansion stroke and improving the efficiency and reducing the fuel consumption. The energy before the turbine is less than that for impulse turbocharging, but as the pressure and temperature before the turbine is nearly constant, the turbine can be adapted to run at peak efficiency, and the blower can supply sufficient air above 50% of engine load. The scavenging air pressure is increased and the compression pressure somewhat decreased compared with impulse turbocharging.

A small auxiliary blower is necessary for satisfactory combustion conditions at loads of up to about 50%. Two blowers of half capacity are used for safety, and even with one of these out of action, the other with half the overall capacity is satisfactory for starting and load increase. The engine will still run at down to 25% load but with a smoky exhaust.

The change from K-GF to L-GF resulted in a 2% improvement in the specific fuel consumption, and the lower speed accounts for a 5% improvement in propeller efficiency. Constant pressure turbocharging adds a further 5% to the improvement, resulting in a 12% reduction in specific fuel consumption between the L-GF and K-GF types. The next engine model of the constant pressure turbocharged type was the L-GFCA (Figs 17.5 and 17.6), shortly succeeded by the L-GB type.

Fig. 17.5 A 12-cylinder L90GFCA engine developing 34800kW at 97 rev/min.

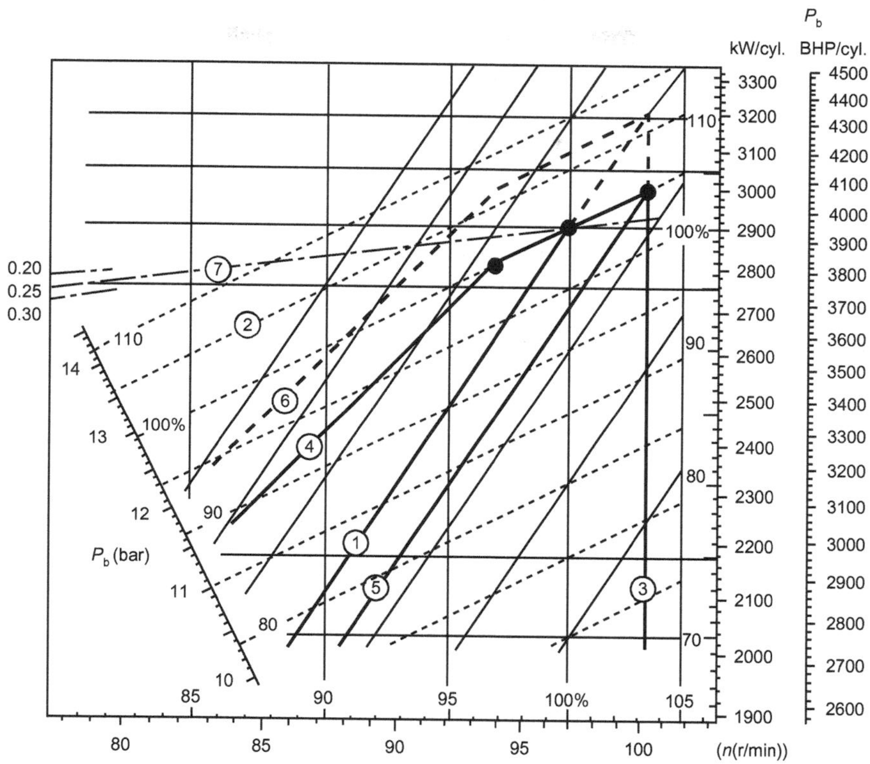

Fig. 17.6 Load diagram for L90GFCA engine.

L-GBE type engines

A further improvement in specific fuel consumption was yielded by the L-GB and L-GBE series engines. By using the optimum combination of longer stroke, increased output and higher maximum pressure, and the newest high-efficiency turbochargers, much lower fuel consumption rates were achieved. The L-GB series had an mep of 15 bar at the same speed as that of the L-GFCA to give an increase in power of 15% and an increase in the firing pressure from 89 to 105 bar. Accordingly, the important P_{max}/mep ratio is almost the same, but the specific fuel consumption is some 4 g/kW h lower than that of the L-GFCA series. A further economy rating is obtained with engines of the L-GBE type by holding the P_{max} at reduced engine output (the so-called derating, as outlined in Chapter 5: the practice of offering propulsion engines in both normal and derated versions was adopted by many engine manufacturers, even for four-stroke medium-speed engines) (Fig. 17.7).

Sulzer RL-type engines

Active in both four-stroke and two-stroke design sectors, Sulzer's links with the diesel engine dated back to 1879 when Rudolf Diesel, as a

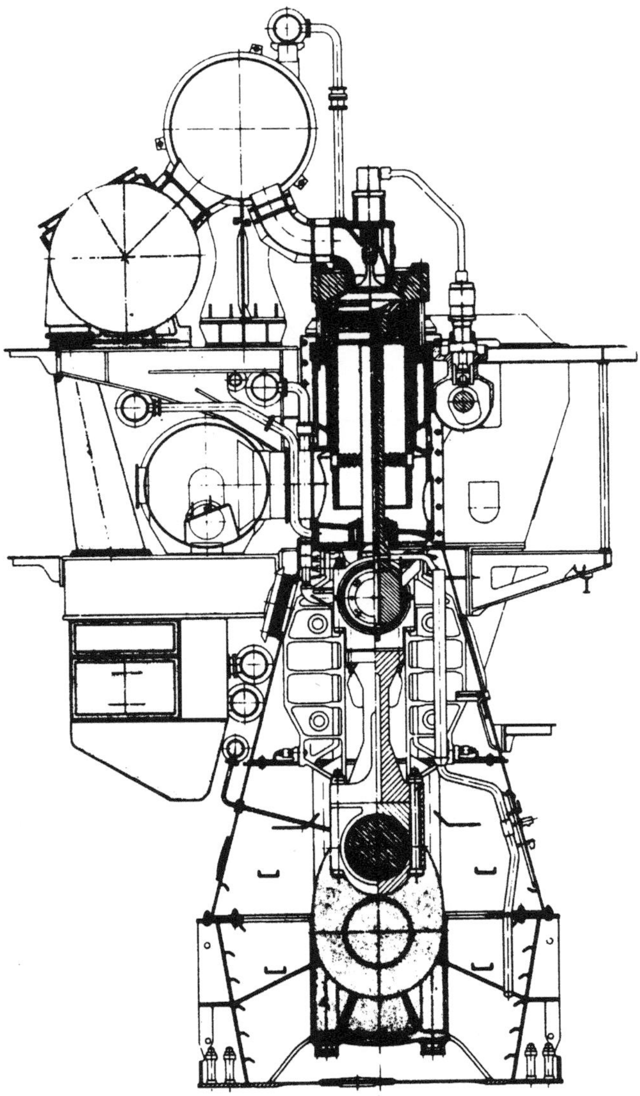

Fig. 17.7 L-GB/GBE engine cross section.

young engineer, followed up his studies by working as an unpaid workshop trainee at Sulzer Brothers in Winterthur, Switzerland. The first Sulzer-built diesel engine was started in June 1898. In 1905, the company built the first directly reversible two-stroke marine diesel engine and, 5 years later, introduced a valveless two-stroke engine with an after-charging system and spray-cooled pistons. Airless fuel injection was applied to production engines in 1932 and turbocharging from 1954.

Low-speed cross head engine designs from Sulzer after 1956 were of the single-acting two-stroke turbocharged valveless type employing loop scavenging and manifested progressively in the RD, RND, RND-M, RLA, and RLB series. A break with that tradition came at the end of 1981 with the launch of the uniflow-scavenged, constant pressure turbocharged RTA series with a single poppet-type exhaust valve (Figs 17.8 and 17.9).The RTA engine's immediate predecessor in the programme was the loop-scavenged RL series of which examples remain in service and merit attention here.

The RLA56, a small-bore two-stroke engine introduced in 1977, incorporated the basic design concept of the successful RND and RNDM series but extended the power range at the lower end. This 560-mm bore engine, of comparatively long stroke design, was the first model in the RLA series.

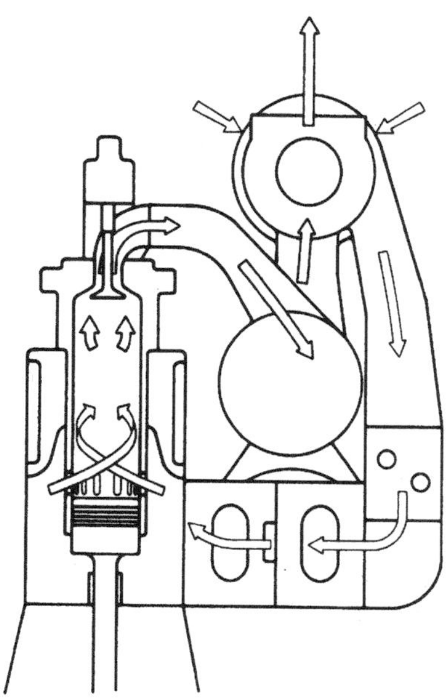

Fig. 17.8 Uniflow scavenging system for RTA engine series.

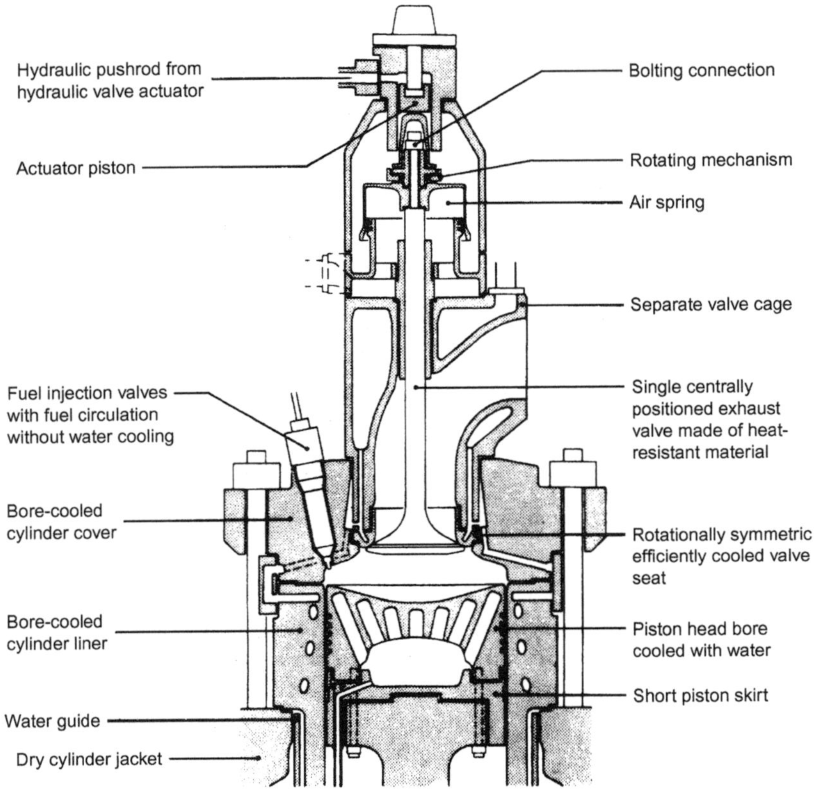

Fig. 17.9 RTA engine combustion chamber components.

It retained many of the design features of the then most recent economical loop-scavenged RLB type (Figs 17.10 and 17.11), both engines using many features of the earlier RND-M series.

A number of RND-M engine features were retained for the RLA and RLB type, namely:

- Constant pressure turbocharging and loop scavenging
- A bore-cooled cylinder liner and one-piece bore-cooled cylinder cover and water-cooled piston crowns
- Double guided cross heads
- New cylinder liner lubrication system with accumulators for the upper liner part and thin-walled aluminium–tin cross head shell bearings.

Major new design features peculiar to the RL types were as follows:

- A new bedplate design with an integrated thrust block; a new box-type column design

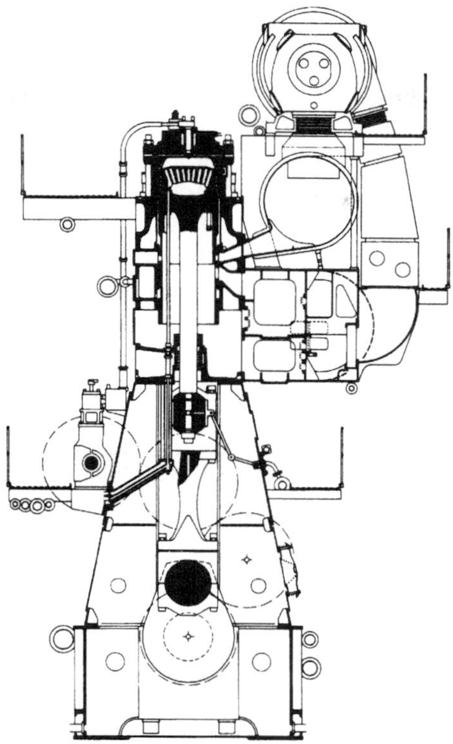

Fig. 17.10 Cross-section of 900-mm bore Sulzer RLB90 engine.

- A semi-built or monobloc-type crankshaft without a separate thrust shaft
- Location of the camshaft gear drive at one end for engines of four to eight cylinders
- Multiple cylinder jackets
- A bore-cooled piston crown
- Modified cross head
- A new design of air receiver.

Bedplate with thrust block

For the RL-type engine bedplate, a new concept of great simplicity was applied. Both the cross girders and the longitudinal structure are of single-wall fabricated design giving very good accessibility for the welded joints. The central bearing saddles are made of cast steel, and only one row of mounting bolts on each side is used to secure the bedplate to the ship's structure.

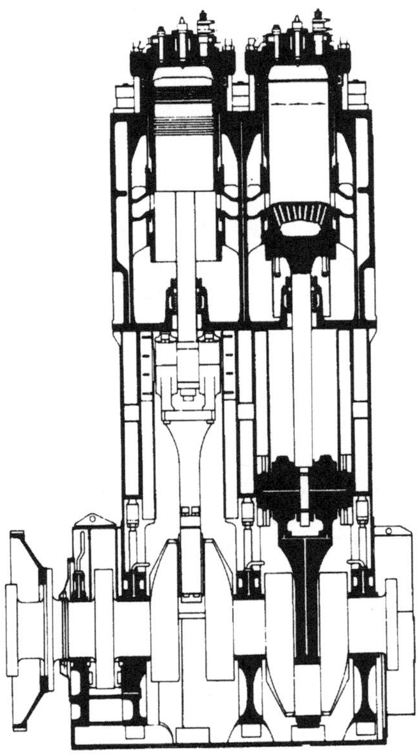

Fig. 17.11 Longitudinal section of RLB90 engine.

The completely new design feature was the method of integrating the thrust block into the bedplate (Fig. 17.12), allowing an extremely compact design and saving engine length. The first crankshaft bearing on the driving end is a combined radial–axial bearing. The bedplate is a one-piece structure for all engines from four to eight cylinders, but, if required, it can be bolted together from two halves.

Thick-walled white-metal-lined bearing shells are used to support the crankshaft and guarantee an optimum safety for the running of the crankshaft.

Crankshaft

In addition to the traditional semi-built-type crankshaft, a monobloc continuous grain flow forged type can be used. From four to eight cylinders, the crankshaft is a one-piece component with an integrated thrust collar section. Only one journal and pin diameter was used for all engines up to eight

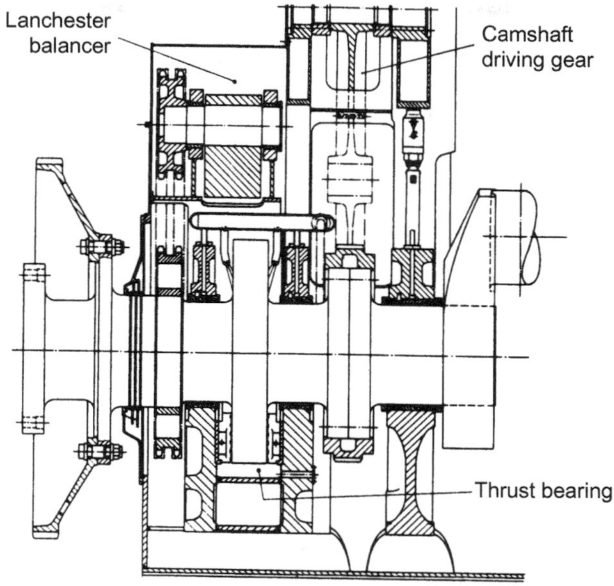

Fig. 17.12 Integrated thrust bearing and camshaft drive for RL engine.

cylinders. For RLA engines larger than the RLA56, the crankshaft was semi-built, and the thrust shaft separately bolted to the crankshaft.

Engine frame

For the small RLA56 engine, a new method of frame construction was used. Cast iron central pieces, on which the cross head guides are bolted, are sandwiched between two one-piece fabricated side-frame girders. This replaced the traditional construction of 'A' frame bolted to the bedplate with side plates attached with access doors forming the enclosure. Larger RL-type engines use 'A' shaped columns of fabricated double wall design, assembled with longitudinal stiffening plates to constitute a rigid structure between the bedplate and cylinder blocks (Figs 17.13 and 17.14).

Cylinder jackets

The cylinder jackets are made of fine lamellar cast iron produced as single block units bolted together in the longitudinal plane to form a single rigid unit, and held on top of the frame section by long tie bolts secured in the bedplate. For the small RLA56 engine, multi-cylinder blocks consisting of two or three cylinders were standard and provide great rigidity of the

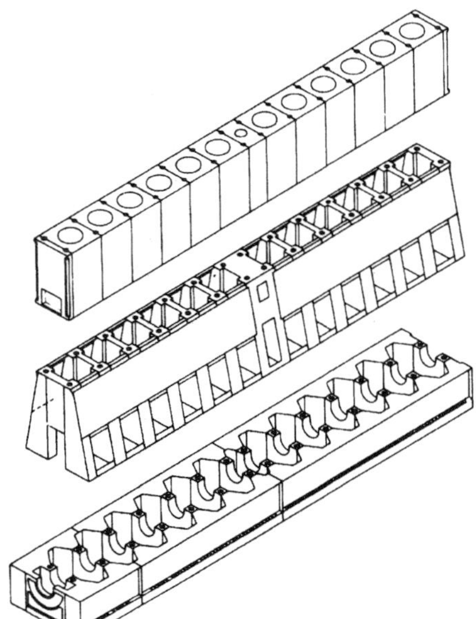

Fig. 17.13 Structural arrangement of bedplate, columns and cylinder jackets for 12-cylinder Sulzer RL engine.

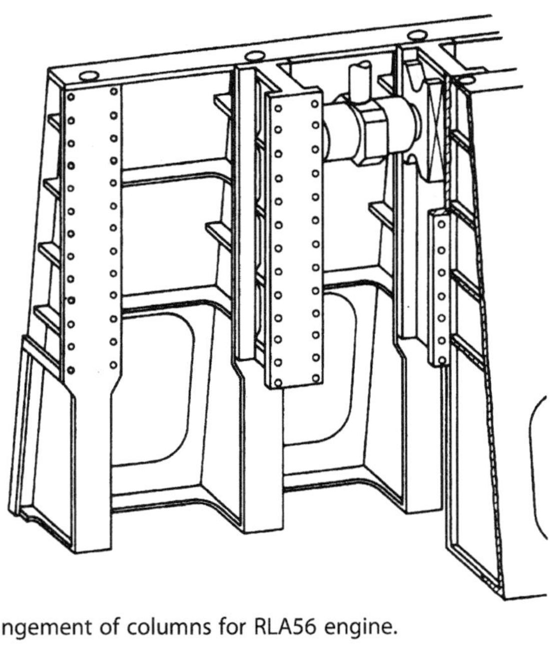

Fig. 17.14 Arrangement of columns for RLA56 engine.

structure. The arrangement of the cooling water passages in the jackets ensures forced water circulation and optimal water distribution around the exhaust canal.

Because of the steadily rising charge air pressures, the design of the air receiver was modified to allow the use of automatic welding techniques. Instead of a rib-stiffened plain side plate, a semicircular pressure containment was fitted with an integral air inlet casing. The auxiliary blower is mounted on the front end of the receiver, thus eliminating inclined ducts.

Combustion chamber components

The combustion chamber of RL-type engines is principally of the same shape as that of the RND-M. The cylinder cover is basically a one-piece steel block with cooling bores, while an identical arrangement of bore cooling is applied to the upper collar of the cylinder liner, which is made from lamellar cast iron, a material of good heat conductivity and wear resistance. Fig. 17.15 shows the arrangement with cover, cylinder liner, and piston.

A new type of piston crown was introduced for the RL series engines. This combustion chamber component similarly uses a bore cooling arrangement (Figs 17.16 and 17.17) as previously only applied to the liner and cover. Water cooling was retained, but the piston bore cooling uses a somewhat different mechanism to the force flow system of the liner and cover.

The cooling space of the piston crown is approximately half filled with cooling water, and, as a result of piston acceleration and deceleration, a 'cocktail shaker' effect is produced to provide excellent heat removal. This effect is capable of ensuring efficient heat transfer under all prevailing load conditions in order to keep the vital temperatures on the piston crown and around the piston rings within suitable limits.

As a result of this new design, the crown temperatures were lowered compared with the previous construction. Forged steel blocks were specified for piston crown manufacture, but cast steel versions were also used.

Cylinder lubrication

Lubrication of the cylinder is through six quills mounted in the upper area of the liner just above the cylinder jacket, as shown in Fig. 17.18. The oil distribution grooves have a very small angle of inclination to avoid blow-by over the piston rings and small but regular quantities of cylinder oil supplied by an accumulator system. Two further lubrication points are provided

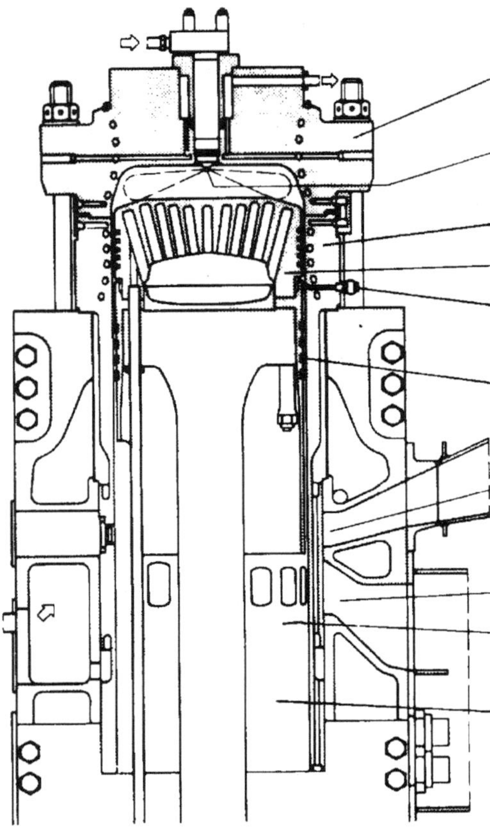

Fig. 17.15 Combustion chamber components for RLA90 engine.

below the scavenge ports on the exhaust side with the necessary oil pumps positioned on the front side of the engine above the camshaft drive.

Cross head design

The cross head is similar to that used for RND engines and has double guided slippers. The pin size was increased for safety, and thin-walled half shells of the aluminium–tin type were used. The pin itself is of forged homogeneous steel of symmetrical design and can therefore be turned around in case of damage. The piston rod is connected to the pin by a single hydraulically tightened nut, while the slippers, made of cast steel and lined with white metal, are bolted to the ends of the pin. The cast iron double guide faces are fitted to the engine columns.

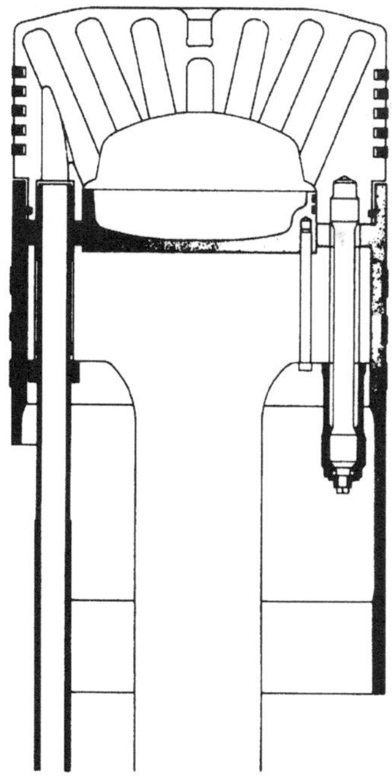

Fig. 17.16 Detail view of RL engine piston with bore cooling.

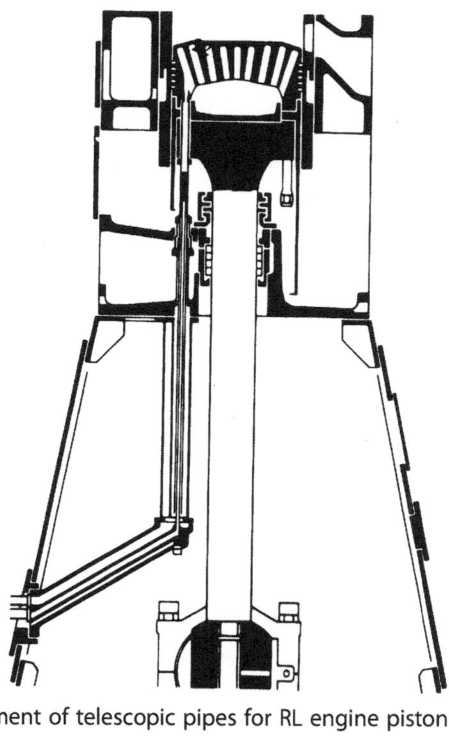

Fig. 17.17 Arrangement of telescopic pipes for RL engine piston cooling.

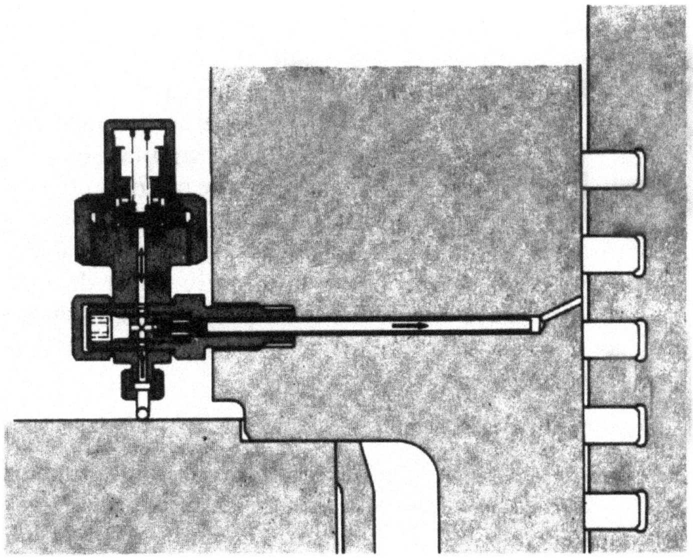

Fig. 17.18 Cylinder lubricator for RL engine.

The connecting rods of traditional marine type have a forged normalized steel bottom end bearing, lined with white metal, held in place by four hydraulically tensioned bolts. Compression shims are provided between the bottom end bearing and the palm of the connecting rod.

Piston assembly

The piston (Fig. 17.16) consists of a water-cooled cast steel crown, a cast iron skirt with copper bandages, and a forged steel piston rod, with the piston rings fitted in chromium-plated grooves.

The water-cooled piston has proved very reliable when running on heavy fuels, and the use of water cooling has resulted in practically negligible system oil consumption. (With oil cooling, oil is consumed usually as a result of thermal ageing on hot piston walls. Oil leaks from oil-cooled pistons may also occur on other engine types.) The two-part gland seals for the piston rod, and telescopic piston cooling pipes can be inspected while the engine is running and can be dismantled without removing the piston. The double-gland diaphragm around the piston rod completely separates the crankcase from the piston undersides, preventing contamination of the crankcase oil by combustion residues or possible cooling water leakages. In addition, the freshwater piston cooling water system is served by an automatic water drain-off when the circulating pumps are stopped. This avoids leakages when the ship is in port.

Turbocharging arrangement

All the Sulzer RL engines employ the constant pressure turbocharging system, and with the high-efficiency Brown Boveri series 4 turbochargers include provision for automatic cleaning. The layout of the turbocharging arrangement is shown in Fig. 17.19.

The use of the piston undersides to provide a scavenge air impulse eliminates the need for large electrically driven auxiliary blowers. At higher loads, a simple flap valve opens to cut out the piston underside pumping effect with a consequent improvement in fuel consumption, whilst a small auxiliary fan incorporated in the scavenging system improves the smoke values at the lowest loads. The piston underside scavenge pump facility allows the engine to start and reverse even with total failure of all turbochargers. The auxiliary fan and engine will even operate at up to 60% load, thus giving a 'take home' facility.

The turbochargers are mounted on top of the large exhaust gas receiver with the scavenge air receiver, which forms part of the engine structure beneath. The charge air is passed down through seawater–cooled intercoolers, which are mounted accessibly alongside the scavenge air receiver. Flap-type valves, which operate according to the scavenge air pressure, direct the air inside the three-compartment air receiver to the piston

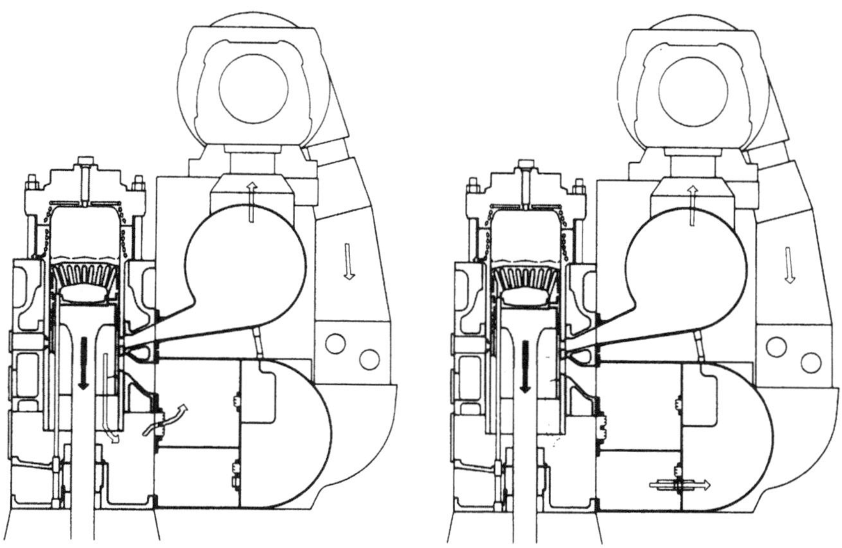

Fig. 17.19 Constant pressure turbocharging and operation of under-piston supercharging (RL engine).

underside (at low scavenge air pressure) or direct to the scavenge ports when the higher air pressure from the turbochargers keeps the underside delivery flap valves shut.

For RL-type engines equipped with only one turbocharger, a separate turbocharger/air intercooler module is available as a standard option. This unit can be located adjacent to the engine at either the forward or aft end, thus considerably reducing the mounting height and width of the main engine. The turbocharger module consists of a base frame onto which the turbocharger, charge air cooler, air ducts, and cooling water pipes are solidly mounted with flexible connections to the engine.

Upgrading RLB engines

Based on field tests carried out between 1991 and 1995, a retrofit package of modifications was introduced for Sulzer RLB90 engines in 1993 and for RLB66 engines in 1995. The main element is the change to loop cooling in the cylinder liners, an improved cooling technique that reduces the thermal and mechanical stresses in the top collar of the liner (Fig. 17.20). By

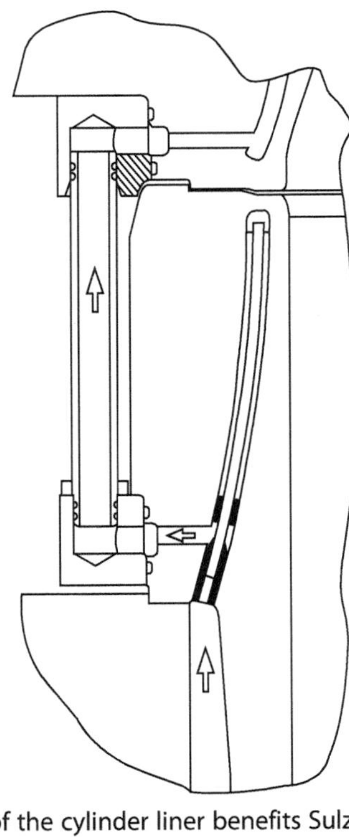

Fig. 17.20 Loop cooling of the cylinder liner benefits Sulzer RLB90and RLB66 engines.

optimizing the running surface temperature, the system also reduces corrosive wear and extends the TBOs up to 2 years.

The retrofit package comprises the loop-cooled cylinder liners as well as:

- A new fully gas-tight, vermicular top piston ring of increased thickness
- A modified top piston ring groove with increased height and increased thickness of chromium layer
- Multilevel lubrication
- A new water guide jacket to suit the new liner
- A modified condensed water drain from the scavenge air receiver.

As RLB engines exploit loop scavenging with both scavenge and exhaust ports in the lower part of the liner, the fully gas-tight top ring significantly reduces the lateral forces at the piston skirt in addition to reducing the blow-by of hot combustion gases—both of which increase the piston skirt life. The package can be applied to the engine either as a full package or step-by-step (intermediate package) depending on service experience with the liners and whether the general piston-running behaviour meets today's expectations. It is also possible to replace individual original liners with loop-cooled units whenever they need to be renewed; there is no problem in running loop-cooled and original design liners together in the same engine.

CHAPTER EIGHTEEN

Medium-speed engines—Introduction

Most of the large ship types, with the exception of cruise ships, choose a low-speed (under 130 rpm) two-stroke main engine in a direct mechanical configuration for the propulsion system. In terms of fuel use and power output, these ships make up 80% of the world fleet. However, in pure number terms they account for just 20%. In most cases, these large ships have a single main engine, although there are a growing number of twin screw vessels which will have two main engines.

The vast majority of the world fleet of ships, tugs, and other craft use medium-speed four-stroke engines (up to around 1200 rpm) for propulsion. A high number will have just a single main engine, but multiple engine configurations are very common. On the larger vessels with low-speed propulsion engines, it is normal to have two or more medium-speed diesel-engined gensets to provide power for the ship's electrical needs.

It is therefore unsurprising that the medium-speed sector supports many more manufacturers than the low-speed sector. Of the low-speed engine makers, only MAN Energy Solutions is also active in medium-speed engine design and production. Wärtsilä was also active in both sectors, but since passing its two-stroke business to WinGD it is now only in the medium-speed sector.

Both companies have a strong global following as does Caterpillar, which now owns MaK. For the other players in the sector, it is noticeable that they tend to attract local buyers rather than selling products on a worldwide basis.

New designs and upgraded versions of established models have maintained the dominance of medium-speed four-stroke diesel engines in the propulsion of smaller ships as well as larger specialist tonnage such as cruise vessels, car/passenger ferries and ro-ro freight carriers. The larger bore designs can also target the mainstream cargo ship propulsion market formed by bulk carriers, container ships, and tankers, competing against low-speed two-stroke machinery, and dual-fuel versions have found a new volume market in LNG carrier propulsion.

Pounder's Marine Diesel Engines and Gas Turbines
https://doi.org/10.1016/B978-0-08-102748-6.00018-9

The growth of the fast passenger and ro-pax ferry sector benefited those medium-speed engine builders who could offer designs with sufficiently high power/weight and volume ratios, an ability to function reliably at full load for sustained periods, and attractive through-life operating costs. Medium-speed engines further enjoy supremacy in the deep-sea vessel genset drive sector, challenged only in lower power installations by high-speed four-stroke engines.

The four stroke engine

The modern medium-speed engine is almost invariably a four-stroke trunk piston engine with significant differences to the low-speed two-stroke cross head engines. The term trunk piston engine derives from the piston skirt or trunk as it is commonly referred to. The trunk acts as the cross head of a two-stroke engine absorbing and transmitting the thrust to the piston. Thus, there is only the single piece connecting rod between the crankshaft and the piston. This makes for a reduced engine height and a correspondingly shorter stroke. Most four-strokes are squarer engines than two-strokes, although a longer stroke has become more common in recent times in the quest for higher power outputs.

A significant difference between the two types is in the lubrication. In medium- and high-speed engines, there is a common system oil lubricating the crankcase and the cylinder liners, and no separate lubrication system is needed for the upper cylinder lubrication as in the low-speed engine.

Significant strides have been made in improving the reliability and durability of medium-speed engines in the past decade, both at the design stage and through the in-service support of advanced monitoring and diagnostic systems. Former weak points in earlier generations of medium-speed engines have been eradicated in new models, which have benefited from finite element method calculations in designing heavily loaded components. Designers now argue the merits of new generations of longer stroke medium-speed engines with higher specific outputs allowing a smaller number of cylinders to satisfy a given power demand and foster compactness, reliability, reduced maintenance, and easier servicing. Progress in fuel and lubricating oil economy is also cited, along with enhanced pier-to-pier heavy fuel burning capability and better performance flexibility throughout the load range.

Completely bore-cooled cylinder units and combustion spaces formed by liner, head, and piston combine good strength and stiffness with good temperature control, which are important factors in burning low-quality fuel oils. Low noise and vibration levels achieved by modern medium-speed engines can be reduced further by resilient mounting systems, a technology which has advanced considerably in recent years.

IMO Tier II limits on NOx emissions in the exhaust gas can generally be met comfortably by medium-speed engines using primary measures to influence the combustion process (in some cases, it is claimed, without compromising specific fuel consumption). Wärtsilä's low NOx combustion technology, for example, embraces high fuel injection pressures (up to 2000 bar) to reduce the duration of injection, a high compression ratio (16:1), a maximum cylinder pressure of up to 210 bar, and a stroke/bore ratio $>$1.2:1. Concern over smoke emissions, particularly by cruise ship operators in sensitive environmental areas, has called for special measures from engine designers targeting that market, notably electronically controlled common rail (CR) fuel injection.

CR designs now dominate production programmes across the sector to the benefit of ship operators and the environment. Significant advantages are delivered in engine operational flexibility, economy, and environmental friendliness by a system in which the generation of fuel pressure and the injection of fuel are not interconnected.

In contrast to a conventional system, the injection pressure in a CR configuration is independent of engine speed, and full pressure is always available at all loads down to idling. Highly efficient and clean combustion is thus fostered across the engine operating range, yielding economic and environmental merits. An optimum injection pressure and timing can be selected for a given operating mode—irrespective of the engine speed—and pilot and postinjection patterns exploited to meet differing demands: for example, invisible exhaust at the lowest loads and NOx emission reductions at medium loads, without undermining fuel economy.

The concept of electronically controlled CR fuel systems had been appreciated for many years before implementation in engine programmes. Feasible solutions, however, awaited the development of fast and reliable rail valves and electronic controls. Advances in materials and manufacturing technology also allowed the creation of systems capable of handling heavy fuel and pressures of 1500 bar and above.

Successful automotive diesel applications of CR layouts, in car and truck power units, were largely driven by stricter emission regulations, which called for flexible fuel injection systems offering injection rate shaping, free adjustment of injection pressure, variable start of injection, and pre- and postinjection patterns. Current and future emission curbs on shipping stimulated the transfer to marine engines. Increasingly stringent regulations on NOx and smoke in the exhaust are difficult to meet without intelligent controls and a flexible injection system if engine efficiency is to remain the same. Part-load operation imposes a particular challenge in satisfying invisible smoke requirements without CR technology.

Ease of inspection and overhaul—an important consideration in an era of low manning levels and faster turnarounds in port—has been addressed by a reduced overall number of components (in some cases, 40% fewer than in the preceding engine generation) achieved by integrated and modular assemblies using multifunctional components. Simplified (often plug-in or clamped) connecting and quick-acting sealing arrangements also smooth maintenance procedures. Channels for lubricating oil, cooling water, fuel, and air may be incorporated in the engine block or other component castings, leaving minimal external piping in evidence. Compact and more accessible installations are achieved by integrating ancillary support equipment (such as pumps, filters, coolers, and thermostats) on the engine. Lower production costs are also sought from design refinements and the wider exploitation of flexible manufacturing systems to produce components.

The cylinder unit concept is a feature of the modern four-stroke designs, allowing the head, piston, liner, and connecting rod to be removed together as a complete assembly for repair, overhaul, or replacement by a renovated unit on board or ashore. This modular approach has been adopted by most of the major producers.

As the attractions of dual-fuel engines have increased, the modular approach has been taken a step further with engines being designed for conversion from oil fuels to dual-fuel operation by way of replacement and addition of a small number of components. In 2017, MAN Energy Solutions converted the MAN 8L48/60B oil-burning engine of the 2011-built feeder container ship *Wes Amelie* to an 8L51/60DF multifuel engine and installed an LNG gas fuel system at the same time.

Before the conversion, the MAN 8L48/60B had an output of 9000kW @500rpm running on heavy fuel oil. After the conversion, the power setting is 7800kW @514rpm. The loss of power was anticipated and considered acceptable because the vessel was previously running at low loads for most of the time.

The conversion involved exchanging the cylinder liners and water-jacket area, piston, piston rings, and the cylinder heads. This was because of the cylinder diameter increase from 48 to 51 cm. In addition, all injection components were exchanged or newly added. The pilot oil system, which is required for the gas use, was a new addition. To realize new timings of the 51/60DF engine, new cams as well as new turbocharger components were installed. The control of the dual-fuel engine 51/60DF is more complex compared to the heavy-fuel-oil engine, meaning engine sensors were either converted or re-instrumentation was necessary.

In newer engine types from different suppliers, more consideration is given to the possibility of conversions, and components are being simplified. This should make future conversions far less extensive.

Compactness and reduced weight remain the key attractions of the medium-speed engine, offering ship designers the opportunity to increase the cargo capacity and lower the cost of a given newbuilding project, and the ability to achieve the most efficient propeller speed via reduction gearing. Medium-speed engine builders can offer solutions ranging from single-engine plants for small cargo vessels to multi-engine/twin-screw installations for the most powerful passenger ships, based on mechanical (geared) or electrical transmissions. Multiengine configurations promote plant availability and operational flexibility, allowing the number of prime movers engaged at any time to match the service schedule. The convenient direct drive of alternators and other engine room ancillaryplant (e.g. hydraulic power packs) is also facilitated via power take-off gearing.

A carbon-cutting ring is now a common feature of medium-speed engines specified to eliminate the phenomenon of cylinder bore polishing caused by carbon deposits and hence significantly reduces liner wear. It also fosters a cleaner piston ring area, low and very stable lubricating oil consumption, and reduced blow-by.

Also termed an anti-polishing or fire ring, a carbon-cutting ring comprises a sleeve insert, which sits between the top piston ring turning point and the top of the cylinder liner. It has a slightly smaller diameter than the bore of the liner, this reduction being accommodated by a reduced diameter for the top land of the piston. The main effect of the ring is to prevent the build-up of carbon around the edges of the piston crown, which causes liner polishing and wear, with an associated rise in lubricating oil consumption.

A secondary function is a sudden compressive effect on the ring belt as the piston and carbon-cutting ring momentarily interface. Lubricating oil is consequently forced away from the combustion area, again helping to reduce consumption: so effectively, in fact, that Bergen Diesel found it necessary to redesign the ring pack to allow a desirable amount of oil consumption. The Norwegian engine designer reports that lubricating oil consumption is cut by more than half and insoluble deposits in the oil reduced dramatically, significantly extending oil filter life. Carbon-cutting rings can be retrofitted to deliver their benefits to engines in service. Removal prior to piston withdrawal is simply effected with a special tool.

Designers now also favour a 'hot box' arrangement for the fuel injection system to secure cleaner engine lines and improve the working environment in the machinery room thanks to reduced temperatures; additionally, any fuel leakage from the injection system components is retained within the box.

Power extremes

The major medium-speed engine builders have long offered 500-mm-bore-plus designs in their portfolios. MAN Diesel still fields its L58/64 series but MaK's 580-mm bore M601 and the Sulzer ZA50S engines have been phased out, as was Stork-Wärtsilä's TM620 engine in the mid-1990s. In the 1970s, MAN and Sulzer jointly developed a V-cylinder 650-mm bore/stroke design (developing 1325 kW/cylinder at 400 rev/min) that did not proceed beyond prototype testing.

Wärtsilä's 64 series, launched in 1996, took the medium-speed engine into a higher power and efficiency territory, the 640-mm bore/900-mm stroke design offering an output of 2010 kW/cylinder at 333 rev/min when last produced. A V12-cylinder model delivered 23,280 kW at 400 rev/min. The range could therefore meet the propulsive power demands of virtually all merchant ship tonnage types with either single- or multi-engine installations. The key introductory parameters were 10 m/s mean piston speed, 25 bar mean effective pressure, and 190 bar maximum cylinder pressure. The Finnish designers claimed the 64 series to be the first medium-speed engine to exceed the 50% thermal efficiency barrier and suggested that overall plant efficiencies of 57%–58% are possible from a combi-cycle exploiting waste heat to generate steam for a turbo-alternator.

The engine was still specified in the company's portfolio in 2007, but with few orders, and an improving 50DF range of engines as well as the 38 and 46 series which provided an overlap in power production for all but the largest 64 engines. The 38 series was also dropped later.

At the other end of the medium-speed engine power spectrum, the early 1990s saw the introduction of a number of 200-mm bore long-stroke designs from leading builders such as Daihatsu, MaK, and Wärtsilä Diesel, contesting a sector already targeted by Sulzer's S20 model. These heavy fuel-burning engines (typically with a 1.5:1 stroke–bore ratio) were evolved for small-ship propulsion and genset drive duties, the development goals addressing overall operating economy, reliability, component durability, simplicity of maintenance, and reduced production costs. Low and short overall configurations gave more freedom to naval architects in planning machinery room layouts and eased installation procedures.

The ~320-mm bore sector is fiercely contested by designers serving a high-volume market created by propulsion and genset drive demands. A number of new designs—including Caterpillar/MaK's M32 and MAN Diesel's L32/40—emerged to challenge upgraded established models such as the Wärtsilä 32.

A Japanese challenger in a medium-speed arena traditionally dominated by European designer/licensors arrived in the mid-1990s after several years' R&D by the Tokyo-based Advanced Diesel Engine Development Company. The joint venture embraced Hitachi Zosen, Kawasaki Heavy Industries, and Mitsui Engineering and Shipbuilding. The 300-mm bore/480-mm stroke ADD30V design, in a V50-degree configuration, developed up to 735 kW/cylinder at 750 rev/min, significantly more powerful than contemporary medium-speed engines of equivalent bore size. A mean effective pressure of around 25 bar and a mean piston speed of 11.5 m/s were exploited in tests with a six-cylinder prototype, although an mep approaching 35 bar and a mean piston speed of 12 m/s are reportedly possible.

In addition to a high specific output, the developers sought a design which was also over 30% lighter in weight, 10%–15% more fuel economical and with a better part-load performance than established engines. Underwriting these advances in mep and mean piston speed ratings is an anti-wear ceramic coating for the sliding surfaces of the cylinder liners and piston rings that is applied by a plasma coating method. A porous ceramic heat shield was also developed for the combustion chamber to reduce heat transfer to the base metal of the piston crown.

A key feature is the single-valve air intake and exhaust gas exchange, contrasting with the four-valve (two inlet and two exhaust) heads of other medium-speed engines. The greatly enlarged overall dynamic valve area and the reduction in pressure losses during the gas exchange period promote a higher thermal efficiency. The system is based on a heat-resistant alloy main poppet valve located over the centre of the cylinder and a control valve placed coaxially against this main valve. The control valve switches the air intake and exhaust channels in the cylinder cover. Both the main and control valves are driven hydraulically.

Fuel injection is executed from the side through multiple injectors arranged around the cylinder periphery instead of a conventional top-mounted central injection system. Combustion characteristics were optimized by raising the fuel injection pressure to around 2000 bar, thus enhancing the fuel–air mixture formation and fostering low NOx emissions without sacrificing fuel economy. A computer-based mechatronics system automatically controls the timing of fuel injection and valve opening/closing to match the operating conditions. The first 6ADD30V production marine engines, built by Mitsui, were specified as the prime movers for the diesel-electric propulsion plant of a large Japanese survey vessel.

Offshore vessel and LNG carrier propulsion market opportunities—and the potential of wider mainstream shipping interest as environmental curbs tighten—have encouraged a number of medium-speed engine builders to develop dual-fuel and gas–diesel designs offering true multifuel capabilities with high efficiency and reliability, and low carbon dioxide emissions. The engines can run on gas (with a small percentage of liquid pilot fuel) or entirely on liquid fuel (marine diesel oil, heavy fuel, or even crude oil). Switching from one fuel to another is possible without interrupting power generation.

The high cost of R&D to maintain a competitive programme and continuing investment in production resources and global support services have stimulated a number of joint ventures and takeovers in the four-stroke engine sector in recent years. Most notable have been Wärtsilä's acquisition of the former New Sulzer Diesel and Caterpillar's takeover of MaK. Earlier, Wärtsilä had acquired another leading medium-speed engine builder, the Netherlands-based Stork-Werkspoor Diesel. This trend towards an industry comprising a small number of major multinational players contesting the world market has continued with the absorption of the British companies Mirrlees Blackstone, Paxman and Ruston (formerly part of Alstom Engines) into the MAN Diesel group. Rolls-Royce inherited the Bergen Diesel interests in Norway through the takeover of Vickers-Ulstein.

Power potential

Considerable potential remains for further developing the power ratings of medium-speed engines, whose cylinder technology has benefited from an anti-polishing ring at the top of the liner, water distribution rings, chrome–ceramic piston rings, pressurized skirt lubrication, and nodular cast iron/low friction skirt designs.

The pressure-lubricated skirt elevated the scuffing limit originally obtaining by more than 50 bar, reduced the piston slap force by 75%, and doubled the lifetime of piston rings and grooves. Furthermore, it facilitated a reduction in lube oil consumption and, along with the simultaneous introduction of the nodular cast iron skirt, practically eliminated the risk of piston seizure. The antipolishing ring dramatically improved cylinder liner lifetime beyond 100,000 h, and lube oil consumption became controllable and stable over time, most engines today running at rates between 0.1 and 0.5 g/kW h. A further reduction in piston ring and groove wear was also achieved, and the time-between-overhauls extended to 18,000–20,000 h. The ring itself is a wear part but is turnable in four positions in a four-stroke engine, fostering a lengthy lifetime for the component.

Such elements underwrite a capability to support a maximum cylinder pressure of 250 bar, of which 210–230 bar is already exploited in some engines today. Leading designers such as Wärtsilä suggest that it may be possible to work up towards 300 bar with the same basic technology for the cylinder unit, although some areas need to be developed: bearing technology, for example, where there is potential in both geometry and materials. A steel piston skirt may become the most cost-effective, and cooling of the piston top will probably change from direct oil cooling to indirect. The higher maximum cylinder pressure can be exploited for increasing the maximum effective pressure or improving the thermal efficiency of the process. Continuing to mould the development of the medium-speed engine will be NOx and carbon dioxide emissions, fuel flexibility, mean time-between-failures, and reduced maintenance.

Designing medium-speed engines

Investment in the development of a new medium-speed engine may be committed for a number of reasons. The engine builder may need to extend its portfolio with larger or smaller designs to complete the available range, to exploit recently developed equipment and systems, or to enhance the reliability, availability, and economy of the programme. As an alternative to a new design, it may be possible to upgrade an existing engine having the potential for improved power ratings, lower operational costs, reduced emissions and weight. An example is the upgrading by MAN Diesel of its 48/60 engine, whose B-version offers a 14% higher specific output with considerably reduced fuel consumption and exhaust emission rates, a lower weight, and the same length and height but a narrower overall width than its predecessor.

Having set the output range of the new engine, a decision must be made on whether it should be built only as an in-line cylinder version or as a V-type engine as well. In the case of the largest and heaviest medium-speed engines, the efforts and costs involved in adding a V-type to the existing portfolio might not be justified if there is a limited potential market for engines above 20,000 kW. Other restraining factors could be the capacity of the foundry for casting very large crankcases or difficulties in transporting engines weighing 400 t or more overland.

Once the power per cylinder of the proposed new engine is known, the first indication of its bore size is determined by the piston load, which relates the specific output to the circular piston surface. Piston loads have increased with the development of better materials: the original 48/60 engine has a piston load of 58 kW/cm^2, while the 48/60B engine achieves 66.4 kW/cm^2.

The bore and stroke of an engine are interrelated by the stroke–bore ratio, which in turn is based on the designer's experiences with earlier engines. Some 25 years ago, it was not uncommon to design medium-speed engines with very similar bore and stroke dimensions (so-called 'square' engines); more recently, the trend has been towards longer stroke designs, which offer clear advantages in optimizing the combustion space geometry to achieve lower NOx and soot emission rates. A longer stroke can reduce NOx emissions with almost no fuel economy penalty and without changing the maximum combustion pressure. The compression ratio can also be increased more easily and, together with a higher firing pressure, fuel consumption rates will decrease. Finally, long-stroke engines yield an improved combustion quality, a better charge renewal process inside the cylinder, and higher mechanical efficiency.

A good compromise between an optimum stroke–bore ratio and the costs involved, however, is an important factor: it is a general rule that the longer the engine stroke, the higher the costs for an engine per kilowatt of output. In the late 1990s, there was a trend for medium-speed engines to have a higher bore to stroke ratio but this has reversed with the wider bores of dual-fuel engines being the main reason.

A decision on engine speed is the next step. The mean piston speed in metres per second can be calculated from the bore and speed of an engine using the following formula: mean piston speed (m/s) = bore (m) × speed (rev/min)/30. The upper limit of the mean piston speed is primarily given by the size and mass of the piston and the high forces acting on the connecting rod and crankshaft during engine running. A mean speed between 9.5 m/s and slightly above 10 m/s is quite common for modern large bore medium-speed engines. Any substantial increase above 10 m/s will reduce operational safety and hence reliability. Since medium-speed engines may be specified to drive propellers and/or alternators, the selection of engine speed has to satisfy the interrelation between the frequency of an alternator (50 or 60 Hz) and the number of pole pairs.

In designing the individual engine components, two main criteria are addressed: reducing manufacturing costs and reducing the number of overall engine parts to ease maintenance.

A new factor that has entered the equation and that never really troubled earlier generations is the choice of fuel types available today, tomorrow, and further into the future. In the period between 2000 and 2010, the medium-speed dual-fuel market was dominated by Wärtsilä and its W50DF.

Developed from Wärtsilä's very successful type 46 diesel engines, the Wärtsilä 50DF engines have cylinder dimensions of 500 mm bore by 580 mm piston stroke. Available in configurations with six, eight, and nine cylinders in line, and 12, 16, and 18 cylinders Vee-form, they develop 950 kW per cylinder MCR at 500 or 514 rev/min for 50 and 60 Hz electricity generation, respectively.

The Wärtsilä 50DF engines can be run alternatively in gas mode or liquid fuel mode. The engines are also fully capable of switching over from gas to liquid fuel (marine diesel oil) automatically should the gas supply be interrupted, while continuing to deliver full power.

Gas fuel is supplied at a low pressure (less than 5 bar) to the engines. In gas mode, the Wärtsilä 50DF engines operate according to the lean-burn Otto process. Gas is admitted into the air inlet channels of the individual cylinders during the intake stroke to give a lean, premixed air–gas mixture in the engine combustion chambers. Reliable ignition is obtained by injecting a small quantity of diesel oil directly into the combustion chambers as pilot fuel which ignites by compression ignition as in a conventional diesel engine.

The Wärtsilä 50DF engines use a 'micropilot' injection with less than 1% of the fuel energy being required as liquid fuel at nominal load. Electronic control closely regulates the "micro-pilot" injection system and air–gas ratio to keep each cylinder at its correct operating point between the knock and misfiring limits.

Whatever the driving factor for development of dual-fuel engines may have been in shore-based power production where medium-speed engines dominate, in marine use the only initial applications were in LNG carriers where a ready supply of fuel in the form of boil off gas (BOG) was available. BOG was already being used to power the steam turbines which all LNG carriers until 2002/3 used for propulsion.

While Wärtsilä were promoting the use of dual-fuel medium-speed engines for LNG carriers as a previously untapped market for diesel engines, MAN pursued a different strategy promoting the concept of HFO-fuelled low-speed engines in a direct mechanical propulsion system and the reliquefaction of the BOG to increase the quantity of cargo delivered.

Increasing interest in the use of LNG as a means of meeting the Tier II and Tier III NOx limits saw most of the leading medium-speed engine makers researching and developing dual-fuel engines of their own.

The slightly lower volumetric power density of LNG compared to oil fuels necessitated a small increase in bore sizes for dual-fuel engines relative to the oil-fuelled engines they would complement. Thus, Wärtsilä's 50DF was based upon its W46 model, and likewise MAN's first dual-fuel engine, the 35/44DF introduced in 2012, was based upon the conventional 32/44 and MaK's M46DF on the oil-burning M43 engines.

One exception to the idea that engine development is based on evolution rather than revolution is the Wärtsilä 31 engine announced in 2015. Intended for multiple roles in power productions as well as marine use, the engine was not, as might be expected, a development of the established W32 series but a radical departure from it in many ways. Arguably, the most innovative development was that it has been designed for complete modular units to be used instead of individual spare parts, similar to the way that cartridges are used in some turbochargers.

The 31 series is available as a dual-fuel engine, an oil-burning engine, or as a spark ignited pure gas engine. Wärtsilä is now achieving sales of the series, but its initial target was the offshore sector which unfortunately suffered a severe economic setback simultaneous with the announcement of the engine. As a consequence, few new ships have been ordered in the sector since 2014, whereas until then it had been the most buoyant shipping sector in what was otherwise a depressed newbuilding market.

CHAPTER NINETEEN

Anglo Belgian Corporation

Anglo Belgian Corporation or ABC, as the company is better known, may be a relatively small player in the overall marine engine market, but it is not insignificant with almost 800 vessels using its engines for main propulsion systems and more for auxiliary purposes. The company is also active in power station and rail engines. In shore power production, some of its engines have been running on alternative fuels produced from animal and vegetable sources for many years.

Following development work with the German technology company Heinzmann, ABC has also developed dual-fuel engines for marine use. It has in its portfolio several dual-fuel variants of its engines and more recently has been working with shipowner CMB in exploring the use of hydrogen in combustion engines.

Three medium-speed engine series have been fielded by ABC for propulsion and auxiliary power applications in recent years. The 242mm bore/320mm stroke DX type was produced in three, six, and eight in-line cylinder versions to cover an output band up to 883kW at 720 or 750rev/min. DX models are naturally aspirated; DXS models are turbocharged; and DXC models are turbocharged and intercooled. This series was dropped from the ABC portfolio by 2018 after the introduction of the new DL engine which was first announced in 2012 to coincide with the company's centenary.

The mainstay of ABC's products, the DZ type, is a 256mm bore/310mm stroke design offered in six and eight in-line and V12- and V16-cylinder form to serve power demands up to 3536kW at 720/750/800/900 and 1000rev/min. The turbocharged/intercooled engine is designed for burning heavy fuel with suitable adaptation. Its nodular cast iron piston features a Hesselman-type head which is oil cooled by the shaker effect, and four rings: a chrome fire ring, two compression rings, and an oil scraper ring located above the floating piston pin.

The cylinder liner is produced from a wear-resistant centrifugal cast iron slightly alloyed with chromium. The slightly alloyed cast iron cylinder head is fitted with two inlet and two Stellited exhaust valves. An intermediate deck assures a forced water circulation around the valve seats and the

Pounder's Marine Diesel Engines and Gas Turbines
https://doi.org/10.1016/B978-0-08-102748-6.00019-0

centrally mounted fuel nozzle. The one-piece crankshaft, forged from 42CrMo4 steel and with induction-hardened running surfaces, is arranged in a box-type crankcase of nodular cast iron. An additional bearing on the flywheel side limits crankshaft deflection and bearing wear.

ABC effectively doubled the power threshold of its programme in 2000 with V-cylinder variants of the in-line DZ design, the new V12 and V16 models offering outputs of 2650 and 3536 kW at 720–1000 rev/min (Fig. 19.1). A 45° V-bank was selected, fostering an overall width of

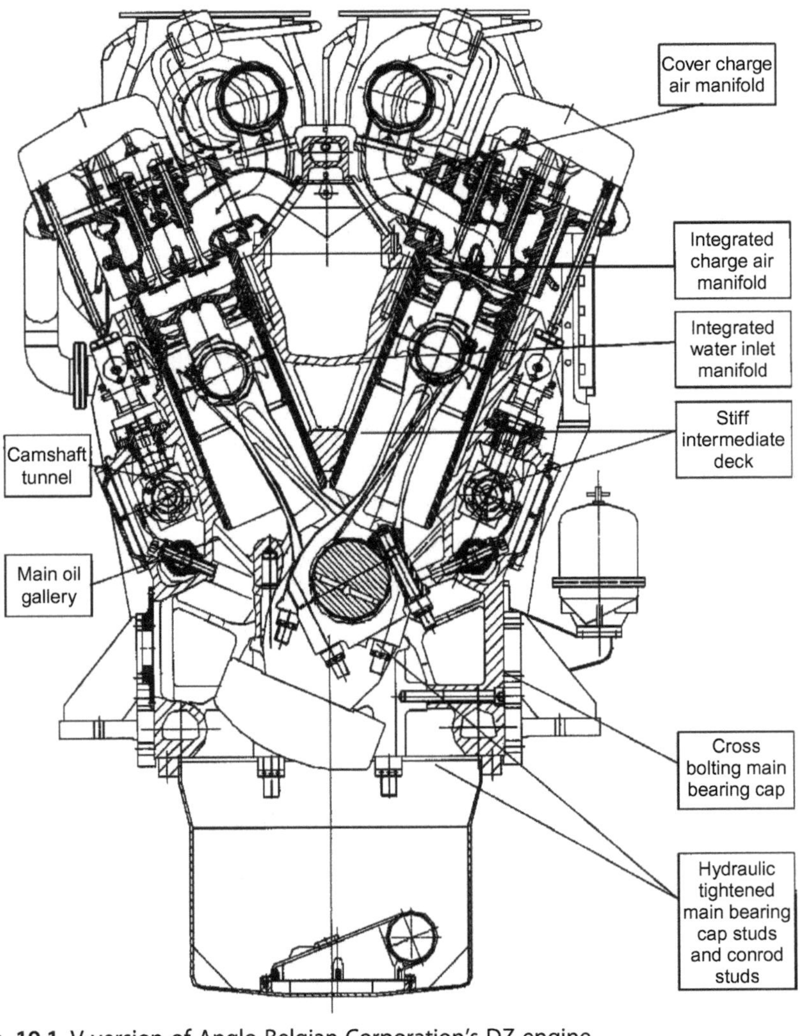

Fig. 19.1 V-version of Anglo Belgian Corporation's DZ engine.

1.65 m; an engine height of 2.5 m is quoted, with lengths of 4.18 and 4.94 m, respectively, for the V12 and V16 models. As much commonality of parts as possible with the in-line cylinder models was sought in the development, but new components were designed to permit a 20% rating rise in the future and to support a peak cylinder pressure of 150 bar. Easy accessibility to parts needing frequent control or regular preventative maintenance was also pursued.

ABC used the occasion of its centenary in 2012 to announce a new range of engines with a 365 mm bore. Initially, the series would contain just two in-line variants with six or eight cylinders. These would be named DL36 engines and would have power outputs of 3900 and 5200 kW. Planned development included adding V12 and V16 variants to the series which would increase the power options up to 10,400 kW.

The increase in power range over the DZ series was aimed at increasing the appeal of the engines to new market sectors. By far, the majority of previous ABC engine types were installed in fishing vessels, tugs, and inland waterways vessels. The D36 range was focused on passenger ferries and coastal ships as well as naval applications.

At 650 kW per cylinder, ABC claimed at the time of the launch that the engines would have the highest power output in their class. Two-stage turbocharging and EGR along with Miller timing and variable valve timing means the engine can comply with IMO NOx Tier III with no SCR system needed. The engines featured a new common rail system and can run on HFO or MDO. Dual-fuel versions were planned for the future.

Main engine data of the D36 engine range.

Cylinder power output	650 kW/cyl
Nominal engine speed	750 rpm
Number of cylinders	6/8 in-line 12/16 in V
Mean piston speed	10.5 m/s
Bore	365 mm
Stroke	420 mm
Swept volume per cylinder	43.9 dm^3
Compression ratio	15.5
Brake mean effective pressure	23.9 bar
Nominal peak firing pressure	210 bar
Engine design peak firing pressure	240 bar
Charging system	2-stage KBB
PTO power	100%
Fuel	MDO HFO (dual fuel)
Main applications	Genset and propulsion

In a paper delivered at CIMAC Shanghai in 2013, three ABC engineers described the development of the D36 series. Particular attention was given to meeting the IMO NOx requirements. In the engines, the Miller timing closes the inlet valve before the piston reaches the bottom dead centre so that the air inside the cylinder can expand and, in this way, cools down. The compression phase now starts from a lower temperature which evidently leads to a lower maximum combustion temperature and lower emission formation.

This measure has also a negative result in that there is less air mass in the cylinder, with the same charging pressure, which results in a lower lambda value, which could lead to higher temperatures and a higher soot creation, which is in contradiction with the desired result. The solution for this issue is an increase of the charging pressure using two-stage turbocharging.

This gives the expected results at nominal load, but at part load and start-up, there is a too low lambda value in the cylinder to maintain good combustion. At the low load points, the charging pressure could not be increased as the turbocharger is almost not working. For this reason, a VVT system has been chosen so that there can be a shift to a more normal inlet valve timing for low loads and to the Miller timing for the higher loads.

To compensate for the negative effects of the Miller timing and to obtain the target lambda of two while introducing additionally EGR, a very efficient charging system was necessary. This implied two-stage turbocharging, but at that time appropriate turbochargers were not readily available. After an intensive investigation, ABC chose to combine an axial low-pressure turbocharger together with a radial high-pressure turbocharger. This gives the advantage that both turbochargers can be built together very short on the exhaust side, which gives the desired compact design as well as the ease for isolation and shielding so that the temperatures are kept below the demanded 220 °C from the SOLAS regulation.

By choosing an axial design for the low-pressure turbocharger, which is also the largest and normally has the biggest inertia, this one will accelerate much faster compared to a radial one. Then the smaller high-pressure turbocharger is of the radial type, but as this one is much smaller, it will not be the limiting factor. Both turbochargers have been newly designed, especially for this two-stage application. The low-pressure turbocharger is the HPA 7000 from KBB, and the high-pressure turbocharger is the HSR 6, also from KBB.

An EGR system is also necessary to complete the means of meeting Tier III NOx levels. In the D36 engines, a part of the exhaust gases coming from the exhaust collector is deviated and cooled in a specially designed exhaust cooler which has a high temperature part and a low temperature part. After the low temperature part, a condensate trap is installed so that the cooled exhaust gases are dry.

These gases are then entered just before the air inlet of the second and high-pressure turbocharger, which makes sure that the exhaust gases are directly mixed in an ideal way. They again increase in temperature by compression so that the eventual residual moisture in the exhaust gas is directly evaporated again. Then this mixture of air and exhaust gas is fed into a second air cooler which cools it down before it is led into the air collector.

Fuel injection is by way of a common rail system from Heinzmann that is able to run on diesel oil, MDO, IFO, and HFO. The HFO will be a logical choice outside ECA areas, where the MDO with a low sulphur content will be mainly used in ECA. A rail pressure of 1800 bar is set, together with an accumulator set-up of the system to have the lowest possible pressure drop during the injection.

The crankcase is cast in nodular cast iron GJS 400 and uses cross bolts to the main bearing cap to ensure a stiff engine block design. The block is of the dry type, and a lot of systems have been integrated in the block design. The camshaft bearing holes are machined directly into the block so that they are perfectly positioned and give a very rigid support to the camshaft. The oil collector is also integrated in the block, and the necessary oil supply lines are drilled directly in the block. On the other side of the camshaft, there is an integrated air collector with a "waved" outer shape so that the sound production of this wall is minimized.

Below the air collector, a special gallery is located to collect the gases from the oil sump so that they can stabilize and so that the oil can already be separated as much as possible from the gases. The crankcase also has two supporting beams so that the engine can be installed with the use of elastic mountings.

By 2010, many engine makers were beginning to produce dual-fuel engines. ABC were to pursue this development in co-operation with Heinzmann. For the project, diesel engines optimized for dual-fuel operation were used. These optimizations enable conversion rates of steady state 95% to be achieved (95% gas/5% diesel).

The key challenges for the engine management system are the variable speed/load and the fact that the torque/power output of the engine is not known. Maintaining a high conversion rate in dynamic operation calls for sophisticated control concepts. In addition, rapid switching functions back to 100% diesel are used, for example, to prevent misfires due to insufficient pilot injection.

A gas metering control unit is used to control the gas mass flow rate. The gas flow rate provides a similar linear relationship to the power output as for the diesel level. This enables the diesel and gas power produced to be calculated and the engine to be protected against overload.

The engine used in the project had been successfully used for a long time as a dual-fuel engine for stationary generator applications. For use in the marine sector, the concept was significantly revised and adapted to reflect the ambient conditions. The diesel side is powered by a mechanical injection system which is connected to an electric actuator. In the dual-fuel mode, this diesel controller is still responsible for the actual speed/power control. An additional control loop then controls the optimum gas quantity. This enables the diesel injection system to respond dynamically to rapid fluctuations or step changes. If the load is disconnected, the engine can quickly be switched back to pure diesel operation. This prevents the diesel level from falling below the minimum required for combustion, thus causing misfires.

A key factor for dual-fuel use is protecting the engine against overload. In a fully optimized and utilized system, both the diesel injection system and the gas system (with a small diesel proportion as pilot fuel) can handle almost the entire engine load. When combined, the engine can easily be overloaded. In generator applications, the load signal from the generator management system is then used as a limit. However, in directly powered ships, there is no such torque/load signal. Therefore, the dual-fuel control unit determines the power generated by the diesel and gas fuels dynamically during operation. To calculate the power generated by the diesel fuel, the power depending on the fill level and speed is recorded on the test bench and plotted in a characteristic map.

To record the power generated by the gas, a gas flow metering unit is used, which uses a differential pressure measurement to determine the gas flow and can control it using a throttle valve. The gas data (density and calorific value) can be used to calculate the gas power from the gas flow. The dual-fuel controller transmits this data along with a gas power set point.

The total power that the engine delivers at the crankshaft is calculated from the actual gas power and the diesel power. As the gas quality will fluctuate according to the gas types in the different ports, a facility has been provided for specifying the gas quality using the visualization unit. A conceivable scenario would be that an engineer receives a measuring report for the gas when refuelling and has to enter the corresponding quality. Otherwise, a corresponding safety factor would have to be maintained.

Following the project with Heinzmann, ABC added dual-fuel versions of its DZ series to its portfolio in 2013 under the title DZD series. All models in the DZ range including the gensets as well as the main propulsion engines now have a dual-fuel equivalent. As with the original DZ engines, they are turbocharged and intercooled. Power outputs range from 720 kW for the 6DZD-720 rpm base engine to 2670 kW for the 16DZD-1000 which is the largest and most powerful model in the range.

CHAPTER TWENTY

Caterpillar

Caterpillar propulsion engines have a strong following in the smaller ship sector and their generating sets and auxiliary engines are found on a great many ship types. Most of their products branded as Caterpillar (the company also owns the German-based manufacturer MaK whose engines are covered in Chapter 21) are high-speed engines, although the company does produce a 280 mm bore medium-speed engine.

For a long time, this engine was marketed as the 3600 series which was a numerical step-up from 3400 and 3500 series of high-speed engines. The 3600 series naming was dropped in the early 2000s when the more logically named C280 series was introduced.

Since its introduction in the early 1980s, the 280 mm bore/300 mm stroke design has earned numerous references in commercial and naval propulsion and auxiliary sectors as well as in power station, locomotive, and industrial applications. Higher-rated, lighter-weight versions were developed in the 1990s and successfully applied to the propulsion of large fast ferries.

There is a strong following for the engine size in the fishing and tug boat sectors and more recently in offshore vessels of different types. A small number are also found as auxiliary engines installed on oil rigs.

The 3600 series programme embraces six- and eight-cylinder in-line and 50° V-configuration 12-, 16-, and 18-cylinder models. An output band from 1490 to 4020 kW at rated speeds of 750–1000 rev/min was initially covered but the upper limit from the V16-cylinder model was raised to 4320 kW and then to 5420 kW through improvements in combustion efficiency. Another 15% rise in specific output over the standard models in the mid-1990s addressed fast ferry propulsion demands, taking the maximum continuous ratings of the V12 and V16 models to 4250 and 5650 kW, respectively. Power-related modifications included a new high-pressure ratio turbocharger, a higher capacity aftercooler, new pistons and cylinder liners, and a new bearing design.

Further refinements for high-performance duty, progressed in conjunction with Bazan Motores of Spain, resulted in the 1998 launch of the

Pounder's Marine Diesel Engines and Gas Turbines
https://doi.org/10.1016/B978-0-08-102748-6.00020-7

V18-cylinder 3618 engine offering up to 7200 kW at 1050 rev/min. The same year brought a rating rise for the 3616 engine, enabling it to be released with an output of 6000 kW

3600 series

A standard specification 3600 series engine (Fig. 20.1) features the following key components:

Cylinder block: A one-piece casting of heavily ribbed, weldable grey iron alloy. The intake plenum runs the full engine length, providing even air distribution to the cylinders.

Crankshaft: A press forging, induction hardened and regrindable. A counterweight for each cylinder is welded to the crankshaft and then ultrasonically examined to assure weld integrity. The crankshaft end flanges are identical to enable full power to be taken from either end of the engine.

Bearings: The main, connecting rod and camshaft bearings are of steel-backed aluminium with a lead–tin overlay copper-bonded to the aluminium.

Connecting rods: These are forged, heat-treated, and shot-peened before machining. The special four-bolt design allows for an extra-large bearing which reduces bearing load, extends bearing life, and improves crankshaft strength and stiffness.

Cylinder liners: High-alloy iron castings, induction hardened, plateau honed, and water-jacketed over their full depth. The flange cooling provided for heavy fuel-burning versions reduces the top piston ring groove temperature.

Pistons: Two-piece components comprise a steel crown and forged aluminium skirt for strength and durability with lightweight. Two rings are arranged in hardened grooves in the crown and another two rings in the skirt. The top compression ring is barrel faced and plasma coated for greater hardness. The two middle rings are taper faced and chrome-plated, while the lower oil control ring is double rail chrome faced with a spring expander. Cooling oil is jet sprayed into passageways within the piston.

Valves: These seat on replaceable induction-hardened inserts. Positive rotators are provided to increase valve life by maintaining a uniform temperature and wear pattern across the valve face and seat. The exhaust valves specified for heavy fuel engines received special attention to extend their lifetime. Caterpillar reports that Vanadium-induced corrosion is significantly minimized by using a Nimonic 80A material in the valves and

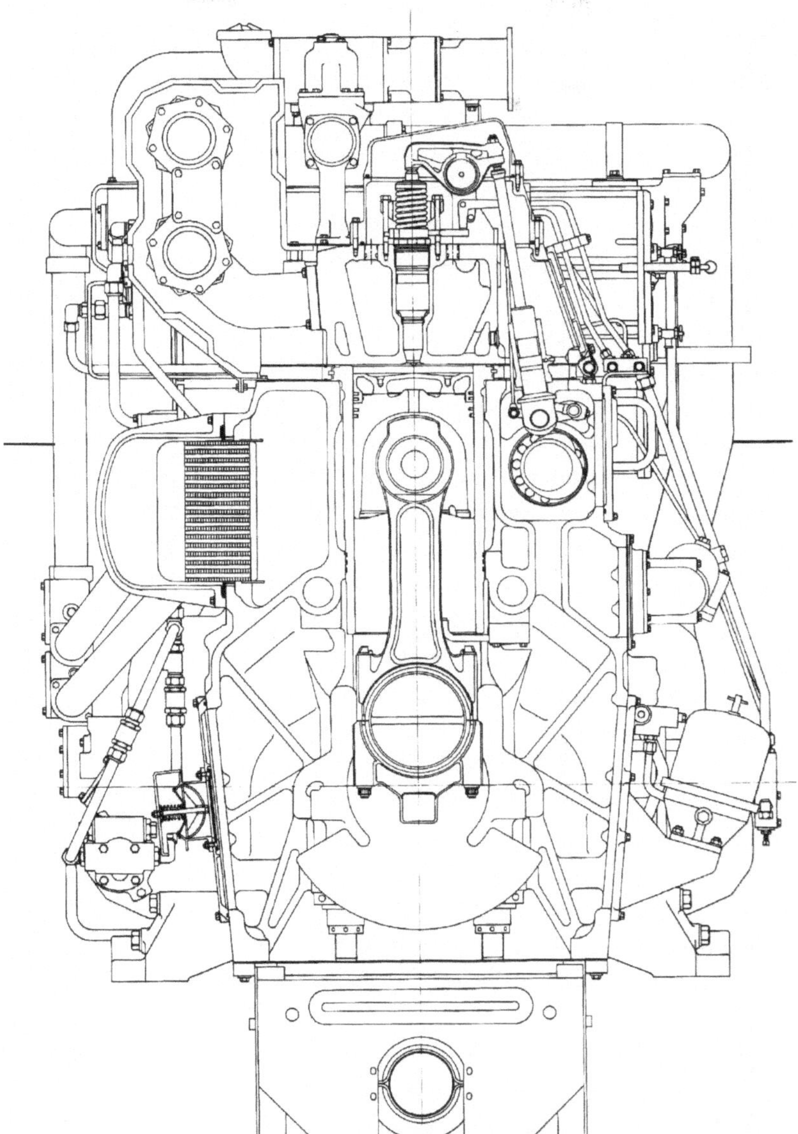

Fig. 20.1 Caterpillar 3600 series medium-speed engine in in-line cylinder configuration.

reducing the exhaust valve temperature to 400°C. A low temperature is maintained by exploiting valve overlap, water-cooling the insert seats, and applying a ceramic coating to the valves.

High-efficiency turbochargers (one set is mounted on the in-line cylinder models and a pair on the vee-models): Equipped with radial flow compressors and axial flow turbines. The axial flow turbine combined with a Caterpillar-supplied washing device is said to be well suited to heavy fuel operation. The turbochargers are water-cooled and their bearings lubricated with engine oil. The turbocharger technology combined with a large cylinder displacement and efficient aftercooling yields a high air/fuel ratio. The results are more complete burning for maximum efficiency and better cooling of the combustion chamber and valves.

Cooling system: A single external cooling system supplies water to the two water pumps. The system sends cool water to the aftercooler and oil cooler, and hot water to the jacket water system to combat fuel sulphur-induced corrosion. The option of separate circuit cooling is offered for heat recovery.

Lubricating system: This features a prelube pump and a priority valve which regulates oil pressure at the oil manifold rather than at the pump. Bearings are thus assured continuous lubrication if the filters plug. Duplex filters with replaceable elements allow servicing without shutdown.

Fuel injection: Caterpillar's unit injector system has reportedly proven itself in distillate, marine diesel oil, and heavy fuel-burning operation. Efficient combustion is promoted by a high injection pressure (1400 bar) and precise timing. The mounting of the unit injector pumps directly in the cylinder head, however, allows the external fuel lines to carry only 5 bar pressure, reducing the danger of line breakage. Injector tip cooling—reducing carbon formation on injectors—is available optionally for heavy fuel-burning engines. Individual control racks for each cylinder foster precise injection timing and minimize fuel wastage.

Exhaust and air intake systems: These contribute to fuel efficiency and response. The air induction system is routed from the turbocharger to the intake plenum with only one right-angle bend to the aftercooler housing. The exhaust manifold system is dry shielded to secure a low surface temperature.

The pumps, filters, and coolers are engine mounted. An accessory module—which can support an expansion tank, heat exchanger, instrument panel, heater, and other support elements—simplifies installation. It can be front-engine mounted or remotely installed up to 8 m from the engine.

Ease of maintenance was addressed by arranging the gear train components for removal or replacement without taking off the front or rear engine cover. The crankshaft damper is mounted outside the engine housing for optimum cooling and accessibility; bolted covers and replaceable nylon bearings permit rebuilding of the damper on site. Four inspection covers for each cylinder allow convenient inspection and service access to the connecting rod and main bearings, the segmented camshaft (one per cylinder), the pushrods and fuel linkage, the injector and valve train.

The 3600-engine family is released for operation on fuels with viscosity and contaminants up to CIMAC class K55 (700 cSt at 50°C) at speeds of 800 rev/min and below, and CIMAC class G35 (380 cSt) ατ 800 rev/min. The following features distinguish the heavy fuel engine from its distillate fuel-burning counterpart: water-cooled valve inserts; ceramic-coated exhaust valves; fuel injector nozzle tips optimized for high viscosity fuel; cooled fuel injector nozzle tips; higher mass flow rate turbocharger components; remotely mounted heated fuel filter; flange-cooled cylinder liners; and a turbocharger water wash system.

Caterpillar 3600 series engine data

Bore	280 mm
Stroke	300 mm
Displacement	18.5 L/cylinder
Cylinders	6/8 L, 12/16/18 V
Compression ratio	13:1
Rated speed	750–1000 rev/min
Mean piston speed	7.5–10 m/s
Mep, mcr	22–23.7 bar
Mep, csr	20–21.5 bar
Output range	1490–7200 kW

3618 (bravo) engine

A V18-cylinder derivative of the 3600 series was jointly developed by Caterpillar with the Spanish engine builder Bazan Motores to target high-speed ferry propulsion opportunities (Figs 20.2 and 20.3). The 3618 engine was initially released in 1998 with a fast commercial vessel rating of 7200 kW at 1050 rev/min; higher sprint ratings were available for military applications. The designers sought improved engine reliability and maintenance-friendliness in addition to a 17% rise in specific output (to 400 kW/cylinder). An improved power-to-weight ratio was also yielded despite increased component robustness, primarily by using nodular cast iron for the block, adding strength without increasing wall thickness and weight.

All critical operating temperatures—coolant, lube oil, and exhaust gases—are lower than previous 3600 series engines, despite the higher power output. An 80°C reduction in the inlet temperature to the

Fig. 20.2 High-performance version of Caterpillar 3600 series engine.

Fig. 20.3 One of four 7200 kW Caterpillar 3618 engines installed in a fast catamaran ferry.

turbocharger was secured, promising improved engine reliability and durability. A new seawater pump provided 33% more flow and 60% more suction lift. A significant contribution to higher performance came from an ABB TPL65E turbocharger, with a compressor bypass system closely matching the amount of intake air to the operating requirements and eliminating mid-speed instability. The new aftercooler was twice as large as before.

Other new elements include: a simplified and improved rear engine gear train (the idler gear removed to form a direct V-drive to the camshafts); new stainless steel fabricated exhaust manifold sections with permanent lagging to reduce maintenance as well as bulk; high strength Nimonic valves for both air intake and exhaust gas flows; a deep bowl steel piston crown connected to a new nodular iron skirt for added strength; interior piston cooling and wrist pin lubrication doubled with two oil jets per cylinder; and a new crankshaft counterweighted to optimize rotary mass balance. Oblique split connecting rods ease servicing: rods are pulled with the piston, allowing cylinder overhauls to be executed without draining the engine.

Some of these features were applied to an uprated 3616 engine, released in 1998 with an output of 6000 kW at 1020 rev/min (equivalent to 375 kW per cylinder) for fast ferry propulsion. A 6% power rise over the standard V16 model's 5650 kW was achieved without increasing peak cylinder pressure. Smoother acceleration, higher efficiency, lower emissions, and more flexible installation requirements were also gained. Optional fore and aft turbocharger mounting was offered, along with off-engine lube oil coolers and a compressor bypass air induction system. The bypass system promotes more efficient engine operation at full power without causing surging at part load, such as during acceleration.

The last of the 3600 series engines was produced not long after the C280 series was introduced.

C280 engine

An updated manifestation of the 3600 series was launched in 2005 in the shape of the C280 series, which benefited from electronic fuel injection to optimize economy and reduce emissions across the operating range. Six and eight-cylinder in-line and V12- and V16-cylinder configurations cover a power range from 1730 to 5420 kW for propulsion plants and from 1900 to 5420 kW for genset and diesel–electric applications (Fig. 20.4). A running speed of 900or 1000 rev/min can be chosen.

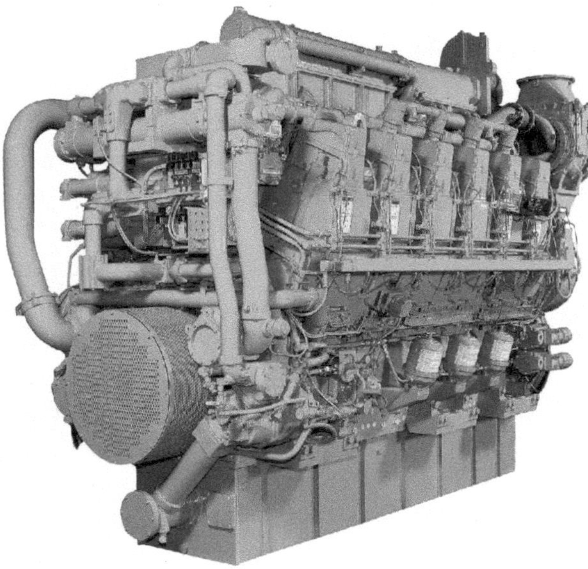

Fig. 20.4 A V12-cylinder version of the C280 engine.

All design parameters and key components—notably the cylinder block, crankshaft, connecting rod, piston, and camshaft—remained unchanged from the 3600 series but an important new feature, Electronic Unit Injection (EUI), replaced the mechanical fuel injection system. Caterpillar's EUI fuel system is based on cam-actuated mechanical pump/nozzle units which have the potential for modelling the course of fuel injection by electronic control. The system combines Caterpillar's ADEM III electronic control system with electronic injection units to secure high precision (Fig. 20.5). Combustion can be optimized for each engine speed and power by variable injection timing and duration. By optimizing the combustion process, the EUI system reduces fuel consumption in part-load operation as well as visible smoke and emissions.

Described as maintenance-free and requiring no fuel cooling, the ADEM III system features watertight electrical connections and a CANJ1939 data link; a back-up module can be configured to provide redundant control.

The electronic fuel system was already proven in Caterpillar's 3500B series high-speed engines. High-pressure fuel lines are not required, allowing threaded connections and double-walled lines to be dispensed within the low-pressure fuel system. The injection pressure is generated by a pump

Fig. 20.5 Electronically activated fuel injection nozzles serve C280 engines.

integrated with the injection units. Magnetic injection valves, replacing mechanical synchronization of the injectors, effect the fuel supply. The ADEM III system controls each injection valve in such a way that the injection timing and duration are optimally set to match the specific power and engine speed requirements for each specified torque.

In addition, the system monitors other engine parameters, such as the boost pressure generated by the turbochargers, and handles power measurement, programmable droop, controllable gain, and adjustable fuel/air ratio. With these features, which can be programmed by the operator, the fuel system can be tailored to the individual application. Histograms of engine operation can also be produced to provide evidence of performance over time.

A number of improvements incorporated in the C280 engine resulted from service experience with the 3600 series in marine installations, among them:

- High-performance aftercoolers, specified as standard, ensuring a lower boost air temperature and extended valve lifetimes.
- An anti-polishing ring in the cylinder liner, preventing the collection of sooty deposits above the piston crown, fostering optimum setting of lube oil quantity, and increasing the liner lifetime.
- Bimetal cylinder liners handling cyclic load changes and contributing to an extended component lifetime.

Field tests and in-service experience have reportedly confirmed the environmental and fuel economy benefits of the engine, along with low noise and vibration.

In 2014 Caterpillar introduced SCR to its marine engines in order to comply with impending NOx Tier III reduction regulations. Although some nonmarine engines were equipped with EGR systems instead of SCR, Caterpillar decided that SCR was simpler and cheaper to use for marine engines. The newer versions have a slightly higher power output at the lower end of the range and have a 1000 rpm speed rather than the 900 rpm of the first generation.

The C280 engines have proved as popular as the older 3600 series and in the same sectors. Unfortunately, the downturn in the offshore market has somewhat limited its sales figures in recent years.

Caterpillar C280-12 engine

Caterpillar C280-12 engine.

CHAPTER TWENTY-ONE

MaK (Caterpillar Marine Power Systems)

The German designer and engine builder MaK have a long history dating back to the early days of motor ships although there is a distinct break at the end of World War II when the company known as Maschinenbau Kiel was founded. In 1959 the company was acquired by Bremer Atlas and changed hands again in 1964 becoming a member of the Krupp group. There are many veteran ships still in operation with MaK or predecessor engines including almost 1500 constructed in the time period mentioned above.

MaK came under the umbrella of Caterpillar Inc. in 1997 as MaK Motoren, adding a valuable medium-speed portfolio to the US-based high-speed engine specialist's programme. The company was renamed Caterpillar Motoren in the year 2000, its Kiel and Rostock production facilities eventually becoming part of Hamburg-based Caterpillar Marine Power Systems. Some MaK models are also produced at a Caterpillar-owned plant in Guangdong, China

At the time of the takeover, the MaK range embraced five designs with bore sizes from 200 to 580 mm offering outputs up to 10,000 kW: the 'new generation' M20, M25, and M32 and the longer-established M552C and M601C models. The latter pair of designs has since been phased out and a fourth new generation series, the M43, was added in 1998. V-cylinder versions of this 430-mm-bore design were launched in 2002, extending the power range of the MaK portfolio to 16,000 kW.

The new generation of long-stroke medium-speed engines was headed in 1992 by the 200 mm bore/300 mm stroke M20 design for small ship propulsion and Genset drive duties. A power ranges from 1020 to 1710 kW at 900/1000 rev/min is covered by six-, eight-, and nine-cylinder models. The M20 was joined in 1994 by the 320 mm bore/480 mm stroke M32 (detailed later in this chapter). The power gap between the M20 and M32 models was bridged in 1996 by a 255 mm bore/400 mm stroke M25 derivative (Fig. 21.1) which covers an output band from 1800 to 2700 kW with six-, eight-, and nine-cylinder models running at 720/750 rev/min. All three engines shared common development goals:

Pounder's Marine Diesel Engines and Gas Turbines
https://doi.org/10.1016/B978-0-08-102748-6.00021-9

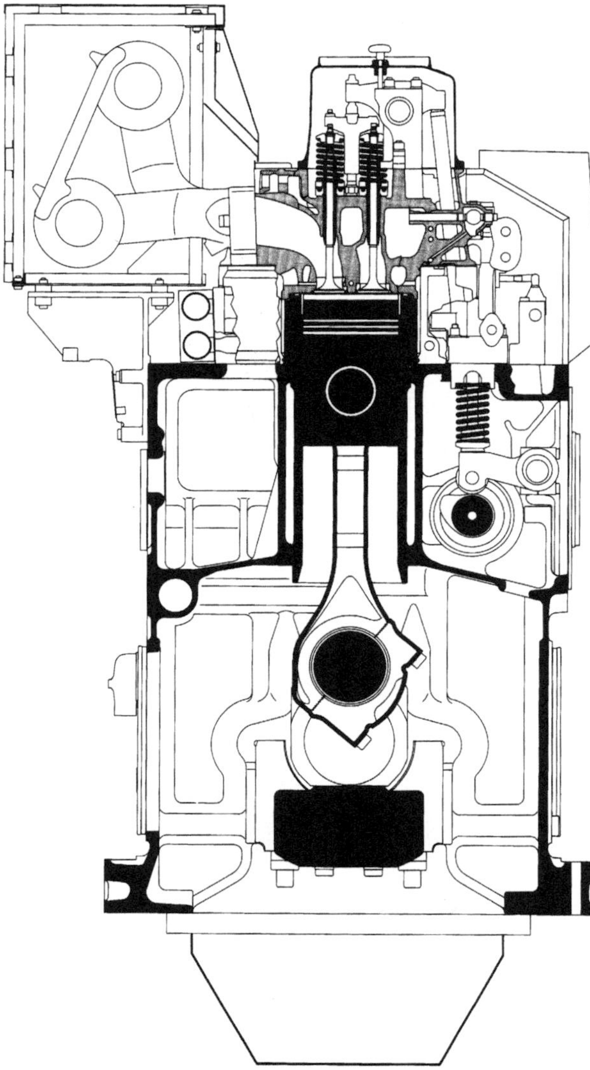

Fig. 21.1 The M25 design typified the new generation of engines from MaK (contrast with Fig. 21.8).

robustness, high reliability, low overall operating costs, simple maintenance, and extended intervals between overhauls.

Operational reliability was sought by avoiding corrosion, reducing mechanical wear and tear, increasing component strengths, and reducing the number of interfaces. Low operating costs were addressed by low fuel and lube oil consumptions, and low-grade heavy fuel-burning capability.

Modest maintenance costs were sought from a reduced number of components, a long service life from components, simple regulation and adjustment procedures, and plug-in connections.

Reduced installation work was achieved by providing simple and easily accessible interfaces for the shipbuilder. All connection points for fuel, lube oil, and cooling water systems are arranged for convenient access at the free end of the engine; and the fuel and lube oil filters are already mounted on the engine. The turbocharger mounting arrangement is variable, either at the flywheel end or free end of the engine.

High functional integration of components achieved engines with 40% fewer components than their predecessors. A reduced amount of piping work in particular (and hence fewer connection points) yielded benefits in assembly, installation, and maintenance. The remaining connections, where possible, are simple plug-in arrangements. Air, water, lube oil, and fuel are guided through bores and plug-in connections inside the individual components, leaving minimal external piping in evidence. The charge air duct, for example, is integrated in the crankcase. The air flows through channels in the crankcase and water guide ring into the cylinder head to the inlet valves. Thus, when maintenance work on the head is necessary, no charge air pipe has to be removed and any danger of leakage or working loose is eliminated. The camshaft is also integrated into the crankcase. Another example is the integration of a slide valve gear into the fuel-injection pump, eliminating starting air distributor, and control air pipes.

Summarizing the common design features, MaK highlights:

- Underslung crankshaft.
- Stiff engine block with integrated charge air and lubricating oil ducts.
- Dry cylinder block; cooling water only where necessary.
- Long stroke (stroke/bore ratio of 1.5 in the case of the M32 in-line cylinder engine).
- Up to 40% fewer components than earlier designs.

A high stroke/bore ratio fosters good fuel injection and combustion in a large combustion space; additionally, the high compression ratio underwrites low fuel consumption and emission figures.

M32 engine

The M32 engine (Fig. 21.2) superseded the well-established M453C series (Fig. 21.8), a 320 mm bore/420 mm stroke design which had been offered as an in-line engine with six, eight, and nine cylinders or as V12

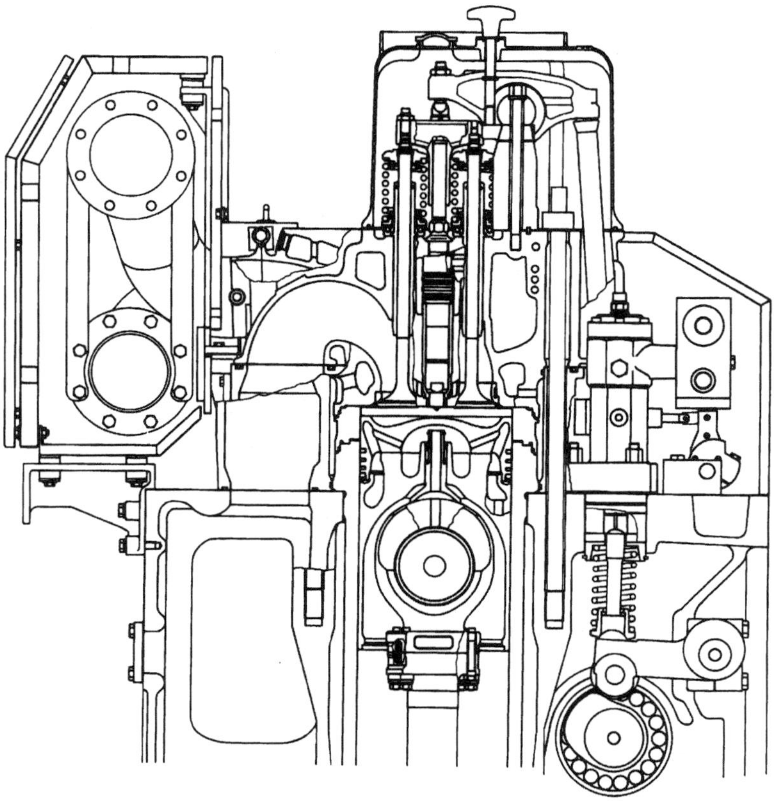

Fig. 21.2 Upper part of M32 engine.

and V16 versions with an output of 370 kW/cylinder. A substantial increase in power was yielded by the 320-mm-bore M32 design whose modular construction also achieved an engine with 40% fewer parts than its predecessor.

The in-line six-, eight-, and nine-cylinder M32 models have a 480 mm stroke and originally yielded an output of 440 kW/cylinder at 600 rev/min, covering an output band from 2400 to 3960 kW at the economic continuous rating (ECR). The rating was raised to 480 kW/cylinder in 1998 to meet market requirements. V-configuration models (12 and 16 cylinders) introduced in 1997 retained the 420 mm stroke of the M453C design and offered up to 480 kW/cylinder at 750 rev/min. The ECR outputs of the V-versions range from 4800 to 7700 kW.

The designer highlighted the following key features of the in-line cylinder M32 models which mainly targeted marine applications:

- *Optimum thermodynamic conditions:* A stroke/bore ratio of 1.5 promotes a favourable air/fuel mixture and combustion in a spacious combustion chamber. Simultaneously, a high compression ratio (14.5:1) fosters low fuel consumption—around 180 g/kW h at full load—and low noxious exhaust emission values.
- *Integral construction:* Advanced machining centres allow components to be fashioned to serve several functions. The engine frame, for example, embodies cast-on boxes for the camshaft drive on the flywheel end, and the vibration damper and secondary pinion gear for auxiliaries on the free end. The frame also conveys the charge air to the cylinders via a cast-in duct. The camshaft runs directly in the frame.
- *Stiff backbone:* The integral engine frame permits the firing and mass forces to flow through the frame, matching the flux of the lines of force and with low deformation. Girder-like cross sections run through the engine housing in a longitudinal direction yielding, together with stiff walls, a system with high bending and torsional rigidity. Rigid as well as resilient foundation work is easily effected, and structure- and airborne sound emissions are kept at a low level.

The 'supporting' and 'cooling' functions are clearly separated in the engine frame which is immune from any corrosion damage because no cooling water flows within it. The cylinder liner, however, is cooled where it is most important: in the upper region outside the engine frame.

- *Robust running gear:* The train from piston to crankshaft is designed for a high load-carrying capacity to secure operational reliability and provide reserves for future development in line with market demands. The integral construction ensures safe bedding for the crankshaft, camshaft, camshaft drive, and cam followers, fostering long service life values (Fig. 21.3).
- *Modular system:* The integrated subassemblies, such as the cylinder head and engine frame, are complemented by support system modules designed with a reduced number of components for easy preassembly and flexibility of application. An example is provided by the modular turbocharging package whose charge air cooler is housed drawer-like in a console screwed onto the frame (Fig. 21.4).
- *Functional groupings:* Arranged to facilitate operation and maintenance procedures. The camshaft side of the engine is easily accessible because there are no obstructions from charge air ducts and other mountings. The exhaust gas pipes are arranged on the opposite side. An optimized flow in the cylinder head ensures low flow coefficients in the tandem conduits

Fig. 21.3 M32 crankshaft installation.

Fig. 21.4 The M32 engine's charge air cooler is housed drawer-like in a console screwed onto the frame.

Fig. 21.5 A six-element hydraulic tool is used to secure or release the M32 cylinder head nuts.

for the inlet air and the exhaust gas. Simplified maintenance was addressed by specifying plug-in type connections throughout, with special tools only required where conventional tools could not guarantee an adequate degree of mounting security.

A six-element hydraulic tool is used to release simultaneously the six nuts securing the cylinder head studs (Fig. 21.5). A single-bolt tightens the clamped joint between the cylinder head and the exhaust gas pipe. The fuel-injection pump and nozzle can be disconnected by releasing just one screwed connection.

The connecting rod (Fig. 21.6) is split just below the lower edge of the piston, promoting a low removal height for the running gear. The bearings do not have to be opened when a piston is drawn.

- *Tough materials:* Essential castings, such as the engine frame and cylinder head, derive stiffness and security against a fracture from high-tensile nodular cast-iron structures. The material inhibits elastic movement of the frame when subjected to firing forces. The circularity of the liners

Fig. 21.6 The connecting rod of the M32 engine is split just below the lower edge of the piston.

Fig. 21.7 M32 cylinder head.

and the bedding of the plain bearings thus remain virtually unaffected and the axes of the gearwheels are maintained in parallel.

The material stiffness and stable design of the cylinder head secure a firm seating for the valves (Fig. 21.7). The high Young's modulus of the nodular cast iron, the thick combustion chamber bottom and the adjacent cooling bores combine to foster operational reliability from the chamber and valves. The piston skirt is also of nodular cast iron, while the steel crown is served by numerous cooling bores.

- *Low thermal load:* The measured exhaust valve temperature of 390°C means that the Colmonoy armouring of the valves is highly resistant to intermetallic corrosion attacks. A temperature of below 300°C measured on the outside of the piston crown permits only soft deposits.

Fuel injection can be adjusted for commencement, duration, and pressure. Specially shaped control edges on the injection pump plunger secure correct commencement of injection as a function of output and speed. The M32 engine can also be provided with eccentric control of the cam follower between the camshaft and the injection pump. This arrangement facilitates simple adjustment of the firing pressure to suit the particular fuel grade or the climatic conditions and achieve optimum economy and operation on heavy fuel. Control of injection pressure is effected by a MaK-patented system which is said to foster part-load running with particularly low emissions and to enhance the ignition of heavy fuels.

Conventional injection pump systems deliver fuel in pulses which can cause vibrations in the associated piping and damage to upstream systems. The M32 engine features a system based on larger volumes with intermediate throttles for damping the delivery process without restraining it. The remaining vibrations in the fuel admission and return pipes are reportedly so low that elastic transition elements can be fitted between the pumps.

Adjustable valve timing is also required for special operating modes or to influence emission behaviour. The pushrods for the inlet and exhaust valves are therefore arranged on cam followers which can be adjusted by an optional eccentric adjustment. This arrangement reportedly allows valve overlap and the turbocharging group characteristic to be adapted to secure good part-load performance.

Like its predecessors, the M32 design is served by a pulse charging system allowing permanent operation on the propeller curve. As an option for particularly difficult applications, the in-line engine can be specified with control of the turbocharger turbine surface by MaK's variable multi-pulse (VMP) system. Faster acceleration, lower temperatures, and higher part-load torques are possible. The designer has also developed systems for air bypass, exhaust gas blow off, and air blow off.

M32 engine data

Bore	320 mm
Stroke, in-line models	480 mm
Stroke, vee-models	420 mm
Cylinders	6, 8, 9L/12, 16V
Speed, in-line models	600 rev/min
Piston speed	9.6 m/s
Speed, vee-models	750 rev/min
Piston speed	10.5 m/s
Output/cyl	480 kW
Mean effective pressure	22.8 bar
Output range	2400–7700 kW

M32C engine

A C-version of the M32 engine was introduced in 2000 with design refinements promising cuts in operating costs, easier maintenance and installation, and performance and fuel economy boosted by a higher efficiency turbocharger (Fig. 21.8). Simple plug-in connections for the cooling water system, improved access to each cylinder unit in the camshaft region, single-pipe exhaust ducting and a newly designed exhaust pipe casing with a smaller installation volume, and a nodular cast-iron crankcase with integrated ducts are among the numerous modifications.

The entire exhaust system was simplified in such a way that only one exhaust pipe (instead of two) is now required for the six-cylinder engine; the number of pipes for the eight- and nine-cylinder models was reduced from three or four to one. Improvements were also made to the cooling water pipework at the supercharge air cooler. The cooler was modified by providing cast elbows as interfaces and simplifying accessibility to the cooler for cleaning. The application of stainless steel components on the 'cold side' also fosters an extended lifetime for these elements. The V12- and V16-cylinder versions of the M32C were introduced with two power ratings: 480 kW/cylinder at 720 rev/min and 500 kW/cylinder at 750 rev/min.

The C-versions were progressively introduced across the MaK engine programme to replace the original designs, the process was completed with the 2007 launch of the M25C.

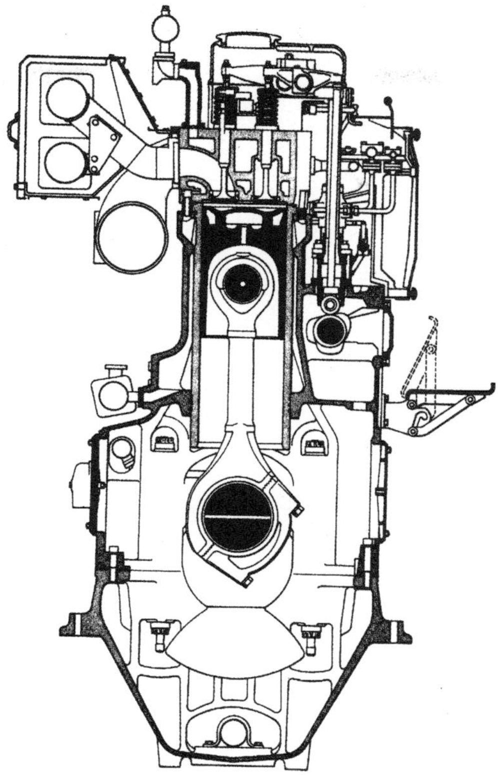

Fig. 21.8 MaK's 320-mm-bore M453C engine was replaced by the M32 design.

M43 engine

The fourth series of new generation long-stroke engines, the M43, was introduced by MaK in 1998, effectively replacing the 450-mm-bore M552C model and exploiting the same design philosophy as that of the M20, 25 and 32 series. With a specific rating of 900 kW/cylinder at 500/514 rev/min, the initial in-line six-, seven-, eight-, and nine-cylinder models covered an output range up to 8100 kW. The power band was extended in 2002 with the launch of V12-, 16-, and 18-cylinder (VM43) models taking the upper output of the series to 16,200 kW (Fig. 21.9).

A total of 80% component commonality is shared by the in-line and vee-cylinder designs, and simplicity of assembly was sought by combining as

Fig. 21.9 A V12-cylinder version of the 430-mm-bore M43 design (The VM43).

many functions as possible into single structural elements, thus reducing the overall number of components in the engine. Common components and component groups are the complete cylinder head; valve drive; cylinder liner with a flame ring; exhaust gas system (without compensator); camshaft bearings; fuel-injection pump; high-pressure connection; injection nozzles; cooling water distributor housing and line (without insertion tube); connecting rod and bearings; and vibration damper.

Cylinder head: A proven double-ended nodular cast iron design ensuring a high level of shape retention, the four-valve head is held on the cylinder liner by six hydraulically tightened studs (all the bolts are tightened simultaneously). Simple dismantling and assembly are fostered by insertion tube connections for all the media. The two exhaust valves are arranged to rotate, and the valve seating rings and ducts are aerodynamically optimized.

Piston: The two-part component comprises a steel crown and steel skirt. The crown is equipped with two compression rings and an oil stripper ring; the first ring is chromium-plated on the flanks. The oil-cooled shaker space of the piston ensures effective cooling at a high mean effective pressure and contributes to a long piston life.

Connecting rod: The split-shaft design allows low piston installation/withdrawal heights. The rod bolts and shaft bolts are tightened hydraulically.

(The size and weight of the hydraulic tools used for tightening bolts on the engine are minimized by operating with 2500 bar pressure technology.)

Special features of VM43 engine

Crankcase is a nodular cast iron design specified for resilient or rigid mounting. Stiffness in bending and in torsion is achieved by strong base strips and by the integration of the boost air and camshaft casings; the transverse bolting of the basic cover also contributes. The multifunctionality of the crankcase is increased by its integration of the gearing space and damper space, which avoids numerous bolted connections and contributes to noise and vibration damping. Integrating the oil pipes in the crankcase also contributes here and facilitates maintenance work. Good accessibility to the main bearings and connecting rod bearings is provided via large space openings. The crankcase is naturally free from cooling water and hence protected against corrosion. The first main bearing is of tandem design, reducing the engine length.

Crankshaft is manufactured from heat-treated alloy steel, with unhardened bearing trunnions and concave fillets. A flange is provided for the vibration damper at the end, opposite to the coupling; a gearwheel for the gear drive is flanged on the coupling end.

Exhaust gas installation is designed as a single-pipe system with gas outlets and pipe cross sections optimized for efficiency. The elements are the same as those used in the in-line cylinder engines. The different distance between cylinders is made possible by compensators of varying lengths.

Turbochargers are configured as a subassembly and equipped with a block aftercooler; the turbocharger group can be arranged at either end of the engine, facilitating electric power generation or propulsion drives. A 10,000 kW power take-off drive can be specified optionally at the front end of the engine. The turbocharger arrangement contributes to a minimized centre distance in twin-engine installations, a low vibration level, reduced engine height, a maintenance-friendly block cooler, and optimized exhaust gas flow. ABB Turbo Systems TPL73 turbochargers serve the V12 engine, and TPL77 models serve the V16 and V18 models.

Contributing to even lower emissions, the optional flexible camshaft technology (FCT) system was applied to a MaK engine for the first time, promising NOx emissions of less than 8 g/kW h and soot emissions below the visibility limit in any operational mode. This is accomplished by adjusting fuel injection and inlet valve timing relative to the given load.

M43 engine data

Bore	430 mm
Stroke	610 mm
Output	900 kW/cyl
Speed	500/514 rev/min
Cylinders	6, 7, 8, 9L/V12, 16, 18
Power range	5400–16,200 kW
Mean piston speed	10.2/10.5 m/s
Mean effective pressure	24.4/23.7 bar
Maximum combustion pressure	190 bar
Specific fuel consumption[a]	175 g/kW h
Specific lube oil consumption	0.6 g/kW h

[a]100% mcr.

M43C engine

The first two M43 engines were commissioned in April 2000 and by end of September 2004 more than 200 examples in service had accumulated an aggregate running time of over 1.4 million hours. Of these, some 90% operated on heavy fuel oil (HFO). By September 2008, the M43 and M43C reference list totalled more than 800 engines with a popular base in a feeder container ship, roro vessel, and ropax ferry propulsion.

In late 2003, Caterpillar Motoren strengthened the competitiveness of the M43 engine by raising the output from 900 to 1000 kW/cylinder at 500/514 rev/min. Uprating to C-status initially benefited the in-line cylinder models but was later applied to the V12- and V16-cylinder models, taking the upper power limit of the M43C programme to 16,000 kW. Apart from seeking a higher power output, the design and development work sought increased reliability, simplified maintenance, reduced noise levels, and lower emissions.

A redesigned cylinder block incorporated features from the established VM43 and M32C engines, such as the modular design lower valve and injection pump drive, the new governor drive and the removal of the mounting plate for the lube oil and cooling water pumps. The changes aimed to improve the maintenance friendliness of these functional groups. New covers for the camshaft drive area also improved noise levels, matching those achieved on the VM43 engines.

The new lower valve and injection pump drive design yielded greater simplicity and eliminated the need for adjustments. It also facilitated the installation of Flexible Camshaft Technology as original equipment or by retrofit, to achieve lower NOx and particulate emissions. NOx levels below 8 g/kW h (30% less than IMO Tier I requirements) and the elimination of visible smoke under part-load and transient conditions could thus be met by primary in-engine measures alone. Access to the camshaft area was also improved significantly.

A new vibration damper for the crankshaft was dictated by the higher power output, a component which can now be removed radially. Easier and safer maintenance was also fostered by removing the plate that formerly mounted the lube oil and cooling water pumps and integrating these assemblies into the cylinder block.

A higher efficiency turbocharger was necessary to maintain the favourable performance data of the original engine, resulting in a redesign of the complete turbocharger assembly group. ABB's TPL71C and TPL76C turbochargers were respectively specified for the six-cylinder model and for the seven-, eight-, and nine-cylinder models. A number of associated improvements were also made, including a new vibration-free charge air cooler, which can now be removed laterally in either direction.

Two other key elements—the aftercooler support and air-box of the turbocharger compressor outlet—were combined into a single housing. One interface was thus removed and potential leakages prevented. The turbine washing device was also improved to make its function more precise and safer; washing is carried out at part load to enhance the efficiency of the process.

Feedback from M43 engines in service was translated into design changes, an example being a new engine turning device designed to be vibration-free, positioned closer to the engine and incorporating a new turning gear and motor.

A higher power rating was also addressed by modifying the exhaust manifold covers for increased durability. Making the whole design more vibration-free was achieved by combining the cast iron blocks for the cooling water brackets with the supports for the manifold covers and detaching the covers from the exhaust pipes themselves. It was also decided to relocate the control cabinet for improved access and adopt the standard cabinet used with all other MaK engines.

Experience with the 900 kW/cylinder M43 engine suggested that improvements to the fuel-injection equipment were necessary. For heavy fuel operation, therefore, the injector was changed to a cooled type, as already used on the VM43 engine. The modification enhanced durability and reliability as well as reducing particulates emissions, especially at part load. Modifications were also made to the cylinder head to accommodate the cooled nozzle.

A feature of modern MaK engines is the minimal use of external pipework; and where a pipe is unavoidable it is fitted with a plug-in connection whenever possible. This principle is applied to the media guides supplying the cylinder heads with injection cooling and rocker arm lubrication and handling the fuel return from the injectors. These connections were unchanged for the M43C but at the far end of the engine, where the media return to the ship's system, plug-in connections were adopted. Overall, efforts were made to reduce the engine's component count by exploiting as many multifunctional parts as possible to promote simplicity and maintenance friendliness.

M43C engine data

Bore	430 mm
Stroke	610 mm
Cylinders	6, 7, 8, 9L/V12, 16
Output	1000 kW/cyl
Power range	5400–16,000 kW
Speed	500/514 rev/min
Mean piston speed	10.2/10.5 m/s
Mean effective pressure	27.1/26.4 bar
Firing pressure	205 bar
Specific fuel consumption	177 g/kW h

C-versions

The M25 engine was upgraded to C-version in 2007 with a slightly increased rating up to 334 kW/cylinder, resulting in a power band extending from 1800 to 3000 kW with unchanged fuel economy from six-, eight-, and nine-cylinder models.

A number of minor changes in the crankcase area fostered smoother engine operation and easier maintenance, the modifications including a modular lower valve drive, improved crankcase covers, and a new speed-governor drive. Well-proven in the larger M32C and M43C engines, the governor drive is a preassembled module with fewer parts and a lower part wear rate than before; its straight-teeth gearwheel drive requires no further adjustment.

An improved casting process for manufacturing the M25C crankcase yields better dimensional precision, and the optional deep oil pan benefits from a strengthened design which is approved for heavy fuel operations.

An unchanged engine footprint for rigid installation and gearbox connection addresses existing engine room designs. Improved damping for resiliently mounted installations further reduces engine vibrations and their transmission to the hull. Resilient mounting is also available for M25C engines with deep oil pans.

Manufacturing quality was raised by preassembling the complete turbocharger group, comprising the turbocharger, charge air cooler, and turbocharger bracket. The turbocharger itself is completely covered to meet the latest Solas requirements, the new cladding preventing hot spots while facilitating the maintenance of the turbocharger and piping. Like other C-engines throughout the programme, the M25C was designed to incorporate advanced emission reduction technologies developed for MaK engines.

After the introduction of the full range of C-versions, the earlier range was gradually dropped resulting in the C-version becoming the base of the MaK portfolio.

Dual-fuel DF-versions

Along with many other engine makers, Caterpillar recognized a growing demand for medium-speed dual-fuel engines around the time of ECAs being established in Northern Europe and looking ahead to the 2016 introduction of the IMO NOx Tier III rules.

The two largest bore sizes were identified as being suited for the development of a dual-fuel version. In line with what other engine makers were doing, Caterpillar adopted a slightly increased bore size for its dual-fuel with the 320 mm bore of the M32-C being increased to 340 mm in the M34 DF and the 430 mm bore of the M43-C increased to 460 mm in the M46 DF.

The first MaK DF engine announced in 2011 was the M46 DF followed the next year by the M34 DF. Both engines shared the same footprint with their diesel counterpart. Designed to operate on gaseous as well as liquid (MDO or HFO) fuels the Mak M34 DF and M46 DF meet IMO III emission limits while operating in gas mode, injecting a small amount of ignition fuel to control combustion. They comply with IMO II emissions regulations while operating in diesel mode.

The M46 DF dual-fuel engine carries a power rating of 900 kW/cylinder at 500 and 514 rpm in diesel and gas modes and is available in an in-line or vee configuration. It was designed to allow retrofitting of existing M43 C engines and therefore shares all the same interfaces as the earlier engine. Another design driver was catering for remote monitoring and engine system diagnostics. The M 34 DF dual fuel engine boasts a power rating of 500 kW/cylinder at 720 and 750 rpm in diesel and gas modes.

The smaller of the dual-fuel engines was intended mainly for use in offshore vessels but its introduction came at the time when that sector suffered badly from a downturn. As of 2019, no M34 DF engines for propulsion use have been contracted. By contrast, the M46 DF has enjoyed some sales success.

A first reference for the M46 DF came with the conversion of the existing M43 C propulsion engine of the 2013-built LPG carrier *Coral Aurelia*. Further sales have been made in this sector but most activity has been in the cruise and ferry arena.

E-versions

After the global economic crash of 2008 which greatly affected world trade and demand for new ships, the only buoyant sectors within shipbuilding were offshore vessels and passenger ships.

Although seemingly very different, these two sectors do share one common feature and that is the diesel-electric propulsion systems are often selected. With this in mind, in 2013 Caterpillar introduced a higher power version of the M 32 series, aimed at diesel-electric installations.

The new M32 E was based on the M32C platform, with the same footprint, but using pistons and other components from the V-format M32C. It featured a power rating of 550 kW/cylinder at 720 and 750 rpm. The M32 E is available in four to nine-cylinder in-line configurations, the nine-cylinder version being the company's first 5 MW engine in the 320-mm-bore sector.

In addition to having HFO capability, the new M32 E featured the same lube oil consumption and maintenance intervals as the M32C product series. The engine is EPA Tier 2 and IMO II compliant and fits in the makers range between the M32C and the more flexible dual-fuel M34 DF engines.

The following year Caterpillar launched a vee variant in two sizes; the VM32 E is available as either a 12 cylinder or 16-cylinder unit setup for mechanical propulsion or as a diesel generator. These two models have power outputs of 6720 and 8960 kW, respectively.

The two VM engines are intended for offshore vessels, they feature a power output of 560 kW/cylinder, 100% PTO capability, and a large portfolio of application-driven options. An optional Part Load Optimization Kit (PLK) for constant speed applications reduces fuel consumption at part load for ships which often operate in DP2/DP3 mode.

Caterpillar developed a wet oil lubricating system to reduce wet engine weight, engine height, and make the installation of the engines simpler. The lube oil systems include a dry lube oil pan, a mechanical lube oil supply pump, and a lube oil storage tank below the engine. With the wet oil lubrication system, Caterpillar offers a solution with a reduced scope of supply but without disadvantages.

The dry oil pan and lube oil storage tank are housed in the wet lube oil pan. The installation is easier and the height of the engine and lube oil system as well as the lube oil volume itself is reduced. The resilient mounting of the wet oil pan concept reduces the transfer of engine vibration into the ship and so increases crew comfort levels.

To make the generator set installation as easy as possible, and to reduce height, weight, and dampen vibration from the engine, Caterpillar developed a new base frame. In this design, the engine is resiliently mounted on the base frame with conical mounts. The base frame itself is reduced in weight and height to make transport and handling during the installation phase easier.

To meet the requirements of the NOx Code, the E-versions make use of SCR.

Low emissions technologies

Three emission levels for MaK engines were envisaged by Caterpillar Motoren in 2000 to address short- to medium-term regulations: a baseline IMO engine fulfilling Marpol 73/78 Annex VI requirements; an IMO-compliant engine with invisible smoke emissions; and a low emission engine (LEE) meeting the expected NOx emission range of IMO II and also having invisible smoke.

Such a strategy favoured the adoption of in-engine measures offering advantages in cost, complexity, and maintenance, with an increased compression ratio for the base engine a key factor for low NOx emissions. A previously typical standard compression ratio of 11–12:1 was raised to 14–15:1 for IMO I, with IMO II calling for ratios of around 17:1.

Another cornerstone of the MaK LEE concept is the Miller cycle, whereby the engine's inlet valve timing is modified to achieve cooler combustion. For IMO I requirements only a small Miller effect of 5% was exploited but IMO II requires a Miller effect of 20%: a challenge for the turbocharger which has to provide boost ratios of around 5 to maintain today's mean effective pressure values.

By combining an increased compression ratio and the Miller effect, NOx emissions can be reduced by around 30% without sacrificing engine efficiency (brake-specific fuel consumption). Such a simple LEE, however, would suffer from poor load pickup at idling and visible smoke emissions at part load. The MaK LEE concept, therefore, exploits Flexible Camshaft Technology (FCT) to secure low NOx emissions, excellent load pickup, and invisible soot at all loads (Figs 21.10 and 21.11).

Fig. 21.10 Mounting a flexible camshaft technology (FCT) unit on a MaK low emission engine.

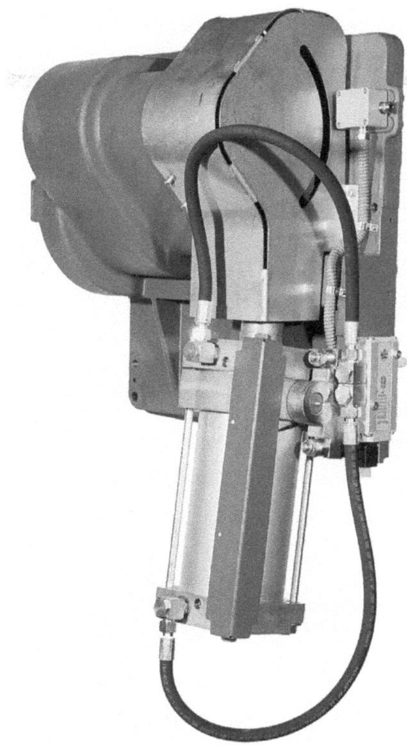

Fig. 21.11 The pneumatic adjusting unit of the FCT system.

Described as a robust mechanical solution, FCT facilitates variation of the fuel and air systems in part-load operations. By advancing the start of fuel injection and increasing the injection pressure, combustion improves and soot emissions reportedly reduces by 50%. Shifted inlet valve timing switches off the Miller cycle and contributes another 25% reduction in soot. Overall, FCT is said to reduce soot emissions by 75% while improving engine performance during transient operation.

FCT essentially comprises a modified lower valve train, a pneumatic adjusting unit and a programmable logic controller. An eccentric lifter lever shaft automatically influences fuel injection timing as well as pressure and valve timing. The engine load required to activate the lever can be flexibly set according to the operator's requirements. Either way, visible smoke is said to be eliminated while IMO NOx emission standards are satisfied at all loads. By the end of February 2008, around 80 MaK M32C and M43C engines had been specified with FCT. All existing MaK M20C,

M25C, M32C, and M43C engines at sea can also be converted to LEE standards, reportedly at around 15%–25% of the cost of a new IMO II-compliant engine.

Common rail fuel system

Over a decade of R&D and comprehensive prefield testing resulted in the Caterpillar Common Rail (CCR) system which was officially launched in 2006 (Although a common rail solution for MaK engines was tested as early as 1988 and unit fuel pumps featuring electronically controlled solenoid valves were developed in 1995.). The CCR system was initially offered for the M32C series but progressively extended to the smaller and larger bore members of the MaK portfolio. The complete system was also designed for retrofit to all C-series engines already in service (Figs 21.12 and 21.13).

Fig. 21.12 The Caterpillar common rail (CCR) fuel injection system made its debut on a six-cylinder MaK M32C propulsion engine.

Fig. 21.13 A CCR system rail segment and injectors.

Caterpillar highlights these merits of CCR technology: high fuel-injection pressure over the entire engine operating range; fully flexible fuel injection fostering optimized emissions and engine performance; and suitability for engines operating on HFO or marine diesel oil. A CCR system incorporates the following main elements:

High-pressure pumps: Two sets deliver the required amount of fuel to the rail and provide the desired rail pressure in closed-loop control. The pumps are based on a proven design but with modifications for heavy fuel operation. A two-pump configuration for all in-line cylinder engines significantly reduces the amount of high-pressure connections and components while securing redundancy; and suction control ensures high pump efficiency (Fig. 21.14).

Rail: The double-walled rails are pressurized and served as an accumulator. One rail segment feeds fuel to several injectors; for example, a nine-cylinder engine has three rail segments, each feeding three injectors, while an eight-cylinder engine has two rail segments, each feeding four injectors. Such an arrangement reduces the overall number of parts and high-pressure contacts required. Flow limiters prevent the cylinder from over-fuelling; and a safety valve acts as pressure relief in the event of an over-pressurized rail. A scavenging circuit enables the complete fuel

Fig. 21.14 Two high-pressure pumps deliver fuel to the common rail of the CCR system.

system (high and low pressure) to be maintained in a preheated condition, which is essential for safe HFO operation.

Injectors: A simple, robust, and compact approach was pursued by using the fuel itself to control the injector, thereby eliminating the need for a separate control fluid. The injector nozzle is cooled with lube oil, a typical feature for HFO operation. For the CCR system, however, the lube oil circuit was extended to provide cooling as well to the electrical components within the injector, thus increasing their lifetime. The injection process is electronically controlled, facilitating flexible injection timing and duration, including the capability for multi-shots. Injection pattern and shaping were optimized using the combustion modelling offered by Caterpillar's own 3D software as well as by comprehensive engine testing.

Injection mapping enables injection characteristics to be fine-tuned for every single engine operating point, reportedly guaranteeing optimum injection pressure and timing at a given load even in transient operation. Both soot and NOx emissions are thereby significantly reduced. There is also the potential through injection mapping to balance the overall wear of an ageing engine using modified control parameters implemented during regular servicing.

Electronics: Caterpillar's proprietary ADEM Electronic Control Module (ECM) manages fuel delivery, air supply, and other fundamental engine processes to achieve an optimum balance of performance and emissions. The software was tailor-made for CCR-equipped MaK engines, addressing the specific demands of HFO operation. The ECM functions can be divided into a core system and a performance system, the former comprising: an A4E4 control device; a crankshaft speed pickup; two camshaft speed pickups; two high-pressure sensors; two suction throttles; and injectors.

Optimum engine settings under all operating conditions are sought by the performance system, using such information as exhaust gas temperature, charge air pressure, ambient conditions, and lube oil temperature. The engine protection system works independently of the engine control system and has its own sensor equipment, ensuring that engine operation is monitored independently of other control systems.

Expert system support

MaK offers intelligent expert systems to lower operational costs, minimize downtime, and increase availability. DICARE is a Windows-based expert system providing real-time engine diagnosis. It describes and diagnoses the actual engine condition and provides references for condition-based maintenance. Remote data transfer to a ship owner's shore base is facilitated by a communication module; additionally, MaK offers a link to its service department for evaluating and interpreting the data.

DICARE was launched in 1987 on an MS DOS platform. A 16-bit MS Windows version in 1997 eventually led to the latest 32-bit variant. Data was initially received from the alarm system but today the engine's existing control bus system is leveraged. Measured values are displayed on engine schematics and gauges, with automatic data recording facilitating black box functions.

DIMOS is a computer-based maintenance and service system covering MaK engines. It delivers service interval supervision and immediate access to maintenance and component information via an electronic operations handbook. Bus technology underwrites safe digital signal transmission and represents a clearly defined connecting point for all ship management systems.

Earlier MaK models

Numerous examples of earlier MaK medium-speed designs are in service in propulsion and auxiliary power installations based on 240-mm-bore M282C/M332C, 320-mm-bore M453C, 450-mm-bore M552C, and 580-mm-bore M601C engines. The 'C' designation models were introduced in the 1980s with refinements enhancing the capability to operate on heavy fuel (Fig. 21.8).

The robust engines feature housings of high-duty cast iron which, in the case of the 320 mm bore and larger models, take the form of multisectional blocks connected by steel tension rods. The high prestressing of the rods, which keep the cast components under compression stress, prevents any tensile stress in the cast material, even under ignition loads. The 240-mm-bore series features an underslung crankshaft.

The cylinder liners of all the engines are salt bath nitrated for high wear resistance and anti-corrosion properties; crankshafts are forged in one piece from wear-resistant, quenched, and tempered alloy steel. Single fuel injection pumps are used throughout the series, the injection system and inlet/exhaust valves being controlled by a one-part camshaft.

CHAPTER TWENTY-TWO

MAN Energy Solutions

In terms of market share in the medium speed sector, MAN Energy Solutions consistently holds the number two spot behind Wärtsilä but enjoys a strong following across many engine sizes. Although a slightly later convert to producing dual-fuel engines it has introduced them to its portfolio and has notched up some success in sales.

Over the years MAN Energy Solutions has gone through several name changes and has acquired other manufacturers whose products have been incorporated into the MAN range of offerings. These include Alpha—MAN's Danish subsidiary, Ruston and Pielstick.

Today MAN Energy Solutions produces medium speed propulsion engines in 11 bore sizes and gensets in 12 bore sizes. The smallest genset is a 160 mm bore engine and the largest propulsion engine is of 510 mm bore. There have been larger bore size engines in the company's portfolio including the 580 mm bore 58/64 range the last of which was installed in 2011. The 58/64 was replaced by the 51/60DF as the most powerful engine in the portfolio with that engine being based upon the 48/60 itself a smaller variant of the 58/64. The following pages describe the evolution of MAN Energy Solutions' engines from the legacy engines of the 1980s through the development of electronic control and common rail and to today's generation of dual-fuel engines.

L58/64 engine

The 'new generation' programme was headed in 1984 by the 580 mm bore/640 mm stroke L58/64 series, whose designers sought a high overall operating economy, reliability, ease of maintenance, component durability, and unrestricted heavy fuel compatibility. The rugged (Fig. 22.1) result can be contrasted with MAN Diesel's previous large bore engine; the 520 mm bore 52/55B (Fig. 22.2).

Among the main features, the L58/64 and its 400 and 480 mm bore derivatives were a stiff monobloc frame casting with continuous tie-rods from the underslung main bearings to the top edge of the engine frame,

https://doi.org/10.1016/B978-0-08-102748-6.00022-0

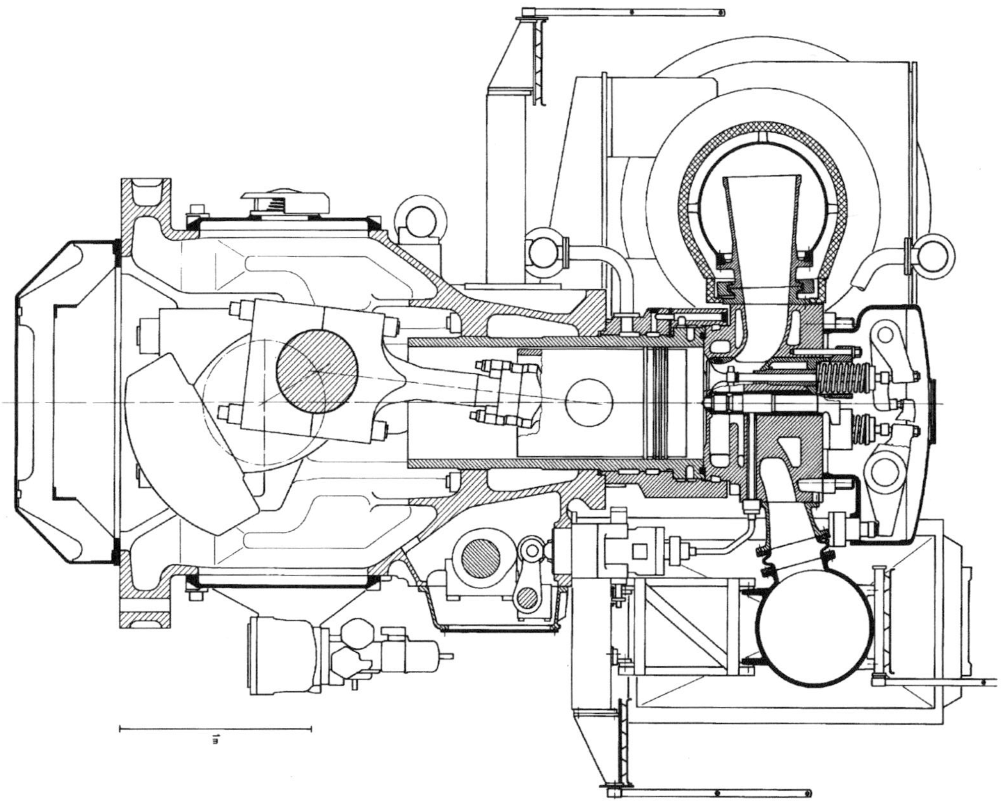

Fig. 22.1 MAN L58/64 design.

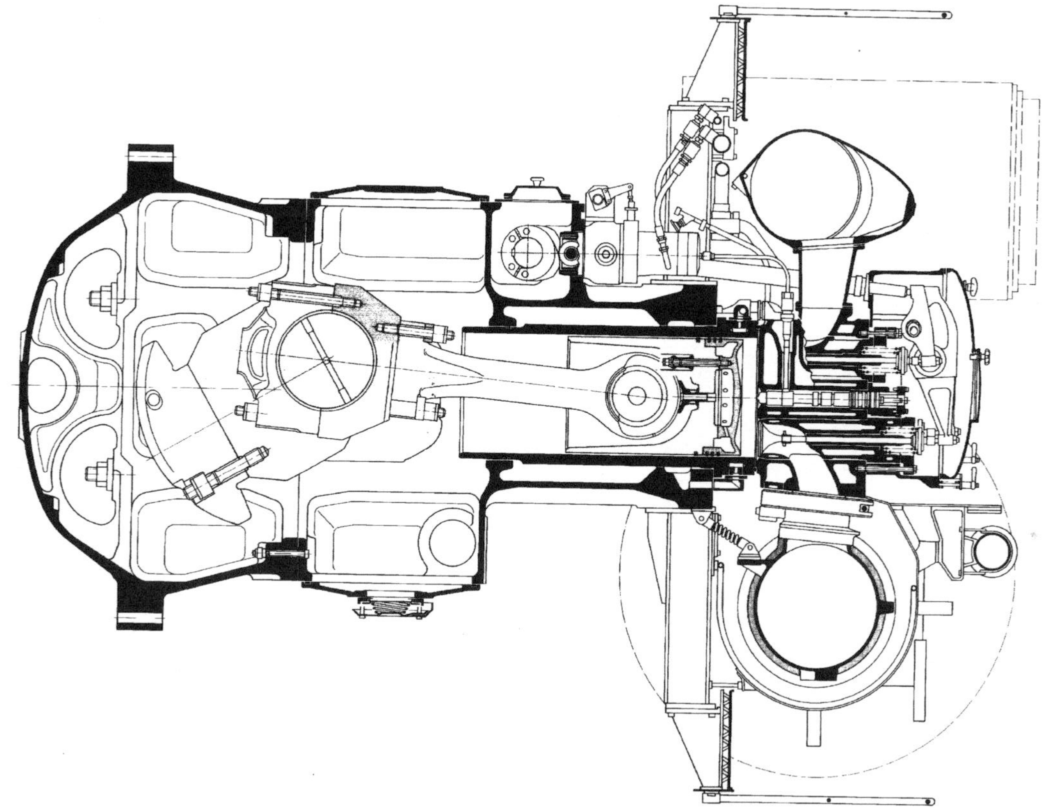

Fig. 22.2 The L52/55B design preceded the 'new generation' range headed by the L58/64 engine.

and tie-rods from the cylinder head to the frame diaphragm plate (Figs 22.3 and 22.4). This arrangement was associated with the following: an optimum flow of forces from the cylinder head down to the crankshaft; minimal deformation of the cylinder liner; more reliable piston performance; all parts of the frame under prestress; cross-braced crankshaft bearings; an amply dimensioned outboard bearing for absorbing high radial forces at the power take off end of the crankshaft; no cooling water passage in the casing (eliminating the potential for corrosion); sturdy landings to prevent plant deformation; and camshaft drive gear and vibration damper integrated into the engine casing.

Minimal deformation of the liner, reliable piston performance, no deformation of the neighbouring cylinder, and reduced deformation over the full length of the engine is further underwritten by individual cylinder jackets.

Cavitation of the cylinder liner (Fig. 22.5) is prevented by thick walls giving high resistance to deformation, while virtually uniform temperatures over the entire liner surface result from intensive cooling of just the upper part of the liner (cooling is not considered necessary elsewhere), thanks to a special arrangement of the cooling water spaces. Good lubricating conditions result and low-temperature corrosion is inhibited.

A thin flame plate for the cylinder head promotes good heat conductivity while high stiffness is imparted by a strong deck plate absorbing the vertically acting gas forces (Figs 22.6–22.8). A long valve life and clean, gas-tight seats are promoted by water-cooled exhaust valve cages which are withdrawable without removing the rocker arms; armoured exhaust valves with gas-driven propeller rotators; and caged inlet valves rotated by a Rotocap device. Pollution of operating media is avoided in the cylinder head by separate spaces for lubricating oil, fuel, and cooling water. A simple connection attaches the exhaust manifold to the head.

A composite piston (Fig. 22.9) is formed from a forged aluminium skirt and a forged thin-walled crown of high-grade steel (the smaller bore engines in the MAN programme feature a nodular cast iron skirt). Effective cooling of the crown is sought from the cocktail shaker method (splash effect) which nevertheless maintains temperatures in way of the ring grooves high enough to prevent wet corrosion. Cooling oil flow arrangements are optimized from inside to outside. A piston pin without cross-bore enhances component reliability. A reduction of mechanical loads on the piston rings was addressed by a narrow piston clearance, keeping out abrasive particles and protecting the lubricating film. A long

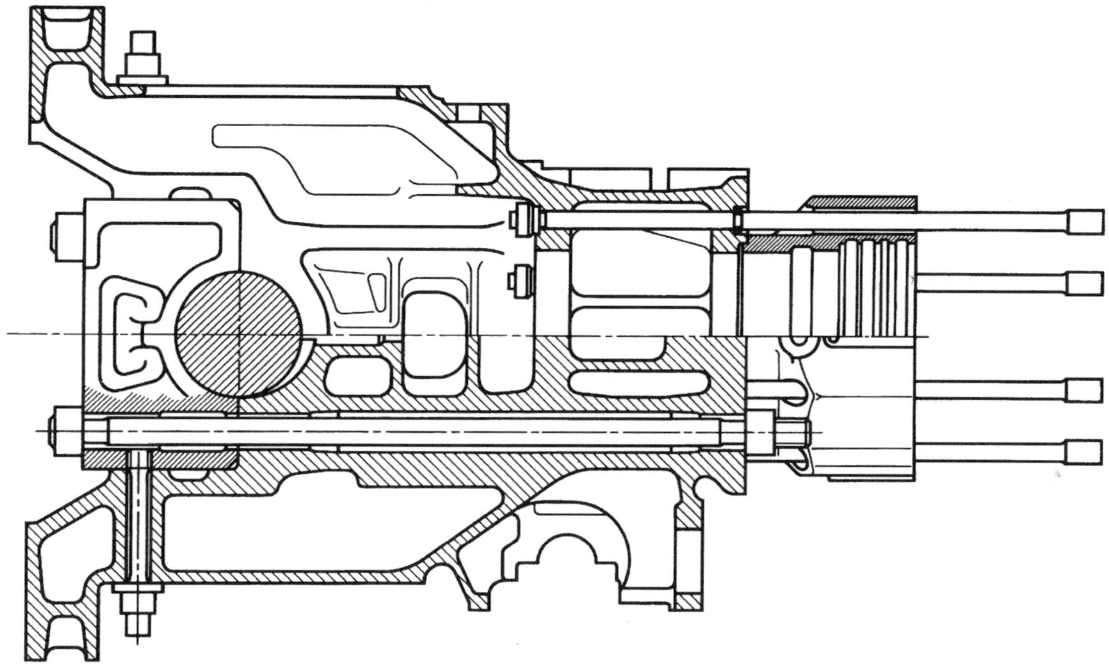

Fig. 22.3 L58/64 engine frame.

Fig. 22.4 Monobloc frame of L58/64 engine.

piston groove life was sought from induction hardening, with the depth of hardening sufficient to allow re-machining several times.

All three compression rings are mounted in the crown, their durability enhanced by plasma coating of the first ring and chrome plating of the second and third rings. A consistently low lubricating oil consumption is reported from the optimized ring set configuration.

The L58/64 engine was reportedly the first to feature a connecting rod with a joint in the upper part of its shaft (Fig. 22.10). A compact and very stiff flange is permitted, and motion of the joint under mass forces (observed in traditional connecting rods) is not encountered. Merit is the very low overhead space required for piston removal, with no need to open the crankpin bearing when pulling the piston (Fig. 22.11).

A fuel-optimized injection system with the economy plunger is installed (Figs 22.12–22.14). A modified control edge on the pump plunger keeps the firing pressure virtually constant between 85% and 100% of maximum

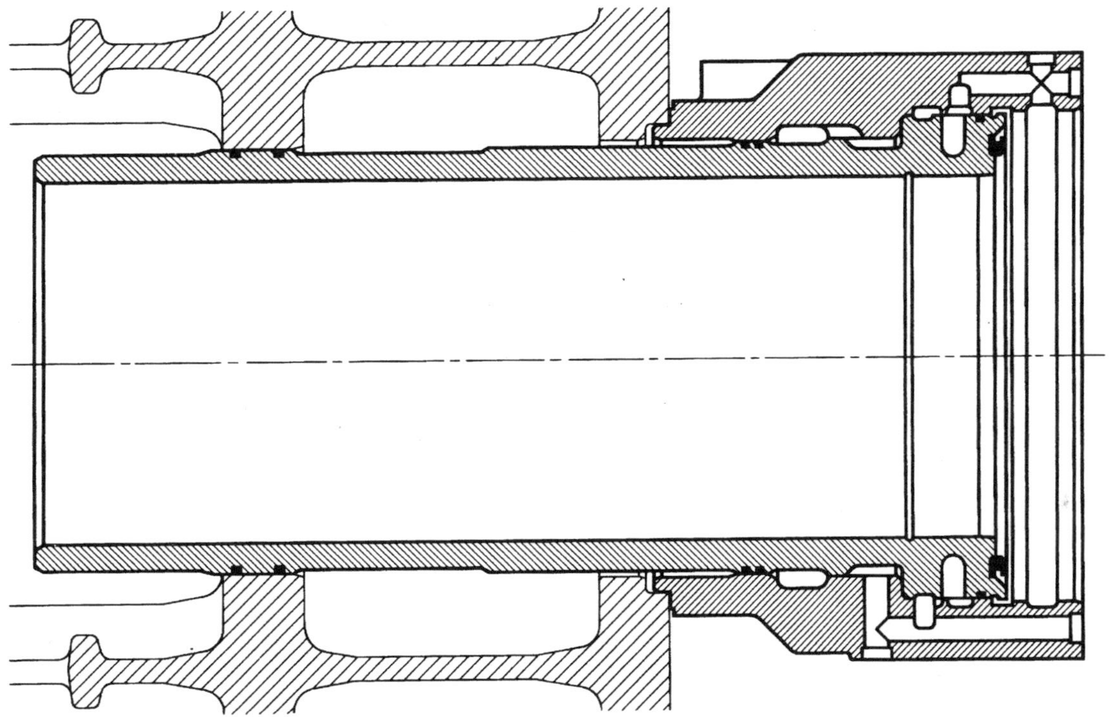

Fig. 22.5 Cylinder liner of original L58/64 engine. A flame ring was specified in later models (see Fig. 22.19).

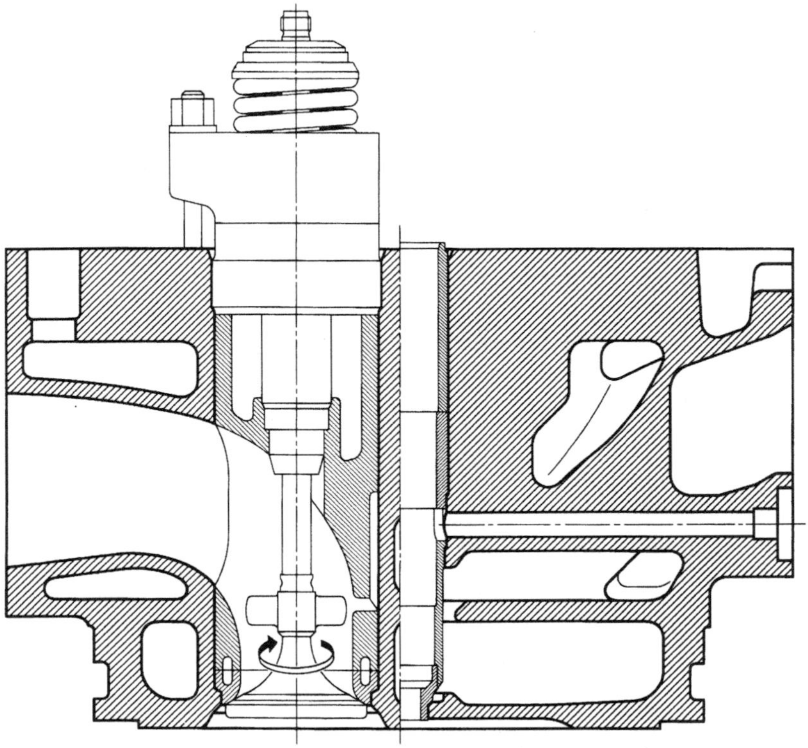

Fig. 22.6 L58/64 engine cylinder head; note the exhaust valve with rotator.

continuous rating, fostering favourable consumption rates throughout the part-load range. A pressure-equalizing valve prevents the formation of partial vacuums (and thus cavitation) as well as pressure fluctuations in the system (and thus fuel dribbling). A high injection pressure yields efficient atomization of the fuel, while an advantageous firing pressure/mean effective pressure ratio underwrites low fuel consumption. A rocking lever between pump plunger and cam—horizontally adjustable by an eccentric shaft—allows simple optimization of injection timing. The injection nozzles can be preheated for operation on heavy fuel oil (HFO).

High-efficiency and high airflow rates at full and part loads are delivered by a constant pressure turbocharging system based on uncooled MAN turbochargers. There is no need to preheat the air for part-load operation.

Fig. 22.7 Cylinder jacket, head, and cover of L58/64 engine.

Fig. 22.8 Cylinder head of L48/60 engine.

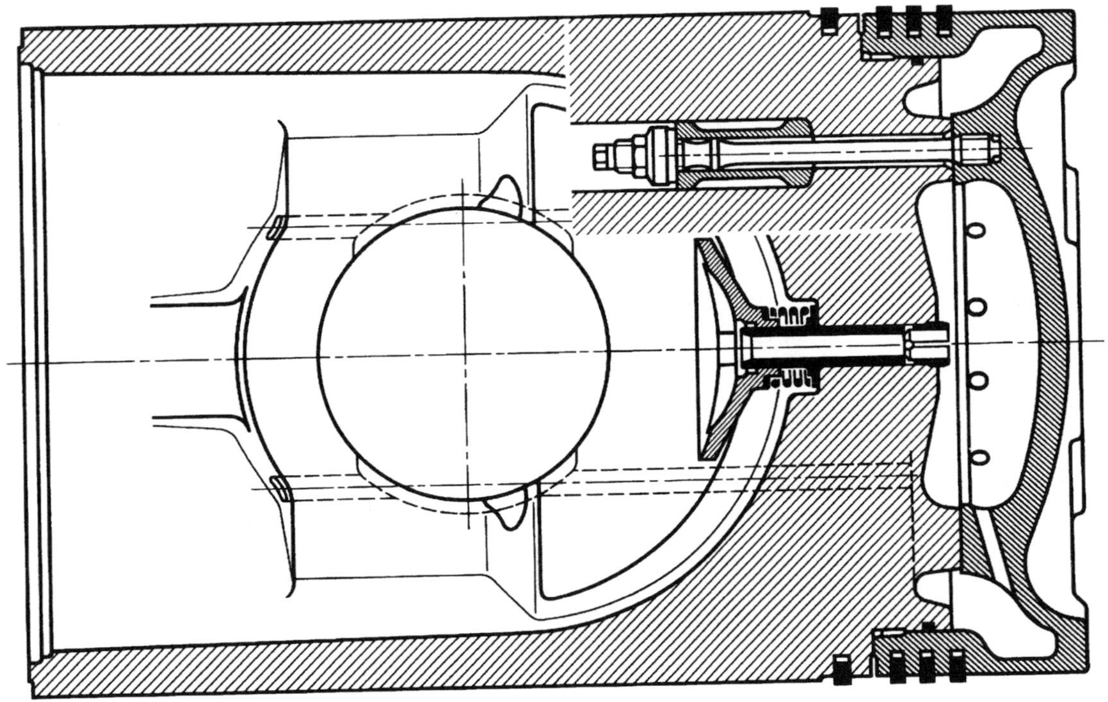

Fig. 22.9 L58/64 engine composite piston.

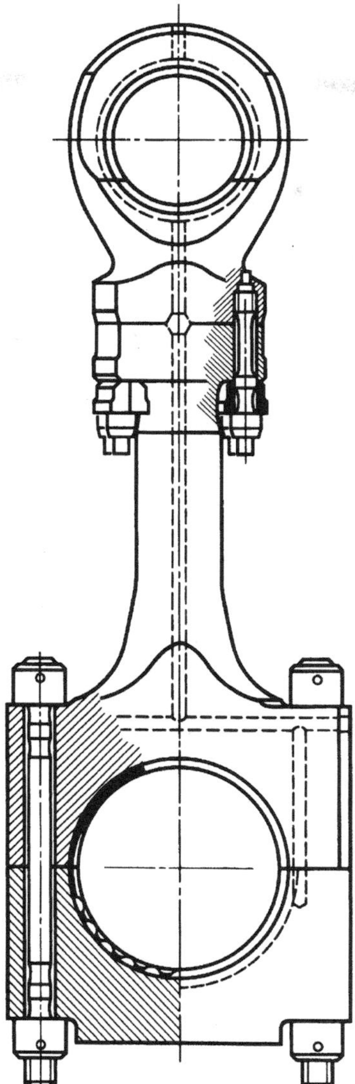

Fig. 22.10 The L58/64 engine connecting rod has a joint in the upper shaft.

Upgraded L58/64 engine

The ageing (although still popular) L58/64 design benefited from an upgrading in 2006, which targeted fuel savings across the operating range, reduced emissions while maintaining the existing power ratings. Measures to enhance reliability and accessibility for servicing were also introduced

Fig. 22.11 A low piston dismantling height is achieved by the joint in the upper part of the connecting rod shaft (L58/64, L48/60, and L40/54 engines).

as well as others seeking to extend maintenance intervals and shorten overhaul times. As an example, the original camshaft housing, which comprised sheeting and countless screws, was replaced by easily removed aluminium covers.

Approximately 70% of the various assembly groups forming the engine were either partly or completely redesigned, the modernization addressing all the components located above the crankcase. Many of the changes had already been implemented on either or both of the smaller bore 32/40 and 48/60B engines. The 48/60B engine was particularly inspirational, some of its concepts applied, tested, and adapted to satisfy the structural conditions of the larger bore model. Four key contributions to the upgrading were:

- Application of MAN Diesel's latest generation TCA turbocharger.
- A new concept for controlling valve actuation.
- A complete cylinder head redesign.
- Extensive redesign of the exhaust system.

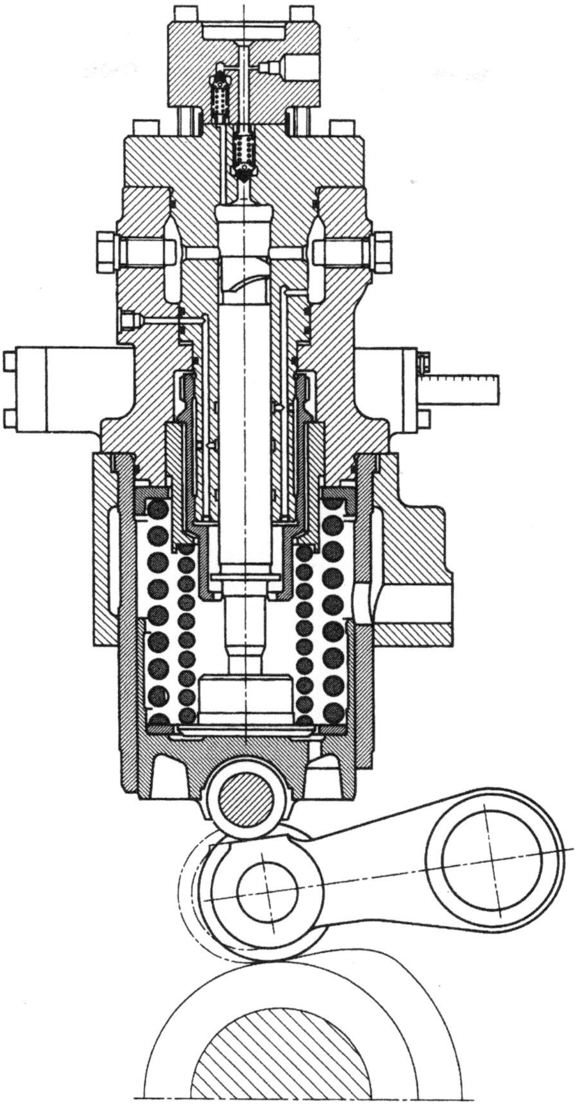

Fig. 22.12 L58/64 engine fuel injection pump.

A final assessment led to changes in the following main areas and assembly groups: turbocharger attachment, new cylinder head and combustion chamber design, new valve control, exhaust gas pipe, and its environment and media supply.

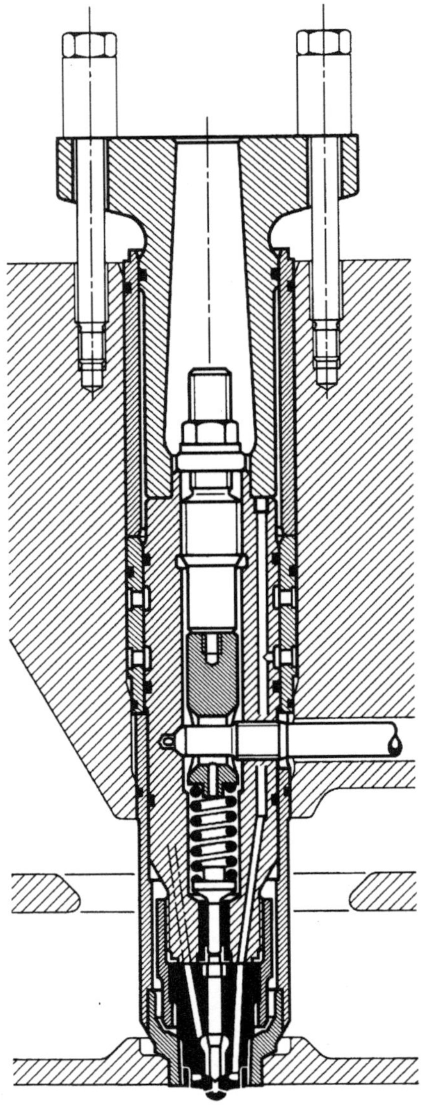

Fig. 22.13 L58/64 engine fuel injection valve.

TCA turbocharger series development by MAN achieved a high specific flow rate, high efficiency, low noise emission, maintainability, uncomplicated mounting to the engine, and high economic efficiency, and reliability. Taking advantage of these advances, TCA55 and TCA66 models displaced the MAN NA48 and NA57 turbochargers previously fitted to the L58/64 engine. Now positioned at the engine's side, the TCA models are also

Fig. 22.14 Main components of L58/64 engine fuel injection valve.

smaller and lighter than the predecessors. Additionally, a new turbocharger bracket was developed, weighing considerably less and requiring less room than the earlier fitting. The resulting free space was used for laying the supply pipes in a compact arrangement. A reduced number of parts involved contribute to lower overall manufacturing costs for the engine.

In developing the new rocker arm concept for the MAN L58/64 engine (Fig. 22.15), the designers fulfilled a brief to create a simple and robust arrangement fostering easy fitting, valve adjustment, and removal. A combination of this valve actuator and new casting technology enabled the integration of the charge air pipe into the rocker arm casing, dispensing with the support system for the old charge air pipe.

Although most of the basic concept for the L58/64 engine's revised cylinder head was derived from that of the 48/60B engine, it became apparent during the design process that adaptation to the structural conditions of the larger bore model dictated the modelling of a new head. By connecting the charge air pipe to the rocker arm casing—thus simplifying the routing of the

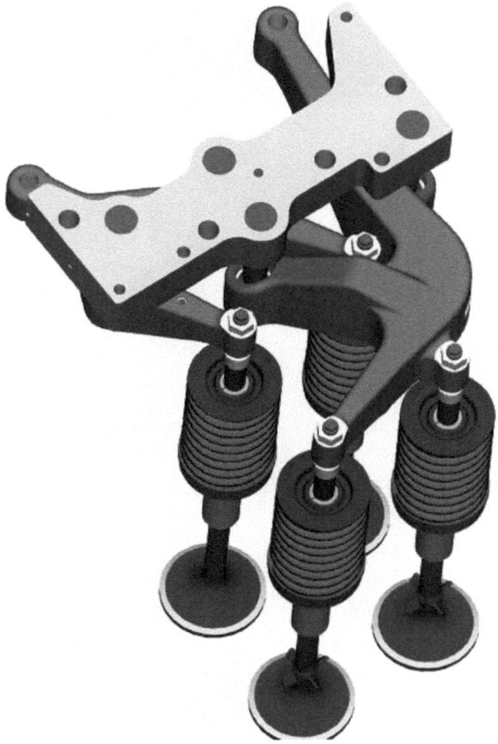

Fig. 22.15 New rocker arm arrangement for the upgraded L58/64 engine, the arms no longer sitting on a shaft but on two ball ends.

airflow—it became possible to design a flatter and more compact cylinder head. This was also facilitated by the operational reliability and wear resistance of the valves; renovating the valves and seats had not been necessary outside regular maintenance intervals.

Both the exhaust pipe and the area around it were redesigned, leading to a reassessment and tailoring of the insulation and covering, exhaust pipe supports, exhaust gas blow-off device, and charge air bypass to meet the demands set by resizing. A flow velocity increase called for a redesign of the gas passage between the cylinder head and exhaust pipe, leading to a 50% reduction in the pipe's cross section. This enhanced exhaust admission to the turbocharger in nonstationary engine operation, resulted in an improved turbocharger response time. As a large number of new components were integrated into the existing engine concept, MAN engineers took the opportunity to redesign the supply pipes, combining the passage of media to the engine at optimized connection points.

L40/54 and L/V48/60 engines

The L58/64 engine was followed on the market in 1987 by the L40/54 derivative, MAN Diesel's experience with 400 mm bore designs at the time extending back over 20 years. The 40/45 design was introduced in the 1970s as an updated shorter stroke version of the RV40/54 series which made its debut in the mid-1960s as a 'first generation' modern medium-speed engine. These older long stroke (540 mm) models were still produced by some licensees after phase-out from the official programme.

A third member of the family, the L48/60 series, became available in in-line cylinder form from end-1989. The prototype V-cylinder version, jointly developed with MAN Diesel's associate company SEMT Pielstick of France, was started up at the end of 1990 and production engines were offered from early 1992. The L/V48/60 series has since matured to B and CR versions, described later.

32/40 engine

The 'new generation' family (Table 22.1) was extended in September 1992 by a smaller member to contest the competitive 320 mm bore propulsion and auxiliary power arenas. The 32/40 series took the MAN medium-speed engine programme down to cover an output band from 2200 to 7920 kW with five, six, seven, eight, and nine in-line and V12, 14, 16, and 18-cylinder models. The five-cylinder model was later removed.

The 320 mm bore/400 mm stroke design (Fig. 22.16) originally yielded an output of 440 kW/cylinder at 720/750 rev/min on mean effective pressures and mean piston speeds at these revolutions per minute ratings of 22.8/21.9 bar and 9.6/10 m/s, respectively. Specific fuel consumption of just under 180 g/kWh was quoted at 85% maximum continuous rating. The 32/40 engine is currently released with a rating of 500 kW/cylinder, the

Table 22.1 MAN's original 'new generation' medium-speed engine line-up.

	Model			
Parameter	**32/40**	**40/54**	**48/60B**	**58/64**
Bore (mm)	320	400	480	580
Stroke (mm)	400	540	600	640
Cylinders	5–9L/12–18V	6–9L	6–9L/12–18V	6–9L
Output (kW/cyl)	480	720	1200	1390
Speed (rev/min)	720/750	550	514	428

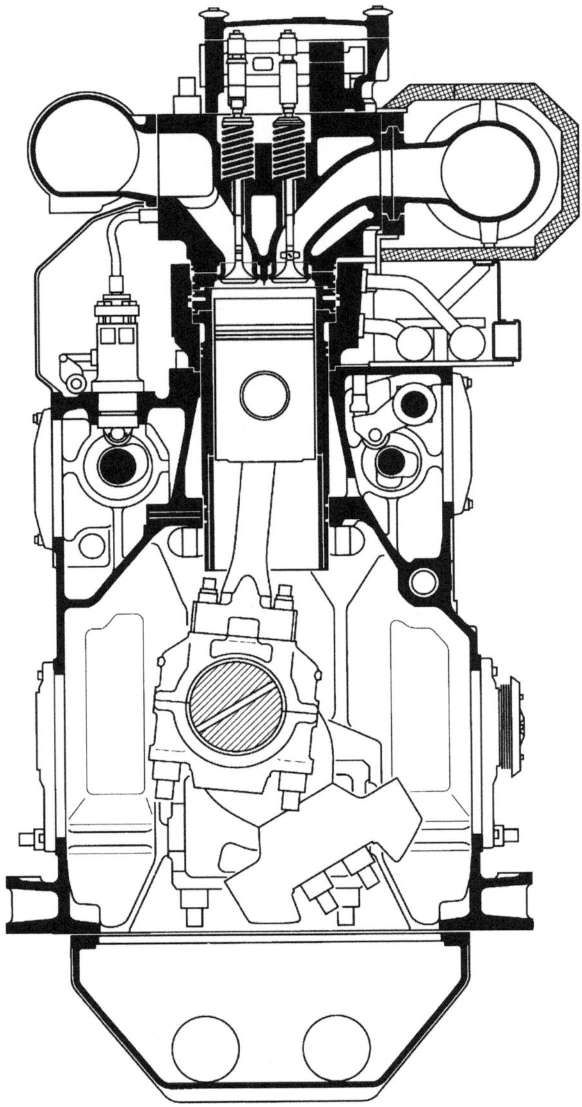

Fig. 22.16 L32/40 engine; note the separate camshafts for fuel injection and valve actuation, respectively, on the left and right sides of this cross section.

series now spanning a power band from 3000 to 9000 kW. The mean effective pressure is 24.9 bar (750 rev/min) and 25.9 bar (720 rev/min).

Proven elements of the larger bore brothers were downscaled for the 32/40 engine: a stiff one-piece frame with tie-rods; cylinder liners in the jackets; load-dependent cylinder lubrication; water-cooled liner fire rings with stepped pistons; and variable injection timing for optimum adaptation of the engine to various fuel grades (for both improved fuel consumption and reduced NOx emissions). Fuel economy was targeted by a high stroke/bore ratio, constant pressure turbocharging, a fuel-optimized injection system with 'economy plunger', pressure-equalizing valves, high-intensity atomization, and high compression ratio.

An innovatory contribution to operating flexibility was made by the adoption of separate camshafts (arranged on either side of the engine) for the fuel injection and valve actuating gear. One camshaft is dedicated to drive the fuel injection pumps and to operate the starting air pilot valves; the other camshaft, on the opposite exhaust side, serves the inlet and exhaust valves. This arrangement allows fuel injection and air charge renewal to be controlled independently, and thus engine operation to be optimized conveniently for either the high economy or low exhaust emissions.

In the larger bore MAN medium-speed engines, fuel injection timing may be influenced by a laterally displaceable rocking lever arranged between the camshaft and pump tappet. The timing of the 32/40 engine, however, is adjusted by turning the camshaft relative to the camshaft driving gear. Such a solution can be applied without restriction because the air charge renewal cams are mounted on a separate camshaft. The relative movement between driving gear and camshaft is via a helically toothed sleeve integrated into the gearwheel, meshing with the corresponding counterpart on the camshaft. By shifting the gear coaxially, the camshaft (which is fixed in the coaxial direction) can be turned infinitely relative to the gear by the required and/or desired angle (Fig. 22.17).

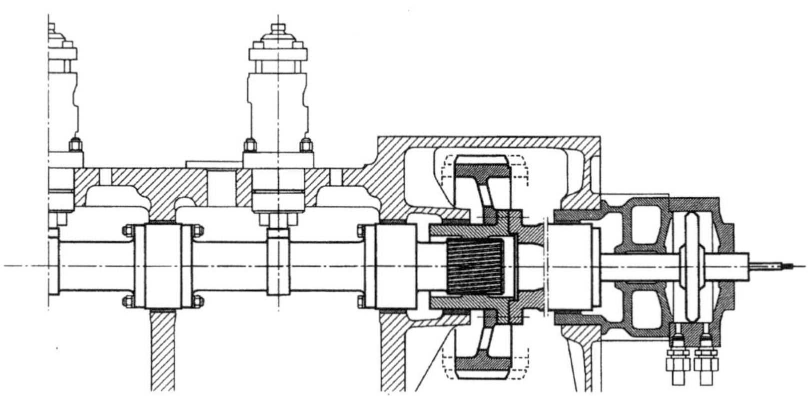

Fig. 22.17 An L32/40 engine camshaft arranged for adjustment of fuel injection timing.

The camshaft-shifting facility is optional; a simpler design is specified for engines whose projected operational profiles do not justify adjustable injection timing. Retrofitting of the facility is possible at a later date, however, if service conditions dictate.

The valve actuating camshaft is provided with different cams for full-load and part-load engine operation, allowing valve timing to be tailored optimally to the conditions. As with a reversible engine, the camshaft can be shifted from one final position to the other to engage the two different cam configurations. Shifting is performed with the engine running. The load point at which switchover from one cam contour to the other takes place can be preset as required via the control system of the shifting mechanism.

Load-dependent adaptation of valve timing promotes low pollutant combustion in both high- and part-load operating modes. Clean combustion with little emission of soot particles is reportedly secured over the full-load range.

The valve camshaft-shifting facility—like its fuel injection counterpart—is an optional device: the standard engine version features just one cam contour. The conversion to an adjustable system for the charge renewal side, however, can be effected at a later date. Independent camshafts, along with the ability to fine-tune the timing and fuel pumps, reportedly underwrite almost constant firing pressures over the 85%–100% rating range, optimized performance down to partial loads, and the ability temporarily to handle fuels with poor ignition qualities. Efficient operation on low-grade fuels with viscosities of up to 700 cSt is also fostered. An optimized cam- and camshaft-shifting facility on the fuel injection side allows IMO requirements on NOx emission levels to be met; and changeover from the open sea to coastal and harbour service (from fuel-optimized to emissions-optimized/SOx-reduced operating mode) is relatively simple.

The 32/40 design is based on a very stiff-cast monobloc single-part frame which is suitable for direct resilient mounting or semi-resilient mounting of the engine, with individual cylinder jackets as well as through bolts for the main bearings and lengthened cylinder cover bolts: a principle applied first in the 58/64 and subsequently in the 40/54 and 48/60 engines. The concept was modified for the smaller bore engine (Fig. 22.18).

Individual cylinder jackets allow the geometry of the cylinder liner to be maintained under all load conditions, avoiding disturbing influences from neighbouring cylinders. The main bearing through bolts and the cylinder cover bolts extending to the middle frame range place the engine frame

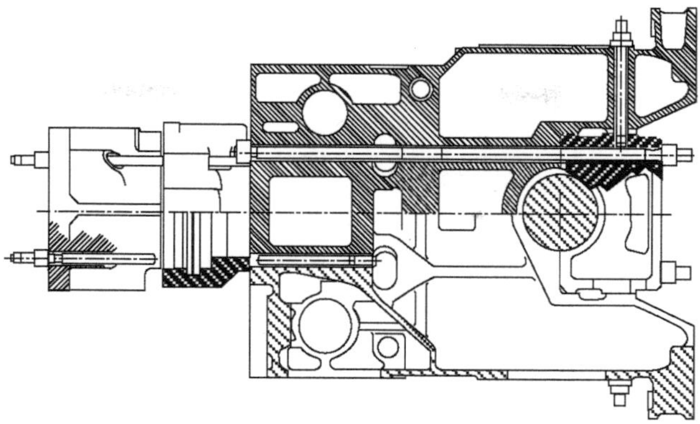

Fig. 22.18 L32/40 engine frame.

under compressive prestress, promoting high component reliability despite the high firing pressures.

The mounting arrangement of the cylinder liner in the jacket and engine frame is shown in Fig. 22.19. Cooling water is supplied to the cylinder jackets only; none circulates in the engine frame structure, eliminating the risk of corrosion, and water penetration of the lubricating oil. Individual cylinder jackets foster low distortion of the thick-walled laser-hardened liners. A uniform temperature pattern over the entire liner surface is reportedly secured through intensive cooling by a special water feed in its upper area.

A separate fire ring (antipolishing or carbon-cutting ring) for the liner in association with an appropriately stepped piston inhibits the formation of combustion residue deposits on the liner; liner surface polishing is thereby prevented and low lubricating oil consumption promoted. The fire ring features jet bore cooling like the previous one-piece liner design. Experience

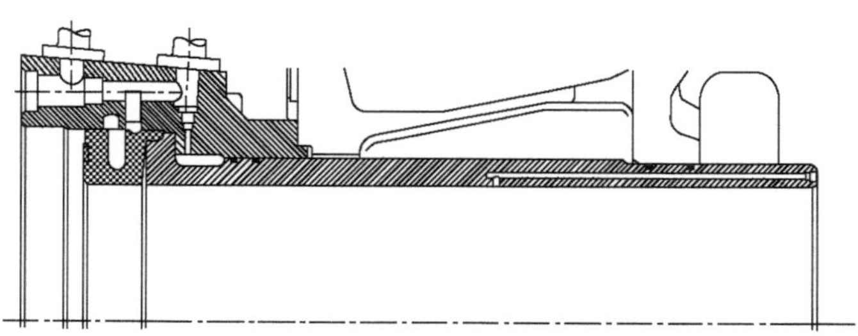

Fig. 22.19 L32/40 engine cylinder liner section; note the flame ring mounted on the top.

has shown that removing the ring—necessary for piston withdrawal—can be carried out quite easily with the simple tools provided.

A four-bolt cylinder cover was designed relatively deep to ensure an excellent flow of power from the nut contact faces of the cover bolts to the seal between the cylinder head and the liner, as well as a very uniform distribution of pressure at the circumference of this seat face (Fig. 22.20). The charge air and exhaust lines are arranged on opposite sides of the cover, the cylinder head depth allowing the flow ducts to be configured to enhance the overall efficiency of the turbocharging system. A bore-cooled flame deck yields low temperatures at the combustion chamber side and a high mechanical strength permitting the high firing pressures to be absorbed with a good safety margin. Valve cages were omitted. Both exhaust and inlet valves are accommodated in cooled seat rings, the exhaust valves featuring a rotator with a gas-driven propeller vane at the valve stem. The cylinder head, rocker arm casing, and a section of the charge air manifold form a unit for maintenance purposes which can be dismantled as a whole when a piston has to be withdrawn.

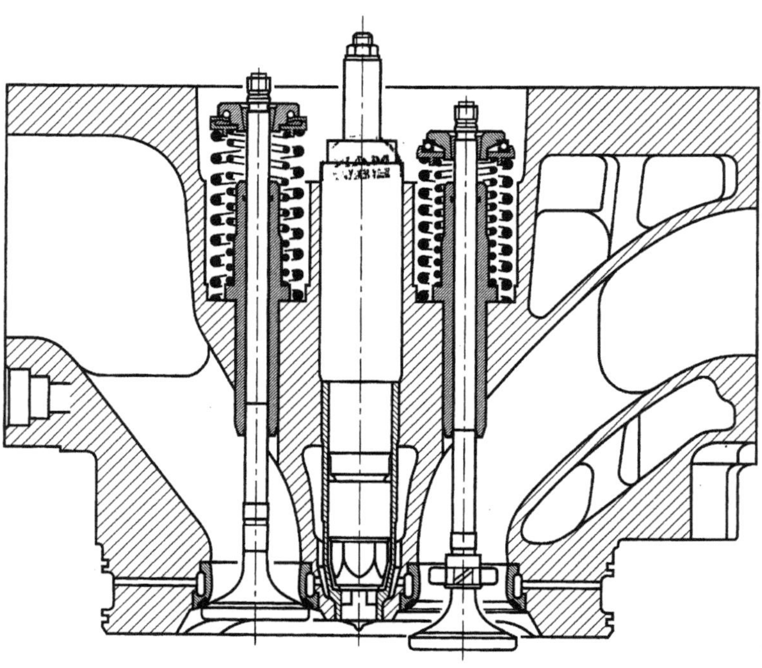

Fig. 22.20 L32/40 cylinder head.

The composite piston, similar to 40/54 and 48/60 engine practice, consists of a nodular cast-iron skirt topped by a forged steel crown whose diameter is reduced in way of the top land to match the fire ring (Fig. 22.21). The crown is cooled by the cocktail shaker method (splash effect) in its inner and outer areas, the latter is also served by bore cooling. The configuration of the cooling chambers ensures a good shaking effect with the cooling oil supplied via the connecting rod.

Chromium/ceramic coatings for the top piston ring and chromium-plated second and third rings are specified to contribute to ring time-between-overhauls of 2 years or more. A nodular cast iron skirt allows a smaller clearance with the rigid liner to be adopted compared with aluminium, this creating favourable preconditions for piston ring running behaviour.

Smaller overall dimensions and more restricted space in the crankcase of the 32/40 engine precluded the use of the connecting rod design used in the larger members of the MAN medium-speed family which features a joint in the upper part of the shank. The traditional marine head design was therefore adopted, with a joint arranged just above the big end bearing box (Fig. 22.22). No drawback is suffered with respect to the required headroom

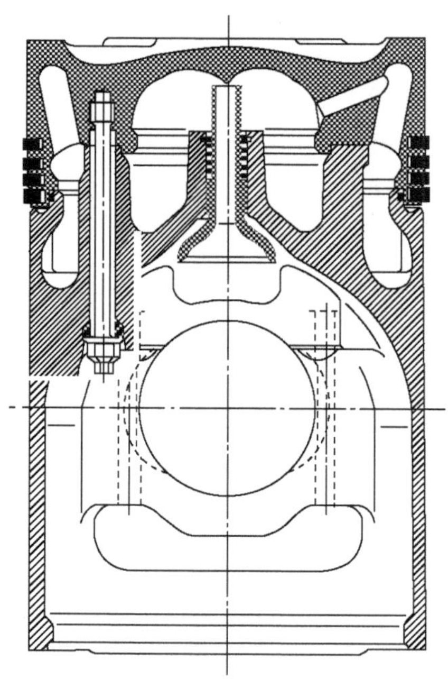

Fig. 22.21 L32/40 engine composite piston.

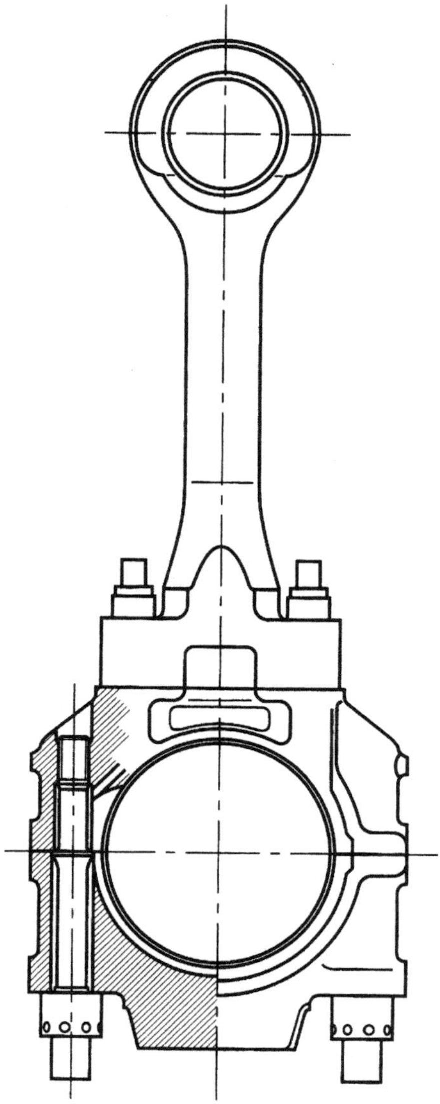

Fig. 22.22 The L32/40 engine connecting rod is a traditional marine head design.

for piston withdrawal since the distance between the four-cylinder cover bolts is sufficient for the connecting rod shank to pass without difficulty.

Design and material specifications of other key wearing components target extended lifetimes and maintenance intervals of at least 2 years for the liners and exhaust valves. A very low wear rate of the valves is promised from the armoured seats and directly water-cooled seat rings. A clean gas-tight seat is maintained by the exhaust valve rotator.

The designer asserts that constant pressure turbocharging is based on a MAN uncooled turbocharger which, along with the charge air and exhaust gas ducting arrangements, promotes a positive charge renewal performance and hence fuel economy. MAN NR26 or NR34 radial-flow turbochargers were typically specified for all cylinder number versions of the engine. Streamlined transitions from the cylinder heads to the exhaust gas line, as well as the double diffuser, arranged downstream of the compressor, help to keep pressure losses in the exhaust gas system at a low level; and part of the kinetic energy of the charge air leaving the compressor at high speed is transformed into pressure on the air side. Both properties contribute to a substantial increase in overall efficiency.

If an engine's load profile dictates, warm charge air can be withdrawn directly after the compressor and supplied to the last section of the exhaust gas line upstream of the turbine via a bypass. The section is provided with a special belt to ensure perfect mixing of the bypassed air with the exhaust gas; a relief valve can also be fitted in the belt if required.

The charge air cooler can be of the single-stage or two-stage type. The latter is specified when the heat contained in the high-temperature stage is to be exploited or if, in the case of frequent part-load operation, the charge air is to be preheated by the engine cooling water in the lower load range to ensure favourable combustion conditions.

Ease of inspection and overhaul—a focus of medium-speed engine designers in an era of low manning—is addressed by the four-bolt cylinder cover with quick-acting seals and clamped and plug-in pipe connections. Only two different types of hydraulic tools are required to handle all the main screwed connections on the engine. The marine-type crankpin bearings allow the connecting rod to be withdrawn up the bore for piston and ring inspection without disturbing the bearings. The exhaust gas line features quick-release couplings that permit swift disconnection from the cylinder head.

MAN Diesel originally anticipated a 550 man-hour annual maintenance workload for an eight-cylinder L32/40 engine running for 6000 h on heavy fuel. The basic regular jobs are: replacing the fuel nozzles every 3000 h; overhauling the exhaust valves (removing the cylinder cover) every 6000 h and overhauling the piston (including the replacement of rings and honing the liner) every 12,000 h. Service lives and maintenance intervals for the key components of current 32/40 engines are indicated in Table 22.2.

Table 22.2 Service lives and maintenance intervals for 32/40 engine components.

Component	Service life (31,000 h)	Maintenance interval (31,000 h)
Cylinder liner	80–100	15–20
Piston crown	60–80	30–40
First piston ring	15–20	–
Inlet valve cone	30–40	15–20
Exhaust valve cone	30–40	15–20
Fuel injection nozzle	6–10	3

32/40 engine refinements

Continual development of the 32/40 engine has pursued reductions in smoke emissions throughout the operating range without undermining fuel economy, easier maintenance, and improved flexibility in setting the engine's parameters for specific projects. Many components were modified in meeting these goals, such as the cylinder head, fire ring, support ring, rocker arms, valve cones, pistons, camshaft, and gas exchange cams. Improved efficiency, particularly at loads less than 50%, was gained from a modified MAN NR-type turbocharger.

Measures to enhance emission values included a new injection nozzle design with a smaller sac volume minimizing the amount of fuel sprayed into the cylinder after the needle is closed. The benefits are reduced dripping and hence reduced smoke formation. A new cylinder head design incorporates an improved combustion chamber geometry promoting unhindered atomization of the fuel jets. The fuel injection pressure was also increased from its base of 1350 bar and closing of the injection needle improved.

Efforts were also directed at reducing engine cost and weight, and facilitating maintenance. A redesigned rocker arm casing, for example, is lighter and allows faster nozzle replacement. A new steel piston was released in 2001. The lubricating oil and cooling water systems were combined into a single unit—the frame auxiliary box—to create a compact, vibration-free package calling for reduced piping.

48/60B engine

The 48/60 engine was originally introduced to series production in 1989 with a rating of 885 kW/cylinder, which was progressively increased to 1050 kW/cylinder. Further development of the 48/60 design was stimulated by the higher propulsion power demands of cruise ships, ferries, and large multipurpose cargo ships, as well as stricter environmental standards. The resulting B-version was launched in 2002 with a 14% rise in specific output to 1200 kW/cylinder at 500/514 rev/min from a lighter and more compact design. The 48/60B programme spans a power band from 7200 to 21,600 kW with six, seven, eight, and nine in-line and V12, 14, 16, and 18-cylinder models. The V-cylinder engines were released first, with the in-line models planned to follow in 2003 (Fig. 22.23).

Retaining the original 480 mm bore/600 mm stroke dimensions, the designers aimed to: reduce fuel consumption; optimize combustion to reduce NOx and soot emissions; boost the specific power output; reduce engine width; lower engine weight; optimize a range of engine components including the big end bearing; and reduce the overall number of engine components. Easier maintenance, higher reliability, and lower manufacturing costs were also pursued. The goals were achieved by developing new subassemblies and/or improving components.

Combustion benefited from: a simplified exhaust gas system with only one manifold and only one turbocharger on the V-cylinder engine; a

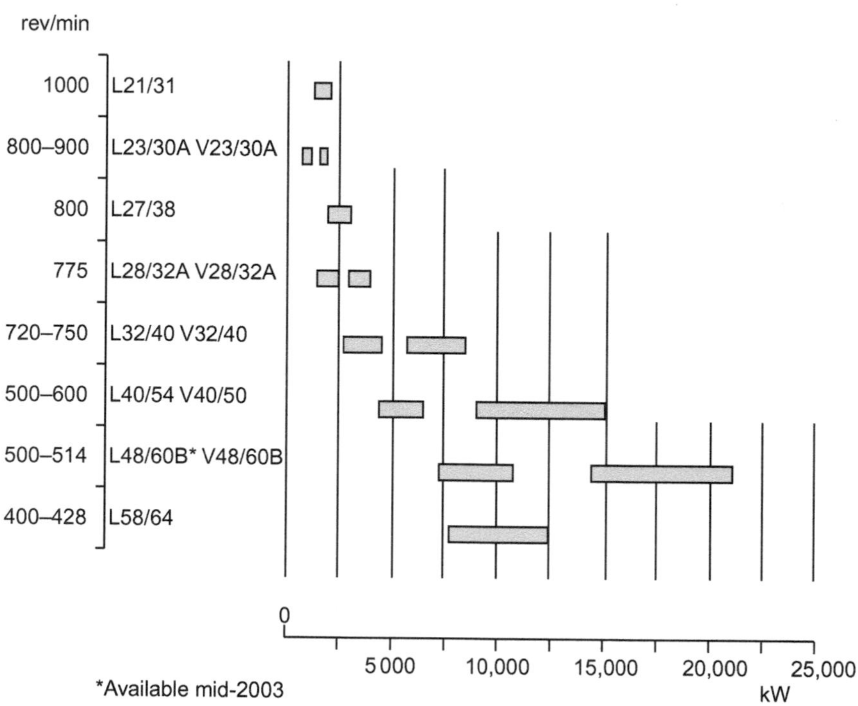

Fig. 22.23 MAN medium-speed propulsion engine programme before the introduction of common rail (CR) fuel injection versions; the L40/54 and V40/50 series have since been dropped.

redesigned combustion space; a new fuel injection system with high injection intensity; modified valve timings; and the new MAN TCA-series turbocharger (see Chapter 10). The compression ratio was increased from 14.4 to 15.3, a rise facilitated by a combustion space optimized by avoiding dead volumes, edges, and ribs in the combustion side of the cylinder cover. The fuel injection pressure was increased from 1300 to 1600 bar, the higher intensity helping to reduce visible flue gas emissions, particularly in the part-load range. Higher NOx emissions were neutralized by slightly delayed injection timing.

A significant contribution to fuel economy came from the new MAN TCA77 turbocharger through an increased flow rate and an efficiency rise of 8%–10% (Fig. 22.24). The minimum specific fuel consumption of 173 g/kWh (for the V-engine at 85% maximum continuous rating) is 7 g/kWh less than that of the original 48/60 design, while the emission values on the IMO NOx test cycle are 8% less at 12 g/kWh (Fig. 22.25).

Accompanying the higher output, fuel economy and environmental friendliness is a lower specific engine weight (reduced by 16% to 11.9 kg/kW). The V12-cylinder 48/60B model, for example, develops 14,400 kW and weighs 181 t, while its predecessor yields 12,600 kW from a weight of 193 t. For the V18-cylinder engines, the reduction in weight is approximately 15 t and the extra power around 2700 kW. The dimensions were also reduced: the height remains the same but the B-engine is 800 mm (15%) narrower, thanks to a new exhaust system concept. A pair of 48/60B

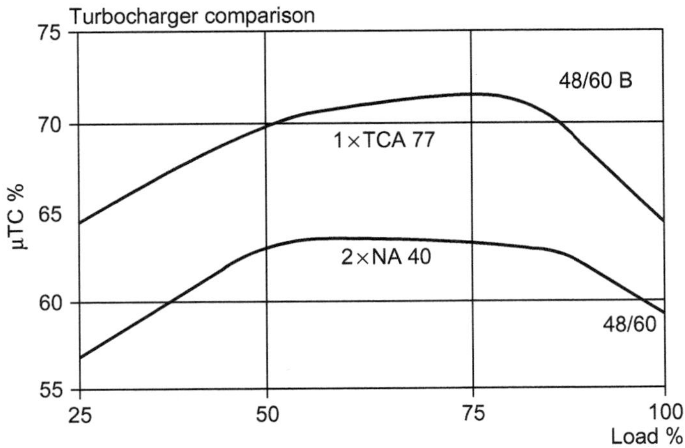

Fig. 22.24 Comparison of turbocharger efficiencies: the MAN TCA77 on the 12V48/60B engine and the MAN NA40 on the original 12V48/60 engine.

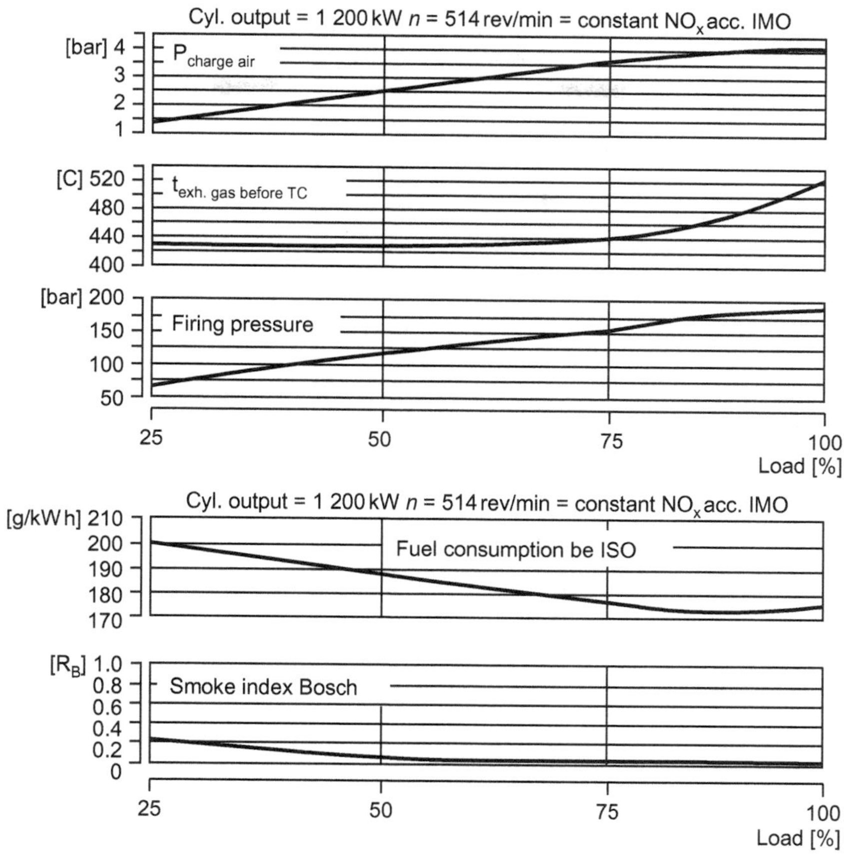

Fig. 22.25 Operating data from the 12V48/60B test engine.

engines can thus be mounted closer together than before, giving an installation width saving of 1 m. The turbocharger and two charge air coolers were combined to form a single module on the V-engine.

Internal refinements included an improved piston, with steel crown and nodular cast iron (or steel, if preferred) skirt. The cylinder head incorporates a reshaped combustion chamber, no valve cages, an innovative rocker arm concept, and a modified rocker arm cover. The connecting rod shank features optimized bearing shells, an increased oil clearance, reduced rotating masses, and a strengthened bearing cover design.

Among the elements retained from the original design, with proven reliability in heavy fuel operation, were the stepped piston with a chrome-ceramic first ring; exhaust valve with a propeller (to ensure a clean and gas-tight seat) and MAN's constant pressure charging system with exhaust gas wasting and bypass control.

Invisible Smoke (IS) variants of the 48/60 became available in early 2001 incorporating measures to achieve low Bosch smoke figures throughout the load range. The development was stimulated by requirements imposed on cruise ships in environmentally sensitive regions such as Alaska. The refinements—including a turbocharger optimized for lower loads, charge air preheating and a new cooled part-load fuel nozzle—can be applied to new engines and retrofitted to engines in service (see Chapter 4 for details). MAN favoured the fuel-water emulsion route to reduce smoke over direct water injection (Fig. 22.26).

Man B&W 48/60B engine data

Bore	480 mm
Stroke	600 mm
Speed	500/514 rev/min
Output	1200 kW/cyl
Cylinders	6, 7, 8, 9L/12, 14, 16, 18V
Power range	7200–21,600 kW
Mean piston speed	10/10.3 m/s
Mean effective pressure	25.8/26.5 bar
Maximum firing pressure	200 bar
Specific fuel consumption[a]	173 g/kWh
Specific lube oil consumption	0.6–0.8 g/kWh

[a]At 85% per cent mcr.

See Chapter 16 for details of 48/60B engine development procedure.

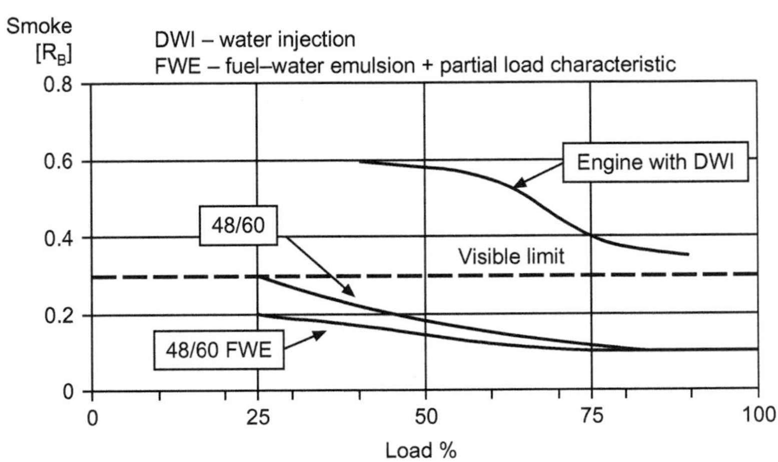

Fig. 22.26 Smoke reduction from a 48/60 engine equipped with fuel-water emulsion (FWE) system.

L27/38 and L21/31 engines

A new generation of smaller medium-speed designs for propulsion and genset drives—the L27/38 and L21/31 series—extrapolated the principles of the successful L16/24 auxiliary engine, including the twin camshaft configuration pioneered by the L32/40 series. The L27/38 series (Fig. 22.27) was introduced in 1997 and the L21/31 series in 2000.

The monobloc cast-iron engine frame is designed for a maximum combustion pressure of 200 bar and direct resilient mounting. A static preloading of the casting is maintained by through-going main bearing tie-rods and deeply positioned cylinder head tie-rods, thus absorbing the dynamic loads from gas and mass forces with a high safety margin. All the rods are tightened hydraulically. Well-supported main bearings carry the crankshaft with generously dimensioned journals. The combination of stiff box design and carefully balanced crankshaft promotes smooth and vibration-free running. Large noise-dampening covers on the frame sides offer good access for inspection and overhaul. A 'pipeless engine' philosophy resulted in integrated oil ducts, with no water flowing through the frame; risk of corrosion or cavitation is thereby eliminated.

An innovation inherited from the L16/24 design, the front-end box, is arranged at the free end of the engine. It contains connecting ducts for cooling water and lube oil systems as well as pumps (plug-in units), thermostatic valve elements, the lube oil cooler, and automatic back-flushing lube oil filter. External pipe connections are arranged on the sides of the front-end box to reduce engine length. A small optional power take off is located on the forward side.

Less advanced but pursuing the same philosophy as the front-end box, an aft-end box is arranged at the flywheel end. This carries the turbocharger and incorporates a two-stage charge air cooler and integrated ducts for high- and low-temperature cooling systems. A regulating valve controls the water flow to the second stage of the cooler, adjusting to the operating conditions to secure an optimum charge air temperature.

High performance from the piston and ring pack depends on the cylinder liner geometry over the entire load range, the temperature in the TDC position of the top piston ring, and measures to avoid bore polishing. The liner is of the thick-walled high collar design, which protects the sealing between cylinder head and liner from the influence of engine frame distortions. Only the collar is water cooled, ensuring a stable geometry under variable load conditions. The temperature in the TDC position of the top piston ring

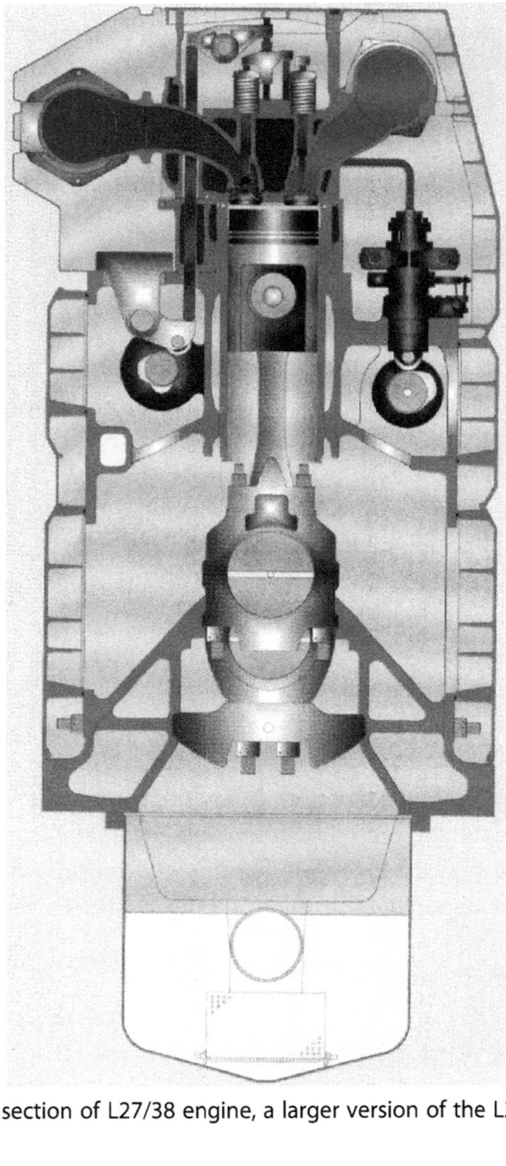

Fig. 22.27 Cross section of L27/38 engine, a larger version of the L21/31 design.

is optimized to prevent cold corrosion, which is especially important in HFO operation.

Years of experience with the flame ring concept successfully applied to the 28/32A and 23/30A engine series was adopted for the L27/38 engine. The flame ring, inserted directly in the top of the cylinder, has a smaller inside diameter than the cylinder bore. The piston has a similar reduction in its top land diameter, allowing the flame ring to scrape away coke deposits and avoid bore polishing of the liner; optimum ring performance and low lube oil consumption over a long period are fostered.

Monobloc pistons in nodular cast iron had been state of the art for a number of years for applications imposing a maximum combustion pressure of around 160 bar. With pressures up to 200 bar, however, a monobloc piston was not suitable for the new engine generation. A composite piston, with bore-cooled steel crown and nodular cast iron body, was therefore specified. The different barrel-shaped profile of the piston rings and the chromium-ceramic layer on the top ring target low wear rates and extended maintenance intervals.

The connecting rod is of the marine head-type, with the joint above the head fitted with hydraulically tightened nuts; the head remains on the crankshaft when the nuts are loosened for removal. A low overhaul height is secured. If engine room space allows, the complete cylinder unit (cylinder head, water collar, liner, piston, and connecting rod) can be withdrawn as a single assembly. A complete unit can thus be sent ashore for overhaul and replaced with a new or overhauled unit. The thin-walled main bearing shells are dimensioned for moderate bearing loads and a thick oil film. The crankshaft is a one-piece forged component provided with counterweights on all crank webs (attached by hydraulically tightened nuts) to yield suitable bearing loads and vibration-free running.

A combustion pressure of 200 bar and a mean effective pressure of 23.5 bar influenced the choice of nodular cast iron as the material for the four-stud cylinder head. The rocker arms for the two inlets and two exhaust valves are mounted on a single shaft supported in the casting. The cylinder segment charge air receiver is integrated into the casting, with the exhaust gas outlet arranged on the opposite side. This cross-flow design, together with flow-optimized inlet and outlet ducts, creates a swirl effect benefiting charge air renewal and combustion.

Inlet and exhaust valve spindles are armoured with heat-resistant hard metal on the seats. The exhaust valve spindles feature integrated propellers for rotation by the gas flow in order to equalize and lower the seat temperature, and to prevent deposits forming on the seat and on the water-cooled valve seat ring. The inlet valve spindles have valve rotators to minimize seat wear.

Two independent camshafts separate the functions of fuel injection and charge air renewal, one operating the injection pumps and one actuating the valve gear. Optimum adjustment of the gas exchange without influencing the injection timing is thus secured. Cam followers on the valve camshaft each actuate two valves via pushrods, rocker arms, and a yoke.

A dedicated camshaft for fuel injection makes it possible to alter the timing, when required, for a low NOx emission mode while the engine is in operation; the engine can also be easily adapted for running on different fuel oil qualities. The injection pump has an integrated roller guide directly activated by the fuel cam (Fig. 22.28). Pipe systems are virtually eliminated, thanks to slide connections between the pumps, promoting easy assembly/disassembly and safety against vibration and leakage. An uncooled injection valve contributes to the simplicity of the injection system, which is designed for a pressure of 1600 bar; effective atomization is thus fostered even at part-load operation with small injection volumes, contributing to a low exhaust smoke value. The complete injection equipment is enclosed behind easily removed covers. A fuel feed pump and filter are standard on propulsion engines.

The constant pressure turbocharging system incorporates a bypass connection from the turbocharger air outlet to the gas inlet side in order to increase the charge air pressure at part load. For certain operating conditions requiring high engine torque at reduced engine speed, a waste gate arrangement is incorporated in the exhaust system.

An entirely closed lube oil system eases installation and avoids the risk of dirt entering the lube oil circuit. A helical gear-type lube oil pump is mounted in the front-end box and draws oil from the wet sump. The oil flows via a pressure regulator through the lube oil plate cooler and the automatic back-flushing filter; this solution eliminates the exchange of filter cartridges as well as the waste disposal problem. The back-flushing filter is drained to the sump and a purifier is connected to maintain the lube oil in a proper condition. An integrated thermostatic valve ensures a constant lube oil temperature to the engine.

A cooling water system based on separate low-temperature (LT) and high-temperature (HT) systems is considered as standard. Both circuits are cooled by fresh water (Fig. 22.29).

Engine speed is controlled by a mechanical or electronic governor with a hydraulic actuator. All media systems are equipped with temperature and pressure sensors for local and remote reading. The number and type of parameters to be monitored are specified in accordance with (but

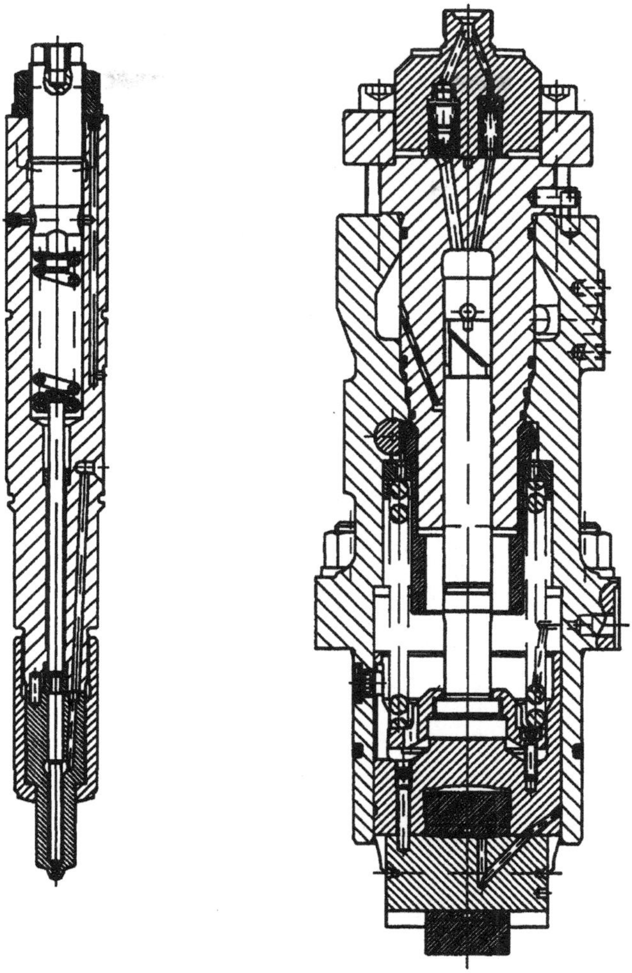

Fig. 22.28 Fuel pump and injector of L27/38 engine.

not limited to) the classification society requirements. Shutdown functions for lube oil pressure, overspeed, and emergency stop are provided as standard. The monitoring, control, and safety system are arranged for compatibility with MAN Diesel's CoCoS system for engine diagnosis, maintenance and planning, and spare parts handling.

L21/31 and L27/38 propulsion engine data

	L21/31	L27/38
Bore	210 mm	270 mm
Stroke	310 mm	380 mm
Cylinders	6–9L	6–9L
Speed (rev/min)	1000	800
Output (kW/cyl)	215	340
Power range (kW)	1290–1935	2040–3060
Mean effective pressure (bar)	24.1	23.5
Mean piston speed (m/s)	10.3	10.1
Combustion pressure (bar)	200	200

Five-cylinder versions are also offered for genset applications.

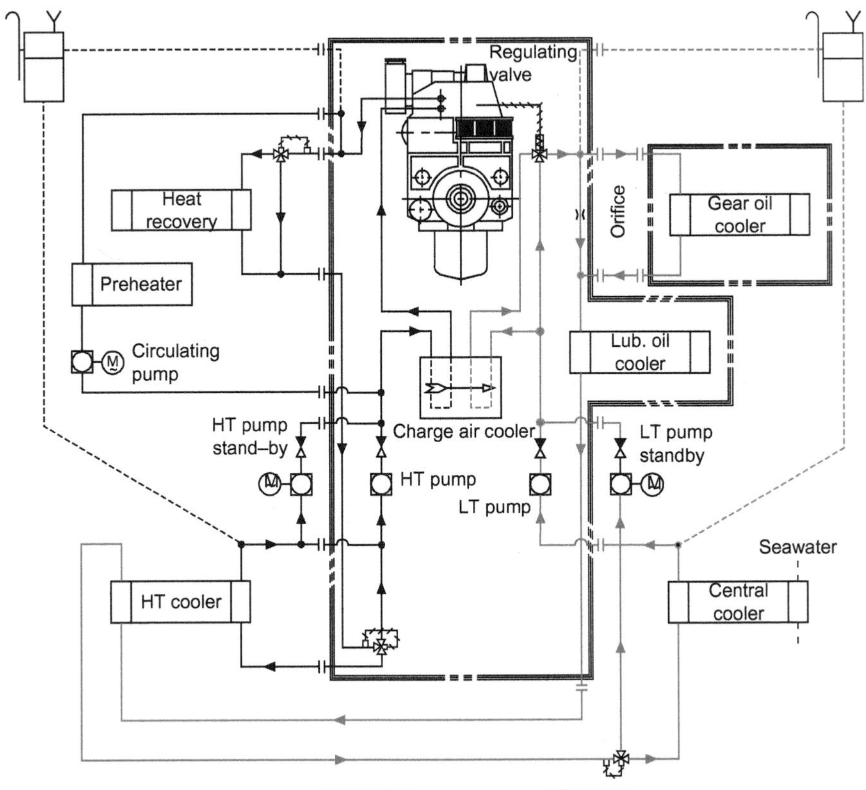

Fig. 22.29 Cooling water system of the L27/38 engine, with the high-temperature (HT) circuit on the left and low-temperature (LT) circuit on the right.

V28/33D engine

Developed by the British designer Ruston, the RK280 engine was inherited by MAN Diesel when the German group acquired the UK portfolio of Alstom Engines and transplanted to Augsburg as the MAN V28/33D series. The first production engines (V20-cylinder models) were delivered at the end-2005 for powering a monohull fast ferry; others in V12, V16, and V20 configurations have since been supplied for high-speed catamaran ferries, a mega yacht and a naval offshore patrol vessel.

A completely new medium-speed engine, the RK280 design had joined the Ruston programme in 2001, offering a considerably higher output than the long-established RK270 series for commercial and naval propulsion and genset drive applications (Figs 22.30–22.32). The 280 mm bore engine was introduced as the world's most powerful 1000 rev/min design, released with an initial rating of 450 kW/cylinder. The programme of V12-, 16-, and 20-cylinder models thus covers a power range from 5400 to 9000 kW. An output of 500 kW/cylinder was specified for the design, however, equating to a future rating of 10,000 kW for the 20RK280 engine.

Fig. 22.30 V16-cylinder version of the 28/33D engine.

Fig. 22.31 Cutaway of V28/33D engine showing the gear train for the camshaft and service pump drives.

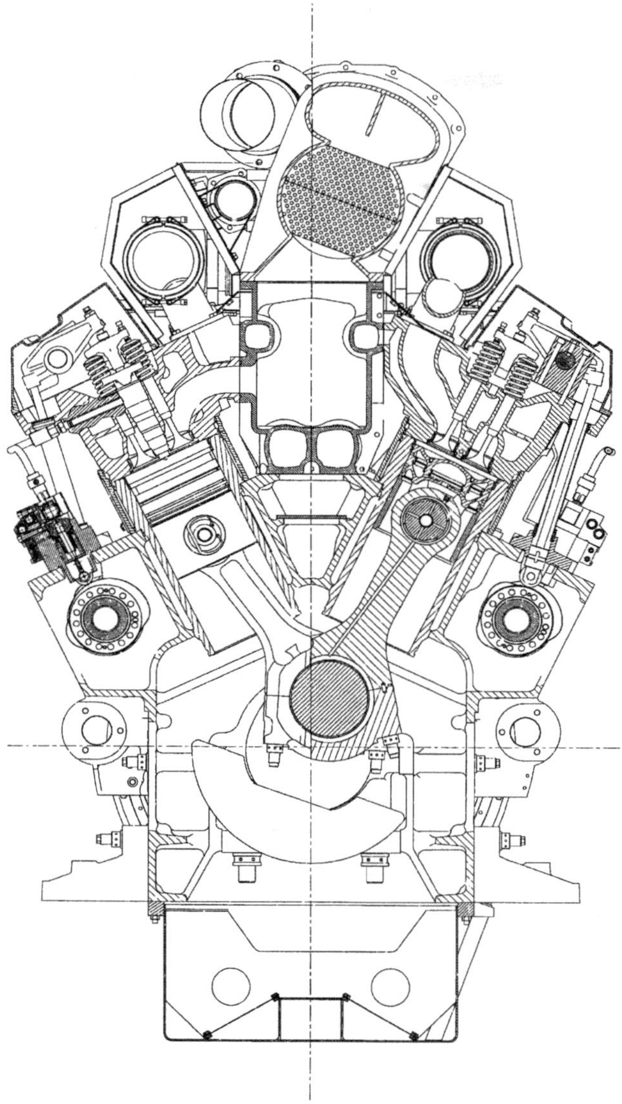

Fig. 22.32 Cross section of V28/33D engine.

An STC (sequentially turbocharged) version of the V28/33D engine features a digitally controlled system programmed to automatically switch off one of the two turbochargers at low revolutions per minute, improving the amount of air reaching the combustion chambers. STC, applied in installations requiring an extended torque envelope, is controlled by the SaCoS electronic control unit.

V28/33D design data

Bore	280 mm
Stroke	330 mm
Cylinders	V12, 16, 20
Output[a]	450 kW/cyl
Rated speed	1000 rev/min
Mean effective pressure	26.6 bar
Mean piston speed	11 m/s
Power range	5400–9000 kW
Power density	5.1 kg/kW
Fuel consumption	190 g/kW h

[a]Continuous rating.

The engine is designed to operate on middle distillate fuels, such as DMA or DMB and equivalent specifications. A competitive specific fuel consumption of less than 190 g/kW h is reported, with NOx emission levels of less than 10 g/kWh from primary reduction measures alone. Compactness and lightness are reflected in a dry weight of 46 t and a power density of 5.1 kg/kW for the V20-cylinder version, which measures 7.33 m long × 2.1 m wide × 3.18 m high. A detailed analysis of all parts was carried out to achieve the high power-to-weight and size ratios, and aluminium alloy materials were used where possible.

The V28/33D (RK280) design benefited from experience with RK270 engines powering fast ferries, as well as feedback from RK215 engine installations. Advantage was also taken of advanced software packages, new development methodologies, and project management techniques. Areas of the RK270 design that had proved particularly successful—such as the integrity of the bottom end in a dynamic fast vessel application exploiting an elastic mounting system—were thus translated into the design parameters of the V28/33D engine. At the same time the opportunity was taken to address other features in eliminating the possibility of problems in service: for example, minimizing overhangs and induced stress in the engine, and reducing pipework on and off the engine.

Among the measures adopted to enhance performance, fuel economy, reliability, and serviceability were:

- Electronically controlled pump-pipe fuel injection system to foster low specific fuel consumption and reduced NOx and particulates emissions throughout the load range.
- Simple, single-stage high-efficiency turbocharging based on ABB Turbo Systems' TPL 65 turbocharger for improved specific fuel consumption

and lower operating temperatures; and a simple exhaust system with rigid heat shields helps to minimize and ease maintenance.

- Integrated pipe design with a four-pipe connection to reduce vibration and ease installation and serviceability.
- Two-stage freshwater-cooled high-efficiency intercooler in the centre of the engine to improve specific fuel consumption and curb NOx emissions while reducing mechanical loading.
- Reduced number of components (around 50% fewer than the RK270 engine) for higher reliability and lower servicing costs. Multifunctional integrated elements include a hollow camshaft doubling as a main lube oil channel. External piping is minimized by the wide use of existing drillings or cast-in passages for fluid transmission; and prewired, modular and encapsulated wiring looms avoid the need for exposed on-engine electrical cables.
- Rigid dry crankcase design for reduced weight and easier servicing.
- Low inertia/low friction two-piece pistons with alloy steel crowns to reduce weight and improve specific fuel consumption; three-ring pack.
- Simplified gear train of robust construction; water and lube oil pumps driven from the free end of the engine through gears housed in the pump drive casing (Fig. 22.31).
- External viscous damper for ease of maintenance.

V28/33D engine details

Crankshaft: manufactured from a high-tensile NiCrMo continuous grain flow steel forging.

Crankcase: machined from spheroidal graphite cast iron and featuring underslung main bearings retained by two vertical studs and two cross-bolts on each side for overall stiffness. The main bearing caps are secured by hydraulically tensioned studs to ensure maximum integrity of the crankcase system. A 52° V-angle minimizes torsional effects and allows the location of the intercooler between the cylinder banks, reducing overhang loadings while minimizing engine height. The inspection covers on both sides of the engine to provide access to internal components, and selected covers carry the explosion relief valves. Engine mounting on either anti-vibration mounting or solid seating is by separate bolt-on feet.

Cylinder liners: individual units incorporating deep flanges, strategically cooled by a separate water jacket enabling a dry crankcase to be used, reducing overall weight. The running surfaces are plateau honed and finished to improve oil retention throughout the liner life; a cutting ring is fitted at the top of the liner to eliminate the buildup of carbon on the piston crowns and minimize lube oil consumption.

Bearings: the generously dimensioned main bearings feature easily replaceable thin wall, steel-backed aluminium-tin shells.

Camshafts: modular design with one cam element per cylinder; they are hollow and form the main lube oil supply to the engine. Optimized cam profiles for electronically controlled fuel injection minimize Hertzian stresses, enhancing reliability and extending component life. The camshaft drive is located at the free end of the engine; the crankshaft gear drives via a compound idler gear for each camshaft.

Piston: two-piece design with a lightweight body and alloy steel crown. A three-ring pack comprises two chrome-ceramic compression rings and an oil control ring. The case-hardened gudgeon pin is fully floating and retained by a circlip at each end. Lube oil is fed from the connecting rod through drillings in the gudgeon pin and piston to a cooling chamber in the piston crown. The oil is then discharged through drillings in the underside of the piston crown back to the sump.

Connecting rods: made of forged high-tensile alloy steel; the rods have obliquely split large ends that carry fully grooved bearings with the cap secured by four hydraulically tensioned studs.

Cylinder heads: the individual heads have a thick combustion face incorporating coolant drillings. The two inlet and two exhaust valves, all with cooled seats, surround the central fuel injector. Twin inlet ports connect directly to the air manifold, while a single tandem exhaust port outlet in the top face allows for ease of maintenance. The heads are held in place by four hydraulically tensioned studs.

Valve gear: each pair of valves, operated via pushrods and rockers, is driven from the camshaft via bucket tappet-type followers mounted in a separate housing bolted to the crankcase.

Air manifold: these are modular castings mounted down the vee of the crankcase and incorporating passages for the lube oil and water systems.

Exhaust system: modular and compact, the system comprises single-cylinder units bolted to the cylinder head and connected to the next unit with expansion bellows. The whole exhaust system is arranged in a lagged enclosure made up of two-cylinder units for ease of maintenance. The exhaust pipes are top-mounted and connected to their respective bellows by vee-band clamps; and quick-release couplings are used on the rigid heat shield.

Charge cooler: the cylindrical two-stage charge cooler contained in a housing includes part of the inlet ducting. The assembly is mounted directly on top of the air manifold to provide good support. Special attention is paid to minimize overhangs on external brackets to reduce the impact of shock loadings in fast commercial vessel or naval applications.

Turbochargers: twin high-efficiency axial turbine turbochargers are mounted on a cast bracket at the free end of the engine.

Fuel system: an electronically controlled pump-pipe injector system is used, with the fuel pump mounted in the cam follower housing which forms part of the body of the pump. Low pressure, modular fuel supply, and return rails connect each pump to the next, and the short high-pressure pipes to the injectors are double skinned. The electronic control for the fuel pump and injector is mounted locally on the engine. The electronic system facilitates control of the fuel quantity and injection timing independent of engine speed, enabling performance to be optimized for the specific application.

Lubricating oil system: all contained on the engine. The lube oil pump is mounted directly on the free end of the crankcase and driven from the camshaft gear drive. The plate-type oil cooler is mounted horizontally on top of the filter housing at the free end of the engine, the duplex filter incorporating an integral oil thermostat.

Cooling system: a twin-circuit cooling system is used, with both pumps mounted on the free end of the engine and driven by the camshaft gear, and provision for a sea-water pump. The charge cooler thermostat is integrally mounted in the turbocharger bracket.

Starting system: the air starting motor incorporates a control valve, pressure regulator and strainer, and engages with a gear ring on the flywheel. A barring motor can be supplied as a service tool or fitted as standard when it is fully protected against inadvertent starting of the engine.

Governor: the engine is served by a digital engine management system which controls its operation and communicates via a CAN bus link to a set of intelligent cylinder control modules that drive the pump and injector solenoids. The system dictates the fuelling, timing, and pressure base on preset mapped information. The CAN bus link is also used to convey diagnostic information back to the master controller for display and action. The cylinder control modules provide emergency governing in the event of a master controller or CAN bus failure. Other similar redundant features are incorporated to promote maximum engine availability.

Common rail engines

Like other major medium-speed engine designers, MAN Diesel has developed a modular CR fuel injection system to foster free selection of injection timing and pressure independent of engine load and speed. The CR system was first announced in 2004 in conjunction with the established 32/40 engine. Fuel pressure generation and injection processes are separated and electronically controlled. A prime advantage is the reduction of smoke emissions, especially at low loads and during starting, and an improvement in the NOx-specific fuel consumption trade-off.

Based on the segmented rail concept, MAN Diesel's CR system is served by high-pressure pumps which compress the fuel to the required pressure and deliver it to in-line accumulator units forming the CR (Figs 22.33 and 22.34). Connections are provided on the accumulators for the injection valves and for the fuel distribution and injection-control components. The CR injection system is fully compatible for use with heavy fuels having viscosities up to 700 cSt at 50°C and temperatures up to 150°C, with all relevant components designed to withstand these conditions; in addition, they meet stiff requirements with respect to resistance to wear by abrasive particles and the aggressive contents of HFO. CR engines can be started and stopped on HFO, and in a load range from 100% down to 20% they can

Fig. 22.33 Key elements of MAN Diesel's CR fuel injection system on a 32/44CR engine showing an accumulator served by a high-pressure pump and supplying two injectors.

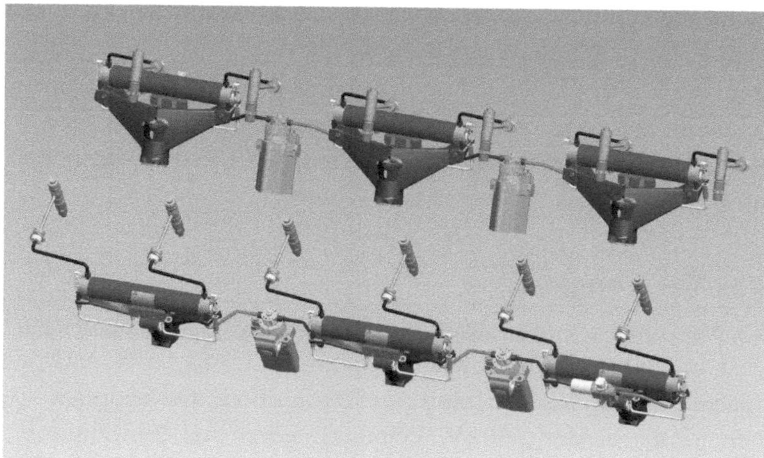

Fig. 22.34 Arrangement of the CR fuel injection system.

be operated without interruption. Below 20% load, the engines can be run on HFO with a time limit.

The CR injection system is modular in design and subdivided into a series of compact pressure reservoirs or accumulators; using multiple accumulators reduces pressure fluctuations in the system and makes rational use of space available on the engine. Each accumulator segment serves one or two fuel injectors featuring conventional nozzles. The solenoid injection-control valves are located on the segmented rails and connected to standard pressure-controlled injectors. No separate servo circuit is required for activating the electronically controlled injection valves which are arranged on the rail away from the hot cylinder heads to secure greater system reliability and easier maintainability. MAN Diesel reports that ease of installation and maintenance is also fostered by the simple assembly procedures.

System safety and redundancy was a priority in development, leading to the adoption of these key features:

- Injection pressure only during injection at the injection valve (no risk of uncontrolled fuel injection because of leaking control or injection valves).
- All high-pressure lines, rails, and units are double-walled (no risk arising from fuel escaping from leaking or damaged lines.
- Fuel-limiting valves on each cylinder (prevent uncontrolled injection).
- Nonreturn valves on each cylinder (ensure that fuel backflow from the fuel low-pressure system into the cylinder is impossible).

- Use of at least three high-pressure pumps (emergency operation possible in the event of failure of one pump).
- Safety valve with additional pressure control valve (emergency operation possible in the event of a fault in rail pressure control).
- Two rail pressure sensors and two engine speed/top dead centre pickups (sustained operation if a sensor fault occurs).

MAN 32/44CR engine

MAN's first all-electronic four-stroke engine, the 32/44CR design, was launched in 2006 to supplement the popular 32/40 series with a CR fuel-injected derivative. Retaining the 320 mm bore, the designers sought a higher specific output (560 kW/cylinder), lower fuel consumption, and reduced emissions from a 10% increase in piston stroke (from 400 to 440 mm). The adoption of CR fuel injection and a considerably more efficient turbocharger also contributed to the power rise and emissions performance.

A higher power density than the 32/40 engine dictated a number of key component modifications. In particular, the cylinder head and valves were modified to cope with the increased cylinder pressure, and the valve guides were extended and the seats modified. In addition, the crank drive was completely revised and optimized using computer design techniques. An all-steel piston incorporating two compression rings as part of an enhanced three-ring package features an enlarged oil cooling gallery and larger piston pin diameter to match the higher rating of the engine. The piston length was also reduced so that, despite these measures, its weight was not higher than the 32/40 engine component.

Additionally, the cross-sectional diameter of the connecting rod shaft was increased and the number of bolts at the big end bearing increased from two to four. The higher number of bolts allowed for a more even distribution of the stresses at the joint line and facilitated a slimmer connecting rod than with the two-bolt design. Two camshafts are featured: a standard full-length unit for actuating the gas exchange valves and a shortened unit serving the high-pressure fuel injection pumps. Shortening the injection camshaft not only reduces its friction characteristics but also the engine manufacturing costs: cited by MAN Diesel as an example of how CR injection can improve not only engine performance but also design and production.

MAN CR engine parameters

	32/44CR	48/60CR
Bore (mm)	320	480
Stroke (mm)	440	600
Cylinders	6–10L/12–20V	6–9L/12–18V
Speed (rev/min)	720/750	500/514
Mean piston speed (m/s)	10.6/11	10/10.3
Mean effective pressure (bar)	26.4/25.3	26.5/25.8
Max. combustion pressure (bar)	230	–
Output/cylinder (kW)	560	1200
Power range (kW)	3360–11 200	7200–21,600
Specific fuel consumption (g/kW h)	177–179	174–180

MAN 48/60CR engine

CR fuel injection technology was extended by MAN Diesel to the 480 mm bore 48/60B design in 2007, the resulting 48/60CR engine retaining the 1200 kW/cylinder rating of its predecessor (Fig. 22.35). Following the precedent of the 32/44CR engine, the solenoid injection-control valves are located on the segmented rails and connected to standard pressure-controlled injectors. No servo assistance of injector opening is required. Such an arrangement keeps sensitive components away from the hot cylinder heads and facilitates retrofitting to engines in service since identical injectors are used on the 48/60B design.

MAN 45/60CR engine

In 2017 the company announced the 45/60CR medium-speed engine as the successor to the 48/60CR with designs on it becoming an engine of choice for the booming cruise ship sector. The new engine combines the proven characteristics of its predecessor—the MAN 48/60CR—including its in-house common-rail injection system, with the latest innovations in diesel-engine technology such as two-stage turbocharging.

With improved power output and specific fuel oil consumption compared to the MAN 48/60CR—the new engine will be available in both in-line and vee configurations.

The 12V and 14V versions, with power outputs of 15,600 and 18,200 kW, respectively, will be launched first towards the end of 2020, with the in-line range of engines from 6 to 10-cylinder models planned to be introduced in 2022. The engine is intended to be the start of a new family, and a dual-fuel version will be introduced at a later stage.

Fig. 22.35 A 48/60CR engine with a cylinder unit exposed for an exhibition.

MAN 45/60CR V12

The timing of the announcement of the new engine was done some 3 years before production was likely to begin so as to generate interest in it and most of cruise ship operators have several projects already under construction with most of the main cruise shipbuilding yards occupied through to 2021/22. If the new engine is specified for new projects after the current crop of newbuildings is delivered, it will be easy to bring into production.

In the development of the 45/60CR, much use was made of computational fluid dynamics to simulate and optimize the combustion process and finite element analysis was used to optimize the engine's mechanical strength and vibration behaviour.

The 12-cylinder vee engine will offer a further 300 kW/cylinder power increase and a −4% decrease in SFOC from 173 to 166 g/kWh compared to the similarly configured 48/60CR, with just an extra 7 t of weight. The extra weight is due to reinforcing the engine base and the addition of a two-stage turbocharger. The turbocharger setup for V-Engines is foreseen as TCX for high-pressure stage and TCA for the low-pressure stage. Despite the turbocharging being two-stage, load pickup behaviour is the same as for the single-stage turbocharged 48/60CR engine.

As well as the two-stage turbocharger, other crucial elements to the new engine include the latest generation safety and control (SaCos) system and a new version of the ECOMAP control software that enables users to optimize operation of the engine.

MAN 45/60CR specifications

Bore	450 mm
Stroke	600 mm
Power output per cylinder	1300 kW
Speed	600 rpm
BMEP	27.2 bar
Specific fuel oil consumption	L: 167 g/kWh; V: 166 g/kWh
6L45/60CR	7800 kW
7L45/60CR	9100 kW
8L45/60CR	10,400 kW
9L45/60CR	11,700 kW
10L45/60CR	13,000 kW
12V45/60CR	15,600 kW
14L45/60CR	18,200 kW

Dual-fuel engines

Like other major medium-speed engine designers, MAN Diesel has developed a range of dual-fuel engines to complement its oil-fuelled range.

The first engine in the MAN portfolio was the 51/60DF introduced in 2007 and covered in Chapter 3.

In 2012 MAN added a new dual-fuel engine in the shape of the 35/44DF marine engine, to serve a broad power band in 10 configurations. Key goals of the technical programme were a high specific output and efficiency and compliance with IMO Tier III emission limits in gas mode and IMO Tier II criteria in diesel mode.

The 350 mm-bore design delivers up to 530 kW/cylinder, the highest power output in the segment for a dual-fuel engine and embodies common rail technology. In a dual-fuel context, it complements the company's larger 51/60DF series, which has made inroads into the LNG carrier dual-fuel electric propulsion sector.

The potential market to be served by the 35/44DF and its 3060–10,600 kW unit power coverage included ferries operating in the European SECAs. One of the aims has been to provide owners with a retrofit option for the 32/44CR-T2 common-rail diesel engine, using a high level of component synergies and the same crankcase, which can be re-machined on board.

Sales have been slow to take off with the first engines in use being a retrofit of a dredger. Two ferries for use in Canadian waters have also been specified with the engines and will be in service in 2021.

The engine is intended mainly for operation in gas mode, and the pilot oil ignition system it employs is independent of the main common rail injection system for running in liquid fuel mode. The pilot injection in the 35/44DF is a high-pressure system derived from technology used for the truck industry.

The main common rail injection arrangements derive directly from the 32/44CR-T2, and provide emergency backup when in gas mode, and allow for deployments in liquid fuel mode. The 35/44DF accordingly offers a very high degree of fuel flexibility, suited to operation on HFO, MDO, and MGO besides natural gas, and has been engineered to ensure a seamless switchover from gaseous to liquid fuel, and vice versa, at any time and load.

Technologies and advanced features combined in the new series include high-efficiency turbocharging, intelligent gas control, double-wall gas piping, air/fuel ratio control, pressure relief valve, on-engine electronics, and knocking control, as well as intelligent, pilot and main common rail oil fuel injection.

CHAPTER TWENTY-THREE

Rolls-Royce/MTU

The diverse shipbuilding and marine engineering interests of Norway's Ulstein group for many years embraced the design and production of Bergen medium-speed engines, after financial problems, the engine business passed to UK Rolls-Royce group's diesel interests in 1999 when Rolls Royce acquired Vickers which had purchased the engine production business from Ulstein earlier in the same year.

In 2011, Rolls-Royce together with Daimler AG of Germany jointly purchased the engine business of Tognum which included the German high-speed engine maker MTU and conducted engine production under the name Tognum until 2014 when it changed the title to Rolls-Royce Power Systems AG. Rolls-Royce subsequently acquired the Daimler stake in the company.

As said most of the MTU engines are high speed but the series 8000 engines introduced in 2000 as the most powerful engines in the MTU range has an engine speed of 1150 rpm. While that could make its inclusion in this chapter as a medium-speed engine, it has been included in the chapter on high-speed engines as its evolution is from the family of MTU high-speed engines and the dividing line in engine speed between medium and high-speed engines is rather indistinct.

In 2018, Rolls-Royce disposed of its entire marine business portfolio to Kongsberg apart from the engine production. However, sales and distribution of engines for marine purposes are managed by Kongsberg.

There were two Bergen engine families for propulsion and genset applications—the 250 mm bore K-series which dates from the early 1970s, and the 320 mm bore B-series introduced in 1986—which were progressively upgraded and were joined in 2001 by a completely new design, the 250 mm bore C-series, developed in co-operation with Hyundai Heavy Industries of South Korea.

K-series

First introduced in 1971, the 250 mm bore/300 mm stroke K-engine was designed mainly to be a genset drive but numerous propulsion plant

Pounder's Marine Diesel Engines and Gas Turbines
https://doi.org/10.1016/B978-0-08-102748-6.00023-2

applications have been logged in tugs, offshore supply ships, and fishing vessels. Engines are available as in-line three-to-nine-cylinder or V12-, 16-, and 18-cylinder models covering an output range from 600 to 4010kW at speeds of 720/750/825/900rev/min.

Design improvements introduced over the years have sought to maintain market competitiveness, the maximum brake mean effective pressure being raised from 20 to 22bar to increase the output to 220kW/cylinder. The later adoption of a 'hot box' arrangement for the fuel injection system followed a trend in larger engine designs. It provides the engine with cleaner lines and improves the working environment in the machinery room: temperatures are reduced and any fuel leakage from the injection system components is retained within the box.

Other improvements have included a revised timing chain, sprockets, and camshaft system to extend service life and accommodate a higher injection pressure. The valve stem rotators have been strengthened and the valve guides and guide seals improved to ensure better operational results running at 900rev/min on heavy fuel.

The provision of a carbon-cutting (or antipolishing) ring at the top of the cylinder unit significantly reduced lube oil consumption and the amount of insoluble deposits in the oil (hence extending filter life). The ring comprises a sleeve insert which sits between the top piston ring turning point and the top of the cylinder liner. It has a slightly smaller diameter than the bore of the liner, this reduction being accommodated by a reduced diameter piston top land. The main effect of the ring is to prevent the buildup of carbon around the edges of the piston crown which causes liner polishing and wear with an associated increase in lube oil consumption.

A secondary function is a sudden compressive effect on the piston ring belt as the piston and carbon-cutting ring momentarily interface; this forces lube oil away from the combustion area and again helps to reduce consumption. The efficiency of the rings was such that Bergen found it necessary to redesign the ring pack to secure a desirable degree of oil consumption. The carbon-cutting ring, refined over a 2-year period, is now standard on all new Bergen engines, and numerous engines in service have benefited from retrofits.

Production of the K series was discontinued in 2009 by which time more than 2000 units had been sold. As of March 2020, the number of vessels still in service with K series engine amounted to around 125. Most of these are offshore vessels or ferries.

B-series

The 320 mm bore/360 mm stroke B-series engine was conceived as a compact, heavy fuel-burning main engine and launched in 1986 with special features facilitating operation on a lower price, poor quality fuel oils. The design's tolerance of fuel quality has been demonstrated by an offshore rig installation running on crude oil direct from the well after degassing. A nominal rating of 360 kW/cylinder at 750 rev/min was initially quoted but some of the first engines sold were contracted at around 400 kW/cylinder, equivalent to a 10% overload.

An uprating in 1991/1992 took the output to 440 kW/cylinder at 750 rev/min which yielded a power band of 1870 to 3970 kW from in-line six-, eight-, and nine-cylinder BR models. A decision to double the upper limit of the range—to 7940 kW—was announced in 1996 through the introduction of V12-, 16-, and 18-cylinder models. These BV engines (see section 'BV engine') were planned to benefit from Bergen's exhaust emissions reduction R&D by adopting faster fuel injection, a change in compression ratio and different valve timing to the existing in-line cylinder engines.

The B-engine was reportedly the first of its size to feature a completely bore-cooled cylinder unit and combustion space (initially, a bore-cooled liner and head and, later, bore-cooled pistons). This arrangement yields good strength and stiffness combined with good temperature control, which is important in heavy fuel operation.

Early in the engine's development, it was discovered that the temperature profile on the liner was not optimum. The resulting design change, benefiting subsequent production engines, has reportedly proved effective in the long term for both propulsion and auxiliary power duties, even at medium load. A drawback to the original bore-cooling concept soon became apparent, however: the use of electrically driven cooling water pumps. Electrical blackouts could not be tolerated, and the fluctuation in cooling water flow away from the optimum resulted in some cracking of liners. An early change, therefore, saw a switch to engine-driven pumps and the incorporation of a new strengthened liner design.

In Arctic areas, it was also found that piston rings were vulnerable to poor ignition qualities and high-temperature gradients, resulting, in some cases, in ring cracking. This problem was solved by optimizing the clearances and the layout of the ring pack.

Other refinements following early installations included an improved bearing arrangement for the timing gears and, in line with other engine builders, a new fuel injection system was introduced in 1991 to counter an industry-wide problem of leakages. The complete injection system was enhanced by bigger pumps, better seals, larger nozzles, and improved high-pressure pipes. The improvements coincided with the 1991/1992 decision to uprate the engine's output to 440 kW/cylinder while retaining the same engine speed of 750 rev/min.

Contributing to this power rise—and a brake mean effective pressure increase to 24.4 bar—was the first application of ABB's VTR-4P high-efficiency turbocharger with a titanium compressor wheel, and a new type of composite piston with a bore-cooled forged steel crown and nodular cast-iron skirt. A carbon-cutting ring (described above in section 'K-series') was introduced to the B-engine in 1994.

The one-piece design cylinder block of nodular cast iron is arranged for an underslung crankshaft and features large crankcase doors. A receiver and supply ducts for oil and water are cast in. The centrifugally cast cylinder liner is bore cooled in the upper part only where cooling is needed; its running surface is specially treated to enhance wear resistance. The bore-cooled cylinder head is provided with a thick bottom for control of mechanical and thermal loads, and six holding bolts for good load distribution. Heavy fuel service is underwritten by exhaust valves with a Nimonic head and welded-on hard seat facing.

A fully forged, continuous grain flow crankshaft has large diameter journals and pins for low bearing loads and is fitted with hydraulically tightened counterweight bolts. The connecting rods are forged in alloy steel and machined all over, and feature an obliquely split big end with hydraulically tightened bolts. The tri-metal thin-walled bush bearings are formed from steel backed with lead/bronze bearing material and soft overlay. The two-piece piston is fitted with three compression rings and an oil scraper ring, all chromium plated to secure low wear.

A L'Orange fuel injection system was developed for an injection pressure of 1400 bar (endurance tested at 1700 bar) with constant pressure unloading for cavitation-free operation at all loads and speeds. A Bergen-developed cleaning/lubricating system cleans the lower pump parts of heavy fuel residues and lubricates the moving parts, helping to reduce the risk of fuel rack sticking problems.

An impulse turbocharging system based on ABB VTR.4 turbochargers is specified for six- and nine-cylinder engines, and a modified pulse converter system for the eight-cylinder model. Operation at low loads is fostered by elevated charge air temperatures through two-stage charge air cooling, high fuel injection pressures, and optimized valve timing.

BV engine

The first BV engines—twin V12-cylinder models, each developing 5294 kW at 750 rev/min—were delivered in 1998 for powering a large anchor handling/tug/supply vessel (Fig. 23.1). The BV design is based on a single-piece engine block cast in GGG500 nodular iron carrying two banks of cylinders in a 55° V-configuration (Fig. 23.2), an underslung crankshaft (Fig. 23.3) and two camshafts; it also incorporates the charge air receiver between the cylinder banks. The camshafts are located outside each bank and housed in open-sided recesses in the block, allowing the complete camshafts to be removed sideways. At the front of the block is an opening for the charge air cooler and another for the auxiliary gear drive; the timing gears are arranged at the rear of the engine. The whole structure is designed

Fig. 23.1 A Bergen B32:40 engine in V12-cylinder form.

Fig. 23.2 The BV engine block is a nodular iron casting.

for firing pressures in excess of 200 bar. An advantage of the block material is that it can be repaired by welding in the event of accidental damage.

A new cylinder liner design specified for the BV engines features a thicker upper wall section than the in-line cylinder BR models and a revised bore-cooling layout (Fig. 23.4). The liner, rated for mean effective pressures up to 32 bar and peak pressures in excess of 220 bar, was subsequently standardized for all B-series engines. Small changes were made to the BV cylinder heads, mainly a new head gasket matching the redesigned liner. The two-piece piston is essentially the same as that used in the BR engines, with a nodular iron skirt and bore-cooled steel crown. The connecting rods were lengthened but the same bearings were used as before.

Shorter fuel injection periods dictated strengthened camshafts to meet the increased loads, with a larger diameter both for the shaft and for the cam base circles. The fuel pumps are the same as those for BR engines but the fuel supply system was modified, chiefly by increasing pipe volumes and changing the layout to avoid cavitation and smooth out vibration caused by pressure pulses from the pumps.

Twin turbochargers mounted above the two insert-type charge air coolers operate on the impulse system, the pipework enclosed in an insulated box between the cylinder banks. A choice of electronic governors is offered,

Fig. 23.3 Crankshaft of the BV engine.

operating in conjunction with a standardized hydromechanical actuator. All electrical transducers on the BV engine are linked to a common electrical rail, one on each side of the engine, in a neat layout enabling faulty transducers to be quickly changed.

B32:40 engine

The B-series benefited from another rejuvenation in the shape of the longer stroke (400 mm) B32:40 design, offering an output of 500 kW/cylinder at 750 rev/min with a mean effective pressure of 24.9 bar.

Fig. 23.4 BV engine cylinder liner with the bore-cooled upper part, also applied to the in-line cylinder BR models.

Production engines became available in the year 2000, supplementing the standard B-series in the programme. The B32:40 was initially offered in six-, eight-, and nine-in-line and V12-cylinder configurations to span a power band from 3000 to 6000 kW; outputs up to 9000 were planned by extending the V-cylinder programme.

A new crankshaft design was incorporated to achieve the longer stroke and provide increased bearing areas, with new bearing technology addressing the larger bearing loads. Modified cylinder liners sought good control over temperatures at all points, and the piston/ring designs were also changed to minimize lube oil consumption and cylinder wear. The cylinder head design and material specification were refined to allow a small rise in the maximum combustion pressure and to handle the greater combustion air throughput from the upgraded turbocharging system. A new fuel injection system provided both an increase in capacity and a higher injection pressure. These changes contributed to specific fuel consumption of 183 g/kW h at the rated speed of 750 rev/min and NOx emissions under the IMO limit. Improved reliability and maintainability were promised by a new pump-end module and turbocharger support structure.

B32:40 details

Cylinder block: one-piece nodular cast-iron design of the rigid structure with an underslung crankshaft, cast-in charge air manifold, and large crankcase doors. The main bearing bolts are hydraulically tightened.

Cylinder liner: centrifugally cast, bore-cooled design with a running surface treated to improve wear resistance; a carbon-cutting ring is incorporated at the top of the liner.

Cylinder head: bore-cooled flame plate; six hydraulically tightened securing bolts to ensure even distribution; new cooled exhaust valve seat.

Crankshaft: continuous grain flow forged; large diameter journals and pins; hydraulically tightened counterweight bolts.

Connecting rod: forged of alloy steel and fully machined; obliquely split and serrated big end; hydraulically tightened big end bolts.

Bearings: steel backed with Sn/Al bearing material.

Piston: improved oil-cooled two-piece design; two compression rings and one oil scraper ring, all chromium plated.

Fuel injection system: pumps designed for 2000 bar injection pressure; totally enclosed in the heat-insulated compartment; constant pressure unloading for cavitation-free operation at all loads and speeds.

Turbocharging system: multipulse converter system based on ABB TPL series turbochargers; easily removable insulation panels for inspection and maintenance.

Clean Design versions of the B32:40 engine were introduced for operators keen to minimize the environmental impact of their tonnage, the first B32:40V12P models to this specification being delivered in mid-2006 for offshore vessel projects. The challenge of reducing NOx emissions to 80% below IMO Tier I requirements was met by Rolls-Royce without undermining power output or significantly increasing specific fuel consumption levels. Among the many refinements adopted were a revised piston bowl shape, along with modifications to the injector nozzles and other fuel system components, and turbocharger matching. Such changes reportedly avoided an increase in smoke output during part-load operation, which is a potential side effect of NOx reduction measures.

A 10% uprating announced in 2008 raised the maximum output to 550 kW/cylinder, the B32:40 series then spanning a power band from 3300 to 8800 kW.

B33:45 engine

Introduced in September 2014, the Bergen B33:45 was designed to offer an increase in power per cylinder compared to existing engines in the Bergen range with power lifted to 600 kW per cylinder. The engine was designed for both mechanical propulsion and electrical generation and in readiness for the 2016 introduction of the IMO Tier III NOx limitations.

The engine is economical with specific fuel consumption of 177 g/kWh at full load and complies with IMO Tier III rules when fitted with Selective Catalytic Reduction.

The Bergen B33:45 has a bore of 330 mm and a stroke of 450 mm and runs at 450–750 rpm as a marine propulsion engine on propeller law or 720/750 rpm for 60/50 Hz generator set drive. In-line engines were the first to be produced, with 6, 7, 8, and 9 cylinders spanning a power range from 3600 to 5400 kW. A 12-cylinder V version followed later extending the power range for the series to 7200 kW. Although 16- and 20-cylinder V models are produced for land use, but could be made available in marinized form if required.

Throughout the B33:45 a modular design approach has been employed which makes for easier maintenance and servicing. The crankcase is designed to incorporate air and oil galleries to minimize the number of external pipes with increased stiffness of the engine block. The design ensures very low levels of vibration. It carries the fully balanced crankshaft which has bolt-on balance weights and is dimensioned so that full power can be taken from either end up to and including the V12 version.

The engine has complete cylinder units comprising the head, liner, piston, and upper conrod which can be withdrawn for an overhaul in one unit. A three-piece conrod allows a piston to be drawn without disturbing the big end bearing.

The strengthened gear-driven camshaft is made up from individual cylinder sections for ease of replacement. Variable valve timing gives optimum response to load changes with reduced visible smoke even at low loads. A Pump Line Nozzle fuel system handles MGO, MDO, and HFO up to

1750 bar injection pressure in an arrangement that minimizes the number of connections and prevents pollution of fuel oil into the lube oil system.

Turbocharging is based on the multipulse system, with exhaust pipes enclosed in an insulated box to reduce heat transfer to the engine room, and two-stage charge air cooling. Turbocharger exhaust outlet and air inlet connections can be rotated in 15 degree steps to simplify duct installation and aid flexibility in routeing. B33:45 engines are designed for 25,000 h between major maintenance when operating within a defined load window.

C-engine (C25:33L)

A completely new design, the C25:33L, was launched in mid-2001 after a joint development project between Bergen and Hyundai Heavy Industries of South Korea, which markets and builds the engine as the HiMSEN H25/33. Flexibility in terms of power and speed ranges was a key development criterion to allow the 250 mm bore engine to target heavy fuel-burning gensets and smaller propulsion plants. Outputs from 1200 to 2700 kW at speeds from 720to 1000 rev/min were initially covered by five- to nine-cylinder in-line models, with V-engines planned to double the power threshold. The C25:33L was expected eventually to replace the popular but ageing K-series, particularly in the genset market. The first four production engines (nine-cylinder models) were due for delivery in spring 2002 as the core of a diesel-electric propulsion plant for an offshore service vessel.

In drawing up the design parameters, the stroke (330 mm) was determined by the piston speed which, in turn, was set by the desired time-between-overhauls of 15,000 h for the top end and 30,000 h for the bottom end, running on heavy fuel oil. The running speed was determined by the frequencies in genset applications, focusing on 900–1000 rev/min for 60 and 50 Hz requirements. These ratings sought the best compromise between an industry preference for a moderate speed in heavy-duty gensets and the potentially lower price per kilowatt from a faster running set. The engines can also be supplied, however, for 720/750 rev/min operation.

The C25:33L engine, Fig. 23.5, is based on a compact and stiff nodular cast-iron frame, with the charge air receiver, lube oil channel, and coolant transfer channel incorporated in the casting to eliminate pipework. A continuous grain flow forged steel crankshaft with steel plate balance weights allowed the cylinder centre distance to be kept the same as the K-series engine in the interests of compactness and rigidity. Full power can be taken

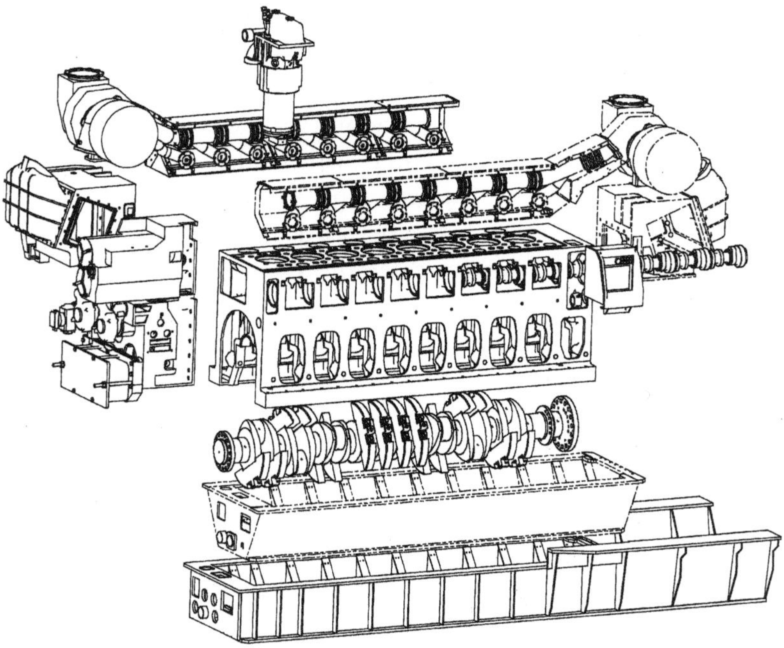

Fig. 23.5 Primary modules of Bergen C25:33L engine.

off at either end of the crankshaft, and an additional main bearing allows single-bearing alternators to be driven.

A key feature is the cylinder unit system, allowing a complete liner, piston, upper connecting rod, water jacket, and cylinder head to be drawn as an assembly for servicing or exchange. The components are clamped together by the cylinder jacket and held down by four-cylinder head bolts. A duct transfers air, exhaust and cooling water to and from the head, each cylinder unit being connected to its neighbour by quick-acting couplings to speed assembly and dismantling. The cylinder liner and water jacket combination is of the 'open deck' type, designed for intensive cooling and high strength without stress raisers in the critical top-end zone.

The piston comprises a steel crown with three rings and a nodular cast-iron skirt. An extended life and low controlled lube oil consumption are fostered by a chrome-ceramic piston ring coating, an anti-polishing ring at the top of the cylinder liner, and special honing of the liner surface. A three-part connecting rod enables its upper part to be detached when drawing pistons without disturbing the big end bearing.

One mechanical fuel pump is provided for each cylinder, with a simple two-step electronic timing feature controlled by the engine management system. (A common rail fuel system was not selected by Bergen because it was not seen as essential for the genset applications envisaged, nor as durable enough for unrestricted heavy fuel operation.) An impulse turbocharging system is based on an uncooled radial turbocharger which can be mounted at either end of the engine to suit the particular installation. A swift response to load changes, with minimal smoke emissions under rapidly changing load and speed conditions, is fostered by the high-efficiency turbocharging, two-stage charge air cooling with electronic temperature control, and a high-speed electronic governor.

Ease of maintenance was addressed by providing good access to key components and a small number of hydraulic tools able to handle all the main bolts. Ancillaries are grouped in a front-end module accommodating pumps, charge air cooler, lube oil cooler, and filters. The components are designed for plugging-in, with a minimum of joints avoiding leakages.

Bergen C25:33 engine data

Bore	250 mm
Stroke	330 mm
Speed	720–1000 rev/min
Output	220–300 kW/cyl
Cylinders	5, 6, 7, 8, 9 L
Power range	1200–2700 kW
Mean effective pressure	22.6–24.7 bar
Mean piston speed	7.9–11 m/s
Firing pressure	190 bar
Specific fuel consumption	182–186 g/kWh

The output of the C25:33 design was raised by 10% to 330 kW/cylinder in 2008, taking the upper rating to just under 3000 kW from the nine-cylinder model, and a Clean Design version introduced with emissions reduced to a minimum of 20% less than IMO limits. The environmental improvement resulted from applying the Miller cycle in combination with an increase in compression ratio. To avoid smoke at low load and poor transient load behaviour in the low load range, which are negative consequences of the Miller cycle, the CD engines are equipped with variable valve timing (VVT) mechanisms. This enables the Miller cycle inlet valve timing to be turned off for low load running, with control of the VVT system exercised by the engine's control logic. Advances in turbocharging technology, delivering higher pressure ratios, contributed to the power uprating in association with some changes in engine parameters.

Gas engines

The growing availability of LNG for gas-fuelled vessels, particularly in Scandinavian and Baltic regions, stimulated Bergen to invest further in spark-ignited engines. The KV-G4 gas engine, now in Norwegian ferry service, was derived from the long-established 250 mm bore K-series medium-speed diesel design. Introduced as a marine genset drive in 2002, it was released in 2004 as a variable speed engine for direct-mechanical propulsion.

Design work on the spark-ignited 350 mm bore B-gas engine (B35:40) began in spring 2002, tapping experience with the successful 320 mm bore B-series diesel engine. A V12-cylinder prototype engine developing 5100 kW made its commercial debut in a Danish combined heat and power station in 2003 after a comprehensive test programme in Bergen. The engine has references in four Norwegian ferries but no longer appears in the marine portfolio.

B36:45 gas engine

Four years after the introduction of the Bergen B33:45 Rolls-Royce announced an LNG variant of the engine with an increased bore of 360 mm. For the B36:45 the main structural components, crankshaft and conrods are the same as those used in the B33:45-series liquid fuel engines.

For the gas version, different liners, pistons, and heads are employed, giving a 30 mm increase in cylinder bore compared with the existing, well-proven, liquid-fuelled B33:45. The stroke remains unchanged, at 45 cm. Other changes involve the fuel system and injectors, which are Bergen Engines' own design.

The power output of the LNG-fuelled B36:45 engine is identical to that of the liquid-fuelled B33:45 at 600 kW per cylinder at 750 rpm. Specific energy consumption is a low 7300 kJ/kWh mechanical ISO including two engine driven pumps; specific lubricating oil consumption is less than 0.4 g/kWh.

The B36:45 is available in six, eight, and nine-cylinder in-line configurations. At the time of the 2018 announcement, Rolls-Royce said a V-12 version was in development to be followed by a 20-cylinder V-engine for very high-power applications. The current portfolio of Kongsberg (March 2020) does not show the V20 model but does indicate a V16 unit could be in development.

CHAPTER TWENTY-FOUR

Wärtsilä

The Wärtsilä Corporation grew from a modest Finnish base in the 1970s to become a world force in four-stroke engine design and production through the development of its own medium-speed programme and international acquisitions. Engine builders absorbed into the group over the years include Nohab Polar of Sweden, Stork-Werkspoor Diesel of The Netherlands, Wichmann of Norway, the French companies Duvant-Crepelle and SACM, Echevarria of Spain, and Grandi Motori Trieste (GMT) of Italy.

In 1997 Wärtsilä Diesel's Finnish parent group Metra took a dominant holding in the newly formed Wärtsilä NSD Corporation, with Fincantieri of Italy taking the remaining minority share. The new group embraced Wärtsilä Diesel, New Sulzer Diesel, and Diesel Ricerche (Fincantieri's R&D arm) as well as GMT, the engine building division of Fincantieri, which became a fully owned subsidiary in January 1999. The New Sulzer Diesel element brought the significant Swiss-designed low- and medium-speed engine portfolio under the Wärtsilä umbrella. Wärtsilä Corporation's medium-speed engine production is now based in Finland and Italy, respectively focusing on small-bore and medium-to-large bore designs. The company is generally acknowledged as being the dominant force in the sector having a share which has varied from 34% to 51% over recent years when measured in terms of total gross power output.

Wärtsilä Diesel laid the groundwork for its international growth in the early 1970s by deciding to develop a 320 mm bore medium-speed engine designed from the outset for operation on heavy fuel. The resulting Vasa 32 engine was tested at end-1976 and the first production model delivered in 1978. The design supplemented the company's established smaller bore marine diesel oil-burning Vasa 22 series and helped to pioneer heavy fuel-burning medium-speed engines for propulsion and auxiliary power.

Underwriting the Vasa 32 design's pier-to-pier heavy fuel-burning capability were a number of innovations which have since become standard medium-speed engine practice: nodular cast iron pistons; forced exhaust valve seat cooling to combat valve burning; an inverse cooling system to foster heavy fuel operation in the low load range; a high injection pressure for

Pounder's Marine Diesel Engines and Gas Turbines
https://doi.org/10.1016/B978-0-08-102748-6.00024-4

improved combustion; and a cylinder liner design with temperature control preventing corrosion and deposit formation.

Wärtsilä Diesel's medium-speed engine programme was progressively extended in the 1980s and 1990s with smaller and larger bore heavy fuel-burning models ranging from bore sizes of 200–640 mm (Table 24.1). As a pioneer of dual-fuel engines, Wärtsilä introduced dual-fuel variants of some of its more popular engine sizes usually with a slightly higher bore.

The current range comprises 11 designs including the Wärtsilä 20 and 20DF, the Wärtsilä 26, the Wärtsilä 32 (a more modern longer stroke derivative of the Vasa 32), the 34DF a dual-fuel version of the Wärtsilä 32, the short-lived Wärtsilä 64 (the world's most powerful medium-speed engine), the Wärtsilä 50DF dual-fuel engine (derived from the Wärtsilä 46 design) and in 2014 the Wärtsilä 46DF which drew on both models. Most recently the Wärtsilä 31 was introduced in 2015 and now comes in several variants including a spark ignited gas engine introduced in 2019. The original Wärtsilä 31 was recognized by the Guinness Book of Records as the most efficient four-stroke diesel engine. The listing is based on the Wärtsilä 31 engine's highest fuel efficiency levels, with its diesel fuel consumption being as low as 165 g/kWh.

The various designs introduced features now widely used by other four-stroke engine builders, Wärtsilä citing: side-mounted one-cylinder camshaft sections; four-cylinder head bolts for large engines; 1500 bar-plus fuel injection pressure; and a 'hot box' for fuel injection pumps and pipes. Advanced machining systems allowed comprehensive integration of fluid systems within the engine block and the creation of multifunctional structural elements, significantly reducing the overall number of components.

Table 24.1 Wärtsilä medium-speed oil-fuelled engines.

Design	Bore/stroke (mm)	Output (kW/cyl)	Speed (rev/min)	Cylinders
W20	200/280	200	720–1000	4, 6, 8, 9L
W26	260/320	340	900/1000	6, 8, 9L/12, 16V
W31	310/430	610	750	8, 10, 12, 14, 16V
W32	320/400	500	720/750	6, 7, 8, 9L/12, 16, 18V
W38B	380/475	725	600	6, 8, 9L/12, 16V
W46C	460/580	1000	500/514	6, 8, 9L/12, 16V
W46F	460/580	1200	600	6, 7, 8, 9L/12V
W64L	640/900	2150	333	6, 7, 8L

Maximum cylinder pressure limits beyond 200 bar have been pursued by the company (Fig. 24.1), exploiting nodular cast iron piston skirts with forced piston skirt lubrication: part of the oil for cooling the piston crown is forced out through four nozzles leading to an annular oil distribution groove, fostering well-controlled lubrication and hence increased reliability and reduced liner wear. Thick-pad bearing technology has benefited engines operating with high cylinder pressures.

A three-ring piston pack is standard for Wärtsilä engines as is an antipolishing ring at the top of the cylinder liner (Fig. 24.2). The function of the latter ring is to calibrate the carbon deposits on the piston top land to a thickness small enough to prevent contact between the liner inner wall and the deposits in any position of the piston. Bore polishing is reportedly eliminated and liner lifetimes can be doubled accompanied by cleaner piston ring areas and significantly reduced lubricating oil consumption.

Low NOx combustion technology (whose principles are detailed in section 'Wärtsilä Vasa 32') can be applied to all current Wärtsilä models, reducing noxious exhaust emissions without undermining fuel economy. The engines can also be arranged for gas–diesel and dual-fuel burning installations (see Chapter 3) and offered as part of group-supplied integrated Propac

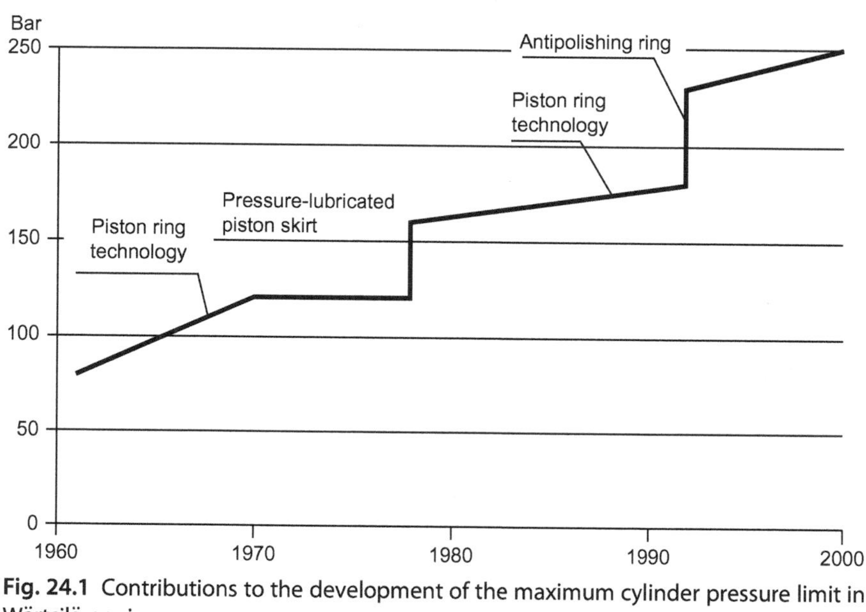

Fig. 24.1 Contributions to the development of the maximum cylinder pressure limit in Wärtsilä engines.

Fig. 24.2 Antipolishing rings have benefited all Wärtsilä engines (W32 engine liner shown).

propulsion packages embracing reduction gearing, propeller, and control systems.

Wärtsilä has pioneered computer-based systems for engine control and monitoring, condition monitoring, and fault diagnosis, as well as multimedia documentation and remote expert communication systems to support servicing and maintenance (Fig. 24.3). Wärtsilä claims to have pioneered the application of computer-controlled CR fuel injection systems operating on heavy fuel, its developments for medium-speed engines starting in 1997 and the first deliveries following in 2000. Numerous installations are now in service and CR fuel injection can be specified throughout the programme for all models except the W64 engine.

Fig. 24.3 The engine computer is the 'brain' of the Wärtsilä Engine Control System (WECS), shown here on a Wärtsilä 32.

Wärtsilä Vasa 32

The 320mm bore/350mm stroke Vasa 32 engine (Fig. 24.4) was designed in the 1970s but benefited from progressive refinements yielding higher power outputs, enhanced reliability and serviceability, and the acceptance of fuels with higher CCAI numbers until the arrival in 1997 of the new-generation Wärtsilä 32 (see section 'Wärtsilä 32'). By that time, around 1900 Vasa 32 engines had been sold for marine service since the seagoing debut of the design in 1978. Sustained demand nevertheless dictated continued production, overlapping with the Wärtsilä 32 engine. The Vasa 32LN was produced in four, six, eight, and nine in-line and V12-, 16-, and 18-cylinder models to cover propulsive power demands from 1480 to 7380kW at 720/750rev/min. The D-rated versions offered 375kW/cylinder while E-rated models yielded 410kW/cylinder with respective mean effective pressures of 21.9 and 24bar.

Fig. 24.4 The Wärtsilä Vasa 32 engine founded a family of heavy fuel-burning designs.

The specification comprises the following main elements:

Engine block: designed for maximum overall stiffness and cast in one piece for all cylinder numbers; arranged for underslung crankshaft; and direct installation on resilient mountings possible.

Crankshaft: forged in one piece; fully balanced; and optional torsional vibration dampers.

Connecting rod: forged in alloy steel and machined, and diagonally stepped split in the big end.

Bearings: designed for maximum wear resistance.

Cylinder liner: special cast iron; and bore cooling for efficient control of liner temperature.

Piston/rings: composite piston with steel top and nodular cast iron skirt; forced skirt lubrication; and ring set comprising three compression rings and one oil scraper ring, all chromium plated.

Cylinder head/valves: grey cast iron; absorbs the mechanical load with a box section formed by an intermediate and an upper deck; mounted on the engine block with hydraulically tensioned studs; twin inlet and twin exhaust valves, all equipped with rotators; and water-cooled exhaust valve seats.

Camshaft: drop-forged one-cylinder shaft segments with integrated cams; and bearing journals fitted to the camshaft segments with flange connections.

Fuel injection: one-cylinder pumps with built-in tappets; throughflow type pumps for heavy fuel operation; uncooled nozzles; and all fuel-carrying equipment located in a drained and insulated space (the hot box) which keeps the system at operating temperature and ensures safety in the event of leakage.

Exhaust pipes: nodular cast iron, the entire system enclosed in an insulating box for safety.

Low NOx Vasa 32

The Vasa 32 engine in 1994 became the first in the Wärtsilä Diesel programme to be released with low NOx combustion technology, a key measure calling for the compression ratio to be increased from 12 to 14:1 to secure a sufficiently high compression temperature. The smaller combustion space dictated reshaping of the piston crown and the cylinder cover flame plate to allow for the fuel jets and good air/fuel mixing.

A new piston was developed for the higher maximum firing pressure (raised by 10–165 bar). Efficient cooling of the alloyed steel crown of the composite piston was secured by applying the cocktail shaker principle. The piston pin diameter was increased by 10% to match the higher cylinder pressure. The three-ring concept (as opposed to the original four-ring set) was also adopted following good experience on the Wärtsilä 46 engine, the two compression rings benefiting from a special wear-resistant chrome-ceramic coating (Fig. 24.5).

The stiff bore-cooled cylinder liner was equipped with an antipolishing ring to keep carbon and ash build-up on the piston top land thin enough to prevent contact between the liner inner wall and the deposits, regardless of the piston's position. The liner lifetime is reportedly doubled and the lubricating oil consumption halved to 0.6 g/kW h or less.

The increased cylinder pressure also required the connecting rod to be changed from the original diagonally stepped two-piece component to a fully machined three-piece design with a horizontally split big end bearing. The change yields greater safety at increased load and eliminates the need to

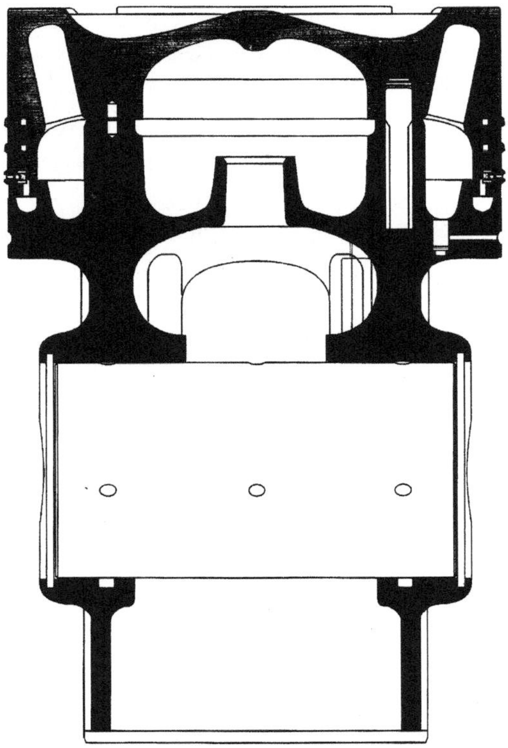

Fig. 24.5 Composite piston of Vasa 32 low NOx engine designed for increased combustion pressure and low friction.

interfere with the big end bearing assembly when overhauling the piston; a disadvantage is the increased number of parts. The load-carrying capability of the big end bearing was raised by switching from an SnSb or SnAl to a BiAl material type.

Only a minor modification to the cylinder head was necessary: a slight reshaping of the flame plate to provide more space for the fuel jets.

The fuel injection system was redesigned to handle demands for an increased injection rate and improved atomization. As a result, the fuel cam is faster and provides higher lift to compensate for the reduced plunger bore/stroke ratio. The injection valve opening pressure was raised from 350 to 600 bar, mainly to improve atomization at the start and stop of injection, and to maintain the higher injection pressure needed for a shorter injection period and reduced ignition delay. Improved part- and low-load performance was apparent, along with reduced NOx formation and fuel consumption. Test results showed it was possible to reach an NOx level of 11 g/kW h, a 30% reduction compared with a standard Vasa 32 engine (IMO's limit is 12 g/kW h NOx at 720/750 rev/min). At the same time, the specific fuel consumption was lowered from 187 to 180 g/kW h while maintaining a reasonable combustion pressure of 165 bar, only 10 bar higher than that of a standard Vasa 32 engine.

Wärtsilä Diesel also developed a planetary gear with controls to replace the standard intermediate gear of the camshaft transmission. The arrangement allowed a retarded fuel injection mode to be temporarily engaged while the engine was in operation to meet modest local NOx emission limits (up to 30% reduction on normal emission levels at the expense of a rise in fuel consumption).

Wärtsilä 32

A new generation 320 mm bore model was introduced to the market in 1997, the Wärtsilä 32 exploiting a longer stroke (400 mm) than its Vasa 32 predecessor and incorporating design features established on the Wärtsilä 20, 26, 38, and 46 engines (detailed later in this chapter). A higher output than its predecessor was offered (460 kW/cylinder at 750 rev/min on a mean effective pressure of 22.9 bar) to give a power range up to 8280 kW from in-line six, eight, nine, and V12-, 16-, and 18-cylinder versions. (The programme now includes a seven-cylinder model.)

Scope for future-specific power rises was demonstrated by the test engine operating comfortably on an mep of 31 bar. A higher rating was subsequently released (500 kW/cylinder at 750 rev/min on an mep of 24.9 bar) to extend the power range to 9000 kW. A specific fuel consumption of 173–180 g/kW h (without engine-driven pumps) is quoted, depending on nominal output or optimal point operation.

Although slightly taller and heavier than the Vasa 32, the Wärtsilä 32 is more compact and has a better power/weight ratio. Integration reduced the number of components in the new engine by 40% compared with its forerunner and contributed to a neater look and lower production costs (Fig. 24.6). Ducts for water and oil are incorporated in the engine block, and low-pressure fuel pipes were replaced by ducts integrated into the injection pump design.

A substantial reduction in the number of joints to be broken during engine overhauls resulted. Further attention was paid to easing maintenance. The multiduct (Fig. 24.7) containing the water system and the exhaust and air piping, for example, stays in place when the cylinder head is lifted; the

Fig. 24.6 Engine-driven pumps are standard for the Wärtsilä 32; note the clean lines achieved by integrated duct systems which eliminate external piping.

joints for starting air, oil, and return fuel are slide-in connections; and, as there is no real jacket water, only a limited amount of water has to be drained when maintaining the cylinder unit. Ergonomic refinements were also pursued: the rocker arm cover is suspended on hinges and stays in place when the rocker arms of the injector are attended. Service pumps, including the prelubricating pump, are arranged on the engine along with a lubricating oil module which incorporates an automatic back-flushing filter (Figs 24.8 and 24.9).

The low-pressure fuel line comprises drilled channels in cast parts clamped firmly on the engine block, these parts consisting of the pump housing, the tappet housing, the fuel transfer housing, and the multicover. For easy assembly/disassembly, the parts are connected to each other with slide connections. Both the whole low-pressure and high-pressure systems are housed in a fully covered compartment. The high-pressure system is designed and endurance tested at 2000 bar, and the injection pressure is around 1800 bar. With a wear-resistant low-friction coating on the fuel plunger, no lubricating oil is required for the pump element. The profiled plunger geometry allows the clearance between plunger and barrel to be

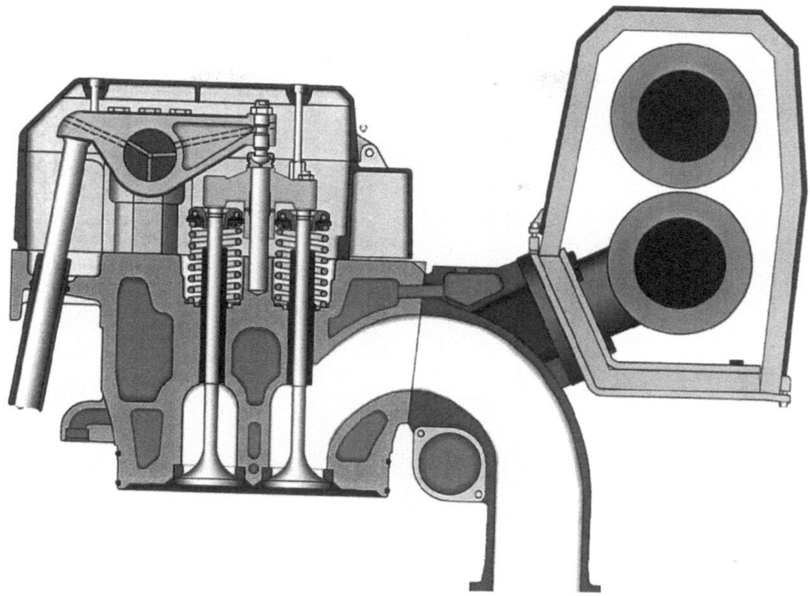

Fig. 24.7 The multiduct remains in place when the cylinder head is lifted (Wärtsilä 32 engine).

Fig. 24.8 Lubricating oil module for mounting on the Wärtsilä 32 engine.

Fig. 24.9 Lubricating oil module mounted on the side of a Wärtsilä 32 in-line cylinder engine.

kept small, thereby allowing only a minimum of oil to pass down the plunger; this small leakage is collected and returned to the fuel system. Any likelihood of fuel mixing with the lube oil is eliminated.

The turbocharging system is specified to suit the requirements of each engine application from these standard options: pulse system, single pipe exhaust (Spex) system, and Spex system with exhaust wastegate and air bypass. The systems are designed for minimum flow losses on both exhaust and air sides, and the interface between engine and turbocharger streamlined to avoid all the adaption pieces and piping often used in the past.

The Wärtsilä Engine Control System (WECS) serving the Wärtsilä 32 measures a range of parameters and undertakes safety monitoring of the main bearing and cylinder liner temperatures, as well as individual exhaust valve monitoring. In the case of the V18-cylinder model, for example, the computer-based system measures 97 temperatures, 10 pressures, four speeds, three positions, and seven levels. It also controls starting and stopping, the low-temperature (LT) thermostatic valve, wastegate and bypass valves, slow turning, and interfaces with external systems.

Wärtsilä 20

Introduced in 1992, the 200mm bore/280mm stroke Wärtsilä 20 engine (Fig. 24.10) was designed primarily as a genset drive to replace the long-established Wärtsilä Vasa 22 but small-ship propulsion installations have also been logged. An output per cylinder of 130–165kW was initially offered, depending on the nominal speed (720–1000rev/min), but subsequent power rises took the rating to 180kW/cylinder (end-1998) and 200kW/cylinder at 1000rev/min (2003). The rating increases were underwritten by designing all key components for a maximum cylinder pressure of 200bar; engines are currently released with a P_{max} of 170bar.

The W20 series now embraces four-, six-, eight-, and nine-cylinder models covering a maximum continuous output range from 720 to 1800kW. Up to 24,000h between overhauls have been achieved by engines running on light fuel oils.

A flat fuel consumption curve, with the lowest point preferably at part load, is desirable for an auxiliary engine which seldom operates at full load and may often run at a very low load. This was addressed in the Wärtsilä 20 design by optimizing the cylinder dimensions in conjunction with modern turbocharging technology; adopting a pulse charging system, and exploiting a high cylinder pressure.

Physically, the Wärtsilä 20 was intended to be shorter and lower than any existing engine in its performance class, with advantages in ease of installation and maintenance. The nodular cast iron engine block was designed for maximum overall stiffness and local stiffness around the upper part of the cylinder liner, and to incorporate a number of cast-in or machined water and oil channels in pursuit of component integration. The camshaft bearings are directly housed in the block as is the camshaft gear train at the flywheel end of the engine. Provision is made for five engine-driven pumps (three are standard). A four-point support configuration underwritten by the overall stiffness of the engine is especially attractive for resilient mounting, and the screwed-on feet arrangement offers considerable freedom in foundation design.

The four-screw cylinder head and box-cone design is another Wärtsilä tradition retained to achieve even sealing pressure and prevent liner deformation under high cylinder pressures. It also secures ample space for large

Fig. 24.10 Three Wärtsilä 6L20-powered gensets in the engine room of a feeder container ship.

inlet and exhaust channels which are necessary for good gas exchange. The head is mounted on the engine block with hydraulically tensioned studs, and is arranged to house two inlet and two exhaust valves, all equipped with rotators; the exhaust valve seats are water-cooled. A starting air motor for all engine cylinder configurations was specified in contrast to traditional direct starting by air injection into the combustion chamber, contributing to optimized cylinder head stiffness and valve performance.

Innovation in the W20 design was a multifunctional duct between the cylinder head and the engine block (Fig. 24.11). The multiduct houses the air duct from the air receiver to the cylinder head (the air duct creates the initial swirl which helps to achieve optimum part load combustion) and incorporates the exhaust gas channel. The cooling water from the cylinder head is led to a cast-in channel in the engine block through the multiduct. The multiduct is also the support for the exhaust pipes, remaining in place when the cylinder head is removed to ease maintenance.

A 'distributed water flow' pattern is used to effect cooling of the exhaust valves and cylinder head flame deck. An even temperature distribution results from increasing the flow velocities on the exhaust side in general and around the exhaust valve seat rings in particular.

Fig. 24.11 The multiduct of the Wärtsilä 20 engine facilitates maintenance of the cylinder components.

The wear-resistant cast iron cylinder liner was designed for minimum deformation thanks to its own stiffness and support from the engine block. It was also important to tune the inner wall temperature to a level just above the sulphuric acid dew point; this was addressed by creating an increased tangential cooling water flow velocity just below the liner collar. The risk of bore polishing is eliminated by an antipolishing ring at the upper part of the liner.

A traditional Wärtsilä concept was retained for the composite piston (Fig. 24.12) but refined for a smaller bore engine operating on heavy fuel: a nodular cast iron skirt, steel crown, and pressurized skirt lubrication. Two compression rings and a spring-loaded oil scraper ring are located in the crown, the efficiency of a three-ring concept having been proven in the Wärtsilä 46 engine. (Most of the frictional loss in a reciprocating combustion engine originates from the piston rings; a three-ring pack is considered optimal by Wärtsilä with respect to function and efficiency.) Each ring is

Fig. 24.12 Wärtsilä 20 engine composite piston with nodular cast iron pressure-lubricated skirt and steel crown.

chromium-plated and dimensioned and profiled for its individual task. The top ring benefits from a special wear-resistant coating.

A connecting rod of forged and machined alloy steel with a stepped split line lower end similar in design to that used in the Vasa 32 engines allows the maximum pin diameter while still making it possible to pull the rod through the cylinder liner (Fig. 24.13). The advantage cited over the conventional straight, oblique split line is an increase in pin diameter of at least 15%, a key factor in an engine designed for high combustion pressures. The fully balanced crankshaft is forged in one piece and amply dimensioned for reliable bearing performance. A small cylinder distance facilitates thin webs and large-bearing surfaces. A low-bearing load and optimized balancing of the rotating masses secure good oil film thicknesses at all engine loads and speeds.

The low-pressure fuel system is an integrated part of the injection pump achieved via a multifunctional housing which acts as a housing for the single-cylinder pump, a fuel supply line along the whole engine, a fuel return line from each pump, a housing for the pneumatic overspeed trip device and guidance for the valve tappets. The high-pressure fuel line is also fully enclosed with an alarm for possible leakages, and all fuel-carrying parts are housed in a fully covered compartment (hot box). The high-pressure side is designed for a 2000 bar injection pressure although the pressure used is around 1500 bar. The pump element itself features a solution to eliminate cavitation erosion. The fuel injection valve is designed for uncooled nozzles which are heat-treated for low wear and high heat resistance.

Fig. 24.13 The connecting rod of the Wärtsilä 20 engine incorporates a stepped split.

A traditional two-circuit cooling water system incorporates a hot circuit for jacket water and a cold circuit for air-cooling and oil cooling. The inverse cooling principle introduced earlier by Wärtsilä on the cold circuit side has reportedly proven valuable for engine operation on low-quality heavy fuels at part load. A stepless electronically controlled version was developed for standard fitment to the W20 engine. The two pumps serving the cooling systems are engine driven. The high-temperature (HT) cooling system is fully integrated into associated engine components, virtually eliminating the use of pipes. Supply and return channels are cast into the engine block while the pump cover and turbocharger support hold the connections between pump and engine block.

The system temperature is kept at a high level (about 95°C) for safe ignition/combustion of low-quality heavy fuels at low loads. An engine-driven pump of the cassette type is standard. The LT cooling system is also fully integrated with the engine components (engine block, pump cover, air, and lubricating oil cooler housings) and served by an engine-driven pump of the same type as in the HT system. The system temperature is electronically controlled as a function of engine load. The charge air temperature is optimized at every operating point by heating the charge air with heat from the lubricating oil according to a preset curve programmed into the electronic board.

The lubricating oil system is grouped compactly into two adjacent modules at the flywheel end: one contains the cooler thermostatic valves and filters, and the other is a pump unit with a prelubricating pump and an engine-driven main pump. The filter arrangement comprises three individual filter chambers which can be switched off one by one when replacing the filter cartridge. A 'safety mesh' is also incorporated in each chamber. The paper-type filter cartridges are horizontally positioned to facilitate proper cleaning of the chamber before a new cartridge is installed. Centrifugal filters fitted for further filtration of the lubricating oil also serve as the oil condition indicator.

A pulse turbocharging system was considered the best choice for an engine designed for installations normally subject to frequent load variation and sudden load application. All cylinder configurations are served by an optimized pulse system; no additional solutions are required for part load operation.

The control and safety system is based on two cabling modules, one on each side of the engine, and sensors of the 'low mass' type to avoid possible vibration damage. The sensors are prefabricated with the correct cable length and a connector for connection to the cabling module. The modules, in turn, are connected with multipin type connectors to the engine control unit (ECU) which is resiliently mounted on the engine. Incorporated in the ECU is instrumentation for the local reading of engine and turbocharger speed, exhaust temperature after each cylinder and after the turbocharger, an operating hour counter and electronic boards for speed measuring and charge air temperature control. Plug-in connectors are provided for connecting the ECU to an external control system.

New W20 layout

Seven years after Wärtsilä 20 engine production started, some design revisions were made, mainly involving the ancillary equipment and with the aim of improving environmental friendliness and adaptability to customer needs. The main components, such as pistons, bearings, and crankshaft, remained untouched. The new layout was introduced at end-2000.

A drawback of the original layout was the paper cartridges used for lube oil filtration. The number of filter cartridges per engine varied between two and four, depending on the cylinder configuration. A typical genset installation of three 9L20 engines running for 5000 h annually on heavy fuel oil would, therefore, generate hazardous waste in the shape of 100 used cartridges a year. The main objective in revising the engine layout was to eliminate this waste and the high cost of new filters for ship owners.

A completely new oil module was designed to enable the use of a fully automatic back-flushing lube oil filter incorporating elements of seamless sleeve fabric with HT resistance. An overhaul interval of 1 year is recommended, and the expected filter lifetime is 4 years. The channels directing the fluids to the thermostatic valves for the LT water and lubricating oil are integrated into the bracket for the filter and the cooler. A horizontal position for the filter was chosen because of space limitations and maintenance benefits. The filter housing drains automatically when the engine is stopped so that any overhaul inside the filter is easier and the risk of foreign particles entering the clean side is minimal.

The oil cooler design was also changed from a tube type to a brazed-plate heat exchanger. The cooler needs no outer casing and its cooling capacity varies with the number of plates (the height of the cooler); this makes it possible to use one bracket for all engine cylinder configurations. The same basic oil module is used on both LF (auxiliary) and LD (main) engines; it is simply rotated 180°.

The number and size of nonmachined parts, after the oil filter, were reduced significantly to minimize the risk of foreign particles entering the system. Since the Wärtsilä 20 cannot be equipped with running-in filters, to prevent damage to the bearings or crankshaft during test runs, this redesign was necessary to allow uninterrupted production. A dual drive, the turbine-driven centrifugal oil filter is used for removing particles from the flush oil coming from the automatic filter. Both dry and wet oil sumps are available. The main oil pump was moved to the front of the engine to optimize the suction arrangement for every installation type.

The charge air cooling module was redesigned to fit both versions of the engine. A self-supporting flange sealing was selected to eliminate the problems with sealing the cooler insert on the original design. The same cooler and the outlet duct are used on both LD and LF engines. The compressor washing arrangement is integrated into the water box for the cooler on the LD engine. Also integrated into the design was the possibility to have a water mist catcher. The water and fuel connections were moved to the sides of the engine to improve the pipe layout and the support for the pipes. No piping disturbs the overhaul space since all the pumps are arranged at the front of the engine.

Among the options are a direct-driven seawater pump, 100% power take-off at both ends, WECS digital control system, fully integrated standby connections for water, fuel and lube oil, and fuel flexibility (light and heavy) possible for all engine types.

The old engine block design with loose feet (to make installation more flexible) had proved unnecessary in practice as it was useful in only a handful of applications. The feet were thus integrated with the engine block casting to reduce cost; the option to use loose feet was nevertheless retained and wider feet were designed to create a more effective resilient mounting. The main oil channel is machined in the engine block to reduce both labour and material costs. The charge air receiver wall was provided with two openings since the charge air duct can be placed at either end of the engine.

A major disadvantage of the Wärtsilä 20 engine for propulsion applications was that the turbocharger could only be mounted at its free end. It was therefore decided to make it possible for the turbocharger to be located at either end on the six-to-nine-cylinder models, the main contenders in the small-ship propulsion market.

In December 2017, Wärtsilä announced an upgraded version of the engine with increased maximum rpms from 1000 to 1200 for mechanical propulsion engines and from 900 to 1200 for 60 Hz generating sets. This technology was passed over from the Wärtsilä 20DF engine, which had been running at this speed from the introduction in 2011.

With an increased cylinder output of 220 kW at 1200 rpms, the upgraded engine is a high-value-added proposition for the yards building merchant ships, and ultimately ship owners looking for lighter and more compact engine configurations to achieve higher payloads with almost the same low fuel consumption. At the introduction, Wärtsilä said it would take 2 years to develop all versions of the engine and also said that in theory, all Wärtsilä 20 engines can be upgraded to the new performance parameters.

The new Wärtsilä 20 is denominated with the 'E' suffix indicating (as in Wärtsilä 20E and Wärtsilä Auxpac 20E) its (alphabetic and other) evolution over previous models, with both models achieving a 10% higher output at 1200 rpms but still based on the classic Wärtsilä 20 engine design. The intention is for the Wärtsilä Auxpac 20E to replace the former Wärtsilä Auxpac 20D3/D4 as an auxiliary generator set which is, for Wärtsilä internally, an enormous reduction in complexity and spare parts handling.

The higher cylinder output will be available in six-, eight-, and nine-cylinder configurations, and due to the new automation system (UNIC 2) including a new main cabinet, now equipped with a touch screen display, will be somewhat longer (30.3–37.2 cm) than previous models and will weigh between 9.3 and 11.6 t. Most of the new engine components are based on the earlier Wärtsilä 20 diesel engine with a few tweaks.

Both new engines—Wärtsilä 20 and Wärtsilä Auxpac 20—have a Wärtsilä UNIC engine control/automation system and includes an engine-mounted control panel showing the engine's operational data on an LED touch screen.

As far as the actual engine block, cylinder heads, and camshaft are concerned, a few new flanges and connecting holes have been machined in to allow for cooling pipes. A new injection system was also installed across the whole Wärtsilä 20 range.

The additional rpms and capacity on the upgraded Wärtsilä 20 stem from the turbocharger and compressor aspects where a new cooling pipe arrangement is needed to reduce the internal temperatures of the turbocharger.

Another major improvement comes in the form of a new AWG (Air Waste Gate) valve, which previously was a pneumatic system and not very accurate. The new Lisk valve is a hydraulic set-up (borrowed from the high-speed engine industry) which has improved functionality. It will also yield a longer lifespan.

Other improvements include a new cable rail for the electrical equipment, new piston rings, a new injection pump and valve. Commonalities between the old and new engine exist because the basic design of the whole engine remains untouched. It is also possible to retrofit a diesel into a dual-fuel unit. In these cases, required changes include pistons, piston rings, cylinder heads (with injectors and valves), camshaft, and electrical equipment including brackets.

Wärtsilä 26

The 260 mm bore/320 mm stroke Wärtsilä 26 engine (Fig. 24.14) was launched in 1995 as an eventual replacement for the SW28 model in the Stork-Wärtsilä Diesel medium-speed programme. The design is characterized by a high power density and the capability to run on heavy fuel up to 730 cSt viscosity. An advanced crankshaft design, significantly reducing the distance between cylinders, and a short connecting rod limiting the height of the engine contributed to a short and low overall configuration.

Fig. 24.14 The W26 engine in V18-cylinder form; it is possible to change the lube oil filter while the engine is running.

Propulsion and genset drive applications are now targeted by six, eight, and nine in-line and V12- and V16-cylinder models covering an output band from around 1950 to 5440 kW at 900–1000 rev/min. Mean effective pressures range from 24 to 25.5 bar, depending on heavy fuel or marine diesel oil operation, with power outputs up to 325 kW/cylinder (900 rev/min) and 340 kW/cylinder (1000 rev/min).

Most of the design work was executed at Stork-Wärtsilä Diesel in The Netherlands. The designers defined five key objectives: high reliability; low fuel consumption and emissions; simple, space-saving installation; multiple applications and fuel choices; and low operating cost per kilowatt. The goals were addressed by combining technology developed specifically for the application with design features of three earlier 'new generation' engines in the Wärtsilä Diesel programme: the medium-speed W38 and W20 series and the high-speed W200 series (Chapter 26).

High reliability was sought from the incorporation of structural and component designs proven in these and other Wärtsilä engines, such as thick-pad bearing technology and pressurized skirt lubrication. Extended times between overhauls and enhanced lifetimes were also targeted.

Compact, more simple and accessible installations are addressed by integrating ancillary support equipment (pumps, filters, coolers, and thermostats) on the engine. Piping is eliminated from outside the engine; the channels for lubricating oil and cooling water are all incorporated in the block and other castings, and the inlet air receiver is also completely integrated into the engine block casting. A multiduct (Fig. 24.15) combines several functions, providing a connection between the inlet air receiver and the inlet port of the cylinder head, a flow channel for exhaust gases and a cooling

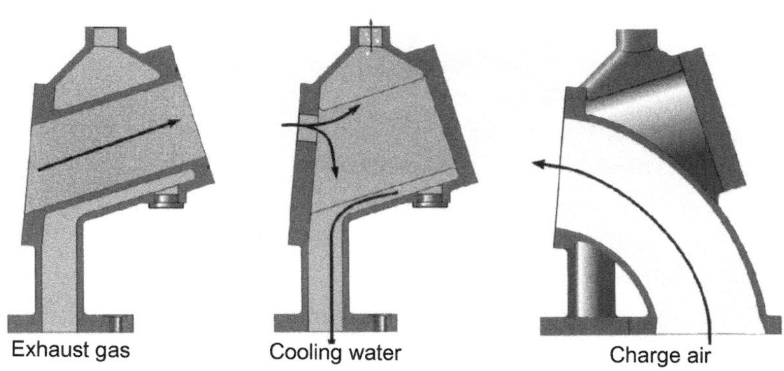

Fig. 24.15 Schematic diagram of multiduct connected to the cylinder head (Wärtsilä 26 engine).

water channel. It also supports the exhaust system and facilitates dismantling of the cylinder head without disturbing that system.

The pursuit of integration reduced the overall number of Wärtsilä 26 engine components by around 30% compared with the SW28 design.

Engine length was saved by giving the crankshaft undercuts. The radii in the transition from the crank web to the crankpin and from the crank web to the journal are included in the web itself. As a result, it was possible to keep the distance between the cylinders short and, at the same time, specify an optimum bearing length. It has a beneficial effect on the oil film thickness of the bearings, and hence their durability was gained. The fully balanced crankshaft, suitable for full power output to be taken off either end, was designed for maximum overall rigidity and moderate bearing loads.

The highest possible rigidity was also sought from the cylinder head design which features a conical intermediate deck and provides optimum support to the flame deck. The flame deck has drilled coolant channels to inhibit large thermal stresses. The exhaust valve seats are water-cooled and the valve temperatures measured by sensors.

Flanged cylinder liners with tangential cooling water flow are installed. The liner is supported symmetrically at the top by the cylinder block and sized so that the temperature on the inner wall remains high enough to prevent LT corrosion yet low enough to ensure good lubrication. This design, proven on the Wärtsilä 20 engine, makes drilled coolant holes unnecessary; as a result, a thinner liner wall can be used. An antipolishing ring at the top of the liner is specified to prevent contact between the piston crown and the liner, thus avoiding any bore polishing by combustion residues. High resistance to wear is further fostered by elements in the alloy material of the liner. The antipolishing ring also contributes to a reported lubricating oil consumption of under 0.5 g/kW h.

The composite piston consists of an oil-cooled steel crown with three rings (two ceramic chromium-plated compression rings and an oil distributor ring) and a nodular cast iron skirt. Crown and skirt are fastened to each other by one central bolt to foster an even, symmetrical load distribution (another feature whose effectiveness had been demonstrated in the Wärtsilä 20 engine). Pressurized skirt lubrication secures excellent tribological behaviour at high cylinder pressures.

A connecting rod with a horizontal split was developed for the Wärtsilä 26 engine, exploiting the finite element method to optimize the component with regard to maximum stiffness and minimum mass. Key merits cited for the resulting design are a short length, high rigidity, reliability, and low mass (and consequently a smaller load on the bearings). The rod is drilled with a single hole without plugs for delivering lubricating oil to the piston.

An injection pressure of 2000 bar, necessary for fine atomization of the fuel and a short injection duration, is attained by closed barrel-type fuel pumps. Fuel feed lines are integrated into the pump housing, and the fuel system is designed to keep pressure pulses low. The high injection pressure dictated a very rigid camshaft made of single-cylinder sections connected to each other by bearing journals.

Design details facilitate easier inspection and maintenance routines. The stud bolts for the cylinder head, connecting rod, and main bearing are hydraulically tensioned. When the head is removed, the multiduct supports the exhaust gas piping and the insulation cover.

The seals of the cooling water pump can be replaced without dismantling the pump itself. Replacement or inspection of the piston rings is possible without disassembly of the liner: the piston can be partially removed from the liner for this purpose.

The fuel pumps can be replaced quickly and simply without taking apart the low-pressure piping and the tappets; the pump can be taken from the engine and overhauled under clean workshop conditions. Both air cooler and lubricating oil cooler are of the insert type, allowing convenient removal for cleaning, and it is possible to change the oil filters while the engine is running.

A Wärtsilä-designed computer-based ECU protects the engine warning against possible problems before damage occurs. Signals from sensors are transmitted to the unit via a CAN bus connection comprising a single cable from one end of the engine to the other.

Wärtsilä 31

The second decade of the 21st century was marked by falling demand for new ships after a near 20-year boom period and increasingly stricter emission and fuel regulations aimed at reducing the shipping industry's environmental impact. A rising fuel cost trend which dictated efficiency improvements came to a sudden and dramatic end in 2014. While this should have helped the shipping industry, its initial impact was to stifle the offshore sector which was effectively the only expanding area of shipping activity.

In this challenging situation, there was an increased interest in the use of different fuels including LNG and other related fuel types and engines capable of dual-fuel operation to give greater flexibility. It was under these circumstances that Wärtsilä developed and introduced its new 31 engine. Other drivers in development were reduced lifecycle cost and increased uptime achieved through an even greater use of modular components.

As mentioned earlier, the Wärtsilä 31 engine was awarded a Guinness World Records title as the world's most efficient four-stroke diesel engine. Its record-breaking fuel efficiency is made possible through the use of technologies, such as two-stage turbocharging, a high-pressure fuel injection system, and adjustable valve actuation, in combination with the next-generation engine control system.

The Wärtsilä 31 can be used both as a main propulsion engine and in diesel-electric configurations, or as an auxiliary engine, and it can be optimized for running either at a constant speed along a propeller curve or with constant torque. It is available only in V form in 8–16 cylinder configurations and has a power output ranging from 4.2 to 9.8 MW, at 720 and 750 rpm.

Designed to serve a wide selection of vessel types, the Wärtsilä 31's application areas include Offshore, Cruise and Ferry, and other marine segments. The engine comes in three alternative versions: diesel, dual-fuel (DF), and spark-ignited gas (SG). The multifuel capabilities that the Wärtsilä 31 brings to the market extend the possibilities for operators to utilize different qualities of fuels; from very light low sulphur fuel to very heavy diesel, and a range of different qualities of gas. It complies with the IMO Tier III regulations when operating on gas or on diesel coupled with SCR without an increase in fuel consumption. While previous engines were typically developed to run on diesel and subsequently adapted for gas, they were never fully optimized for the gas or dual-fuel modes. In contrast, the Wärtsilä 31 delivers maximum fuel efficiency and performance across all its fuel variants.

A new-generation, common rail fuel injection system was developed specifically for the Wärtsilä 31. The diesel and DF engines both use identical fuel injection systems with multiple injection capability, thereby facilitating future conversion needs, as no separate pilot fuel needs to be added. The flexibility of the fuel injection results in improved performance—especially at low loads which many ships tend to operate at. The Wärtsilä 31 can operate on low sulphur fuels, marine diesel oil, and HFO, without any restrictions on low load operation, while the DF engine can switch instantly between the different fuel modes independently of the fuel that is utilized.

Another key technology is the two-stage turbocharging system. The Wärtsilä 31 utilizes the second-generation, two-stage system with a pressure ratio capability above 10 bar and turbocharger efficiency of more than 75%. This improves upon the typical efficiency level of around 65%–70% for a single-stage turbocharger. By enabling earlier inlet valve closing, resulting in lower combustion temperature, the increased pressure ratio capability enables increased engine output, as well as improving fuel economy.

The two-stage concept also offers benefits in terms of operational flexibility, because the reduced inertia of the high-speed turbocharger improves loading performance and load acceptance from low loads. Furthermore, the second-generation, two-stage system has been optimized for minimum downtime through the incorporation of an extractable cartridge concept that facilitates turbocharger overhaul without touching any engine connections.

In order to make optimal use of the common rail fuel system and two-stage turbocharging system, adjustable valve actuation is necessary. With flexible valve actuation, the correct air-fuel ratio is ensured at all engine-operating conditions. Meanwhile, control of the inlet valve closure can be set individually on each cylinder—a feature that also can be used to improve cylinder balancing. The uptime and availability of the engine are enhanced further due to the fact that the new hydraulic valve actuation does not require periodic adjustment of inlet and exhaust valve clearance. On most other modern engines, this has to be carried out at 1000-h intervals, as the valve actuation is mechanically controlled.

With regard to operational flexibility, the Wärtsilä 31 maintains excellent performance across its complete operating range. This optimal operating performance is the result of the unique way in which the advanced UNIC engine control system has been combined with state-of-the-art technologies. A similar improvement in fuel consumption is achieved in both low-load running and high load operation. Meanwhile, fast starting and load transient response are improved as a result of the flexibility of the fuel injection and air-fuel ratio control, along with the benefits of the two-stage turbocharging system.

Maintenance operations and frequencies have been drastically reduced, resulting in significantly improved uptime. The first major overhaul is scheduled at 32,000 h or 5 years of operations when operating on LSFO or gas. While most engines of a similar output require their first maintenance interval after about 2000 operating hours, the Wärtsilä 31 has extended this by a factor of 4–8000 h.

Extensive rig testing was used to ensure reliability which has also been ensured through the minimization of maintenance operations. For instance, valve clearance has been eliminated, and new service concepts using exchange units are introduced. One such example is the power pack unit, which consists of a connecting rod, piston, cylinder liner, and cylinder head with related pipes combined in one exchange unit, as opposed to the traditional single spare part concept. Maintenance intervals have also been extended.

The key factor facilitating the increased uptime of the Wärtsilä 31 is its new modular design. The modular design also makes it possible to carry out fast, efficient engine conversions as the diesel, DF, and SG engine uses the same technologies and components. Furthermore, the standardized component interfaces allow engines to be converted to run on different fuels, without requiring any machining. Modular design does not only facilitate quick repairs but it also supports future upgrades. Shipowners will be able to instal the latest, state-of-the-art technologies simply by replacing the module containing the upgrade.

Wärtsilä 38

The 380mm bore/475mm stroke W38 design was developed by Wärtsilä Diesel's Dutch subsidiary Stork-Wärtsilä Diesel as a replacement for its ageing TM410 filling a power gap in the group portfolio between the Vasa 32 and W46 designs. The W38 series was introduced in 1993 with a specific rating of 660kW/cylinder at 600rev/min to cover a powerband from 3960 to 11,880kW with six, eight, and nine in-line and V12-, 16-, and 18-cylinder models (Fig. 24.16).

The design was distinguished by a new combustion philosophy applied to achieve a high thermal efficiency and reduced NOx emissions (down by 50%–70%) without compromising fuel economy. The key parameters are a stroke/bore ratio of 1.25:1, a maximum cylinder pressure of up to 210bar and a maximum fuel injection pressure of 1500bar. Such parameters provide the flexibility to select a compression ratio, injection timing, and injection rate that allow the desired low NOx values to be attained.

A completely new fuel injection system was developed with the following characteristics: suitable for continuous 1500bar injection pressure; closed barrel; optimized constant pressure valve design for short injection period and high brake mean effective pressure; no mixing of lubricating oil and fuel between plunger and barrel to avoid lacquering and sticking; separation of leakage fuel from the oil sump to avoid lubricating oil contamination; integrated low-pressure fuel supply lines; fuel injection valves with uncooled

Fig. 24.16 V18-cylinder version of the Wärtsilä 38 engine.

nozzles; and shielded high-pressure lines and connections. The fuel injection system is fully covered by the hot box, and possible low- and high-pressure leakages are detected by alarms.

The fuel pump has an integrated low-pressure channel which means there is no low-pressure pipe to fit the pump because the fuel pipes are fitted directly to the fuel pump foundation. Integrated fuel pulse dampers in the fuel supply lines ensure low-pressure pulses on the system side. Each fuel pump is equipped with a flow distribution throttle. The high-pressure fuel pipe is shielded and fitted with a leak alarm. Maintenance is facilitated by a locking device flange on the cylinder head for the high-pressure pipe.

The engine starting system incorporates a starting fuel-limiter to secure a smooth and safe start. Each fuel pump is equipped with an overspeed trip cylinder to ensure that the fuel rack is pushed to zero in the event of overspeed. Additional safety is provided by an extra stop cylinder turning the regulating shaft to zero in the event of overspeed. Overall compactness and lightness were prime targets for the W38 engine designers. Component integration resulted in 40% fewer parts than the older generation TM410 engine, with many of the traditional bolted-on elements now incorporated in the main block casting (e.g., lines and piping for lubricating oil, cooling water, and fuel).

The simplicity of maintenance and extended overhaul intervals were other design goals, with only a modest envelope required for servicing and dismantling. Engine-driven cooling and lubricating oil pumps as well as engine-mounted coolers and filters simplify installation and release machinery space. Integrated brackets are provided for resilient mounting systems. The bending stiffness and torsional rigidity of the engine block and crankshaft allow savings in the specification of flexible couplings, torsional vibration dampers, and elastic elements for resilient mounting.

W38A specification

Engine block: underslung crankshaft; integrated air receiver; single-piece casting for all cylinder numbers; and arranged for resilient mounting with screwed-on feet.

Crankshaft: fully balanced and designed for space-saving, maximum overall rigidity, and moderate bearing loads. The crankshaft of the V-cylinder models has undercuts for both the main and pin fillet, the design combining a short cylinder distance and maximum bearing length. Normal fillets are applied in the in-line cylinder engines.

Connecting rod: marine head design, completely machined and dimensioned to achieve the lowest dismantling height (Fig. 24.17).

Bearings: Thick-pad bearing technology and moderate main bearing load.

Cylinder liner: high-collar design with optimized operational roundness and straightness; bore cooling for effective control of liner temperature; and antipolishing ring.

Piston and rings: composite piston with forged steel crown and nodular cast iron skirt; pressurized skirt lubrication; two ceramic chromium-plated directional barrel-shaped compression rings in the crown; a ceramic chromium-plated taper-faced compression ring and an oil distributor ring in the skirt.

Cylinder head: stiff box with four screws; flame plate supported by the intermediate deck, allowing no deformation of valve seat environment.

Camshaft: single-cylinder shaft segments with integrated cams; and built-in valve tappet modules screwed onto the engine block.

Turbocharger/charge air cooling: modular exhaust gas pipes in an insulated hot box; two-stage air cooler; and integrated, rigid turbocharger support/charge air cooler.

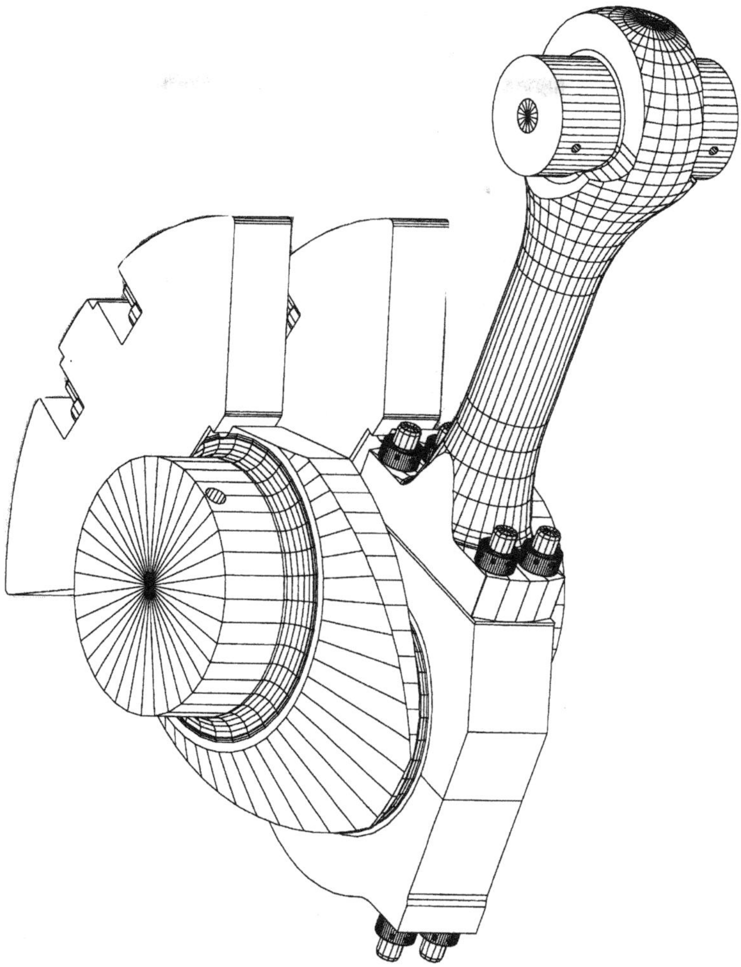

Fig. 24.17 Wärtsilä 38 engine connecting rod.

W38B version

The popularity of the Wärtsilä 38 encouraged the development of a B-version offering higher powers with lower fuel consumption and reduced emissions. The new model also benefited from an advanced control and monitoring system and full modularization to ease assembly. The B-series was fully released in 2001 with in-line six, eight, and nine and V12-, V16-, and V18-cylinder models covering a powerband from 4350 to 13,050 kW at 600 rev/min. The V18-cylinder model was subsequently dropped, lowering the upper output of the programme to 11,600 kW.

Two specific ratings are now offered: 675 kW/cylinder (25.1 bar mep) and 725 kW/cylinder (26.9 bar mep). The original engine, redesignated the Wärtsilä 38A, remained in the programme for a while, both versions assigned for production at Wärtsilä Italia in Trieste. The W38 engine had disappeared from the company's portfolio by 2017 in which year three vessels including two Japanese ferries were delivered with engines from the range.

Careful selection of thermodynamic processes and progress in component design contributed to the achievement of the development goals. The refinements included a higher maximum cylinder pressure, compression ratio and mean effective pressure, and advanced fuel injection timing. Special attention was paid to optimizing the combustion chamber characteristics, the fuel injection profiles and the piston shape. The airflow system was fully revised (inlet bends to the cylinder head, flow channels at the head, port design, exhaust bends, and turbocharger). State-of-the-art valve timing (including the application of the Miller cycle) was also introduced. New-generation turbochargers were adopted to further boost overall performance and economy, with the integration of the bypass and wastegate valves. The position of the valves and their controls was selected to secure low vibration levels, low local temperatures, and short pipework.

These measures resulted in a significant reduction in specific fuel consumption (6 g/kW h below the 38A engine) and opened the way to further improvements in fuel economy. The limits on pollutant emissions were also comfortably met; the NOx emission levels can be lowered further by up to 50%–60% by incorporating direct water injection (see Chapter 4).

Critical components were given greater strength to support the higher power levels and raise the safety margins for further rating growth; examples include additional strengthening of the connecting rod and geometrical modifications at the interface of the cylinder liner with the cylinder head (area of the gas sealing ring). All bearing loads are kept low and an ample oil film thickness ensures safe bearing operation.

A new three-piston ring pack and an antipolishing ring incorporated at the top of the cylinder liner foster low lube oil consumption. The 'hot box' environment was completely redesigned to secure high reliability and functional integration; externally, there are swinging hot box covers and new cylinder head covers enclosing the hot box area; internally a new LP and HP fuel system, which is based on a new generation pump.

The lubrication module incorporates a cooler, an automatic back-flush filter, a centrifugal filter in the back-flush line and thermostatic valves. The overall system includes main and prelubrication pumps and the regulating valve. Lubricating oil flow through the engine was revised and based on optimized bearing clearances to allow large safety margins. Running-in filters of the spin-on type can be mounted on the nonoperating side of the engine and removed after the factory approval test. The LT and HT cooling water systems embrace pumps and thermostatic valves built on the engine.

Compactness and easier installation were promoted by 'all on' engine support systems, with complete lubrication and cooling water modules mounted on the engine (Fig. 24.18). All piping connections are located at the free end to allow simplified interfacing with external systems. Among the advanced automation features is Wärtsilä's WECS 7000 engine-integrated system, which is tailored for the group's medium-speed designs.

Wärtsilä 38A and 38B differences

	38A	38B
Bore	380 mm	380 mm
Stroke	475 mm	475 mm
Output/cylinder	660 kW/cyl	725 kW/cyl
Cylinders	6, 8, 9L/12, 16, 18V	Same
Speed	600 rev/min	600 rev/min
Compression ratio	12.7	15
Maximum combustion pressure	180 bar	210 bar
Mean effective pressure	24.5 bar	27 bar
Piston speed	9.5 m/s	9.5 m/s
Injector holes	11	10
Injector position	Ref	6 mm higher
Fuel cam	Ref	Low speed
Cylinder head	Ref	Optimized air flow
Piston shape	Ref	Optimized
Inlet valve timing	Ref	Miller cycle
Injection timing	Ref	Advanced
Exhaust system	Ref	Optimized flow
Turbocharger efficiency	65%	701%
Specific fuel consumption	Ref	26 g/kW h
NOx emissions	IMO	IMO-50%[a]

[a]Prepared for direct water injection.

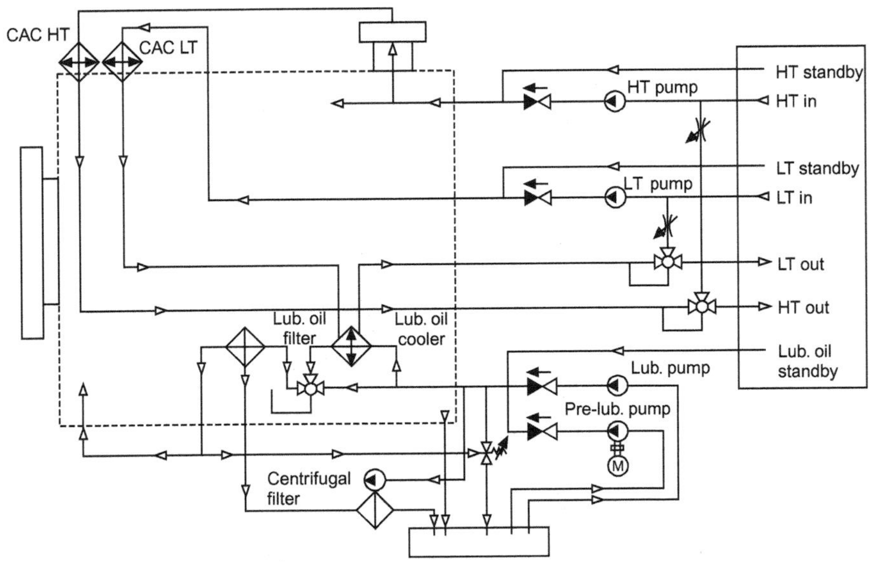

Fig. 24.18 Integrated cooling water and lubrication oil systems of W38B engine.

Wärtsilä 46

Several innovative features were introduced by Wärtsilä Diesel with the launch of the W46 engine in 1988 which became popular in diverse passenger and cargo ship propulsion sectors (Figs 24.19 and 24.20). The designer's prime development goal in creating a large bore medium-speed trunk piston engine to compete with low-speed cross head machinery was the highest reliability. A significant contribution was sought from thick-pad bearing technology, exploiting large bearings with thick oil films, and pressure-lubricated piston skirts.

During its first 4 years of production, the 460 mm bore/580 mm stroke design was offered with an output of 905 kW/cylinder at 450/500/514 rev/min as the Wärtsilä 46A. Experience with pilot installations encouraged the release in 1992 of a B-version with an output of 975 kW/cylinder at 500/514 rev/min. The C-output version followed in autumn 1995, yielding 1050 kW/cylinder at 500/514 rev/min with reduced thermal load and lower emissions thanks to advances in turbocharging and fuel injection systems and the adoption of low NOx combustion principles.

The three ratings allow the 46 series to cover a powerband from 5850 to 15,600 kW with six, eight, and nine in-line and V12- and V16-cylinder

Fig. 24.19 Six-cylinder version of the Wärtsilä 46 engine.

models. The C-output versions operate with mean effective pressures up to 28.8 bar and the engine is designed for a maximum combustion pressure of 210 bar.

A nodular cast iron engine block (Fig. 24.21) is configured to achieve a rigid and durable construction for flexible mounting (the W46 engine was perhaps the first of its size designed for elastic and superelastic mounting, Fig. 24.22). The main bearings are of the underslung type with hydraulically tightened bolts; side bolts add further rigidity to the main bearing housing. In-line cylinder engines are equipped with an integrated air receiver fostering rigidity, simplicity, and cleanliness. A welded block of steel castings and plates is offered as an alternative.

A rigid box-like cylinder head design aims for even circumferential contact pressure between the head and cylinder liner, with four fixing bolts simplifying maintenance procedures. No valve cages are used, improving reliability and increasing the scope to optimize exhaust port flow characteristics. Water-cooled exhaust valve seat rings are specified. Both inlet and exhaust valves receive a forced rotation from Rotocaps during every opening cycle, fostering an even temperature distribution and wear of the valves, and keeping the sealing surface free from deposits, resulting in good heat conduction.

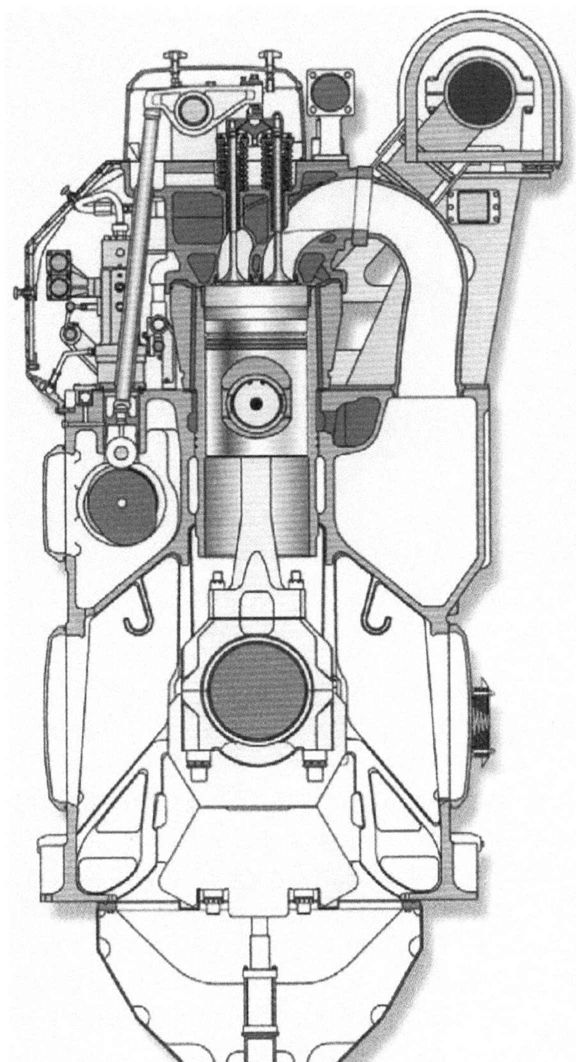

Fig. 24.20 Cross-section of Wärtsilä 46 engine.

Cylinder liner deformations are normally caused by cylinder head clamping and thermal and mechanical loads. A special liner design with a high collar-to-stroke ratio for the W46 engine minimizes deformation, the round liner bore in combination with efficient lubrication enhancing conditions for the piston rings and reducing wear. The liner material is a special grey cast iron alloy developed for wear resistance and high strength. An

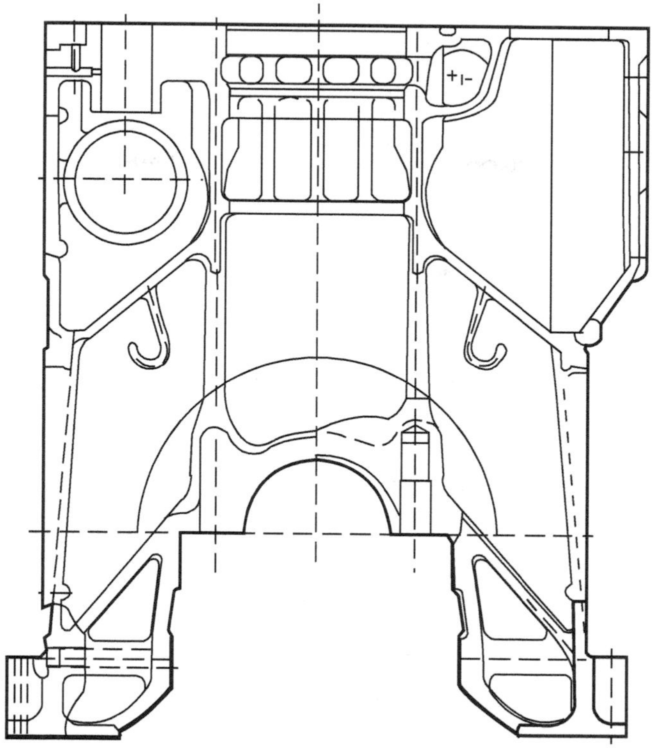

Fig. 24.21 Engine block of Wärtsilä 46 engine.

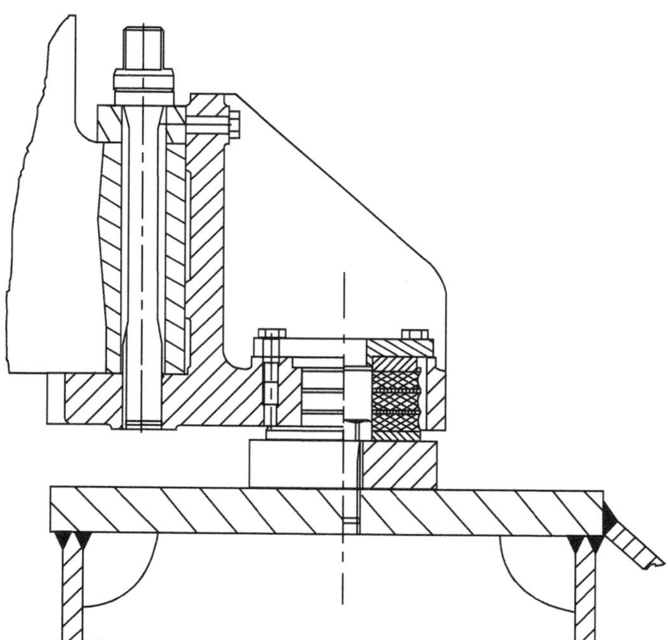

Fig. 24.22 Flexible mounting system for Wärtsilä 46 engine.

antipolishing ring in the upper part of the liner prevents the bore polishing that leads to local liner wear and increased lubricating oil consumption. The simple yet highly effective device was introduced as standard on later engines and retrofitted to existing installations.

A composite low-friction piston with a nodular cast iron skirt and steel crown incorporates a special cooling gallery configuration to secure efficient cooling and high rigidity for the piston top. It is designed to handle combustion pressures beyond 200 bar. A long lifetime is sought from hardened top ring grooves. The three-ring set includes a top ring with a special wear-resistant coating; all the rings are dimensioned and profiled for maximum sealing and pressure balance. Low friction is addressed by a skirt lubrication system delivering a well-distributed clean oil film that eliminates the risk of piston ring scuffing and reduces the wear rate and fostering cleaner rings and grooves free from corrosive combustion products. Noise and wear are reduced by hydraulically damped tilting movements provided by an oil pad between liner and piston.

Hard skirt contact against the liner experienced with the first seagoing W46 engine installation was solved by introducing a piston with increased compression height, a slightly increased skirt length, and an optimized skirt form.

A three-piece marine design connecting rod was designed for distribution of the combustion forces over a maximum bearing area with relative movements between mating surfaces minimized. Piston overhaul is facilitated without touching the big end bearing, and the bearing can be inspected without removing the piston. The three-piece design also reduces the piston overhauling height.

The one-piece forged crankshaft is designed to accept a high combustion pressure while maintaining a conservative bearing load; rigidity is underwritten by a moderate bore/stroke ratio and large pin and journal diameters. Counterweights fitted on every crank web provide 95% balancing. Reliability is sought from the thick-pad bearing design and bearing loads reduced by increasing crankshaft journal and pin diameters as well as length. Low-bearing loads allow for softer bearing materials with greater conformability and adaptability, underwriting a virtually seizure-free bearing, according to Wärtsilä. A thick corrosion-resistant overlay of tin–antimony is used in the big end bearings.

The key elements of Wärtsilä's thick-pad bearing technology are ample oil film thicknesses achieved by proper overall dimensioning; corrosion-resistant bearing materials with excellent embeddability properties; radially rigid bearing assemblies during both mass and gas force loading conditions; conformable bearing materials and axially conformable housings, which minimize edge pressure; and oil grooves and oil holes located where no harm can be caused to the oil film.

The camshaft is built of single-cylinder sections with integrated cams. The shaft sections are connected through separate bearing journals which make it possible to remove the sections sideways from the camshaft compartment. The valve follower is of the roller tappet type, the roller profile being slightly convex for good load distribution. The valve mechanism includes rocker arms working on yokes guided by pins.

W46 fuel injection

An innovation on the original W46 engine was a twin injection fuel system exploiting a pilot valve and the main injection valve (Fig. 24.23). The initial aim of the system was to secure safe combustion with the poorest fuels on the market; the very fast main injection process facilitated by the system also contributed to low fuel consumption. It became evident too that the twin injection configuration yielded a positive effect on NOx emissions, stimulating further refinement of the system.

The first seagoing W46 engine installation was later optimized with a reduction of the pilot fuel amount to below 3%. Both pilot and main injection valves were designed with a small heat-absorbing surface facing the combustion space and arranged for efficient heat transfer to the cooling

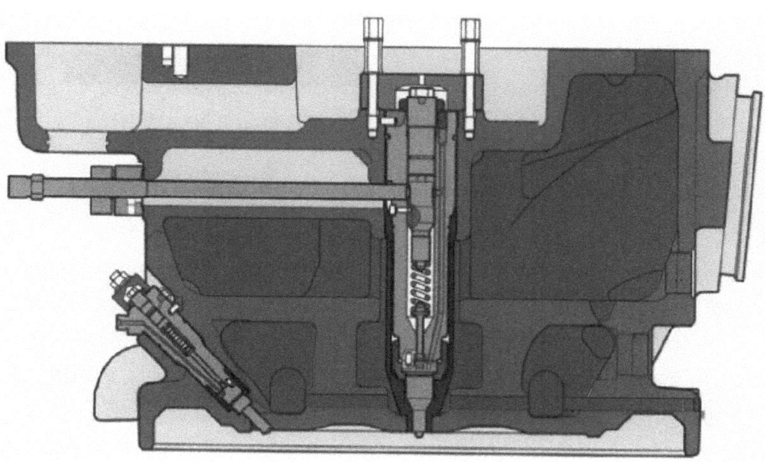

Fig. 24.23 Wärtsilä 46 twin injection system, showing the pilot fuel valve (left) and the main fuel valve (centre).

water, eliminating the need for a separate nozzle temperature control system. Fuel was transported over the shortest distance from the pump to the valve via a high-pressure pipe in the cylinder head.

Wärtsilä cited the following merits of twin injection through individually optimized nozzles: excellent pilot fuel atomization and combustion over a wide load and speed range; reasonable injection rates even under idling conditions when the combustion process is often a poor compromise; facility to burn fuels of low ignition quality with a minimum of ignition delay; low combustion noise level thanks to controlled preburning, resulting in a lower rate of increase in cylinder pressure and consequently lower mechanical load; reduced delay in the preburning phase, which lowers peak pressure and temperature and hence curbs NOx emission formation; improved starting ability; and increased thermal efficiency.

The mono-element injection pump (Fig. 24.24) was designed to serve the twin injection system. A constant pressure relief valve eliminates the risk of cavitation erosion by maintaining a residual pressure at a safe level over the full operating field. A drained and sealed-off compartment between the pump and tappet prevents leakage fuel from mixing with lubricating oil. The precalibrated pumps are interchangeable.

Fuel system pipes and main components are located in a hot box to secure safety at high preheating temperatures; fuel pipes outside the hot box are also carefully shielded. The fuel feed pipes on the engine are equipped with pressure pulsation dampers. Any fuel leakage from pipes, injection valves, and pumps is collected in a closed piping system which keeps the hot box and the engine dry and clean.

The twin injection system was later replaced by a single injection configuration as higher pressure traditional layouts became available.

Spex turbocharging

Swirl-Ex turbocharging—a constant pressure system with an in-built diffuser—was specified for early Wärtsilä 46 engines. It demonstrated high efficiency but service experience showed up a disadvantage due to the large volume in the exhaust pipe which made the load pickup slower than the standard in other Wärtsilä engines. A faster turbocharging system based on a single pipe with a much smaller volume was therefore developed, resulting in the Spex system. This offers a swift load pickup and almost the same efficiency at high load as the Swirl-Ex system.

The aim of an exhaust system is to transfer the energy released from the cylinder, after the exhaust valve has opened, to the turbine of the

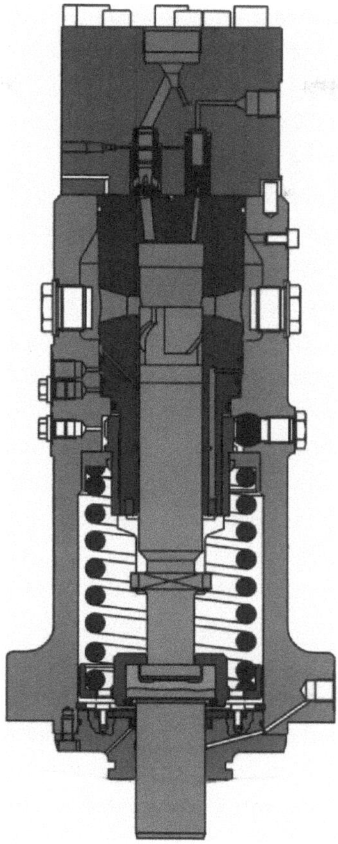

Fig. 24.24 Wärtsilä 46 engine fuel pump designed to serve the original twin injection system.

turbocharger in the most effective way. This could easily be done by making the pipes as short as possible and the areas wide without restrictions. There are, however, other parameters that must be taken into account, such as uniform scavenging. Since all the cylinders are connected to the common system there is an obvious risk that a pulse coming from one cylinder at the beginning of the blowdown phase disturbs the scavenging of another cylinder. To avoid this, the design has to be suited to all the possible cylinder configurations and speed ranges that the engine type will cover.

Another parameter affecting scavenging is the pressure drop between inlet and exhaust manifolds. The higher the pressure difference, the better resistance the cylinder has against the disturbing pulses in the exhaust manifold; the pressure difference is largely a function of the turbocharger efficiency. Thus there are many parameters affecting the performance of the exhaust system, and the final design is a compromise between various factors.

Turbocharging of the W46 engine is now optimized for the given application exploiting the Spex system as standard (original Swirl-Ex engine installations in service were reconfigured). For special applications, where low load running accounts for a large proportion of the total operating time, a three-pulse system is also offered for appropriate cylinder numbers. Pulse charging is reportedly excellent at partial load operation and for fast load response, while Spex charging is said to be excellent at high and steady loads, with the load response remaining good. Spex is based on an exhaust gas system that combines the advantages of both pulse and constant pressure charging. Compared with a constant pressure system, the ejector effect of the gas pulses provides better turbine efficiency at partial loads. Very small deviations in the scavenging between the cylinders result in an even exhaust gas temperature.

Modular-built exhaust gas systems on the W46 engine are designed to handle high-pressure ratios and pulse levels while being elastic enough to cope with thermal expansion in the system. Exhaust wastegate, air bypass and air wastegate arrangements can be specified to satisfy specific operating requirements, such as load response or partial load performance.

A load-dependent cooling water system is used for the LT water circuit. At low load, the water temperature is elevated by transferring heat from the lubricating oil to the charge air. This eliminates the risk of cold corrosion in the combustion space and also improves the ability to burn fuels of low ignition quality at low loads. The HT cooling water system operates constantly at an HT to make the temperature fluctuations in the cylinder components as small as possible and to prevent corrosion due to undercooling. The charge air cooler is split into HT and LT sections to maximize heat recovery. Engine-driven pumps are available as an option.

A wet or dry sump lubricating oil system can be specified. On its way to the engine the oil passes through a full flow automatic filter unit and a safety filter for final protection. A small proportion of the flow is diverted to a centrifugal filter which acts as a lubricating oil condition indicator and warns the operator of excessively dirty oil and wear. Both the safety filter and the indicator filter are designed for cleaning while the engine is running. An engine-driven lubricating oil pump can be provided optionally.

Wärtsilä 46F

Market drivers, particularly environmental regulations, dictated not just an upgrade but the development of a new 46 engine design based on the original cylinder dimensions but exploiting novel solutions for key components and systems (Fig. 24.25). The resulting 46F engine, launched in 2004, was described as more powerful without compromising reliability and offering attractive power-to-weight and space ratios, competitive fuel and lube oil consumptions, and low emissions regardless of fuel quality.

The original 46 engine series remained on offer alongside the 46F programme, which covers a powerband from 7200 to 14,400 kW with six-to-nine in-line and V12-cylinder models. (A V20-cylinder version introduced in 2007 with an output of 23,000 kW was initially intended for land-based

Fig. 24.25 An in-line six-cylinder version of the 46F engine.

power installations.) Installation, operational and maintenance benefits flowed from the increased specific rating since an engine with fewer cylinders (say, seven instead of nine) can be specified to deliver the same power. More compact machinery spaces and easier installation procedures are further fostered by built-on support system modules.

A higher specific output (1250 kW/cylinder) was achieved by increasing engine speed (from 500/514 to 600 rev/min) instead of mean effective pressure (25.9 bar) and applying the latest developments in turbocharging to make wider use of the Miller valve timing concept. Under full load operation, early closure of the inlet valves fosters a low effective compression ratio and hence comparatively low temperatures at the end of the compression stroke. The charge air, both somewhat expanded and cooled on its way through the receiver into the cylinders, contributes to creating the initial conditions for environmentally friendly combustion: a low global temperature that is still high enough to guarantee reliable and stable ignition of the fuel-air mixture.

An optional variable inlet valve closure (VIC) facility—first designed for the Sulzer ZA40S medium-speed engine—controls the closing time of the inlet valves. With VIC it is possible to arrange an early Miller cycle or remove it completely, control being possible while the engine is operating.

Wärtsilä's common rail fuel injection system, standard fitment on the 46F engine, maximizes the scope for adjusting the injection process to the prevailing operating conditions, fuel characteristics, and emission limits (see Chapter 11). The engine is optionally offered with a conventional fuel injection system based on twin-plunger injection pumps (see section 'Wärtsilä 64').

In the CR system updated for the W46F engine, the high-pressure pumps are mounted on a tappet housing module which is interconnected between cylinders. The module carries all low-pressure fuel system supply and returns, including drains and sealing oil systems, fostering a low level of piping density. The pumps are driven by a camshaft incorporating two cam lobes, with one pump assigned to every two cylinders of the engine; pump filling is suction controlled by linear solenoid valves. The suction control valve is built as a module together with the solenoid.

Each pump primarily feeds into an accumulator, each of which in turn serves two injectors. The accumulators are interlinked by small-bore piping to form a CR. The piping between the accumulators undertakes three functions: equalization of the pressure; reduction of pressure pulsations; and system connectivity during engine warm-up circulation with heavy fuel oil.

Fuel flow to the two-stage injector's nozzle is controlled by a shuttle valve actuated by control oil which, in turn, is controlled by a solenoid valve. The control oil pump is engine driven for an easy black start. Nozzle needle movement is controlled by the fuel pressure in the nozzle gallery, and pressure acting on the press-down piston which, under certain safety conditions, keeps the nozzle shut.

A Spex system is standard, with options of exhaust gas wastegate or air bypass depending on the application. The turbocharger can be located at either the free or the driving end of the engine, transversal alignment allowing the exhaust gas outlet to be inclined in the longitudinal direction. The charge air receiver is integrated into the engine block, and the two-stage self-supporting charge air cooler comprises separate HT and LT water sections (an advantage for heat recovery applications).

A continuous bearing temperature monitoring system covers not only the main bearings (as in some earlier engines) but for the first time also the big end bearings, the latter benefiting from a new installation system patented by Wärtsilä. The bearing temperatures are recorded at a point roughly halfway between the crank webs and a few millimetres beneath the crankpin surface. A stationary antenna mounted on the intermediate wall in the crankcase records the bearing pin temperature every time the crank throw passes the registration unit (several times a second). The temperature pickup is passive and requires no electrical supply. Kongsberg Maritime's Sentry wireless temperature monitoring system, specified as standard for the big end bearings, can be linked to Wärtsilä's condition-based maintenance system.

Nodular cast iron was retained for the engine block but the new component, with an attractive stiffness-to-weight ratio, is both lower and lighter than that of the 46 engine. The rigid and self-supporting block is designed for resilient mounting applications, hydraulically tightened main bearing screws and side screws, and to provide easy internal access for service personnel.

Wärtsilä 64

The most powerful medium-speed design ever brought to the market—and reportedly the first to exceed the 50% thermal efficiency barrier (Fig. 24.26)—the Wärtsilä 64 engine was prototype-tested in a six-cylinder form in September 1996. Production models became available from autumn 1997 but the engine had a relatively short production life for marine

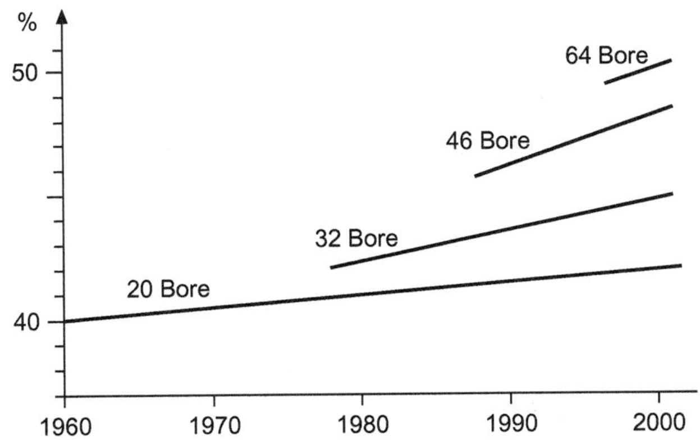

Fig. 24.26 Rise in the thermal efficiency of successive Wärtsilä medium-speed engine designs.

purposes finding applications in container vessels and general cargo vessels. Today just 17 vessels are sailing with this engine, 15 of which are the S-type general cargo vessels built by Dutch operator Spliethoff's between 2000 and 2005.

A nominal output of 2010 kW/cylinder at 333 rev/min was originally quoted for the 640 mm bore/900 mm stroke in-line cylinder design on a mean effective pressure of 25 bar and a maximum cylinder pressure of 190 bar (Fig. 24.27). The rating was later raised to 2150 kW/cylinder with meps up to 27.2 bar.

Power demands up to 17,200 kW were offered by a programme of six-, seven-, and eight-cylinder in-line models, the 9L- and V12-cylinder versions in the original programme having been dropped. Only six and seven-cylinder variants remain in service in 2020.

The V-engine had the same bore but a shorter stroke (770 mm) and lower specific power rating (1940 kW/cylinder at 400–428 rev/min on an mep of 22–23.5 bar) than the in-line design. The performance was boosted by a high-efficiency TPL80E turbocharger, one of ABB Turbo Systems' latest TPL series (see Chapter 10).

Reliability was sought from traditional Wärtsilä solutions, notably a pressure-lubricated piston skirt, large bearings with thick oil films, thick collar cylinder liners fitted with an antipolishing ring, high fuel injection pressures for optimized combustion, and camshafts with high torque capacity and low Hertzian pressures. Easier installation was addressed by built-on cooling water pumps and lubricating oil system (including automatic filters).

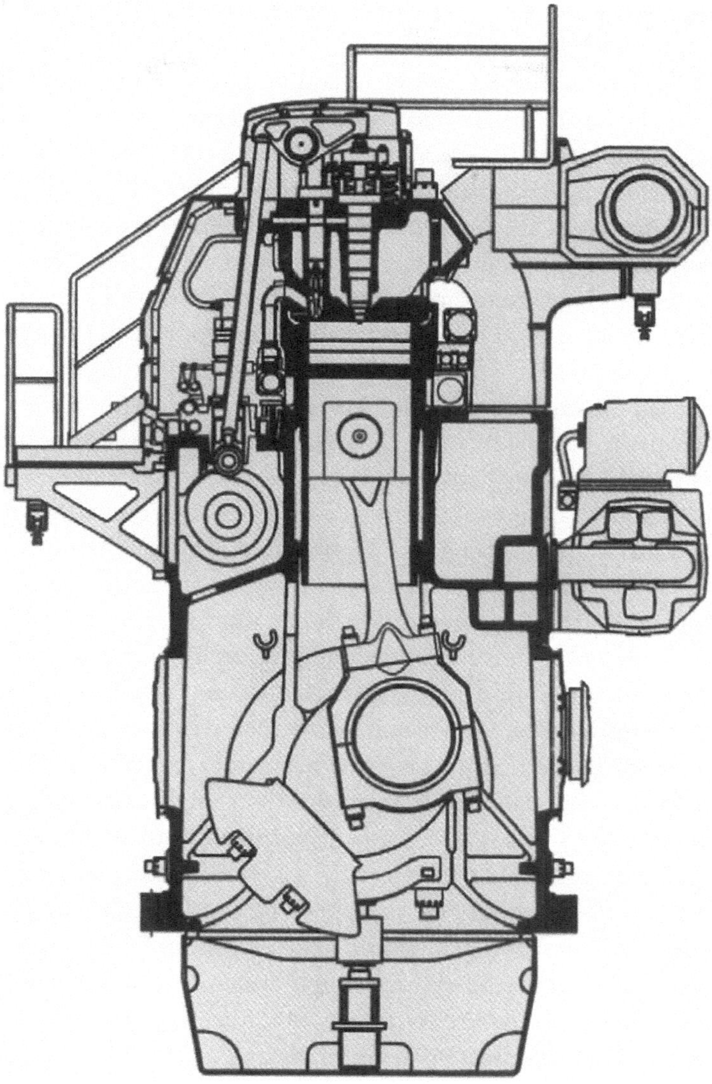

Fig. 24.27 Cross-section of Wärtsilä 64 design.

W64 engine details

Engine block: Wärtsilä suggests that nodular cast iron was the natural choice for modern engine blocks because of its strength and stiffness properties and the freedom that casting offers. Optimum use was made of current foundry technology to integrate most oil and water channels, resulting in a virtually pipe-free engine with a clean outer exterior. Resilient mounting, now common, calls for a stiff engine frame; integrated channels designed with this in mind thus serve a dual purpose.

Crankshaft and bearings: advances in combustion development require a crank gear which can operate reliably at high cylinder pressures. The crankshaft must be robust and the specific bearing loads kept at an acceptable level; this was achieved by optimizing the crank throw dimensions and fillets. The specific bearing loads are conservative and the cylinder spacing (important for the overall length of the engine) is minimized. Apart from low-bearing loads, the other crucial factor for safe bearing operation is oil film thickness. Ample film thicknesses in the main bearings are ensured by the optimized balancing of rotational masses and in the big end bearing by ungrooved bearing surfaces in the critical areas. All these features ensure a free choice of the most appropriate bearing material. The other thick-pad bearing technology concepts proven on the Wärtsilä 46 engine (see p. 698) are also applied.

Piston and rings: a rigid composite piston with a steel crown and nodular cast iron skirt has been adopted for highly rated diesel engines for years to secure reliability in high cylinder pressure and combustion temperature environment. Wärtsilä's patented skirt lubrication is applied to minimize frictional losses and ensure appropriate lubrication of piston rings and skirt. Each ring of the three-ring pack is dimensioned and profiled for a specific task. The pressure balance above and below each ring is crucial in avoiding carbon deposits in the ring grooves of a heavy fuel engine (Fig. 24.28).

Cylinder liner and antipolishing ring: the thick high-collar type liner is designed with the stiffness necessary to withstand both pretension forces and combustion pressures with virtually no deformation. Its temperature is controlled by bore cooling of the upper part of the collar to achieve a low thermal load and avoid sulphuric acid corrosion. The cooling water is distributed around the liners with simple water distribution rings at the lower end of the collar. At its upper end, the liner is fitted with an antipolishing ring to eliminate bore polishing and reduce lube oil consumption. The ring's function is to calibrate the carbon deposits formed on the piston top land to a thickness small enough to prevent any contact between the liner wall and the deposits at any piston position. When there is no contact between liner and piston top land deposits, no oil can be scraped upwards by the piston; at the same time, liner wear is also significantly reduced.

Connecting rod: a three-piece rod with all highly stressed surfaces machined is the safest design for engines of this size intended for continuous operation at high combustion pressures, according to Wärtsilä. For easy maintenance and accessibility, the upper joint face is placed right on top

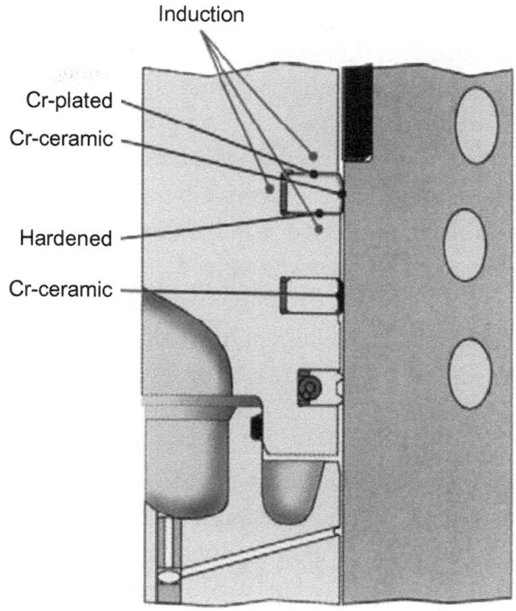

Fig. 24.28 Three-ring pack for Wärtsilä 64 engine piston; note the antipolishing ring incorporated in the upper cylinder liner (top right).

of the big end bearing housing. A special hydraulic tool was developed for simultaneous tensioning of all four screws. An intermediate plate with a special surface treatment is arranged between the main parts to eliminate any risk of wear in the contact surfaces.

Cylinder head: high reliability and ease of maintenance were sought from a stiff cone/box-like design able to cope with the high combustion pressure and secure both cylinder liner roundness and even contact between the exhaust valves and their seats. The head design is based on the four-screw concept developed by Wärtsilä and applied for over 20 years. Such a design also provides the freedom needed for designing inlet and exhaust ports with minimal flow losses. The port design was optimized using computational fluid dynamics (CFD) analysis in conjunction with full-scale flow measurements. Wärtsilä's extensive heavy fuel-burning experience contributed to the exhaust valve design, the basic criterion for which is correct temperature; this is achieved by carefully controlled cooling and a separate seat cooling circuit to secure long lifetimes for valves and seats.

Fuel injection system: the split-pump technology first introduced on the W64 engine offers advantages in terms of operating flexibility, mechanical strength, and cost-effectiveness. Fuel injection timing can be freely adjusted independently of the injected quantity, and tuning of the injection parameters according to the engine's operational conditions improves engine performance and reduces exhaust emissions. Smaller closed-type pump elements—derived from high volume production of smaller engines—reduce mechanical stresses and enhance reliability, while lower loads on rollers, tappets, and cams improve pump drive reliability.

This new solution was dictated when injection pump manufacturers suggested that for such a large medium-speed engine it would be very difficult to produce pump plungers of the size and accuracy required to secure the reliability associated with smaller engine designs. Since the output of the Wärtsilä 64 is approximately double that of the established Wärtsilä 46, it was decided to use two plungers (each of roughly W46 size) per engine cylinder.

The two plungers have slightly different functions (Fig. 24.29). Both pump fuel at each stroke and are connected to the same line, from where fuel is led to the nozzle via a single high-pressure line. Although both plungers pump the fuel in a similar way, only one of them needs to be controlled to adjust the fuel quantity. This made it possible to reserve the other plunger for another task: turning it to control the injection timing during engine operation. New possibilities thus opened up for controlling different load modes and fuel qualities, even allowing injection retardation when lower NOx emission values are demanded.

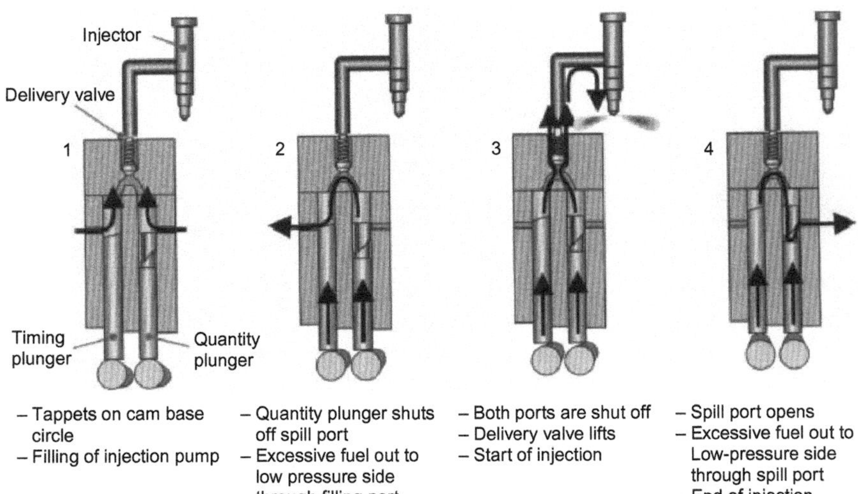

Fig. 24.29 Functions of twin plungers of the fuel pump for the Wärtsilä 64 engine.

A contribution to reliability by the fuel pump design comes from splitting the plunger loading between two cams and rollers, thus lowering the loading on these components and underwriting safe operation at injection pressures up to 2000 bar. The associated tappets for these components are both integrated into the same housing as the tappets for the inlet and exhaust valves.

The high-pressure fuel system was designed and endurance tested at 2000 bar; an actual injection pressure of around 1400 bar thus represents a substantial safety margin. No lubricating oil is required for the pump element because the plunger has a wear-resistant low-friction coating. A profiled plunger geometry keeps the clearance between the plunger and barrel small, allowing only minimal oil to pass down the plunger; the small leakage is collected and returned to the fuel system. Any chance of fuel mixing with the lube oil is eliminated. Both the nozzle and nozzle holders are made of high-grade hardened steel to withstand the high injection pressures and, combined with oil cooling of the nozzles, to foster extensive nozzle lifetimes.

Low-pressure fuel system safety is underwritten by the Wärtsilä-patented multihousing concept. The fuel line consists of channels drilled in cast parts which are clamped firmly on the engine block and connected to each other by simple slide-in connections for ease of assembly and disassembly. The pumps are connected together to form the complete low-pressure fuel line with both feed and return channels; any need for welded pipes is eliminated. Safety is further enhanced by housing the entire low-pressure and high-pressure systems in a fully covered compartment.

Turbocharging system: based on noncooled turbochargers with inboard plain bearings lubricated from the engine's lube oil system. The Spex turbocharging system is standard, with the option of exhaust wastegate or air bypass according to the application. Spex, which exploits the pressure pulses without disturbing the cylinder scavenging, is described in section 'Wärtsilä 46'. The interface between the engine and turbocharger is streamlined, eliminating all the adaptation pieces and piping formerly used.

Cooling system: split into separate HT and LT circuits (Fig. 24.30). The cylinder liner and cylinder head temperatures are controlled through the HT circuit; the system temperature is kept at a high level (around 95 °C) for safe ignition/combustion of low-quality heavy fuels, including operation at low loads. An additional advantage is maximum heat recovery. To further increase the recoverable heat from this circuit, it is connected to the HT part of the double-stage charge air cooler. The HT water pump is integrated into

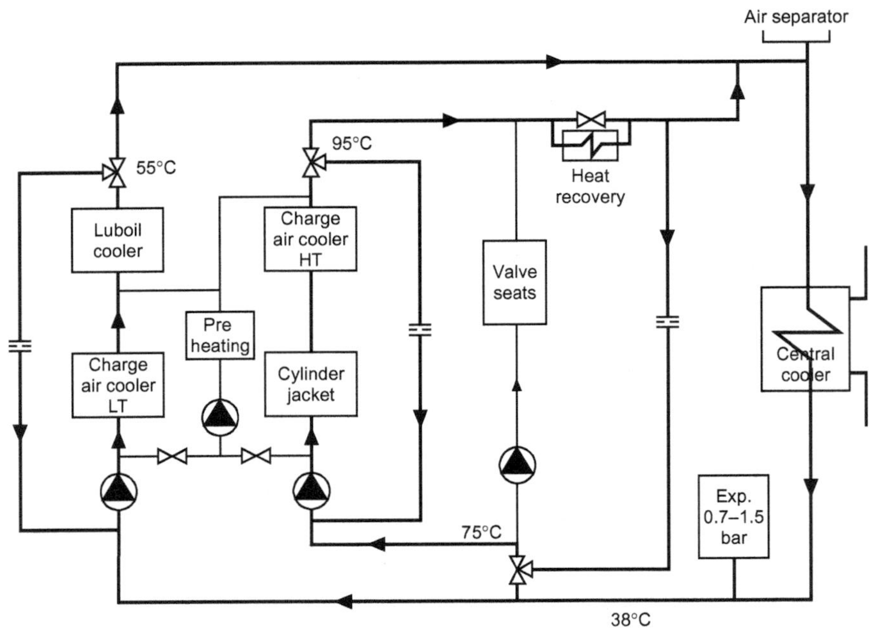

Fig. 24.30 Cooling water system of Wärtsilä 64 engine.

the pump cover module at the free end of the engine; the complete HT circuit is thus virtually free of pipes.

The LT circuit serves the LT part of the charge air cooler and the built-on lube oil cooler. It is fully integrated with engine parts such as the LT water pump with pump cover module, the LT thermostatic valve with the lube oil module and transfer channels in the engine block. In addition, the LT circuit provides separate cooling of the exhaust valve seats and a lower seat/valve temperature, thus promoting long lifetimes for these components. Directly driven pumps ensure safe operation even during a short power cut.

Lubricating oil system: all W64 engines are equipped with a complete built-on lube oil system comprising:

- Pump cover module: engine-driven main screw pump with built-in safety valve; prelubricating module; electrically driven prelubricating screw pump; pressure regulating valve; and centrifugal filter for lube oil quality indication.
- Lubricating oil module: lube oil cooler; oil thermostatic valves; full flow automatic filter; and special running-in filters before each main bearing, camshaft line, and turbocharger.

On in-line cylinder engines, the lubricating oil module is always located at the backside of the engine, while on V-engines it can be built on the engine at the flywheel or free end, depending on the turbocharger position. The lube oil filtration is based on an automatic back-flushing filter requiring minimal maintenance and no disposable filter cartridges.

Automation system: an engine-integrated system, WECS, is standard and has these main elements:

- The main control unit (MCU) cabinet, which comprises the MCU itself, a relay module with backup functions, a local display unit (LDU), control buttons and backup instruments. The MCU handles all communication with the external system.
- The distributed control unit (DCU) handling signal transfer over a CAN bus to the MCU.
- The sensor multiplexing units (SMU) transferring sensor information to the MCU.

Software loaded into the system is easily configured to match the instrumentation and the safety and control functions required for each installation. The MCU cabinet is well protected and built into the engine; most of the remaining hardware is housed in a special electrical compartment alongside the engine.

Wärtsilä dual-fuel engines

As described in Chapter 3, Wärtsilä took a lead role in the development of dual-fuel engines, thus it is not surprising that there should be several within its range of medium-speed engines.

The first engine developed was the 32DF based on the oil-burning Wärtsilä 32 but it was short-lived to find commercial applications in only a handful of ships before being replaced in 2008 with the 34DF. After the 32DF, the next product developed was the Wärtsilä 50DF. The company drew on its proven Wärtsilä 32DF and Wärtsilä 46 engines to design the Wärtsilä 50DF. In all, about 40% of the Wärtsilä 50DF components were identical to the Wärtsilä 46, a further 40% were based on the Wärtsilä 46, 10% were identical to the Wärtsilä 32DF, and 10% upscaled from the Wärtsilä 32DF. The 50DF has become a popular engine for LNG carriers.

The series of DF engines was further expanded in 2009 with the Wärtsilä 20DF for use as a propulsion engine for smaller vessels or as a genset and in 2014 by the Wärtsilä 46DF. This engine drew upon experience with an updated 34DF introduced in 2011 and the 50DF and 46 oil-fuelled engines.

Although not the largest in the bore size of the DF family, the 46DF is more powerful than its larger 50Df sibling.

The 31DF version was announced in 2017 as part of a new family of engines.

Across the ranges mentioned above, there are 26 different engines available with power outputs from 1110 kW for the smallest engine the 6L20DF, to 18,320 kW for the 16V46DF.

Wärtsilä 20DF

The Wärtsilä 20DF design is the smallest of the company's dual-fuel offerings and is based on the well-proven and reliable Wärtsilä 20 diesel engine which was introduced on the market in the early 1990s. It can run on HFO, distillate fuels or LNG. It was initially intended to complete the multifuel engine room concept as generating set next to Wärtsilä 31DF, 34DF, 46DF, or 50DF main propulsion machinery but also suits as a mechanical drive prime mover for smaller applications, such as small cargo vessels, ferries, or tugs.

It has a bore of 200 mm and a stroke of 280 mm. Three versions are offered all in line units with six, eight, or nine cylinders. As a 60 Hz genset the engine produces 185 kW/cyl at 1200 rpm and with a mep of 22.0 bar. The 50 Hz version produces 160 kW/cyl at 1000 rpm and 21 bar mep. As the main propulsion engine, the data is as for the 60 Hz genset allowing a power range from 1056 kW from the six-cylinder engine running on LNG through to 1665 kW for the nine-cylinder engine running on HFO.

For the engine to be able to run 100% load in gas the methane number must be 80 or above, it is, however, possible to run the engine on gases with lower methane number at a reduced output. IMO NOx Tier III requirements can be met by the engine running in gas mode but SCR is needed when running on oil fuels.

The engine block, made of nodular cast iron, is cast in one piece for all cylinder numbers. It has a stiff and durable design to absorb internal forces and enable the engine to be resiliently mounted without any intermediate foundations. The engine has an underslung crankshaft held in place by main bearing caps. The crankshaft is forged in one piece and the journals are of the same size regardless of the number of cylinders. The crankshaft is fully balanced to counteract bearing loads from eccentric masses by fitting counterweights in every crank web. This results in an even and thick oil film for all bearings. If necessary, the crankshaft is provided with a torsional vibration damper.

The main bearing caps, made of nodular cast iron, are fixed from below by two hydraulically tensioned screws. They are guided sideways by the engine block at the top as well as at the bottom. Hydraulically tightened horizontal side screws at the lower guiding provide a very rigid crankshaft bearing.

A hydraulic jack, supported in the oil sump, offers the possibility to lower and lift the main bearing caps for inspecting the bearings. Lubricating oil is led to the bearings and piston through this jack. A combined flywheel/thrust bearing is located at the driving end of the engine.

The cylinder head is made of grey cast iron, the main design criteria being high reliability and easy maintenance. The mechanical load is absorbed by a strong intermediate deck, which together with the upper deck and the side walls form a box section in the four corners of which the hydraulically tightened cylinder head bolts are situated.

The cylinder head features two inlet and two exhaust valves per cylinder. All valves are equipped with valve rotators. No valve cages are used, which results in very good flow dynamics. The basic criterion for the exhaust valve design is the correct temperature by carefully controlled water cooling of the exhaust valve seat. The thermally loaded flame plate is cooled efficiently by cooling water led from the periphery radially towards the centre of the head. The bridges between the valves cooling channels are drilled to provide the best possible heat transfer.

There is one campiece for each cylinder with separate bearing pieces in between. The cam and bearing pieces are held together with flange connections. This solution allows removing of the camshaft pieces sideways. The drop-forged completely hardened camshaft pieces have fixed cams. The camshaft bearing housings are integrated into the engine block casting and are thus completely closed. The bearings are installed and removed by means of a hydraulic tool.

The camshaft covers, one for each cylinder, seal against the engine block with a closed O-ring profile. The valve mechanism guide block is integrated into the cylinder block. The valve tappets are of the piston type with self-adjustment of the roller against cam to give an even distribution of the contact pressure. Double valve springs make the valve mechanism dynamically stable.

Wärtsilä 31DF

This is the newest of the Wärtsilä dual-fuel engine family and while it has many of the characteristics of the other engines in terms of basic construction, it has been designed for reduced lifecycle costs and increased longevity due to greater use of modular components. As a relatively new engine, it has yet to achieve any in-service references, but it has been specified for seven new buildings including, one ferry, one cruise vessel, four chemical tankers, and a fishing vessel. The ferry being built for Balearia Lines will have the largest engines with four 16V31DF units. The cruise vessel being built for Ponant has six 14V31DF engines and all of the other vessels mentioned will feature a single 10V31DF engine in their propulsion systems.

The DF version of the engine has a lower mep of 27.2 bar compared to the 30.1 bar of the oil-fuelled variant and its power output is lower at 550 kW/cyl as opposed to 610 kW/cyl. This difference means that the rated power output of the DF versions run from 4400 kW for the 8-cylinder model to 8800 kW for the 16-cylinder version or 10% below that of the oil-fuelled engine.

Wärtsilä 31DF engines are usually installed for dual fuel operation meaning the engine can be run either in gas or diesel operating mode. The operating mode can be changed while the engine is running, within certain limits, without interruption of power generation. If the gas supply fails, the engine automatically transfers to diesel mode operation.

In gas operating mode the main fuel is natural gas which is injected into the engine at low pressure. The gas is ignited by injecting a small amount of pilot diesel fuel. The Gas and pilot fuel injection are solenoid operated and electronically controlled common rail systems.

In diesel operating mode the engine operates only on liquid fuel oil. MDF or HFO is used as fuel with a common rail system/Electronic fuel injection rate optimized nozzle system. The pilot fuel injection remains active in order to avoid clogging of the pilot nozzle.

The fuel gas can typically be contained as CNG, LNG at atmospheric pressure, or pressurized LNG. The design of the external fuel gas feed system may vary, but every system should provide natural gas with the correct temperature and pressure to each engine.

Before the gas is supplied to the engine it passes through a Gas Valve Unit (GVU). The GVU include a gas pressure control valve and a series of block and bleed valves to ensure reliable and safe operation on gas.

The unit includes a manual shut-off valve, inerting connection, filter, fuel gas pressure control valve, shut-off valves, ventilating valves, pressure transmitters/gauges, a gas temperature transmitter, and control cabinets. The filter is a full flow unit preventing impurities from entering the engine fuel gas system. The fineness of the filter is 5 μm absolute mesh size. The pressure drop over the filter is monitored and an alarm is activated when the pressure drop is above permitted value due to a dirty filter.

The fuel gas pressure control valve adjusts the gas feed pressure to the engine according to engine load. The pressure control valve is controlled by the engine control system. The system is designed to get the correct fuel gas pressure to the engine common rail pipe at all times.

Readings from sensors on the GVU as well as opening and closing of valves on the gas valve unit are electronically or electro-pneumatically controlled by the GVU control system. All readings from sensors and valve statuses can be read from Local Display Unit (LDU). The LDU is mounted on control cabinet of the GVU.

The two shut-off valves together with gas ventilating valve (between the shut-off valves) form a double-block-and-bleed function. The block valves in the double-block-and-bleed function effectively close off the gas supply to the engine on request. The solenoid operated venting valve in the double-block-and-bleed function will relieve the pressure trapped between the block valves after closing of the block valves. The block valves and inert gas valve are operated as fail-to-close, i.e. they will close on current failure. Venting valves are fail-to-open and will open on current failure. There is a connection for inerting the fuel gas pipe with nitrogen.

Wärtsilä 31DF engine presents a cleaner less cluttered appearance than older engine models.

Wärtsilä 34DF

The 34DF engine was introduced in 2008 and upgraded in 2014. It features advanced dual-fuel technology and high efficiency. For marine applications, it is manufactured in configurations from 6 to 16 cylinders covering a power range of 2880–8000 kW. Based on the well-proven and reliable Wärtsilä 32 diesel engine introduced in the mid-1990s and also from the 50DF engines, the Wärtsilä 34DF's fuel flexibility means that the engine can be optimized for constant speed generating sets, as well as variable speed mechanical drives for main engine applications.

When first introduced, it came in six and nine-cylinder in-line versions and three V engines of 12 or 16 cylinders. A larger 20 V model was developed for power industry use. The Wärtsilä 34DF operates on the lean-burn principle, whereby the mixture of air and gas in the cylinder has more air than is needed for complete combustion. Lean combustion reduces peak temperatures and, therefore, NOX emissions. It also reduces heat flow to the walls of the combustion chamber, as well as the tendency for knocking. Because of the reduced heat loss and less likelihood of knocking, efficiency is increased and a higher output is attained.

Combustion of the lean air-fuel mixture is initiated by injecting a small amount of fuel oil pilot fuel into the cylinder. The pilot fuel is ignited in a conventional diesel process, providing a high-energy ignition source for the main fuel charge, which is a mixture of natural gas and air. To obtain the best efficiency and lowest emissions, the main fuel flow to each cylinder is individually controlled to ensure operation at the correct air-fuel ratio, and with the correct amount and timing of pilot fuel injection.

The engine functions are controlled by an advanced automation system that allows optimum running conditions to be set, independent of the ambient conditions or fuel used. The electronic control system is designed to cope with the demanding task of controlling the combustion in each cylinder and to ensure optimal performance in terms of efficiency and emissions, under all conditions, by keeping each cylinder within the operating window.

All fuel ignition parameters are controlled automatically during operation. Incorporated into the system is a cylinder pressure-based control. As this control utilizes the measurement of cylinder pressure for combustion optimization, cylinder pressure sensors have been added as standard in each cylinder.

The Wärtsilä 34DF can be started in diesel mode, using both main diesel and pilot fuel, or in gas mode. If the engine is started in diesel mode, gas admission is activated when combustion is stable in all cylinders. When running the engine in gas mode, the pilot fuel, which is always present, amounts to less than 1% of full load fuel consumption. The amount of pilot fuel is controlled by the engine control system. In the event of a gas supply interruption, the engine switches from gas to fuel oil operation at any load instantaneously and automatically. Furthermore, the separate backup fuel system makes it possible to switch from LFO to HFO without load reduction.

When running the engine in backup fuel mode, the pilot is also in use to ensure nozzle cooling and to avoid clogging of the injector tip. The engine can also be started without using the backup fuel system, in which case, the engine is started on pilot fuel with gas admission activated.

The natural gas is supplied to the engine through a gas regulation unit. The gas is first filtered to ensure a clean supply. The gas pressure, which depends on engine load, is controlled by a valve located in the valve station. On the engine, the gas is supplied through large common-rail pipes running along the engine. Each cylinder then has an individual feed pipe to the gas admission valve on the cylinder head. Gas pipes on the engine have a double-wall design.

The fuel oil supply on the engine is divided into two separate systems: one for the pilot fuel, and the other for backup fuel. The pilot fuel is elevated to the required pressure by a pump unit. This includes duplex filters, a pressure regulator, and an engine-driven radial piston-type pump. The high-pressure pilot fuel is then distributed through a common-rail pipe to the injection valve at each cylinder.

The pilot fuel is injected at a pressure of approximately 900 bar, and the timing and duration are electronically controlled. The backup fuel is separated from the pilot fuel system and is fed to a normal camshaft-driven injection pump. From the injection pump, the high-pressure fuel goes to a spring-loaded injection valve.

The Wärtsilä 34DF has a twin-needle injection valve with two separate nozzles. The larger needle and nozzle are used in diesel mode for LFO or HFO operation, and the smaller one for pilot fuel oil—when the engine is running in gas mode—and in backup fuel operation to ensure nozzle cooling. The pilot injection is electronically controlled, and the main diesel injection is hydraulically controlled. The individually controlled solenoid valve allows optimum timing and duration of the pilot fuel injection into every cylinder when the engine is running in gas mode. Gas is admitted to the cylinders just before the air inlet valves.

The gas admission valves are electronically actuated and controlled by the engine control system to give the precise amount of gas needed to each cylinder. In this way, the combustion in each cylinder can be fully and individually controlled. Independent gas admission ensures the correct air-fuel ratio and optimal operating point with respect to efficiency and emissions. The gas admission valves have a short stroke and are made of specially selected materials, thus providing low wear and long maintenance intervals.

Having the correct air–fuel ratio under any operating conditions is essential to optimum performance and emissions. For this function, the Wärtsilä 34DF is equipped with an exhaust gas wastegate valve. Part of the exhaust gases bypasses the turbocharger through this wastegate valve. The valve adjusts the air–fuel ratio to the correct value, depending on the varying site conditions, under high engine loads.

The Wärtsilä 34DF is equipped with the modular-built Spex (single pipe exhaust) turbocharging system, which combines the advantages of both pulse and constant pressure charging. The interface between engine and turbocharger is streamlined with a minimum of flow resistance on both the exhaust and air sides. High-efficiency turbochargers with inboard plain bearings are used, and the engine lubricating oil system is used for the turbocharger. The wastegate is actuated electro-pneumatically.

The 34DF was given an upgrade in 2013 to produce 11% more power than the original version. In diesel mode, it now consumes 7 g/kWh less fuel, while in the gas mode it features increased efficiency of 1.6 percentage points. The performance improvements were achieved without altering the physical dimensions of the engine.

The main changes included new turbochargers capable of boost pressures of more than 5.5. This higher boost pressure provides a larger window over which to optimize the engine, thus giving increased output with added efficiency. The upgraded engine is also now based on the Wärtsilä 32 diesel engine platform, which has a design firing pressure that is 20 bar higher.

A number of components have been modified for the higher firing pressure and pressure ratio. The pistons have been reinforced to give higher firing performance. Furthermore, the geometry of the top of the piston and compression ratio have been optimized to give the optimal combustion space. The cylinder head, which is a common design for diesel, DF, and SG engines, has been reinforced. So has the crankshaft, for the higher firing pressure. The stronger camshaft increases output in both diesel and gas mode.

Reinforcement of the main engine block was also necessary to carry the new higher capacity turbocharger. In addition to a stiffer engine block, the main bearing caps and large studs that hold the engine together have been reinforced to accommodate the higher firing pressure. The newer engines feature advanced Miller timing and variable inlet valve closing. Variable inlet valve closing provides optimum load response at low loads. This is very beneficial on ships where there the propulsion system experiences fluctuating loads in rough seas.

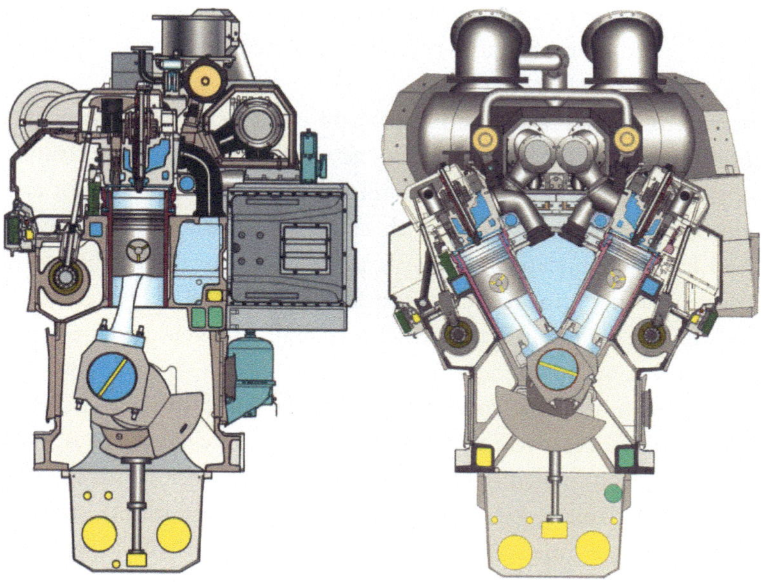

Cross-sections of the 34DF in line and V-engines.

Wärtsilä 46DF

After the economic crash of 2008, construction of merchant cargo ships declined dramatically and only the offshore and passenger ship sectors retained any buoyancy. In the offshore sector, there was a rapid development of large construction vessels with much higher power requirements than the ubiquitous PSVs and AHTS. This lead Wärtsilä to develop a dual-fuel engine specifically for these vessel types bearing in mind the impending introduction of the IMO NOx Code Tier III level on 1 January 2016.

In March 2014, the company announced the introduction of its new Wärtsilä 46DF engine based upon the Wärtsilä 46F engine but bringing in aspects of its other dual-fuel offerings, notably the 50DF. The fuel flexible engine is capable of operating on natural gas, HFO, or MDO. It extended Wärtsilä's dual-fuel engine family by covering the power range from 6.8 to

18.3 MW at 600 rpm. The engine has a bore of 460 mm and a stroke of 580 mm and can produce 1145 kW/cyl. It is offered in 6-, 7-, 8-, and 9-cylinder in-line versions and 12, 14, and 16 cylinder Vee versions.

All of the 46DF models produce more power than equivalent cylinder versions of the 50DF. The largest model in the range—the 16V46DF puts out 18,320 kW which compares very favourably with the larger 18V50DF (not used in marine applications) which has a maximum power output of 17,550 kW.

The Wärtsilä 46DF is offered in two different versions to meet specific customer needs. The high-efficiency version offers lower fuel consumption with a cylinder power of 1045 kW, while the high power version is capable of a cylinder power of 1145 kW with excellent engine thermal efficiency. For LNG carrier applications, the Wärtsilä 46DF can offer fuel savings of as much as 20 t/day compared to the first introduced DF engines. With up to 14 fewer cylinders installed, the overall lifecycle installation costs are significantly and positively impacted.

When operating in gas mode, the Wärtsilä 46DF engine is compliant with IMO Tier III regulations without any secondary exhaust gas purification systems. When fuelled by gas, the SOx and CO_2 emissions are notably reduced, and smokeless operation is attained. In liquid fuel oil mode, the Wärtsilä dual-fuel engines are fully compliant with the IMO Tier II exhaust emission regulations and can make use of SCR for meeting Tier III levels.

The switch between fuel types is made without loss of power or speed. The engine automation adapts automatically to the relevant fuel selection, both in normal and emergency modes.

In gas mode, the natural gas is fed to the engine at low pressure. This facilitates a simpler and space-saving engine room configuration while providing easier and faster maintenance activities. The engine's gas piping is double-walled as standard, and the advanced integrated automation system enables enhanced safety and local monitoring. This leads to safer and more reliable operations, under all conditions. The complete built-in automation minimizes the need for external controls, thus saving engine control room space.

The first in-service engines were installed in a crane vessel delivered in late 2019. As of February 2020, some 34 engines have been specified for nine vessels. Most of the engines are 14 or 16 cylinder types with cruise vessels being the most numerous applications.

Wärtsilä 46DF engine introduced in 2014.

Wärtsilä 50DF

Intended originally mainly for LNG carriers, the Wärtsilä 50DF was among the earliest dual-fuel engines on the market having been announced in 2001 and with the first engines coming into service in 2006. The engine was developed mainly with LNG carriers in mind with dual-fuel engines considered as being more efficient than the steam turbine systems normally found on LNG carriers of the period. The vessel concerned was an LNG carrier constructed at the French shipyard Chantiers de l'Atlantique for Gaz de France. The ship was the first LNG carrier to be powered by electric propulsion. The four six-cylinder in-line engines each developed 5700 kW at 514 rpm.

Development of the engine drew on the already existing 32DF and the 46 oil-fuelled engine. For marine applications, the Wärtsilä 50DF is manufactured in configurations from 6 L up to 16 V giving 950/975 kW per cylinder and a total maximum mechanical output of 15,600 kW. The bore is naturally 500 mm and the stroke 580 mm.

The engine has proved popular for LNG carriers with this type of ship and related gas storage vessels accounting for all but a handful of the 200 ships equipped with the 50DF engine. In most applications, the engine is used in a genset configuration providing power for electric motors. In a number of recent Japanese new buildings, the engines have been combined with a steam turbine in a hybrid configuration known as STaGE (Steam Turbine and Gas Engine).

The Wärtsilä 50DF remains in the company's portfolio although in terms of efficiency and compactness it has been superseded by the newer Wärtsilä 46DF. This is understandable since many operators will be specifying the engine for new builds in their fleets because of ensuring crew familiarity and for maintaining stocks of spare parts.

CHAPTER TWENTY-FIVE

Other medium-speed engines

There are many more manufacturers that produce or have produced medium-speed marine engines, than just the main players in this sector covered in earlier chapters. Some are relative newcomers although they may have a long history of local production. Meanwhile, some of the older brands have been absorbed by other manufacturers or have become defunct. Where the number of engines still in service warrant the engines from both categories are covered in this chapter.

Daihatsu

A Japanese contender in the domestic and international small-bore medium-speed engine arenas, Daihatsu evolved its current programme from the PS series which was developed in 1950 and helped establish the company as a major supplier of auxiliary diesel engines. Some 10,000 examples of the PS-26 model, for example, were sold worldwide.

The PS was succeeded in 1967 by a second-generation engine, the DS series, whose development goals focused on high output, low-specific fuel consumption, and compactness. Market demands for higher fuel economy with the capability to burn lower-grade bunkers led to the introduction of a third-generation engine, the longer stroke DL series, which includes eight medium-speed models with bore sizes ranging from 190 to 400mm and covering an output band up to 4415kW. The DL series remained in the production programme alongside the fourth generation DK series, designed to address requirements in the 1990s for high reliability, durability, and extended service intervals.

The DK series (Fig. 25.1) embraces 200, 260, 280, 320, and 360mm bore medium-speed models covering power demands up to 6325kW at 600/720/750/900rev/min from in-line and V-cylinder models.

The engine is based on a one-piece cast iron frame designed and manufactured for high rigidity to minimize noise and vibration. Large crankcase windows are arranged on both sides to ease access for maintenance. The thick bore-cooled flame plate of the cylinder head seeks improved

https://doi.org/10.1016/B978-0-08-102748-6.00025-6

Fig. 25.1 Daihatsu DK-28 engine in six-cylinder form.

mechanical and heat resistance, and extended service life. The head is secured by four hydraulically tightened studs. The crankshaft incorporates hardened large diameter pins and journals forged in one piece and supported by heavy-duty bearings for durability. The connecting rod is formed from three pieces with horizontal joints to reduce piston overhauling height and yield high rigidity.

A gap between the DK-20 and DK-28 models was filled by the 260 mm bore/380 mm stroke DK-26, intended as a genset drive or propulsion plant for small ferries, cargo vessels, and tugs. Produced as in-line five- and six-cylinder models, the design develops up to 1710 kW at 720/750 rev/min.

Similar design principles to the previous DK models are exploited by the DK-26, notably a high stroke/bore ratio of 1.46:1, but the engine features an improved combustion chamber profile and higher compression ratio. Combustion is claimed to be stable over the entire load range when operating on heavy fuel, without compromising NOx emission and fuel consumption performance. The engine frame is fitted with deep mounting feet with I-contour sections so that rubber elements for resilient supports can be attached directly to the feet. The charge air manifold, lubricating oil gallery, and cooling water passages are all integrated into the frame, creating a double-wall configuration to help reduce noise and vibration.

A simple yet efficient exhaust layout designated Daihatsu single-pipe (DSP) exhaust system recovers gas energy effectively by allowing a good match to the turbocharger. An NOx emission level of 9.4 g/kW h and specific fuel consumption of 185 g/kW h at full load is claimed, with temperatures of key combustion chamber components well within the optimum range.

Supplied with matching gearboxes, the DKM series is produced in 200, 250, 260, 280, and 360 mm bore sizes in six- and eight-cylinder models covering an output band from 880 to 4413 kW at speeds from 900 to 600 rev/min.

A new auxiliary engine, the DC-17, was introduced in 2001 in five- and six-cylinder versions offering generator outputs of 440 and 560 kW at 900/1000 rev/min. The 170 mm bore/270 mm stroke design is released for heavy fuel-burning applications.

Blending experience from the DK series and the DC-17 engine, Daihatsu subsequently developed a 320 mm bore version of the latter. Six- and eight-cylinder variants of this DC-32 design deliver outputs of 2905 and 3884 kW at 720/750 rev/min. A significant reduction in the number of overall parts contributed to the designers' goals of compactness, simplicity and lightweight, ease of installation and maintenance, high reliability and durability, low operating costs, and environmental friendliness. Swift withdrawal of a complete unit assembly (cylinder head, piston, connecting rod, and liner) is facilitated, although conventional dismantling of the individual elements is possible.

A new generation of engines was launched in 2012 with the DE range (see Table 25.1). The E representing Environment, Endurance and Reliability, Easy Maintenance, Economy, and Ease and Safety according to publicity material. Initially launched as just two models the DE-18 with a bore and

Table 25.1 Daihatsu DE engine range.

Model	Cylinders	Output (kW)	rpm	Bore (mm)	Stroke (mm)
6DEM-18	6	810	900	185	280
6DEM-23	6	1200	750	230	320
		1440	900		
6DEM-28	6	2240	750	285	390
6DEM-33	6	3390	750	330	440
8DEM-33	8	4520	750	330	440

stroke of 185 and 280 mm and the DE-23 with a 230 mm bore and a stroke of 320 mm. Both were only available as six-cylinder in line models with the smaller engine designated for use as a genset and the larger engine for both genset and main propulsion. Used as gensets the engines could be run at either 720 or 900 rpm producing outputs of 360–808 kWe for the DE-18 and 760–1425 kWe for the larger engine. When used in propulsion mode, the DE-23 offered two power outputs one of 1200 kW based on a speed of 750 rpm and the other of 1500 kW for the engine running at 900 rpm.

The DE engines were smaller and lighter than earlier engines of similar output. The 6DEM-23 eclipsing the maximum output of the 8DKM-20e by some 250 kW with a maximum output of 1500 kW against 1250 kW.

Later introductions were of a DE-28 with a bore of 285 mm and stroke of 390 mm and the DE-33 with 330 mm bore and 440 mm stroke. Both of these are also available as six-cylinder in-line models and there is also an eight-cylinder version of the largest engine (Fig. 25.2).

The DE type incorporated much of the tried and tested technology of the DK-type engine and was further complemented by new technologies aimed at meeting the IMO NOx Tier II rules. During the design phase,

Fig. 25.2 Cross section of the DE engine.

extensive use was made of new computer analysis including CAD, CAE, CFD, and performance simulation.

Engine frame: The frame has a monobloc structure made of cast iron. Each of the flow paths is formed integrally with the frame, which includes the air intake path, the lubricating oil path, and the cooling water inlet. Since the hanging metal method is adopted, the main bearing is firmly mounted by the mounting bolts and the side bolts hydraulically tightened.

Crankshaft and bearings: the crankshaft is integrally forged from alloyed steel and features increased crankpin and journal diameters for optimized surface pressure and oil film thickness. Large-capacity weights are applied for better rotational balance adjustment and reduced vibration. The flywheel is attached to the rear of the shaft, and the gears for the camshaft and pumps are mounted at the other end. The main and crankpin bearings are of special corrosion- and wear-resistant aluminium alloy with a thrust metal attached to the flywheel side No.1 bearing.

Camshaft and timing gears: air inlet valve, exhaust valve, and fuel injection cams are arranged on a large diameter unit camshaft built-in two-cylinder sections. The camshaft timing gear is attached to the front end of the engine, positioned between the crank gear and cam gear. Each gear is surface hardened for durability.

Cylinder liner: for the DE-18, the cylinder liner is made of special cast iron. The high-temperature cooling water comes from the frame water chamber and flows from under to over the water chamber formed with the frame and the liner, effectively cooling the liner. After this, the cooling water goes into the water chamber of the cylinder head. On the DE-23, the cylinder liner is of a thick-wall bore cooling type. The high-temperature cooling water is led from the cooling water path in the frame to the bore-cooling hole of the cylinder liner, effectively cooling the upper part of the liner. After this, the cooling water goes into the water chamber in the lower part of the cylinder head.

Piston and connecting rod: the composite piston features an alloyed steel crown and a ductile iron skirt, with the grooves for the top and second rings induction surface hardened for durability. Effective forced piston cooling is performed by oil passing through the connecting rod and piston pin. The horizontally cut three-piece rod design allows the piston to be overhauled without removing the crankpin bearing section; it also reduces the height necessary for piston withdrawal.

Cylinder head and valves: a special high-strength cast iron improves the durability of the head which is fastened to the engine frame by four hydraulically tightened bolts. The twin inlet and twin exhaust valves are fitted with rotators. The water-cooled exhaust valve seats are attached directly to the cylinder head.

The fuel injection valve and the surrounding area at the centre of the head are cooled by the tornado cooling method (Daihatsu patent), allowing noncooled valves to be used even for heavy fuel operation. Piping on and around the head is eliminated to ease maintenance.

Fuel injection: Bosch-type high-pressure injection pumps with built-in tappets and a closed-type plunger barrel are used, with a pressure valve incorporated in the fuel outlet valve. The noncooled heat-treated injectors are connected to the pumps by a forged high-pressure pipe.

Turbocharging and exhaust system: the turbocharger includes a radial turbine with high efficiency and high-pressure ratio, which is not cooled but forcibly lubricated as standard. Each of the turbocharger and the air cooler is independently fixed on the gear case in this structure in consideration of maintenance workability.

A pulse turbocharging method is adopted for the exhaust pipe, which enhances the turbocharging efficiency and is excellent in the transient response characteristics. The exhaust pipe is arranged on the same side as the air intake duct integrated with the frame, improving the easiness in handling.

Freshwater cooling system: split into separate high-temperature and low-temperature circuits; a two-stage intercooler system benefits combustion when burning heavy fuel at low loads. The main cooling water outlet pipes for the engine cylinders are connected by sliding flanges for ease of dismantling.

Lube oil filter: an automatic, continuous back-flushing filter serves both engine and turbocharger, eliminating the need for an extra filter for the latter. All equipment related to the lubrication system is arranged on a side surface on the exhaust pipe side. The lubricating oil cooler, the thermostat, the pressure regulator, and the automatic backwash filter are connected by the block unit. This module allows accessibility and eases maintenance. To effectively remove foreign substances and foreign particles, a large-capacity centrifugal filter is standard (Fig. 25.3).

Daihatsu has developed a common-rail fuel injection system for the DK series and has applied two-stage turbocharging systems in conjunction with the Miller cycle for applications calling for very low NOx emissions. Pursuing compliance with IMO Tier II NOx emission requirements, Daihatsu developed a complete range of standard e-Diesel engines, headed by

Fig. 25.3 Lubricating equipment of the DE engine. The automatic backwash filter is mounted behind the large oil cooler and the oil pump and a centrifugal filter positioned below and forward.

shipments of the first 6DK-20e models in June 2010. Compliance was affected by optimizing the compression ratio, valve timings, combustion chambers, and fuel injection timings, and adopting high-pressure turbochargers without undermining safety, reliability, overall performance, and fuel economy. For meeting Tier III NOx rules, the company has opted for SCR treatment.

General Electric

Four-stroke medium-speed trunk piston engines from US designer GE Marine have found wide appeal in the international marine propulsion and genset drive market the current programme embracing V228 and L/V250 designs. GE Marine was part of the wider GE Group that includes aviation and marine gas turbine manufacture but the diesel engine production was moved to GE Transportation and in 2019, this division was acquired by Wabtec an industrial organization mostly involved in rail and locomotive production.

The marine engine business continues to produce and offer the engine ranges mentioned.

GE V228 series

A power range from 1044 to 3355 kW at 900–1050 rev/min is covered by V8-, 12-, and 16-cylinder versions of the 228.6 mm bore V228 (formerly 7FDM) design. High-strength nodular cast iron is used to create a rigid monobloc mainframe with a V45° cylinder configuration and all water, fuel, and exhaust piping externally mounted to eliminate the potential for internal contamination. Large access doors are provided to ease inspection and maintenance, while casing and ribbed cross-members dampen vibration.

Forged from high-quality steel and nitrided for long life, the one-piece crankshaft features generous journals and crankpins to minimize surface loads; large radii fillet-welded counterweights also reduce stress to foster a longer lifetime. The tri-metal main bearings are oversized to reduce pressure while providing an oil-wedge formation for low wear, and a sacrificial running-in overlay promotes rapid conformity in the start-up. Superior heat resistance is sought from a nickel barrier, while a lead–tin overlay serves as a high-tolerance contaminant trap for particles and degraded oil.

Four-stud mounting of the unitized power assembly facilitates the removal and replacement of key components. Electron-beam welding of the steel cylinder liner to the forged cylinder head prevents leakage, and stainless steel valve seats, Inconel exhaust and chromed inlet valves and valve rotators are specified to extend overhaul intervals and component lifetimes. High-strength composite pistons incorporate forged steel crowns with oil splash cooling. A three-ring arrangement distributing pressure more evenly and a cutback design contribute to lower lube oil consumption. The lightweight aluminium skirt offers high heat resistance. The master and articulated rod share a common journal, minimizing the engine length. Forged and hardened cams and journals on the segmented camshaft foster reduced wear, and oversized end bearings decrease loads for a longer camshaft life.

Precise fuel management and control are provided by GE's electronic fuel injection system, which helps to secure compliance with regulatory emissions standards. GE group expertise in gas turbine technology was applied in designing the turbocharger, a high compression ratio improving efficiency across the load range. Dual modular pulse piping preserves exhaust pulse energy for maximizing thermal efficiency, the dual pipe made of stainless steel for extended life.

A GE PowerStar controller houses the electronic governor unit, speed reference, and load core, protecting the engine against overspeed, overload, low oil and water pressures, and positive crank pressure.

GE L/V250 series

GE Transportation started developing its Evolution-series locomotive engines in 1998, targeting a platform that would meet or exceed US EPA Tier II emissions requirements without compromising fuel economy, reliability, and durability. The resulting medium-speed design, well proven in locomotive duty, was refined in 2006 and launched as the V250 series for marine applications.

Compact and high power-to-weight installations are promised from the 250mm bore/320mm stroke design, which is released in V12- and V16-cylinder versions to cover an output range from 2732 to 4661 kW at speeds from 900 to 1050 rev/min (Table 25.2). An electronic fuel injection system based on an electronically controlled unit pump serving a standard configuration side entry injector fosters flexible operation in a range of applications as well as optimized combustion at different engine speeds and loads.

GE's PowerStar II controller drives the fuel injection system, exploiting advanced closing point detection to secure better fuel economy and emissions control. The system features improved diagnostics and display of all engine parameters and faults without the need for a laptop computer. Engine protection algorithms can predict problems early to avoid damage.

Turbochargers based on GE's established design were developed in partnership with GE Aviation to boost the V250 engines, all aerodynamic features being upgraded to satisfy combustion and engine-efficiency requirements (Fig. 25.4).

A new articulated piston with three rings meets weight requirements and exceeds peak firing pressure demands; the skirt and pin boss are supported by the piston pin and are optimized to provide excellent lubrication and guidance. A cylinder head with significantly increased safety factors is achieved with a wall supported by the head gasket, a thicker flame deck, optimized

Table 25.2 GE L/V250 engine data.

GE L/V250 data	
Bore (mm)	250
Stroke (mm)	320
Cylinders	6/8 L/12/16 V
Compression ratio	16.8
Speed, rev/min	900–1050
The power range (kW)	1498–4661

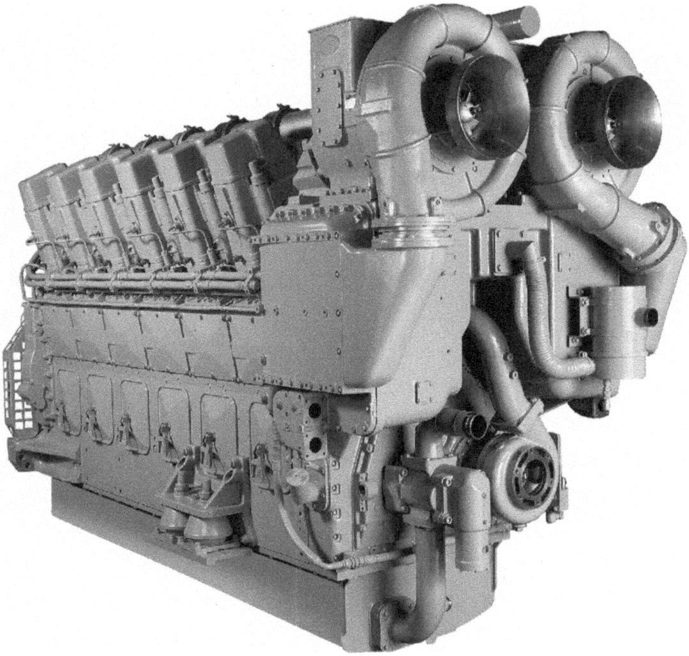

Fig. 25.4 GE Marine's V250 engine.

placement, larger exhaust seat ring cooling bores, and smooth transition areas to avoid stress concentrations.

The connecting rod benefited from considerable design changes addressing reliability. Requirements for oil film thickness and pressure were met through a combination of an increased bearing width and a smaller oil groove in the bearing shell. The angle between the shank and big end split line maximizes stiffness while allowing disassembly through the crankcase bore. Advanced simulation tools developed by GE corporate R&D were used to verify the rod design's low-fretting risk.

High levels of reliability and durability were sought from the main bearing, and the block, crankshaft, and drive end were repackaged to offer sufficient room for bearing size increases. The same hardening process used on the established FDM/FDS engines was applied for improved crankshaft fatigue strength.

Easier maintenance is addressed by a segmented camshaft, sectional exhaust manifold, unitized power assembly, and other features to simplify component removal and replacement. The cylinder head, water jacket, cylinder liner, piston, and connecting rod can be withdrawn from the engine as one assembly by removing a small number of connections and fasteners. Both inlet and exhaust manifolds are compactly arranged between the V-cylinder banks to contribute to a narrow overall engine width.

The V250 marine engine shares most components with its locomotive parents, such as connecting rods, power assemblies, cams, pistons, high-pressure fuel equipment, mainframe, exhaust system, and crankcase. A number of elements, however, had to be developed or refined for marine applications: rigid and isolation mounting systems; dual-impeller cooling water pump; mechanical fuel transfer pumps; deep lube oil sump; class-certifiable integrated wire harness; explosion-relief crankcase doors; heat shielding on the exhaust system; flywheel and air starting system; and seamless steel shielded fuel lines.

An in-line cylinder derivative of the V250 family was introduced in 2008, the L250 engine offered in six- and eight-cylinder models with respective maximum ratings of 1922 and 2564 kW and continuous ratings from 1498 to 2330 kW. While based on the V-design, the engine featured a number of refinements and new components, including the mainframe with optimized ribs and an integrated air manifold, and a crankshaft with wide webs and tangent fillets. The air and oil filters, fuel pumps, low- and high-temperature thermostatic valves, and seawater pumps were also new to the in-line engine.

In addition, the L250 was designed with engine-mounted ancillaries for the ease of installation and maintenance, and an integrated power take-off system. The GE-designed and manufactured turbocharger can be flexibly mounted at either end of the engine, a facility (along with a narrower engine frame) that eases retrofit projects.

By 2012, the L/V250 and V228 ranges were upgraded to meet IMO Tier II rules and in 2016, the L/V250 was again improved to meet IMO Tier III standards and designated 250MDC. The new engines employ exhaust gas recirculation rather than SCR to meet Tier III standards and the maker considers this a strong selling point for the engines.

Current ratings of 250MDC engines start at 1700 kW, for the 6L250 unit running at 900 rpm (190 kW is available at an engine speed of 1000 rpm), with 3500 kW and 4660 kW being available from 12 and 16 cylinder products, respectively, at the higher speed. The new engines have improved response times to step load increases and can be matched for optimum performance with either fixed or controllable pitch propellers.

Grandi Motori Trieste

A wide range of medium-speed engines were designed and produced over many years by Grandi Motori Trieste (GMT), which was founded in the late 1960s to pool the diesel engine expertise and resources of Fiat, Ansaldo, and CRDA and rationalize the Italian engine-building industry. GMT later became part of the state-owned Fincantieri shipbuilding group's diesel engine division and in 1997 entered a new era as a member of the Wärtsilä Corporation (see Chapter 24), after which its own design portfolio was phased out.

GMT's last large engine, the A55 series, was the final evolution of a 550mm bore model first designed and produced in Torino by Fiat Grandi Motori in the early 1970s. The original A550 design with a 590mm stroke and a rating of 880kW/cylinder was developed into a B550 version with an output of 1080kW/cylinder at 450rev/min. The BL550 derivative (Fig. 25.5), launched in 1986, retained the same bore size and running speed but exploited a longer stroke (630mm) to yield a 10% higher output and a reduction in specific fuel consumption on low-grade fuel. The power

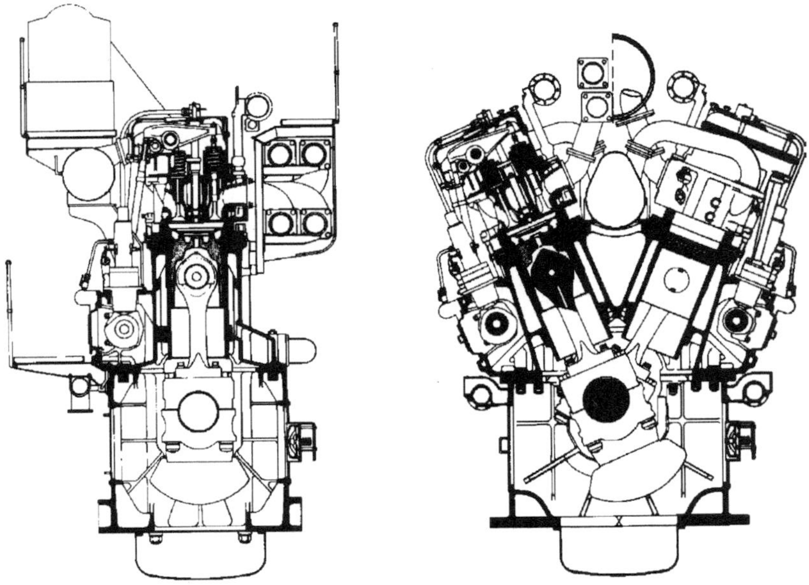

Fig. 25.5 GMT's BL550 design in in-line and V-cylinder form.

uprating—to 1213 kW/cylinder—was achieved with a modest increase in mean effective pressure (from 20.6 to 21.6 bar) and a rise in mean piston speed from 8.85 to 9.45 m/s.

Higher mechanical and thermal loads dictated some changes in the design of combustion chamber elements and running gear but the engine frame and bedplate remained separate units of nodular cast iron for the in-line cylinder engines. Unlike the original design, the eight- and nine-cylinder bedplates and frames are monobloc castings secured by tie rods; however, the V-type engines feature a welded steel bedplate and a special cast iron frame for which tie rods are unnecessary and replaced by stud bolts.

The cylinder head for the BL550 engine (Fig. 25.6) was redesigned to assure a more even distribution of temperatures between the exhaust valves and the fuel nozzle, and additional bore cooling introduced in the gas-side walls for better distribution. A new exhaust valve cage (Fig. 25.7) with a thinner seat wall and more efficient cooling from improved water circulation was specified to reduce the valve seat temperature and extend valve life when the engine burns fuels with high sodium and vanadium contents. Better cooling of the cage seat and effective pulse turbocharging lowered the temperature of the continuously rotated composite exhaust valves which comprise a Nimonic 80A alloy head and anticorrosion coated steel stem.

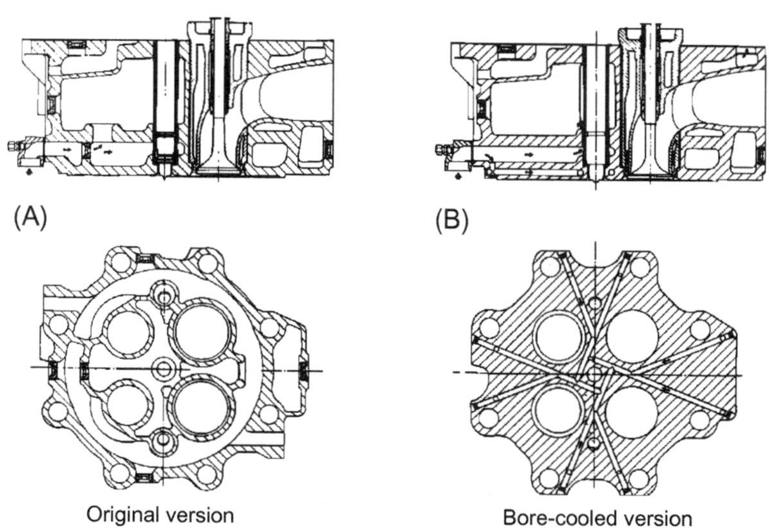

Fig. 25.6 Sections through GMT B550 cylinder head (left) and the redesigned BL550 component.

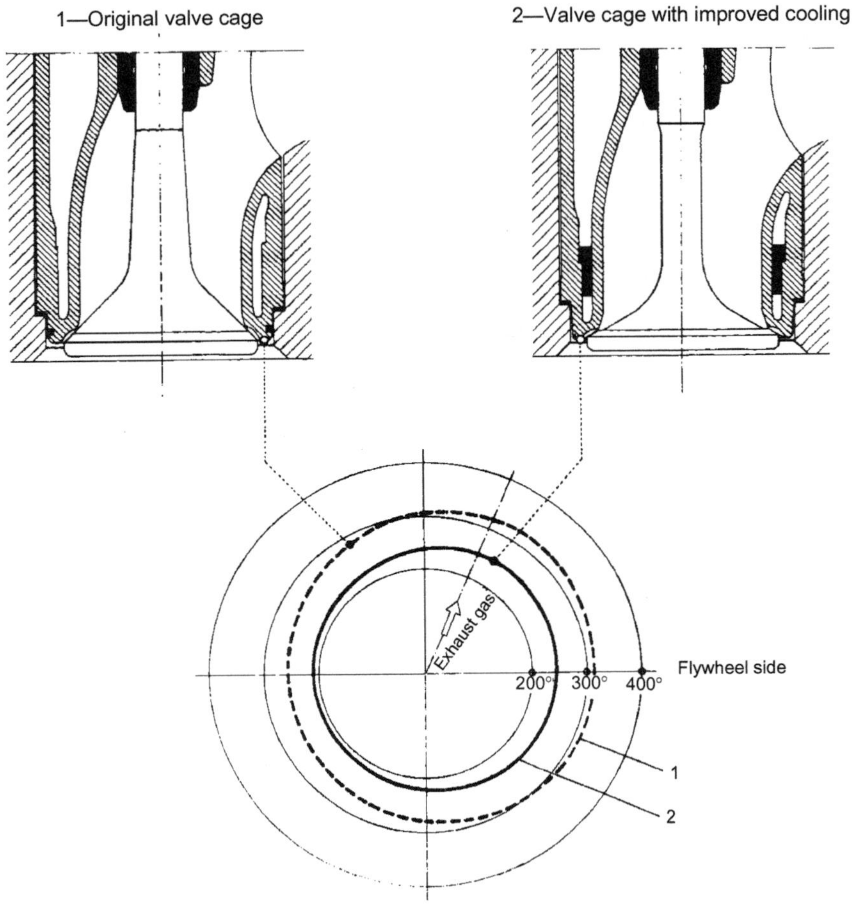

Fig. 25.7 Temperature distributions of the GMT B550 valve cage and the improved design for the BL550 engine (21 bar mep and 450 rev/min).

A much stiffer cylinder liner unit was designed, without a heat shield and bore cooled at the top near the gas-side wall. The cooling bores are of the blind type in which the inlet and outlet water flows are separated by diaphragms, allowing the radial outlet holes to be positioned in the lower part of the liner and thus eliminate a concentration of stresses in those areas most affected by dynamic load fluctuations below the cylinder head gasket. Tests showed that the bore-cooling arrangement secures a lower peak temperature and a more uniform distribution on the piston ring rubbing surface. Load-dependent cooling was specified as standard.

The composite pistons (steel crown and aluminium alloy skirt) carry three compressions and one oil scraper ring in chromed grooves. The upper

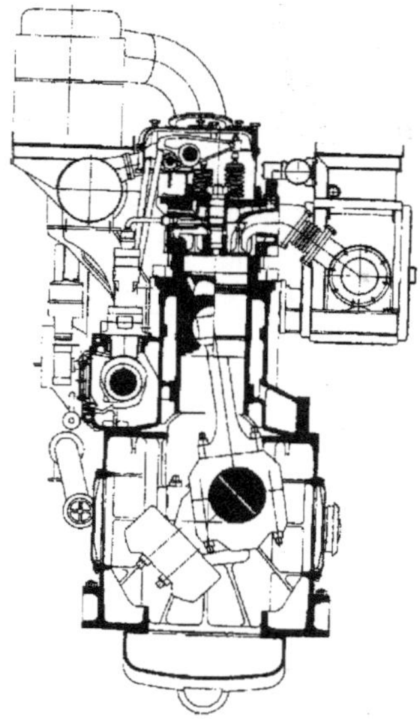

Fig. 25.8 GMT's A55 design.

two compression rings were increased in thickness to withstand the higher combustion pressure and reduce torsional deformation.

A subsequent development programme in 1992/93 resulted in the A55 design (Fig. 25.8) which retained the 550 mm bore of its precursors but featured an extended stroke (680 mm). An 8% power increase over the BL550 series was derived—taking the specific output to 1250 kW/cylinder at 425 rev/min—along with enhanced fuel economy. The redesign also sought simplified maintenance procedures and higher reliability from a range comprising six, eight, and nine in-line and V12-, 14-, 16-, and 18-cylinder models covering a power band up to 22,500 kW. Special propulsion market targets included cruise ships and high-powered ferries, GMT citing the low height of the A55 engine for its power output: 5.765 m for the in-line version and 5.67 m for the V-cylinder model.

The A55 benefited from GMT's technical collaboration with New Sulzer Diesel, permitting elements of the Swiss designer's successful ZA40S medium-speed engine to be exploited. The longer stroke took the stroke/bore ratio from 1.15 to 1.24:1, giving a better thermal efficiency. This was achieved without raising the cylinder head; a more compact piston allowed the same head casting to be used. A reduction in engine speed compared with the BL550 limited the mean piston speed and yielded the increased output with virtually the same brake mean effective pressure as the earlier model.

Fuel injection system improvements were also made. Fuel enters the valve laterally rather than vertically, underwriting a reduced fuel pipe length and a shorter fuel valve. The injection itself takes place over a shorter period, covering around 30° of the cycle, and at a higher pressure (around 1200 bar).

The crank mechanism and other components were also redesigned, mainly to simplify maintenance. The connecting rod, for example, is in three parts (a shank and two caps) connected by hydraulically tightened studs. Despite the longer stroke, it was possible to house the crank mechanism in the same crankcase as the BL550 by redesigning the crankshafts and connecting rods. The bedplate and cylinder block remained virtually unchanged.

Most notable of the A55 refinements was the adoption of Sulzer's unique rotating piston concept, previously unique to the ZA40S engine and its predecessors. A bore-cooled steel crown is mated with a cast-iron skirt. Cast iron rather than aluminium alloy was specified to allow the operating clearance of the skirt to be reduced; this, together with the inertia from the rotating piston, reduces impact velocities with the liner resulting from piston slap. The rotating piston underwrites reliability and lower lubricating oil consumption. Simpler and more efficient turbocharging was introduced. In-line and V-cylinder propulsion engines are served by a single-pipe exhaust system associated with high efficiency and compact configuration; an air/gas bypass is also provided.

GMT was also active in the popular 320 mm bore class, its A32 design having emerged in the 1990s as a longer stroke derivative of the A320 engine introduced in 1986. The 320 mm bore/390 mm stroke A32 series had a maximum continuous rating of 440 kW/cylinder at 750 rev/min and was produced in six, eight, and nine in-line and V12-, 14-, and 16-cylinder versions to cover a power range up to 7040 kW.

A monobloc casting of high-strength iron arranged for an underslung crankshaft forms the structure of both in-line and V-cylinder engines. The high-strength lamellar iron cylinder head is secured by four stud bolts for ease of dismantling and has a thick bore-cooled bottom plate for low thermal loading and high stiffness. The inserted thin-walled exhaust valve seat rings are directly cooled by water circulation to achieve a low seat temperature and extended time-between-overhauls on heavy fuel.

High-strength cast iron cylinder liners have a thick bore-cooled collar for low mechanical and thermal stresses. The composite pistons feature thin-forged steel crowns and aluminium alloy skirts. The connecting rods are drop forged in two pieces and withdrawable from the top along with the piston.

Reliable operation on residual fuels was addressed by basic design parameters (mean piston speed, compression ratio, combustion pressure, and air–fuel ratio) and attention to key component and system design, GMT citing:

- A facility to provide automatic load-dependent control of jacket cooling water and charge air temperatures, gradually raising them with diminishing load to improve combustion, reduce fouling, and maintain the combustion chamber walls above the acidic condensation temperature to prevent cold corrosion.
- Exhaust valves of Nimonic 80A material with corrosion-resistant coated stems and rotating devices to prevent hot spots.
- Forged steel piston crown carrying corrosion- and wear-resistant plasma-coated rings.
- High-pressure fuel injection system ensuring fine atomization for better ignition; and availability of a remote-controlled facility for recirculation of fuel through the injectors at engine standstill for starting on fuels of the highest viscosities.
- Bore cooling of the hot components of the combustion chamber (cylinder liner, head, and exhaust valve seats) through machined passages, allowing precise temperature control in inhibiting both hot and cold corrosion.

GMT also developed 420 mm bore engines for a number of years. The A420 had a stroke of 500 mm and ran at 500 rev/min to deliver 515 kW/cylinder, while the shorter-stroke (480 mm) A420H ran at 600 rev/min for an output of 588 kW/cylinder.

HiMSEN (Hyundai Heavy Industries)

A prolific South Korean builder of two-stroke and four-stroke diesel engines under licence, Hyundai Heavy Industries developed its first own-design medium-speed engine, the 210 mm bore HiMSEN H21/32 series, for launching in 2001. Since then the HiMSEN brand has expanded to embrace 170, 250, and 320 mm bore models based on common design principles and covering power demands up to 10,000 kW for genset and propulsion plant applications. Home and export market penetration were swiftly achieved, supported by installations for the HHI group's substantial newbuilding commitments.

Initially targeting marine genset drives, the H21/32 engine was also intended for propulsion markets (Fig. 25.9). A heavy fuel-burning

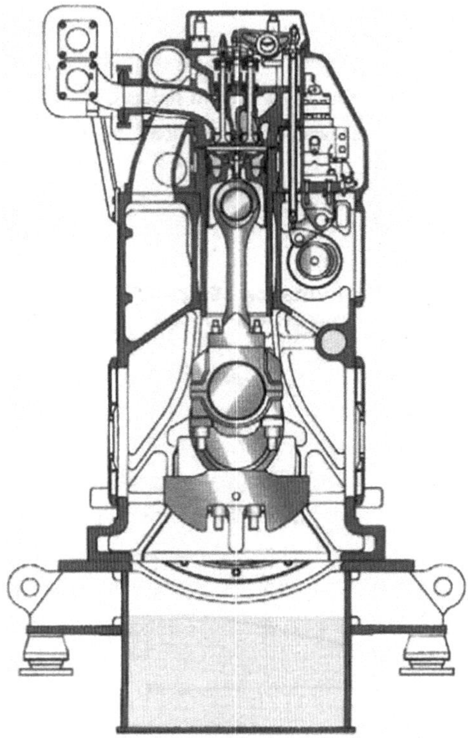

Fig. 25.9 Cross section of HiMSEN H21/32 engine, whose design principles were extended to 170, 250, and 320 mm models.

210 mm bore/320 mm stroke design was selected for development; the high stroke/bore ratio (1.52) and high compression ratio (17) were considered appropriate for fuel economy and low NOx emissions. A simplified configuration with a minimal number of components was pursued to enhance reliability and to ease production and maintenance. Two sets of prototypes were built and tested to confirm engine performance and reliability. The current programme covers a power band from 1200 to 1800 kW at 900 rev/min with in-line six-, eight-, and nine-cylinder models; specific ratings from 150 to 200 kW per cylinder at 720/750/900 and 1000 rev/min are delivered.

Key structural and running elements of the design were optimized for Hyundai's production facilities, with generous dimensions underwriting moderate fatigue safety factors. Ease of maintenance was addressed by adopting a cylinder unit concept, allowing the cylinder head, valve train, fuel injection equipment, cylinder liner, piston, and connecting rod to be withdrawn as an assembly for servicing onboard or ashore (Fig. 25.10).

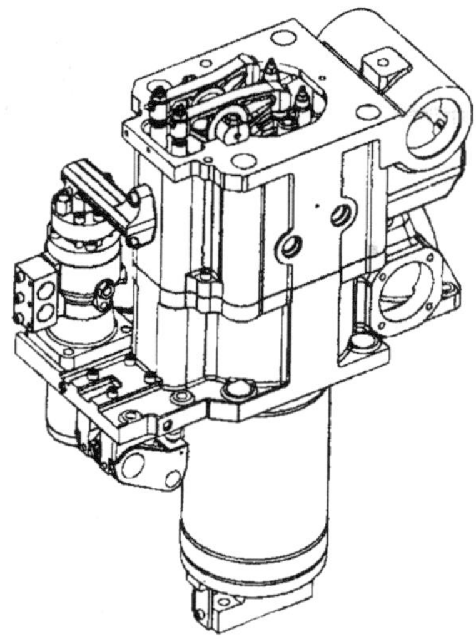

Fig. 25.10 H21/32 engine cylinder unit.

The engine block is a simple and robust structure of nodular cast iron. A large volume of the combustion air chamber and lubricating oil channel are incorporated but water spaces are excluded to avoid any risk of corrosion and flooding into the oil chamber. The nodular cast iron cylinder head features a V-type fire deck support to combat the high maximum firing pressure (200 bar); the inlet and exhaust ports were evolved in-house by Hyundai using its own flow test rig and computational fluid dynamics (CFD) analysis. The structural design of the head and the cylinder liner was optimized based on the CFD analysis of coolant flow as well as finite element analysis.

The two-piece piston comprises a steel crown and box-type steel skirt, delivering lower deformation under high cylinder pressure. Two compression rings and one oil scraper ring are fitted, the top ring coated with chromium-ceramic material on its running surface to curb wear. The connecting rod is a three-piece marine head-type component, shot-peened to prevent any fretting from relative movement of the mating surfaces of rod and big end. The crankpin bearing is an aluminium tri-metal Rillenlager type, and the main bearing an aluminium bimetal component.

A special chromium–molybdenum steel camshaft incorporates a fuel cam and intake and exhaust cams arranged as a one-piece element serving each cylinder. The cam profiles were optimized using the Hermite Spline Curve method developed by Hyundai. The robust fuel injection system is designed for pressures up to 2000 bar. The injection pump has a roller tappet with generous dimensions, the injection pipe is of the short block type and the injection valve has a high opening pressure capability with an oil-cooled nozzle body for heavy fuel operation.

An innovative design of air cooler cover controls the cooling water flow and hence the combustion air temperature at low load; the air cooler is arranged to function as a heater at low loads. ABB Turbo Systems' compact TPS turbocharger is specified to boost engine efficiency, power density, and durability, offering the potential to support future rating rises.

The 250 mm bore HiMSEN H25/33 design emerged from a joint development project with the Rolls-Royce group's Bergen Diesel and was independently updated to suit Hyundai's specific requirements. Rolls-Royce offers the engine as its C series. The H25/33 programme embraces six-, eight-, and nine-cylinder models covering an output band from 1740 to 2610 kW at 900 rev/min

Hyundai stepped up its challenge to established medium-speed engine builders in 2007 with the launch of a contender in the popular 320 mm bore propulsion and genset drive arena. With a specific rating of 500 kW/cylinder at 720/750 rev/min, the H32/40 series covers power demands up to 4500 kW from in-line six- to nine-cylinder propulsion models. The first prototype, an eight-cylinder model, was type approved by the major classification societies in December 2006 for commercial availability in the following year. In 2008, the V series was also type approved.

In preparation for the IMO NOx Tier III rules, Hyundai launched a new generation of engines under the acronym CLEAN (Customer, Liability, Environment, Acceptable Technology, No Defect). The first engine chosen for this adaptation was the H21/32 with the new version being labelled HiMSEN H21C. The changes allowed for an improved power output by 20% and fuel efficiency up to 2% compared to the existing same class four-stroke engines, H21/32, while making its physical size and weight 10% smaller and lighter, respectively, by means of high-pressure combustion technology.

A similar upgrade was carried out in 2019 to another engine in the portfolio. The H32C(V) engine with a bore of 320 mm covers a power range from 3.6 to 10.8 MW is an upgraded version of the H32/40(V) engine. The H32C(V) improves power output by 20% and fuel efficiency by 6% compared to the existing same class four-stroke engines, while weight per power is 10% lower, thanks to a variable fuel injection timing device and two-stage turbocharging system.

Another new engine, the HiMSEN H46/60V was given its type approval in 2013. The type approval test completion of the IMO Tier III compliant H46/60 expanded Hyundai Heavy's medium-speed marine engine line-up and covered outputs from 7500 to 22.500 kW. The engine is available for propulsion, genset use, and as a stationary engine. The early commercialization was of the stationary engines but by 2020 the marine propulsion range was available as four in-line engines of six, seven, eight, or nine cylinders under the designation H46/60P, and as three V models of 12, 16, and 18 cylinders designated as H46/60VP.

As the designation suggests that the engines have a bore of 460 mm and a stroke of 600 mm. Both types run at 600 rpm. They meet IMO NOx Tier II as stand-alone engines and NOx Tier III with SCR treatment of exhaust gases.

Engine block: the engine block is made of monobloc nodular cast iron and has a lubricating oil channel. The main bearing and main bearing cap for the underslung crankshaft are carried in high duty supports tightened hydraulically by two vertical main bearing studs and two horizontal side studs. There is easy access for maintenance and access to the crankshaft including connecting rod and camshaft for each cylinder. In accordance with class requirements, relief valves are fitted to some of the crankcase doors.

Crankshaft: the crankshaft is made of continuous grain flow die-forged special alloy steel with high tensile strength. The crankshaft has a counterweight for balancing the mass forces on each crank web, which is fastened with two hydraulically tensioned studs. At the free end, there a gear wheel drives lubricating oil and cooling water pumps and a vibration damper will be mounted depending on the result of the torsional vibration calculation for each installation. At the flywheel end is the camshaft driving gear and the transmission coupling flange.

Connecting rod: is a die-forged steel type with a three-piece marine head allowing easy maintenance without the removal of other components.

Camshaft: the camshaft is of a split type for each cylinder. Each piece of the camshaft has a fuel cam, an intake cam, and an exhaust cam. Extreme miller timing for the intake cam has been applied for the reduction of NOx emissions.

Piston and piston rings: the piston is of composite type with steel crown and nodular cast-iron skirt. Cooling is achieved by the engine's lubricating oil. There are two compression rings and one spring-loaded oil scraper ring. The piston ring grooves are hardened and the first compression ring is coated for excellent wear resistance and for low fuel consumption.

Cylinder liner: the cylinder liner is made by gravity cast of special cast iron for good wear resistance and high strength and fitted in a bore in the engine block. The liner is fixed by the cylinder head and seated on the water jacket. The wall thickness of top part is optimized for both the resistance from the combustion force and for effective cooling. The lower part of the liner is uncooled in order to maintain a suitable temperature for preventing cold corrosion. There is no water in the engine block. To reduce bore polishing and lubricating oil consumption, a flame ring is fitted on the top of the cylinder liner.

Cylinder head: the cylinder head has been designed taking the thermal load distributions and the maximization of flows for charge air and exhaust gas into considerations. It is constructed of ductile cast iron and has cast-in passages for cooling water, intake air, and exhaust gas. It also has drilled holes for supplying lube oil to valve drives from the engine block.

Charge air and exhaust: a high compression ratio turbocharger. The MPC (modular pulse converter) exhaust system allows better performance within the limited space of the Vee-type engine. A compact two-stage charge air cooler with a large cooling surface and the charge air receiver is designed to ensure minimum pressure variation.

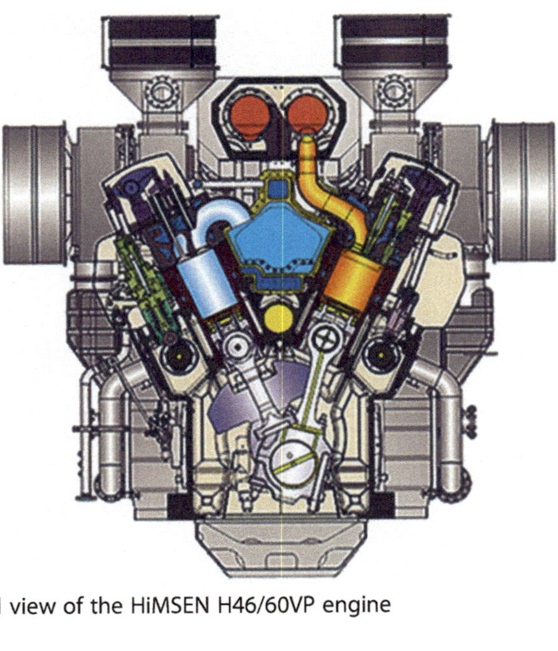

Cross-sectional view of the HiMSEN H46/60VP engine

Mirrlees Blackstone

An impressive medium-speed engine-building tradition was established by Mirrlees Blackstone of the United Kingdom which traced its roots back to the first British licence arranged with Dr. Diesel at the end of the 19th century. The Stockport-based company was acquired by the MAN B&W Diesel (now MAN Energy Solutions) group from Alstom Engines in June 2000, along with Paxman and Ruston.

Mirrlees Blackstone's final programme—now phased out—was headed by the 800kW/cylinder MB430L engine, a 430mm bore/560mm stroke in-line cylinder derivative of the MB430 design which was launched in V45° cylinder form in 1985 with a shorter stroke of 480mm. The two types were, respectively, produced in six, eight, and nine in-line and V12-, V16-, and V18-cylinder versions to cover power demands up to 13,144kW, the in-line engines running at 514rev/min, and the V engines at 600rev/min.

The MB430 (Fig. 25.11) was introduced with a rating of 665kW/cylinder (later increased to 730kW) and the following key component specification.

Crankshaft: underslung, thin web, and high overlap.

Cylinder liners: centrifugally cast grey iron of heavy section with intensively cooled flanges.

Cylinder heads: individual, with six holding-down bolts, housing caged exhaust valves, and air inlet valves seated on pressed-in seat inserts; valve operation is by Y lever-type gear.

Connecting rods: three-piece construction with a palm-ended shank secured to the large end block with four alloy steel studs, and the halves of the large end block secured together by alloy steel bolts. Large end bearing shells are of steel-backed lead–bronze design. Oilways are provided in the connecting rod to facilitate the lubrication of the piston pin bearings and deliver a copious flow of cooling oil to the piston.

Pistons: two-piece construction with heat-resistant alloy steel crown and aluminium alloy skirt.

Camshafts: driven by a crankshaft gearwheel via compound intermediate gears; all meshing points are lubricated and cooled by oil sprays; separate cylinder casings house the camshafts and high-level fuel pumps.

Engine-driven lubricating oil and water pumps are mounted at the free end of the engine; air manifolds are mounted on the outside of the vee; and turbochargers and intercoolers can be located at either end of the engine.

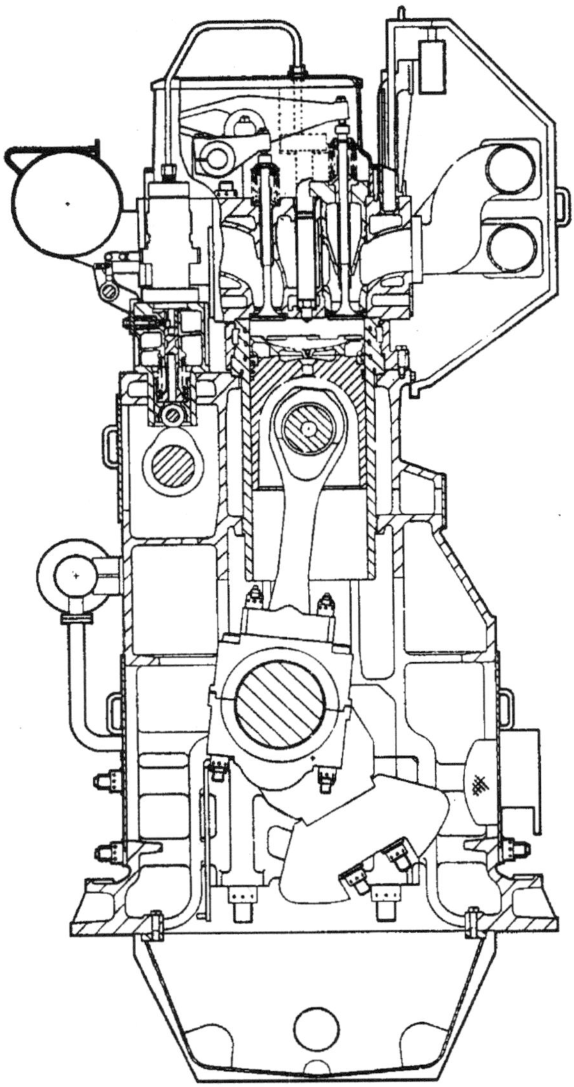

Fig. 25.11 Mirrlees Blackstone MB430 design.

The MB430 retained many basic features of Mirrlees Blackstone's long-established 400 mm bore/457 mm stroke K Major series which remained in the programme in Mark 3 form, delivering an output of 545 kW/cylinder at 600 rev/min. The design matured from the K-engine of the 1950s, one of the first large medium-speed engines in the world, which in its previous Mark 2 form featured a 381 mm bore (Fig. 25.12).

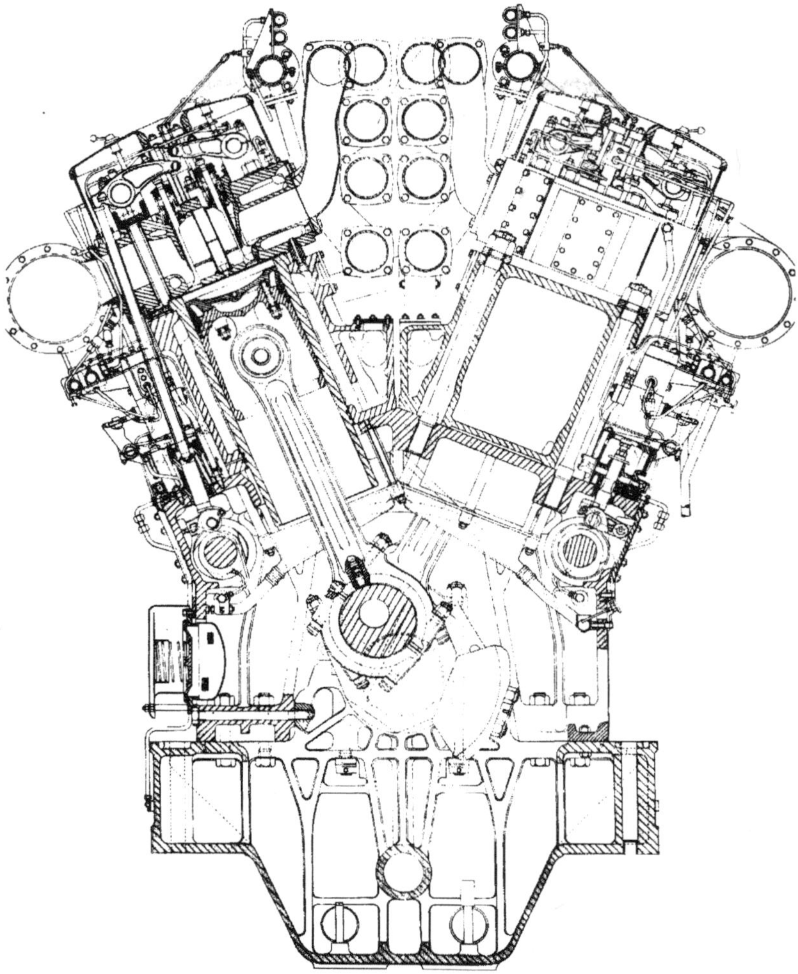

Fig. 25.12 Cross section of Mirrlees Blackstone KV Major Mk 2 engine.

The MB430 design also exploited some concepts introduced for the MB275 series in 1979. This 275 mm bore/305 mm stroke engine was available in six and eight in-line and V12- and 16-cylinder versions covering a power range up to 5070 kW at 1000 rev/min.

Lower power requirements were served by Mirrlees Blackstone's 222 mm bore/292 mm stroke ESL series whose upper output limit was doubled in 1986 by V12- and V16-cylinder models, the latter developing 2985 kW at 1000 rev/min. The programme also embraced five, six, eight, and nine in-line cylinder models.

The V-versions tapped the pedigree of the former E range but incorporated several interesting features, including an underslung crankshaft and a belt drive for the lubricating oil and water pumps. The V-models have a 45° cylinder configuration, deep main bearing caps, side bolts for added structural stiffness, and a central air intake. A simplified rocker arrangement operates four valves (two inlets and two exhaust) through two levers and bridge pieces enclosed within an aluminium cover. The valve gear is actuated by side-loaded camshafts driven from the flywheel end through nitride-hardened gears. The three-piece construction of the connecting rod allows piston withdrawal without disturbing the large end bearing.

A pulley and flat belt arrangement for the service pumps, mounted on the engine front, represented a departure from the traditional gear drive used by most engine builders. Its adoption was considered beneficial in lower production costs, pump speed flexibility, easier installation and maintenance, and weight and power savings. The belt comprises a core of endless polyester cords helically wound with a chrome leather friction face covering. A minimum belt life of 10,000 h was anticipated.

Mitsui

A major licensee of MAN Diesel in the production of MC/ME low-speed engines, Mitsui also once offered its own 420 mm bore/450 mm stroke medium-speed design whose last version was designated the L/V42MA series. The launch of the Japanese group's earlier venture in the four-stroke sector, the 600 mm bore 60M design with a rating of 1120 kW/cylinder, unfortunately, coincided with the 1973 oil crisis which undermined its market.

The smaller bore L/V42M design, however, benefited from that ill-fated project and became available in in-line and V-cylinder form with an output per cylinder of 558 kW at 530 rev/min. The frame is in two parts, comprising a welded steel crankcase structure made from cast sections and plate and a cast-iron cylinder housing held together with tie-bolts. The underslung crankshaft bearings are secured by vertical and lateral bolts, all hydraulically tightened. The designers sought a stress-free structure, controlling the lateral fit between bearing keep and frame, for example, by tapered adjusting liners before tightening to avoid stressing the frame unduly.

For the crankshaft bearings, Mitsui was able to keep within the stress capacity of thin white metal (on a lead–bronze backing in the case of the crankpins) with an overlay. While this is a function of the control of

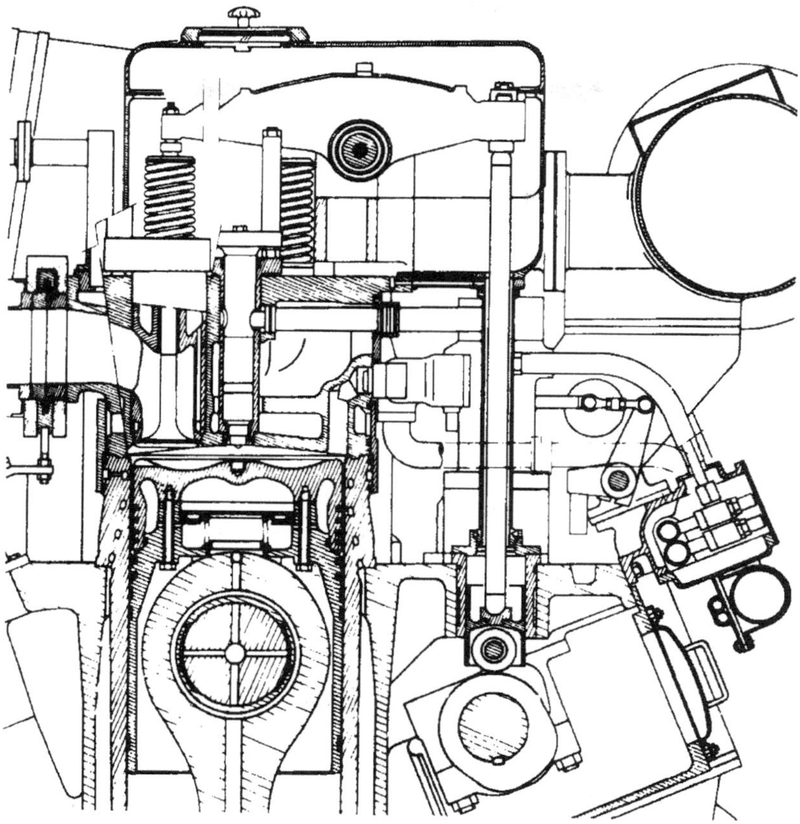

Fig. 25.13 Cross section through cylinder head and piston of Mitsui 42M engine.

maximum bearing pressures, it also means that damage to a bearing does not necessarily lead to shaft damage.

The combustion area and cylinder head configuration can be appreciated in Fig. 25.13. Note the composite piston, bore-cooled cylinder liner, three-deck head, caged exhaust valves, and rigid high-pressure fuel line, as well as the camshaft bearing arrangement. In addition to the four compression rings in the piston crown (the second is Ferrox-filled to improve running-in), there are two conformable SOC rings at the top of the cast iron piston skirt and two broad lead–bronze bands above the gudgeon and one below; these also assist during the running-in phase.

A three-part marine-type connecting rod was selected with serrations at the big end joint. The exhaust valve cage can be removed easily by taking out the pushrod and lifting clear the rocker. Rotators are applied on the exhaust valves, and guide vanes are formed in the inlet passages of the head to foster swirl in the charge air. A mist catcher is installed after the inter-cooler to reduce the water loading of the charge air in humid ambient conditions.

The fuel injector sleeve is provided with small water passages at the seat. These passages communicate with the cylinder head water passages and foster a convection flow which gives adequate cooling to the nozzle for operation on all fuels without the complication of water cooling the nozzles.

Subsequent refinements led to the Mitsui L/V42MA design with an output raised to 625 kW/cylinder at 600 rev/min, allowing 6L- to V18-cylinder models to cover a power band up to 11,250 kW.

Niigata

A wide range of own-design high- and medium-speed engines from Niigata Engineering of Japan was supplemented by licences to build SEMT-Pielstick PA and PC series and MAN B&W Holeby 23/30 and 28/32 series engines. Niigata's own MG (Marine Geared) medium-speed programme embraces designs with 190, 220, 250, 260, 280, 320, 340, 400, 410, and 460 mm bore sizes; the high-performance FX series includes 165 and 260 mm bore versions (*see Chapter 26 for Niigata high-speed engines*).

Its largest medium-speed engine, the MG46HX, is a 460 mm bore/600 mm stroke design with an output per cylinder of 823 kW at 450 rev/min and the capability to operate on 700 cSt fuel. Power demands up to 13,180 kW are met by 6/8L and V12/16-cylinder models, the targeted propulsion plant applications including medium-to-large ferries, container ships, and RoRo freight vessels. The crankcase is built as a box structure to secure rigidity in handling a maximum combustion pressure of 180 bar; engine noise and vibration levels are also claimed to be low. Easier maintenance was addressed by simplified component designs and hydraulic tools.

Niigata also developed the HLX series, which embraces 220, 280, and 340 mm bore sizes covering a power band from 1392 kW to just under 10,000 kW from in-line and V-cylinder models.

Nohab

The Nohab engine business was acquired by Wärtsilä, which focused on the Swedish designer's Nohab 25 series before phasing out the programme. This 250 mm bore/300 mm stroke design was offered in six-in-line and V12- and 16-cylinder versions covering output demands up to 3680 kW at 720/750/825/900 rev/min and 1000 rev/min for auxiliary and propulsion plant markets. The engine proved particularly popular in the offshore power arena. A cross section of the 6R25 model is shown in Fig. 25.14.

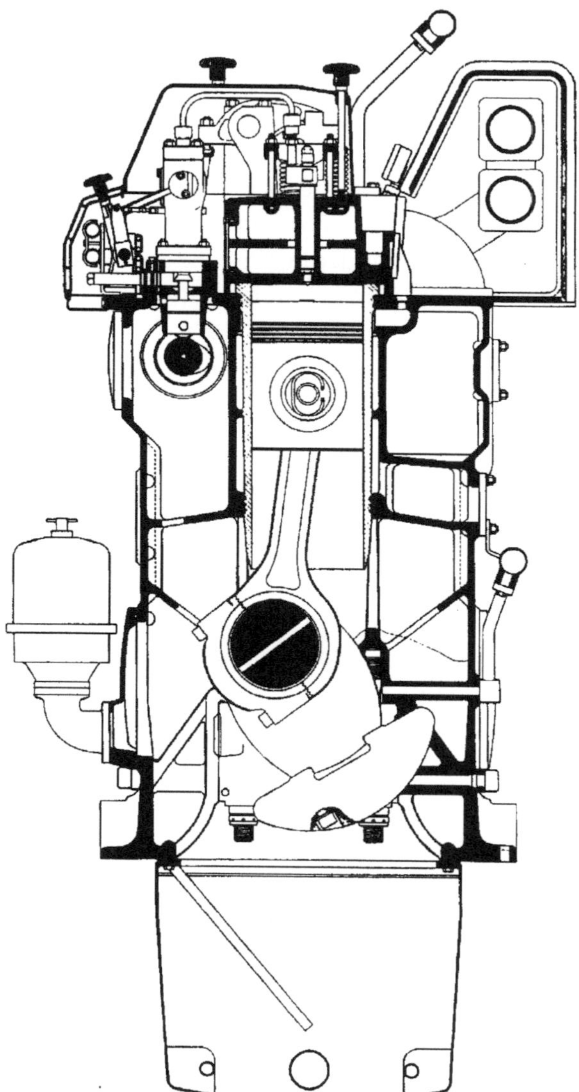

Fig. 25.14 Wärtsilä Nohab R25 design.

SEMT-Pielstick

The Paris-based designer SEMT-Pielstick fielded three long-established high-speed programmes, the PA4, PA5, and PA6 series, before and after its acquisition first by MTU and MAN B&W Diesel and then wholly by MAN Diesel. Some engines are still produced for special land-based and submarine applications but not for mainstream marine markets.

The PA4 series, produced in 185 and 200 mm bore versions (both having a stroke of 210 mm), featured a variable geometry (VG) precombustion chamber; while the PA5 is a 255 mm bore/270 mm stroke design.

Development benefited the 280 mm bore PA6 design, introduced in 1971 with a stroke of 290 mm and a maximum continuous rating of 258 kW/cylinder at 1000 rev/min. The output was raised to 295 kW/cylinder in 1974 and to 315 kW/cylinder in 1980. Extensive service experience in naval vessels moulded progressive refinements over the years, including the creation of a BTC version in 1980 which yielded 405 kW/cylinder at 1050 rev/min through the adoption of a reduced compression ratio in conjunction with two-stage turbocharging. This PA6 BTC model was released in 1985 with a higher output of 445 kW/cylinder. The series was extended in 1983 by a longer stroke (350 mm), slower speed PA6 CL variant with a rating of 295 kW/cylinder at 750 rev/min.

A sequential turbocharging system (STC) was introduced to the 290 mm stroke engine in 1989, this PA6 STC model developing 324 kW/cylinder at 1050 rev/min. The performance was enhanced in a 330 mm stroke B-version from 1994, an STC system and a cylinder head with improved air and gas flow rates contributing to a nominal maximum continuous rating of 405 kW/cylinder at 1050 rev/min. A maximum sprint rating of 445 kW/cylinder at 1084 rev/min equated to an output of 8910 kW from the V20-cylinder PA6B STC engine which targeted high-performance vessels such as fast ferries.

At the nominal MCR level, the V12-, 16-, and 20-cylinder models, respectively, offered 4860, 6480, and 8100 kW for fast ferry propulsion, being released for sustained operation at these ratings. The weight/power ratio of the 20PA6B STC model at 8100 kW is 5.2 kg/kW.

PA6B STC design

High rigidity without compromising overall engine weight was achieved by a stiff one-piece crankcase of nodular cast iron specially treated for shock resistance, with transverse bolt connections between both crankcase sides through the underslung-type bearing caps. Integrated longitudinal steel piping supplies lubricating oil to the main bearings.

Large dimension main journals yield a particularly large bearing surface and conservative pressures which underwrite prolonged bearing life, as do the large surface area connecting rod bearing shells. The alloy steel one-piece forged crankshaft has high-frequency hardened crankpins and journals and is bored to feed lubricating oil to the connecting rods.

A nodular cast-iron cylinder head was configured to achieve improved air and gas flow rates for the PA6B STC version. The composite piston (steel crown and aluminium skirt) is fitted with five rings. The forged steel connecting rod is fitted with a large surface bearing shells to yield high durability under high firing pressures. Special low-wear characteristics were sought from the centrifuged cast-iron material specified for the cylinder liner. The camshaft is formed from several sections for ease of dismantling.

Operational flexibility and power output are fostered by the single-stage sequential turbocharging system which is based on two turbochargers and delivers a large combustion air excess at partial loads. Supercharging is affected by one turbocharger for engine loads up to 50% of the nominal power, and its effort boosted by the second identical turbocharger at higher loads. Switching from single to twin turbocharger mode is performed automatically by opening two flap valves. Engine performance at prolonged low load is thereby improved with respect to fuel consumption, smoke emissions, fouling resistance, and transient performance. In addition, engine utilization is expanded towards the high torque/low revolutions per minute area.

Developing the high-performance derivative

Refining an established engine for high-speed vessel propulsion duty must focus on achieving a lighter and more powerful package within a more compact envelope, while also enhancing operating flexibility. The resulting PA6B STC design was released with a nominal maximum continuous rating of 405 kW/cylinder at 1050 rev/min and a sprint rating of 445 kW/cylinder at 1084 rev/min.

SEMT-Pielstick's development goals aimed to:

- Increase the power rating by 25%.
- Use an operating field allowing operation following a double propeller law.
- Maintain, and if possible reduce, the specific fuel consumption.

- Reduce the weight/power ratio and enhance engine compactness.
- Reduce installation costs.
- Retain the maximum number of existing components.

Power can be raised in two ways: by increasing the brake mean effective pressure and by increasing the piston speed. Increasing piston speed can be affected by increasing the revolutions per minute and by lengthening the stroke. Increasing the revolutions per minute was not attractive because of a need to maintain an acceptable synchronized speed for land power plant applications but also because of these drawbacks: increased specific fuel consumption; higher wear rate; and higher noise levels.

Raising piston speed by lengthening the stroke is considered preferable because it reduces specific fuel consumption and fosters optimum performance at starting and low load (the compression ratio is secured without resorting to a combustion chamber that is too flat). This route was therefore selected but with the piston speed limited to 11.5 m/s (representing an increase of 14% with a stroke of 330 mm). To reach the required power rating, the brake mean effective pressure had to be increased by 10%; this was achieved by adopting high-performance turbochargers. Attaining the desired specific fuel consumption called for the maximum cylinder pressure to be raised to 160 bar.

Coping with these new parameters dictated redesigning some key components. The existing cylinder head, for example, was incompatible with the targeted specific fuel consumption level. This was partly because its mechanical strength was insufficient in relation to the necessary peak combustion pressure, and partly because the pressure loss through the inlet and exhaust ports would be unacceptable in relation to the intervening gas flow dictated by the increased power. A new cylinder head was therefore designed with a reinforced bore-cooled fire plate and incorporating inlet and exhaust ports with large dimensions. In addition, cooled seats were specified for the inlet valves as well as the exhaust valves to maximize seat reliability and avoid risks of burning.

A redesigned connecting rod was also necessary to address the increased peak compression pressures and the inertia efforts linked to the lengthening of the stroke. The bevel-cut design of the original component was abandoned in favour of a straight-cut rod to avoid weak points, such as the bevel-cut serrations and the threading anchorage in the shank. Piston cooling by jet replaced the traditional oil supply through the connecting rod from the crankpin, simplifying machining and allowing the bearing shell grooves to be eliminated and yield these benefits: the increasing capacity of the bearing shells; stopping cam wear of the crankpin (differential wearing of the pins between the side areas of the plain bearing shell and the central area; that is, the area including the groove); and cutting out the risk of cavitation erosion on the bearing shells at the end of the grooves. The new connecting rod is 10% lighter than its bevel-cut forerunner. The weight reduction, along with an improved bearing shell capacity, partly compensates for the increased inertia efforts.

Temperature measurements on the jet-cooled piston head indicated similar levels to those of the original piston, and even lower in some areas. Critical points of the piston were modified to cope with the higher peak combustion pressure: a spherical shape was given to the support spot face used for the piston head/skirt tightening spacers, and the radius under the skirt vault was increased.

Crankshaft dimensions were modified to target the same reliability from the component as before, despite the increased stroke and peak combustion pressure. Finite element analyses of the crank webs and hydrodynamic calculations led to an increase in the journal diameter from 230 to 250 mm, and in the crankpin diameter from 210 to 230 mm.

High-performance turbochargers were necessary to secure the brake mean effective pressure increase with the required efficiency, a model from MAN B&W's then-new NA series—the NA 34S—being selected to meet the performance and compactness parameters. Sequential turbocharging was applied, based on the principle of reducing the number of turbochargers in operation as the engine speed and load fall. The speed of the turbochargers still operating consequently rises and significantly larger quantities of air are thus delivered to the engine.

A simple system was adopted using only two turbochargers, one being switched off at below approximately 50% of the nominal engine power rating. This was affected by closing two flap valves located at the compressor outlet and at the turbine inlet of one of the turbochargers. The designer cited the following benefits from the PA6 engine's sequential turbocharging (STC) system:

- High torque and power ability at reduced engine speed.
- A gain in fuel economy at low and part loads.
- Capability to run the engine at very low loads for extended periods with minimal fouling (the light deposits can be cleaned out by running for half an hour at 50% load).
- Invisible smoke emissions over a wide working range.
- Reduced exhaust temperature.

- Lower thermal stresses in the combustion chamber components at part loads.

A higher output rating naturally reduced the engine's weight/power ratio but other measures were pursued to trim overall weight. The scope for using aluminium alloy was explored for all components where operating stresses (particularly thermal) and class rules allowed, leading to the engine supports, turbocharger support, air manifolds (after the air cooler), lube oil and water cooler support, and lube oil filter support being designed in cast aluminium. Studies assessed other components for which aluminium could not be considered, either to use alternative materials with higher mechanical properties and so reduce thickness or simply to optimize existing shapes and thicknesses. As an example, specifying high yield point steel sheet for the manifolds connecting the turbocharger to the air cooler allowed a reduction in thickness from 10 to 4 mm. The lube oil sump plate was also modified by reducing the material thickness.

A 10% reduction in the original engine weight, along with the increased power output, contributed to a weight/power ratio of 4.8 kg/kW, including all ancillaries. In parallel with the weight trimming studies, SEMT-Pielstick focused on reducing the overall dimensions of both engine and ancillaries.

A key element in this regard was the combi-cooler, integrating one lube oil and one freshwater plate cooler circulated by a common seawater system. The combi-cooler's support is used as the rear plate of the cooler and includes an as-cast part of the connections to the HT freshwater and lube oil systems. Its front plate incorporates as-cast the connections to the LT freshwater loop and the water thermostatic valve. The main self-cleaning lube oil filter is incorporated axially in cast support located under the combi-cooler and includes the lube oil thermostatic valve as well as the centrifugal oil filters.

Such solutions fostered compactness, a simple pipeless configuration, and good access to the main subjects of maintenance. The integration of the ancillaries on the engine further eases shipboard installation procedures. All pumps (water, oil, fuel make-up), as before, are driven by the engine upon which the fuel filter is also mounted.

A potential for burning an intermediate fuel oil grade such as IF30 was addressed in the development programme and reflected in the specification of cooled valve seats and exhaust valve rotators. The 75°C temperature necessary to reach an adequate viscosity for its injection could be derived by taking heat from the engine's HT freshwater system. In such installations, however, the engine was derated to 360 kW/cylinder and the time-between-overhauls reduced.

SKL

Four-stroke engines in a number of bore sizes—160, 200, 240, and 320 mm—were produced by SKL whose pedigree dates back to 1905 when the Magdeburg, German-based company's forerunner (Lokomobilenfabrik R. Wolf), took out a licence to build diesel engines. SKL Motor was taken over in 2008 by MTU Friedrichshafen, which has invested in a remanufacturing centre for its own high-speed engines.

Several of SKL Motor's medium-speed engines are still in service including the last produced VDS 29/24 AL series for propulsion and genset duties which was offered in six, eight, and nine in-line cylinder versions to cover an output range up to 2350 kW at 750/900 rev/min and 1000 rev/min. Its brake mean effective pressure rating is 23.9 bar (Fig. 25.15).

Stork-Werkspoor diesel

Formed in 1954 by the merger of Stork and Werkspoor, along with other interests in the Dutch engine-building industry, Stork-Werkspoor Diesel (SWD) was acquired in 1989 by the Finland-based multinational Wärtsilä Diesel group, now the Wärtsilä Corporation. SWD's own-design medium-speed engines were gradually phased out of the production programme and replaced by the Wärtsilä 26 and 38 series largely designed by SWD staff and subsequently manufactured in Zwolle, The Netherlands. These models were later assigned for production by Wärtsilä Italia in Trieste (Figs 25.16–25.18).

Yanmar

Like Daihatsu above, Yanmar Diesel is a prolific Japanese producer of high-speed and smaller medium-speed engines for genset and propulsion drives. Its medium-speed propulsion engine programme has included the S185, M200, M220, T240, T260, N260, Z280, N280, and N330 series, the designation signifying the bore size in millimetres. Most are built in six-cylinder in-line form, with the larger bore models also offering an eight-cylinder version. The 8N330-EN engine delivers 3310 kW at 620 rev/min.

Yanmar added to an extensive genset engine portfolio in 2006 with the six-cylinder EY18 model, which covers power demands from 360 to

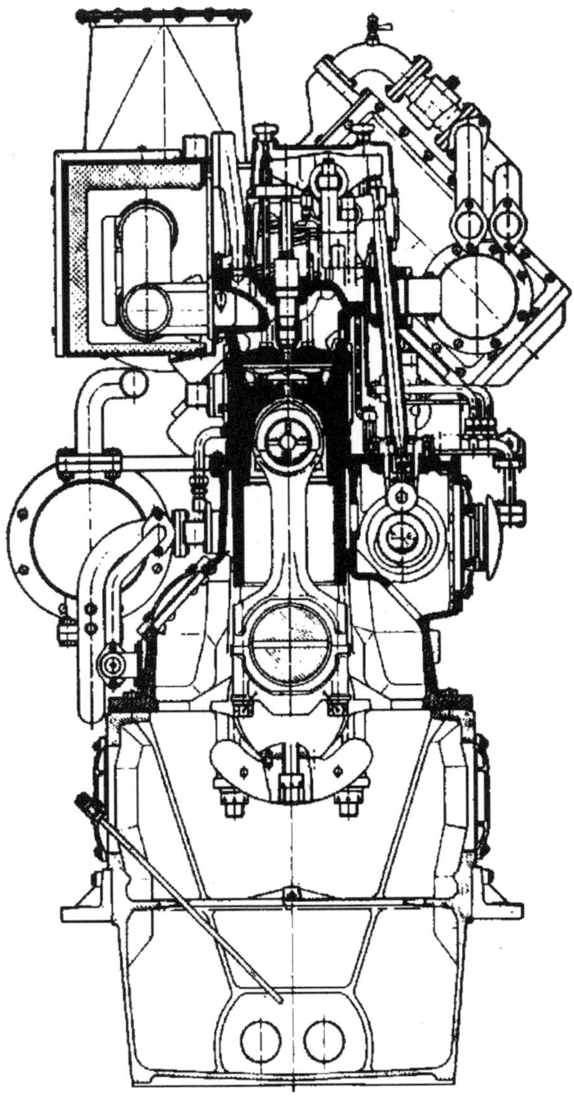

Fig. 25.15 SKL's VDS 29/24 AL design.

750kWe with 720 and 900rev/min variants. This model complemented the 1840kW output 6EY26 propulsion engine launched 2years previously.

The EY18's compact 180mm bore heavy fuel-burning design features low-sac fuel valves, variable jacket water temperature control, and a valve stem seal with a back-pressure lip. Fuel economy is enhanced and NOx emissions reduced by a staggered injection hole nozzle and deep combustion chamber, and the latest generation turbocharger. The designer's ASSIGN combustion

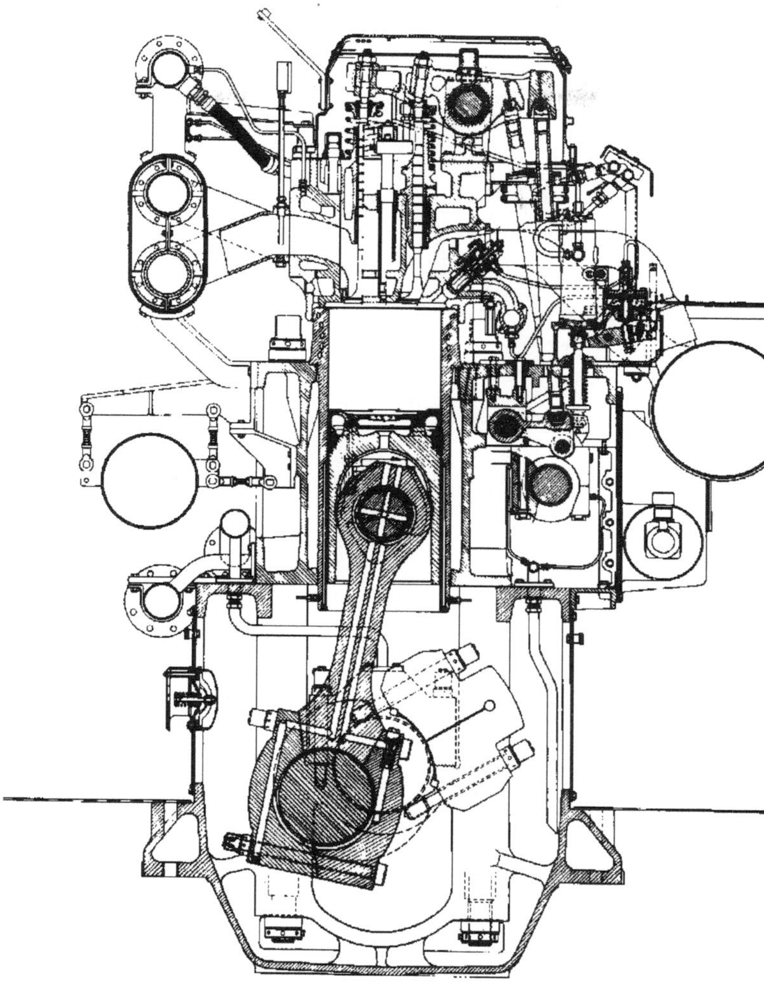

Fig. 25.16 Cross section of in-line SWD TM410 engine.

technology and ECODiesel SAVETEN measures are said to ensure a balance between fuel consumption and emissions of NOx and smoke.

Development goals pursued by Yanmar to benefit the designs have included enhanced compatibility with heavy fuels and strengthened but lighter cylinder blocks, bedplates, and cylinder heads to lower overall engine weight. A compact and lightweight reverse-reduction gear and compact arrangement of turbocharger and air cooler contribute to reduced overall engine length and height.

Measures to ease engine installation and servicing are also cited. Large inspection windows on both sides of the cylinder block facilitate dismantling

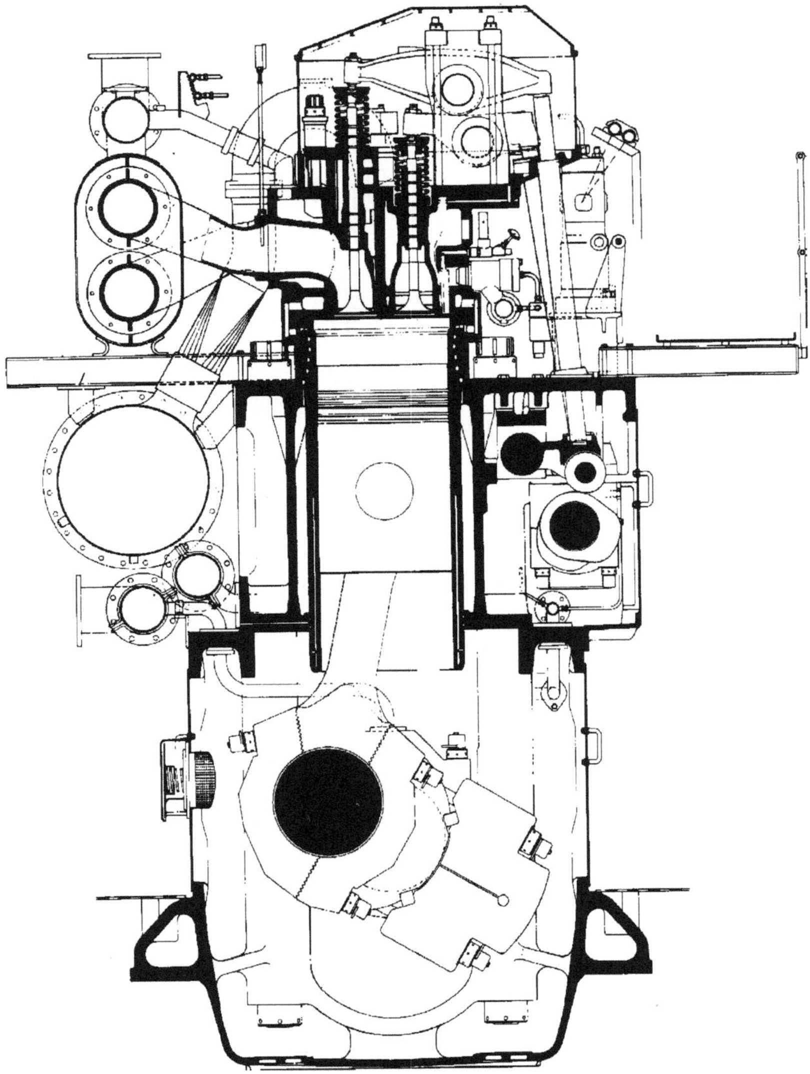

Fig. 25.17 Cross section of in-line SWD TM620 engine.

and re-assembly of the main bearing and connecting rod. All pumps and filters are arranged above the engine installation level to simplify access. The full power take-off is available from the front of the engines.

Since the introduction of the EY18, Yanmar has added to this range with two new bore sizes (220 and 330 mm). Each bore size is now represented by a six-cylinder in-line engine and the 260 and 330 mm bore sizes also feature an eight-cylinder model. In order to meet IMO NOx Tier II rules, Miller timing is employed.

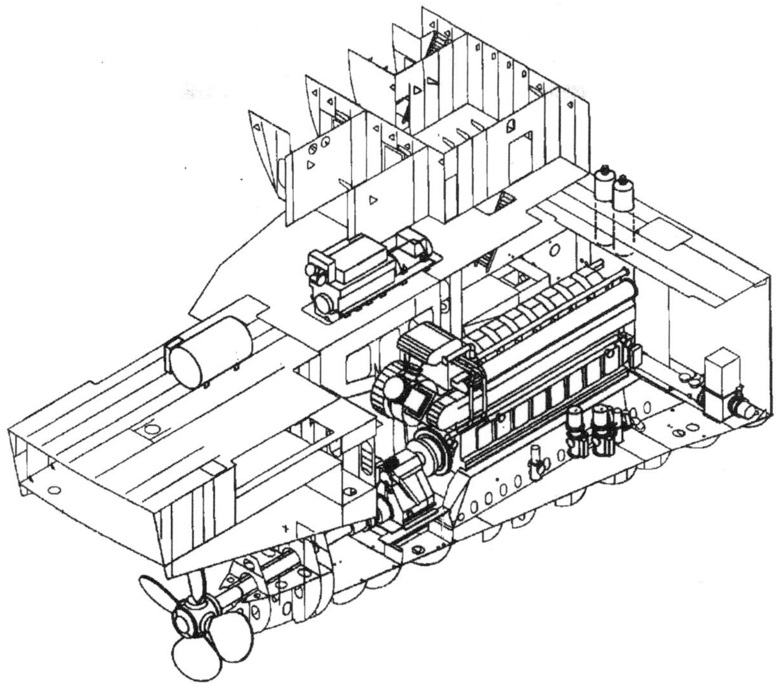

Fig. 25.18 An eight-cylinder Stork-Werkspoor Diesel TM620 engine installed to power a container ship.

The 6EY22AW was launched in 2010 and has a stroke of 320mm and can be configured for five power outputs all at 900rpm but with mean effective pressure ratings between 1.62 and 2.50MPa. The power outputs span 885–1370kW.

Two years later in 2012, the 6EY18 was replaced by the 6EY17 which with an engine speed of 1450rpm could be considered as a high-speed engine.

For the next larger bore size (and the oldest engine in the series) the 6EY26W the stroke is 385mm and only three power outputs are offered with the engine running at 750rpm. The highest rating at 2.50MPa is 1920kW. There is also an eight-cylinder model in this bore size, the 8EY26W. With a choice of four power outputs, this engine also runs at 750rpm and can output 2560kW. In 2014, a gas-burning version of this engine was released and in 2016 a dual-fuel model followed.

In 2015, the 6EY33 was also released and later the 8EY33. This bore size has a stroke of 440mm and between them, the two engines span a power range of 2500–4500kW. The EY33 type was first installed in a ship in 2018 and there are further vessels under construction which will feature one or other of the variants available.

Two-stroke medium-speed engines

The medium-speed engine market has long been dominated by four-stroke, uniflow-scavenged designs but at one time a number of two-stroke designs enjoyed popularity, notably the Polar loop-scavenged type—which ceased production when Nohab introduced its four-stroke F20 range—and Sulzer's uniflow-scavenged ZH40.

Wichmann's commitment to the two-stroke loop-scavenged trunk piston concept was renewed in 1984 with the launch of its 295 kW/cylinder WX28 engine (Fig. 25.19). The simple 'valveless' approach had been proven in service by the Norwegian company's earlier AXE, AXG, and AXAG designs. The 280 mm bore/360 mm stroke WX28 covered an output band from 1180 to 4735 kW at 600 rev/min with four, five, and six in-line and V8-, 10-, 12-, and 16-cylinder models.

Development focused on low fuel and maintenance costs with high reliability. The engine was also claimed to be one of the lightest and most compact in its power class. The ability to operate on heavy fuel (180 cSt) under all conditions was another goal. Specific fuel consumption of 188 g/kW h resulted from enhanced scavenging and fuel injection systems, and a maximum combustion pressure of 140 bar is underwritten by rugged construction. The mean effective pressure is 13.5 bar.

The valveless cylinder cover is of simple construction, the lack of ducts for hot exhaust gas promoting uniform temperature distribution and low stress. Fastened by eight hydraulically tightened nuts, the cover can be removed in a few minutes and the piston withdrawn in 10 min. The connecting rod can be disconnected while leaving the big end bearing on the crankshaft; this feature reduces the necessary removal height.

Wichmann stressed the overall simplicity of the engine and the impact on reliability and serviceability, citing fewer moving parts, and hence fewer wearing parts. Separate cylinder lubrication—a standard feature—permits matching of the lubricating oil total base number to the fuel sulphur content. The oil is distributed over the cylinder surface by a hydraulic lubricator via four bores and quills.

Wichmann engines—including the earlier 300 mm bore/450 mm stroke AXAG design—found particular favour in the Norwegian fishing and offshore vessel propulsion sectors. A/S Wichmann became part of the Finland-based Wärtsilä Diesel group in 1986 and changed its name from Wärtsilä Wichmann Diesel in January 1994 to Wärtsilä Propulsion A/S (now Wärtsilä Propulsion Norway A/S).

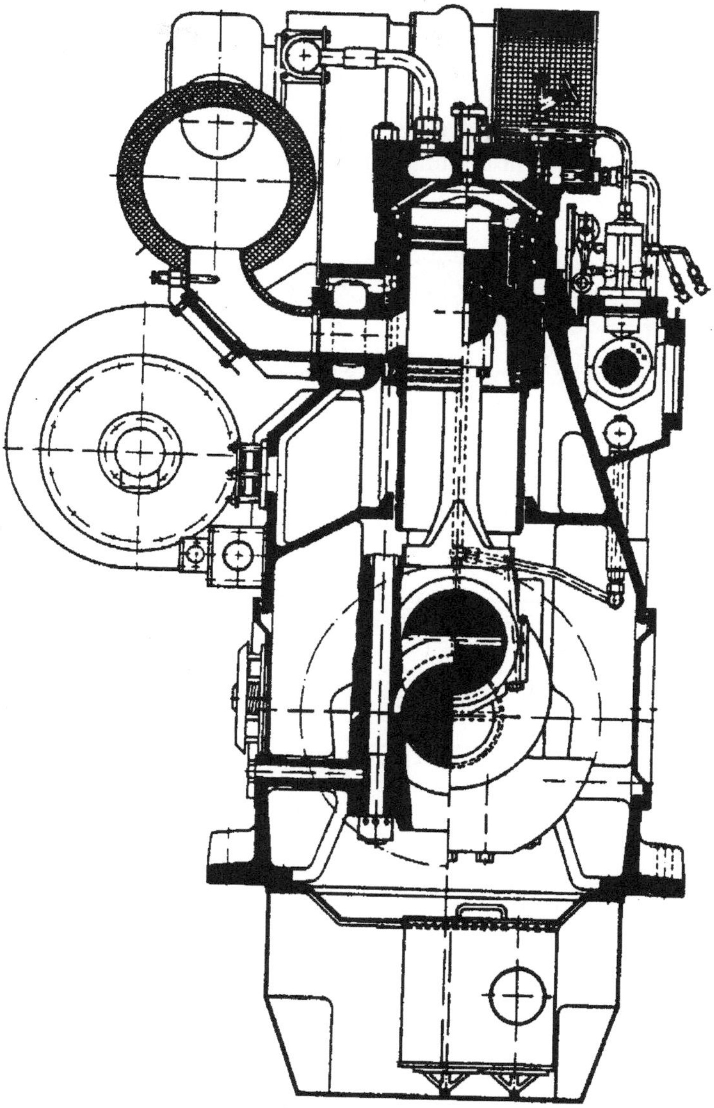

Fig. 25.19 Wichmann WX28L design.

The Wichmann 28 engine, which remained in production until 1997, was released with the following specification.

Cylinder block: cast-iron monobloc design with the integrated crankcase, scavenging air receiver, water manifold, and camshaft box; underslung type of crankshaft support.

Crankshaft: fully forged and machined in Cr—Mo steel; dimensionally laid out for 50% power growth potential.

Cylinder liner (Fig. 25.20): wear-resistant cast-iron alloy; bore cooled with strong-backed top section; balanced cooling water flow for efficient temperature control; and separate cylinder lubrication through four quills.

Cylinder head: cast iron, valveless, simple design; bore cooled with strong backing to secure efficient cooling and low-stress level.

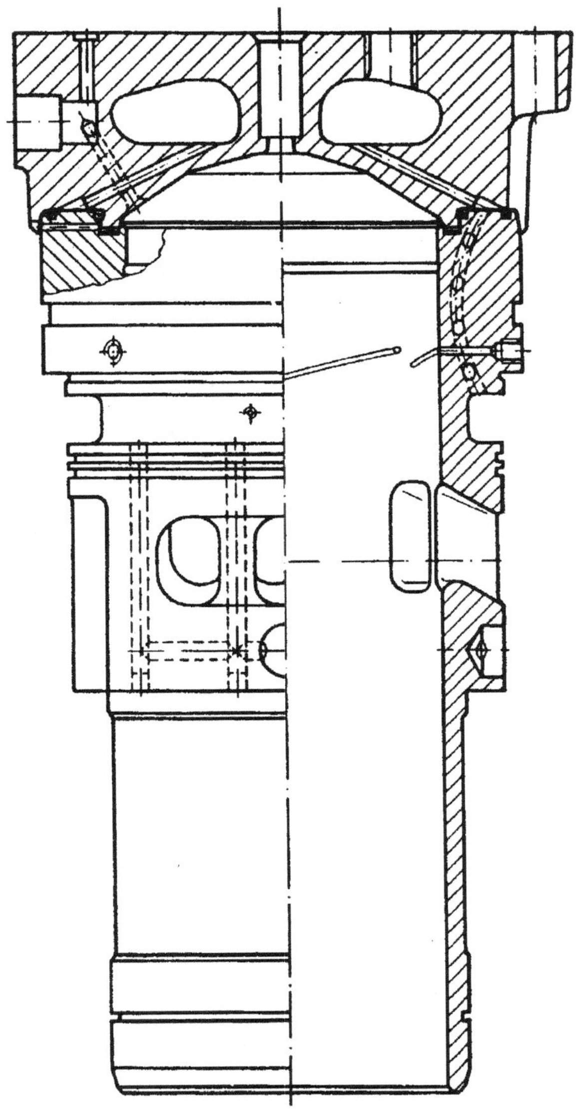

Fig. 25.20 Bore-cooled cylinder liner and cover of Wichmann WX28 engine; separate cylinder lubrication is standard.

Piston (Fig. 25.21): oil-cooled composite design with cast-iron skirt and steel crown; ring grooves hardened for low wear rate in heavy fuel operation; and integrated small end bearing in full gudgeon pin length.

Connecting rod: drop forged and fully machined; separate large-end bearing unit for easy piston withdrawal and low removal height.

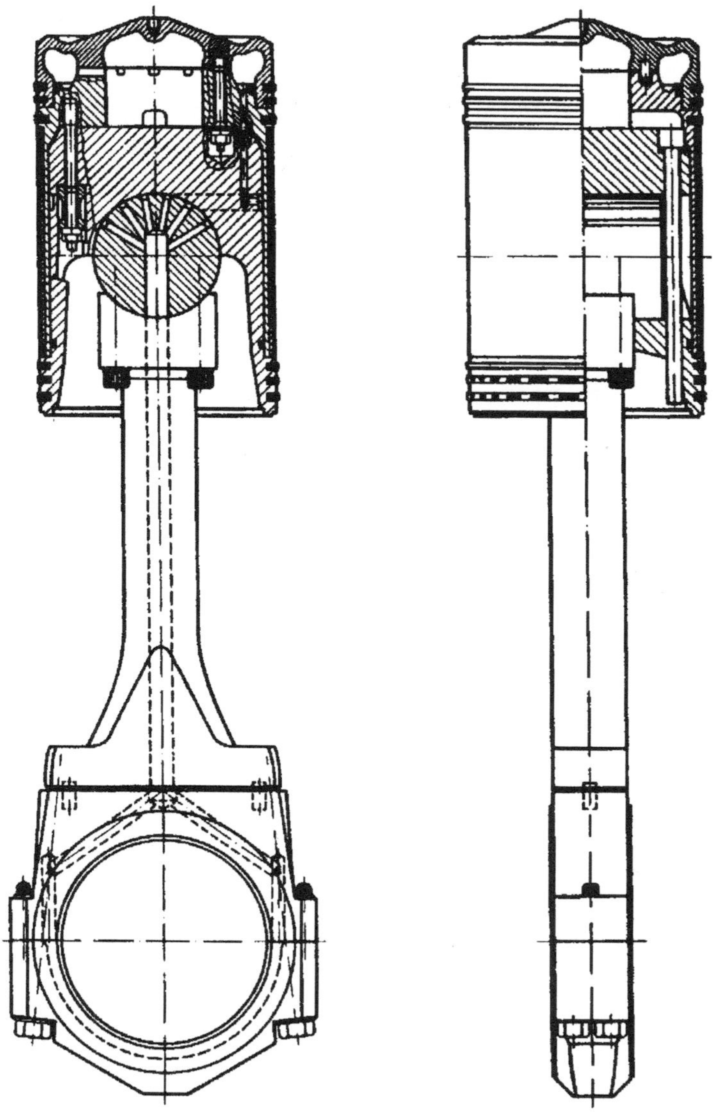

Fig. 25.21 Composite piston (high alloy steel crown, cast iron skirt, and light alloy gudgeon pin support) and connecting rod of Wichmann WX28 engine.

Bearings: three-metal steel-backed type, interchangeable with main and crank journal.

Turbocharging: constant pressure system with auxiliary blower in series; the moderate speed auxiliary blower boosts the turbocharger effort to ensure an adequate air supply under all load conditions; the blower is engine driven via low-pressure hydraulics using the engine lubricating oil and pump.

Fuel injection system: individual high-pressure monobloc pumps with built-in roller tappet; short high-pressure pipes and temperature-controlled nozzles for heavy fuel operation.

Auxiliary pumps: engine gear-driven units for lubricating oil, freshwater, and seawater.

Loyal to the two-stroke uniflow-scavenged principle for medium-speed trunk piston engines—exploiting air inlet ports in the cylinder liner and exhaust valves in the head—is the former General Motors' **Electro-Motive Division (EMD)**, now the only exponent of the concept still serving the market. The US designer argues better component wear life, reliability, and serviceability over four-stroke designs serving the same power range. Established in 1922, EMD was fully incorporated into General Motors in 1941 and sold to a US-based private equity partnership in 2005. The company claims to be the world's largest builder of diesel–electric locomotives, a rugged pedigree that has benefited marine and offshore propulsion and genset engine sectors.

EMD's current 710G series covers an output band from 1249 to 3730 kW at 720/750/800/900 rev/min from V8-, 12-, 16-, and 20-cylinder models. The 230.2 mm bore 3279.4 mm stroke design (Figs 25.22 and 25.23) was launched in 1986 as a longer stroke derivative of the established 645FB range. A more advanced turbocharger (yielding a 10% increase in the overall air–fuel ratio) and larger fuel pump plunger diameter contributed to the higher power rating and lower fuel consumption. Increasing the stroke raised the cylinder displacement by 10% from 645 to 710 $in.^3$ (hence the model nomenclature).

A number of improvements benefited the 710GB, GC, and GC-T2 models which offer outputs up to 187 kW/cylinder at 900 rev/min.

- An L-11 liner design for enhanced durability and performance, reduced scuffing, and higher wear resistance and better fuel economy.
- A Duracam camshaft, extending valve train component life and lowering valve vibration.
- A Diamond six-cylinder head with tangent flow fireface, yielding improved cooling and better valve sealing, eliminating core plugs and

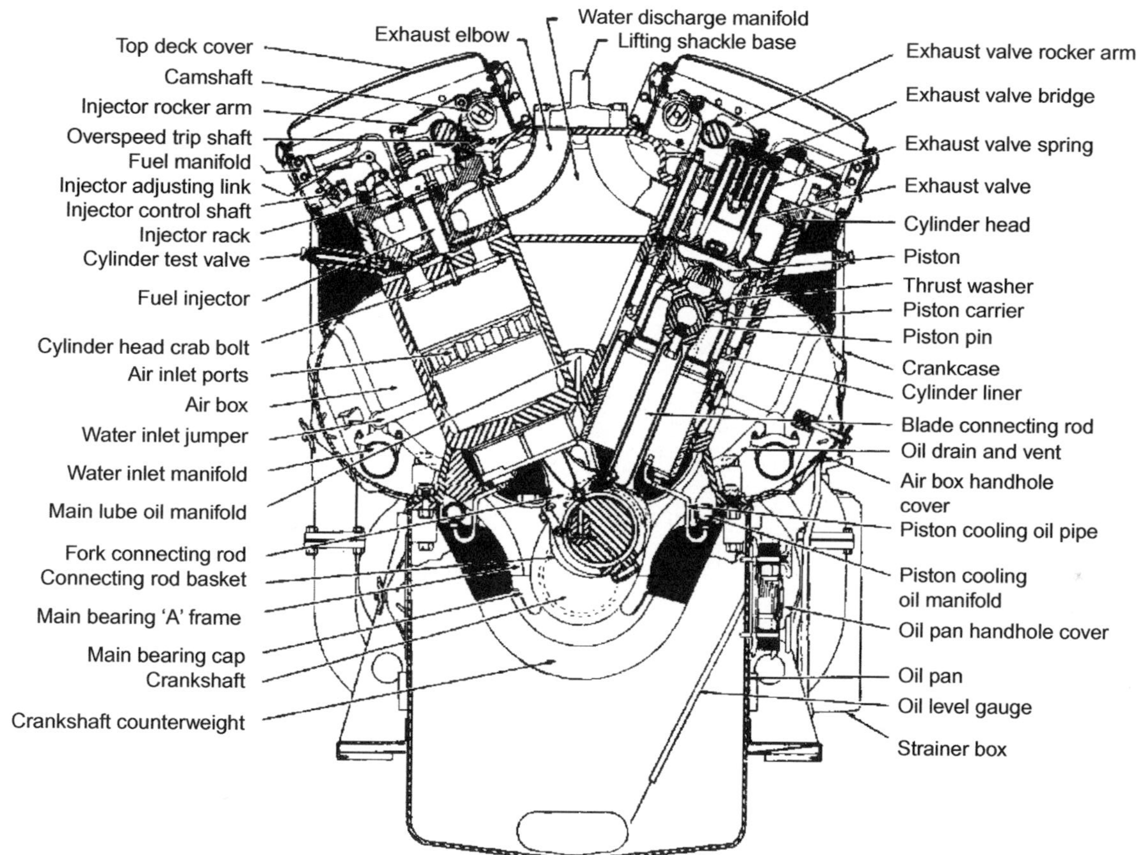

Fig. 25.22 General Motors EMD 710G two-stroke medium-speed engine with overhead camshafts and unit injectors.

Fig. 25.23 An EMD 710G two-stroke medium-speed engine in V20-cylinder form.

hence water leaks, and providing hardened valve guides for improved valve and valve guide life.

- Improved unit fuel injector, with new seal design, enhanced check valve, and stiffer follower spring.
- A four-pass aftercooler achieving improved thermal efficiency and hence fuel economy, and exhaust emissions reduction.
- A new turbocharger incorporating an external clutch for easier maintenance.
- Lower vibration levels from a new crankshaft and coupling disc-balancing technique.

With cylinders arranged in a V45° configuration, the two-stroke uniflow-scavenged 710G design features overhead camshafts actuating the four exhaust valves per cylinder and the unit fuel injectors. A unitized cylinder power assembly with jacketed liner and minimal bolted and gasketed water connections simplifies maintenance and strengthens structural integrity.

A higher displacement than the 645FB series called for changes to the crankcase, a larger diameter crankshaft, a new camshaft, and larger diameter plunger injectors. Longer piston rod assemblies and cylinder liners accommodated the extended stroke, but overall the 710G engine was only fractionally higher and longer than its predecessor. The 10% increase in overall air–fuel ratio was secured by the G turbocharger, significantly reducing thermal loading on critical components. A unit fuel injector with a larger (14.29 mm) plunger diameter than the 645FB engine (12.7 mm) delivered a higher injection rate and shorter injection period. In addition, a spray tip incorporating seven orifices yielded better combustion and higher thermal efficiency.

A piston crown design incorporating a toroidal-shaped combustion bowl further fostered higher combustion efficiency when combined with the established air swirl and well-atomized fuel spray. The larger cylinder displacement enabled a 10% larger combustion chamber, helping to minimize quenching effects and promote lower emissions.

The capacity of the camshaft/valve train system was more than doubled over the 645FB engine, allowing a reserve for future growth in injector size. The camshaft base circle was increased from 2.5 to 3.25 in. diameter; the cam roller follower diameter was increased by 0.375 in., and the injector and exhaust rocker arms were strengthened.

A larger and more efficient turbocharger achieved a 15% increase in airflow for reduced thermal loading. Transferring the over-running clutch from the turbocharger housing to the engine gear train improved accessibility of the clutch for inspection and replacement without disassembling the turbocharger.

An interesting feature of the 710G series is the single turbocharger with an externally mounted clutch which enables, in the lower load ranges, the turbocharger to be driven by the engine gear train in supercharger mode. Once the engine load reaches around 65%, the clutch disengages and allows the exhaust gas to drive the turbocharger turbine. Full load can be accepted swiftly and smoothly at any engine speed, says EMD, and a 10% overload capability can be exploited for two out of every 24 h of running.

The electronically controlled engines are said to offer precise control and facilitate tuning refinements, with full feedback of operating parameters to customer control and monitoring systems. Continuing efforts to extend the period between overhauls has resulted in a current recommendation of 30,000 h for engines with an average duty cycle. Ease of inspection and maintenance is also claimed, a unit power assembly being exchangeable in 4 h. Emissions performance of 710G-T2 engines is reflected in IMO, US EPA Tier II, and EU Stage IIIA certification.

Another medium-speed two-stroke design philosophy was pursued for many years by **Bolnes** of The Netherlands until its acquisition by the Wärtsilä group and the subsequent cessation of engine production. The company produced the world's smallest two-stroke crosshead engine, the last manifestation being the 190/600 series. The earlier 150/600 design is illustrated in Fig. 25.24.

The 190 mm bore/350 mm stroke design delivered a maximum continuous output of 140 kW/cylinder at 600 rev/min on a mean effective pressure of 14.1 bar. The range embraced 3–10 in-line models (excluding a four-cylinder version) and V10–20 cylinder models covering an output band from 400 to 2800 kW. The crosshead took the form of a lower piston/scavenge pump, with uniflow scavenging arranged via a single valve in the head. Air is drawn in by the turbocharger, passes through a first-stage air cooler to the scavenge pump and then fed to the cylinder for the compression stroke via a second-stage air cooler.

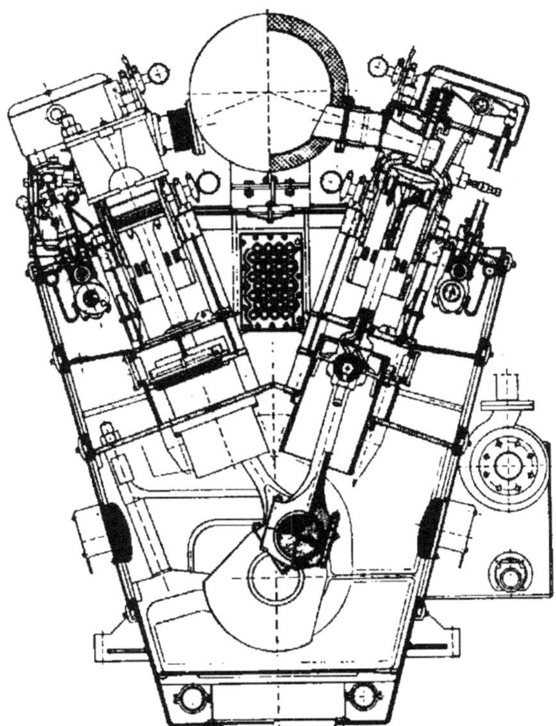

Fig. 25.24 Cross section through Bolnes VDNL 150/600 two-stroke engine.

Bolnes cited the following merits of the design:

- Very low lubricating oil consumption due to the complete separation of combustion space and crankcase.
- Unique air control due to the design of the 260 mm bore crosshead scavenge pump which was claimed to secure completely smokeless combustion under all load conditions.
- High performance at low speed (e.g. 110% torque at 70% rev/min).
- Separate lubrication systems promoting good heavy fuel-burning operation.
- Simple maintenance.

Bolnes engines had a faithful following in fishing, dredger, coaster, and survey vessel propulsion sectors. The design was also valued as a test engine in fuel and lubricating oil research laboratories.

CHAPTER TWENTY-SIX

High-speed engines

High-speed four-stroke trunk piston engines are widely specified for propelling small, generally specialized, commercial vessels and as main and emergency genset drives on all types of tonnage. The crossover point between high- and medium-speed diesel designs is not sharply defined but for this chapter engines running at 1200 rev/min and over are reviewed.

High-speed engines tend to be found in smaller craft such as tugs, workboats, and similar types as well as military vessels and fast ferries. They are found in numerous propulsion configurations including powering vessels by way of conventional propellers, niche propulsion systems such as Schottel Rudder Propellers and Voith Schneider systems. They are also popular for powering waterjets.

Marine high-speed engines traditionally tended to fall into one of two design categories: high-performance or heavy-duty types. High-performance models were initially aimed at the military sector, and their often complex designs negatively affected manufacturing and maintenance costs. Applications in the commercial arena sometimes disappointed operators, the engines dictating frequent overhauls and key component replacement.

Heavy-duty high-speed engines in many cases were originally designed for off-road vehicles and machines but have also found niches in stationary power generation and locomotive traction fields. A more simple and robust design with modest mean effective pressure ratings compared with the high-performance contenders yields a comparatively high weight/power ratio. But the necessary time-between-overhauls and component lifetimes are more acceptable to civilian operators.

In developing new models, high-speed engine designers have pursued essentially the same goals as their counterparts in the low- and medium-speed sectors: reliability and durability, underwriting extended overhaul intervals and component longevity and hence low maintenance costs; easier installation and servicing; compactness and lower weight; and enhanced performance across the power range with higher fuel economy and reduced noxious emissions.

Pounder's Marine Diesel Engines and Gas Turbines
https://doi.org/10.1016/B978-0-08-102748-6.00026-8

Performance development progress over the decades is highlighted by considering the cylinder dimension and speed of an engine required to deliver 200 kW/cylinder (Fig. 26.1). In 1945, a bore of 400 mm-plus and a speed of around 400 rev/min were necessary; in 1970, typical medium-speed engine parameters resulted in a bore of 300 mm and a speed of 600 rev/min, while typical high-speed engine parameters were 250 mm and 1000 rev/min to yield 200 kW/cylinder. By 1995, that specific output could be achieved by a 200 mm bore high-speed design running at 1500 rev/min.

Modern production methods such as modular design and flexible manufacturing systems (FMS) have allowed a different approach to engine design. The reduced cost of machining has made possible integrated structural configurations, with more functions assigned to the same piece of metal. The overall number of parts can thus be reduced significantly over earlier engines (by up to 40% in some designs), fostering improved reliability, lower weight, and increased compactness without compromising on ease of maintenance. FMS also facilitates the offering of market-adapted solutions without raising cost: individual engines can be optimized at the factory for the proposed application.

Fig. 26.1 Cylinder dimensions and speeds for medium- and high-speed engines delivering 200 kW/cylinder (1945, 1970, and 1995). *(Source: Wärtsilä.)*

Bore (mm)	420	300	(250)	200
Stroke (mm)	500	360	(300)	240
Speed (rpm)	428	600	(1000)	1500
Year	1945	1970		1995

Widening market potential for small high-speed engines in propulsion and auxiliary roles encouraged the development in the 1990s of advanced new designs for volume production. The *c.*-170 mm bore sector proved a particularly attractive target for leading European and US groups which formed alliances to share R&D, manufacture, and marketing—notably Cummins with Wärtsilä Diesel and MTU with Detroit Diesel Corporation (DDC). Not all of these alliances prospered, however, the Cummins/Wärtsilä connection eventually being cut and Wärtsilä withdrawing from own-design high-speed engine production.

High-speed engine designs have benefited from such innovations as modular assembly, electronically controlled fuel injection systems, common rail fuel systems and sophisticated electronic control/monitoring systems. Some of the latest small-bore designs are even released for genset duty burning the same low-grade fuel (up to 700 cSt viscosity) as low-speed cross head main engines. On large vessels, high-speed engines will frequently be found in use as emergency generators and also for powering lifeboats and fast rescue craft.

Evolving a new design

An insight into the evolution of high-speed engine design for powering fast commercial vessels is provided by MTU of Germany with reference to the creation of its successful 130 mm bore Series 2000 and 165 mm bore Series 4000 engines, which together cover an output band from 400 to 4300 kW. These engines have proven immensely popular with more than 2000 vessels of different types including, ferries, ro-paxes, fishing vessels, yachts, tugs, patrol craft, research vessels, and different types of inland cargo vessels.

MTU notes first that operators of fast tonnage place a high value on service life and reliability, with fuel economy and maximized freight capacity also important. In the fast vessel market, conflicting objectives arise between key parameters such as low-specific fuel consumption, low weight/power ratio, and extended engine service life. If one parameter is improved, at least one of the others is undermined. The engine designer's aim is therefore to optimize the coordination of the parameters to suit the application.

Knowledge of the anticipated service load profile is vital for determining the specific loads that must be addressed during the engine design stage so that the required maintenance and major overhaul intervals can be established.

Load acceptance characteristics and performance map requirements have a strong influence on turbocharging and the maximum possible mean pressures.

Specifying performance map requirements is simultaneously connected with the selection of the lead application, in this case, high-speed tonnage. The maximum possible mean pressures are determined on the basis of the power–speed map requirements of various vessel types (e.g. air cushion, hydrofoil, and planing hull types) and the form of turbocharging (sequential or nonsequential, single or two-stage). With increasing mean pressures (higher power concentration), the weight/power ratio of the engine can be improved.

The maximum mean piston speed is derived from a service life requirement (time-between-overhaul) and the target for the weight/power ratio. With increasing mean piston speed, a greater power concentration in a given volume is achieved, and thus improving the weight/power ratio. For fast vessel engines, mean piston speeds of 11–12 m/s and mean effective pressures up to 22 bar (single-stage turbocharging) or 30 bar (two-stage turbocharging) are typical. Fig. 26.2 shows the correlations of four-stroke engines for determining bore diameter, stroke, and speed. The output per cylinder (Pe) is known from the power positioning of the proposed new engine, and the maximum mean pressure and maximum mean piston speed have already been established. The required minimum bore diameter (D) can thus be determined.

The appropriate stroke (s) is determined using the specified stroke/bore ratio (s/D). If large s/D ratios are selected, a large engine height and width results (V-engine); small s/D ratios are associated with somewhat reduced fuel efficiency. For relevant MTU engines with good weight/power ratios, the stroke/bore ratio lies within a range of 1.1–1.25. The engine speed

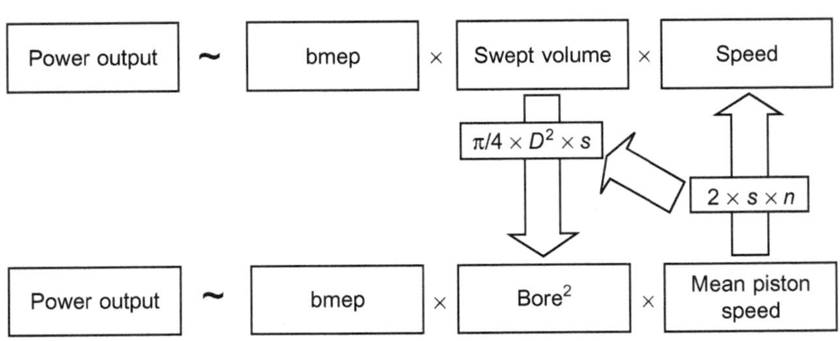

Fig. 26.2 Calculation of power output.

appropriate to the established stroke is determined via the resulting mean piston speed. Engine speed is an important factor for the customer as the size of the gearbox required is based on the speed and torque.

The peak firing pressure or peak firing pressure/mean pressure ratio is the most important factor influencing the specific effective fuel consumption. For high efficiency, MTU suggests that peak firing pressure/mean pressure ratios of around 8 should be targeted. With the mean pressure already specified, the peak firing pressure can be established. If, for example, a mean pressure of 22 bar (single-stage turbocharging) is selected, the peak firing pressure should lie in the 160–180 bar range. For two-stage turbocharged engines with correspondingly high mean pressures, peak firing pressures of above 200 bar must be targeted. With two-stage turbocharging, the potential for fuel consumption reduction can be increased using charge-air cooling.

The 4000 series was further evolved in 2016 to bring the engines in conformity with the IMO Tier III NOx emission requirements by using MTU's inhouse SCR systems. The engine was displayed at the SMM exhibition in Hamburg in 2016 when it was announced that it would be available as 12, 16, and 20-cylinder versions for a power range extending from 1920 to 3900 kW. The injection and turbocharger technology was also upgraded and the engine control system new compared with the preceding models. At the exhibition, MTU revealed that the first order had already been placed for a 16-cylinder unit for a Turkish-built mega yacht.

High-speed engine manufacturers

There are more manufacturers of high-speed marine diesels than of low- and medium- speed models. The following are details of some of the major players and their products.

Baudouin

Moteurs Baudouin is a long-established French engine manufacturer building small engines for fishing vessels and naval craft. TA various times, it had cooperative ventures with Renault and John Deere but after a foray into building engines for fast leisure craft and powerboats, the company returned to its roots in the 1980s.

In 1991, it developed a new engine the M26S. It is available in six- and eight-cylinder in-line versions as well as in a 12-cylinder Vee engine.

The engine range had power outputs up to a maximum of 671 kW. A supercharged version of the M26SR was added later. After introducing the engine, the company changed hands several times and was finally acquired by Chinese engine manufacturer Weichai Power in 2009. The acquisition was marked by the introduction of a new version of the M26 as the M26.2 which lifted maximum power output to 1014 kW.

The new engine featured a cast-iron cylinder block one inspection door per cylinder and cast-iron cylinder liners of the wet type. There was a separate cast-iron cylinder heads equipment for valves and replaceable valve guides and seats.

A follow up to the M26.2, the M26.3 was launched in 2016 and the M26.3 engine had more power and with a common rail structure, it allowed Baudouin to meet the IMO Tier III NOx regulations using SCR. The older engine was retained in the portfolio. The new range of six-cylinder in-line engines and a V12 at 90° deliver from 441 to 1214 kW in the 1800–2300 rpm spread. With the same bore and stroke of 150 mm, the engine had featured in all its evolutions.

The engine is Baudouin's first with electronic common rail and features twin-engine control units (ECUs), for redundancy. These new ECUs have been developed jointly with Bosch specifically for the M26.3 and allow the 12V engine to operate on one bank of six cylinders only; the company believes this is a unique feature for this size of marine diesel and will be very useful in a get-you-home situation.

In 2017, launched a longer stroke version of the engine with a stroke of 185 mm. The M33.2 extends Baudouin's portfolio into the heavy-duty sector with powers of 485–1120 kW range. There are two engines in the range, the 12M33.2 V12 39.2 L and 6M33.2 straight six 19.6 L. Its high engine capacity, relative to the power, is designed to conserve engine life in high strain, load-bearing applications such as fishing trawlers, offshore support, workboats, and ferries.

Caterpillar

A wide programme of high-speed engines from the US designer Caterpillar embraces models with bore sizes ranging from 100 to 170 mm. The largest and most relevant to this review is the 170 mm bore/190 mm stroke 3500 series which is produced in V8-, V12-, and V16-cylinder versions with standard and higher B-ratings to offer outputs up to around 2237 kW. The engines, with minimum/maximum running speeds of 1000/1200/1925 rev/min, are suitable for propelling workboats, fishing vessels, fast commercial craft, and patrol boats. Genset applications can be covered with ratings from 750 to 2281 kVA.

Fast vessel power ratings for the Series II high displacement (HD) versions of the 3512B and 3516B engines enable them to target the mega-yacht propulsion sector, with respective outputs of 1895 and 2525 kW at 1800 rev/min. Among the hardware, refinements are a one-piece steel monotherm piston, modular pulse exhaust manifold, ABB turbochargers, and an optimized fuel injector spray angle and combustion crater. Cat 3500C series engines are offered in V8-, V12-, and V16-cylinder versions for propulsion installations, with HD ratings up to 2525 kW at 1800 rev/min.

The series-B engines (Fig. 26.3) benefited from a number of mechanical refinements introduced to take full advantage of the combustion efficiency improvement delivered by an electronic control system (ECS). Electronically controlled unit fuel injectors combine high injection pressures with an advanced injector design to improve atomization and timing. Outputs were raised by 17%–30% above earlier 3500 series models.

A special high-performance variant of the V16-cylinder 3500 series model was introduced to target niche markets, the refinements seeking increased power, enhanced reliability, and lower fuel and lubricating oil consumptions without undermining durability. This Phase II high-performance version

Fig. 26.3 Caterpillar 3512B engine with ECS.

of the 3516 has an upper rating of 2237 kW at 1925 rev/min. It was released for fast passenger vessels with low load factors with a standard maximum continuous rating of 1939 kW at 1835 rev/min and a 'two hours out of 12' rating of 2088 kW at 1880 rev/min. Optional higher ratings up to 2205 kW at 1915 rev/min can be specified for cooler climate deployment, with the revised turbocharger, fuel injector, and timing specifications.

Key contributions to higher performance came from high-efficiency ABB turbochargers, a seawater aftercooler to supply colder air to the combustion chambers, larger and more aggressive camshafts, and a new deep crater piston design. The fuel is delivered through strengthened unit injectors designed and manufactured by Caterpillar to secure injection pressures of 1380 bar.

An optimum air–fuel mixture which can be burned extremely efficiently is fostered by the combination of denser air intake and the high injection pressure. The reported result is a specific fuel consumption range at full load of 198–206 g/kW h with all fuel, oil, and water pumps driven by the engine. Modifications to the steel crown/aluminium skirt pistons and rings lowered lubricating oil consumption to 0.55 g/kW h.

A particularly desirable feature for fast ferry propulsion is underwritten by the high-efficiency combustion and low crevice volume pistons which help to eliminate visible exhaust smoke at all steady points along the propeller demand curve. The rear gears were widened and hardened to serve the higher-pressure unit injectors. New gas-tight exhaust manifolds with bellow expansion joints and stainless steel O-rings improved engine room air quality by eliminating exhaust gas leaks.

A longer-stroke variant of the Cat 3500 series-B engine was introduced after marine field tests undertaken from early 1998, these 3512B and 3516B models offering as much as 13% higher powers than their standard counterparts, with respective maximum commercial ratings of 1380 and 1864 kW at 1600 rev/min. Seven per cent improvements in power-to-weight ratio and fuel economy were reported, along with lower emission levels than the standard engines. The higher output was achieved by enlarging the cylinder displacement (increasing the stroke by 25–215 mm) and without raising cylinder pressure or undermining bearing life or the durability of other key components.

A new single-piece forged crankshaft has more mass and is made from a stronger steel alloy than before to handle the higher loads. The connecting rods are longer and feature stronger shaft geometry, and a more robust rod pin end enhances the durability required for the increased piston speeds and higher inertia loads. The pistons are of the same two-piece design proven in standard Cat 3500 series-B engines, a steel crown, and aluminium skirt securing high strength and reduced weight. The engine footprint of these more powerful variants remained unchanged; only the dimensions of the higher capacity aftercooler and turbochargers were increased.

All Caterpillar 3500 series-B engines are controlled by a microprocessor-based electronic control module (ECM). Information is collected from engine sensors by the ECM which then analyses the data and adjusts injection timing and duration to optimize fuel efficiency and reduce noxious exhaust emissions. Electronic control also supports onboard and remote monitoring capabilities, the ECM reporting all information through a two-wire Cat Data Link to the instrument panel. The panel records and displays faults as well as operating conditions. An optional Customer Communications Module translates engine data to standard ASCII code for transmission to a PC or via satellite to remote locations.

Caterpillar's Engine Vision System (EVS) is compatible with the high-performance 3500 series-B engines and the company's other electronically controlled engines. The EVS displays engine and transmission data, vessel speed, trip data, historical data, maintenance intervals, diagnostics, and trouble-shooting information. Up to three engines can be monitored simultaneously, the system transferring between the vision display and individual ECMs via the two-wire data link.

An upgrade announced in 2002 introduced the 3500B series II engines with enhancements to their electronic control, monitoring, display, and cooling systems as well as new derating and operating speed options designed for specific applications. New electronics included the latest Caterpillar ADEM III control system, allowing more engine parameters to be controlled and monitored, with more accuracy and fault-reporting capability. A new 'programmable droop' capability allows precise governor control for load-sharing applications. A combined cooling system, rather than two separate circuits, became an option. A higher maximum continuous rating of 2000 kW from the 3516B series II engine was offered to yield more power and bollard pull capacity for larger harbour tugs; the higher rating also addressed some types of ferries and offshore service vessels.

ACERT and WOSR: Caterpillar's ACERT technology and Wide Operating Speed Range (WOSR) capability, respectively, underwrite compliance with emissions regulations and boost fuel economy and operating flexibility. An ACERT fuel injection and combustion management system enables an engine to comply with US EPA Tier II requirements and provides a foundation for compliance with even stricter future limits. Higher ratings with lower peak cylinder pressures exhaust temperatures, and smoke levels are reported without undermining fuel consumption in cruise mode.

The WOSR facility allows operators to run their vessels at full power over a wide speed range, thus enhancing productivity. It is designed to exploit a nominal 1600 rev/min rated engine and allow power to be maintained at a high level up to 1800 rev/min in the governor over-run before dropping off the high idle speed. An excellent reserve power for quick vessel acceleration is also reportedly fostered. Full advantage of the technology can be taken by specifying a propeller sized as normal at the rated speed of 1600 rev/min.

Caterpillar asserts that WOSR engines promise a cost-saving alternative to torque converters, multi-speed transmissions, and CP propellers. The real advantage is derived when the vessel is running free, without passengers or cargo or downriver. In this mode, the governor over-run maintains a high power level up to 1800 rev/min; as a result, the engine and thus the propeller are electronically allowed to run at a higher speed, yielding an increased vessel speed.

WOSR is particularly useful for tugs that are not running under tow but need to move quickly between locations for salvage and rescue operations. The performance curve is also beneficial in the dredging market, where higher productivity can be achieved by dredgers. Advantages for fast ferries, crew boats, and other high-speed craft include the ability to increase speed when passenger levels vary seasonally or in cyclical markets.

Cummins

A new range of Quantum System high-speed engines for propulsion and auxiliary power generation from US-based designer Cummins was headed in 2005 by the six-cylinder/19-L QSK19 model. A similar specification—based on a common rail fuel injection system and advanced electronics—was subsequently applied to create V12- and V16-cylinder derivatives, with respective displacements of 38 and 50 L (Fig. 26.4). These QSK19, QSK38, and QSK50 models share a common 159 mm bore/159 mm stroke cylinder unit and cover an output range up to 1567 kW at 1500/1800/1900 rev/min. Higher powers (up to 1716 kW) are offered by the QSK60 engine, a V16-cylinder/60.2-L design with the same bore size but a longer stroke (190 mm) than the other QSK models.

Fig. 26.4 Cummins V12-cylinder QSK38 Tier II engine.

A significant contribution to meeting stricter US EPA Tier II emission standards is made by Cummins' own modular common rail fuel system (MCRS), which allows multiple injection events to precisely control the combustion process. A constant high injection pressure, regardless of engine speed or load conditions, improves overall performance and idling stability as well as delivering more low-end torque. In addition, the system is said to ensure no visible smoke across the operating range and to considerably reduce noise and vibration. The MCRS exploits electronically actuated injectors featuring an integrated fuel accumulator that facilitates connection in series, thus eliminating the need for a large forged common rail. Injector load on the camshaft and overhead rocker levers are also eliminated, reducing wear on these key components. Higher safety and reliability are cited for the system in conjunction with double-wall fuel lines.

Proven base components from the established K19, K38, and K50 engines were used in the new QSK models but with some refinements. A new ductile cast-iron single-piece piston with nitride-coated rings and a hardened cylinder liner were specified to increase durability and reduce maintenance. Improvements also benefited the exhaust system, which incorporates a triple-wall water-cooled manifold with an air gap between the exhaust gas and jacket water to maintain thermal efficiency while cooling engine surface temperatures. A single-piece fabricated construction prevents exhaust leakage and reduces installation costs.

Turbocharging is effected by a water-cooled unit optimized for the projected operating conditions. A twin-pump, two-loop low-temperature (LT) aftercooling arrangement is installed to maximize efficiency and enhance performance. An engine-mounted titanium plate heat exchanger is incorporated for durability and minimal maintenance demands.

A number of options are offered to ease servicing and reduce downtime and maintenance. The Centinel oil management system, for example, can extend the time between, or even eliminate, oil changes, while the Eliminator filter replaces disposable lube oil filters with a self-cleaning centrifuge. Working either together or individually, says Cummins, these systems foster cost-effective maintenance and reduce the potential for lube oil contamination during oil-related servicing.

Minimal changes to the footprint, mountings, ratings, and optional equipment of the KV engines were sought to ease the installation of the QSK engines for newbuilding and retrofit projects.

Quantum System electronics are driven by an ECM proven in rugged applications across previous Cummins platforms. The C-Command instrument panels include a selection of modular monitoring and display options that assist in protecting and enhancing engine operations. These facilities also help to manage operating costs by logging critical parameters such as engine load, duty cycle, speed, and fuel consumption as well as providing diagnostic and prognostic capabilities. A common point of connection in the customer interface box simplifies the link between engine electronics and ship systems and reduces installation complexity.

A comprehensive history of alarms and faults can also be logged for efficient trouble-shooting and service scheduling, the data downloadable via an Ethernet connection. Direct access to the latest Cummins parts and service information is accessible via the Internet from the panel itself. All C-Command systems can help operators manage the engine life cycle by pinpointing the optimum speed for the best possible fuel efficiency under the given conditions, says Cummins.

Earlier models in the Cummins programme include the KTA50-M2 model, which became available from 1996. The 159 mm bore/159 mm stroke design is produced in V16-cylinder form with ratings of 1250 and 1340 kW for medium continuous duty and 1030 and 1180 kW for continuous duty applications. The running speeds range from 1600 to 1900 rev/min, depending on the duty; typical applications include fishing vessels, tugs, crewboats, and small ferries.

The KTA50-M2 engine benefited from a new Holset turbocharger, LT aftercooling, and gallery-cooled pistons. Cummins's Centry electronics system contributes to enhanced overall performance and fuel economy, providing adjustable all-speed governing, intermediate speed controls, dual power curves, a built-in hour meter, and improved transient response. Diagnostic capabilities are also incorporated.

Customer requests stimulated Cummins in 2007 to upgrade the KTA38-MO engine to meet EPA Tier II emission levels as well as EU Stage IIIA regulations. The resulting V12-cylinder/38-L design was initially offered with a rating of 634 kW at 1800 rev/min for continuous marine duty. Direct retrofits for Tier I engine installations are facilitated.

In 2011, Cummins announced the development of a new larger 16-cylinder engine with ratings from 2386 to 3132 kW at 1700 rpm. The bore is 190 mm and the stroke 210 mm. The QSK95 achieves a power output previously available only from larger medium-speed marine engines, while bringing the advantages of a lower capital cost and a more compact installation.

The QSK95 capacity of 95 L is compact taking up the same space as a typical 78-L package. Nested cylinders and a 60-degree V enable a short, narrow engine block relative to other engines of comparable displacement. Even with the compact package size, Cummins claims that the QSK95 provides best in class power density. In addition, the QSK95 weighs in just over 13,000 kg this is between 25% and 70% less than medium-speed platforms of similar power output.

The QSK95 delivers faster transient response through a unique turbo arrangement and a dry system. By using one turbo per four cylinders, the QSK95 can utilize a small turbo model. The dry turbo housings and dry exhaust manifold maximize the available energy to the turbos, allowing them to spool up quickly, resulting in fast engine response. First in-service models were brought into operation in 2017.

Deutz

A long tradition in high-speed engine design was maintained by Deutz of Germany (formerly Deutz MWM) whose last marine programme focused on the 616 and 620 series with an upper output limit of 2336 kW for propulsion and genset drives. Other high-speed engines produced under the Deutz banner covered a power range down to 14 kW.

The 616 series is a 132 mm bore/160 mm stroke design covering propulsion plant applications from 320 to 1360 kW at speeds up to 2300 rev/min by V8-, 12-, and 16-cylinder models (Fig. 26.5).

The 170 mm bore/195 mm stroke 620 series design also embraces V8-, 12-, and 16-cylinder versions covering a power band from 829 to 2336 kW at speeds up to 1860 rev/min for propulsion duty (Fig. 26.6). Good exhaust

Fig. 26.5 V12-cylinder block of Deutz 616 series engine.

gas exchange and optimum combustion with minimal emissions are fostered by cross-flow cylinder heads and four valves per cylinder, while a relatively high compression ratio promotes excellent cold-start characteristics and low fuel consumption. High-pressure (HP) fuel injection further contributes to these operating qualities. Correct swirl under both low-load and full-load conditions is secured by the boost pressure-controlled throttle of the HALLO-Swirl system, a variable swirl-supporting air intake system.

Outputs from 1035 to 3600kW at 1000rev/min are delivered by six, eight, and nine in-line and V12- and V16-cylinder versions of the 240mm bore/280mm stroke 628 series, which can be operated on heavy fuels (RMK 35) as well as marine diesel oil.

HiMSEN

A relatively new name to the marine diesel sector, HiMSEN is the name given by Hyundai Heavy Industries to its own in-house developed engine

Fig. 26.6 V16-cylinder version of Deutz 620 series engine.

family. Hyundai is a long-time licensee of the two main two-stroke engine makers but was determined to develop its own brand of engine. The name HiMSEN is a contraction of Hi-Touch Medium Speed Engine which was the title under which the company's first medium-speed engines were marketed in 2000.

After 2008, when construction of most commercial ship types was declining rapidly, HiMSEN turned its eyes towards the offshore propulsion and auxiliary markets which were the only growth sectors. To do this, it determined the need for its own high-speed engine to challenge the likes of Caterpillar, Cummins, and the other makers competing in that arena. In 2011, the first—and so far only—high-speed engine under the HiMSEN brand was introduced.

The H17/21V engine was developed for marine propulsion and power generation, stationary generators, and emergency generators. The new engine boasted a speed twice that of the company's medium-speed range at 1800 rpm. Experience from the medium-speed engine designs—particularly the H17/28 engine—benefited a development that pursued easier production and maintenance from a minimized number of components.

With a bore of 170 mm and a stroke of 210 mm, the H17/21V design had a power range from 1500 to 3500 kW at 1500/1800 rpm from V12, 16, 18, and 20-cylinder models. Three engines were built to support the development of the H17/21V series as a compact, high-performance engine with low emissions. A V12-cylinder model was first assembled for testing purposes.

A nodular cast-iron engine block was designed to achieve a lightweight from minimal thickness without compromising structural strength. Oil and water passageways are incorporated for reinforcing the structure and reducing pipework. Torsional vibration analysis of the high tensile Ni-Cr-Mo alloy steel crankshaft was applied to secure a sufficient fatigue safety factor for the journal, pin, and fillet. Provision was also made for fitting a power take-off at the crankshaft's free end. The alloy steel connecting rods are manufactured by stamp forging, with an oblique split-type and serrated joint configuration adopted between the shaft and big end cap for the ease of maintenance.

Fatigue strength evaluation via full-cyclic nonlinear structural analysis was performed to confirm structural strength integrity, and fretting wear risk analysis carried out for the serrated contact surfaces of the connecting rod. An optimized rod design was pursued by refining the shape and dimensions and adjusting the number of serration teeth.

Heat flow and structural analysis were essential in designing the cylinder head and liner, based on measurements of actual temperatures at principal points during testing. Low-cycle fatigue testing was also effective in investigating the endurance reliability of the combustion chamber components.

Swirl flow in the combustion chamber, necessary to improve the fuel-air mixing rate, is influenced by the intake port and valve seat design. The effects on engine performance for different port geometries were investigated by 3D in-cylinder combustion analysis. Combustion chamber geometry plays a significant role in optimizing combustion. Four parameters—piston bowl diameter, bowl height, bowl depth, and fuel injection angle—were examined in optimizing the piston bowl shape for reducing fuel consumption.

Hyundai was able to tap broad experience with mechanical fuel injection system technologies to satisfying environmental regulations for its medium-speed HiMSEN engines. The pump-line-injection system is typically used for such engines burning heavy fuel oil but a mechanical unit injector was adopted for the H17/21V engine as it is arranged to burn distillate fuels. Mechanical unit injectors are seen as a first stage before electronic engine control development.

Unit fuel injection was optimized for the application in terms of increased injection pressure and nozzle specification refinement (e.g. exit area, coefficient, and injection angle).

A high-efficiency turbocharger, an air cooler, and Miller cycle technology enable IMO Tier II NOx emission regulations to be met while securing favourable specific fuel consumption figures. Addressing stricter Tier III emissions regulations in the future, Hyundai was said to be developing appropriate solutions as well as planning common rail injection and gas engine derivatives of the H17/21V diesel design. However, these appear not to have materialized and the 2020 engine catalogue has also dropped the propulsion engine leaving just the gensets apparently in the market.

The range of models for marine propulsion systems previously included four models all of which are V configuration. Power outputs range from 1920 to 3200 kW. For use as a genset, the H17/21 range also includes in-line 6, 8, and 10 cylinder models as well as the 12, 16, 18, and 20V models. For 50 Hz use, these run at 1500 rpm and produce from 722 to 2941 kWe and for 60 Hz service running at 1800 rpm the output increases to a range from 825 to 3362 kWe.

Isotta Fraschini

A family of high-speed engines from Isotta Fraschini, part of Italy's Fincantieri group, is headed by the 170 mm bore/170 mm stroke 1700 series which is available in different versions for light, medium, and heavy propulsion duties in commercial and military vessels, as well as for genset drives. A nonmagnetic variant was developed for mine warfare vessels.

The 1700 series embraces V8-, 12-, and 16-cylinder models, all with a 90° configuration arranged on a high-tensile alloy iron crankcase and featuring a direct fuel injection system and four valves per cylinder. Upgrades raised the power output, the largest model—the V1716 T2—then offering up to 2595 kW at 2100 rev/min, compared with its previous maximum rating of 2350 kW at the same speed.

In 2009, a new V16 cylinder model was introduced as the VL1716C2 MSD type. It still has the standard bore of 170 mm but an increased stroke of 185 mm, giving a displacement of 67.2 L. It uses common rail fuel injection and has three sequential turbochargers. Power output is 2545 kW at

2030 rpm at the special fast craft rating and 2800 kW at 2100 rpm for high performance.

MAN Energy Solutions

MAN's forte is the large two-stroke engines and more recently its medium-speed engines rather than high-speed engines. However, in mid-2000, the company then trading as MAN B&W Diesel acquired UK-based Paxman Diesels and in doing so the VP185 originally launched in 1993.

Over 60 years' experience in high-speed diesel design was exploited by Paxman Diesels in creating the VP185 engine, launched to join the company's established 160 mm bore Vega and 197 mm bore Valenta series which, respectively, offered specific outputs of 107 kW/cylinder and 206 kW/cylinder. The new 185 mm bore design was introduced initially in V12-cylinder form with outputs up to 2610 kW and complemented in 1998 by a V18-cylinder version extending the power limit of the series to 4000 kW at a maximum speed of 1950 rev/min.

In designing the 12VP185 engine (Fig. 26.7), Paxman sought improvements in compactness and lightness over its earlier models, goals dictating a smaller swept volume, slightly higher ratings, and a review of piston speeds. The first parameter to be fixed was the stroke which was set at 196 mm; this, linked to a crankshaft speed of 1800 rev/min, gave a mean piston speed of 11.8 m/s for continuous duty. An improved speed platform for marine applications was derived, the edge of which is bounded by a maximum speed of 1950 rev/min with a mean piston speed of 12.8 m/s.

The cylinder bore was fixed at 185 mm which, coupled with a design maximum mean effective pressure of 25.3 bar, delivered a maximum power of 2610 kW at 1950 rev/min from the V12-cylinder configuration: a rating primarily targeting high-speed military and commercial marine markets, as well as mega-yacht propulsion. An unrestricted rating of 2180 kW at 1800 rev/min was quoted (at 45°C air and 32°C seawater conditions) for continuous marine duties (e.g. fast ferry propulsion) where the engine's high torque capability over a broad speed range was considered especially attractive for hydrofoil service.

The 18-cylinder VP185 model offers a maximum output of 4000 kW at 1950 rev/min, a 3300 kW continuous power rating for fast ferries and patrol craft, and 3000 kW at 1770 rev/min for unrestricted marine duty.

Fig. 26.7 Paxman 12VP185 engine; note the compact turbocharging arrangement along the top.

A smaller bore and stroke than the Valenta design, combined with a switch from a 60° cylinder bank angle to a 90° configuration, achieved a layout which was short and having width under 1.5 m while providing an adequate platform for the charge-air system mounted above the engine. The 90° bank angle itself also lowered the profile of the charge-air system to foster space efficiency.

A higher mean effective pressure normally leads to a lower compression ratio which in turn promotes poor starting characteristics, reduces thermal efficiency, and can increase risk of ignition delay damage. Some of these problems can be addressed by the use of intercylinder charge-air transfer arrangements but Paxman decided to adopt a compression ratio in excess of 13:1. The resulting VP185 engine was claimed to be simple and easy to start under cold conditions, with good thermal efficiency and the potential to satisfy low NOx emission limits. The combination of high boost levels and compression ratios demands a robust construction, but Paxman sought to avoid a heavy and cumber-some engine by securing robustness within a small and hence a stiff envelope.

The reciprocating assembly of the engine was designed for high strength and rigidity in handling a high maximum cylinder pressure and securing high reliability with a long lifetime. The backbone is a compact crankcase cast in high-strength spheroidal graphite iron to provide stiff and solid support for the underslung crankshaft. Crankcase doors along each side of the engine give access to the connecting rod large ends for in situ servicing and piston removal.

The crankshaft is a machined steel forging, fully nitrided to yield strength and durability. It is secured by main bearing caps which are drawn internally against the deep-fitting side faces of the crankcase by high-tensile set screws and hydraulically tensioned main bearing studs. Such a configuration, the designer asserted, delivered good strength, and stiffness characteristics to the bottom end of the crankcase and provided solid support for the crankshaft. A generous overlap between the crankpin and main bearing journals added to stiffness and strength. Both main and big end bearings are of steel-backed aluminium–tin type with thrust washers controlling axial location. The crank-shaft is provided with a viscous torsional damper totally enclosed within the gearcase.

Connecting rods of side-by-side design are forged in high-tensile steel, fully machined, and ferritic nitro-carburized. The large end is obliquely split to allow the rod to pass through the cylinder liner and the joint faces are serrated. The one-piece nodular cast-iron pistons incorporate a large oil-cooling galley above the ring belt, oil being supplied by accurately aligned standing jets mounted in the crankcase. The pistons run in wet cylinder liners made of centrifugally cast high-grade iron.

The cast-iron cylinder heads are designed to withstand the high firing pressures with an internal configuration ensuring maximum flame face stiffness and high cooling efficiency. They incorporate two exhaust valves, two inlet valves, and a centrally mounted unit fuel injector (combined pump and injector) which is fixed with a single screw. The valves and injectors are actuated through a pushrod arrangement from a single large diameter camshaft mounted centrally in the vee of the engine. The stiffness of the actuating system reportedly achieves valve control similar to that of an overhead camshaft, while the specially designed unit injectors yield very high rates of fuel injection with clean cut-off, contributing to low NOx generation and good fuel efficiency. The fuel cam design was based on an optimized width for long life and a large base circle, coupled with high rates of lift to secure the key injection characteristics.

A maximum injection pressure of 1400 bar promotes good combustion characteristics with low emissions and fuel consumption. The unit injector is rack controlled, the rack itself being controlled by a linkage operated from twin shafts in the centre vee of the engine. The shafts in turn are coupled to the governor via an overspeed protection assembly which closes the fuel racks independently of the governor in the event of the engine overspeeding.

Safety merit of the unit injector (which combines fuel pump and injector in one unit) is that it eliminated the need for HP fuel pipes. The LP fuel oil is delivered to the injectors by a gear-driven lift pump mounted on the main gearcase. The fuel is fed to a subassembly—comprising cooler, filter (duplex optional), reservoir, and solenoid-operated shut-off valve—mounted high at the free end of the engine for convenient access.

A single camshaft reflects the policy adopted by the designers of minimizing the number of components. This, allied to the 90° cylinder bank angle, allowed the camshaft gear to mesh directly with the crankshaft gear and so eliminate the need for idlers in a critical area and reduce the component count even further.

Charge air is delivered by a 'valve-less' two-stage turbocharging system designed for simplicity and ease of maintenance. Preliminary explorations of single-stage HP ratio turbocharging showed poor low-speed torque characteristics due to surge limitation, along with poor acceleration and load acceptance. Sequential turbocharging was considered but the necessary valve gear and associated control system conflicted with the primary aim of keeping the engine as simple as possible, Paxman explains. Another factor was that few turbocharger designs in the smaller size range were capable of meeting the pressure ratio required.

A two-stage turbocharging arrangement with intercooling and aftercooling was finally adopted for the VP185 engine, based on six Schwitzer automotive-type turbochargers with broad and stable operating characteristics: four turbochargers provide low-pressure charge air and two provide the HP air.

MAN 175D

The VP185 remained the sole high-speed MAN engine other than those used by its trucking division until 2014 when the new 175D was introduced. With the MAN 12V175D displayed at SMM in Hamburg. The 12-cylinder model, developed especially for use in the marine industry, is part of a product initiative aimed at providing MAN customers with a portfolio that covers every power requirement, from high to low speed.

The new engine with a 175 mm bore and stroke of 215 mm was initially offered with an output spectrum from 1500 to 2200 kW and optimized for propelling ferries, offshore supply vessels, tugs, and workboats. Other central areas of application such as the markets for superyachts and marine applications are served by additional specialist model versions. The current range in 2020 includes versions for light, medium, and heavy-duty running at speeds from 1600 to 2000 rpm with power outputs of 1499 to 2400 kW for the 12-cylinder model, 2000 to 3200 kW for the 16-cylinder variants, and 2500 to 4000 kW for the 20 cylinder version. The power-to-weight ratio measured at MCR varies from 3.8 to 4.9 kg/kW depending upon model and rating. The engine has common rail fuel injection and features engine mounted control devices and operating panels.

The MAN 175D scores highly in terms of its eco-friendliness. Its compact and modular exhaust gas after-treatment system uses the selective catalytic reduction (SCR) method and is based on the MAN Ad Blue technology from road vehicles. Fuel consumption from 195 to 204 g/kWh when conforming to NOX Tier II levels increasing to 197–20 g/kWh when the SCR system is used to meet Tier III NOx rules.

MAN 175D showing SCR NOx reduction system.

MAN 175D showing SCR NOx reduction system.

Mitsubishi

Fast ferry propulsion business potential stimulated the development in the early 1990s of the Mitsubishi S16R-S high-speed engine by the Japanese group's Sagamihara Machinery Works. The higher-performance V16-cylinder model was evolved from the established S16R design which had been in production since 1989 as a general-purpose marine engine. A constant pressure turbocharging system based on a newly developed turbocharger and a revised fuel injection system contributed to a 20% rise in the power output. The 170 mm bore/180 mm stroke design has a maximum continuous rating of 2100 kW at 2000 rev/min with overload ratings up to 2300 kW.

The weight was reduced to 89% of the original engine, primarily through the adoption of aluminium alloy components optimized in size for the duty. An overall weight of 5500 kg underwrites a power-to-weight ratio of 2.62 kg/kW.

An upgraded model the S16R2-T2MPTAW engine was introduced in 2012 to meet IMO Tier II NOx emission rules. The engine had a much longer 220 mm stroke and a 79.9 L displacement. Its bore, stroke, and engine speed are the same as Mitsubishi's popular S6R2.

As a propulsion engine, it comes with three ratings for light, medium, or heavy duty representing fast vessels with low load, fishing vessels and tugs, and cargo vessels, respectively. The HD engine has a power output of 1450kW at 1350rpm, the MD engine 1600kW at 1400rpm, and the LD engine 1885kW at 1500rpm.

The engine's cylinder heads are individually divided by cylinder, and large inspection covers are available on the crankcase and the oil pan. Its original Mitsubishi fuel injection pump and turbocharger bring combustion matching, and a hydraulic or electronic governor is available to provide optimum control of the engine.

The S6R2 engine on which the S16R2-T2MPTAW was based is a smaller six-cylinder in-line engine for fishing vessels and workboats. This engine was itself upgraded in 2015 with the newly released S6R2-T2MPTAW3M. The new engines are a compact designed 30L in-line six cylinder engine that is available in four versions designed for various operational uses in workboats, fishing vessels, tugs, and ferries. Power output ranges from 324 to 759kW depending upon model and usage.

MTU

A portfolio of high-performance high-speed designs with an upper output limit of 9100kW is offered by MTU (Motoren- und Turbinen-Union) of Germany, which was created in 1969 when Daimler-Benz and MAN consolidated the development and production of relevant engines from MAN and Maybach Mercedes-Benz. MTU Friedrichshafen became part of the Deutsche Aerospace group in 1989 and subsequently a DaimlerChrysler company before coming under the umbrella of the Tognum Group in 2006. Rolls-Royce and Daimler jointly owned Tognum from 2011 to 2014 when Rolls-Royce became the sole owner. Today MTU is part of the Rolls-Royce Power division.

Series 396

Addressing lower power demands, MTU's Series 396 engine was long established as a propulsion and genset drive, the 165mm bore/185mm stroke design delivering up to 2560kW at 2100rev/min from V90° configuration 8-, 12-, and 16-cylinder models (Fig. 26.8).

Production was discontinued in the early 2000s but several vessels mostly ferries, yachts, and patrol craft fitted with the engines are still in service. Engines were available in three different versions:

- TB04, with external charge air cooling (intercooler in raw water circuit).

Fig. 26.8 MTU 16V 396 TE94 engine with the split-circuit cooling system.

- TC04, with internal charge air cooling (intercooler in engine coolant circuit).
- TE04, with internal charge air cooling (intercooler in engine coolant circuit; split-circuit coolant system).

TE split-circuit cooling system: The Series 396 engine—and other MTU designs—could be supplied with the TE split-circuit coolant system with a power-dependent subcircuit to cool the combustion air. The optimum performance was fostered throughout the engine's power range: at idling speed the air supply is heated to achieve complete fuel combustion; in the medium power and full-load-range conditions are optimized for high output while keeping thermal stress on engine components at a low level.

The coolant flow from the engine is split into two. Approximately two-thirds of the flow passes through an HT circuit and returns directly to the engine inlet, while the remainder is fed into a thermostatically controlled LT circuit. During engine idling or low load operation, the thermostat allows heated coolant to bypass the re-cooler on its way to the intercooler in order to warm up the combustion air and prevent white smoke in the exhaust.

An annular slide valve in the thermostat remains in its initial position until increasing power raises the coolant temperature, causing the wax pellet in the thermostat to expand. The gradual closing of the bypass line now directs the coolant stream through the re-cooler. As a result, coolant entering the intercooler is at a low temperature which, in turn, underwrites a high-combustion air volume and, consequently, maximum engine power. After flowing through the intercooler and the oil heat exchanger, the 'cold' coolant rejoins the uncooled HT circuit, thereby cooling the total volume flow before it re-enters the engine.

Sequential turbocharging is exploited for propulsion engines required to deliver high power in the lower- and medium-speed ranges. The system incorporates two or three turbochargers with automatic on/off control as a function of engine speed, power demand, and turbocharger maximum efficiency. In addition to increased torque, the system yields reduced fuel consumption and lower exhaust temperatures.

Cylinder cut-out (no fuel injection into selected cylinders) is adopted for engines operating under varying speed conditions (low idling speed). The system enhances combustion at idle, thus shortening the warm-up phase and avoiding white smoke from unburnt fuel, and generally contributing significantly to environmental compatibility.

Series 595

Another engine now discontinued but with several examples still in service. The 595 series featured a power range from 2000 to 4320 kW at 1500–1800 rev/min covered by the 190 mm bore/210 mm stroke engine, introduced in 1990 in V12- and V16-cylinder versions for propulsion and genset applications.

Contributing to a compact design yielding power-to-weight ratios of under 3 kg/kW and a power-to-space ratio of 250 kW/m^3 (including all ancillaries) are the following features: a V72° cylinder configuration with all ancillaries arranged for space saving within the engine contour; nodular cast-iron crankcase extending below the crankshaft centreline to maximize rigidity; crankshaft with hardened main and connecting rod bearing radii, designed to withstand firing pressures of up to 180 bar; individual fuel injection pumps designed for injection pressures of up to 1500 bar; plate-type coolant heat exchanger integrated with the engine; split-circuit coolant system; and two-stage sequential turbocharging system with charge-air cooling for boost pressures of up to 4.8 bar. Turbocharging sequencing is electronically controlled.

Components requiring regular servicing are located to provide good access at the engine's auxiliary power take-off end. Maintenance is also smoothed by the modular arrangement of functionally interlinked components, plug-in connections, and the omission of complex pipework, hose couplings, and cabling. A new ECS was purpose-developed by MTU for the Series 595 engine to provide information for the operator, promote easier operation, and enhance safety, reliability, and economy. The safety functions embrace engine and plant monitoring, overload protection, diagnostics, automatic start-up, and load control (Fig. 26.9).

Series 1163

The highest power outputs from the MTU programme were offered for many years by the Series 1163 'frigate' engine (Fig. 26.10). The 230 mm bore/280 mm stroke design is produced in V60° configuration 12-, 16-, and 20-cylinder versions delivering up to 7400 kW at 1300 rev/min for commercial and naval propulsion installations. The key elements of the design are summarized as follows.

Crankcase: nodular cast-iron structure with access ports on both sides and a flange-mounting facility for alternators or other driven machinery; a welded sheet steel oil pan is provided.

Fig. 26.9 V16-cylinder MTU Series 595 engine.

Fig. 26.10 MTU Series 1163 engine in V20-cylinder form.

Cylinder liner: replaceable, wet-type centrifugally cast.

Cylinder head: cast-iron component arranged with two inlet and two exhaust valves, all equipped with Rotocap rotators, a centrally located fuel injector separated from the rocker area, and a decompression valve.

Valve actuation: by two camshafts, roller tappets, pushrods, and rocker arms. Valve clearance adjustment is performed via two screws located in the rocker arms.

Crankshaft: a single-structure forged component finished all over and featuring bolted-on counterweights; axial crankshaft alignment is affected by a deep-groove ball bearing.

Bearings: thin-walled, two-piece, replaceable steel-backed tri-metal sleeve bearings for the crankshaft and connecting rod, with cross-bolted bearing caps.

Connecting rods: forged and finished all over and grouped in pairs to serve two opposite cylinders.

Piston: composite-type with light-alloy skirt and bolted-on steel crown, cooled by oil spray nozzles; three compression rings are fitted in the crown, with one oil control ring between crown and skirt; all rings replaceable after piston crown removal.

Fuel injection: direct injection by individual pumps and via short HP lines; pump replacement requires no readjustment; gear-type fuel delivery pump;

two duplex fuel filters; cylinder cut-out facility available (see section Series 396 Engine).

Governor: hydraulic MTU unit mounted on the gearcase, with the linkage between the governor and fuel injection pumps accommodated in the gearcase and camshaft space. Engine shutdown solenoids act on the fuel rack; in addition, independent emergency air shut-off flaps are arranged to block the engine's air supply.

Lubricating oil system: self-contained forced feed system, with oil flow progressing through gear-type pumps, heat exchanger, and filters to the engine lubrication points. Oil for piston cooling is tapped off after the heat exchanger and filters; the engine-mounted oil heat exchanger is integrated into the engine coolant circuit (Series 1163-02 model) or raw water circuit (Series 1163-03 model).

Cooling system: two-circuit system; closed engine coolant circuit using two centrifugal pumps; thermostatic control; recooling provided by raw water heat exchanger or fan cooler.

Turbocharging: the Series 1163-02 model exploits pulse charging based on two turbochargers (one turbocharger for the 12-cylinder engine) with intercoolers incorporated in the raw water circuit. The Series 1163-03 model exploits constant pressure turbocharging with two-stage air compression and interstage cooling; four or five charger groups (depending on the application) are under sequential turbocharging control; the constant pressure exhaust manifold is located in the V-bank, and charge-air pipework mounted externally; high- and low-pressure intercoolers are incorporated in the raw water circuit. For start-up and low load operation, the Series 1163-03 model features a charge-air preheater, fed with engine coolant, in each HP intercooler outlet. Coolant jackets for the exhaust manifolds and turbocharger turbines of this model secure reduced heat rejection.

Like the Series 595 engine, the Series 1163 design features inboard HP fuel injection lines and the enclosure of all hot exhaust components in water-cooled and gas-tight casings: valued for unmanned engine rooms by helping to reduce the possibility of fire in the event of fuel or lubricating oil leakages. A triple-walled insulation design also maintains surface temperatures well within the limit of 220°C dictated by classification societies, as well as considerably reducing radiant heat in the engine room.

Sequential turbocharging (output-dependent control of the number of turbochargers deployed) fosters high torque at low revolutions per minute (wide performance band) and hence good acceleration. Optimization during sea trials, with assistance from an ECS, can eliminate black smoke emissions.

An uprated Series 1163 engine for fast ferry propulsion benefits from:

- An improved fuel injection system to optimize consumption and exhaust emissions reduction.
- A higher cylinder head fatigue strength, thanks partly to bore cooling.
- Improved power and torque characteristics achieved by modifying the two-stage sequential turbocharging system.
- Adaptation and optimization of the turbochargers to match the increased cylinder output (325 kW).
- Intelligent electronic engine management.

Series 8000 engine

MTU's thrust in commercial and naval propulsion sectors was strengthened in 2000 with the introduction of its most powerful engine ever designed, the advanced 265 mm bore Series 8000 design (Fig. 26.11). Innovative fuel injection, turbocharging, and electronic management systems proven on earlier MTU engines are exploited, while a new modular power unit-based structure helps to ease maintenance and minimize life-cycle costs.

The Series 8000 engine was conceived with the aim of consolidating the company's position in the market for fast ferries and extending opportunities to mainstream commercial arenas, such as cruise liners. Outputs up to

Fig. 26.11 MTU Series 8000 engine in V20-cylinder form. Note the fuel pumps (lower right) for the common rail injection system.

9100 kW for naval vessels and mega yachts were initially sanctioned from the V20-cylinder model which launched the series, with an 8200 kW rating at 1150 rev/min approved for high-speed commercial tonnage. The engine speed could allow this engine to be described as medium speed but its heritage incorporates many features from the 4000 series of high-speed engines so it is apt to include here.

A potential for firing pressures up to 230 bar underwrites future power rises, projected after successful long-term testing of engines in service. The first four productions 20V 8000 engines were installed as a mega-yacht propulsion plant, subsequent orders calling for units to power a large catamaran fast ferry, frigates, and naval supply vessels.

A V48° cylinder configuration contributes to a narrow engine (1.9 m wide), well suited for fast catamaran and monohull ferries where machinery space is at a premium. The specific power output related to volume and weight is reflected in figures of 200 kW/m^3 and just over or under 5 kg/kW, depending on the application.

Common rail fuel injection, proven on MTU's 1996-launched 165 mm bore Series 4000 engine, allows all injection parameters affecting combustion to be independently controlled; these include such variables as the timing, period, and pattern of injection as well as the injection pressure. Independent control fosters reduced fuel consumption and exhaust emission levels across the entire engine power curve. The common rail system also benefits engine noise and vibration levels, substantially lowering them at idling and mid-range running speeds compared with conventional systems.

Two HP fuel pumps mounted at the free end of the engine develop pressure levels of up to 1800 bar. The fuel is pumped through double-walled delivery lines arranged longitudinally down the engine and directed to the cylinders via distributors into individual accumulators. The accumulators are large enough to prevent any drop in pressure before the fuel injector when maximum fuel quantities are being injected and to prevent pressure oscillations in the system. The fuel is injected via electronically controlled injectors arranged centrally in the cylinder head.

Experience gained from numerous Series 4000 engines in service moulded the common rail system created by L'Orange for the new larger engine, which is detailed in Chapter 11. Considerable scope for further development of the system promised further advances in fuel economy, emission levels, and noise reduction in future production engines.

The Series 8000 design's sequential turbocharging system had been successfully applied by MTU to its Series 396, 538, 595, and 1163 engines for

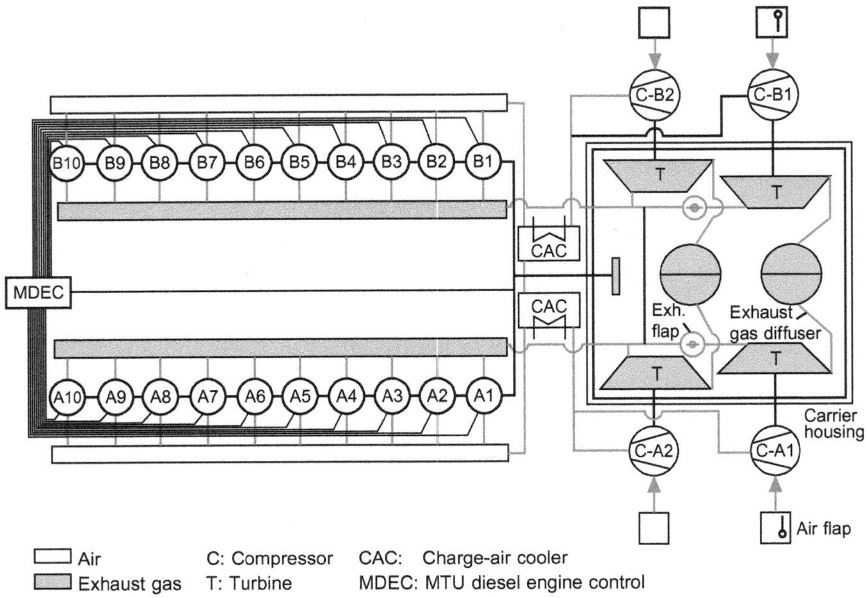

Fig. 26.12 Thermodynamic model of the MTU 20V 8000 M70 engine and its sequential turbocharging system.

many years to secure a large performance map width. First fitted on production models in 1982, the system marries a series of individual turbochargers that can be switched-in and -out while the engine is running to match the load. High-brake mean effective pressures can thus be sustained even at low engine speeds. Among the merits cited for the sequential turbocharging system are excellent acceleration and substantially lower fuel consumption and smoke emission levels at low- and medium-power rates. In addition, in twin-engine plants with common transmission and fixed pitch propellers or waterjets, the system can be run on one engine virtually up to its rated power. Quadruple MTU ZR 265 turbochargers are specified for the single-stage turbocharging system serving the 20V 8000 engine (Fig. 26.12).

Both common rail fuel injection and sequential turbocharging systems contribute to a claimed specific fuel consumption of below 195 g/kW h across a broad power range, and less than 190 g/kW h at the most economical points. These figures are achieved while maintaining NOx emissions within IMO limits. Smoke indices are reportedly 0.4 Bosch throughout the entire engine performance map and 0.2 Bosch at full load.

Electronics are increasingly exploited by engine designers, MTU addressing the full potential from the conceptual stage of the Series 8000.

The resulting MDEC (MTU Diesel Engine Control) engine management system allows all functions and parameters to be precisely monitored and controlled for optimum effect. The system also incorporates the basic functions for trend analysis and diagnostics and the facility for integrating the engine in a comprehensive control and monitoring package for the entire propulsion system as well as other ship systems. MTU offers these advanced options in the shape of its RCS (remote control system) and MCS (monitoring and control system) products.

Structurally, the Series 8000 is the first MTU engine to feature modular 'power units', which can be quickly removed and replaced (Fig. 26.13). Each integrated module comprises:

- A cylinder head, including valve gear, fuel injector, exhaust T-piece, and coolant line segment
- Cylinder liner
- Piston and connecting rod
- Spacer (for channelling coolant to the upper section of the liner and to the coolant pipes)

Each power unit is attached to the crankcase by four hydraulically tightened stud bolts which have to withstand the high ignition pressures but do not perform a sealing function. A good seal is affected by securing the cylinder liner directly to the cylinder head from below with 24 bolts. By separating the functions of sealing and retention, MTU explains that wider design scope was created for optimizing the channels, valve gear, and cylinder head cooling.

High design rigidity and endurance strength were sought from the transverse-flow nodular grey cast-iron cylinder head. Its exhaust ports and particularly the inlet ports are said to yield excellent flow coefficients which establish the fundamental conditions for fuel economy and low emission levels. The circularity of the cylinder liner when the engine is running is sustained by the 24 evenly spaced bolts around its perimeter and by which it is attached to the cylinder head. That circularity underwrites highly efficient sealing by the piston rings and, combined with a special liner honing process developed by MTU, promotes smooth running and durability. The high sealing efficiency of the rings and liners is also the basis for the 'exceptionally low blow-by levels' demonstrated by the Series 8000 engine, the designer reports. This significantly extends the life of the lubricating oil by reducing the amount of dirt escaping past the rings.

A robust nodular cast-iron crankcase extending well below the centreline of the crankshaft is described as particularly rigid, the structure

Fig. 26.13 All the cylinder components of the MTU Series 8000 engine are combined in a functional module termed a 'power unit' for ease of withdrawal and replacement.

offering large inspection ports to ease checking and maintenance. Integrated into the crankcase is large lateral charge-air ducts to the cylinders, the central bore for the camshaft operating the valve gear and a centrally located main oil gallery. The cylinder liner is cooled only in its upper region and the crankcase is therefore completely coolant free.

Wide, cross-bolted bearing caps retain the crankshaft with its bolt-on counterweights. Both big end and crankshaft main bearings are of the micro-groove type, and the latter can be replaced in situ. Other key elements of the running gear are fully machined connecting rods with diagonally split big ends, and composite pistons comprising a forged steel skirt (for the first time in an MTU engine) bolted to a steel crown carrying two compression rings and an oil control ring. The pistons are cooled by nozzles spraying cooling oil.

Exceptional engine rigidity imparted by the power unit concept and consequently low-wear characteristics were expected to help keep lube oil consumption at a constantly low level throughout the period between major overhauls. Confidence in a promised overhaul interval of up to 24,000 running hours—depending on the engine's load profile and power rating—was supported by trials during which the test engines were subjected to an extremely severe alternating load sequence as well as continuous operation at full and overload powers.

Minimizing the number and complexity of necessary interfaces, the pumps for engine coolant, raw water, fuel (HP and LP), and lube oil, as well as the associated filters and oil coolers, are mounted on the engine. The electronic MDEC engine management unit is also arranged here. Locating all these components together as a 'service block' on the free end of the engine fosters accessibility and simplifies the integration of the engine with the ship's systems. Installation work is further reduced by an engine mounting concept proven with the Series 1163 engine in many high-speed ferries, the torsionally resilient, offset-accommodating shaft couplings, and a special arrangement of the air intake and exhaust connections.

Engine starting is affected by an air starter, the sequence incorporating an initial slow-turn phase which allows engine functioning to be quickly checked without having to open the decompression valves. This is followed immediately by the main starting sequence.

Contributing to the trouble-free, low-maintenance operation are an automatic oil filter and two oil centrifuges in the secondary oil circuit (an indicator filter is fitted in the return line of the automatic filter for monitoring purposes); fuel filter cartridges with paper elements that can be replaced while the engine is running; and an engine coolant filter. The circulation systems are also fitted with running-in filters which trap any dirt that has found its way into the engine fluids during the installation process.

Adopted from previous MTU engines is the TE cooling system. Raw water never comes into contact with the engine components and the inlet temperature of the coolant at the engine can be controlled to suit the operating conditions. At low ambient temperatures and/or low engine loads, for example, the recooler is bypassed so that the coolant remains at a high temperature and heats up the combustion air. 'White smoke' emissions produced by cold engines can thus be eliminated. At full power, however, the system provides highly efficient cooling and creates the optimum conditions for combustion, both in terms of fuel economy and NOx emission levels, according to MTU.

MTU 20V 8000 M70 engine data

Bore	265 mm
Stroke	315 mm
Cylinders	V20
Output/cylinder	410 kW
Power, max	9000 kW
Power, continuous	8200 kW
Rated speed	1150 rev/min
Idle speed	380 rev/min
Mean effective pressure	24.6 bar
Mean piston speed	12.1 m/s
Weight/power ratio	5.3 kg/kW
Weight	43 t
Specific fuel consumption	190–195 g/kW h
Dimensions (L 3 W 3 H)	7.4 m 3 1.9 m 3 3.3 m

Uprated Series 8000 design

The V20-cylinder Series 8000 engines have a continuous rating of 8200 kW in M70 form and peak power of 9000 kW in the M90 form. Customer interest in higher performance as well as anticipated stricter emission limits stimulated the development of a 20V 8000 M71 variant of enhanced design with rated outputs up to 9100 kW, which was unveiled in 2007. A challenge to the designers was to raise power while improving the engine's characteristic qualities.

An 11% power increase was achieved with the same bore and stroke dimensions and rated speed by raising the brake mean effective pressure to 27.3 bar. To avoid increasing the thermal load on the engine, the charge-air pressure was raised by developing the single-stage turbocharging system. At rated power, the charge-air pressure reaches an absolute value of 4.1 bar. The high maximum permissible combustion pressure of 230 bar affords sufficient possibilities for a high compression ratio, thus ensuring that the engine's starting and idling characteristics with the increased power are as favourable as the original design.

Classifying the M71 application called for the engine to be run at 10% overload, thereby achieving an output of 10,000 kW. The resulting 20V 8000 M71 engine was released in both consumption-optimized IMO and emissions-optimized EPA Tier II variants.

A new combustion process was developed to achieve an improvement in emissions performance meeting future requirements at all the power output options without compromising fuel economy. The configuration of the cylinder crown recess in conjunction with the injection pattern and fuel injector design was the most important variable in combustion engineering. Computer-aided 3D combustion modelling made it possible to refine the design in fewer steps. Development of the sequential turbocharging system, based on four MTU ZR265 turbochargers, also contributed to the achievement of low fuel consumption and minimal smoke levels at low engine speeds.

Continual feedback from 20V 8000 engines in diverse service contributed to design refinements seeking to maximize reliability and safety, MTU citing these problems and solutions:

- Wear on the connecting rod axial bearing: coating of the bearing area reduced the friction and significantly reduced the wear.
- Vibrations on the HP fuel pump were reduced by a new screw design and optimized piping arrangement.
- Leakages in exhaust pipe gaskets between cylinders and turbochargers: a modified vee-clamp, improved gasket, and E-ring were able to compensate for the higher axial deformation.
- Leakages due to the vibration of the coolant water connection pipes to the turbocharger carrier housing: design changes reduced the vibration level and eliminated the leakage.

MTU/DDC designs

An alliance formed by MTU in 1994 with DDC resulted in the creation of two advanced high-speed designs for joint marketing by the German/US partners, the Series 2000 and Series 4000 engines, which were launched in 1996. MTU was responsible for the basic design development of both engines and the variants for propulsion applications. Both series have proven highly successful in propulsion and genset drive sectors, benefiting from progressive refinements.

Series 2000

The Series 2000 design, with a 130 mm bore/150 mm stroke, is based on the Mercedes-Benz 500 commercial vehicle engine (Fig. 26.14). Models with V90° 8, 10, 12, and 16 cylinders cover a propulsion band from 500 to 1790 kW, while genset drive applications start from 270 kW. The most powerful version, intended for yachts, has a maximum speed of 2300 rev/min. A single-stage turbocharging system is specified as standard but, when special requirements regarding power band width and acceleration characteristics are dictated, Series 2000 engines can benefit from MTU's sequential turbocharging system.

Fuel injection, adapted from the commercial vehicle engine, was originally a solenoid valve-controlled pump/fuel line/injector system, each cylinder having its own separate injection pump. MTU's third-generation common rail injection system was introduced from 2004 to create the more compact and higher performance Series 2000CR engine, which reportedly set new standards in power density in its class. The V16-cylinder version delivers 1790 kW from a package weighing 3380 kg.

Marine versions are fitted with MTU's TE twin-circuit cooling system (detailed earlier in the section Series 396 Engine) which fosters an optimum

Fig. 26.14 An MTU V16-cylinder Series 2000CR engine.

temperature for all operating conditions: intake air can be heated at idling speeds or at partial throttle and cooled at full throttle.

Series 4000

A number of notable design and performance features were highlighted by MTU for the 165 mm bore/190 mm stroke Series 4000 engine whose designers focused on high reliability and ease of maintenance without compromising compactness (Figs 26.15 and 26.16). A propulsion power band from around 840 to 2720 kW at 2100 rev/min is covered by the V90° 8-, 12-, and 16-cylinder versions.

Testing confirmed fuel economy figures that reportedly set a new standard for compact engines in the design's performance class and speed range. Specific fuel consumption of 194 g/kW h was considered exceptional for an engine with a power-to-weight ratio of between 2.7 and 3.5 kg/kW.

Fig. 26.15 MTU's upgraded Series 4000 programme includes a V20-cylinder model delivering up to 4300 kW.

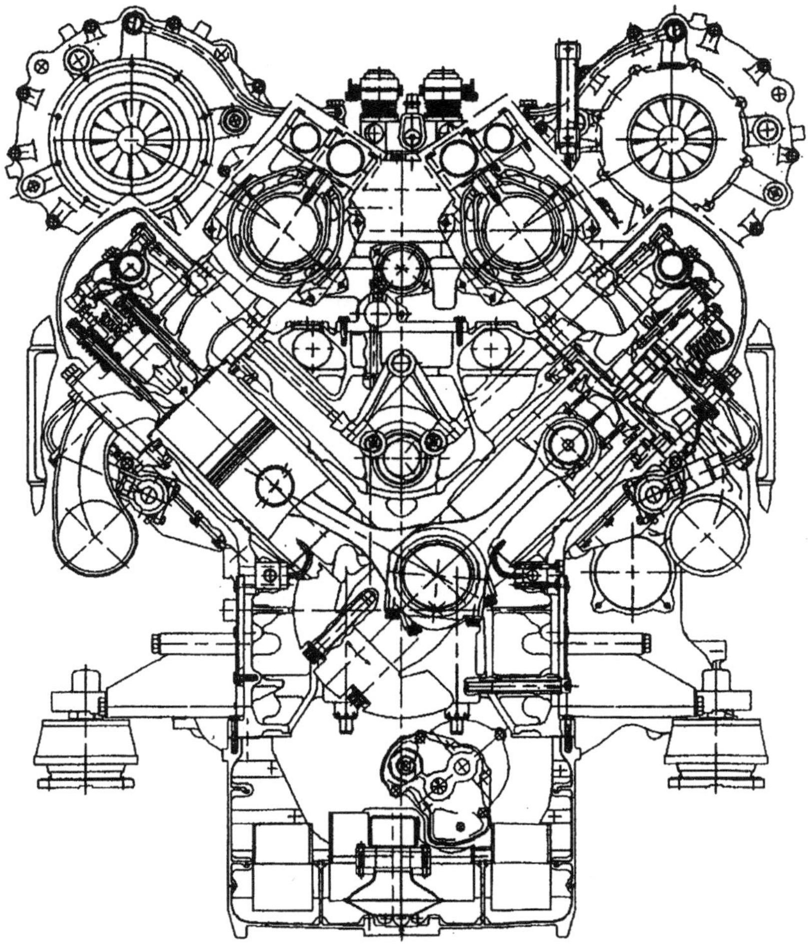

Fig. 26.16 Cross section of MTU Series 4000 engine.

A key contribution, along with a high peak firing pressure potential, is made by the common rail fuel injection system—an innovation in this engine category—whose development was carried out in conjunction with MTU's specialist subsidiary company, L'Orange (see Chapter 8).

The system, which allows infinite adjustment of fuel injection timing, volume, and pressure, embraces an HP pump, a pressure accumulator, injectors, and an electronic control unit. The tasks of pressure generation and fuel proportioning are assigned to different components: pressure generation is the job of the HP pump while metering of the fuel relative to time is the job of the injectors.

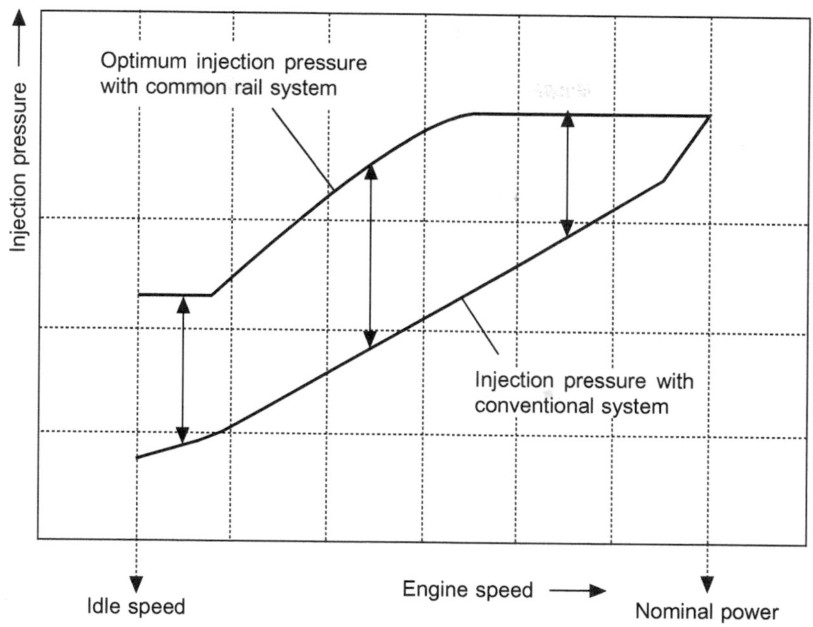

Fig. 26.17 Performance of common rail fuel injection system of MTU's Series 4000 engine.

The flexibility of the common rail system across the engine power band enables it to deliver the same injection pressure (around 1200 bar) at all engine speeds from full-rated speed down to idling (Fig. 26.17). The HP fuel pump and electronically controlled injectors are fully integrated into the ECS. Peak pressures with the common rail system are around 20% lower than with conventional systems, reducing stress on the HP components. The pump is also simpler and its plunger has no helices. Since the pressure-relief requirement of conventional systems is eliminated, the mechanical effort is reduced, which further enhances fuel economy.

One fuel rail is provided for the injectors and their respective solenoid valves on each bank of cylinders. The fuel pressure is generated by the HP pump driven via a gear train at the end of the crankcase. The electronic system controls the amount of fuel delivered to the injectors by means of the solenoid valves, while the injection pressure is optimized according to the engine power demand. Separate fuel injection pumps for each cylinder are thus eliminated, and hence the need for a complicated drive system for traditional pumps running off the camshafts. A reduced load on the camshaft and gearing system is therefore realized, and the removal of the mounting holes for separate injection pumps enhances the rigidity of the crankcase.

Maximum strength and optimized rigidity were sought from the crankcase, with the integration of the main lubrication and coolant circulation channels reducing the number of separate components (Fig. 26.18). The crankshaft machined all over and with bolted-on counterweights, is designed to withstand maximum combustion pressures of up to 200 bar. Wear-resistant sleeve bearings contribute to longer service intervals. The cylinder head design, with integrated coolant channels and stiffened base, also addresses a 200 bar pressure. Provision is made for two inlets and two exhaust valves, and a fuel injector located in the centre of the head (Fig. 26.19).

A piston comprising an aluminium skirt and a bolted-on steel crown is provided with chrome-ceramic coated rings which, in combination with the plateau-honed cylinder liner, foster low lubricating oil consumption, and extended service life. An optimized combustion chamber shape and the bowl of the composite piston promote low fuel consumption and low emission levels.

Exhaust and turbocharging systems on the Series 4000 engine benefit from established MTU practice. A triple-walled exhaust system, with an outer water-cooled aluminium casing, ensures that the surface temperatures do not exceed permissible levels at any point while also securing gas tightness. Heat dissipation via the cooling system is reduced. The exhaust pipework is positioned centrally between the V-cylinder banks. The

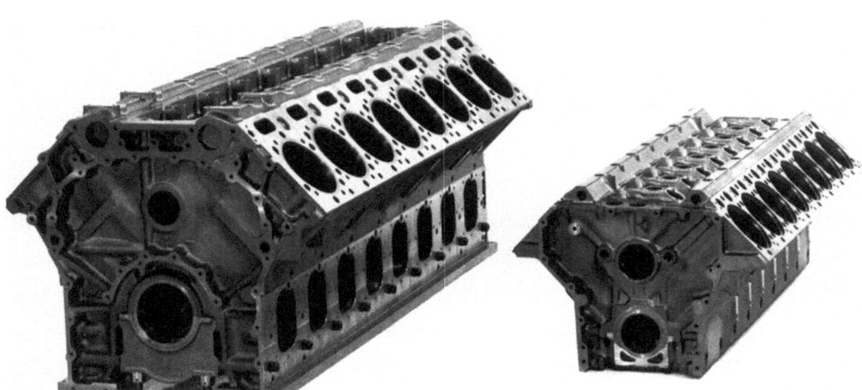

Fig. 26.18 The crankcases of MTU's Series 2000 (right) and Series 4000 engines.

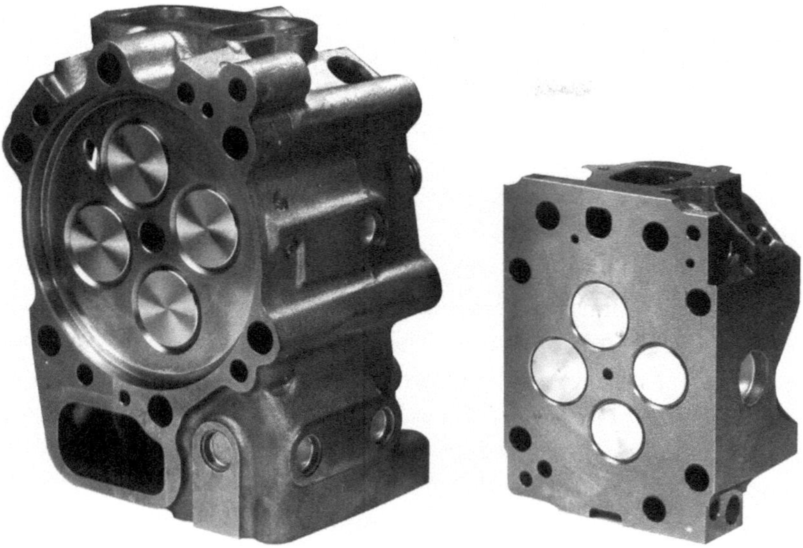

Fig. 26.19 The four-valve cylinder heads of MTU's Series 2000 engine (right) and Series 4000 engine (left) with the central fuel injector.

turbochargers are mounted in a water-cooled housing. A sequential turbocharging system is exploited to deliver high torque at low engine speeds. A wide powerband at a high fuel economy is also claimed, along with excellent acceleration capabilities without black smoke emission.

A split-circuit cooling system for the engine and intake air complements the sequential turbocharging system. It acts as an 'intelligent' cooling system to maintain engine coolant, lubricating oil, and intake air at an optimum temperature for all operating conditions. When the engine is idling or running at a low load, the temperature of the intake air is raised in order to ensure smooth and complete combustion without generating white smoke emissions.

On the front end of the engine is an integral service block providing ease of inspection and maintenance of the sea-water cooler, oil cooler, oil filter, oil centrifuge, and fuel filter. The turbochargers and charge-air coolers are located at the flywheel end.

Access openings in the crankcase are said to be large enough to allow all running gear servicing to be performed without removing the engine, even under restricted machinery space conditions. The application-tailored service block integrated into the auxiliary power take-off end also simplifies routine maintenance tasks. A variety of ancillary mounting options enable the engine to be matched to specific customer requirements.

An early indication of necessary maintenance, based on the actual duty profile, is provided by an electronic engine management system designed to foster reduced downtime periods and lower servicing costs. A time-between-major-overhauls of 18,000 h was anticipated by MTU.

Uprated Series 4000 engines

Four enhanced-power Series 4000 models developing 6%–10% more output than earlier versions were introduced in 2004 in response to the demands of workboat and ferry operators. These V12- and V16-cylinder models benefited from extensive service experience with the original design, which remained in the programme. Two V16 M61R and M61 models with respective ratings of 1520 and 2000 kW were added to the range for workboat propulsion, while V12 and V16 M71 models were uprated to 1850 and 2465 kW, respectively, for the ferry propulsion market.

Even higher power rises along with improved shock resistance, electromagnetic compatibility, and acoustic signatures were offered for Series 4000 models launched in 2006 for naval applications. These engines fulfil the strictest military requirements, such as the latest NATO standards, and comply with US EPA Tier II emission regulations. Output per cylinder was increased by over 26% (from 170 to 215 kW) at a rated speed of 2170 rev/min. A V12-cylinder model can thus deliver 2580 kW and a V16 model 3440 kW. In addition, the upper power limit of the Series 4000 programme was proposed to be significantly extended by the introduction, for the first time, of a V20-cylinder version offering outputs up to 4300 kW.

Advances in fuel injection, turbocharging, and electronic management systems (all MTU in-house specialties) contributed to the higher performance and lower emissions. A new generation common rail fuel injection system exploits an HP in-line pump and LEAD injectors with individual fuel accumulators to secure a virtually constant pressure of 1800 bar across the entire engine operating range. MTU's latest generation ADEC (Advanced Diesel Engine Control) electronic management system facilitates triple injection modes (pilot, main, and after-injection) which promote optimum control of the injection process for a low pollution, high-efficiency combustion.

Progress in own-design turbocharger development was also tapped. The new engines are fitted with two turbochargers (V12- and 16-cylinder models) or four turbochargers (for the V20-cylinder model) arranged in a single-stage sequential charging system, the turbochargers being switched-in or -out depending on the engine speed.

In 2016, the engine was again modified to allow compliance with IMO Tier III NOx requirements. This was achieved by a variety of technological advances to the turbocharging system, the combustion process, and the injection system of the engines combined with MTU's new SCR system, and an additional diesel particulate filter was not required to meet the requirements.

Niigata

Niigata has been producing marine engines since 1919 and is today part of Ishikawajima Heavy Industries.

Compact and highly rigid engines were sought by the Japanese designer from a monobloc structure fabricated from nodular cast iron, with the extensive use of light alloy elements contributing to a modest overall weight. Engine width was minimized by adopting a V60° cylinder bank and overall length reduced by minimizing the distance between cylinders while maintaining the required bearing width (Figs 26.20–26.22).

Niigata's current range of high-speed FX series offers 165 and 205 mm-bore designs with power outputs ranging from 1545 kW at 1950 rpm to 4000 kW at 1650 rpm with V12- and V16-cylinder models. The compact and highly rigid engines are based on a monobloc structure of nodular cast iron, with extensive use of light alloy elements contributing to a modest overall weight: a 12V16FX model has a dry mass of 4.5 tons and a 16V20FX model developing 4000 kW weighs 13.8 tons. Engine width was minimized by adopting a V60-degree cylinder bank, and overall length reduced by minimizing the distance between cylinders while maintaining the required bearing width.

A one-piece nodular cast-iron piston was specified for the V16FX engine but the higher mechanical and thermal stresses imposed by the cylinder pressure parameters of the V20FX (Blue Arrow 6000) engine influenced the selection of a component with a steel crown and aluminium skirt for the larger design.

Marketed as the Blue Marine 1100, Niigata's 6MG17HX engine was developed to provide smaller vessels with a high power output from a compact and lightweight installation. The six-cylinder 165 mm-bore design develops a maximum rating of 809 kW at 1703 rpm.

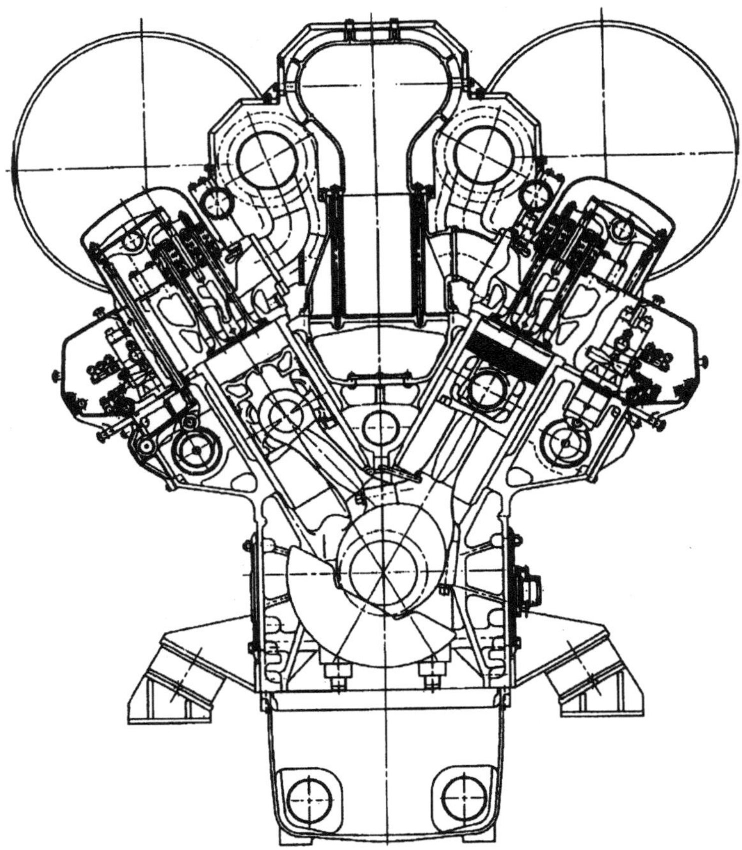

Fig. 26.20 Cross section of Niigata V26FX engine.

A twin-turbocharger system contributes to engine acceleration performance and favourable smoke characteristics in the low load and transitional operational ranges. An antipolishing ring at the top of the cylinder liner helps to secure consistently low lube oil consumption and to protect the lubricant from fouling and degradation.

A one-piece nodular cast-iron piston was specified for the V16FX engine, but the higher mechanical and thermal stresses imposed by the cylinder pressure parameters of the V20FX 'Blue Arrow' engine influenced the selection of a component with a steel crown and aluminium skirt for the larger design. A bore cooling-type oil gallery structure on the crown reduces the temperature of the piston at higher mean effective pressures. Higher cylinder pressures also foster increased piston blowby, calling for a high sealing

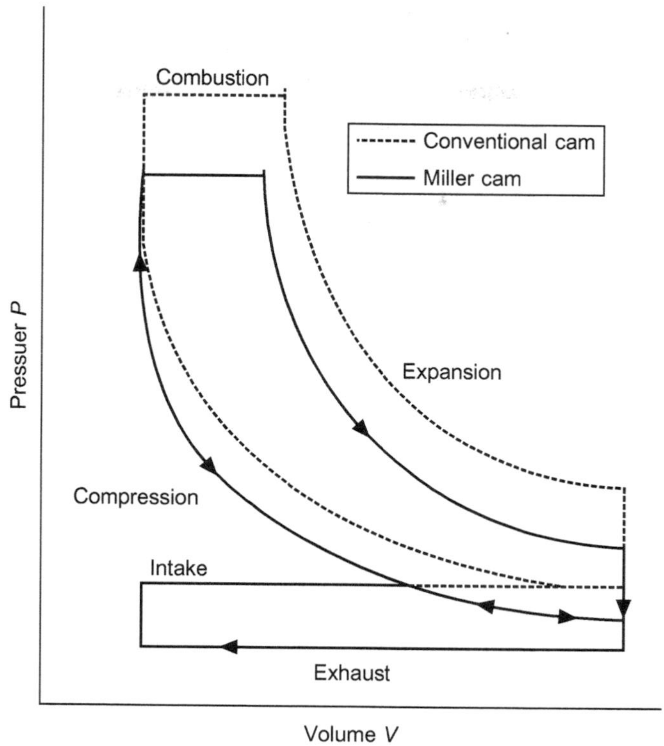

Fig. 26.21 P–V diagram for the Miller system (Niigata).

capability and wear resistance from the piston rings. A chromium-ceramic coated top ring contributes to these properties, while an antipolishing ring in the top of the cylinder liner yields a lower lube oil consumption rate.

Niigata 16V20FX engine data

Bore	205 mm
Stroke	220 mm
Cylinders	V16
Output, mcr	4000 kW
Speed	1650 rev/min
Mean piston speed	12.1 m/s
Specific fuel consumption	210 g/kWh
Weight	12,100 kg
Weight/power ratio	3.15 kg/kW

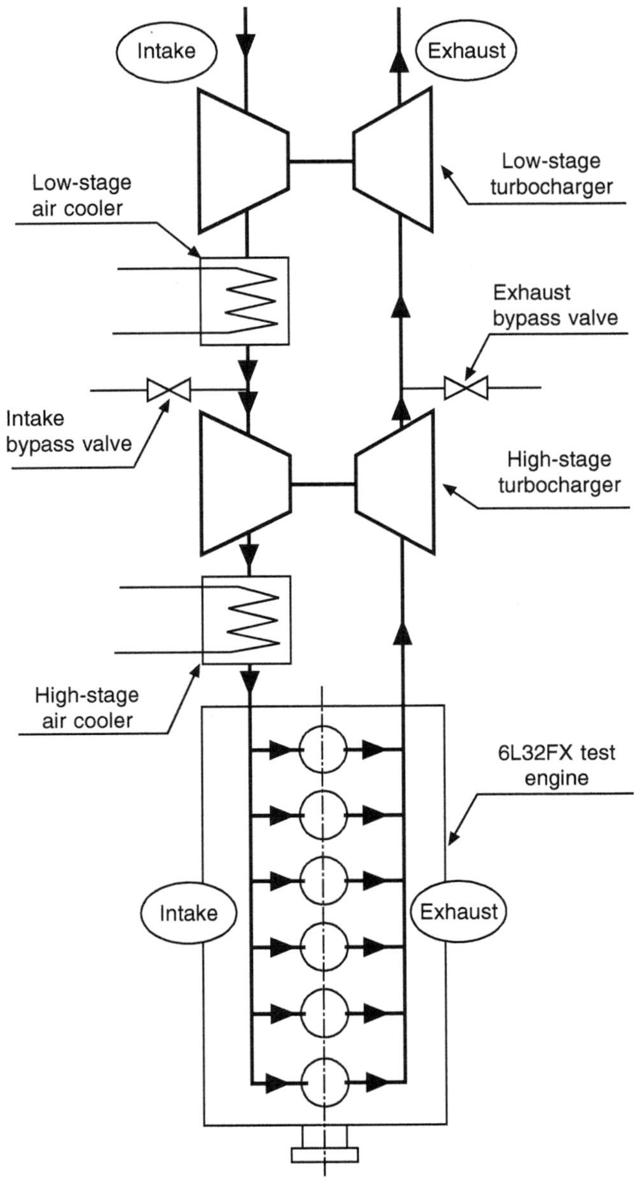

Fig. 26.22 Two-stage turbocharging system (Niigata).

Wärtsilä

The medium-speed specialist Wärtsilä added high-speed designs to its four-stroke engine portfolio with the acquisition in 1989 of SACM Diesel of France. The Finnish parent group saw the market potential for a new generation engine blending the best features of high- and medium-speed designs for continuous duty applications, resulting in the 1994 launch of the Wärtsilä 200 series (Figs 26.23 and 26.24).

The 200mm bore/240mm stroke engine was produced by Wärtsilä France for propulsion and genset drive applications, with an output band from 2100 to 3600kW at 1200 or 1500rev/min covered by V12-, 16-, and 18-cylinder models. High reliability in continuous duty (defined as 24h a day operation with an annual running period of over 6000h) was sought from an engine structure and main components designed for a maximum cylinder pressure of 200bar. Wärtsilä's medium-speed engine technology was exploited to achieve a high power density, low emissions and fuel consumption, and ease of maintenance. The engine was comparatively short lived as the company concentrated on developing dual-fuel medium speed gas engines (Figs 26.25–26.28).

It was not until 2018 that the first Wärtsilä-branded high-speed engine was introduced. The Wärtsilä 14 is a high-speed, compact engine designed to fit requirements for limited space and weight, lower capital expense, to

Fig. 26.23 Wärtsilä W200 engine.

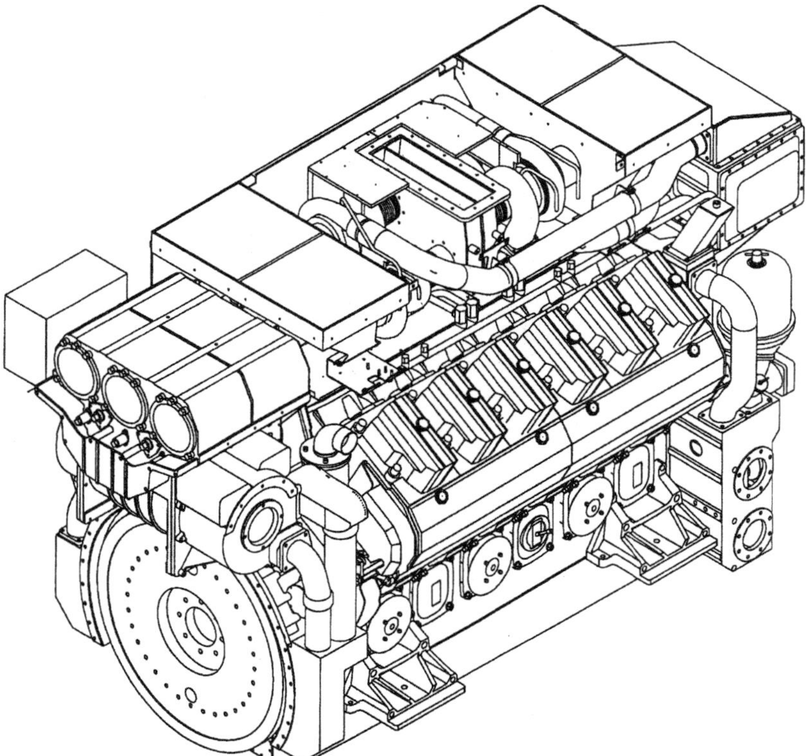

Fig. 26.24 Wärtsilä W200 engine. The original automotive-type turbochargers shown here were later replaced by ABB or Holset models.

Fig. 26.25 The nodular cast-iron block of the Wärtsilä W200 engine incorporates fluid channels.

Fig. 26.26 The Wärtsilä W200 engine crankshaft has gas-nitrided surfaces.

Fig. 26.27 The fuel injection pump of Wärtsilä's W200 engine is integrated into the multi-housing.

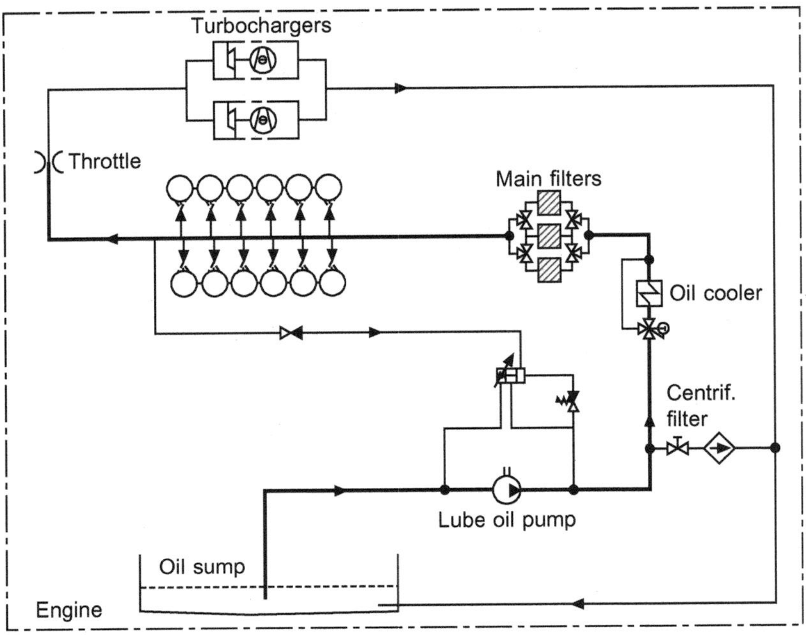

Fig. 26.28 Lubricating system of W200 engine.

meet current and future global emissions regulations, and to provide customers with improved efficiency, safety, and environmental sustainability.

Highly versatile and engineered to support a wide range of applications within marine and offshore markets, the Wärtsilä 14 is intended to serve both as main propulsion and auxiliary genset and is ideal for hybrid installations. This small but powerful high-speed engine is available in 12- and 16-cylinder configurations, delivering a power output of 755–1340 kW in mechanical propulsion, and 675–1155 kWe in auxiliary and diesel–electric configurations. It is aimed at vessels where the power to weight ratio, fuel-type, efficiency, safety, and environmental compliance are key considerations. Examples of such operating profiles include tugs, fishing vessels, offshore service vessels, small ferries, and as an auxiliary on larger commercial vessels.

Developed in close cooperation with Liebherr, one of the largest construction machine manufacturers in the world and a leading technology provider for a number of other industry verticals, first deliveries of the Wärtsilä 14 were planned for the latter part of 2019. Liebherr is responsible for product development, classification, and manufacturing of the engines. At the same time as launching the new engine in its genset form, Wärtsilä announced the introduction of its HYTug 40 hybrid tug as a suitable platform.

A smaller version of the HYTug launched in 2017 and designed for shallow draft operation, the HyTug 40 would feature two Wärtsilä 14 engines along with a hybrid propulsion solution which delivers various flexible operating settings, including a zero-emissions mode.

The Wärtsilä 14 has a bore of 135 mm and a stroke of 157 mm. The 12V version can output between 755 and 1005 kW at speeds of 1500 and 1900 rpm, respectively. For the 16V model, the power outputs are 1005–1340 kW at the same speed ranges.

Wartsila 14 engine.

Yanmar

A long-established Japan-based engine maker, Yanmar was one of the earliest Asian makers in the marine field. Today it has medium and high-speed offerings in its portfolio some of which are based upon engines for land use. The range of high-speed engines is extensive but many are too small for mainstream commercial work.

The AY series is the most powerful and comes in six cylinder in line and V12 engine variants with the highest output of 1340 kW being delivered by the 12AYM-WGT, running at a shaft speed of 1940 rpm. The engines both have a bore of 155 mm and a stroke of 180 mm. This product, in its latest generation, has been developed as an 'Eco Diesel' using exhaust gas recirculation to satisfy IMO Tier II NOx rules. Fuel consumption has also been improved and smoke generation has been reduced. The EGR is facilitated through control of the lift and timing of intake and exhaust valves, with the partial reverse flow of exhaust gases air back into the cylinder from the common exhaust manifold.

The AY engine has been developed based on long-term experience gained from older 12LA and 16LA series engines, first put into service in the 1980s. Many performance improvements have been derived from a new combustion system, which uses an all-new design of fuel injector with multiple micro-sized injection holes to produce both a finer fuel mist and a staggered injection pattern. Deeper combustion chambers have been introduced, along with re-profiled cylinder heads, and the air intake port geometry produces a swirl or vortex flow into the combustion chamber, which promotes better fuel-air mixing. Combustion efficiency is therefore improved, as is engine specific power capability. Compared with the previous equivalent engine, typical fuel savings are estimated as being 10 g/kWh, or around a 5% reduction in fuel burned for the same power output.

The AY series has low lubricating oil consumption levels and long overhaul intervals due to the use of special ceramic surface treatment on cylinder liners. The process embeds Silicon Carbide onto the wear surface increasing its hardness and gives the liners superior wear resistance and durability. Combined with the use of nitrided stainless steel piston rings and tight clearances between the piston and liner, leakage past the piston rings is minimized. The engine also has an optimized flywheel weight, water-cooled exhaust manifold and is designed to be leak-free; an increasingly important aspect with new rulings on oil spills now introduced in some waters. From a maintenance perspective, the engines are again designed for in situ piston replacement and cylinder heads are individual. Yanmar has also retained a mechanical engine management system to avoid risks of failure of delicate electronic equipment in often restricted and hot marine engine rooms.

Automotive-derived engines

Marinized automotive-derived high-speed engines are popular for small craft-propulsion and genset drives on larger ships, where their ruggedness and reliability (based on numerous truck applications) are appreciated.

Swedish truck engine specialist **Scania** launched its most powerful-ever marine engine in 2001, the DI16M design replacing the long-established DI14M series. The new 16-L engine exploited an extra 2 L of displacement in a package that was both shorter and lower than its forerunner. Increased torque and lower emissions were also delivered by the DI16M, a twin-turbocharged V8-cylinder design offered with commercial power ratings from 338 to 590 kW at speeds from 1800 to 2200 rev/min. The highest rating addressed light-duty applications such as patrol boat propulsion; ratings from 440 to 590 kW (medium-duty/high-power commercial) served pilot, police, and rescue vessels; and the lower output range was applied for heavy-duty commercial propulsion (tugs, workboats, fishing vessels).

The DI16M was the first Scania marine engine to use electronic unit injectors (EUIs) and the company's new S6 fully electronic engine management control module. EUI allows fuel injection timing to be adjusted to suit all engine running conditions, from cold start to idling and throughout the power range. The system fosters improved power output, fuel economy, and emission control, with reduced exhaust smoke.

Based on Scania's 16-L truck engine, the 127 mm bore/154 mm stroke DI16M shared the same separate four-valve cylinder heads, cylinders, and valve mechanism as its smaller sibling, the DI12M. New composite pistons featured aluminium bodies and steel crowns to handle the high combustion pressure. The engine measures 1.25 m long 3 1.178 m wide 3 1.15 m high overall and has a dry weight of 1500–1600 kg, depending on its equipment.

Scania's current 12- and 16-L engines, respectively, in-line V6- and V8-cylinder versions of a 127 mm bore/154 mm stroke design, are offered in a range of ratings for propulsion duties. The power band extends from 221 kW at 1800 rev/min for continuous service to 588 kW at 2200 rev/min for patrol craft/short service applications. Intermittent service and patrol craft/long service ratings between these limits can also be specified for the engines which are popular for small ferries, rescue boats, fishing vessels, and pilot and patrol craft.

An electronic engine management system in combination with EUIs fosters optimum combustion and enables data to be collated and processed every microsecond to regulate injection timing and fuel quantity injected. Precision in controlling these parameters, together with high injection pressure, promotes lower fuel consumption and emission levels, including compliance with IMO, EU Stage IIIA, US EPA Tier II, and CCNR Stage II regulations.

Low emissions were a high priority for another Swedish specialist, **Volvo Penta**, when designing its 16-L TAMD 165 automotive-based marine engine. The in-line six-cylinder model benefited from an upgraded fuel injection pump and a new injector as well as a new charge-air cooling system, which reduces the inlet air temperature by around 10°C compared with earlier engines. New pistons and rings reportedly reduced lube oil consumption by 50%. High reliability and low noise and vibration were also

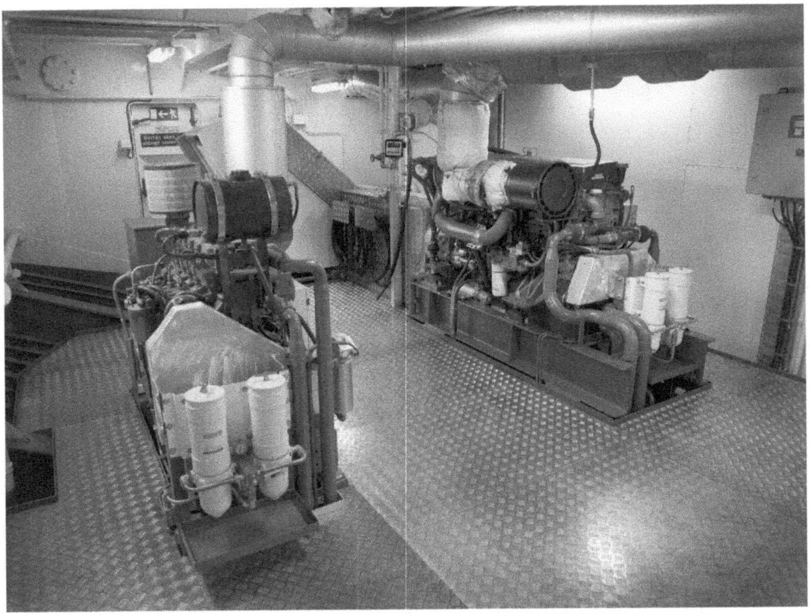

Fig. 26.29 Two Volvo Penta D16 MH engines arranged to power a double-ended road ferry.

targeted, relevant measures including the cast-iron engine block, trapezoidal connecting rods, and a crankshaft with seven main bearings.

With a peak output of 551 kW at 1900 rev/min in its Rating 2 form, Volvo Penta's latest D16 engine (Fig. 26.29) offers an output 110 kW higher than its predecessor but with similar installation dimensions and weight. The 16-L design is available for propulsion (MH) and genset (MG) applications. The torque at low revolutions per minute is also 50%–90% higher thanks in part to a centrally mounted twin-entry turbocharger facilitating pulse charging.

Improved manoeuvrability and load pick-up and reduced load sensitivity result from the higher torque, while a new charge-air cooling system designed for lower charge-air temperatures contributes to the torque performance, reliability, and lower emissions. The Volvo Penta-developed EMS 2 electronic engine management system fosters performance and fuel efficiency as well as improving engine protection. Unit fuel injectors operating with injection pressures up to 2000 bar further contribute to higher fuel economy and lower emissions.

CHAPTER TWENTY-SEVEN

Gas turbines

Gas turbines have dominated warship propulsion for many years but their use in the commercial shipping sector has never reached the expectations of the few specialists in this niche technology sector. Breakthroughs in container ships, a small gas carrier and the Baltic ferry *Finnjet* during the 1970s promised a deeper penetration that was then thwarted by the rise in bunker prices and the success of diesel engine designers in raising specific power outputs and enhancing heavy fuel burning capability.

In more recent years, gas turbine suppliers with suitable designs secured propulsion plant contracts from operators of large cruise ships and high-speed ferries, reflecting the demand for compact, high-output machinery in those tonnage sectors, rise in cycle efficiency, and tightening controls on exhaust emissions. A new generation of the marine gas turbine—superseding designs with roots in the 1960s—may eventually benefit from the massive investment in aero engine R&D strengthening competitiveness in commercial vessel propulsion, particularly as more stringent emission limits favour the adoption of very low sulphur-content distillate fuels.

The main candidates for gas turbine propulsion in commercial shipping are as follows.

Cruise ships: the compactness of gas turbine machinery can be exploited to create extra accommodation or public spaces, and the waste heat can be tapped for on-board services. In a large cruise ship project some 20–100 more cabins could be incorporated within the same hull dimensions, compared with a diesel-electric solution, depending on the arrangement philosophy. Compactness is fostered by the smaller number of prime movers and minimal ancillary systems (roughly around 50% fewer than a diesel-based plant).

Large fast passenger ferries and freight carriers: the extremely high-power levels demanded for such vessels is difficult to satisfy with diesel machinery alone; coastal water deployment favours the use of marine diesel oil in meeting emission controls.

On 15 June 2013, *Francisco* a 1000-passenger catamaran ferry built by Australian shipyard Incat for Argentina's Buquebus achieved 58.21 knots—making it the fastest commercial ship in the world. The ship is powered by two 25 MW GE LM2500 marine gas turbines.

Pounder's Marine Diesel Engines and Gas Turbines
https://doi.org/10.1016/B978-0-08-102748-6.00027-X

LNG carriers: an ability to burn both cargo boil-off gas and liquid fuel at higher efficiency than traditional steam turbine propulsion plant should be appreciated. Between 2014 and 2018, a number of gas turbine-powered LNG carrier designs have been given Approval in Principle by classification societies. All of these have involved turbines from GE Marine.

Fast container ships: the compact machinery allows space for additional cargo capacity. However, current thinking at the IMO does not include fast cargo ships and indeed the pressure being applied by environmental groups would rather see ships subject to maximum speeds well below what would once have been considered typical for container ships.

Among the merits cited for gas turbine propulsion plant in commercial tonnage are:

- High power-to-weight and power-to-volume ratios; aero-derived gas turbines typically exhibit power-to-weight ratios at least four times those of medium-speed diesel engines; compactness and weight saving releases machinery space for extra revenue-earning activities; a General Electric LM2500 aero-derived gas turbine unit delivering 25,000 kW, for example, measures 4.75 m long × 1.6 m diameter and weighs 3.5 t.
- Low noise and vibration.
- Ease of installation and servicing fostered by modular packages integrated with support systems and controls.
- Modest maintenance costs, low spare part requirements, and ease of replacement.
- Environmental friendliness (lower NOx and SOx emissions than diesel engines).
- Reduced manning levels facilitated by full automation and unmanned machinery space capability.
- Operational flexibility: swift start-up: no warm-up or idling period required; idle can typically be reached within 30 s, followed by an acceleration to full power; deceleration can be equally as rapid, after which the turbine can be immediately shut down; subsequent restarts, even after high power shutdown, can be instant with no 'cool down' restriction.
- High availability, underwritten by excellent reliability and rapid repair and/or turbine change options.

Plant configurations

Optimizing the combination of low-power manoeuvring and high-power operation is accomplished in many naval applications by using a combined diesel or gas turbine propulsion system (CODOG). Other arrangements can be configured to suit the power demands and/or operational flexibility required for a project: combined diesel and gas turbine (CODAG—Fig. 27.1); combined gas turbine and gas turbine (COGAG); combined gas turbine or gas turbine (COGOG); and combined diesel and gas turbine electric propulsion (CODLAG or CODEG).

A notable COGAG plant powered the Stena HSS 1500-class high-speed passenger/vehicle ferries, whose service speed of 40 knots was secured by twin General Electric LM2500 and twin LM1600 gas turbines arranged in father-and-son configurations with a total output of 68,000 kW. All four turbines were deployed for the maximum speed mode, with the larger or smaller pairs engaged alone for intermediate speeds; this enabled the turbines to operate close to their optimum efficiency at different vessel speeds, with consequent benefits in fuel economy (Fig. 27.2). Three ships of the type were built, two of which were scrapped and the sole survivor is being used for office accommodation in Turkey. Whether its turbines are still in situ is unknown.

An example of a CODEG plant is provided by Cunard's cruise ship *Queen Mary 2*, whose 117,200 kW power station combines twin General Electric LM2500+ turbine sets with four Wärtsilä 16V46 medium-speed diesel engines, all driving generators.

Combined-cycle gas turbine and steam turbine electric (COGES) plants embrace gensets driven by gas and steam turbines. Waste heat recovery boilers exploit the gas turbine exhaust and produce superheated steam (at around 30 bar) for the steam turbine genset. Such an arrangement completely changes the properties of the simple-cycle gas turbine: while gas turbine efficiency decreases at low load the steam turbine recovers the lost power and feeds it back into the system. The result is a fairly constant fuel consumption over a wide operating range. Heat for ship services is taken directly from steam turbine extraction (condensing-type turbine) or from the steam turbine exhaust (back-pressure turbine), and there is thus normally no need to fire auxiliary boilers.

Installations supplied by General Electric for large Royal Caribbean Cruises' liners pioneered the COGES plant at sea, the first entering service in mid-2000 (Figs 27.3 and 27.4). Each 59,000 kWe outfit comprises a pair of GE LM2500+ gas turbine-generator sets, rated at 25,000 kWe apiece, and a 9000 kWe noncondensing steam turbine-generator. Heat recovery steam generators located in the exhaust ducts of the gas turbines produce the steam to drive the steam turbine and feed auxiliaries such as evaporators and

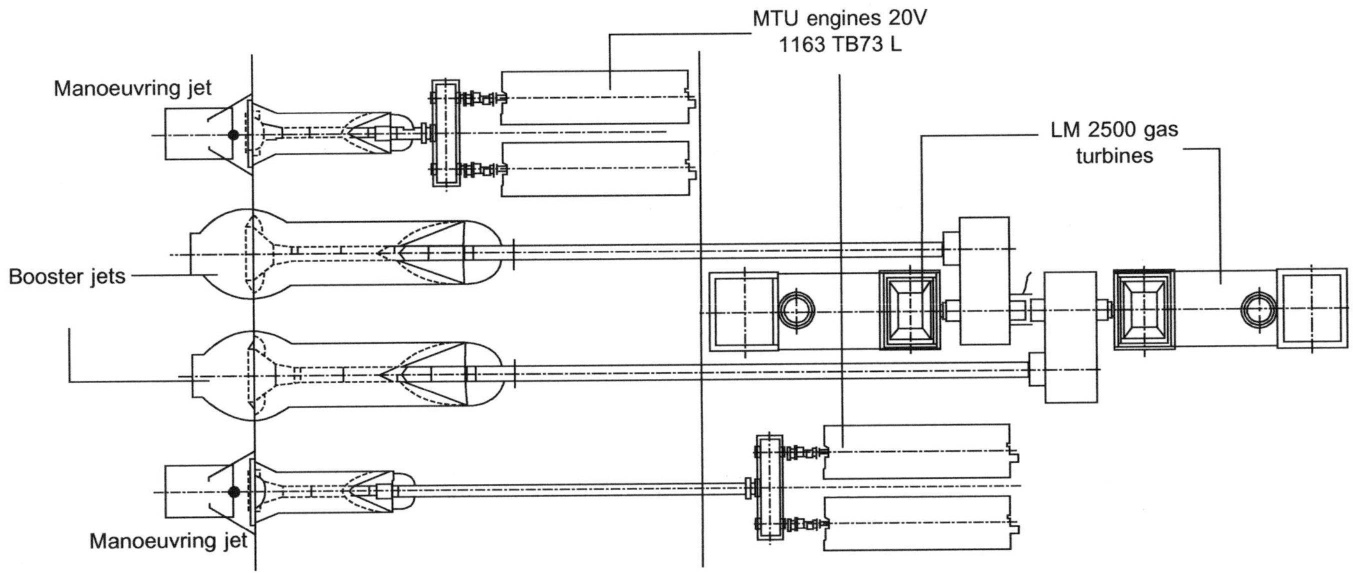

Fig. 27.1 CODAG propulsion plant configuration for a large monohull fast ferry.

Fig. 27.2 Final preparation of a General Electric LM2500 propulsion module for a large high-speed ferry.

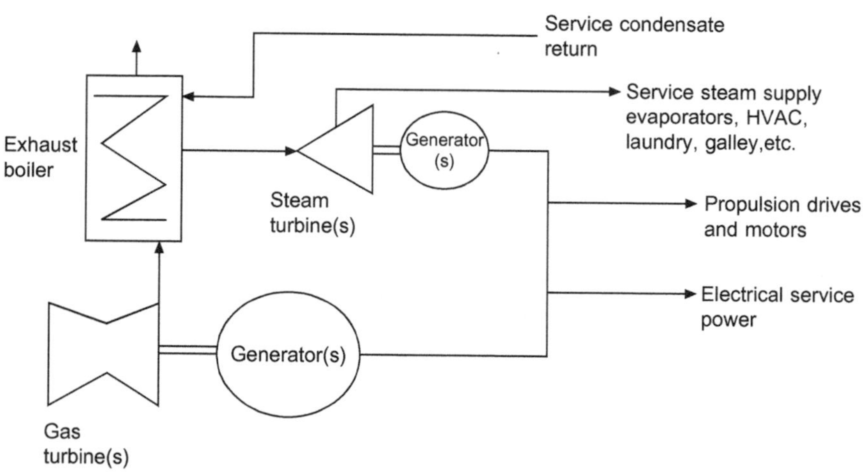

Fig. 27.3 Schematic layout of COGES plant for a large cruise ship (GE Marine).

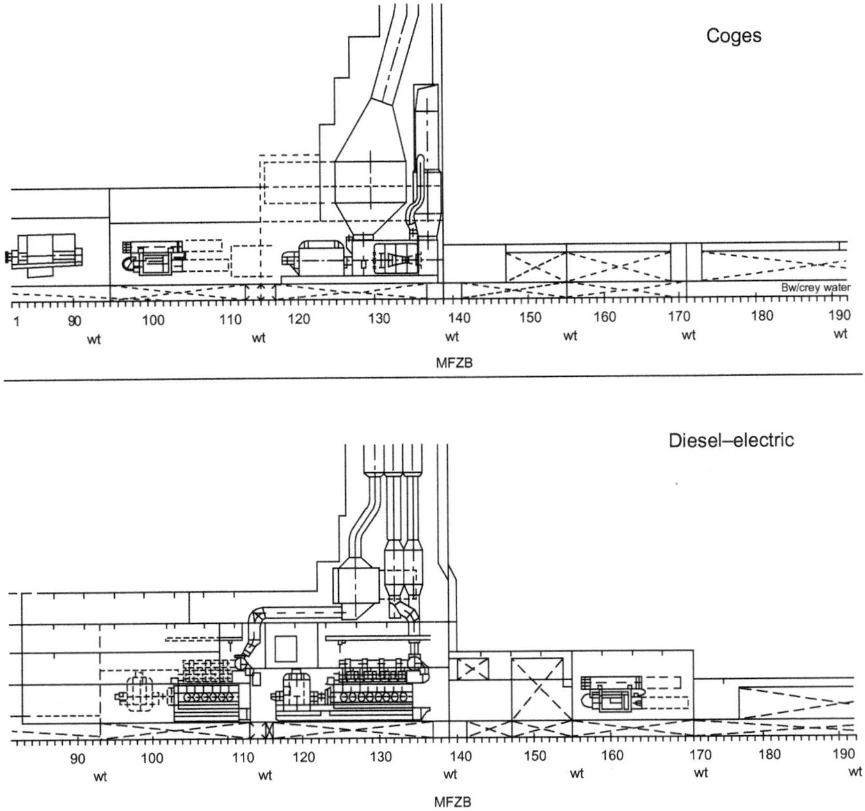

Fig. 27.4 Space-saving of a COGES plant compared with a diesel–electric installation for a cruise ship (Deltamarin).

heating systems. As no additional fuel is consumed to drive the steam turbine, the additional power it generates represents a 15%–18% increase in efficiency with the gas turbines operating at rated power. The auxiliary steam represents a further efficiency improvement. The GE models are currently installed in just under 20 passenger and cruise vessels of various sizes.

Lightweight gas turbine-generator sets (weighing approximately 100 tons) have allowed naval architects to locate them in the base of a cruise ship's funnel. Such an arrangement replaces the gas turbine inlet and exhaust ducts normally running to the engine rooms with a smaller service trunk housing power lines, fuel and water supplies to the gas turbine package. A significant area on every deck between the funnel and the machinery spaces is thus released for other purposes.

Cycles and efficiency

Significant progress has been made in enhancing the thermal efficiency of simple-cycle gas turbines for ship propulsion over the years, R&D seeking to improve the part-load economy and reduce the fuel cost penalty compared with diesel engines. In 1960 marine gas turbines had an efficiency of around 25% at their rated power, while second-generation aero-derivatives were introduced in the 1970s with efficiencies approaching 35%. Subsequent advances—design refinements, new materials and cooling techniques, and the appropriate matching of higher compressor pressure ratios—have resulted in some large simple-cycle turbines achieving efficiencies of over 40% (Fig. 27.5).

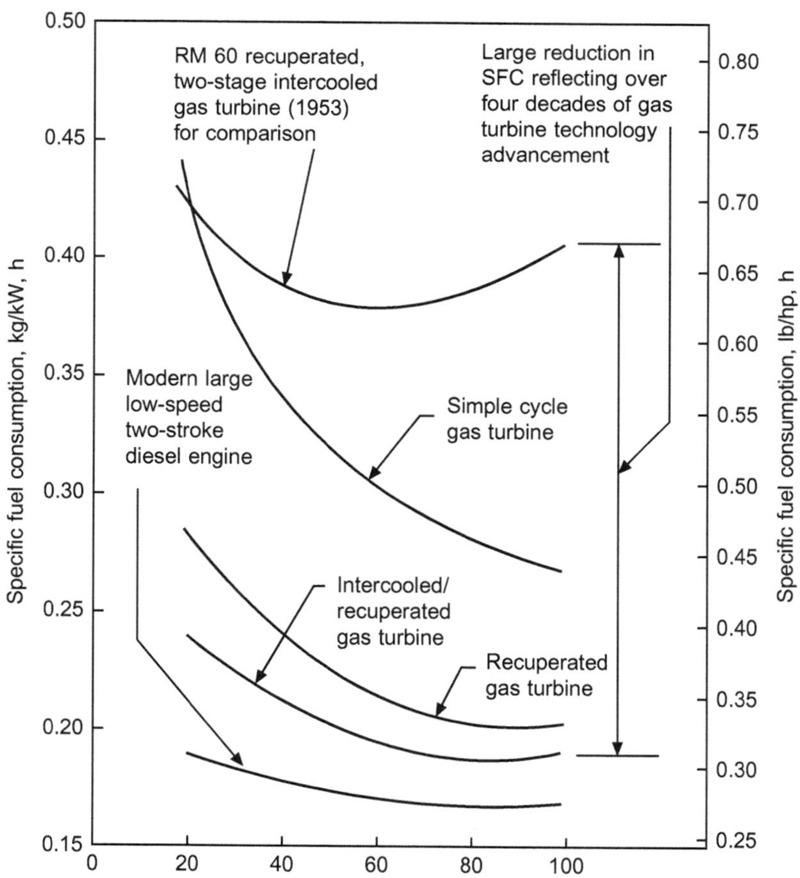

Fig. 27.5 Comparison of sfc curves against load for various gas turbine cycles and a low-speed two-stroke diesel engine.

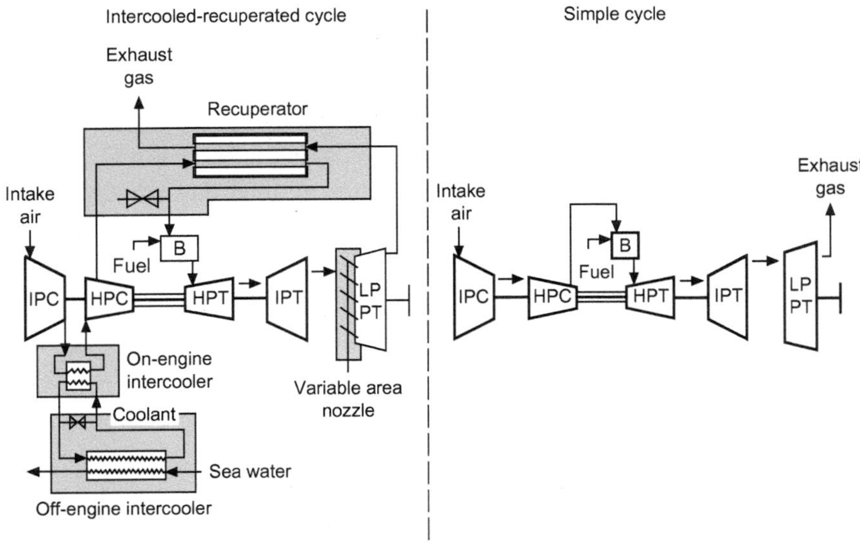

Fig. 27.6 Schematic layout of ICR cycle and simple-cycle gas turbines (Rolls-Royce).

Gas turbines, particularly in the lower range of outputs up to 5 MW, are significantly less efficient than diesel and also less flexible as diesel has a wider load range before efficiency falls to suboptimal levels.

More complex gas turbine cycles can deliver specific fuel consumptions (sfcs) closely approaching the very flat curve characteristics of larger diesel engines. Part-load efficiency can be improved in a number of ways, notably through the intercooled recuperated (ICR) cycle which uses the exhaust gas to heat the combustor inlet (Fig. 27.6). This method was chosen for the WR-21 gas turbine (detailed in section 'Rolls-Royce') which has achieved a 42% thermal efficiency across 80% of the operating range. (The first advanced-cycle gas turbine, the Rolls-Royce RM60, was operated in *HMS Grey Goose* for 4 years from 1953. Unfortunately, the technical complexity of the intercooling and recuperation was far ahead of the contemporary production techniques, thwarting commercial success, but the turbine proved reliable and efficient during its comparatively brief service.)

The ICR cycle and other performance-/efficiency-enhancing solutions were examined by Mitsubishi Heavy Industries, taking as an example, its own MFT-8 aero-derived gas turbine developed for marine use. The Japanese designer investigated various methods of improving the thermal efficiency and power output to similar levels as those of the diesel engine, focusing on six gas turbine cycle configurations (Fig. 27.7):

– *Simple cycle*: the standard layout comprising compressor, turbine, and combustor.

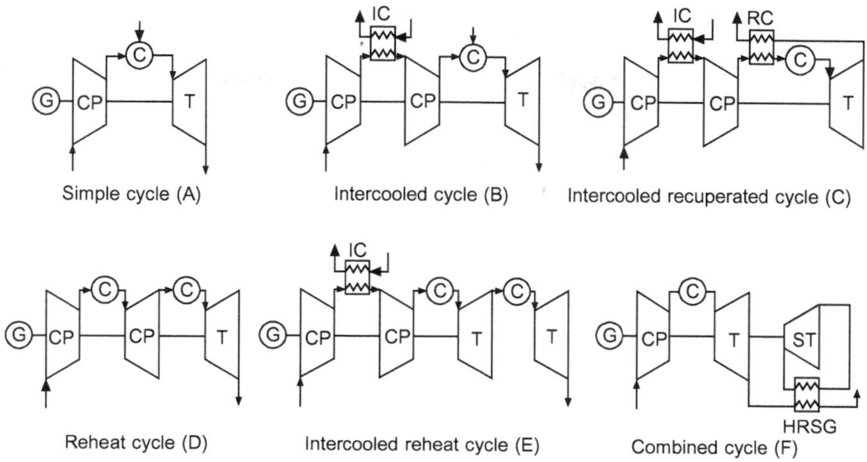

CP: Compressor IC: Intercooler C: Combustor T: Turbine RC: Recuperator
ST: Steam turbine HRSG: Heat recovery steam generator

Fig. 27.7 Various gas turbine cycle configurations (Mitsubishi).

- *Intercooled cycle*: as for the simple cycle but with an intercooler added to increase performance through a reduction of high-pressure (HP) compressor power.
- *ICR cycle*: as for the intercooled cycle but with a recuperator installed to recover heat from the gas turbine exhaust to the combustor inlet.
- *Reheat cycle*: as for the simple cycle but with a reheat combustor added downstream of the HP turbine.
- *Intercooled reheat cycle*: a combination of intercooled and reheat cycle configurations.
- *Combined cycle*: a bottoming steam turbine cycle added to a simple-cycle gas turbine.

The main conditions that are changed by varying the cycle, MHI explains, are the turbine inlet temperature and the pressure ratio (1150°C and 20:1, respectively, in the case of the MFT-8 simple-cycle gas turbine). Parametric studies were carried out to establish the optimum performance from each cycle, within the bounds of what is practically achievable in terms of temperatures and pressures. With this in mind, a maximum turbine exhaust temperature of 600°C was specified, and pressure ratios for maximum efficiency and maximum power were determined for each cycle.

The highest efficiency and power was obtained with the combined cycle option, which had the further advantage of a pressure ratio lower than even the simple cycle. The major disadvantage of the combined cycle is the fact that the heat recovery steam generator is physically large and contains very heavy components. Although this is not necessarily a problem in industrial applications, where the combined cycle is commonly used, it is a considerable drawback in marine installations.

MHI's investigations favoured the ICR cycle, which showed high efficiency and in power output was second only to the combined cycle. Moreover, maximum efficiency and maximum power were identical, at the achievable pressure ratio figures of 20 in each case. In the other nonsimple cycles, the two pressure ratio values were far apart, with one or other of the values being higher than 20, a ratio that incurs many practical disadvantages. The only problem with the ICR cycle, according to MHI, is that both the cost and size of the recuperator need to be reduced to make it more applicable to a marine gas turbine.

Although atmospheric pressure variation is an important consideration for aero gas turbines, it is relatively small for marine units operating at sea level both temperature and humidity can vary significantly. Both of these parameters influence the specific heat of the air as a working fluid within the gas turbine. Atmospheric variations can have a significant effect on thermal efficiency and sfc. Increased air temperature, in particular, reduces efficiency and lowers fuel economy; relative humidity has a less significant influence though at high inlet temperatures an increase in humidity will have an adverse effect on sfc.

Emissions

A lower operating temperature and more controlled combustion process enable gas turbines to deliver exhaust emissions with significantly lower concentrations of nitrogen oxides (NOx) and sulphur oxides (SOx) than diesel engines. Gas turbines typically take in three times the amount of air required to combust the fuel and the exhaust is highly diluted with fresh air. Along with a continuous combustion process, this yields very low levels of particulate emissions and cleaner exhaust.

Combustion in gas turbines is a continuous process, with average temperatures and pressures lower than the peak levels in diesel engines that foster NOx emissions. The fundamental characteristic of continuous combustion in a gas turbine is that residence time at high flame temperatures (a key cause of NOx formation) is capable of being controlled. In a gas turbine, a balance between smoke production and NOx generation can be easily secured.

Staged premixed combustion allows NOx emission levels to be reduced without the need for expensive selective catalytic reduction systems typically required by diesel engines to meet the most stringent controls (Fig. 27.8).

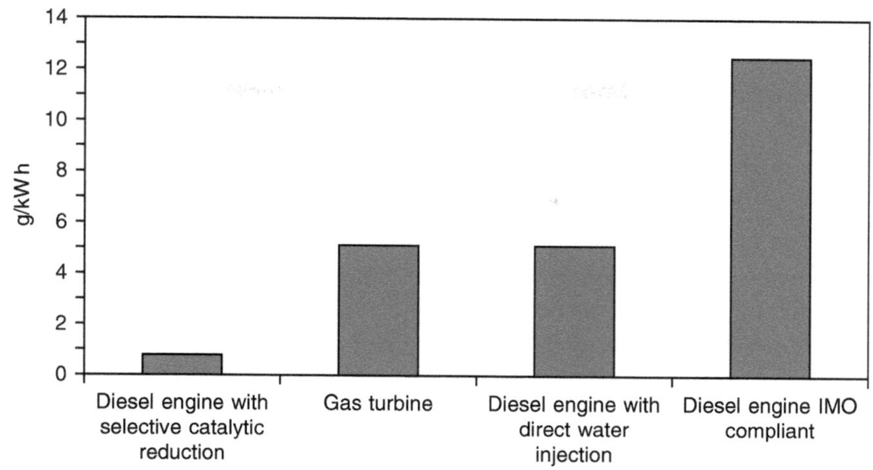

Fig. 27.8 Specific NOx emissions for prime movers operating on marine gas oil.

Proven technologies are available to reduce NOx emissions from gas turbines even further:

- Wet technology based on steam or water injection into a standard combustor can lower NOx emissions to less than 1 g/kW h, exploiting a concept proven in land-based industrial applications. The availability of clean water can restrict operation at sea but the technology may be suitable for limited use in coastal areas.
- Dry low-emissions (DLE) technology yields much lower emissions than current marine engine requirements, well below 1 g/kW h. Lean premix DLE combustion technology maintains a near-optimum fuel-air distribution throughout the combustion zone and the flame temperature in a narrow band favourable both to low NOx and low carbon monoxide production.

SOx emissions are a function of fuel sulphur content. The heavy fuel typically burned by diesel engines may have sulphur content up to 4.5% (3%–3.5% on average). In contrast, gas turbines fired by marine distillate fuels have a maximum 1%–2% sulphur content and the average level in gas turbine fuels is less than 0.5% (see section 'Rolls-Royce', WR-21 gas turbine).

Lubrication

The quantity of lubricating oil in circulation in a gas turbine is much smaller than in a diesel engine of equivalent output, and similarly, the oil consumption is significantly less. (The Rolls-Royce Marine Spey, for example, has a recorded in-service average lube oil consumption of 0.1 L/h.) The smaller charge, however, means the oil is subjected to far greater stress than in a diesel engine, and the reduced consumption means it is not refreshed as often.

Apart from lubricating the bearings and other key components of a gas turbine, the bulk of the oil performs an intensive cooling function. Unlike its diesel engine counterpart, the lubricant does not come into contact with the combustion process and does not have to remove products of combustion or neutralize acids formed by burning sulphur-containing fuels.

In aero-derived gas turbines the concentration of power and heat, coupled with the comparatively small quantity of lubrication oil in circulation, results in peak oil temperatures typically over 200°C. Mineral oils are not suitable for this type of gas turbine because they would quickly oxidize at such temperatures: gums, acids, and coke deposits would form and the viscosity of the oil would rapidly increase. Synthetic lubricating oils are therefore favoured because of their intrinsic ability to withstand much higher temperatures than mineral oils.

There are two types of lubricating oil generally available for gas turbine applications: 'standard' or 'corrosion inhibiting'. Most turbine operators have moved towards premium corrosion inhibiting (C/I in US military specifications) oils to take advantage of the higher protection afforded to the bearings and other key components.

A low-coking propensity, good swell characteristics, and compatibility with seals are desirable properties of a synthetic lube oil for aero-derived gas turbines. Coking is the formation of hard, solid particles of carbon due to high temperatures and can result in the blocking of oil ways. The tendency of an oil to break down and produce coke may be exacerbated by turbine operating procedures commonly encountered in fast ferry services, such as rapid acceleration and sudden shutdowns, resulting in the oil being subjected to excessive temperature rises.

Swell is caused by synthetic seals coming into contact with synthetic oils and absorbing the oil. Some swelling is desirable to ensure good sealing but too much can damage the seal and result in leakage. In addition to swelling characteristics, the oil and the elastomers which come into contact must be compatible in all other respects so that degradation is avoided.

Most of the marine lubricant manufacturers produce synthetic lubricants for marine gas turbines that conform to the commonly required viscosity of most synthetic oils for gas turbines, nominally 5 cSt at 100°C.

Air filtration

The use of air filtration systems tailored to the individual application can significantly enhance the reliability and efficiency of a marine gas turbine installation. A typical filtration system comprises three stages:

- *Vane separator*: a static mechanical device that exploits inertia to remove liquid droplets from an airflow passing through it. The intended purpose of this first stage is to remove the bulk of any large quantities of water and coarse spray that may otherwise overload the remaining components of the filtration system. This type of device can remove droplets down to around 12 μm in size, beyond which inertia has little effect. Unfortunately, the damaging salt aerosol experienced within this environment is predominantly below this size and so will pass to the next filtration stage.
- *Coalescer*: this filter-type device—usually of the 1-in depth pleated variety—is specifically designed to coalesce small water (and particularly salt aerosol) droplets; in other words, to capture small liquid droplets and make them form larger droplets. This comparatively easy task for a filter can be achieved with a relatively open filter material which, although a good coalescer, only has a very limited efficiency against dust particulate.

A low-efficiency open media filter has a low-pressure (LP) loss and thus for this application can be run at much higher airflow face velocities than traditional filters in other applications before it has an impact on gas turbine performance. Velocities are typically between 6 and 8 m/s, compared with 2.5 and 3.5 m/s for other applications. This enables a much smaller filter system to be designed for naval vessels, where space is at a premium.

High velocity does not come without a price, however. A coalescer operating at low velocity is likely to be able to drain all of the liquid it collects—through its media—with carryover into the duct beyond. But a high-velocity coalescer will re-entrain the liquid captured as large droplets, and so a third stage is required, particularly as this liquid is heavily laden with salt.

- *Vane separator*: this third-stage device is very similar to the first-stage and in many systems is exactly the same. Its purpose is to remove droplets being re-entrained into the airflow by the second-stage coalescer; as such, it must be totally reliable as this liquid has the potential to cause significant damage to the gas turbine because of its salt content.

A successful filtration system serving the gas turbine plant of a fast ferry should have the following features and characteristics:

- It should be a three-stage (vane, coalescer/dust filter, and vane) system.
- The filter element should have an appropriate degree of dust filtration performance.
- The filter should have a suitable lifetime.
- The system should be of the high velocity, compact type.
- The system should have an LP loss that is not significantly affected by wet conditions.
- The aerodynamics of the intake should be carefully considered.

Marine gas turbine designs

Most marine gas turbines are derived from aircraft jet engines where billions of dollars are spent on development and upgrading designs. Such an investment cannot be justified for the maritime industry but the extensive testing and service experience accumulated by aero engines provides an ideal basis for marine derivatives. Marine gas turbines are available only in specific sizes and ratings, unlike a given diesel engine design which can cover a wide power range with different in-line and V-cylinder configurations. The economical application of gas turbines to marine propulsion, therefore, dictates matching the unit to the project and hence calls for technical and economic analysis of the vessel's proposed deployment.

Operational costs include availability for service and maintenance requirements, cost of consumables (fuel and lube oil) and manning costs. All these factors have to be considered for military ship designs but not to the extent that is called for in commercial shipping. Trade-offs among the various parameters can be conducted to determine the right fit of the propulsion system to the application.

All marine gas turbines incorporate the same fundamental components (Fig. 27.9):

- A compressor to draw in and compress atmospheric air.
- A combustion system into which fuel is injected, mixed with the compressed air and burned.
- A compressor turbine which absorbs sufficient power from the hot gases to drive the compressor.
- A power turbine which absorbs the remaining energy in the hot gas stream and converts it into shaft power.

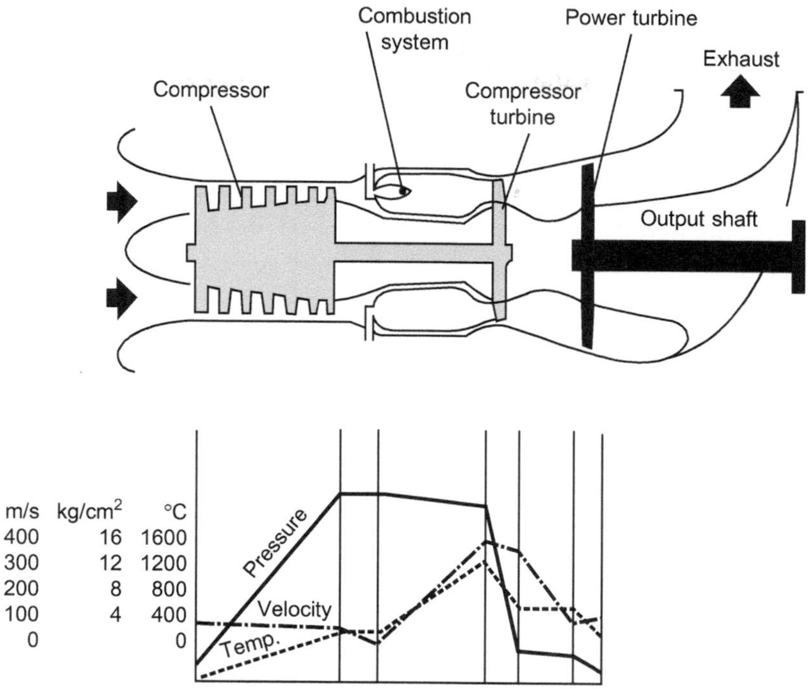

Fig. 27.9 Main elements of an aero-derived gas turbine, and the pressure, temperature, and air/gas speeds during the cycle.

Lightweight aero-derived, rather than heavyweight industrial-derived, gas turbine designs are favoured for marine propulsion applications. The aero jet engine extracts only sufficient energy from the gas stream to drive the compressor and accessories and releases the surplus gas at high velocity through a convergent nozzle to propel the aircraft by reaction.

When converting the jet engine into a shaft drive machine it is necessary to provide an additional free power turbine to absorb the energy left in the gas stream and to transmit that energy in the form of shaft power. The original jet engine is then termed the *Gas Generator* and the whole assembly becomes a *Gas Turbine*. Using a free power turbine reduces the starting power requirement as only the gas generator rotating assemblies have to be turned during the starting cycle. The resulting prime mover benefits from the intensive development and refinement of the original aero engine, both before and during service.

Certain aero engines are already fitted with power turbines and termed turboprop or turboshaft machines; these are readily adaptable to industrial and marine use. Special technical and production techniques are applied to the redesigned gas turbine to ensure its suitability for sea level operation at high powers and for the marine environment.

A basic gas generator has one rotating assembly: the compressor and its turbine coupled together. The characteristics of axial-flow compressors vary considerably, however, over the operating range from starting to full power. On some high-compression ratio compressors, it is necessary to fit automatic blow-off valves and to alter the angle of the inlet guide vanes and first stages of stator blades to ensure efficient operation.

To achieve the necessary stability in larger gas generators, the compressor is divided into two separate units: the LP and HP compressors, each driven by its own turbine through coaxial shafts. Each compressor is able to operate at its own optimum speed, giving flexibility of operation and efficient compression throughout the running range. Only the HP rotor needs to be turned during starting, and therefore even the largest gas generator can be started by a battery.

Although such a two-spool compressor is very flexible, it is still necessary in some cases to adjust the inlet guide vanes to deal with the changing flow of air entering the compressor. This is effected automatically by pressure sensors acting upon rams which alter the angle of the guide vanes.

An axial-flow compressor consisting of alternate rows of fixed and rotating blades draws in atmospheric air through an air intake and forces it through a convergent duct formed by the compressor casing into an intermediate casing, where the compressed air is divided into separate flows for combustion and cooling purposes. A typical annular combustion chamber receives about 20% of the airflow for combustion into which fuel is injected and burned.

Initial ignition is executed by electrical igniters, which are switched off when combustion becomes self-sustaining. The resulting expanded gas is cooled by the remainder of the airflow which enters the combustion chamber via slots and holes to reduce the temperature to an acceptable level for entry into the one or more axial-flow stages of the turbine. The turbine drives the compressor, to which it is directly coupled. The remaining high-velocity gases are exhausted and are available for use in the power turbine.

A power turbine of the correct 'swallowing' capacity is required to convert the gas flow into shaft power (Fig. 27.10). Typically of one or more axial-flow stages, the power turbine may be arranged separately on a base frame designed to mount the gas generator, to which the turbine is linked

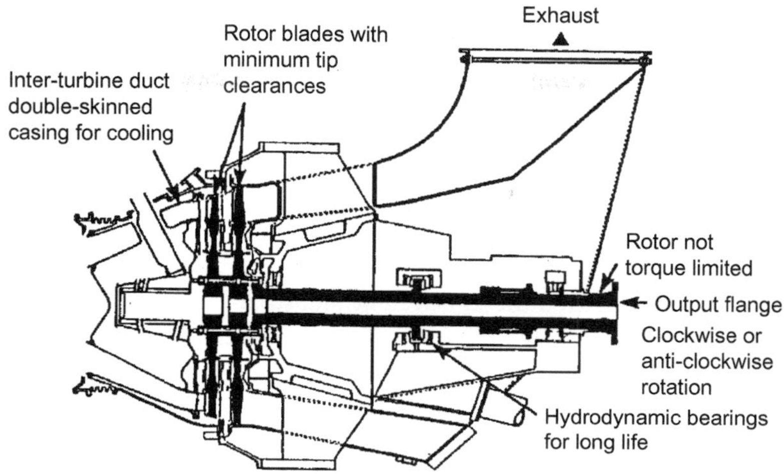

Fig. 27.10 Power turbine of a Rolls-Royce Spey SM1C engine.

by a bellows joint to avoid the need for very accurate alignment and to allow for differential expansion. In some cases, however, expansion is allowed for in the design of the power turbine and the gas generator is mounted directly on to the inter-turbine duct. The power turbine, usually designed to last the life of the plant, is surrounded by an exhaust volute which passes the final exhaust to the atmosphere through a stack.

Most marine gas turbines incorporate a free power turbine, which is free to rotate at whatever speed is dictated by the combination of the power output of the gas turbine change unit (GTCU) and the drive train behind it. The gas turbine is designed so that there are no critical speeds in the running range, and it can operate continuously with the power turbine at any speed from stop to the full power revolutions per minute of the power turbine. This facility is especially important under manoeuvring conditions where the combination of a free power turbine and reversing water jets provides a flexible and effective propulsion system.

Aero- and industrial-derived marine gas turbines are available with unit outputs ranging from around 2000 kW to over 50,000 kW. At the smaller end, the most popular for yachts and hovercraft Vericor's TF40 marine gas turbine with a power output between 3000 and 3500 kW is an aero-derivative engine used for naval and fast commercial vessel propulsion. With more than 50 years of heritage originating with the Lycoming Engine Company, the rugged TF series gas turbines have accumulated hundreds of thousands of hours in military and commercial service around the world. Vericor Power Systems is headquartered in Alpharetta, Georgia, US.

Larger designs (with ratings of 17,000 kW upwards) are reviewed here.

Siemens

Lower operating costs through burning cheaper fuels than the light gas and diesel oils normally required by aero-derived gas turbines are promised by Siemens, which inherited the lightweight industrial-based turbine interests of Swedish designer ABB Stal from Alstom Power in 2003. Successful marine service experience on IF30 intermediate fuel is cited for the SGT-500 (formerly designated GT35) gas turbine, supported by a capability to burn IF180 fuel demonstrated in land-based plants.

The GT35 emerged from a postwar aero engine development programme in Sweden involving Stal Laval and Volvo Flygmotor. Small turbojets were designed and built for testing in the late 1940s/early 1950s but never entered commercial production. Stal Laval saw potential in transforming the engine into an industrial gas turbine for power generation and mechanical drive duties. The first GT35, rated at 9000 kW, was tested in 1955 and installed during the following year in a Swedish power station.

Performance and reliability were subsequently demonstrated in land-based and offshore platform power generating installations. The design has since benefited from two upgradings to attain its present output rating of 17,300 kW at 3450 rev/min for mechanical drive applications, with an efficiency of 33% (Fig. 27.11).

The gas generator of the GT35 is a two-shaft engine equipped with a 10-stage LP compressor, an eight-stage HP compressor, seven can-type combustion chambers, a single-stage HP turbine and a two-stage LP turbine (Fig. 27.12). The power turbine, developed in parallel with the gas generator, is a three-stage design. A refinement made in the 1990s enabled the power turbine to operate at variable speed and enhanced its suitability for mechanical drive and marine applications.

Refining the GT35 for marine applications called for the following measures: modifying the lube oil system to allow for vessel movements; modifying the support arrangement to protect the turbine from hull deflections; and modifying the power turbine by introducing a second combined journal and thrust bearing, as well as a modified active tip clearance control mechanism (achieving a self-supported power turbine and thus eliminating the need for a thrust bearing in the gearbox).

Fig. 27.11 One of two 17,000 kW ABB Stal GT35 (now designated Siemens SGT-500) gas turbines supplied to drive the water jets of a Stena HSS 900-class fast ferry.

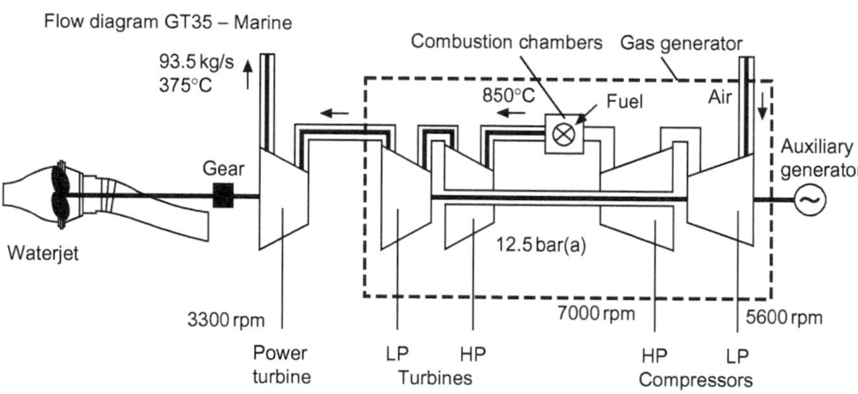

Fig. 27.12 Flow diagram of ABB Stal GT35 (now designated Siemens SGT-500) gas turbine in a marine propulsion installation.

High reliability from the marinized GT35 was sought by exploiting as many standard components as possible from the power generation and mechanical drive versions. An extensive recalculation of low-cycle fatigue and creep properties of certain critical elements of the turbine was also undertaken, leading to a subsequent redesign of some parts. High availability and maintainability in arduous marine duty were addressed by adapting the installation to the specific vessel, with the aim of reducing the time for unit replacement and facilitating servicing. The maintenance programme was also adapted to operating conditions imposing only short shutdown times for routine inspections and unit change-outs.

Four key factors contribute to high fuel flexibility, underwriting operation of a GT35 turbine set on IF30 fuel (normally produced from 75% IF180 fuel/25% marine gas oil) after suitable treatment:

- A high-viscosity tolerance due to the spill flow fuel system in which the amount of fuel injected into the turbine is governed by a regulating valve mounted in the return pipe from the injectors. The system configuration makes it possible to circulate the fuel through the injectors prior to ignition and hence warm them up sufficiently to facilitate starting on fuels with a viscosity of up to 30 cSt/50°C (equivalent approximately to 10 cSt/95°C, the temperature to which the fuel is heated on the vessel).
- Maximum inlet temperature of 880°C (fairly low compared with other gas turbine designs) makes it unnecessary to use cooled blades and vanes, yielding such benefits as insensitivity to high-temperature corrosion from sodium sulphate; no need for blade coating; and no risk of plugging cooling holes and channels with ash deposits and reducing the lifetime of blading.
- Large combustion chambers (Fig. 27.13) secure low sensitivity to variation in radiation heat depending on the type of fuel that is burned. The size and design of the chambers ensure their lifetime matches that of the complete turbine unit.
- A simple online 'soft blast' turbine cleaning procedure—with ground nutshells injected directly into the turbine section via inspection windows in the combustion chambers—results in sustained high turbine efficiency and power output combined with low-outage time for cleaning. Cleaning can be effected with the turbine operating at close to full load; depending on the fuel properties, the procedure is necessary at intervals varying between 50 h and several months' operation.

Retaining a high output-to-weight ratio dictated special attention to the marine GT35's auxiliary systems, the designers reportedly achieving a reduction in weight of approximately 25% over a land-based unit. Among the measures was a redesign of the starting system from a pneumatic to an electric system and a reconfigured base frame. Quiet running and high safety

Fig. 27.13 Seven large can-type combustion chambers (foreground) contribute to the lower grade fuel burning capability of the ABB Stal GT35 (now designated Siemens SGT-500) gas turbine.

are fostered by solid turbine casings which allow sets to be installed without an enclosure. Packages can thus be better tailored to individual hulls, while the free-standing auxiliary systems can be located in the most convenient position.

Tests at Siemens' Finspong facility in Sweden aimed to determine the limits for good combustion when using commercially available heavy fuel oil, with regard to viscosity and carbon residue. The fuel system was modified (pressurized since it would be subjected to heating in excess of 100°C) to accept higher viscosity fuels. All the test targets were achieved and, in some cases, exceeded. The operation was problem-free and combustion efficiency was high, with no significant effect on turbine maintenance indicated. Furthermore, no visible smoke was emitted down to loads below 20% maximum rating.

After reviewing the results of heavy fuel operation, classification society DNV signed a statement of preliminary approval that made the SGT-500 (formerly GT35) the only industrial gas turbine offered for marine applications burning such fuel. The following ISO 8217 marine fuels can be handled: DMX, DMA, DMB, DMC, RMA 30, RMB 30, RMD 80, RMG 180, RMF 180, and RMG 380. Siemens also reports a seamless transfer between gas and liquid fuels and the capability to run on virtually any combination of liquid and gas.

Since acquiring the former Alstom Power portfolio, Siemens has supplied dual-fuel SGT-500 sets for offshore floating production storage offloading (FPSO) vessels.

A dry low NOx combustor system maintains emissions below 42 ppmV on gas fuel. This means a NOx level 10 times lower than conventional burner techniques. Water or steam is injected into the combustor section to reduce the NOx level when the gas turbine is running on liquid fuels. The SGT-500 satisfies stringent noise requirements and regulations for NOx emissions.

General electric

General Electric (GE) introduced its first aero-derivative gas turbine, the LM100, in 1959 and in the same year the LM1500, derived from the successful J79 aircraft engine. The first LM1500 installation, serving the hydrofoil vessel *H.S. Denison*, was followed by over 160 applications in land-based catapults, pipeline pumping systems, marine propulsion, and power generation. The mainstay of the US designer's programme, the LM2500, arrived in 1969 and has since benefitted from developments which have doubled its initial rating of around 15,000 kW. Other models subsequently joined the LM-series portfolio, which now embraces six simple-cycle designs with maximum power ratings from around 4500 to 42,750 kW and thermal efficiencies up to 42%.

Valuable experience from LM-series gas turbines in naval propulsion applications ranging from patrol craft to aircraft carriers was tapped by GE Marine Engines in targeting commercial vessel projects during the 1990s, earning a number of large cruise ship and fast ferry references.

GE Marine Engines' aero-derived LM gas turbine family

Model	Maximum rating	Thermal efficiency
LM500	4470 kW	32%
LM1600	14,920 kW	36.7%
LM2500	25,060 kW	37.1%
LM2500 +	30,200 kW	39%
LM2500 + G4	35,300 kW	41.3%
LM6000	42,750 kW	42%

The LM500 is a two-shaft gas turbine derived from GE's TF34 aircraft engine; the LM1600 is a three-shaft gas turbine derived from the F404 engine, and the LM6000 is a two-shaft gas turbine derived from the CF6-80C2 engine. The most successful models in the commercial and military marine markets are the LM2500 and LM2500+ models, described in the next section.

LM2500 and LM2500+

Derived from the GE military TF39 and commercial turbofan aircraft engines, the LM2500 marine gas turbine is a simple-cycle, two-shaft engine comprising a gas generator, a power turbine, attached fuel and lube oil pumps, a fuel control and speed-governing system, associated inlet and exhaust sections, lube and scavenge systems, and controls and devices for starting and monitoring engine operation. The four main elements (Fig. 27.14) are:

- A 16-stage compressor, with an 18:1 pressure ratio; the first seven stages feature variable stators and inlet guide vanes promoting easy starting, good part-load performance, high efficiency, and a high stall margin over the entire operating range.
- The fully annular combustor, with externally mounted fuel nozzles supporting liquid fuel combustion; virtually smokeless operation is assured through the complete power range even when burning heavy distillate fuels.
- Two-stage HP turbine, air-cooled for a long life, which drives the compressor and accessory-drive gearbox.
- Six-stage LP power turbine; a low-speed, low-stress machine aerodynamically coupled to the gas generator and driven by its high energy release exhaust flow.

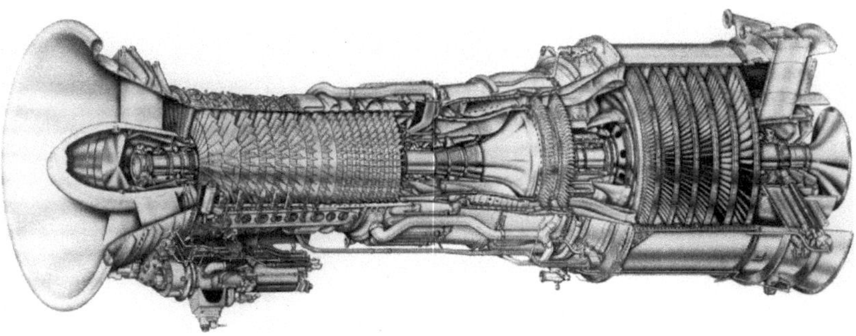

Fig. 27.14 Cutaway of GE Marine Engines' LM2500 aero-derived gas turbine showing (from left to right) the compressor, annular combustor, HP turbine, and LP power turbine.

Lower installed and life-cycle costs per unit kilowatt were sought from the LM2500+ derivative released in the late 1990s with a 25% higher power rating than its precursor. An introductory rating of 27,600 kW was achieved with a thermal efficiency of more than 37%; release at the design rating of 29,000 kW and thermal efficiency of 38% followed early sets in service demonstrating the traditional reliability and availability of the LM2500. The current rating is 30,200 kW with a thermal efficiency of 39%. The design refinements applied to the LM2500+ focused on:

- The compressor rotor: airflow was increased by approximately 20%; Stage 1 blades were redesigned to a wide chord configuration eliminating the mid-span dampers; an LM6000 rotor airfoil design was added to Stages 2 and 3; LM6000 stages were incorporated to improve compressor efficiency, and a new inlet guide vane assembly was specified.
- HP turbine rotor and stator: redesigned to reduce maintenance costs and new materials exploited for improved oxidation life; Stage 1 and 2 contours optimized for higher flows.
- Power turbine: redesigned in line with the higher output; Stage 1 and 6 blades optimized for aerodynamic efficiency, and the rotor strengthened for the higher torque and potential energy.

No changes were made to either the standard or DLE combustion system.

Only slightly longer and heavier than the LM2500, the LM2500+ offers dual-fuel capability (distillate and gas), rapid start-up and loading, variable speed operation, and excellent part-load efficiency, GE claims.

LM2500+G4

Further investment in the LM2500 gas turbine resulted in 2005/2006 in the LM2500+G4 derivative with a rating of 35,300 kW at ISO day conditions and thermal efficiency of 41.3%. This fourth-generation design offers 17% more power with a 6% higher airflow than the established LM2500 +model.

Essentially an LM2500+with increased flow capacity in the HP compressor, HP turbine and LP turbine, the G4 version has a pressure ratio of 24.2 compared with the original 23.6. Design changes were limited to the minor blade and stationary vane airfoil adjustments that provide the mass flow increase. HP turbine modifications included minor blade-cooling improvements and proven material upgrades from aircraft technology which enhance higher temperature capability. The effective area of the compressor discharge pressure seal was adjusted in order to maintain the optimum rotor thrust balance.

Structurally, all frames—such as front, compressor rear, turbine mid- and rear frames—remained unchanged, as did the sump hardware and the number of main shaft bearings. Among the internal changes benefiting the G4 model were:

- Redesigned inlet guide vanes and re-staggered Stage 0 blisk airfoils.
- Increased HP compressor life design through rotor material changes.
- Redesigned Stage 1 HP turbine blade with improved cooling and oxidation coating.
- Application of thermal barrier coatings to HP turbine Stage 1 nozzle.
- Material change and improved cooling for the Stage 2 nozzle.
- Redesigned Stage 2 HP turbine rotor for extended life.
- Redesigned Stages 2 and 3 nozzles for the six-stage low-speed (nominal 3600 rev/min) power turbine for increased output.

GE also offers a two-stage high-speed power turbine with a design speed of 6100 rev/min and an operating speed range of 3050–6400 rev/min for mechanical drive and other applications where continuous shaft output speeds of 6400 rev/min are desirable. Both six-stage and two-stage power turbines can be operated over a cubic load curve for mechanical drive duties.

With a single annular combustor, the LM2500+G4 has the same fuel flexibility as the LM2500+, allowing it to operate on natural gas, liquid fuel, or dual fuel. Water or steam injection can be adopted for NOx emission abatement.

Improvements in base load power over the LM2500+ were focused across a wide range of ambient conditions. Specifically, during off-ISO conditions, such as cold day/hot day, the LM2500+G4 yields an 11/12% power increase. In combined-cycle mode, it was expected to have 8.5% power and 0.75% heat rate advantages.

A naval propulsion debut for the LM2500+G4 gas turbine was secured in 2006 from the FREMM frigate programme for the French and Italian navies.

Mitsubishi

A Japanese challenge in the aero-derived marine gas turbine market was spurred by the country's Techno-Superliner (TSL) fast freight carrier project of the early 1990s which called for an indigenous high output prime mover with a very low weight. In response, Mitsubishi created the MFT-8 from the marriage of its own power turbine and the GG-8 gas generator from Turbo Power and Marine Systems, a subsidiary of Pratt & Whitney (Fig. 27.15).

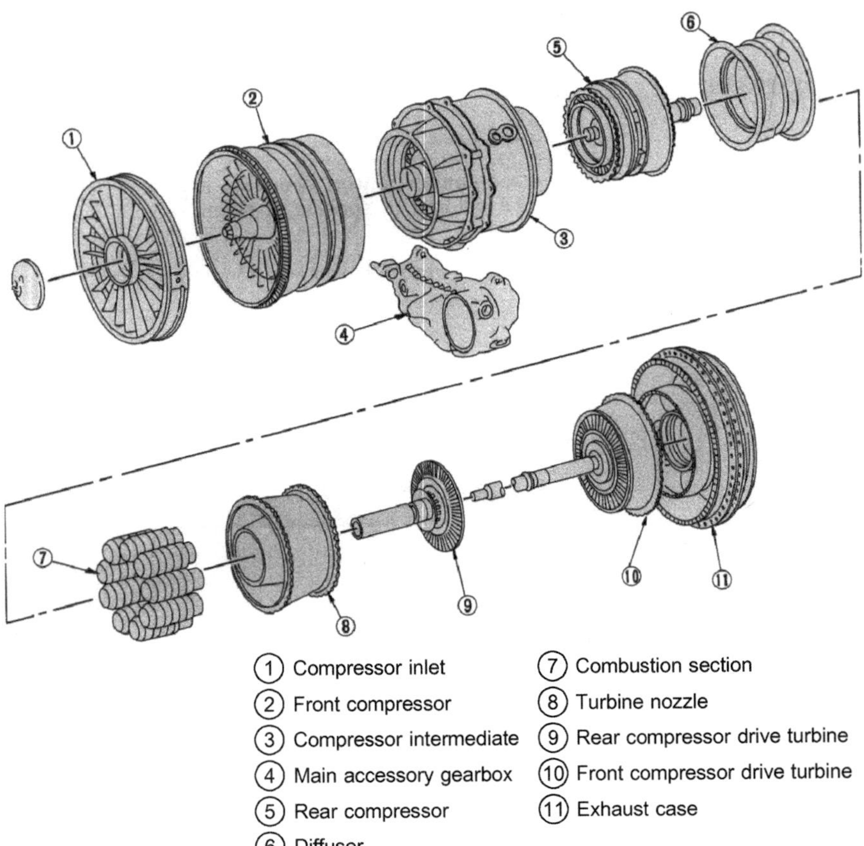

Fig. 27.15 Major engine groups and sections of Mitsubishi MFT-8 aero-derived marine gas turbine.

The GG-8 is a land and marine derivative of the US group's JT8D turbofan aero engine, the design modified to replace the front fan with the forward part of the front compressor; otherwise, it is fully compatible with the core engine. The key elements of the gas generator are an eight-stage LP compressor; seven-stage HP compressor; nine burner cans; a single-stage air-cooled HP turbine; and a two-stage LP turbine. Gas is fed to a high-performance power turbine developed by Mitsubishi's project team as a three-stage design with a lightweight, overhung ball, and roller bearings.

The main modules and components are designed for ease of removal and replacement during inspection and maintenance, and provision is made for borescope inspection of the hot parts and rotating parts of the gas generator and power turbine without disassembly.

After the prototype achieved a full power rating of 24,300 kW during shop trials, two production sets were delivered for installation in the TSL craft built by Mitsubishi. The Japanese designer subsequently offered the MFT-8 turbine to the wider market with a propulsion plant rating of 25,790 kW and thermal efficiency of just under 39%.

Rolls-Royce

Marine Spey

The Marine Spey first went to sea with the UK Royal Navy in 1985, and the latest C-rated version entered service in 1990. Numerous warship propulsion references have been logged and commercial opportunities targeted, notably for high-speed ferry and freight vessel projects (Fig. 27.16). The twin-spool design is released with respective commercial and military marine ratings of 18,000 and 19,500 kW, and sfcs of 230 and 226 g/kW h.

Gas generator, power turbine, and all ancillary systems are mounted on a common bedplate to form a self-contained unit for ease of installation

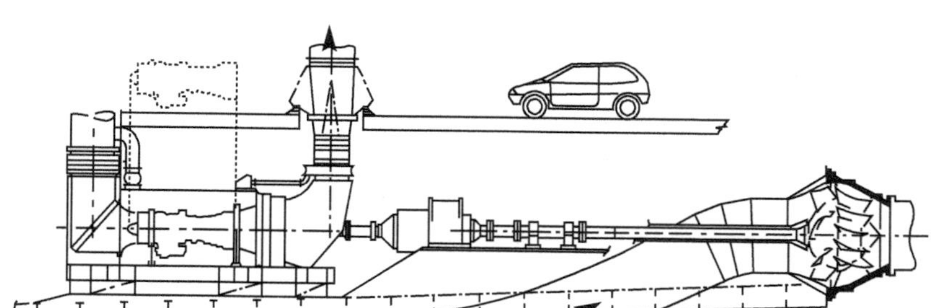

Fig. 27.16 Rolls-Royce Marine Spey machinery arrangement for a water jet-powered fast ferry.

Fig. 27.17 A fully packaged Rolls-Royce Marine Spey propulsion module.

(Fig. 27.17). Interfaces for the shipbuilder are limited to the intake and uptake flanges, the mountings to the hull and the output flange. A complete propulsion module, including power turbine, has a length of just over 7.5 m and weighs 17.5 t in its commercial vessel configuration; the gas generator weighs only 1.8 t. An 80% weight saving and 60% space saving over a medium-speed diesel installation of equivalent output are reported.

The gas generator is a high-efficiency, second-generation unit of twin-spool design designated the SM1C, which is a development of the SM1A and incorporates technology from the Rolls-Royce Tay and RB211 aero engines. The core achieves a 20:1 pressure ratio and good overall efficiency from a compact unit with relatively few stages of blading. The designers sought excellent operational flexibility, rapid acceleration, and good aerodynamic stability, with no visible exhaust smoke at any power level. The modular construction allows gas generator sections to be exchanged on-site, the engine comprising five maintenance assembly change units (MACUs).

Specifically designed for the marine environment, the power turbine is a high-efficiency, two-stage axial-flow design with short and rugged shrouded blades capable of withstanding significant foreign body damage. Its rotor system is supported on hydrodynamic bearings housed in a rigid centre body and is available with clockwise or anticlockwise output shaft rotation. The power turbine is designed for the life of the ship but is removable in the event of serious damage (Fig. 27.10).

Marine Spey data

Power, commercial service rating		18,000 kW
Intake mass flow		64.3 kg/s
Exhaust mass flow		65.5 kg/s
Exhaust temperature		480°C
Compressor stages	LP	5
	HP	11
Turbine stages	LP	2
	HP	2
	PT	2
Shaft speeds	LP	7900 rev/min
	HP	12,000 rev/min
	PT	5000 rev/min
Combustion system	Cannular	10 combustors
Number of shafts		21 free power turbine

MT30

A simple-cycle marine gas turbine with a thermal efficiency exceeding 40%, the Rolls-Royce Marine Trent 30 was scheduled for commercial availability from early 2004 offering outputs up to 36,000 kW from a package weight of 25 t. In fact it was 2008 before the entry into service for the MT30 when it powered the US Navy's first Littoral Combat Ship, *USS Freedom*. The propulsion system features twin MT30 engines and two diesel engines driving waterjets in a sophisticated combined diesel and gas turbine (CODAG) mechanical arrangement. The MT30 is the eleventh engine type derived from the British designer's core aero technology (Fig. 27.19) and has gained references in UK Royal Navy and US Navy projects including the Royal Navy's new aircraft carriers *HMS Queen Elizabeth* and *HMS Prince of Wales* which each feature twin MT30 turbines as the main propulsive power units.

The aero Trent 800 was selected from a diverse Rolls-Royce engine family as the parent for this new generation of marine gas turbine because it best matched current and emerging customer requirements with margins in hand. The three-spool marine derivative allows variable speed operation, and its 'turndown' facility secures low-power harbour manoeuvring capability. Furthermore, the aero engine's track record (99.9% dispatch reliability) gave confidence for commercial marine applications with high annual usage.

Availability, reliability, and maintainability were prime goals in setting the design parameters for the MT30, with an overhaul target of 12,000 h for the hot end and 24,000 h between major overhauls. The design reportedly features some 50%–60% fewer parts than other aero-derived turbines in its class. A sound basis is provided by an 80% commonality of parts with the aero Trent. Specialized coatings are applied where necessary for protection against the marine environment. Internal temperatures and pressures are substantially lower than those at the aero engine take off rating to foster component longevity.

A rating of 36,000 kW is available at the power turbine output shaft at ambient air temperatures up to 26 °C, with a corresponding sfc of 207 g/kW h;

MT30 gas turbine data

Power	36,000 kW @ 25°C ambient 30,000 kW @ 35°C ambient
Specific fuel consumption	207 g/kWh
Exhaust mass flow	113 kg/s
Exhaust temperature	466°C
Compressor stages	IPC–8 HP–6
Turbine stages	IP–1 HP–1
Power turbine stages	PT–4
Power turbine shaft speed	3600 rev/min

under tropical conditions (32°C air temperature), the output is 34,100 kW. Competitive efficiency is sustained down to 25,000 kW and thermal efficiency is similar to that of high-speed diesel engines, Rolls-Royce claims. The MT30 is designed to burn the widely available distillate marine fuel-grade DMA.

Essentially the core unit of an aero Trent engine minus the fan section, the gas generator of the MT30 benefited from the latest design techniques, materials, and production technology during its refinement for marine applications. An eight-stage variable geometry axial intermediate pressure (IP) compressor driven by a single-stage IP turbine feeds air to the HP spool, which comprises a six-stage compressor and a single-stage HP turbine. An annular combustion chamber is arranged between the HP compressor and the HP turbine. The IP and HP spools rotate in opposite directions. A two-spool compressor design yields reduced start time, good balance retention, and no lock-up problems, says Rolls-Royce, resulting in greater operational flexibility and no need for in situ balancing.

Hot gas from the gas generator section is supplied to a four-stage free power turbine whose full speed of 3300 rev/min was selected for mechanical drives to yield the lowest sfc. The design addresses the demands of water jet operation, notably sudden large load changes if the air is drawn into the jet unit in severe sea conditions.

The MT30 is supplied as a package with an acoustic enclosure housing auxiliaries and inlet and exhaust diffusers, and incorporating automatic fire detection and extinguishing systems (Fig. 27.18). A complete unit—measuring 9.1 m long × 3.8 m wide × 4 m high—weighs around 22 t dry, depending on the options selected. The GTCU, including power turbine, weighs 6200 kg dry and is air-freightable. Installation of the main package can be effected in a single lift at the shipyard, while the control panel, starter and lube oil modules can be mounted where convenient in the engine room (Fig. 27.19).

A key element of the package—the integrated control and monitoring system—is housed in a free-standing cabinet with an uninterrupted power supply for protection in the event of mains failure. An engine health monitoring system is also offered. Condition-based maintenance is facilitated by internal condition sensors; routine servicing is limited to checking fluid levels and visual examinations. Maintenance contracts covering servicing, overhauls, and spares supply, long practised in the airline industry, are available to support operators and foster competitive through-life costs.

WR-21

The only advanced-cycle marine gas turbine currently on the market, the WR-21 was developed by Northrop Grumman in the USA, Rolls-Royce, and the French Naval Group DCN which supplied specialist marine input during the 1990s with funding from the US Navy, UK Royal Navy, and French Navy. It is the first aero-derived gas turbine to incorporate compressor intercooling and exhaust heat recuperation (Fig. 27.20) to foster a low sfc across the operating range: a fuel burn reduction up to 30% over simple-cycle turbines is reported.

Rated at 25,000 kW, the WR-21 is based on the successful Rolls-Royce aero RB211 and Trent engines, with modifications to marinize the components and effectively integrate the heat exchangers and variable geometry.

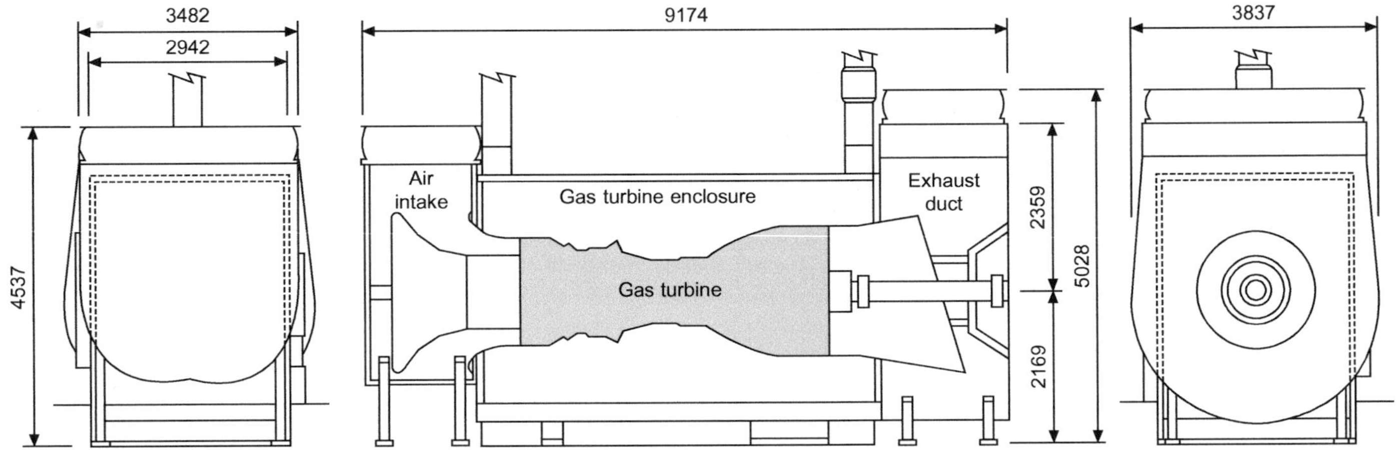

Fig. 27.18 Rolls-Royce MT30 in mechanical drive package (dimensions in millimetres).

Fig. 27.19 Rolls-Royce MT30 gas turbine.

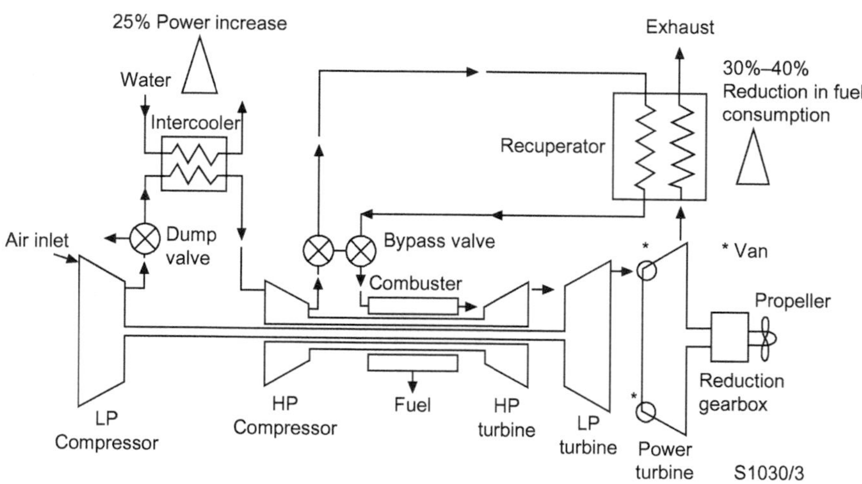

Fig. 27.20 ICR system applied to the Rolls-Royce WR-21 gas turbine.

The first seagoing installations (featuring twin sets) entered service from 2008 in the UK Royal Navy's new Type 45 class destroyers—a series of six headed by *HMS Daring*—but commercial propulsion opportunities, particularly cruise ships, are also targeted. The fuel consumption characteristics—approximately 205 g/kWh from full power down to around

30% power—enable the WR-21 to fulfil the role of both cruise and boost engines.

Compared with simple-cycle turbines, the benefits cited for the WR-21 include an improvement in fuel efficiency over the entire operating range (with a radical improvement at low power), easier maintenance through enhanced modularization and the facility to retrofit ultralow-emission reduction systems. The ICR propulsion system package was designed to occupy the same footprint as the existing power plant (Fig. 27.21).

A significant fuel saving over simple-cycle turbines is achieved by using heat exchangers to improve the part-load cycle efficiency. Recuperation alone improves the thermal efficiency of LP ratio cycles where the exhaust temperature of the turbine is significantly higher than that of the air leaving the compressor. Heat is transferred to the compressed air before it enters the combustion system, reducing the amount of fuel required to attain the cycle turbine entry temperature.

The recuperator thus preheats the combustion air by recovering waste energy from the exhaust, improving cycle efficiency and reducing fuel consumption. Low-power efficiency is improved still further by power turbine variable area nozzles (VAN); these maintain a constant power turbine entry temperature, which in turn maintains recuperator gas side entry conditions and improves recuperator effectiveness as power reduces.

The intercooler cools the air entering the HP compressor, reducing the work required to compress the air; the intercooler also reduces the HP compressor discharge temperature, which increases the effectiveness of the recuperator.

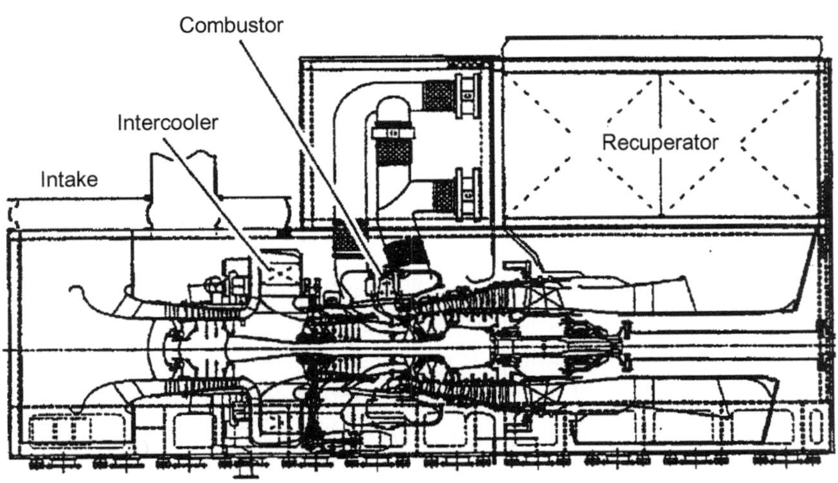

Fig. 27.21 Layout of WR-21 gas turbine package.

In 2009 it was discovered that the Northrop Grumman intercooler as fitted in the WR-21, on Type 45 destroyers, had a major design flaw, failing to operate in water temperatures beyond 30°C. The intercooler of *HMS Daring*, failed in the mid-Atlantic in 2010 and had to be repaired in Canada, with further repairs for intercooler failure in 2012 in Bahrain.

To resolve the issue, the UK government initiated Project Napier which in a two-phase programme will replace the two existing diesel generators on each Type 45, add a third one, and modify the high voltage systems on the ships to handle the additional power. This is on top of the Equipment Improvement Plan (EIP), in which Rolls-Royce is working to upgrade certain components of the gas turbines. The UK Ministry of Defence expects the first vessel with all the new modifications will be ready by the end of 2021 and the entire the fleet will have received the upgrades by sometime in the early 2020s.

Air enters the compressors via a composite radial intake designed to maintain uniform circumferential velocity at the entry to the gas generator. The IP compressor (so-called because it reflects commonality with the parent three-spool turbofan aero engine) includes six stages of compression; stages two to six are in common with the RB211 engine, while the first stage is modified to suit the increased flow requirements of the ICR cycle.

The intercase between the compressors transmits structural load from the engine through two support legs to the subbase; it also houses the five intercooler segments in an outer casing and the internal gearbox within an inner casing. Both casings combine to form the airflow path between compression stages. Incorporating an intercooler between the compression stages on a twin-spool cycle increases the specific power of the engine. The amount of work needed to drive the compressor is reduced, thereby increasing the net power available. Bypass valves on the intercooler are modulated depending on the pressure and relative humidity of the air so that condensation formation can be avoided.

Cycle thermal efficiency is approximately the same as that of a simple-cycle engine, as additional fuel is required to offset the drop in compressor exit temperature. Combining an intercooler with a recuperator, however, is attractive for higher pressure ratio engines, leading to high specific power outputs and good thermal efficiency.

The HP compressor also has six compression stages and is aerodynamically identical to its aero origin. Compression is split 30:70 between IP and HP compressors, and both stages incorporate additional borescope holes to allow greater flexibility for inspection.

The Marine Spey SM1C combustor was adopted as the basis for the WR-21 combustor design as a proven system in use worldwide. Although otherwise conventional in its construction, the combustor features a Reflex Airspray Burner method of fuel injection developed specifically for the marine versions of the Spey. This achieves a controlled mixing of fuel and air, allowing a high burner exit air–fuel ratio (AFR) to be maintained with adequate flame stability. Based on previous experience, high (lean) AFR was considered an important factor in reducing visible smoke when burning diesel fuel.

Preservation of the proven aero RB211 HP and IP spool lengths, which are characterized as short, rigid high-integrity structures, was a principal design objective. The requirement to remove compressor delivery air and return recuperated air within the length constraint dictated that the annular RB211 combustor be replaced by a radially orientated turbo-annular system.

The manifold designs underwent extensive analysis and testing to confirm structural integrity, aerodynamic flow distributions, ease of manufacture, and maintainability. The inner casing carries the structural load from the HP compressor outer casing to the turbine casing; the component is designed so that it provides the minimum of blockage for HP compressor exit air entering the combustion manifold and also transfers HP compressor Stage 4 air to the power turbine for sealing and cooling. In-situ removal of the combustor and burner is a key element of the maintainability strategy. The design addresses ease of life monitoring and timely repair or replacement of hardware (without engine removal) in the event of premature failure.

The HP nozzle guide vanes and rotor blade airfoils maintain commonality with the RB211-524 parent engine, with slight modifications to provide a smooth gas path from the radically swept discharge nozzles. The disc seals and bearing arrangements are essentially unchanged from their aero origins. The IP nozzle guide vane is skewed relative to the aero RB211-535 vane and incorporates the addition of a cast boss to facilitate a borescope inspection hole. The blade is uncooled and manufactured from a single nimonic to extend creep life.

Consistent with the pedigree of the gas generator, the power turbine also originates from an aero parent, the Trent. Stages two-through-five incorporate three-dimensional orthogonal blade geometry to maximize turbine efficiency, but the main difference is the incorporation of VANs which control flow area. The VANs are hydraulically actuated via a single-geared ring designed to maintain VAN-to-VAN throat areas within a specified

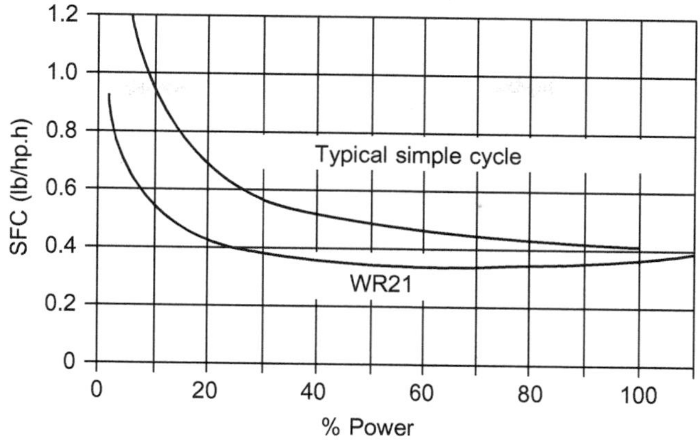

Fig. 27.22 Fuel savings from ICR cycle WR-21 gas turbine over the power range compared with a simple-cycle unit.

tolerance. The VAN is fully open at full power and closes at part power; this has the effect of retaining the efficiency benefits across the whole power range by maintaining the high exhaust temperature at part powers. The recuperator can thus be exploited fully to give the characteristic flat fuel consumption curve for the WR-21 (Fig. 27.22).

The gas generator and power turbine comprise 12 interchangeable, prebalanced modules, whose small size and low weight facilitate removal via simple routes; new or leased modules can be fitted in situ, reducing maintenance costs and downtime.

The enclosure panels are structurally designed to meet blast overpressure, shock and airtightness, with insulation and acoustic damping treatment applied to satisfy thermal and noise requirements. All the panels are removable, except those welded to the recuperator housing. The engine can be withdrawn sideways by removing vertical supports at its front. The enclosure is cooled by ship's ventilation, air entering it via a connection above the recuperator ducts.

An electronic engine controller (EEC), incorporating a full authority digital engine control (FADEC) system, is responsible for controlling the system in response to commands from either the ship's monitoring and control system or from the local man–machinery interface (MMI). The controller covers sequencing, steady state and transient control, surveillance, fault detection, and protection.

Easy control of NOx emissions when entering controlled waters is fostered by the flexibility of the ICR cycle. Punitive charges can be avoided by switching to recuperator bypass mode, lowering the combustion temperature beneath the threshold for NOx generation. Additionally, the high combustor inlet temperatures over the power range in recuperated operation ensure complete combustion with extremely low levels of carbon monoxide and unburned hydrocarbons.

Gas turbine maintenance

Gas turbines have nominal power ratings established for marketing purposes. A power rating at particular ambient conditions determines the internal engine temperatures and the resulting expected service life of components exposed to these temperatures. These 'hot section' components include the combustor and the HP turbine blades and vanes. The life of the hot section components will determine the interval between major maintenance actions, resulting in an estimated average maintenance cost.

The normal relationship between internal parts life and a representative internal gas turbine temperature, in this case, the power turbine inlet temperature, is indicated in Fig. 27.23. At higher internal temperatures the component life is determined by creep-rupture and decreases rapidly with increased temperature; at lower internal temperatures the component lives are determined by corrosion and do not vary as rapidly. (Turbine blades operate at high temperature and rotate at high speed. The blade material has an overall creep life which is proportional to time at a given power level.)

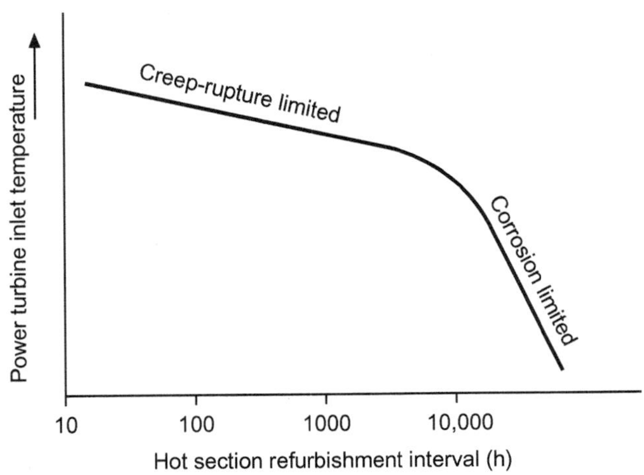

Fig. 27.23 Relationship between internal parts life and power turbine inlet temperature.

The nature of the curve in the figure illustrates the selection of power rating for a particular application. The operation for large percentages of time at power levels in the creep-rupture regime will severely limit component lives in the gas turbine and increase the unit's average maintenance cost. Conversely, the best advantage of a gas turbine's capability can be taken with respect to maintenance near the top of the corrosion-controlled regime. Establishing a rating around the 'knee' of the curve, therefore, tends to be a good starting point with respect to a trade-off between optimum power and reasonable maintenance costs.

Vessel operations will generally involve some representative profile with a mixture of high- and low-power elements, and a range of ambient operating temperatures. The optimized gas turbine rating makes the best use of the engine at the predominant mode of operation while ensuring that the maintenance costs combined with other operating costs produce a minimum total life-cycle cost.

A second factor affecting maintenance costs is the number of cycles of operation per unit time. Since some internal components may be limited in life to a maximum number of thermal cycles, operation with very short mission lengths will result in higher average maintenance costs.

Fig. 27.24 illustrates the nominal relationship of average maintenance cost to the length of the mission. For example, a fast ferry operation with mission lengths of 1 h will experience a higher average annual maintenance cost than for 5- or 6-h runs at the same power rating and ambient conditions. As the length of the mission increases, the maintenance requirements approach the level determined by the power level and ambient conditions

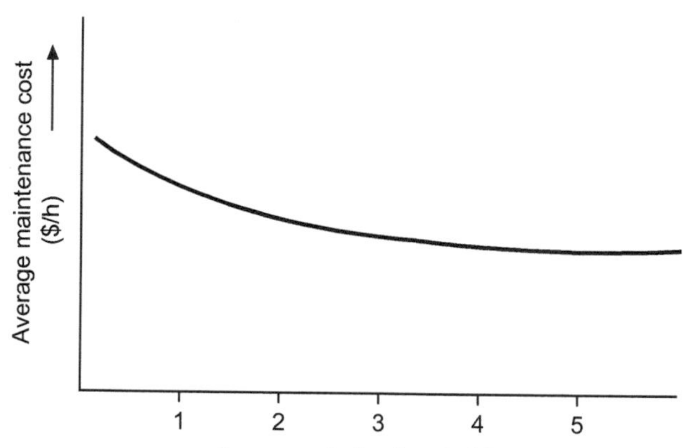

Fig. 27.24 Average gas turbine maintenance cost vs length of the mission.

resulting from the hot section life. For short missions, however, the replacement of cyclic-limited components can have a significant effect on average maintenance costs. Total cycle times for very short applications are usually determined by the turnaround time at each end of a route. The additional speed is less beneficial for the short cycle because a smaller percentage of time is spent in transit. Gas turbine propulsion generally yields greater benefits in longer cycle applications.

High reliability is a requirement imposed by military ship applications because of the potentially critical nature of missions, even so, issues do occur, for example, with the intercoolers of the WR-21 mentioned above. In commercial operations, however, time in service translates into revenue production; factors crucial to the commercial success of the vessel design are therefore minimal downtime and short maintenance intervals.

High availability results from two design considerations. First, the maintenance requirements must be minimized through an appropriate rating of the gas turbine, and mission planning must allow for maintenance that minimizes disruption of the operating schedule. Second, appropriate installation design must foster quick removal and replacement of modular components, minimizing time out of service.

Commercial vessels are usually designed with a particular service profile in mind, but the design must be flexible enough for diverse deployment if resale to other operators is planned later. The propulsion plant must, therefore, allow flexibility in operation over a range of power levels while maintaining operating efficiency with respect to fuel and lube oil consumptions.

Gas turbine engine maintenance may be considered broadly in two groups: routine tasks to be carried out on-board (Fig. 27.25) and routines that call for service engineer support or scheduled removal of the GTCU.

Routine on-board maintenance is limited, consisting mainly of daily visual checks of the running engine. An internal water wash is executed after a fixed number of operating hours to remove any salt deposits in the compressor which would otherwise eventually result in degraded performance.

The engine itself can be divided into two main elements: the GTCU and the power turbine with ancillaries package. The GTCU can be readily removed from the vessel, within 6–12h, when dictated by condition monitoring or scheduled removal. The power turbine and ancillaries are designed to remain for longer periods, up to the life of the vessel.

In general, the operator will regard the repair of the GTCU as repair by replacement; it will be removed and replaced, and the vessel returned to

Fig. 27.25 Routine maintenance of the Rolls-Royce MT30 gas turbine (shown here on the test bed) is limited to checking fluid levels and visual examinations.

service in hours rather than days. An owner will therefore normally choose to own a calculated number of spare engines or major engine components (MACUs: maintenance assembly change units) to support a particular operation with a required availability.

A maintenance regime can be provided through a power-by-the-hour arrangement directly with the gas turbine manufacturer. An indicative figure for such an arrangement is often requested by operators, the figure typically including scheduled removal and overhaul of the engine and its replacement by a spare during the overhaul cycle. It is invariably misleading to quote such a figure, which is not a feature purely of the engine but highly dependent on the usage. The calculation will vary strongly with the following factors, notes Rolls-Royce:

- The operating profile of the vessel.
- The number of starts, stops, and idling periods during the route.
- The ambient operating temperatures.
- The maximum and minimum powers required from the engine.

These factors are fundamental in deciding engine life and therefore the maintenance costs. The resulting calculations are based on creep and cyclic performance. With knowledge of the operating profile, an assessment can be made of the GTCU life at which the blading must be replaced. At one extreme, on a longer ferry route with operation at high power and high ambient temperatures, engine life may be creep limited.

The multiple compressors and turbine stages of the GTCU comprise individual discs with blades attached. Any metal disc that is continuously and repeatedly accelerated from rest to a high speed then decelerated, will eventually crack. All the discs have a declared cyclic life, a full cycle being defined as from rest to full speed and back to rest. The declared life falls well within that at which a disc might be expected to crack, but when this life is reached the disc must be replaced.

Cyclic life may become a limiting factor on short ferry crossings, particularly if this is carried out at high engine speeds; it may dictate whether the engine can be shut down or should be left running at idle in harbour, largely for economic reasons based on engine lifting.

It is the combination of creep and cyclic life calculations in conjunction with condition monitoring of the engine that decides the periodicity of the required GTCU changes, and therefore impacts directly on any fired-hour maintenance support arrangement. By working with a prospective operator, the engine manufacturer is then able to make an accurate assessment of the maintenance costs. The combination of engine characteristics and selected operation determine these costs rather than the engine alone.

Gas turbine compressor washing is an important maintenance routine in sustaining power output. Since the compressor consumes up to two-thirds of the total engine power even slight blade fouling significantly reduces the net available output and hence the drive power available or increases fuel costs. Offline washing is an effective countermeasure in restoring performance and overcoming the efficiency drop caused by compressor fouling. Online washing is available when engines cannot easily be shut down for an offline wash (as with cruise ships during long voyage legs).

Acknowledgements

GE Marine Engines, Mitsubishi, Rolls-Royce, and Siemens.

CHAPTER TWENTY-EIGHT

Engine-room safety matters

Safety and training

As in all areas, onboard ship engine-room safety is a combination of basic training, drills, familiarity with equipment, maintenance, a well-defined company safety system, and personal attention to safety matters.

Basic training requirements are laid out in the STCW Convention and apply to all SOLAS ships. Flag states can impose stricter requirements but are left to their own choices in the case of non-SOLAS vessels. It is the duty of the ship operator and the senior officers on the ship to ensure that crew are qualified for their role on board and medically fit.

An engine room and the associated machinery spaces are hazardous areas and even in the best-run vessels, great care must be taken to avoid accidents. As well as being a noisy, hot, and vibration affected workspace, engine and machinery layouts mean constricted passageways, stairs, obstructions and exposed pipes, cabling, and other hazards.

All staff should have access to personal protective equipment (PPE) including hard hats, ear protection, safety goggles, appropriate footwear, gloves, and overalls. More importantly, senior staff must ensure that appropriate PPE is being used by each crewman at all times. Good management practices will ensure that crew instinctively safeguard their own safety and those around them.

Safety matters on ships and within shipping companies are the subject of the IMO's International Safety Management Code adopted in 1993 by resolution A.741(18) and which entered into force on 1 July 1998. This is a mandatory system whereby each ship operator is obliged to establish and maintain a safe management system (SMS). The Code is regularly reviewed and amended by the IMO.

The SMS must identify key items of equipment which would include the engine and peripherals and establish procedures to ensure their availability. With regard to the engine, these would normally be linked to the manufacturer's recommendations for servicing and repair.

Pounder's Marine Diesel Engines and Gas Turbines
https://doi.org/10.1016/B978-0-08-102748-6.00028-1

The SMS also requires that hazards are identified and means to avoid them taken. Often shipping companies make use of proprietary training materials produced and distributed by equipment makers and third parties such as Videotel has been producing materials for many years.

In its engine rooms safety training products, Videotel has identified 10 major hazards that are common in engine rooms. These are:

1. Garbage—there can be a lot of waste produced in an engine room and too often there is a garbage can overflowing with oily rags that may catch on clothes and tools and be deposited near a hot running engine. Oily rags are a well-known fire hazard whereby the substances on the rags can ignite due to an exothermic reaction from the nearby hot engine, causing a fire. So housekeeping is of the utmost importance.
2. Working at height—often engineers may not think of themselves working at height. As a machinery space can be at a significant height, any missing or damaged sections of handrail, or chains left hanging can be extremely dangerous. Such thoughtless, simple, and careless acts as leaving the chain unhooked is a fall hazard for anyone passing.
3. Safe Working Loads—there can be many heavy lifts in the engine room, large pieces of equipment or spares being moved into position or taken out of the machinery space. It can be all too common for loads on chain blocks to exceed the safe working load, which may result in the catastrophic failure of the equipment and injury to the crew.
4. Bilge and Tank Alarms—when bad things happen in the engine room there can be little time to act, so alarms are hugely important. Unfortunately, especially on older vessels, there can often be instances when alarms have been blocked, or lights with bulbs removed. There can be an assumption that all is well. However, alarms are there for a reason and need to be activated and working.
5. Insulation—there are many extremely hot pieces of equipment and piping in the engine room, and insulation is a vital component of safety. It can protect people from coming into contact with hot spots, and it can also reduce the dangers of ignition where other pieces of equipment or items may come into contact.
6. Wiring and electricity—the dangers of high voltage are clearly marked around an engine room, but that does not always mean that the right actions are taken. Inspectors often report fuse boxes being left open with exposed wires. There is a real danger of electrocution if wires are left exposed, so these need to be insulated and where necessary electrical switchboards should always be properly closed, and insulation matting provided and in place.

7. Deck plates—in a distracting place, perhaps with some degree of slipperiness underfoot, and especially if a vessel is moving, then keeping footing can be a challenge. This is more so if there are any loose or missing floor plates. It is vital that anywhere people are meant to tread is safe and secure. Any missing, damaged or loose floor plating may result in a crew member falling into the bilge spaces below the plating, causing serious injury.
8. Lighting—it is vital that engineers and especially visitors to the machinery space are able to see where they are going or working. Lighting is hugely important, and it needs to be sufficiently bright, but without causing areas of excessive shadow in which dangers can lurk. Where lighting is improper, ill-positioned or damaged, then this reduces the ability of the crew to work effectively and safely in the engine room.
9. Blocked escape routes—incredibly surveyors and inspectors often report that escape routes are found to be impeded, or escape trunks not illuminated, which will obviously hinder safe evacuation in an emergency.
10. Doors open—while there are often escape routes blocked off, locked, or closed. Conversely, and just as dangerously, there are often reports of fire doors being tied, or even wedged open. While the most ironic use of fire extinguishers is to have them used to stop fire doors from shutting.

Fire safety

Statistics reveal that a shipowner operating a fleet of 20 vessels can expect one major engine-room fire every 10 years. Avoiding fires, or at least preventing engine-room incidents from developing into serious emergencies, is therefore paramount in ship management. A research study by the Oslo-based classification society Det Norske Veritas (DNV) carried out before it merged with Germanischer Lloyd (GL) to form DNV GL in 2013 found that two-thirds of 165 fires on board the DNV-classed fleet started in the engine room. Of these engine-room fires, some 56% (excluding shipyard repairs) originated from oil leaking onto a hot surface. The direct cost of repairing engine-room fire damage is compounded by the off-hire penalty and loss of goodwill. Similar studies carried out by DNV GL and others since show similar results.

Regulations referring to the fire protection of engine compartments are covered by SOLAS. The requirements regarding the fire-extinguishing systems are included in Part C: Suppression of fire, Regulation 10.5 Fire-extinguishing installations in machinery spaces, SOLAS 1974 II-2/10.5. The provisions concerning fire-extinguishing systems and fire-extinguishing appliances that are on board are incorporated into the International Code for Fire Safety Systems (FSS Code) developed by IMO.

Machinery spaces, in which oil-fired boilers or oil-fuel units are placed, should be equipped with one of the following fixed fire-extinguishing systems:

- a fixed gas fire-extinguishing system,
- a fixed foam fire-extinguishing system,
- a fixed pressure water-spraying fire-extinguishing system, and
- fixed aerosol fire-extinguishing systems equivalent to fixed gas fire-extinguishing systems required by SOLAS.

From 1 July 2003, all ships of 500 gt and above in international trading have had to comply with IMO SOLAS regulations covering the protection of hot surfaces in the engine rooms as well as fuel oil systems. The requirements were made retroactive from that date for tonnage built before 1 July 1998. Water-based firefighting systems have been required for all new cargo and passenger tonnage from 1 July 2002 and for existing passenger ships from 1 October 2005.

Most fuel oils will ignite spontaneously when hitting surfaces with temperatures above 250°C, DNV explains, and class rules, therefore, require that all surfaces above 220°C should be shielded or insulated. However, it is surveyors note that such protection is often impaired in service.

The DNV study referred to above recommended a number of methods that can be used to detect hot surfaces: surface/contact thermometers, laser-based infrared heat tracers, and infrared thermo-scanning video equipment.

Fuel leakage sources include damaged flexible hoses, couplings, filters and fractured pipes, calling for attention to the installation, location, and condition of all such components. Operators are urged to inspect engine-room fuel systems, which are also subject to class inspections.

Oil mist can be formed by minute leaks in oil lines which, under pressure, produce a very fine atomized spray; or it can be formed by oil boiling off in contact with a hot surface, producing a fine vapour. In the first case, the danger occurs when the particle size formed is between 3 and 10 μm and allowed to form a mist in the atmosphere. When oil vapour reaches the range of flammability the condition can be classed as truly hazardous and, if no action is taken, a fire may result. The ignition temperature for this type of oil mist can be extremely low, depending on the type of oil being atomized.

Oil mists generated by being boiled off can produce particles between 3 and 10 μm. The mist is visible and known as 'blue smoke'. The temperature and area of surface contact affect the rate of oil mist generation; at this stage, a temperature as low as 150°C could result in ignition. Among the sources of oil mist are pump seals, leaking injectors, loose or incorrectly fitted pipe elements, weld fractures, and poor machinery maintenance. Possible heat sources causing ignition include heat exchangers, exhaust pipes, turbochargers, electrical contacts, static electricity, and faulty wiring.

Acknowledging that most engine-room fires result from the formation of oil mist, the IMO's Maritime Safety Committee in 2003 approved a *Code of Practice for Atmospheric Oil Mist Detectors*, giving advice on their location, setting alarm levels, test procedures, and maintenance.

Detection systems can be single sampling units or multiple sampling systems. In each case, the number of detectors or sampling points depends on the size and layout of the application. Sample lines in a multiple sampling system (fitted in a suitable location away from the application) are fed to a common manifold with a control unit allowing alternative samples to be taken from continuously flowing sample streams. One unit can thus be used to monitor several points, the oil mist is drawn into the unit by a built-in fan or an independent blower.

To determine suitable positions for mounting detectors or for fitting sampling lines, a smoke test is required to verify air movements in relation to the application. In general, air will move towards ventilation extractors and turbochargers, so any detector or sampling line should be positioned as closely as possible to the machinery. Similarly, detectors or sampling lines should not be sited next to ventilation blowers as these will prevent mist formation from being drawn into the unit.

On installation, the IMO *Code of Practice* advises, a smoke test should be carried out with all engines, ventilation, and machinery fully operational to ensure that detectors/sampling lines are correctly positioned. If the detector units are to be located close to the source of the application, care is necessary to avoid locations where: vibration is excessive; extremes of temperature may be experienced; access for maintenance personnel is difficult; high levels of humidity and water may arise; and there is a risk of electromagnetic interference. Locating any detector in an explosive atmosphere should not be undertaken unless the unit is certified as intrinsically safe for the hazardous area.

Preventing crankcase explosions

Explosions caused by overheating of surfaces within the engine crankcase or by leakage of hot gases past the piston have always posed a danger. Among the components identified as the cause of crankcase explosions have been overheated main bearings, gudgeon pin bearings, chain wheel bearings, cylinder gland rings seizing on the piston skirt, and seized pistons and piston rods. An overheated component can convert the normal crankcase atmosphere of air and oil droplets into an explosive oil mist, the same hot spot then acting as a source of ignition of the mist. The primary explosion, often of limited severity, may be followed by a more severe secondary explosion.

Serious crankcase explosions on the British passenger ship *Reina del Pacifico* in 1947—killing 28 personnel and seriously injuring another 23—highlighted the dangers and stimulated regulatory action and technical developments to combat the problem. The four main engines were 12-cylinder four-stroke trunk piston designs with a combined output of 11,800 kW. The subsequent court of inquiry held that the primary cause was the piston of No. 2 port outer engine overheating and igniting a flammable mixture present in the crankcase of that engine.

Prior to the explosions, the engines had been running at full power for some six-and-a-half hours when it was reported that No. 2 cylinder liner was hot. The control lever on the engine was brought to the stop position and, after a few minutes, was moved to the starting position. Shortly afterwards, the first explosion occurred, spreading throughout the crankcase, blowing off the crankcase doors, and igniting the whole of the lower engine-room atmosphere.

It was not clear how the flame from the port outer engine reached the crankcases of the other three engines. The consensus was that the pressure produced in the engine-room atmosphere forced the unsupported lower ends of certain crankcase doors of each engine inwards; and then, when the flammable vapour inside the crankcase was ignited, the pressure blew the doors outwards.

SOLAS regulations introduced in the 1960s aimed to improve the safety of marine engines through monitoring by oil mist detection (OMD). Until the mid-1990s, the adopted standards appeared to be effective as there were no further incidents of crankcase explosions reported with severe loss of life.

Ongoing research intensified during the 1990s, however, when several major ship owners and engine builders began to express concerns that the number and scale of explosions appeared to be increasing. Estimates suggested that a minimum of six crankcase explosions were occurring every month worldwide. Between 1990 and 2001 Lloyd's Register recorded 143 crankcase explosions in its classed fleet that caused damage in one form or another. Of these incidents, 21 occurred in two-stroke engines and 122 in four-stroke engines. The causes in two-stroke engines were identified as due to bearings (43%), pistons (38%), and others (19%); the causes in four-stroke engines were due to bearings (39%), pistons (47%), and others (14%). The analysis showed that both engine types were affected in almost the same way. Higher specific power ratings and inferior fuel qualities were cited among the contributory factors.

The International Association of Classification Societies' (IACS) Machinery Panel was tasked with reviewing the applicability and suitability of the two Unified Requirements—UR M10 and M9—covering crankcase explosions.

In January 2005 IACS agreed on major revisions to both URs (M10: *Protection of internal combustion engines against crankcase explosions*; and M9: *Crankcase explosion-relief valves for crankcases of internal combustion engines*). These URs set uniform standards of installation and functionality for the equipment—OMD and alarm equipment (M10) and relief valves (M9)—to be applied by all IACS members.

Statistics from incidents reported to DNV GL show that main bearing and journal problems are the most frequent cause of damage in medium-speed main engines and high-speed engines, with crank bearings and pins the second major contributor. Damages caused by bearing failure in low-speed engines are considerably lower in percentage terms. An investigation of the causes of bearing damage identified the following:

- Lubrication oil supply failure or contamination (from particles or water).
- Connecting rod bolt shear (material fatigue, over-stressing).
- Abnormal wear (ovality, pounding, corrosion).
- Overloading (revving, piston seizure, abnormal vibration).
- Faulty bearing material.
- Incorrect bearing assembly.
- Nonoptimum engine construction.

Three different basic approaches to detecting and preventing bearing damage are available: OMD; oil temperature measurement; and measurement of metal temperatures.

OMD is the most well-known and applied system. Overheating of a sliding surface under friction can be recognized by detecting the developing oil mist. When the dry friction continues and the component reaches ignition temperature the oil mist concentration can rapidly increase to the level of creating a powerful explosion if not detected early.

Oil mist (sometimes erroneously termed 'oil smoke') consists of very fine oil droplets—colloidal oil—that are suspended in the air, their size in the range of approximately 1–10 μm. Oil mist that has arisen through mechanical atomization has particles lying more in the upper size range; oil mist arising from oil vapour that has condensed again has particle sizes more in the lower size range. The atmosphere of the crankcase compartment contains both oil mist from re-condensed oil vapour and from mechanical atomization, though the latter can be ignored because of the far greater amount of mist that is generated even at the start of a damage event through vaporization and recondensation.

An important feature of OMD systems is that they cover all heated surfaces inside the engine. It is questionable, however, if such systems—being an indirect bearing temperature measurement method—have the capability to give the fast response necessary in the event of evolving bearing damage, suggests Norwegian specialist Kongsberg Maritime.

The fact that engines develop oil mist during normal operation (condensation from evaporating cooling oil returning to the crankcase from pistons, bearings—including those of the turbocharger—and the blowby gases) presents a difficulty in identifying the start of friction damage. If the setting of the oil mist detector is too sensitive it can lead to false alarms. The other extreme—when the detector is set to react at a higher mist concentration—is also problematic: engine components may reach an advanced stage of damage and become costly to repair.

An alternative approach is to measure bearing temperatures by measuring the temperature of the associated lubricating oil using probes (also known as the 'splash oil method') or by direct measurement of the bearing metal temperatures.

The compromise here, according to German OMD pioneer Schaller Automation, is that the temperature probe (such as a PT100) must be very close to the heat source. In the case of crankshaft bearings in large engines, this is difficult due to the fact that the bearing area is wide; and in the event of friction, the lubricating oil first cools and the heat sink effect of the voluminous metal parts (bearing cap or shaft) around the probe reduces the efficiency of this measurement method. Ultimately, says Schaller, there is an economic restriction on the number of temperature probes that can be installed in an engine, and not every sliding surface can be provided with a temperature sensor.

In the splash oil method, the temperature of the lube oil splashing out of each bearing is measured. Critics argue that a measurement point located at some distance from the engine wall will not deliver any useful values for the timely detection of bearing damage. Furthermore, in high- and medium-speed engines piston cooling oil can be mixed with the splash oil from the crank bearing, making the temperature measurements less reliable.

By comparing the results, Kongsberg Maritime concludes that a system based on directly measuring the bearing metal temperatures has a better performance both in terms of response time and measurement accuracy. It, therefore, pursued a system designed to be permanently installed inside an engine to perform real-time monitoring of bearing metal temperatures (Fig. 28.1).

An innovative wireless telemetry technology was exploited to create the Sentry GB-100 sensor-based system capable of real-time temperature monitoring of crosshead and crankpin bearings. A combination of specialized radar technology and acoustic wave technology facilitates the use of highly

Fig. 28.1 Continuous big end bearing temperature monitoring is performed by Kongsberg Maritime's Sentry system on Wärtsilä's 46F medium-speed engine.

reliable surface acoustic wave (SAW) sensors without wiring or internal power sources. The passive GB-100 sensors are reportedly capable of working at a range of 1–50 mm between sensor and antenna as well as in harsh environments.

Sentry embraces a signal processing unit (SPU); a wireless temperature sensor; and a stationary antenna. The SPU generates a low-energy, high-frequency radar pulse which is transmitted to the wireless sensor via the stationary antenna. When the wireless sensor passes the stationary antenna the radar pulse is transferred. The sensor will then immediately reflect a pulse signal, which is received by the SPU via the stationary antenna. The shape and characteristics of the received signal are then used to determine the temperature of the sensor, this information will then be communicated to the engine control and monitoring system.

A key merit cited by Kongsberg Maritime is the system's high flexibility in arranging the sensor and antenna with respect to the gap, angle, and lateral position between these units. In addition, there is no need to calibrate the temperature sensors and there is no degradation of sensor accuracy over time.

The final development of the Sentry system was carried out in close co-operation with Wärtsilä, which performed a series of tests including a crank bearing seizure on a W4L20 medium-speed engine. The seizure was affected by cutting the supply of lube oil to one of the cylinders while the engine was running at approximately 1050 rev/min. The main conclusion from the test was that the response time of the system was fast enough to save the crankpin from being damaged.

Several long-term performance and reliability tests were also conducted, and the first installation made on a large two-stroke engine to measure crosshead-bearing temperatures. Wärtsilä adopted the Sentry system as a standard fit for temperature monitoring the big end bearings of its new 46F medium-speed engine. The SPU can be linked to the group's condition-based maintenance system or any alarm and monitoring system.

Approved by classification societies as an alternative to traditional oil mist detectors, the Sentry system can be installed to monitor the big end and crosshead bearings of two-stroke engines, such as MAN Diesel and Wärtsilä designs, at newbuild stage or by retrofit.

In the early 1980s, Schaller Automation found that interacting machine parts in a state of friction and made of different metals (e.g. a steel shaft running in a bronze bearing) produce a thermoelectric effect which can be exploited to produce instant measurable signals. With an appropriate method, this effect can be measured to provide a diagram showing whether there is a normal operational condition or serious damage is evolving.

Schaller's Bearing Overheating Monitoring System (BEAROMOS) was developed primarily for protecting sliding bearings and surfaces in combustion engines. A sensor is incorporated to obtain information on signals that become noticeable when the rotating crankshaft is touching the metallic surface of the main bearings. Applied to engines, the basic principle of the system is to measure the thermoelectric effect produced by a thermocouple formed by the crankshaft (steel) when sliding on the soft metal alloy of bearing shells. This physical effect takes place for a few seconds at each start and stop of the engine when the crankshaft is still making contact with the bearing shell at the instant that the oil film is partly interrupted.

As soon as the oil film is complete, the thermoelectric current is interrupted and BEAROMOS registers normal operation. Should the film be interrupted again at any moment (e.g. due to dirt or metal particles within the bearing clearance, or lack of pressure or overheated oil) the sensor will at an early stage show signals which can be used for taking operational precautions: a prealarm or shutdown procedure. An important aspect of such a measurement, says Schaller, is that when integrated into a system it is free from false alarms. This is due to the fact that the signal is only generated when the oil film is interrupted and frictional heat begins to develop in a particular bearing.

The XTS-W bearing wear sensor system from UK-based Amot addresses two-stroke engine applications, promising the capability to detect the onset of plain bearing problems before expensive secondary damage can occur. The simple-to-install proximity-based system is claimed to indicate problems well in advance of temperature sensors or OMD systems.

Wear is detected by the system in the main, crankpin and crosshead bearings by accurately measuring the vertical position of the guide-shoe at bottom dead centre: wear occurring in any of these bearings causes the guide-shoe position to drop. The relevant data is processed by the system, which provides the operator with a graphic display of current status and trends. A bar chart display shows the current status: a small green bar is normal; a larger amber bar is a warning that calls for further monitoring; and a red alarm bar is the final stage.

The alarm levels are user-configurable but the factory-set limits of 0.3 mm warning and 0.5 mm alarm avoid the potential for nuisance alarms while retaining sufficient headroom on a typical 1-mm-thick bearing shell to allow time for corrective action. Trend monitoring provides the rate of degradation data that allows the operator to determine whether immediate action is required or whether the issue can be addressed at the next port of call.

Delivering this level of functionality makes it necessary to compensate for normal engine conditions that can affect measurement accuracies, such as engine load and temperature. The algorithms in the Amot system were refined during extensive engine testbed and sea trials to provide repeatability of, typically, 50 μm. A typical XTS-W system may comprise: two 4–20 mA output proximity sensors per cylinder mounted via brackets; and an SPU with capacity for up to 14 sensors (seven cylinders) mounted on the engine. Each sensor can be calibrated individually or simultaneously, and each microprocessor-based SPU generates continuous signals proportional to the wear detected, compensating for environmental and engine load conditions.

Testbed trials reportedly confirmed successful operation of the system under real bearing failure conditions. When the main bearing on the test engine was machined to minimize contact area, the system clearly indicated the problem: the level of wear detected correlated to the actual value within 0.05 mm, and no significant rise in temperature was recorded.

Amot worked closely with MAN Diesel during the development of the system, the two-stroke engine designer also partnering sea trials and subsequently adding the XTS-W bearing condition monitoring system to the extent of the delivery list for all its engines.

Crankcase explosions and relief valves

Classification society requirements for OMD, bearing temperature monitoring and crankcase pressure relief valves have contributed significantly to reduce the risk of life-threatening crankcase explosions.

Explosion-relief valves are used on internal combustion engines to protect the crankcase in the event of an explosion of the oil mist, the inlet manifold in the case of backfire, and the exhaust system in the event of a misfire (Fig. 28.2). A classic sequence of events in a crankcase explosion can be summarized as follows:

- A mechanical fault develops with consequent rubbing of moving parts and the creation of a hot spot somewhere in the crankcase.
- Above around 200°C lubricating oil begins to vaporize where it is in contact with the hot spot.

Fig. 28.2 Hoerbiger explosion-relief valves installed on a medium-speed engine.

- Movement of engine components within the crankcase causes the oil vapours to circulate through the engine; as the vapours pass to cooler areas they re-condense and a white mist of oil droplets is formed; the droplets are around 0.005–0.01 mm in diameter.
- The oil mist concentration steadily increases until it reaches the lower explosive limit of around 50 mg/L, equivalent to an oil mist air ratio of 13%.
- Ignition of the oil mist is possible if the temperature of the hot spot is higher than 500°C.

Apart from a hot spot (typically a bearing), there are other causes of lube oil vaporization in a crankcase, such as a severe external fire or the products of cylinder combustion finding their way into the crankcase via a failed piston.

The process of ignition is complex, explains Austria-based explosion relief-valve specialist Hoerbiger. Essentially, it is a function of the composition and uniformity of the oil mist mixture within the crankcase, where explosions can range from puffs to violent, damaging events. Following the ignition of an oil-rich mist, there will be expulsion of pressure through the relief valves and a reduction in crankcase pressure. The reduced crankcase pressure causes atmospheric air to enter the crankcase via the vent pipes, thus creating a further flammable mixture and the possibility of a secondary explosion in the crankcase.

Tests have shown that ignition of oil mist vapour within a crankcase is only achievable over a very narrow mixture range: weaker or richer mixtures than around 50 mg/L will not ignite. There is no question of self-ignition of the vapour, either theoretically or practically, as there must always be a cause of ignition, such as a hot spot from rubbing components or a solid, loose object in the crankcase causing sparks. Fundamental for an explosion to occur are three factors: an ignitable substance (oil mist); oxygen (air in the crankcase); and a source of ignition (hot spot).

Effective crankcase explosion-relief valves are vital in a holistic approach to crankcase safety, asserts Hoerbiger, with a design featuring no flame escape being the ultimate because of its proven capability and practical viability (Fig. 28.3). A number of key factors must be addressed in designing such valves, the company adds the valve size that determines the free valve

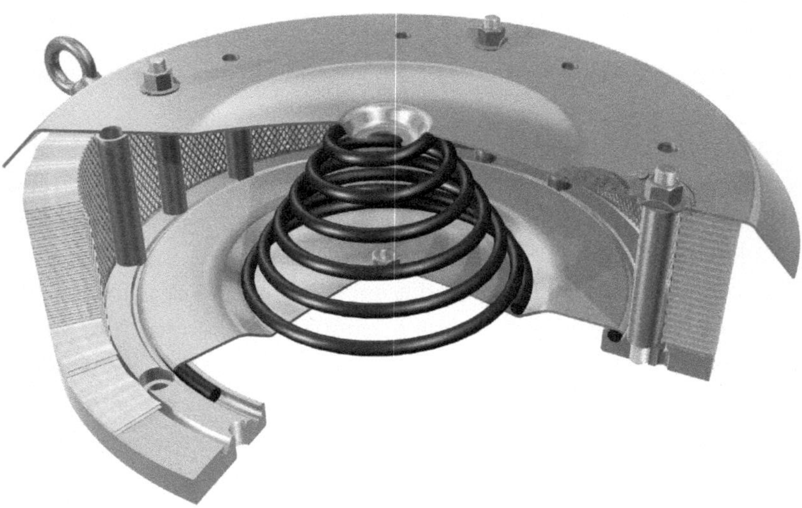

Fig. 28.3 The large relief cross section of a Hoerbiger valve allows explosion waves to disperse freely and rapidly; the low mass, frictionless guided valve plate opens in fractions of a second and then closes tightly to prevent the inrush of air (oxygen); and an integrated flame absorber prevents the escape of blast flames.

seat area; the valve opening pressure and the relieving capacity; the ratio of free valve seat area to crankcase volume; the number of valves; the sealing of the valve plate; and the zero flame escape flame arrester.

It is important that the functionality of crankcase relief valves is demonstrated to an agreed standard by an independent recognized testing institute, assessing:

- The effectiveness of the flame arrester arrangement in preventing flame transmission outside the engine during a crankcase explosion.
- The effectiveness of the valve in limiting the pressure rise during an explosion and preventing damage to the engine.
- That the valve closes and is gas/airtight before and after a crankcase explosion.

Explosion-relief valves from Hoerbiger have reportedly passed extensive tests, its EVN series produced in sizes covering a wide range of relief areas. Excellent sealing and swift frictionless opening and closing (avoiding an inrush of air) are claimed, with a metallic flame arrester design fully absorbing the flames and effectively protecting personnel and plant from secondary explosions. A specific opening pressure of approximately 0.05 bar is quoted for the standard series, with absolute tightness maintained after opening and closing. Low maintenance demands are promised and the valves do not need replacing after an explosion.

Acknowledgements

DNV GL, Lloyd's Register, Kongsberg Maritime, Schaller Automation, Amot, Wärtsilä, and Hoerbiger.

Index

Note: Page numbers followed by *f* indicate figures, and *t* indicate tables.

D

F

M

S

T

U

Y

FSC
www.fsc.org
MIX
Papier | Fördert gute Waldnutzung
FSC® C083411

Kolibri 360 GmbH
Ferdinand-Jühlke-Straße 7
99095 Erfurt, Deutschland
produktsicherheit@kolibri360.de